TROPICAL FOREST COMMUNITY ECOLOGY

TROPICAL FOREST COMMUNITY ECOLOGY

Editors

Walter P. Carson
University of Pittsburgh
Department of Biological Sciences
Pittsburgh, PA
USA

Stefan A. Schnitzer
University of Wisconsin – Milwaukee
Department of Biological Sciences
Milwaukee, WI
USA

and

Smithsonian Tropical Research Institute
Apartado 0843-03092, Balboa
Republic of Panama

WILEY-BLACKWELL

A John Wiley & Sons, Ltd., Publication

This edition first published 2008
©2008 by Blackwell Publishing Ltd

Blackwell Publishing was acquired by John Wiley & Sons in February 2007. Blackwell's publishing program has been merged with Wiley's global Scientific, Technical and Medical business to form Wiley-Blackwell.

Registered office
John Wiley & Sons Ltd, The Atrium, Southern Gate, Chichester, West Sussex, PO19 8SQ, UK.

Editorial offices
9600 Garsington Road, Oxford, OX4 2DQ, UK
The Atrium, Southern Gate, Chichester, West Sussex, PO19 8SQ, UK
111 River Street, Hoboken, NJ 07030-5774, USA

For details of our global editorial offices, for customer services and for information about how to apply for permission to reuse the copyright material in this book please see our website at www.wiley.com/wiley-blackwell.

The right of the Walter P. Carson and Stefan A. Schnitzer to be identified as the author of this work has been asserted in accordance with the Copyright, Designs and Patents Act 1988.

Library of Congress Cataloging-in-Publication Data
Data available

ISBN: 978-1-4051-1897-2 (paperback)
ISBN: 978-1-4051-8952-1 (hardback)

A catalogue record for this book is available from the British Library.

Set in 9/11 pt Photina by Newgen Imaging Systems Pvt. Ltd, Chennai, India
Printed in Singapore by Ho Printing Pte Ltd

1   2008

To my parents Walter and Alice Carson: thank you for years of unconditional love and support, and to my son Chris Carson: you have changed the way I see the forest.

Walter P. Carson

CONTENTS

PREFACE

It is not hyperbole to say that there has been an explosion of research on tropical forest ecology over the past few decades. The establishment of large forest dynamics plots in tropical forests worldwide, in and of itself, has led to a near revolution in our understanding of forest change. In addition, there has been a substantial increase in the use of models and experiments to test long-standing theories developed to explain the striking patterns found in tropical forests and the putative mechanisms that underlie these patterns. When we started this project, we felt that a comprehensive synthesis of tropical forest community ecology was necessary in order to help the field move forward. Of course, no single volume could do this. Nonetheless, this book is our attempt to make a significant contribution to the field, and to ask anew: What are the main theories in tropical ecology, and which ones are supported or refuted by empirical data? Thus, we have attempted to assemble a volume that describes the most up-to-date findings on the important theories of tropical forest community ecology. We hope that this book accomplishes this goal to the degree possible, while at the same time providing a road map of what we know, what we think we know, and where future research is most needed.

The focus of the chapters in the volume is at the community level because this is where some of the most fundamental questions in tropical ecology exist. Indeed, perhaps the greatest challenge to community ecologist is to explain what processes account for the maintenance of the staggering diversity of plants and animals common in tropical forests around the globe. Still, our emphasis on communities definitely reflects our bias as community ecologists. While we have focused on communities, we certainly recognize the important contributions to tropical ecology that have come from those who study different levels of ecological organization. Indeed, it is difficult to understand communities without understanding the ecology of populations and individuals. We decided to focus on forest communities because, to date, that is where the bulk of research on tropical community ecology has been conducted. We acknowledge that our focus has forced us to omit many important studies. Nonetheless, the emphasis on tropical forest community ecology provides enough material to fill multiple edited volumes, and thus we have attempted to focus on the areas that have received the most empirical attention, along with some topics that are currently nascent, but are rapidly becoming key areas in tropical ecology.

Each chapter in this book was reviewed by at least two relevant experts. We thank these reviewers for their efforts and we are indebted to all of them. We will not list them by name, thus allowing them to remain anonymous. We also thank the production team at Newgen Imaging Systems, and our editors at Blackwell for guiding us through the publication process.

This book, as with all edited volumes, would not have been possible without the dedicated contributions of the authors, each of whom is an expert in his or her respective area of study. For their hard work, truly top-notch contributions, and their patience throughout this process, we owe them a great deal of gratitude. This book is a tribute to their research, along with the research of all of the other scientists whose work is cited in this volume.

Walter P. Carson
Pittsburgh, Pennsylvania
2007

Stefan A. Schnitzer
Gamboa, Republic of Panama
2007

FOREWORD

The present volume captures the excitement generated by an explosion in tropical forest research. When I was a graduate student in the late 1970s, it seemed to be possible to read every new article published on tropical forests. The ISI Web of Science© confirms this schoolboy memory. Just 289 articles published between 1975 and 1979 included the words "forest*" (for forest, forested or forests) and the name of a tropical country (or tropic*) in their titles. By reading just one or two articles a week, I was able to keep abreast of the entire literature on tropical forests. This would be nearly impossible today. Between 2002 and 2006, 2593 articles met the criteria described above reflecting a nine-fold increase in the rate of publication of tropical forest articles since the late 1970s. This explosion has been driven by new discovery; new theory; new technology; new challenges posed by global change, deforestation and other threats to tropical biodiversity; and ongoing interest in theory posed in the 1970s and earlier. This volume illustrates each of these developments.

In the 1970s, we all "knew" that ants were predatory with the exception of an insignificant few observed at extrafloral nectaries. No one guessed that plant exudates supported most of the great biomass of ants (Chapter 6). Likewise, no one guessed that plants consisted of a mosaic of plant plus endophytic fungi and that the endophytic fungi were hyperdiverse with tens to hundreds of species inhabiting each leaf in the forest (Chapter 15). The roles of herbivorous ants and endophytic fungi are only beginning to be explored, and their implications for forest biology are potentially profound. New theories of chance, dispersal and seed limitation (Chapters 2, 8 and 14) and new tradeoffs postulated between fecundity and habitat tolerance (Chapter 11) also hold the potential to change our understanding of how tropical forest communities are structured and are only now beginning to be explored.

In the 1970s, we would have been mystified by functional (Chapter 10) and phylogenetic (Chapter 20) approaches to plant community ecology and the knowledge base in physiology, morphology and molecular genetics that makes these approaches possible today. Both approaches have the potential to reduce the immense number of species of tropical forest plants to a manageable number of ecologically distinct groups or crucial relationships among species' traits. Today, we are striving to bring functional, phylogenetic and ecological approaches together for 6000 plus tropical tree species found in the network of large Forest Dynamics Plots maintained by the Center for Tropical Forest Science (Chapter 7).

A graduate student in the late 1970s would have been familiar with the plant favorableness (Chapters 3 and 4), regeneration niche (Chapter 6), Janzen–Connell (Chapter 13) and bottom-up versus top-down hypotheses (Chapters 16–19 and 21) addressed by one third of the chapters in this volume and would be delighted to read the progress summarized here. I was also familiar with the potential of large forest plots – Robin Foster and Steve Hubbell were busy generating excitement for a grand new plot when I was a graduate student on Barro Colorado Island – and it is also a delight to see that potential realized (Chapter 7). Likewise, Phyllis Coley and I were contemporaries as graduate students on BCI as she revolutionized the study of herbivory (and I muddled about with island communities of birds and lizards), and it is a delight to see many of her ideas extended to a new framework to explain herbivory gradients across tropical rainfall gradients (Chapter 5) and to bioprospecting for new pharmaceuticals (Chapter 25).

The final section of this volume (Chapters 22–28) would shock a 1970s graduate student. A potential tropical deforestation crisis was only first publicized in the early 1970s (Gómez-Pompa *et al.* 1972 *Science* 177, 762–765). The severity of deforestation in 2007 and the many exacerbating problems (Chapters 24, 26 and 27) would be entirely unexpected. The potential for solutions through natural secondary succession on abandoned agricultural land (Chapters 22 and 23) and conservation action (Chapter 25) proposed, in some cases, by my peers from the late 1970s on BCI would be equally surprising and heartening.

Where do we go from here? What might a graduate student do in 2007 to have the greatest future impact? There are many answers. Spectacular new data sets are being made available by the Angiosperm Phylogeny Group, by several new efforts to assemble global plant and animal trait data, and by the new remote sensing technologies mobilized in global change research. Those trained to capitalize on these and other similar data sets will make many important contributions.

Simultaneously, we are still in the age of discovery in tropical forest ecology. No one suspected that there might be millions of species of endophytic fungi in tropical leaves until Elizabeth Arnold looked starting in 1996. We are equally ignorant of the roles of myriad other organisms. Even the local point diversity of herbivorous insects remains an unknown. Basic discovery will continue to make many crucial contributions to tropical forest ecology.

Finally, I will return to the nine-fold explosion in tropical forest publication rates mentioned in the first paragraph. The publication rate for extra-tropical forests increased just 4.3-fold over the same time interval. This latitudinal difference has been driven by a 15-fold increase in publication rates for authors from tropical countries. The increase in tropical forest publication rates falls to 5.8-fold when authors with tropical addresses and unknown addresses are excluded. The rapid increase in publication rates for authors from tropical countries is very uneven. Scientists from Brazilian and Mexican institutions increased their rate of tropical forest publications by 71-fold between 1975–1979 and 2002–2006 (from just 9 to 644 articles). Perhaps not surprisingly the authors of this volume include one Brazilian (Chapter 21) and two Mexicans (Chapter 5). Increasingly, scientists from Brazil, Mexico, and other tropical countries will formulate the tropical forest research agenda and determine what research has the greatest future impact. This is a positive development.

S. Joseph Wright
Smithsonian Tropical Research Institute
Apartado 0843-03092, Balboa
Republic of Panama

LIST OF CONTRIBUTORS

Gregory H. Adler
Department of Biology and Microbiology
University of Wisconsin – Oshkosh
Oshkosh, WI
USA

Jill T. Anderson
Department of Ecology and Evolutionary Biology
Cornell University
Corson Hall, Ithaca, NY
USA

A. Elizabeth Arnold
Division of Plant Pathology and Microbiology
Department of Plant Sciences
University of Arizona
Tucson, AZ
USA

Karina Boege
Department of Biology
Stanford University
Stanford, CA
USA
and
Instituto de Ecología
Universidad Nacional Autónoma de México
Ciudad Universitaria
México

Nicholas Brokaw
Institute for Tropical Ecosystem Studies
University of Puerto Rico
San Juan
Puerto Rico

Robyn J. Burnham
Department of Ecology and Evolutionary
    Biology
University of Michigan
Ann Arbor, MI
USA

and
Museum of Paleontology
University of Michigan
Ann Arbor, MI
USA

Charles H. Cannon
Department of Biology
Texas Tech. University
Lubbock, TX
USA

Todd L. Capson
Smithsonian Tropical Research Institute
Apartado 0843-03092, Balboa
Republic of Panama

Walter P. Carson
Department of Biological Sciences
University of Pittsburgh
Pittsburgh, PA
USA

Jérôme Chave
 Laboratoire Evolution et Diversité Biologique
UMR 5174 CNRS/UPS, Bâtiment 4R3
Toulouse
France

Robin L. Chazdon
Department of Ecology and Evolutionary Biology
University of Connecticut
Storrs, CT
USA

Phyllis D. Coley
Department of Biology
University of Utah
Salt Lake City, UT
USA
and
Smithsonian Tropical Research Institute
Apartado 0843-03092, Balboa
Republic of Panama

Steven C. Cook
Department of Biology
University of Utah
Salt Lake City, UT
USA
and
Department of Entomology
Texas A&M University
College Station, TX
USA

Richard T. Corlett
Department of Ecology &
    Biodiversity
The University of Hong Kong
Hong Kong

Luis Cubilla-Rios
Department of Chemistry
University of Panama
Panama City
Republic of Panama

James W. Dalling
Department of Plant Biology
University of Illinois
Urbana, IL
USA

Diane W. Davidson
Department of Biology
University of Utah
Salt Lake City, UT
USA

Stuart J. Davies
Center for Tropical Forest Science
Smithsonian Tropical Research
    Institute
Apartado 0843-03092
Balboa
Republic of Panama

Julie S. Denslow
Institute of Pacific Islands
    Forestry
USDA Forest Service
Hilo, HI
USA

Saara J. DeWalt
Department of Biological Sciences
Clemson University
Clemson, SC
USA

Rodolfo Dirzo
Department of Biology
Stanford University
Stanford, CA
USA

Lee A. Dyer
Department of Ecology and Evolutionary
    Biology
Tulane University
New Orleans, LA
USA

Daniel A. Emmen
Department of Zoology
University of Panama
Republic of Panama

Kenneth Feeley
Center for Tropical Forest Science
Harvard University – Arnold
    Arboretum
Cambridge, MA
USA

Paul V.A. Fine
Department of Integrative Biology
University of California
Berkeley, CA
USA

Catherine A. Gehring
Department of Biological Sciences
Northern Arizona University
Flagstaff, AZ
USA

William Gerwick
Scripps Institution of Oceanography
La Jolla, CA
USA

Mahabir P. Gupta
Centro de Estudios Farmacognósticos de la Flora
    Panameña (CIFLORPAN)
Faculty of Pharmacy
University of Panama
Republic of Panama

Maria V. Heller
Secretaría Nacional de Ciencia
Tecnología e Innovación (SENACYT)
Clayton, Ancon
Republic of Panama

Stephen P. Hubbell
Department of Ecology and Evolutionary
    Biology
University of California
Los Angeles, CA
USA
and
Center for Tropical Forest Science
Smithsonian Tropical Research
    Institute
Apartado 0843-03092, Balboa
Republic of Panama

Robert John
Smithsonian Tropical Research Institute
Apartado 0843-03092, Balboa
Republic of Panama

Kaoru Kitajima
Department of Botany
University of Florida
Gainesville, FL
USA
and
Smithsonian Tropical Research Institute
Apartado 0843-03092, Balboa
Republic of Panama

Thomas A. Kursar
Department of Biology
University of Utah
Salt Lake City, UT
USA

and
Smithsonian Tropical Research Institute
Apartado 0843-03092, Balboa
Republic of Panama

William F. Laurance
Smithsonian Tropical Research Institute
Apartado 0843-03092, Balboa
Republic of Panama
and
Biological Dynamics of Forest Fragments
    Project
National Institute for Amazonian
    Research (INPA)
C.P. 478 Manaus
Brazil

Egbert G. Leigh Jr
Smithsonian Tropical Research Institute
Apartado 0843-03092, Balboa
Republic of Panama

Joseph Mascaro
Department of Biological Sciences
University of Wisconsin-Milwaukee
Milwaukee, WI
USA

Kerry McPhail
College of Pharmacy
Oregon State University
Corvallis, OR
USA

Helene C. Muller-Landau
Department of Ecology, Evolution and
    Behavior
University of Minnesota
St. Paul, MN
USA

Eduardo Ortega-Barría
Institute of Advanced Scientific
    Investigations and High
    Technology Services (INDICASAT)
Clayton, Ancon
Republic of Panama

John R. Paul
Department of Biological Sciences
University of Pittsburgh
Pittsburgh, PA
USA

Carlos A. Peres
School of Environmental Sciences
University of East Anglia
Norwich
UK

Chris J. Peterson
Department of Plant Biology
University of Georgia
Athens, GA
USA

Lourens Poorter
Forest Ecology and Forest
    Management Group
Wageningen University
Wageningen
The Netherlands

Richard B. Primack
Biology Department
Boston University
Boston, MA
USA

Francis E. Putz
Prince Bernhard Chair for International
    Nature Conservation
Utrecht University
Utrecht
The Netherlands
and
Department of Botany
University of Florida
Gainesville, FL
USA

Dora I. Quiros
Department of Zoology
University of Panama
Republic of Panama

Richard H. Ree
Department of Botany
Field Museum of Natural History
Chicago, IL
USA

Luz I. Romero
Institute of Advanced Scientific
    Investigations and High
    Technology Services (INDICASAT)
Clayton, Ancon
Republic of Panama

Stefan A. Schnitzer
Department of Biological Sciences
University of Wisconsin-Milwaukee
Milwaukee, WI
USA
and
Smithsonian Tropical Research Institute
Apartado 0843-03092, Balboa
Republic of Panama

Pablo N. Solis
Centro de Estudios Farmacognósticos de la Flora
    Panameña (CIFLORPAN)
Faculty of Pharmacy
University of Panama
Republic of Panama

John Terborgh
Center for Tropical Conservation
Nicholas School of the Environment and
    Earth Sciences
Duke University
Durham, NC
USA

Tad C. Theimer
Department of Biological Sciences
Northern Arizona University
Flagstaff, AZ
USA

Jill Thompson
Institute for Tropical Ecosystem Studies
University of Puerto Rico, San Juan
Puerto Rico

Stephen J. Tonsor
Department of Biological Sciences
University of Pittsburgh
Pittsburgh, PA
USA

Campbell O. Webb
Arnold Arboretum of Harvard University/Center
    for Tropical Forest Science
Kotak Pos 223, Bogor 16002
Indonesia

Jess K. Zimmerman
Institute for Tropical Ecosystem Studies
University of Puerto Rico, San Juan
Puerto Rico

Pieter A. Zuidema
Prince Bernhard Chair for International Nature
    Conservation and Section of Plant Ecology and
    Biodiversity
Utrecht University, Utrecht
The Netherlands

INTRODUCTION

Chapter 1

SCOPE OF THE BOOK AND KEY CONTRIBUTIONS

Stefan A. Schnitzer and Walter P. Carson

Tropical forests are vastly complex systems with a myriad of interactions that ecologists are now just beginning to understand. Thus, for many years tropical forest ecology was, by necessity, largely a descriptive- and demographic-based science. More recently, however, tropical ecologists have begun to test more sophisticated ecological theory. Steve Hubbell has called this a time in tropical ecology where the theoretical rubber finally meets the empirical road. Tropical ecologists are now beginning to unite theory and long-term empirical studies to address a broad array of questions and theories that are of particular importance to tropical systems. These questions include the mechanisms responsible for large-scale patterns of species abundance and distribution, species coexistence and the maintenance of the vast species diversity, trophic interactions, and the dynamics of secondary forest succession, to name a few. These issues are not only important for the advancement of tropical ecology, but are crucial for our overall understanding of basic ecology in any system.

This volume represents a comprehensive synthesis of recent and significant advances in tropical forest community ecology. We have divided the book into five main sections: (1) Large-Scale Patterns in Tropical Communities; (2) Testing Theories of Forest Regeneration and the Maintenance of Species Diversity; (3) Animal Community Ecology and Trophic Interactions; (4) Secondary Forest Succession, Dynamics, and

Invasion; and (5) Tropical Forest Conservation. These broad categories encompass some of the most active areas of tropical forest community ecology. We acknowledge that we have omitted some active and important areas of tropical forest research. For example, more chapters in this book were devoted to plants than to animals and some traditionally important areas of tropical ecology (e.g., mutualisms) were not explicitly addressed (but see Arnold Chapter 15, Theimer and Gehring Chapter 17). This bias towards plants, large-scale patterns, and mechanisms for the maintenance of diversity reflects, to some degree, our own expertise as plant ecologists, as well as the abundance of these studies and their impact on tropical forest ecology. The chapters within each of the five major sections of this book represent some of the most recent advances in the field. Below we highlight the importance of each of these chapters.

LARGE-SCALE PATTERNS IN TROPICAL COMMUNITIES

In this section, Chave (Chapter 2) re-examines traditional studies of patterns of vegetation change and diversity at multiple spatial scales (beta-diversity) using new advances in both remote sensing techniques and statistical approaches. He examines theories that inform ecologists about the underlying causes for contrasting patterns of beta-diversity among regions. For example, Chave points out how new approaches can "partition beta-diversity into deterministic and stochastic processes." Chave's final conclusion is bold: "The debate over the validity of the neutral theory is now behind us" because there is now little doubt that unpredictability and dispersal limitation play crucial roles in structuring plant communities, as does environmental determinism.

Fine, Ree, and Burnham (Chapter 3) revive and expand the geographic area hypothesis, which predicts that tropical latitudes will have more species because of greater total land area. Although the hypothesis was recently dismissed in the literature (Schemske 2002), Fine *et al.* demonstrate convincingly that it explains significant variation in latitudinal patterns of tree species richness but only when biome area is integrated over time to include land area fluctuations over millions of years. Future tests of this hypothesis with other growth forms will now have to account for historical shifts in land area.

Paul and Tonsor (Chapter 4) explore the sticky issue regarding the relationship between a species' age and its range size. This idea can be traced back to Willis (1922) and his studies in Sri Lanka. In the first test using tropical plants (*Piper* spp.), Paul and Tonsor found a significant positive linear relationship, where species age explained 25% of the variation in geographic range. They call for further research that fully evaluates the shape of the age–area relationship (linear, unimodal, etc.) among many taxa so that its underlying causes can be fully elucidated.

Dirzo and Boege (Chapter 5) develop a new conceptual model of plant defense allocation. This model uses contrasting patterns of foliage availability in strongly seasonal tropical dry forests versus tropical rainforests to predict variable patterns of herbivory and defense. Dirzo and Boege point out that past theoretical frameworks failed to consider how seasonality in rainfall creates highly episodic resource availability. This water availability–phenology hypothesis will likely promote the development of additional models designed to predict patterns of herbivory and damage in other community types that have sharply contrasting patterns of resource availability.

Webb, Cannon, and Davies (Chapter 6) use a phylogenetic approach to explain the taxonomic and ecological composition of tropical trees at multiple scales. By examining the evolution of ecological characters among species within a community, Webb *et al.* provide the tools to evaluate the relative roles of the biotic and abiotic mechanisms that together act to filter local species composition. Ultimately they seek models that will predict the taxonomic and ecological composition of tropical forest communities; if successful this will be a huge step forward.

Zimmerman, Thompson, and Brokaw (Chapter 7) bring to bear the power of large forest dynamic plots (while confessing their limitations) to tackle the issue of the relative role of neutral dynamics, negative density dependence (NDD), and gap dynamics in explaining high species diversity in tropical forests around the world. They argue persuasively that NDD is pervasive at these sites, thereby weakening the "value of neutral theory as a general explanation" for high species diversity. They reject a major role for gap dynamics for tall-statured tree species, a view that is now well supported, but which contrasts strongly with views from the 1970s and 1980s.

TESTING THEORIES OF FOREST REGENERATION AND THE MAINTENANCE OF SPECIES DIVERSITY

In an engaging, unique, and wide-ranging chapter in the second major section, Leigh (Chapter 8) addresses the relationship between theory and what we really need to know to understand tropical forests. Leigh weighs in on issues ranging from the limitations of neutral theory to how theory addresses the limits on gross primary production. He tells us "what mathematical theory has done" for tropical ecology and why it remains in relatively "crude" form.

Hubbell (Chapter 9) argues that understanding complex ecological systems, such as tropical forests, can best be accomplished using empirical studies to test simple theoretical models that use few free parameters. Hubbell uses the Neutral Theory to illustrate how simple theoretical models make predictions that are consistent with patterns of tree species abundance and diversity that have been observed in tropical forests worldwide. The chapter begins with an entertaining and candid account of the history and development of the Neutral Theory. Hubbell then provides an excellent overview of the theory, beginning with the model in its most simple terms and subsequently adding complexity. Along the way, he explains the key components of the model and emphasizes their unique attributes and importance. Hubbell concludes with some recommendations for the advancement of ecology, including the value of simple, approximate theoretical models, as well as the need for honesty, not advocacy, in testing theory in ecology.

In Chapter 10, Kitajima and Poorter tackle the concept of "the niche" by evaluating the functional basis of resource specialization of tropical trees. They demonstrate that light is partitioned among tree species at all developmental stages and trade-offs exist between growth, survival, and reproduction for many species. The next logical question then becomes, are there similar trade-offs along other niche axes? If so then there will likely be far more niche opportunities when additional environmental gradients (e.g., fertility and soil moisture) are considered. These findings would appear to challenge the viability of neutral theory.

Muller-Landau (Chapter 11) addresses important trade-offs that are putatively responsible for the maintenance of species diversity. She focuses mostly on the trade-off between competition and colonization, which has garnered much theoretical and empirical attention over the last 50 years. Muller-Landau argues that the available empirical evidence does not support this trade-off for tropical trees, and that other important trade-offs (dispersal–fecundity and tolerance–fecundity) are much more likely candidates to explain plant species coexistence.

Schnitzer, Mascaro, and Carson (Chapter 12) revisit the long-held belief that gaps promote the maintenance of plant species diversity in tropical forests. They concur with Zimmerman *et al.* (Chapter 7) that gaps do not maintain the diversity of tall-statured shade-tolerant trees. Nonetheless, they argue that gaps may be critical for the persistence of lianas, pioneer trees, and small-statured species trapped in the understory, groups that comprise more than 50% of most tropical floras. In addition, nearly all studies have failed to evaluate the degree to which gaps enhance the fecundity of any life-form, including trees – a potentially important oversight.

The Janzen–Connell hypothesis is one of the most widely accepted explanations for the maintenance of species diversity in tropical forests. Carson *et al.* (Chapter 13) compile the available literature to evaluate Janzen–Connell. They conclude that there are many examples of distance- and density-dependent effects on survival, growth, and recruitment. There remains, however, a paucity of evidence that these effects maintain diversity at the community level. Additionally Carson *et al.* argue that falsifying Janzen–Connell is extremely challenging and suggest that the Janzen–Connell effect could be strongest in the least common species.

Dalling and John (Chapter 14) examine the critical role that seed, dispersal, and recruitment limitations play in structuring plant communities. Using simulations based on seed and dispersal traits of pioneers on Barro Colorado Island, Panama, they evaluate whether these limitations minimize competitive interactions, thereby reducing the probability of competitive exclusion. They find that pioneer species appear to be strongly seed limited. However, even the most seed-limited species can become relatively abundant, suggesting that other processes also structure pioneer tree communities.

In Chapter 15, Arnold describes the nascent but increasingly important study of endophytic fungi. To date, endophytes have been found in the photosynthetic tissue of every tropical plant ever examined, and a single tree may harbor thousands of species. The ecological role of endophytes in tropical forests is substantial and complex. Endophytes may act as "environmental acquired immune systems" for plants or bolster a plant's own defense system against pathogens and herbivores. Arnold points out that elucidating the role of natural enemies in structuring plant communities may rest on understanding plant–endophyte–pathogen interactions.

ANIMAL COMMUNITY ECOLOGY AND TROPHIC INTERACTIONS

Although community-level theories in tropical ecology are most commonly tested with plants, they can also be addressed using animals. For instance, in this section, Dyer (Chapter 16) explores the complexity of tritrophic interactions and argues compellingly that the empirical basis for much conventional wisdom within tropical community ecology remains "largely untested." This includes such standards as tropical consumers are more specialized and that predation is more intense in tropical habitats. Dyer reviews the diversity of trophic cascades and identifies the shortcomings of previous studies. He provides a clear roadmap to how future research will need to integrate solid natural history, phylogenetics, modeling, and experimental approaches.

Theimer and Gehring (Chapter 17) examine the tritrophic interaction among terrestrial vertebrates, tree seedlings, and mycorrhizal fungi using a vertebrate exclusion experiment in an Australian tropical forest. They report that after nearly 5 years, vertebrates reduced seedling species richness via increased rates of density-independent mortality, and concomitantly increased arbuscular mycorrhizal fungi richness via spore dispersal. The authors propose a conceptual model to address how these complex opposing but interrelated effects can alter forest community dynamics and diversity.

Terborgh and Feeley (Chapter 18) exploit an excellent model system of predator-free fragmented forests on small islands in Venezuela to explore the role of complex trophic cascades among plants, herbivores, and their predators. Their results "strongly supported the hypothesis of Hairston, Smith, and Slobodkin," which posits that regulation by predators prevents herbivores from decimating plant populations. Nevertheless, Terborgh and Feeley ultimately conclude that trophic cascades are "far more complex than implied by simple [tritrophic] models" and that "plant composition is established and maintained by ... numerous interaction links ... between plants and animals."

Adler (Chapter 19) also examines forest fragments on small islands to evaluate top-down versus bottom-up forces in central Panama. Adler finds that when predators are absent, herbivores (spiny rats) can be resource limited, even in times of resource abundance. He argues that fragmentation will increase conditions where predators are absent, leading to strong trophic cascades. His take-home message: "attempts to categorize herbivore populations as being limited solely by either top-down or bottom-up processes are likely to fail" because both processes operate, but their relative strengths vary seasonally.

Why are arboreal ants the most dominant arthropods of tropical forest canopies in terms of abundance and biomass? Davidson and Cook (Chapter 20) address this and other sizeable questions using the unique approach of ecological stoichiometry, which is the elemental balances (and imbalances) between an organism and its food. The authors use this framework, combined with knowledge of ant digestive anatomy and function, to examine interactions among different ant functional groups and between ants, plants, and trophobionts.

Utilizing an extensive neotropical dataset, Peres (Chapter 21) provides one of the first large-scale tests of the theory that mammalian biomass is directly correlated with soil fertility, which drives plant productivity and food quality (Janzen 1974). Peres's data support this theory and he provides a predictive model for estimating primate biomass, abundance, and diversity along gradients of soil fertility. He then extends the model to other continental vertebrate communities, urging ecologists to continue to link "soil processes to vertebrate populations ... at large spatial scales."

SECONDARY FOREST SUCCESSION, DYNAMICS, AND INVASION

In this section, Peterson and Carson (Chapter 22) review and identify the major constraints on woody species colonization into pastures and call for studies that test broad general hypotheses of species turnover. They find that most temperate models of succession fail to apply in tropical regions because these models place too little emphasis on propagule limitation and facilitation. Nonetheless, they propose that with refinements, and with the addition of quantitative models of dispersal, a temperate model of succession from the 1970s (the nucleation model) may accurately explain early patterns of succession in tropical pastures.

Chazdon (Chapter 23) provides a thorough introduction to and review of succession, with a focus on secondary forests. She applies the stages of succession in the tropics developed by Oliver and Larson (1990) for temperate forests: stand initiation, stem exclusion, understory reinitiation, and old-growth phases. This framework is important because it unites conceptual patterns of succession in the tropics with those found in temperate forests. Chazdon questions the notion that a stable climax would ever be reached, thus forcing us "to view all forests as points along a successional continuum."

Denslow and DeWalt (Chapter 24) examine four likely hypotheses to explain how continental tropical forests resist invasion. Using recently published studies, they conclude that high functional group diversity, high rates of competitive exclusion, and high pest loads may all confer resistance to exotic invasion. Contrary to conventional wisdom, however, high species diversity alone is unlikely to deter invasion. The authors emphasize that the data to test these hypotheses are still relatively weak, and they provide a strategy for future research on this important topic.

TROPICAL FOREST CONSERVATION

The conservation of tropical forests can be promoted by demonstrating their direct value in terms of human services. In this section, Kursar et al. (Chapter 25) outline how basic research can guide bioprospecting, and thus promote tropical forest conservation. The authors show how knowledge of plant species and life-history traits can increase the probability of finding novel active secondary compounds for drug discovery. This exciting approach has resulted in technology transfer, tropical forest conservation, and advance

in combating some of the most devastating human diseases of our time.

Corlett and Primack (Chapter 26) take a global perspective on tropical forest conservation and conclude that there are "many rainforests" and "many threats" and that the conservation of the world's richest ecosystems needs to be a global effort yet reflect clear regional differences. They outline how threats vary among the world's major forests: Asia, Africa, Madagascar, New Guinea, Central and South America, Australia, and island rainforests. They conclude that "the single most important strategy for protecting intact rainforest communities is to establish – and effectively manage – protected areas."

While Corlett and Primack took a worldwide focus, Laurance (Chapter 27) hones in on threats and promise for conservation in the Amazon. The outlook is sometimes bleak; in a reference to the biblical book of Revelation, Laurance argues that the four horsemen of the future tropical apocalypse will be uncontrolled agriculture, logging, wildfires, and widespread fragmentation. Annual deforestation is staggering and additional threats include burgeoning immigration and massive economic development. Thankfully, Laurance finds a silver lining in the figurative dark cloud hanging over tropical forest conservation. He suggests that this is also a time of "unparalleled opportunity for conservation" due to expanding networks of reserves, corridors, and other conservation units.

Putz and Zuidema (Chapter 28) argue passionately that ecologists need to examine conservation within a much broader social, economic, and political context. They suggest that in many instances (though not all) "expertocratic approaches," such as creating a system of walled-off protected reserves, are inappropriate to local, cultural, and political realities. The authors call for expanded research into processes that promote and maintain biodiversity in human-altered landscapes that vary in size from small fragments to large plantations. Are we as ecologists sequestering ourselves away in pristine forests while secondary forests are starved of inquiry?

Overall, we believe that these chapters serve to not only synthesize the current state of knowledge on the ecology of tropical forest communities, but they also point out some of the long-standing but yet unresolved issues in the field. We hope that this volume stimulates additional research in those critical areas.

REFERENCES

Janzen, D.H. (1974) Tropical blackwater rivers, animals, and mast fruiting by the Dipterocarpaceae. *Biotropica* 6, 69–103.

Oliver, C.D. and Larson, B.C. (1990) *Forest Stand Dynamics*. McGraw-Hill, Inc., New York.

Schemske, D.W. (2002) Ecological and evolutionary perspectives on the origins of tropical diversity. In R. Chazdon and T. Whitmore (eds), *Foundations of Tropical Forest Biology: Classic Papers with Commentaries*. University of Chicago Press, Chicago, IL, pp. 163–173.

Willis, J.C. (1922) *Age and Area*. Cambridge University Press, Cambridge.

LARGE-SCALE PATTERNS IN TROPICAL COMMUNITIES

SPATIAL VARIATION IN TREE SPECIES COMPOSITION ACROSS TROPICAL FORESTS: PATTERN AND PROCESS

Jérôme Chave

OVERVIEW

Understanding the causes of spatial variation in floristic composition is one of the overarching goals of plant ecology. This goal has been challenged by the difficulty of unfolding the spatial component of biodiversity, and of interpreting it biologically, especially in the tropics. Hence until recently, virtually nothing was known about the real impact of land-use change on tropical biodiversity, in spite of the rapid rates of tropical deforestation and habitat loss. This picture has changed dramatically over the past few years, with the development of large-scale inventory projects and the implementation of methods for quantitative analysis of floristic data. Here, I provide an overview of the definitions of spatial floristic turnover, or beta-diversity, and a statistical toolkit for the analysis of beta-diversity. I also contrast ecological theories which underlie the statistical tests. I then review recent empirical studies on plant beta-diversity in tropical forests. This panorama shows that a consensus on field and analytical methods is now being reached. There is a need for careful reinterpretations of published ecological patterns in light of well-formulated ecological hypotheses. Only through ambitious field studies and collaborative approaches will further progress be achieved in this fascinating research area.

INTRODUCTION

If the traveller notices a particular species and wishes to find more like it, he may often turn his eyes in vain in every direction. Trees of varied forms, dimensions and colours are around him, but he rarely sees any one of them repeated. Time after time he goes towards a tree which looks like the one he seeks, but a closer examination proves it to be distinct. He may at length, perhaps, meet with a second specimen half a mile off, or may fail altogether, till on another occasion he stumbles on one by accident.

Ever since Wallace's (1895) description of the diversity in tropical tree species, this outstanding variety of form has been regarded as a curiosity and a scientific challenge. Over the past decade, record-setting levels of tree diversity have been reported, where, in one hectare of old-growth forest, every other tree represents a new species (Valencia *et al.* 1994, de Oliveira and Mori 1999). Recent diversity mapping projects have also demonstrated the great variability of tree species richness in the Amazon (ter Steege *et al.* 2003) and in Borneo (Slik *et al.* 2003). What explains these striking floristic changes in tropical forests at both local and regional scales? To answer this question, it is necessary to examine changes in biological diversity at all scales, for they are caused by processes that, themselves, operate at various spatial scales (Levin 1992, Huston 1999, Mouquet and Loreau 2003, Ricklefs 2004). Historically, however, tropical plant ecology has focused almost exclusively on the mechanisms of local species coexistence, placing less emphasis on intermediate-scale patterns of diversity.

Our limited knowledge of the scaling properties of biodiversity is partly due to the scarcity of datasets available to quantify these patterns. This, of course, is a result of the difficulty of gathering large-scale and consistent diversity data in species-rich ecological communities (Ashton 1964, Gentry 1982). A second cause for the limited interest in documenting patterns of spatial species turnover is the complexity of statistical measurement procedures, and also the lack of a consistent theoretical framework for testing ecological hypotheses. While phytosociologists predicted that dispersal, together with biotic and abiotic factors, played a prominent role in spatial plant turnover (Braun-Blanquet 1932), they lacked statistical approaches to test appropriate biological hypotheses. A number of studies have contributed to the recent revival of interest in tree beta-diversity patterns (Tuomisto and Ruokolainen 1994, Duivenvoorden 1995, Tuomisto *et al.* 1995, Terborgh *et al.* 1996, Hubbell 1997, Ruokolainen *et al.* 1997, Pitman *et al.* 1999, Condit *et al.* 2002), and beyond (McKnight *et al.* 2007, Woodcock *et al.* 2007). By using networks of plots established in several neotropical forests, these studies tested the theoretical expectation that among-plot species similarity might be predicted by abiotic environment or that it should decrease predictably with between-site geographical distance. Their effort benefited from improvements in methods to analyse spatial turnover in diversity. For example, rapid and normalized techniques for chemical soil analysis are now available, with modern instruments such as inductively coupled plasma-mass spectrometers (Lucas *et al.* 1993, Clinebell *et al.* 1995) now being routinely used to measure the concentration of macro- and micronutrients in the soil. Major advances have also been made in the long-term, cross-scale, prediction of climatic variables (New *et al.* 2002, Hijmans *et al.* 2005, Flikkema *et al.* 2006), in topography (worldwide mapping at a 90 m resolution from the Shuttle

Radar Topography Mission), and in mapping spatial envelopes by remote sensing (Tuomisto 1998, Clark *et al.* 2004).

The conjunction of these conceptual advances, technological progress, new empirical work, and social demand make this field a very exciting one, and recent achievements are evidence for this claim. My goal here is to convey the message that even more remains ahead of us. I will review available tools for measuring spatial variation in floristic diversity, and statistical and dynamic models. I will then discuss evidence for and against the role of environmental variation in predicting beta-diversity. New methods are available to partition beta-diversity into deterministic and stochastic processes, and these approaches should be used more consistently across a broad array of tropical forest landscapes.

DOCUMENTING PATTERNS OF SPATIAL VARIATION IN SPECIES DIVERSITY

There has been a tremendous wealth of research on the statistical measurement of biodiversity. General discussions of these techniques can be found in Pielou (1975), Engen (1978), Gaston (1994), Colwell and Coddington (1994), Krebs (1999), and Magurran (2004). I restrict the present section to a selective introduction of common measurements of diversity across scales in the context of tropical tree communities.

Whittaker (1960, 1967, 1972) pioneered the study of spatial diversity. He offered a conceptual spatial diversity partitioning scheme by distinguishing four scales at which diversity could be measured: samples (point diversity), habitats (alpha-diversity), landscapes (gamma-diversity), and biogeographic provinces (epsilon-diversity).

Although intuitively appealing, this classification has been interpreted differently among different authors. For instance, point diversity and alpha-diversity are often confused, given the difficulty of delimiting objectively habitats for plants. Further, many measures of diversity depend on sampling effort (they are "biased"). This makes it difficult to compare sampling units of unequal size. To resolve this problem, one may choose to compare only equal-sized subsamples, a method called "rarefaction" (Hurlbert 1971), or to assume an underlying species abundance distribution. For instance, if the species abundance distribution follows Fisher's logseries, then an unbiased index of alpha-diversity is Fisher's α. This assumption has been tested in several tropical tree communities (Condit et al. 1996), but it would be interesting to test it further in other forests.

Many biological questions relate to species turnover, or changes in species composition from one community to another, rather than just local diversity as defined above. In such cases, one can define a relationship between alpha- and gamma-diversities, coined beta-diversity by Whittaker (1972). Beta-diversity is useful for studying ecological processes such as habitat specialization and dispersal limitation, but also large-scale patterns of abundance, rarity, and endemism. Two extreme cases may occur: alpha-diversity may be much smaller than gamma-diversity, when most species are spatially clumped; in this case, beta-diversity is large. Conversely, alpha-diversity may be on the same order as gamma-diversity, in which case most species would be represented in any local sampling of the region, and beta-diversity would be low. More precisely, Whittaker (1972) defined beta-diversity as the ratio of gamma-diversity over alpha-diversity:

$$\beta^W = \frac{D_\gamma}{D_\alpha} \tag{2.1}$$

where D_α and D_γ are the expected species diversities at local and regional scales, respectively. A statement equivalent to Equation (2.1) is that gamma-diversity is equal to the product of alpha- and beta-diversity, that is, there exists a multiplicative partition of gamma-diversity into a strict local contribution and a spatial turnover contribution. Lande (1996) pointed out that an additive partition of diversity into alpha- and beta-diversity is more natural than Whittaker's multiplicative partition. He defined beta-diversity as the difference of gamma-diversity minus alpha-diversity:

$$\beta^L = D_\gamma - D_\alpha \tag{2.2}$$

This additive partitioning scheme simply results from the exact definition of the diversity indices.

Lande (1996) chose to define local diversity as the probability of two random chosen individuals to belong to different species, a quantity also known as Simpson diversity. Gamma diversity may be defined using exactly the same statistical interpretation but at the regional scale. Because probabilities are additive, the scheme of Equation (2.2) make more sense than that of Equation (2.1). For further details on this additive diversity partitioning scheme, the reader is referred to Lande (1996), Crist et al. (2003), and Jost (2006).

Wilson and Shmida (1984) reported six different measures of beta-diversity based on presence/absence data. More recently, Koleff et al. (2003) reported a literature search of 60 papers quantifying beta-diversity, in which they found no fewer than 24 different beta-diversity measures based on presence/absence data. While all of these measures are increasing functions of the number of shared species, as intuitively expected for a measure of species overlap, their mathematical behavior differs broadly, and this contributes to obscuration of the discussion on patterns of beta-diversity. Wilson and Shmida (1984) and Koleff et al. (2003) proposed a terminology for these indices (which I here follow), together with a useful interpretation in terms of two sampling units with overlapping species lists. If a and b are the number of species restricted to samples 1 and 2 respectively, and c the number of shared species (Krebs 1999, Koleff et al. 2003; Figure 2.1), then the total number of species is $D_\gamma = a + b + c$, and the local diversity can be defined as the average number of species in the two samples: $D_\alpha = (a + c)/2 + (b + c)/2$. Thus it is easy to see that $\beta^L = (a + b)/2$. Many other overlap measures have been used in the literature, but I shall mention just two. The Sørensen index $\beta^{Sørensen}$ is the number of shared species c divided by the average number of species in the two samples: $\beta^{Sørensen} = 2c/(a + b + 2c)$. The Jaccard index $\beta^{Jaccard}$ is the number of shared species divided by the total number of species: $\beta^{Jaccard} = c/(a + b + c)$. Evidently this generalizes to more than two sites; beta-diversity indices are then defined for any pair of sites, $\beta_{i,j}$. The diagonal terms of this diversity matrix compare any plot with itself, so that, for instance, $\beta_{i,i}^W = 1$ and $\beta_{i,i}^L = 0$.

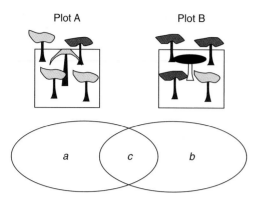

Plot A   Plot B

Figure 2.1   Various ways of measuring the species overlap between two sites, plot A and plot B. In this example, each plot contains five individuals and three species, two of which are shared between the two plots (defined as c in the main text). The number of species only in plot A, a, is equal to one, and the number of species only in plot B, b, is also one. Thus, for instance, the Sørensen index is $\beta^{\text{Sørensen}} = 2/3$, and the Jaccard index is $\beta^{\text{Jaccard}} = 2/5$.

Koleff *et al.*'s approach provides a consistent framework for comparing previously published measures of species overlap. However, it veils a number of sampling issues: the true number of species in a landscape is usually larger than $a + b + c$ unless the two samples are very large and effectively contain most species (Colwell and Coddington 1994). In addition, alpha-diversity tends to be underestimated. A simple argument can be used to estimate the influence of these biases on the calculation of beta-diversity indices. Species accumulation curves suggest that for larger samples, species richness is less underestimated than for smaller ones, hence D_γ should in general be less underestimated than D_α. Thus, both indices defined in Equations (2.1) and (2.2) should be overestimated. Chao *et al.* (2000) devised statistically unbiased estimates for species overlap between two communities, when sampling is accounted for (see also Magurran 2004). This approach has seldom been used in tropical plant ecology (but see Chazdon *et al.* 1998). Plotkin and Muller-Landau (2002) addressed the closely related question of how well similarity indices

are estimated given that only small fractions of the total landscape can be sampled, and given that one knows the local species abundance distributions and aggregation patterns. They developed an exact formula for the expected Sørensen index, and provided methods for estimating the model's parameters. Much of this recent work remains to be introduced into the community ecologist's toolbox, and freely available statistical software may help serve this goal (e.g., EstimateS, http://viceroy.eeb.uconn.edu/EstimateS, developed by R.K. Colwell).

Another approach for estimating beta-diversity is based on species abundance and produces measures that are generally less biased than those based on presence/absence data. Local diversity may be measured as the probability that two individuals taken at random from the community belong to different species, a quantity known in ecology as the Simpson index. Among-site overlap may be defined as the probability that two individuals taken from two communities belong to different species. Such a measure of species overlap has been used in the literature (Wolda 1981, Leigh *et al.* 1993, Chave and Leigh 2002) and is similar to the Morisita–Horn index. Defining N_{ik} as the number of individuals of species i in site k, then $x_{ik} = N_{ik}/\sum_i N_{ik}$ is the relative abundance of species i in site k. The probability that two individuals, one from site k, the other from site l, both belong to species i is $x_{ik}x_{il}$, hence the probability that the two individuals belong to different species is

$$\bar{D}_{kl} = 1 - \sum_i x_{ik} x_{il} \qquad (2.3)$$

These indices cannot be simply deduced from species numbers a, b, c as above. Also, this measure places the emphasis on abundant species rather than on rare species and is asymptotically unbiased for large samples. This measure of diversity has a simple probabilistic interpretation, and it is formally equivalent to a universally used measure of local and spatial diversity in the closely related discipline of population genetics, which leads to formulas for the additive partitioning between local and landscape diversity. Nei (1973) showed that a measure of diversity between two populations, excluding intra-population diversity, would be (see also Lande 1996):

$$\beta_{k,l}^{\text{Nei}} = \overline{D}_{kl} - \frac{\overline{D}_{kk} + \overline{D}_{ll}}{2} = \frac{1}{2} \sum_{i=1}^{S} \left(x_{ki} - x_{li} \right)^2$$

(2.4)

This measure of strict beta-diversity is proportional to the squared Euclidean distance between site i and site k, a quantity already used in the ecological literature (Ricklefs and Lau 1980). The total diversity across a total of K populations can then be defined as

$$\overline{D}_T = \frac{1}{K} \sum_{k=1}^{K} \overline{D}_{kk} + \frac{1}{K^2} \sum_{k=1}^{K} \sum_{l=1}^{K} \beta_{k,l}^{\text{Nei}}$$

The first term is the contribution of local diversity, while the second term is the contribution of strict beta-diversity (Nei 1987, Lande 1996, Chave et al. 2007).

The Steinhaus index provides an index of beta-diversity that is based on species abundance. This index was also called the Renkonen index or the complement of the Bray–Curtis index (i.e., Steinhaus index equals one minus the Bray–Curtis index). Note that historically, the first reference to this index was due to O. Renkonen and it would be more appropriately named after this author (Renkonen 1938, Plotkin and Muller-Landau 2002). This index of similarity between sites k and l reads:

$$\beta_{k,l}^{\text{Steinhaus}} = \frac{\sum_{i=1}^{S} \min \left(N_{ki}, N_{li} \right)}{\left(N_k + N_l \right)/2}$$

where $N_k = \sum_{i=1}^{S} N_{ki}$ is the total number of individuals in sample k. The corresponding index of diversity would then be $1 - \beta_{k,l}^{\text{Steinhaus}}$. The Steinhaus index of similarity can be rewritten approximately in terms of relative abundances where the sample sizes are not too dissimilar: $\beta_{k,l}^{\text{Steinhaus}} \approx \sum_{i} \min(x_{ki}, x_{li})$. The Nei index is more intuitive than the Steinhaus index, because a probabilistic interpretation of the latter is less obvious. Recently, Green and Plotkin (2007) have offered a sampling theory for betadiversity

including species abundance, thereby generalizing results of Plotkin and Muller-Landau (2002).

Species are but one way of measuring biodiversity. This implicitly assumes that all species are independent units. Any evolutionary biologist knows that this is far from true: species are organized according to a definite structure, and this structure is defined by their evolutionary history. Indeed, two species in the same genus tend to share a larger amount of evolutionary history than two species in different genera. Pavoine et al. (2005) called the amount of unshared evolutionary history of a species, its 'originality', and Nee and May (1997) and Purvis et al. (2000) discussed this effect in light on conservation biology, as a means for evaluating the potential loss of evolutionary history link to species extinction. Based on this reasoning, Faith (1992) proposed to measure the biological diversity of a species assemblage by the amount of evolutionary history in this assemblage. If a dated phylogenetic hypothesis is available for a given species assemblage, then one may implement Faith's biodiversity index by measuring the total branch length in the phylogenetic tree, measured in millions of years. One recent example of measuring the 'phylogenetic' diversity of plant species assemblage is due to Forest et al. (2007). Nei's (1973) measure of local diversity based on a probabilistic interpretation can also be generalized to account for the amount of shared evolutionary history among species. This fact has been formalized mathematically by Rao (1982), but it is only recently that it has received some further scrutiny (Pavoine et al. 2005, Chave et al. 2007).

SEARCHING FOR ENVIRONMENTAL CORRELATES OF SPATIAL VARIATION IN DIVERSITY

Plant ecologists have long sought to predict species occurrence from environmental characteristics, both soil and climate (Warming 1909, Braun-Blanquet 1932, Whittaker 1956). There is a vast literature reporting correlations between floristic variables and environmental or geographical variables using a large number of different statistical methods. Unfortunately, these methods rely heavily on complex statistical concepts and it is very easy to get lost along the way. Here I try to avoid technicalities, while pointing the interested reader towards the relevant literature.

The most intuitive approach in performing species association analyses is where the abundance or occurrence of one species is correlated with environmental descriptors, independently of all other species in the community. This approach has been used largely for temperate plants. Because tropical tree species are usually rare and infrequent, however, analyses for tropical tree assemblages often lack statistical power and can be applied only to abundant species (Newbery and Proctor 1984, Baillie *et al.* 1987, Swaine 1996, Pitman *et al.* 1999, Svenning 1999, Clark *et al.* 1999a, Webb and Peart 2000, Pyke *et al.* 2001, Phillips *et al.* 2003, Svenning *et al.* 2004).

This omission for rare species, which compose the majority of species in most tropical forests, may bias species association analyses because rare species may behave very differently from abundant ones (Condit *et al.* 2000).

A measure of environmental dissimilarity may be defined by the absolute value of the difference in the environmental variable between sampling units (e.g., rainfall, nutrient concentration). Before applying many of these statistical methods, it is important to make sure that environmental dissimilarity data are normally distributed, so a non-linear transform may be advisable (Phillips *et al.* 2003, Tuomisto *et al.* 2003a, cf. Figure 2.2). The environmental variable may also be qualitative: soil types may be classified as fertile or infertile, dry or wet (Swaine 1996), well drained or poorly drained, sandy or

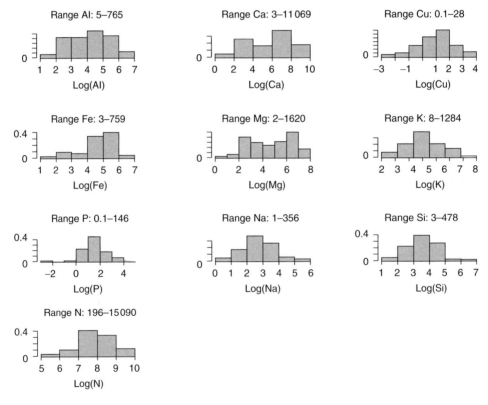

Figure 2.2 Histograms of the log-transformed soil chemical concentrations for 69 plots reported in Gentry (1988) (see Clinebell *et al.* 1995). All variables are measured in ppm (or $mg\,kg^{-1}$).

clayey (Sabatier *et al.* 1997, Clark *et al.* 1999a), ridge, valley, or mid-slope (Harms *et al.* 2001).

The full environmental variability among several sites is described by a collection of environmental distance matrices. Correlations are then directly performed between the floristic distance matrix, as defined in Equation (2.4) for instance, and environmental dissimilarity matrices. Alternatively, several environmental dissimilarity measures can be combined into one compound variable, such as Gower's index (Gower 1971, Legendre and Legendre 1998, Potts *et al.* 2002). Another important predictive variable when comparing the floristics of two sampling units is their geographical distance. As we shall see below, a correlation between diversity and geographical distance should be consistent with a prominent role of dispersal limitation. In the case of a fragmented habitat, it may be advisable to use an "effective distance" rather than the straight-line distance among sites, a distance that takes into account the habitat heterogeneity among plots.

Ordination methods provide a powerful framework for detecting environmental correlations, and illustrating them graphically (ter Braak 1987). These methods enable the visualization of beta-diversity over several sites, on a two-dimensional plot. Observed patterns are qualitatively interpreted in terms of the environmental variables (Whittaker's 1967 indirect gradient analysis), or are directly regressed against these environmental variables (direct gradient analysis). The ordination axes often represent either linear combinations of species abundance data or linear combinations of environmental variables. A disadvantage of ordination is that it is difficult to provide a biological interpretation of these axes. Another disadvantage of ordination is that it is difficult to grasp the relative merit of the numerous existing ordination methods. As a result, most users are led to treat statistical techniques as cooking recipes rather than intuitively interpretable statistical methods, and consequently do not make full use of the data, or interpret these methods inappropriately (for an illustration, see the controversy between Legendre *et al.* 2005 and Tuomisto and Ruokolainen 2006). The different ordination techniques are compared

and contrasted by, for example, ter Braak (1987), ter Braak and Prentice (1988), and Legendre and Legendre (1998).

Floristic diversity may also be modeled by multiple regressions on environmental dissimilarity and geographical distance matrices (Borcard *et al.* 1992, Legendre and Legendre 1998, Ohman and Spies 1998). The significance of correlations is then measured using Mantel tests for simple regressions, or partial Mantel tests for multiple regressions. This approach has been widely used in the recent literature (Potts *et al.* 2002, Phillips *et al.* 2003, Tuomisto *et al.* 2003a), but it has also yielded a remarkable amount of controversy over which an appropriate method should be used to partition the causes of variation of diversity into geographical and environmental distance. Legendre *et al.* (2005) reviewed these methods and listed the potential pitfalls related to them. In particular, they emphasized that partitioning on distance matrices should not be used to study the variation in community composition among sites, because the variance of a dissimilarity matrix among sites cannot be interpreted as a measure of beta diversity. The direct result of using this method is that the amount of explained variation is underestimated, and tests of significance had less power than the tests associated with the canonical ordination method. Tuomisto and Ruokolainen (2006) contended that both approaches have merits, and especially that methods based on distance matrices were more appropriate to test the neutral theory of biodiversity (Hubbell 2001). On this controversy, it should be simply stated that (1) Legendre *et al.* (2005) provide compelling and indisputable evidence for why ordination methods are superior to distance methods, (2) in contrast with Tuomisto and Ruokolainen (2006)'s main statement, ordination methods can be used to test the neutral theory (Chust *et al.* 2006), (3) given that many published results have already made use of distance matrices, it should be possible to use these results and gain biological insight, but these results should not be over-interpreted, (4) in any case, users are advised at least to publish analyses based on correlation of distance matrices together with analyses based on ordination methods. I will further discuss at length these approaches below, as they provide an evaluation of the relative importance of dispersal and of environment in the shaping of beta-diversity patterns.

UNDERSTANDING THE CAUSES OF SPATIAL VARIATION IN SPECIES DIVERSITY

The goal of community ecology is to explain patterns, not just to document them. Theoreticians have now developed models that can be used to derive predictions for beta-diversity patterns. Hubbell (2001) provided a good overview of these theoretical attempts and classified them into two groups, referred to as niche-assembly theories and dispersal-assembly theories (Figure 2.3). Before delving into the neutral theory's predictions of beta-diversity patterns, I shall first discuss the more traditional theory, that of habitat specialization, and its predictions for patterns of beta-diversity.

Niche-assembly theories

Niche-assembly theories posit that species are distributed not randomly but as a result of environmental constraints and competitive displacement. Two types of niche-related mechanisms may act upon a species and determine its presence at a given site, namely a *physiological filter* and a *biotic filter*, a distinction that is far from new even in the tropical ecology literature (see questions (a) and (c) in Poore 1968, p. 144).

By physiological filter (or "stress," *sensu* Grime 1977), I mean that certain environmental features prevent the establishment of plants that do not present specific adaptations to the chemical composition of soils, water availability, or light availability. Soils may contain metals toxic for certain plants, either naturally in acidic soils or as a result of human contamination (aluminum, lead, cf. Baker 1987). Water availability, including length of the dry season, is another crucial environmental feature for plant species vulnerable to embolism and cavitation in the xylem water column (Tyree and Sperry 1989, Engelbrecht and Kursar 2003). Other species may be unable to establish in the anoxic soils prevailing in periodically flooded forests or peat swamps, where other plants thrive (Ashton 1964, Webb 1969, Newbery *et al.* 1986, Tuomisto and Ruokolainen 1994). Finally, plants may develop protections against herbivores that are specific to the habitats

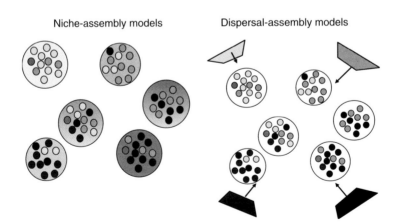

Niche-assembly models   Dispersal-assembly models

Figure 2.3 Theoretical distributions of plant species. Four species are represented and they occupy six patches. According to the niche-assembly theory (left) the distribution of the species is mostly due to the species' preference for environmental conditions (the patterning of the patch matches the pattern of the locally most adapted species). According to the dispersal-assembly theory (right) the environment is homogeneous across patches, but species may have dispersed from different regions, which is reflected in their current local distribution.

they live in, a mechanism that would indirectly lead to habitat specialization (Fine *et al.* 2004). In all these cases, species that have not developed particular adaptations to cope with the specific environmental conditions will be unable to establish a population.

The biotic filter operates through interspecific competition for available resource. In the resource use theory developed by Tilman (1980, 1982), individuals of each species immobilize resources, for instance nutrients. One species may draw these resources to below a level at which potential competitors would survive (Tilman's $R*$ theory). This process of resource-driven competitive exclusion is formally equivalent to that predicted by the Lotka–Volterra competition theory (Levin 1970). This theory is compatible with the assumption that soil environments are a major cause of the landscape-scale variation in plant distributions (Ashton 1964, 1976). However, according to Tilman's theory, the presence of a species in a landscape is not fully determined by its environmental requirements: the failure to encounter the species locally may be attributable to either competitive displacement or sampling limitation. Thus, environmental variation may determine the spatial contours of the *fundamental* niche for any species, but their *realized* niche is a smaller area that also depends on competitive displacement by other species. One important limitation of this theory is that it assumes perfect mixing in resources, a reasonable assumption in marine environments, but likely incorrect in terrestrial environments (Huston and DeAngelis 1994, Loreau 1998). In the presence of limited resource mixing, the competitive exclusion principle holds only in a small area around each individual, and the biological filter has only a limited influence on the presence/absence of a species.

What predictions do niche-assembly theories make about beta-diversity patterns? The capacity of a species to exclude other species from its niche depends on its status within a competitive hierarchy. At a very small scale (micro-sites), there is one and only one competitively dominant species, provided it is present in the local community. If such a clear competitive hierarchy among species does exist, two environmentally similar sites will tend to be floristically similar. More precisely, the

higher the environmental dissimilarity across the landscape, the higher the beta-diversity index, a statement that can be tested with standard statistical approaches. However, this theory makes no prediction regarding the quantitative relationship between biological and environmental diversities.

Dispersal-assembly theories

Dispersal-assembly theories place an emphasis on demographic and seed dispersal processes. The presence/absence of a species within a landscape is due not to its preference for a specific environmental condition, but to the ability of the plant to reach maturity at one site, having dispersed from another one. Moreover, few of the produced seeds are dispersed, and few of the dispersed seeds will ever produce a mature plant. The fundamental premise of this theory is therefore that the limited ability of species to colonize remote habitats should be a major explanatory factor for their heterogeneous distributional patterns. Species should then be relatively insensitive to environmental differences, so long as the resources are not too limiting. Soil and climate maps show that huge expanses of tropical land are relatively homogeneous, and covered by forests growing typically on oxisols, with rainfall of more than 1500 mm year^{-1}, and experiencing a dry season of less than 3 months. Under such conditions, most plant species should be able to persist in tropical forests irrespective of the abiotic environment. Wong and Whitmore (1970) made a similar proposal in their study of tree species distributions in a rainforest in Peninsular Malaysia. To examine how well a model with no environmental constraints could fit to biodiversity patterns, Hubbell (1979) and Hubbell and Foster (1986) constructed a neutral theory by assuming that all individuals in all species have the same prospects of reproduction and death, irrespective of the environment they grow in. They also assumed that seed rain is not homogeneous at the landscape scale, but clustered around the parent plants.

Hubbell's neutral theory makes a number of quantitative predictions about biodiversity patterns. Chave and Leigh (2002) produced a spatially explicit version of the neutral theory for beta-diversity of tropical trees using theoretical results from population genetics (Malécot 1948). Let us consider that a community is saturated with ρ individuals per unit of area, such that any dying individual is immediately replaced by a young individual, not necessarily of the same species. New species may appear in the system as a result of point-wise speciation or long-distance dispersal (immigration from outside the community). Thus, a dying individual is replaced at rate ν by an individual belonging to a species not yet represented in the landscape. Recruitment will occur through seed dispersal from existing individuals to neighboring sites, some of which may be empty. The probability that a seed falls r meters away from its parent is defined as $P(r)$, and is commonly called a *dispersal kernel* in the theoretical literature (Kot *et al.* 1996, Clark *et al.* 1999b, Chave *et al.* 2002). A crucial feature of this model is dispersal limitation, that is, seeds are more likely to fall close to the parent than far from it. The dispersal kernel may take a broad array of mathematical forms. The Gaussian dispersal kernel is defined by $P(r) = (1/(2\pi\sigma^2)) \exp(-r^2/\sigma^2)$, hence the variance σ^2 is the only parameter of the dispersal model. In general the Gaussian dispersal function provides a poor fit to empirical data (Clark *et al.* 1999b, Nathan and Muller-Landau 2000). The so-called "2Dt" dispersal kernel (Clark *et al.* 1999b), defined as

$$P(r) = \frac{p}{\pi u \left(1 + r^2/u\right)^{p+1}} \qquad (2.5)$$

provides a better fit (this kernel is parametrized by p and u, both positive). The 2Dt kernel is similar to a Gaussian dispersal function if both u and p become large, such that $2\sigma^2 = u/p$. For r^2 larger than u, elementary calculations on Equation (2.5) show that the dispersal function is approximately equivalent to a power law: $P(r) \approx p u^p/\pi r^{2p+2} \sim 1/r^{2p+2}$. Hubbell (2003) proposed the use of another "fat-tailed" class of dispersal kernels, namely Lévy-stable dispersal kernels (for a technical definition, see Gnedenko and Kolmogorov 1954). Lévy-stable dispersal kernels have not been used in the literature because they are difficult to manipulate both mathematically and numerically. They behave as power laws at large values of r, but so does the 2Dt dispersal kernel. Results obtained

with the 2Dt dispersal model should therefore be qualitatively similar to those obtained with any dispersal kernel with a power-law tail, and the mathematically more tractable kernel should always be preferred.

In a spatially structured neutral model, it is possible to find how the species overlap index in Equation (2.3) varies with the geographical distance between pairs of sites. Chave and Leigh (2002) call $F(r)$ the similarity function, the probability that two individuals taken from two different sites belong to the same species. Since the only parameters in the Gaussian neutral model are the speciation/immigration rate ν, the dispersal parameter σ, and the density of individuals ρ, the similarity function can be exactly expressed as a function of ν, σ, and ρ (for an exact expression, see Chave and Leigh 2002). A simple approximation is

$$F(r) \approx -\frac{2}{2\pi\rho\sigma^2 + \ln(1/\nu)} \ln\left(\frac{r\sqrt{2\nu}}{\sigma}\right) \qquad (2.6)$$

The similarity function decreases logarithmically with increasing distance, and this approximation is valid as long as $1 \le r/\sigma \ll 1/\sqrt{2\nu}$. Assuming values of 50 m for σ and 10^{-8} for ν (Condit *et al.* 2002), the range of validity of this equation would be $50 \le r \ll 7000$, in meters. Two remarks should be made at this point. First, Equation (2.6) is not valid in the range $r/\sigma \ge 1/\sqrt{2\nu}$, and it should be replaced by an exponentially decreasing function (Chave and Leigh 2002). Large-scale analyses confirm an exponentially decaying pattern at larger scales (Nekola and White 1999, Qian *et al.* 2005). Second, in the case of a 2Dt kernel, the similarity function is parametrized by u and p, rather than σ only. Here again an exact formula and useful approximations are available (Chave and Leigh 2002). For instance, it can be shown that for large r, $F(r) \sim 1/r^{2p}$.

TESTING THEORIES

Landscape-scale patterns of tree diversity

The confrontation of niche-based and dispersal-based theories in community ecology has turned into an active field of research since the mid-1990s. Terborgh *et al.*'s (1996) test of Hubbell's (1979) non-equilibrium hypothesis for tropical forests played an important role in this progress. Unlike previous works, Terborgh *et al.* (1996) addressed the relevance of the non-equilibrium hypothesis at the landscape scale, not just locally as had previously been done. They identified all trees greater than 10 cm diameter in several plots along the río Manu, southeast Peru, and showed that a few species were both abundant and widespread (i.e., "oligarchic species," *sensu* Pitman *et al.* 1999). Thus Hubbell's non-equilibrium theory, even if valid at the local scale, cannot be true at a larger scale. Hubbell (1997) then extended his local neutral model to a "two-scale" model where the local community interacts with a regional species pool in which diversity is maintained by evolutionary processes (see also Hubbell 2001, 2003, Ricklefs 2003, Chave 2004). Thus, by allowing for predictions at a larger scale, the neutral theory was significantly improved. Hubbell's (1997) theory viewed each local community as one sample of a regional species pool, hence allowing for regionally abundant species to be abundant in local samples. Hubbell (2001) further expanded this idea, showing that the mechanism of seed dispersal limitation can be included in a neutral theory and may contribute to explain the decrease of species similarity with distance across a landscape. This aspect was formalized in greater mathematical detail by Chave and Leigh (2002).

Condit *et al.* (2002) brought theory together with empirical data using permanent tree plot censuses previously assembled by three independent research groups (Terborgh *et al.* 1996 in Peru, Pitman *et al.* 2001 in Ecuador, Pyke *et al.* 2001 in Panama). More precisely, they tested one of the predictions of Chave and Leigh's (2002) model described above, that the similarity function $F(r)$ should decrease logarithmically with increasing distance (Equation (2.6)). Using simple regression approaches, Condit *et al.* found that a logarithmic function indeed provided a correct fit for $F(r)$ across several decades for r, from 0.1 to 100 km (Figure 2.4). They were also able to estimate the value of the model's parameters, and found values of the dispersal parameter σ of about 50 m, consistent with known estimates for the dispersal of tropical trees. The value for

the speciation rate ν was also estimated, but with rather poor accuracy (on the order of 10^{-14} to 10^{-8}). This was the first explicit test of the influence of distance in the shaping of tree communities at the landscape scale (see also Hardy and Sonké 2004 for a similar analysis in the tropical forests of Cameroon).

Interestingly, the Chave and Leigh model failed to fit the observations at both short and large distances. The large-distance discrepancy can be easily explained by the fact that Equation 2.6 is only an approximation of the exact formula predicted in Chave and Leigh (2002). The short-distance discrepancy is more troublesome: in most cases the similarity at short distance is far greater than that predicted by the neutral model. At least two, non-mutually exclusive, explanations can be provided for this discrepancy. First, the neutral model analyzed by Chave and Leigh (2002) assumes a Gaussian dispersal kernel which is convenient but may not provide the best fit of real dispersal kernel at short distance. Second, other biological processes may be acting at this local scale, for example density-dependent factors (T. Zillio *et al.* unpublished results, and S.P. Hubbell personal communication) or habitat specialization. It is premature to decide which explanation is the more likely, and more research on this topic is necessary.

Another surprising feature of the Condit *et al.* (2002) study is the striking difference in the similarity functions in the large plot data from Panama and western Amazon (Peru and Ecuador). The Panamanian similarity function decreased steadily with increasing logarithmic distance. In contrast, the western Amazon similarity function decreased rapidly for distances less than 100 m, and much more slowly at larger distances. One interpretation for this pattern is that the histories of the two forests are different: tree species may have had more time to disperse in the western Amazon than in Panama (Condit *et al.* 2002).

Partitioning the effects of dispersal and environment

Duivenvoorden *et al.* (2002) pointed out that the differences between Panama and the western

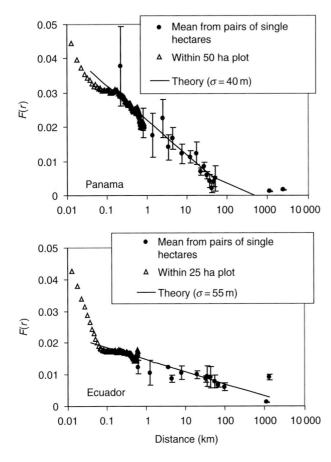

Figure 2.4 The probability F that randomly selected pairs of trees are the same species, as a function of distance r, on a semilogarithmic scale, in Panama (top) and Ecuador (bottom), and a best fit of the dispersal model to the data for $r > 100$ m. Modified from Condit *et al.* (2002).

Amazon may be explained by differences in habitat heterogeneity. They quantified the amount of floristic variation attributable to space and environmental variation in the Panama dataset and found that most of the floristic turnover was caused by the environment, in particular rainfall. Ruokolainen and Tuomisto (2002), then Legendre *et al.* (2005) had noticed flaws in Duivenvoorden *et al.*'s approach and we recently reanalyzed the Panama dataset using both ordination and distance matrix approaches (Chust *et al.* 2006). We found that environment alone (i.e., rainfall, topography, and soil properties) explained 10–12% of the floristic variation, space alone (logarithmically transformed geographical distance) 22–27%, and the interaction between the two 13–18%, depending on the statistical method we employed. The unexplained fraction varied between 46 and 49%. Phillips *et al.* (2003) performed a similar variation partitioning approach for a network of 88 permanent plots around the city of Puerto Maldonado, southeast Peru (ca. 50 km east of Terborgh *et al.*'s study plots) and found that about 10% of the tree floristic variation could be explained by space and 40% by abiotic habitat conditions. A preliminary comparison of the two studies therefore shows that floristic similarity is better explained by distance in Panama than in the western Amazon, as suggested in Condit *et al.* (2002), even when habitat differences are accounted for. This conclusion is further supported by an independent smaller-scale study in a Mexican tropical forest by Balvanera *et al.* (2002), who reported that the fractions of deviance explained by either distance or habitat variables alone were 15 and 43%, respectively, strikingly similar to the results of Phillips *et al.*

In contrast, two additional studies (Duque *et al.* 2002, Potts *et al.* 2002) reported a small and non-significant correlation between floristic similarity (as measured by the Steinhaus index) and linear geographical distance, and a very high and significant correlation with environment (Mantel tests). For example, Duque *et al.* (2002) reported another elegant field study where all trees greater than 2.5 cm diameter at breast height (dbh) were identified in thirty 0.1 ha plots along the río Caquetá, southern Colombia (see also Potts *et al.* 2002). Unfortunately, neither study tested the expectation that floristic similarity should decrease with the log-transformed geographical distance. For this reason, the conclusion in both papers that habitat specialization plays a far more important role than distance in structuring tropical tree communities remains unconvincing.

Similar conclusions were reached by two studies aimed at examining large-scale biodiversity patterns in selected plant groups. Tuomisto, Ruokolainen and colleagues have worked on the distribution of ferns and fern allies (Pteridophyta) and shrubby plants in the family Melastomataceae (henceforth melastomes). They have assembled an unrivaled dataset comprising ca. 300,000 ferns and 40,000 melastomes in numerous neotropical forest sites (Tuomisto and Ruokolainen 1994, Tuomisto *et al.* 1995, 2003a). Vormisto and colleagues have worked on neotropical palm species (Vormisto *et al.* 2004). Because ferns, melastomes, and palms have such distinct biological features (habitat specialization, dispersal syndromes, growth form), it is particularly relevant to contrast these studies. Both Vormisto *et al.* (2004) and Tuomisto *et al.* (2003b) used similar floristic diversity indices (Sørensen and Jaccard, respectively). They found that space alone explained 22, 31, and 22% of the floristic variation (ferns, melastomes, and palms, respectively), environment alone 34, 33, and 8%, and the interaction between the two 15, 16, and 38%.

All of the abovementioned studies are based on very large field inventories, and they show remarkably convergent patterns. The contribution of space to floristic variation appears roughly constant across several groups of understory and canopy plants, at around 20–30%. The contribution of the environment appears much more variable, perhaps because of the important differences in habitat characteristics measured across studies, but there is little doubt that the environment explains 10–40% of the floristic variation (Phillips *et al.* 2003, Tuomisto *et al.* 2003b). Thus, any forthcoming mechanistic theory of biodiversity aiming at predicting patterns of biodiversity at the landscape scale should take into account both dispersal and environmental factors. The above results hold even though some of the studies have employed statistical methods that do not really partition beta-diversity (Legendre *et al.* 2005). This is an obvious limitation of the above comparison, but should also be taken as an evidence that the method based on distance matrices does grasp the main trends in biological datasets.

Measuring the environment of plants

At present, there is no comprehensive review on the quantitative relationships between soil features and beta-diversity, but Sollins (1998) provided an excellent critique of studies focusing on the detailed influence of soil on plant species occurrence in tropical forests, which may serve as a background in this discussion. He suggested that the following four soil factors may influence species occurrence, in decreasing order of importance: (1) available P content; (2) free Al content; (3) soil physical properties; and (4) availability of base-metal cations. It is unclear whether the influence of soil on floristic diversity would be similar to that on species occurrence. To my knowledge, only Phillips *et al.* (2003) mentioned a significant Mantel correlation between the floristic diversity matrix and log-transformed P content (reported in their figure 7). However, this correlation could be spurious because of the contrast between rich Holocene soils and poor Pleistocene soils, which also show strikingly different floristic composition. Phosphorus is generally limiting in tropical soils (Vitousek 1984), it may be an important driver of beta-diversity (Gartlan *et al.* 1986). Soil Al content may also prevent the establishment of some species. Tuomisto *et al.* (2003a) reported that the large-scale distribution of melastomes was positively correlated with soil Al content ($r = 0.26$, $P < 0.05$, Mantel test), as expected given the status of most melastomes as Al-accumulators (Jansen *et al.* 2002). Soil texture (fraction of sand, silt, and clay) and the availability of base-metal, or exchangeable, cations (Ca^{2+}, Mg^{2+}, Na^+, K^+) have been investigated in other studies. Soil texture was found to be a significant predictor for understory plants (Tuomisto *et al.* 2003b, Vormisto *et al.* 2004) but not for canopy trees (Phillips *et al.* 2003). Although this finding deserves further scrutiny, it also confirms the naïve prediction that large-statured plants are more tolerant to variation in soil texture, perhaps because this variable primarily controls the water holding capacity. On the last correlation emphasized by Sollins (1998), a significant partial correlation between floristic diversity and base-metal cations was indeed supported by data from Phillips *et al.* (2003), Tuomisto *et al.* (2003a), and Vormisto *et al.* (2004).

DISCUSSION

Recent progress in the study of tropical plant beta-diversity has been greatly facilitated by the establishment of ambitious field sampling protocols (Gentry 1988, Duivenvoorden 1995, Terborgh *et al.* 1996, Pitman *et al.* 2001, Pyke *et al.* 2001, Balvanera *et al.* 2002, Tuomisto *et al.* 2003a, Vormisto *et al.* 2004). These works have sought general explanations for the observed patterns in beta-diversity, and have made extensive use of correlative approaches, served by the recent biostatistical literature (Legendre and Legendre 1998). All too often, however, patterns are described and post-hoc explanations are proposed without explicit reference to a theoretical framework. This has led to a tension between results and their interpretation and calls for a tighter connection between empirical work and ecological theory (Hubbell 2001, Chave *et al.* 2002, Chave and Leigh 2002, Ricklefs 2003, 2004). Before asking whether a correlation should be sought between tree floristic diversity across sites and geographical distance between these

sites or, say, the difference in the soil concentration of base-metal cations, one should provide a conceptual focus with which a priori hypotheses can be tested. In a model of isolation by distance, Chave and Leigh (2002) have shown that a correlation was expected between one measure of floristic diversity and the logarithm of the geographical distance. Similar models should be developed to justify searching for correlation between floristic diversity and abiotic environmental features.

There are other reasons why it might be difficult to relate the results of correlative approaches to a theoretical framework. Ideally, one environmental variable would predict the variation in floristic composition, and this variation would then be interpretable physiologically. However, an existing correlation between plant species occurrence and environmental variation may also fail to be detected, due to dispersal-related processes: trees may be present in places where the species is not perfectly adapted to the local conditions, just because a large population is present nearby (Shmida and Wilson 1985, Pulliam 1988, Cannon and Leighton 2004). This source–sink phenomenon may cause correlative approaches to overestimate the importance of dispersal over niche-assembly processes. Alternatively, a species adapted to the environmental conditions of a site may fail to be encountered in this site because it may have been unable to disperse there, or unable to invade in the absence of a facilitating species (Law and Morton 1996). This leaves room for less-adapted species, but also confuses the interpretation of any correlative analysis. Of course, one might argue that simple correlative approaches are better than mechanistic theories that make no, or patently false, predictions (Currie 1991). However, it is unlikely that simple correlative approaches will predict patterns of biodiversity at all scales (Latham and Ricklefs 1993, Qian *et al.* 2005). Although Francis and Currie (2003) demonstrated a strong correlation between family-level richness in angiosperms worldwide and environmental variables (temperature, potential evapotranspiration, rainfall), the abovementioned studies at smaller spatial scales show that historical factors and complex environmental gradients also play an important role, not captured in a simple energy-based theory. At this finer scale, the one of most interest for land-use planners and conservationists, it is likely that the simple correlative model predicted by Francis and Currie (2003) will fail.

Niche-assembly and dispersal-based theories highlight different constraints on the distribution of plant species. The niche-assembly theory emphasizes the physiological constraints on plants, constraints that cannot be ignored in the study of plant distributional patterns. Many of the most common tree species of the Amazon forest never occur in swampy areas, or in areas with intense dry seasons. Species in the Melastomataceae and Vochysiaceae are tolerant to soils rich in free aluminum (Al^{3+}), while most other tropical tree families are not (Jansen *et al.* 2002); some species are capable of ectomycorrhizal associations which puts them at an advantage in phosphorus-limited environments. Moreover, it is likely that distance will not limit the spread of plants with small propagules; the minute spores of ferns are known to be dispersed great distances, thereby making the fern community effectively panmictic on large scales (Wolf *et al.* 2001). On the other hand, history should also be an important predictor of the distribution of many plant species. The Central American and South American tree floras remain clearly differentiated 3 million years after the closure of the Panama land-bridge (Gentry 1982, Dick *et al.* 2005). A look at almost any distribution map of congeneric species in the Flora Neotropica shows that species with similar functional features and broad environmental niches occur in clearly distinct and restricted areas (consider, e.g., the case of the neotropical palms in Henderson *et al.* 1995), despite the fact that they have had millions of years to spread across the continent. These few examples show that spatial species mixing at the regional scale is not a particularly efficient mechanism. Thus, niche-assembly theories can be criticized because they tend to minimize the role of dispersal limitation, while the dispersal-assembly theories can be criticized for ignoring species' physiological peculiarities.

The debate over the validity of the neutral theory is now behind us. The fundamental ingredients of the neutral theory, namely the crucial role of unpredictability and of dispersal limitation – in

short, of historical factors – can hardly be debated. The signature of this effect, the space dependence of floristic diversity, has been demonstrated in various studies in different sites and with different plant groups. The fundamental premise of classical community ecology, the influence of the abiotic environment on floristic turnover, has also been confirmed in all these studies. Future unifying theories should take into account both historical fluctuations and environmental determinism. Yet much remains to be done to achieve this goal, simply because we still know so little about how plants respond to their abiotic environment. It is known that some chemical elements of the soil are necessary while others are toxic, and that trade-offs exist among physiological functions, including nutrient use efficiency (Reich *et al.* 1999), but the balance for each species, or across species lineages, remains poorly documented. It may still be useful to explore the two theories independently, as long as it is understood that they represent only part of the larger picture (Ricklefs 2004). One must not forget that most theories in ecology have an illustrative purpose. In May's (1973) words, ecological models "are at best caricatures of reality, and thus have both the truth and the falsity of caricatures."

ACKNOWLEDGMENTS

I thank Guillem Chust, Jim Dalling, Nicolas Mouquet, and Oliver Phillips for useful correspondence, and Christophe Thébaud, Stefan Schnitzer, Kalle Ruokolainen, and an anonymous reviewer for pointing out a number of deficiencies in previous versions. Finally, I thank S Schnitzer for offering an opportunity to update this work in January 2008, some 2 1/2 yrs after the manuscript was first written.

REFERENCES

Ashton, P.S. (1964) Ecological studies in the mixed dipterocarp forests of Brunei State. *Oxford Forestry Memoirs* 25, 1–75.

Ashton, P.S. (1976) Mixed dipterocarp forest and its variation with habitat in the Malayan lowlands: a re-evaluation at Pasoh. *Malayan Forester* 39, 56–72.

Baker, A.J.M. (1987) Metal tolerance. *New Phytologist* 106(Suppl.), 93–111.

Baillie, I., Ashton, P., Court, M. *et al.* (1987) Site characteristics and the distribution of tree species in mixed dipterocarp forest on tertiary sediments in central Sarawak, Malaysia. *Journal of Tropical Ecology* 3, 201–202.

Balvanera, P., Lott, E., Siebe, C. *et al.* (2002) Patterns of β-diversity in a Mexican tropical dry forest. *Journal of Vegetation Science* 13, 145–158.

Borcard, D., Legendre, P., and Drapeau, P. (1992) Partialling out the spatial component of ecological variation. *Ecology* 73, 1045–1055.

Braun-Blanquet, J. (1932) *Plant Sociology. The Study of Plant Communities.* McGraw-Hill Book Company, New York and London.

Cannon, C.H. and Leighton, M. (2004) Tree species distributions across five habitats in a Bornean rain forest. *Journal of Vegetation Science* 15, 257–266.

Chao, A., Hwang, W.-H., Chen, Y.-C. *et al.* (2000) Estimating the number of shared species in two communities. *Statistica Sinica* 10, 227–246.

Chave, J. (2004) Neutral theory and community ecology. *Ecology Letters* 7, 241–253.

Chave, J. and Leigh, E.G. (2002) A spatially explicit neutral model of beta-diversity in tropical forests. *Theoretical Population Biology* 62, 153–168.

Chave, J., Muller-Landau, H.C., and Levin, S.A. (2002) Comparing classical community models: theoretical consequences for patterns of diversity. *American Naturalist* 159, 1–23.

Chave, J., Chust, G., and Thébaud, C. (2007) The importance of phylogenetic structure in biodiversity studies. In D. Storch, P. Marquet, and J.H. Brown (eds), *Scaling Biodiversity*, Santa Fe Institute. Blackwell.

Chazdon, R.L., Colwell, R.K., Denslow, J.S. *et al.* (1998) Statistical estimation of species richness of woody regeneration in primary and secondary rainforests of NE Costa Rica. In F. Dallmeier and J. Comisky (eds), *Forest Biodiversity in North, Central, and South America and the Caribbean: Research and Monitoring.* Parthenon Press, Paris, pp. 285–309.

Chust, G., Chave, J., Condit, R., Aguilar, S., Lao, S., and Perez, R. (2006) Determinants and spatial modeling of beta-diversity in a tropical forest landscape in Panama. *Journal of Vegetation Science* 17, 83–92.

Clark, D.B., Palmer, M.W., and Clark, D.A. (1999a) Edaphic factors and the landscape-scale distributions of tropical rain forest trees. *Ecology* 80, 2662–2675.

Clark, D.B., Read, J.M., Clark, M.L. *et al.* (2004) Application of 1-M and 4-M resolution satellite data to ecological studies of tropical rain forests. *Ecological Applications* 14, 61–74.

Clark, J.S., Silman, M., Kern, R. *et al.* (1999b) Seed dispersal near and far: patterns across temperate and tropical forests. *Ecology* 80, 1475–1494.

Clinebell, H.R.R., Phillips, O.L., Gentry, A.H. *et al.* (1995) Prediction of neotropical tree and liana species richness from soil and climatic data. *Biodiversity and Conservation* 4, 56–90.

Colwell, R.K. and Coddington, J.A. (1994) Estimating terrestrial biodiversity through extrapolation. *Philosophical Transactions of the Royal Society of London Series B* 345, 101–118.

Condit, R., Hubbell, S.P., LaFrankie, J.V. *et al.* (1996) Species-area and species-individual relationships for tropical trees: a comparison of three 50-ha plots. *Journal of Ecology* 84, 549–562.

Condit, R., Ashton, P.S., Baker, P. *et al.* (2000) Spatial patterns in the distribution of tropical tree species. *Science* 288, 1414–1418.

Condit, R., Pitman, N., Leigh, J.E.G. *et al.* (2002) Beta-diversity in tropical forest trees. *Science* 295, 666–669.

Crist, T.O., Veech, J.A., Gering, J.C. *et al.* (2003) Partitioning species diversity across landscapes and regions: a hierarchical analysis of a, b and g diversity. *American Naturalist* 162, 734–743.

Currie, D.J. (1991) Energy and large-scale patterns of animal- and plant-species richness. *American Naturalist* 137, 27–49.

Dick, C.W., Condit, R., and Bermingham, E. (2005) Biogeographic history and the high β-diversity of rainforest trees in Panamá. In R. Harmon (ed.), *Rio Chagres, Panama: A Multi-Disciplinary Profile of a Tropical Watershed.* Springer, Dordrecht, The Netherlands, Chapter 6, pp. 259–270.

Duivenvoorden, J.F. (1995) Tree species composition and rain forest – environment relationships in the middle Caquetá area, Colombia, NW Amazonia. *Vegetatio* 120, 91–113.

Duivenvoorden, J.F., Svenning, J.-C., and Wright, J. (2002) Beta diversity in tropical forests. *Science* 295, 636–637.

Duque, A., Sánchez, M., Cavelier, J. *et al.* (2002) Different floristic patterns of woody understorey and canopy plants in Colombian Amazonia. *Journal of Tropical Ecology* 18, 499–525.

Engelbrecht, B.M.J. and Kursar, T.A. (2003) Comparative drought-resistance of seedling of 28 species of co-occurring tropical woody plants. *Oecologia* 136, 383–393.

Engen, S. (1978) Stochastic abundance models, with emphasis on biological communities and species diversity. Chapman & Hall, London.

Faith, D.P. (1992). Conservation evaluation and phylogenetic diversity. *Biological Conservation*, 61, 1–10.

Fine, P.V.A., Mesones, I., and Coley, P.D. (2004) Herbivores promote habitat specialization by trees in Amazonian forests. *Science* 305, 663–665.

Flikkema, P.G., Agarwal, P.K., Clark, J.S. *et al.* (2006) Model-driven dynamic control of embedded wireless sensor networks. In V.N. Alexandrov *et al.* (eds), ICCS 2006, Part III, LNCS 3993, 409–416. Springer-Verlag, Berlin and Heidelberg .

Francis, A.P. and Currie, D.J. (2003) A globally consistent richness-climate relationship for angiosperms. *American Naturalist* 161, 523–536.

Gartlan, J.S., Newbery, D.M., Thomas, D.W. *et al.* (1986) The influence of topography and soil phosphorus on the vegetation of Korup Forest Reserve, Cameroun. *Vegetatio* 65, 131–148.

Gaston, K. (1994) *Rarity*. Chapman and Hall, London.

Gentry, A.H. (1982) Patterns of Neotropical plant species diversity. *Evolutionary Biology* 15, 1–84.

Gentry, A.H. (1988) Changes in plant community diversity and floristic composition on environmental and geographical gradients. *Annals of the Missouri Botanical Garden* 75, 1–34.

Gnedenko, B.V. and Kolmogorov, A.N. (1954) *Limit Distributions for Sums of Independent Random Variables*. Addison Wesley, Reading, MA.

Gower, J.C. (1971) A general coefficient of similarity and some of its properties. *Biometrics* 23, 623–637.

Green, J.L. and Plotkin, J.B. (2007) A statistical theory for sampling species abundances. *Ecology Letters*, 10, 1037–1045.

Grime, J.P. (1977) Evidence for the existence of three primary strategies in plants and its relevance to ecological and evolutionary theory. *American Naturalist* 111, 1169–1194.

Hardy, O.J. and Sonké, B. (2004) Spatial pattern analysis of tree species distribution in a tropical rain forest of Cameroon: assessing the role of limited dispersal and niche differentiation. *Forest Ecology Management* 197, 191–202.

Harms, K.E., Condit, R., Hubbell, S.P. *et al.* (2001) Habitat associations of trees and shrubs in a 50-ha Neotropical forest plot. *Journal of Ecology* 89, 947–959.

Henderson, A., Galeano, G., and Bernal, R. (1995) *Field Guide to the Palms of the Americas*. Princeton University Press, Princeton.

Hijmans, R.J., Cameron, S.E., Parra, J.L., Jones, P.G., and Jarvis, A. (2005) Very high resolution interpolated climate surfaces for global land areas. *International Journal of Climatology* 25, 1965–1978.

Hubbell, S.P. (1979) Tree dispersion, abundance and diversity in a dry tropical forest. *Science* 203, 1299–1309.

Hubbell, S.P. (1997) A unified theory of biogeography and relative species abundance and its application to tropical rain forests and coral reefs. *Coral Reefs* 16, S9–S21.

Hubbell, S.P. (2001) *The Unified Neutral Theory of Biodiversity and Biogeography*. Princeton University Press, Princeton.

Hubbell, S.P. (2003) Modes of speciation and the lifespans of species under neutrality: a response to the comment of Robert E. Ricklefs. *Oikos* 100, 193–199.

Hubbell, S.P. and Foster, R.B. (1986) Biology, chance, and history and the structure of tropical rain forest tree communities. In J. Diamond and T.J. Case (eds), *Community Ecology*. New York, Harper and Row, pp. 314–329.

Hurlbert, S.H. (1971) The non-concept of species diversity: a critique and alternative parameters. *Ecology* 52, 577–586.

Huston, M.A. (1999) Local processes and regional patterns: appropriate scales for understanding variation in the diversity of plants and animals. *Oikos* 86, 393–401.

Huston, M.A. and DeAngelis, D.L. (1994) Competition and coexistence: the effects of resource transport and supply rates. *American Naturalist* 144, 954–977.

Jansen, S., Broadley, M.R., Robbrecht, E. *et al.* (2002) Aluminium hyperaccumulation in angiosperms: a review of its phylogenetic significance. *Botanical Review* 68, 235–269.

Jost, L. (2006) Entropy and diversity. *Oikos*, 113, 363–375.

Koleff, P., Gaston, K.J., and Lennon, J.J. (2003) Measuring beta-diversity for presence–absence data. *Journal of Animal Ecology* 72, 367–382.

Kot, M., Lewis, M.A., and van der Driessche, P. (1996) Dispersal data and the spread of invading organisms. *Ecology* 77, 2027–42.

Krebs, C.J. (1999) *Ecological Methodology*. Addison-Wesley, Menlo Park, CA.

Lande, R. (1996) Statistics and partitioning of species diversity, and similarity among multiple communities. *Oikos* 76, 5–13.

Latham, R.E. and Ricklefs, R.E. (1993) Global patterns of tree species richness in moist forests: energy-diversity theory does not account for variation in species richness. *Oikos* 67, 325–333.

Law, R. and Morton, R.D. (1996) Permanence and the assembly of ecological communities. *Ecology* 77, 762–775.

Legendre, P. and Legendre, L. (1998) *Numerical Ecology.* 2nd English Edition. Elsevier, Amsterdam.

Legendre, P., Borcard, D., and Peres-Neto, P.R. (2005) Analyzing beta-diversity: partitioning the spatial variation of community composition data. *Ecological Monographs* 75, 435–450.

Leigh, E.G., Wright, S.J., Putz, F.E. *et al.* (1993) The decline of tree diversity on newly isolated tropical islands: a test of a null hypothesis and some implications. *Evolutionary Ecology* 7, 76–102.

Levin, S.A. (1970) Community equilibria and stability, and an extension of the competitive exclusion principle. *American Naturalist* 104, 413–423.

Levin, S.A. (1992) The problem of pattern and scale in ecology. *Ecology* 73, 1943–1967.

Loreau, M. (1998) Biodiversity and ecosystem functioning: a mechanistic model. *Proceedings of the National Academy of Sciences of the United States of America* 95, 5632–5636.

Lucas, Y., Luizao, F.J., Chauvel, A. *et al.* (1993) The relation between biological activity and mineral composition of soils. *Science* 260, 521–523.

Magurran, A.E. (2004) *Measuring Biological Diversity.* Blackwell Publishing, Oxford.

Malécot, G. (1948) *Les Mathématiques de l'Hérédité.* Masson, Paris.

May, R.M. (1973) *Stability and Complexity in Model Ecosystems.* Princeton University Press, Princeton.

McKnight, M.W., White, P.S., McDonald, R.I. *et al.* (2007) Putting beta-diversity on the map: broad-scale congruence and coincidence in the extremes *PLoS Biology,* 5, 2424–2432.

Mouquet, N. and Loreau, M. (2003) Community patterns in source-sink metacommunities. *American Naturalist* 162, 544–557.

Nathan, R. and Muller-Landau, H.C. (2000) Spatial patterns of seed dispersal, their determinants and consequences for recruitment. *Trends in Ecology and Evolution* 15, 278–285.

Nee, S. and May, R.M. (1997). Extinction and the loss of evolutionary history. *Science* 278, 692–694.

Nekola, J.C. and White, P.S. (1999) The distance decay of similarity in biogeography and ecology. *Journal of Biogeography* 26, 867–878.

Nei, M. (1973) Analysis of gene diversity in subdivided populations. *Proceedings of the National Academy of Sciences of the United States of America* 70, 3321–3323.

Nei, M. (1987) *Molecular Evolutionary Genetics.* Columbia University Press, New York.

New, M.G., Lister, D., Hulme, M. *et al.* (2002) A high-resolution data set of surface climate for terrestrial land areas. *Climate Research* 21, 1–25.

Newbery, D.M. and Proctor, J. (1984) Ecological studies in four contrasting lowland rain forests in Gunung Mulu National Park, Sarawak: IV. Associations between tree distribution and soil factors. *Journal of Ecology* 72, 475–493.

Newbery, D.M., Renshaw, E., and Brunig, E.F. (1986) Spatial pattern of tree in kerangas forest, Sarawak. *Vegetatio* 65, 77–89.

Ohman, J.L. and Spies, T.A. (1998) Regional gradient analysis and spatial pattern of woody plant communities of Oregon forests. *Ecological Monographs* 68, 151–182.

Oliveira, A.A. de and Mori, S.A. (1999) A central Amazonian terra firme forest I. High tree species richness on poor soils. *Biodiversity and Conservation* 8, 1219–1244.

Pavoine, S., Ollier, S., and Dufour, A.-B. (2005) Is the originality of species measurable? *Ecology Letters* 8, 579–586.

Pennington, R.T. and Dick, C.W. (2004) The role of immigrants in the assembly of the South American rainforest tree flora. *Philosophical Transactions of the Royal Society of London* 359, 1611–1622.

Phillips, O.L., Nuñez, P., Monteagudo, A.L. *et al.* (2003) Habitat association among Amazonian tree species: a landscape-scale approach. *Journal of Ecology* 91, 757–775.

Pielou, E.C. (1975) *Ecological Diversity.* Wiley Inter-Science, New York.

Pitman, N.C.A., Terborgh, J.W., Silman, M.R. *et al.* (1999) Tree species distributions in an upper Amazonian forest. *Ecology* 80, 2651–2661.

Pitman, N.C.A., Terborgh, J.W., Silman, M.R. *et al.* (2001) Dominance and distribution of tree species in upper Amazonian terra firme forests. *Ecology* 82, 2101–2117.

Plotkin, J.B. and Muller-Landau, H.C. (2002) Sampling the species composition of a landscape. *Ecology* 83, 3344–3356.

Poore, M.E.D. (1968) Studies in the Malaysian rain forest. I. The forest on Triassic sediments in the Jengka Forest Reserve. *Journal of Ecology* 56, 143–196.

Potts, M.D., Ashton, P.S., Kaufman, L.S. *et al.* (2002) Habitat patterns in tropical rain forests: a comparison of 105 plots in Northwest Borneo. *Ecology* 83, 2782–2797.

Pulliam, H.R. (1988) Sources, sinks, and population regulation. *American Naturalist* 132, 652–661.

Purvis, A., Agapow, P.-M., Gittleman, J.L. and Mace, G.M. (2000). Nonrandom extinction and the loss of evolutionary history. *Science* 288, 328–330.

Pyke, C.R., Condit, R., Aguilar, S. *et al.* (2001) Floristic composition across a climatic gradient in a neotropical lowland forest. *Journal of Vegetation Science* 12, 553–566.

Qian, H., Ricklefs, R.E., and White, P.S. (2005) Beta diversity of angiosperms in temperate floras of eastern Asia and eastern North America. *Ecology Letters* 8, 15–22.

Rao, C.R. (1982). Diversity and dissimilarity coefficients: a unified approach. *Theoretical Population Biology* 21, 24–43.

Reich, P.B., Ellsworth, D.S., Walters, M.B. *et al.* (1999) Generality of leaf trait relationships: a test across six biomes. *Ecology* 80, 1955–1969.

Renkonen, O. (1938) Statistisch-Ökologische Untersuchungen über die terrestriche Kaferwelt der finnischen Bruchmoore. *Archivum Societalis Zoologicae Botanicae Fennicae* 6, 1–226.

Ricklefs, R.E. (2003) A comment on Hubbell's zero-sum ecological drift model. *Oikos* 100, 185–192.

Ricklefs, R.E. (2004) A comprehensive framework for global patterns of biodiversity. *Ecology Letters* 7, 1–15.

Ricklefs, R.E. and Lau, M. (1980) Bias and dispersion of overlap indices: results of some Monte Carlo simulations. *Ecology* 61, 1019–1024.

Ruokolainen, K., Linna, A., and Tuomisto, H. (1997) Use of Melastomataceae and pteridophytes for revealing phytogeographic patterns in Amazonian rain forests. *Journal of Tropical Ecology* 13, 243–256.

Ruokolainen, K. and Tuomisto, H. (2002) Beta diversity in tropical forests. *Science* 297, 1439a.

Sabatier, D., Grimaldi, M., Prévost, M.-F. *et al.* (1997) The influence of soil cover organization on the floristic and structural heterogeneity of a Guianan rain forest. *Plant Ecology* 131, 81–108.

Shmida, A. and Wilson, M.V. (1985) Biological determinants of species diversity. *Journal of Biogeography* 12, 1–20.

Slik, J.W.F., Poulsen, A.D., Ashton, P.S. *et al.* (2003) A floristic analysis of the lowland dipterocarp forests of Borneo. *Journal of Biogeography* 30, 1517–1531.

Sollins, P. (1998) Factors influencing species composition in tropical lowland rain forest: does soil matter? *Ecology* 79, 23–30.

Svenning, J.-C. (1999) Microhabitat specialization in a species-rich palm community in Amazonian Ecuador. *Journal of Ecology* 87, 55–65.

Svenning, J.-C., Kinner, D.A., Stallard, R.F. *et al.* (2004) Ecological determinism in plant community structure across a tropical forest landscape. *Ecology* 85, 2526–2538.

Swaine, M.D. (1996) Rainfall and soil fertility as a factor limiting forest species distribution in Ghana. *Journal of Ecology* 84, 419–428.

Terborgh, J.R., Foster, R.B., and Nuñez, V.P. (1996) Tropical tree communities: a test of the non-equilibrium hypothesis. *Ecology* 77, 561–567.

ter Braak, C.J.F. (1987) Ordination. In R.H.G. Jongman, C.J.F. ter Braak, and O.F.R. van Tongeren (eds), *Data Analysis in Community and Landscape Ecology*. Pudoc Wageningen, The Netherlands, Chapter 4, pp. 91–173.

ter Braak, C.J.F. and Prentice, I.C. (1988) A theory of gradient analysis. *Advances in Ecological Research* 18, 271–313.

ter Steege, H., Pitman, N., Sabatier, D. *et al.* (2003) A spatial model of tree α-diversity and -density for the Amazon. *Biodiversity and Conservation* 12, 2255–2277.

Tilman, D. (1980) Resources: a graphical-mechanistic approach to competition and predation. *American Naturalist* 116, 686–692.

Tilman, D. (1982) *Resource Competition and Community Structure*. Princeton University Press, Princeton.

Tuomisto, H. (1998) What satellite imagery and large-scale field studies can tell about biodiversity patterns in Amazonian forests. *Annals of the Missouri Botanical Garden* 85, 48–62.

Tuomisto, H., Poulsen, A.D., Ruokolainen, K. *et al.* (2003a) Linking floristic patterns with soil heterogeneity and satellite imagery in Ecuadorian Amazonia. *Ecological Applications* 13, 352–371.

Tuomisto, H. and Ruokolainen, K. (1994) Distribution of Pteridophyta and Melastomataceae along an edaphic gradient in an Amazonian rain forest. *Journal of Vegetation Science* 5, 25–34.

Tuomisto, H., Ruokolainen, K., Kalliola, R. *et al.* (1995) Dissecting Amazonian biodiversity. *Science* 269, 63–66.

Tuomisto, H., Ruokolainen, K., and Yli-Halla, M. (2003b). Dispersal, environment, and floristic variation of Western Amazonian forests. *Science* 299, 241–244.

Tuomisto, H. and Ruokolainen, K. (2006) Analyzing or explaining beta diversity? Understanding the targets of different methods of analysis. *Ecology* 87, 2697–2708.

Tyree, M.T. and Sperry, J.S. (1989) Vulnerability of xylem to cavitation and embolism. *Annual Review Plant Physiology and Molecular Biology* 40, 19–38.

Valencia, R., Balslev, H., and Paz Y Mino, G. (1994) High tree alpha-diversity in Amazonian Ecuador. *Biodiversity and Conservation* 3, 21–28.

Vitousek, P.M. (1984) Litterfall, nutrient cycling, and nutrient limitation in tropical forests. *Ecology* 65, 285–298.

Vormisto, J., Svenning, J.-C., Hall, P. *et al.* (2004) Diversity and dominance in palm (Arecaceae) communities in terra firme forests in the western Amazon basin. *Journal of Ecology* 92, 577–588.

Wallace, A.R. (1895) *Natural Selection and Tropical Nature*. Macmillan, London.

Warming, E. (1909) *Oecology of Plants*. Oxford University Press, Oxford.

Webb, C.O. and Peart, D.R. (2000) Habitat associations of trees and seedlings in a Bornean rain forest. *Journal of Ecology* 88, 464–478.

Webb, L.J. (1969) Edaphic differentiation of some forest types in Eastern Australia: II soil chemical factors. *Journal of Ecology* 57, 817–830.

Whittaker, R.H. (1956) Vegetation of the Great Smoky Mountains. *Ecological Monographs* 26, 1–80.

Whittaker, R.H. (1960) Vegetation of the Siskiyou Mountains, Oregon and California. *Ecological Monographs* 30, 279–338.

Whittaker, R.H. (1967) Gradient analysis of vegetation. *Biological Reviews* 42, 207–264.

Whittaker, R.H. (1972) Evolution and measurement of species diversity. *Taxon* 21, 213–251.

Wilson, M.V. and Shmida, A. (1984) Measuring beta diversity with presence–absence data. *Journal of Ecology* 72, 1055–1062.

Wolda, H. (1981) Similarity indices, sample size and diversity. *Oecologia* 50, 296–302.

Wolf, P.G., Schneider, H., and Ranker, T.A. (2001) Geographic distributions of homosporous ferns: does dispersal obscure evidence of vicariance? *Journal of Biogeography* 28, 263–270.

Wong, Y.K. and Whitmore, T.C. (1970) On the influence of soil properties on species distributions in a Malayan lowland dipterocarp rain forest. *Malayan Forester* 33, 42–54.

Woodcock, S., van der Gast, C.J., Bell, T. *et al.* (2007) Neutral assembly of bacterial communities. *FEMS Microbiology Ecology* 62, 171–180.

Chapter 3

THE DISPARITY IN TREE SPECIES RICHNESS AMONG TROPICAL, TEMPERATE, AND BOREAL BIOMES: THE GEOGRAPHIC AREA AND AGE HYPOTHESIS

Paul V.A. Fine, Richard H. Ree, and Robyn J. Burnham

OVERVIEW

According to theoretical models, larger land areas should experience higher speciation rates and lower extinction rates and thus contain higher species richness than smaller areas, all else being equal. This idea has been applied to explain the latitudinal gradient in species diversity, and has been named the geographic area hypothesis (GAH). Although putative differences in the geographic area between tropical and non-tropical biomes within continents have been linked to the disparity in species richness between biomes, no one has tested the GAH with a global dataset. Using estimates of tree diversity for 11 biome areas on six continents, we evaluated the importance of geographic area in explaining patterns in tree diversity at the largest spatial scales. We found that the tree diversity of a biome was not correlated with its geographic area. However, because area is predicted to influence *in situ* speciation and extinction rates within a biome, we considered changes in a biome's size over tens of millions of years, a time period appropriate for those processes. We found a significant correlation between current tree species richness and biome size integrated over time since the Miocene, the Oligocene, and the Eocene. These results suggest that both the wet lowland tropics' larger area and their longevity have played a significant role in generating and maintaining the extraordinarily high tree diversity. In addition, minimum biome area during the Pleistocene and current tree diversity were positively correlated, suggesting that extinction due to contraction of available habitat during glaciation in temperate and boreal areas may have been important factors reducing diversity at high latitudes. These results support the predictions of a related hypothesis, the tropical conservatism hypothesis (TCH), and may explain why most tree lineages have arisen in the tropics, and why tropical forests contain such high tree species richness compared with extra-tropical forests. One of the implications from our results is that conservation of large areas of tropical forests should be given the highest priority because these forests should be the most sensitive to extinction due to habitat loss.

INTRODUCTION

In the mid-1990s, in his book *Species Diversity in Space and Time*, Rosenzweig championed area as the primary factor producing the latitudinal gradient in species diversity (Terborgh 1973,

Rosenzweig 1992, 1995). The geographic area hypothesis (GAH) posits that, all else being equal, larger areas should promote speciation and reduce extinction; thus a large biome adjacent to a smaller biome should contain more species. If the GAH is true, the latitudinal distribution of biome area should match that of species richness. However, despite considerable disagreement on the methods used to define biome boundaries and calculate the area of the world's biomes (Terborgh 1973, Rosenzweig 1992, 1995, Rohde 1997, Rosenzweig and Sandlin 1997, Gaston 2000, Fine 2001, Hawkins and Porter 2001), a continent's biome areas do not generally increase in size towards the equator, matching the latitudinal gradient of increasing species diversity. Moreover, attempts to match biome area with the species richness of trees (Fine 2001) or birds (Hawkins and Porter 2001) did not confirm the prediction that the area of a biome should correlate with the total number of species that have distributions within it. Because of these inconsistencies, the GAH has not been embraced by most biologists concerned with understanding the latitudinal diversity gradient. Indeed, in a recent review of hypotheses to explain the latitudinal gradient, the GAH is only briefly mentioned, and ultimately rejected as unimportant (Schemske 2002).

Should the GAH be dismissed? Here we make the case that doing so would be premature because it has not been adequately tested. First, due to the difficulty of defining and determining the spatial extent of biomes, and to the lack of reliable data on global distributions of most organisms, the two studies that sought to match biome area with species totals were not conducted with complete datasets: Fine (2001) considered only North American trees north of Mexico and Hawkins and Porter (2001) excluded tropical birds from their study. Second, the mechanisms by which geographic area is thought to affect species richness involve speciation and extinction, processes that operate over large time scales. When comparing the sizes of a continent's biomes with species richness, one must consider how the biomes may have changed in area during the time that its lineages have been undergoing speciation and extinction (McGlone 1996, Pennington et al. 2004, Ricklefs 2004). The GAH was advanced to explain diversity patterns at the largest spatial scales. Its mechanistic basis lies in differential rates of speciation and extinction, both of which occur over time scales of 10^5–10^6 years (Magallón and Sanderson 2001, Whittaker et al. 2001, Ricklefs

2003, 2004). Below, we explore a new way to incorporate time into an evaluation of the GAH.

In this chapter, we first review the theory and evidence for population genetic mechanisms by which geographic area may influence the species diversity of a biome. Second, we discuss the difficulties inherent in testing the GAH, and present recommendations for defining biomes, calculating their areas, and evaluating the number of species found within them. Next, we present data on tree diversity in forest biomes, and evaluate the GAH in the context of current biome area for the world's moist forests. Then, we correlate current tree diversity of biomes with estimates of the areas of past forest biomes integrated over the last 55 million years. Finally, focusing on trees, we discuss our results within the context of two alternative hypotheses which have been advanced to explain the latitudinal gradient of species diversity: the tropical conservatism hypothesis and the species–energy hypothesis.

MECHANISMS

Why are larger areas predicted to include more species? Specifically, what role does area play in the population-level processes influencing rates of speciation and extinction? These questions are important in determining whether large, diverse areas like the tropics act as cradles of biodiversity from which new species arise, or as museums that preserve existing species from extinction (Stebbins 1974, Moritz et al. 2000).

In the cradle-versus-museum debate (reviewed by Chown and Gaston 2000), an important premise has been that larger areas allow species to have larger ranges, and much attention has been given to how range size relates to speciation and extinction rates. The expectation of a positive, peaked, or negative relationship between range size and the probability of speciation seems to depend on which parameters (extrinsic or intrinsic) are emphasized. For example, consider the extrinsic effect of geographic barriers, such as mountain ranges, rivers, etc. If large ranges are more likely to be subdivided by such barriers, disrupting gene flow and causing allopatric speciation, we might predict a positive relationship between range size and speciation rate. On the other hand, if barriers are small relative to the largest ranges and less likely to subdivide them completely, then the probability of speciation will decrease with range size. These models are not mutually exclusive: the former may operate at smaller range sizes, and the latter at larger range sizes, which may result in the probability of allopatric speciation peaking at intermediate range sizes (Rosenzweig 1995).

Intrinsic effects of range size on population genetic processes affecting speciation have generally been proposed in the context of larger ranges being commonly associated with greater dispersal ability and higher population abundances (see Chown 1997). The effect of geographic area on speciation may thus depend on the extent to which large areas promote selection for life-history traits such as high vagility, short generation time, and good colonization ability, as these enable species to increase their range (Marzluff and Dial 1991).

Some modes of speciation are thought to be more common if range size is large. For example, peripatric speciation (isolation of small populations at the periphery of a range) will be accelerated if peripheral populations experience local selective regimes that differ from those across the rest of the range (Mayr 1954). This is a more likely circumstance within large ranges, which usually encompass more habitats than small ranges. This hypothesis predicts a faster rate of evolution and divergence in small peripheral isolates. In centrifugal speciation (Brown 1957), isolation between a smaller peripheral population and its parent is driven by accelerated evolution in the parent. Because centrifugal speciation is dependent on population size, and population size increases with range size (Gaston 1996), centrifugal speciation also supports a positive relationship between range size and the probability of peripatric speciation (Rosenzweig 1995).

There are two ways that large range size may impede speciation. First, if high vagility increases gene flow in addition to range size, population cohesion will be maintained and result in a lower probability of speciation (Mayr 1963, Stanley 1979). However, genetic evidence

in marine species shows that dispersal is not always associated with gene flow (Palumbi 1992). Second, higher local population densities and dispersal rates characterize large ranges, preventing stochastic divergence. Gavrilets et al. (2000) conducted simulations in one- and two-dimensional systems to study how range size, population density, mutation rate, and migration influence the timing and mode of speciation. Their results supported the centrifugal model if new genetic variation was the limiting factor, and conversely supported the "centripetal" model (rapid evolution and isolation at the periphery) if genetic variation was sufficient but gene flow between populations was low. They concluded that large ranges are not more likely to undergo speciation. The question of whether a large range promotes speciation remains largely unresolved, as there appear to be viable theoretical arguments that predict both higher and lower rates of speciation in large-ranged species.

In contrast to speciation, the expected negative relationship between range size and probability of extinction is less controversial, and is supported by empirical studies of the fossil record (e.g., Gaston and Blackburn 2000, p. 120). A common view is that species with large ranges are more likely to have broader habitat tolerances and more genetic variation than species with small ranges, and are thus more resilient to stochastic extinction caused by changing environments (Rosenzweig 1995). As a result, even if species with large ranges have lower speciation rates over the short term, over the long term they may generate a disproportionate number of new species merely by persisting longer than small-ranged species (Chown and Gaston 2000) and thus increase the diversity of large areas like the tropics. The expectation that long-persisting, widespread species are significant in situ sources of diversity in large areas highlights how being a museum may, in fact, also mean being a cradle; the two models are not mutually exclusive.

Study of the evolutionary dynamics underlying the signal of geographic area in the latitudinal diversity gradient is challenging because several relevant parameters (range size, speciation rate and mode, extinction, mutation rate, dispersal) interact and modulate each other's effects in a complex way. Thus, it is difficult to make robust inferences about process from current distributions of range size. However, ecological modeling of lineage and population dynamics over spatial and temporal gradients is a topic that is ripe for progress. A good theoretical starting point might be Chown's (1997) modified "fission" model in which the probability of speciation peaks at relatively low values for range size, dispersal rate, and abundance, but declines slower than the probability of extinction as those variables increase. The importance of area in driving this pattern might become clearer if population genetic models incorporating those variables were used to test the GAH. Furthermore, combining theoretical models with population genetic data is also likely to be productive. In an empirical study, Martin and McKay (2004) surveyed geographical patterns of genetic diversity over a wide range of animal species and found that at low latitudes, genetic divergence between populations (and hence potential for speciation) is greater than at high latitudes. Similar studies that focus on genetic patterns within and between tropical and extra-tropical biomes are needed to more adequately test the speciation models underlying the GAH.

TESTING THE GAH

In theory, testing the GAH involves simply calculating a biome's surface area and counting the number of species found within it, and then comparing those figures across biomes. This is difficult to do in practice for several reasons. Biome boundaries may be different depending on which organisms are considered. Biomes are traditionally defined by climatic variables (temperature and rainfall), but these vary in how they affect the ranges of different kinds of organisms, for example plants versus animals. Birds, for instance, have an array of behavioral and physiological adaptations to cold and drought that are not analogous to the adaptations of plants. As a consequence, dozens of bird species have ranges that span the North American continent, including boreal, temperate, and subtropical

biomes – in contrast to North American tree species, which rarely cross even one biome border (Fine 2001).

A second problem in testing the GAH is the pervasive lack of data on species distributions. Once a biome is carefully defined for a group of organisms, it must be inventoried. For no group of organisms, not even birds, are comprehensive species distribution maps available for the entire globe. Moreover, even where species lists exist, caution should be exercised in comparing species richness among continents. For example, North American birds have been intensively studied for centuries, and in some cases very closely related taxa are considered distinct species (Zink 2004). In contrast, some Amazonian bird species are morphologically uniform, but analyses of genetic data reveal substantial geographic structure, suggesting a plethora of undiscovered cryptic species in the Amazon (Bates *et al.* 1999).

A third challenge in testing the GAH is that it is an equilibrium hypothesis – that is, the areas of biomes should correlate with species totals only if the dynamics of species turnover are at equilibrium. However, there is little reason to presuppose equilibrium at any given time for all biomes because paleoclimates have not been stable (Ricklefs 2004). Current diversities may reflect the climatic history of biome areas as much as they reflect current climatic conditions. Indeed, since climate change and glaciations disproportionately affect biomes that are closer to the poles compared with those closer to the equator, it is possible that the biota of northern biomes are depauperate and below their equilibrium values, especially for long-lived and slowly dispersing taxa like trees (Svenning and Skov 2004). Similarly, tropical biomes once covered a much larger area than they do today, and species totals of tropical rainforests may be "above" equilibrium diversity. Thus, an adequate test of the GAH must incorporate the size of biome areas over time. In the rest of this chapter, we estimate the geographic extent of boreal, temperate, and tropical biomes for each continent over the last 55 million years and correlate these areas integrated over time with current tree species richness to test the geographic area and age hypothesis (GAAH).

TESTING THE GAAH

We chose three biomes defined on the basis of temperature parameters that are biologically meaningful for trees. In our view, tropical, temperate, and boreal biomes should not be defined by latitudinal boundaries; instead, they should be defined by the physiological boundaries important for trees (Fine 2001). We define boreal biomes as areas that experience $-40°C$ temperatures in winter. This is the threshold of spontaneous nucleation of supercooled water, which requires a specific adaptation in plants to avoid death by xylem cavitation (Woodward 1987). We define temperate biomes as those delimited by the $-40°C$ isocline on the high-latitude border, and by the frost isocline on the low-latitude border. The frost-line is important because the lack of frost tolerance limits tropical trees from expanding into temperate areas (Sakai and Weiser 1973, Woodward 1987, Latham and Ricklefs 1993, Fine 2001). Finally, tropical biomes are defined as areas that never experience $0°C$ temperatures. Extra-tropical trees are likely limited from crossing into tropical areas by the trade-off in growth that accompanies frost tolerance, giving temperate trees a competitive disadvantage in tropical areas (MacArthur 1972, Loehle 1992, Fine 2001).

We divided each of six continents into boreal, temperate (including subtropical), and tropical moist/wet and dry forests using the World Wildlife Fund Ecoregions data tables (Olson *et al.* 2001). We were not able to obtain estimates for the world's tropical dry and temperate Mediterranean tree floras, so we ignored the dry forest areas and present only moist/wet forest data. Eurasia and North America each include two separate temperate moist forest biomes, geographically separated by more than 1000 km and with almost no overlap in species composition (Petrides and Petrides 1992, Petrides 1998). Therefore, each of the temperate biomes in these continents is treated here as a separate entity.

We searched for estimates for tree species richness for 11 biome areas: two boreal areas (North America and Eurasia), six continental temperate areas (Europe, East Asia, Eastern North America, Western North America, South America, and Australia), and three tropical areas (Neotropics,

African Tropics, and Asian Tropics [including India, Malayan Peninsula, and Borneo]). The Australian tropics and Papua New Guinea (PNG) were not included because we could not find reliable estimates on the species richness of the tree flora of PNG, nor of the amount of overlap between PNG and Asia and/or Australia (to decide whether Australia/PNG warranted a separate designation from the Asian Tropics). Estimates for tree diversity per biome area are admittedly speculative (see Table 3.1 for sources), especially for tropical forests. Significance of the relationships between biome area and species richness was tested by pairwise correlation of the log-transformed variables.

ESTIMATING HISTORIES OF BIOME AREAS

Testing the GAH while taking into account past fluctuations in biome area requires some knowledge of the tempo of plant diversification. Magallón and Sanderson (2001) estimated speciation rates for angiosperms and proposed an average overall rate of 0.0893 net speciation events per million years (maximum rate of 0.32 events per million years in the most rapidly diversifying clade, Asteraceae). From these values, we deduced that the recent history of biome size on the order of tens of millions of years was a reasonable window over which fluctuations in biome area could be expected to have an effect on extant diversity levels.

To estimate the size of biome areas through time, we used paleoclimatic and paleovegetation maps that estimated lowland moist/wet tropical, temperate, and boreal biomes from five recent sources (Dowsett *et al.* 1999, Morley 2000, Beerling and Woodward 2001, Willis and McElwain 2002, and C.R. Scotese's PALEOMAP project [www.scotese.com]; see also Parrish *et al.* 1982, Scotese 2004) (Figure 3.1). Eocene, Oligocene, and Miocene reconstructions were largely drawn from reconstructions by Willis and McElwain (2002), Pliocene reconstructions were almost entirely based on Dowsett *et al.* (1999), and mid-Holocene and Last Glacial Maximum reconstructions were based on Beerling and

Table 3.1   Area and number of tree species for the 11 biomes included in the present study.

Biome	Area (in 1000 km^2)	Estimated no. of tree species	Sources
North American Boreal	5,117	61	Petrides and Petrides (1992), Petrides (1998)
Eurasian Boreal	10,010	100	Hytteborn *et al.* (2005), A. Shvidenko (personal communication)
North American Eastern Temperate	3,396	300	Petrides (1998)
North American Western Temperate	1,698	115	Petrides and Petrides (1992)
European Temperate	6,374	124	Latham and Ricklefs (1993)
East Asian Temperate	4,249	729	Latham and Ricklefs (1993)
South American Temperate	413	84	Rodriguez *et al.* (1983)
Australian Temperate	735	310	Francis (1981)
Neotropics	9,220	22,500	R. Condit (personal communication), R. Foster (personal communication)
Asian Tropics	5,903	14,000	R. Condit (personal communication), J. LaFrankie (personal communication)
African Tropics	3,471	6,500	R. Condit (personal communication)

Notes: Current biome area is estimated pre-human impact, and comes from the World Wildlife Fund (see map of biomes in Olson *et al.* 2001). Sources for estimated numbers of species per biome are listed in the table.

Woodward (2001). We synthesized these five estimates to provide a "best guess" of tropical, temperate, and boreal biomes for these particular slices in time. These methods for estimating biome area through time are crude and approximate, given the coarse temporal resolution of the underlying maps, the method used to calculate areas over time in Figure 3.2, and the uncertainties associated with paleoclimatic reconstruction.

We traced biome boundaries based on reconstructions of the Eocene (55 million years ago [Ma]), Oligocene (30 Ma), Miocene (11.5–5 Ma), Pliocene (3.5 Ma), Last Glacial Maximum (21,000 years ago), and Mid-Holocene (6000 years ago) onto equal area projections and used ImageJ for Mac OSX to calculate the area of each of the 11 biome areas at each time period.

For each biome area, geographic area size was plotted against time (Figure 3.2), the area under each curve was traced, and this area estimated using ImageJ. Then, these area–time measures were log-transformed and tested for significance by pairwise correlations with log-transformed current tree diversity. This was performed for the area under the curve since the Miocene, the Oligocene, and the Eocene (Figure 3.2).

The extant species richness of biomes may also be driven by extinction caused by contractions in biome area during Pleistocene climate change. To test this hypothesis, we correlated the minimum size for each of the 11 biome areas at either the Last Glacial Maximum or the Mid-Holocene with log current tree diversity.

EMPIRICAL TESTS OF THE GAH AND GAAH

We found no significant relationship between current biome area and tree diversity ($R^2 = 0.13$, $P = 0.3$; Figure 3.3). Although trends are evident associating larger biomes with higher species richness within the two boreal biomes, the three tropical biomes, and the six temperate biomes,

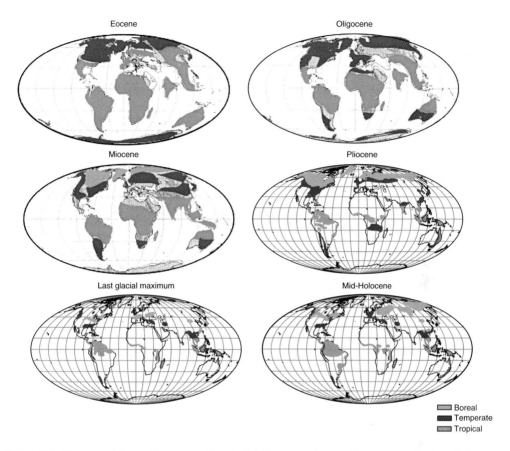

Figure 3.1 The maps of the past biomes, used to calculate the composite area–time measures in Figure 3.2. Paleocoastlines from Smith *et al.* (1994) are shown for the Eocene, Oligocene, and Miocene maps. Equal area maps of present-day coastlines are shown for the Pliocene, Last Glacial Maximum, and Mid-Holocene maps.

the relationship disappears when all biome areas are considered together. Thus, the current size of biomes does not explain tree species richness totals, as predicted by the GAH.

The integral of biome area over time (log biome area × age) exhibited a significant and positive correlation with current tree species richness, a result that holds for cumulative time periods since the Miocene ($R^2 = 0.35$, $P < 0.05$; Figure 3.4a), the Oligocene ($R^2 = 0.51$, $P < 0.01$; Figure 3.4b), and the Eocene ($R^2 = 0.67$, $P < 0.001$; Figure 3.4c). The R^2 value of the correlation increases as time increases.

Extant tree diversity is also significantly and positively correlated with minimum biome size during the Pleistocene ($R^2 = 0.48$, $P < 0.02$; Figure 3.5). This suggests that extinction via range contraction during Pleistocene climate change may also help explain the disparities in species richness among the 11 biome areas we tested.

DISCUSSION

We found that current geographic area size and species richness are not positively correlated across biomes, a result that does not support the GAH as a primary explanation for the latitudinal diversity gradient. However, because the GAH is based on factors influencing *in situ* speciation

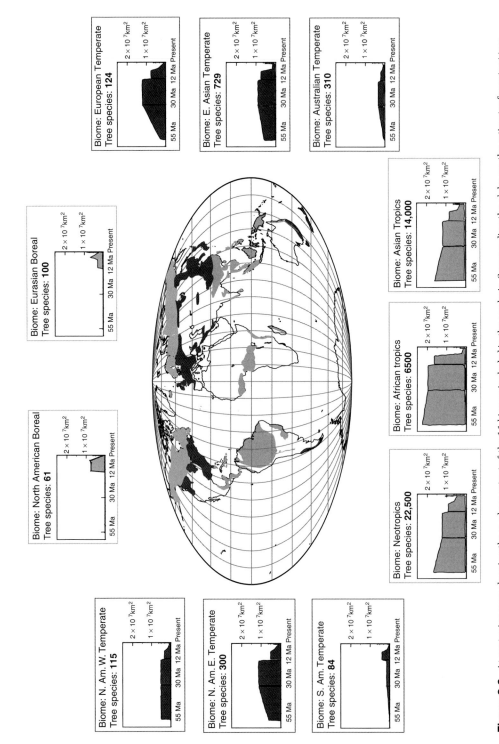

Figure 3.2 At center is a map showing the modern extent of the 11 biomes included in the analysis. Surrounding the globe are the estimates for extant tree species richness and composite area–time measures for each biome. The area of each area–time plot was quantified, log-transformed and then correlated with log tree diversity to test for the time-integrated species–area effect.

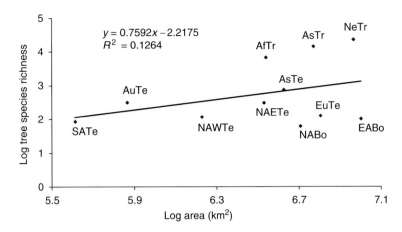

Figure 3.3 Log of extant tree species richness plotted against the log of 11 extant biome areas. AfTr, African Tropics; AsTr, Asian Tropics; NeTr, Neotropics; SATe, South American Temperate; AuTe, Australian Temperate; NAWTe, North American Western Temperate; NAETe, North American Eastern Temperate; AsTe, Asian Temperate; EuTe, European Temperate; NABo, North American Boreal; EABo, Eurasian Boreal.

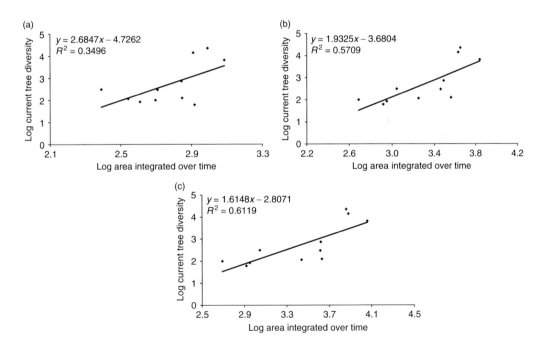

Figure 3.4 Log current tree diversity plotted against log area integrated over time since (a) the Miocene to present (11.2 Ma), (b) the Oligocene to present (30 Ma), and (c) the Eocene to present (55 Ma). Each point corresponds to the amount of area under the curve in each biome history plot in Figure 3.2. In (a), only the black areas (from 11.2 Ma to the present) were included, in (b) the black and dark gray areas (from 30 Ma to the present) were included, and in (c) the entire area under the curve was included (from 55 Ma to the present).

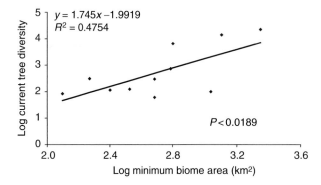

Figure 3.5 Log current tree diversity plotted with the minimum size of each biome during the Pleistocene. The smallest areas for each of the 11 biomes in the two Pleistocene maps (21,000 and 6000 years ago) were log-transformed and plotted with log current tree species richness.

and extinction rates, this conclusion is unsatisfying because it ignores the possibility that current biome diversities are not at equilibrium – that is, at levels attained if current areas were stably maintained for tens of millions of years.

By incorporating the history of biome size into the analysis, we found significant correlations between the integral of area over time and current tree diversity, whether evaluated from the Miocene, Oligocene, or Eocene. Fine and Ree (2006) reported similar results when they correlated the same tree species diversity estimates with area–time composites based on five separate interpretations of past climates based on five sources. Together, these results suggest that the combined size and longevity of a biome are important factors in explaining its current species richness. Examined alone, however, neither factor produced a significant trend. The idea that time is important is hardly new: Wallace (1876), Willis (1922), and others have claimed that the extraordinarily high diversity of tropical rainforests is due to the stability or greater age of these forests. Many have also noted that most plant lineages appear to have originated in the tropics (Crane and Lidgard 1990, Latham and Ricklefs 1993, Judd et al. 1994, Ricklefs 1999, 2004, Wiens and Donoghue 2004). Other large-scale studies have noted that area and species diversity are positively related. Tiffney and Niklas (1990) found that fossil plant species richness correlated with the overall land area of the northern hemisphere at 12 slices in time between 410 Ma and 10 Ma, and a recent study of palynological data by Jaramillo et al. (2006) found that neotropical tree

diversity peaked in the Eocene when we propose that tropical forests covered the largest amount of area.

Extinction

It is difficult to assess the relative importance of speciation versus extinction as underlying causes of the result that area size integrated over time correlates with extant tree diversity (Figure 3.4). The significant correlation between minimum biome area during the Pleistocene and extant tree diversity is suggestive that extinction caused by glacial cooling and drying is one important factor in explaining extant tree diversity patterns. However, another important consideration might be the amount of latitudinal shift in each biome during glacial periods. For example, the tropics also decreased in area during the Pleistocene, but tropical refugia remained within earlier tropical biome borders (Bush 1994). In contrast, temperate and boreal refugia were located closer to the equator during glacial advances, and largely outside of their pre-glacial period borders (Figure 3.1). If trees disperse more slowly than their moving refugia (cf. McLachlan et al. 2005), temperate tree species extinction may have been higher than that attributable solely to the overall reduction in biome area.

The tropical conservatism hypothesis

The GAAH that we test here is in some ways similar to the tropical conservatism hypothesis

(TCH) recently proposed by Wiens and Donoghue (2004). The TCH is based on three basic ideas. First, if a clade originated in the tropics, it is expected to include more tropical species because of the longer temporal duration of the tropics, and hence greater opportunity for diversification in tropical regions. Second, if tropical areas have covered larger areas for longer durations than extra-tropical areas, a correspondingly higher proportion of extant lineages should have originated in the tropics. Third, if adaptations to survive freezing temperatures are necessary to invade extra-tropical regions and these adaptations are difficult to acquire and maintain, then niche conservatism within tropical lineages will maintain the disparity in species richness over time (Latham and Ricklefs 1993, Ricklefs 1999, Wiens and Donoghue 2004).

The GAAH and the TCH both predict peaks in species richness in tropical areas, but they approach the disparity of species richness between biomes from different angles. The GAAH focuses on the intrinsic properties of biomes that influence *in situ* diversification of resident species. The TCH focuses on lineages, particularly the phylogenetic distribution of tropical and non-tropical taxa, as this bears directly on general inferences about the geographic history of diversification across many clades. The two hypotheses intersect in their emphasis on the tropics being larger and older than extra-tropical regions.

However, the GAAH provides a general rule for why most lineages can be traced to the tropics and why tropical lineages have undergone greater diversification overall than temperate and boreal lineages. For example, the TCH argues that because frost tolerance is a difficult physiological barrier for angiosperms to overcome, relatively few lineages were able to colonize the temperate zone. But phylogenetic niche conservatism explains only why so few lineages cross into the temperate zone – it does not address why the lineages that do acquire frost tolerance have not diversified to the same degree as their tropical relatives. It is not likely that those lineages that did cross the frost-line are inherently constrained in their potential for diversification. We suggest that if temperate areas were large and stable through time, an interval on the order of 55 million years

might be sufficient for the development of a comparably diverse temperate flora. However, climate changes and glaciation at higher latitudes during the last 55 million years have resulted in much smaller effective areas for temperate and boreal biomes than tropical ones. During this time, tropical biomes overall have been larger than temperate biomes (Figure 3.2), and are therefore expected to be characterized by lower extinction rates (and perhaps higher speciation rates as well), leading to higher species richness in the tropics (Fine and Ree 2006).

If the assembly of forest communities has been characterized by evolutionary responses to physiological thresholds that exist between biomes (e.g., frost tolerance), then boreal lineages should tend to be phylogenetically nested within temperate lineages, and temperate lineages within tropical lineages (Wiens and Donoghue 2004). The data currently available support this prediction (e.g., see Judd *et al.* 1994, Hoffmann 1999, Scheen *et al.* 2004). If diversification within a lineage could be dated and mapped onto reconstructions of past biomes (similar to Figure 3.1), it would allow for a powerful test of how the area of a biome over time affects speciation rates (Ricklefs 2004). We caution, however, that a large number of independent lineages would need to be studied to avoid sampling bias in detecting any general relationship. In addition, it is important to note that some tropical lineages have crossed the frost-line, but disappear from our analysis of temperate areas because they include trees in the tropics but only herbaceous plants in the temperate zone (i.e., Clusiaceae). Thus, our non-phylogenetic focus on "trees" rather than monophyletic groups underscores the different approaches needed by clade-based (like the TCH) versus biome-based (like the GAAH) analyses of variation in species richness.

Another exciting avenue to follow would be to estimate the rates at which lineages cross biome boundaries. For this, likelihood-based inference methods for historical biogeography would be useful (e.g., Ree *et al.* 2005). Using data on the location, frequency, and timing of lineage expansions across biome boundaries, one could ask: Do tropical and extra-tropical lineages diversify at similar rates? Are some boundaries between biomes more frequently crossed than others, and if so, is there a relationship with the length of the boundary line? Does lineage diversification (and expansion into a new biome) coincide with increases or decreases in the area of a biome over time?

Species–energy hypothesis

The relationship between extant log biome area and log biome species richness is not linear largely because of the two boreal biome areas (Figure 3.3). These boreal sites cover disproportionately large areas, yet are depauperate in tree species (Table 3.1). Rosenzweig (1992, 1995) accounted for this discrepancy by adding productivity as a corollary to the GAH. Indeed, leaving aside area, many other proponents of the "species–energy hypothesis" have linked global diversity patterns to productivity (or correlates of productivity: Currie and Paquin 1987, Adams and Woodward 1989, Wright *et al.* 1993). All of these studies found strong correlations between tree diversity and actual evapotranspiration (AET). What is the mechanism by which productivity affects speciation and extinction? The common view is that higher productivity enables more individuals to inhabit an area, increasing population density and leading to higher speciation and lower extinction rates compared with areas with low productivity (Gaston 2000). This argument is strikingly similar to that underlying the GAH. Ricklefs (1999) argued convincingly that explanations involving environmental determinism (like productivity) and explanations involving history and regional effects should be disentangled. Although productivity (or energy) may be important at the local level, regional processes and historical events contribute to species richness patterns and can override local effects (Ricklefs and Schluter 1993, Ricklefs 1999).

Boreal biomes have been in existence for only the past 4–10 million years (Graham 1999, Willis and McElwain 2001). Because trees require specific adaptations to survive boreal climates (Woodward 1987), it would require a radiation several orders of magnitude faster than the fastest known plant radiation (Hawaiian silverswords, Baldwin and Sanderson 1998) for boreal biomes to have species richness totals similar to tropical biomes. In other words, while productivity may slightly mediate the effect of area (by influencing the number of individuals that can share space in a biome), the effect of productivity on current tree species richness patterns must be negligible compared with the effect of the size of biomes through time (McGlone 1996, Fine and Ree 2006). As a thought experiment (cf. ter Steege *et al.* 2000), let us imagine a world where moist tropical areas were small and periodically reduced in size, perhaps by extreme dryness, while large extra-tropical areas with climates similar to today's boreal biomes stayed the same size for tens of millions of years. In such a world, would we find highly diverse tropical rainforest and a low-diversity boreal forest, or the reverse?

CONCLUSION

We evaluated the importance of geographic area in explaining tree diversity patterns at the largest spatial scales. Because area is predicted to influence *in situ* speciation and extinction rates in a biome, we considered changes in a biome's size over time periods appropriate for those processes. We tested the GAH with empirical data on global tree diversity and estimates of biome extent over three large slices of time within the last 55 million years to the present, finding a significant relationship between biome size integrated over time and current species diversity. Although other explanations may also be valid for the latitudinal gradient in tree diversity, our analysis suggests a significant role for the size and age of a biome area in determining its species richness. Under this explanation, tropical forests simultaneously represent both a museum that preserves and a cradle that generates new lineages. In addition, the differential reduction of habitable area in tropical, temperate, and boreal zones, which likely caused differential increases in extinction during the Pleistocene in each of the 11 biome areas, may have been an important factor affecting current tree diversity patterns.

Because different explanations for the latitudinal diversity gradient may be valid for different kinds of organisms, some have recommended that we cease trying to find a universal cause (Gaston 2000). We are sympathetic to this view. For example, the mechanisms causing the famous reverse latitudinal gradients, such as in salamanders, are unlikely to be the same as the mechanism causing forward latitudinal gradients (Willig *et al.* 2003). But it also seems appropriate to consider that organisms like trees take on a greater importance because of their role in providing habitat, food, and shelter for so many other organisms (e.g., Huston's 1994 structural versus interstitial organisms). If we understand the most important causes of global diversity gradients in trees, we will also understand one element controlling gradients of organisms that are dependent on them for survival, such as specialist herbivorous insects, which may represent most of the world's biodiversity.

Human activity is currently causing both massive losses of habitat and rapid climate change (Corlett and Primack Chapter 26, this volume, Laurance Chapter 27, this volume). Understanding the consequences of fluctuations of biome area through time on global tree diversity is critical as we strive to develop effective conservation strategies. If we consider conservation of natural habitats at the largest scales, the results reported here suggest that tropical areas should be more sensitive to habitat loss than high-latitude areas because extinctions during the last glacial period have likely set temperate and boreal biomes well below their equilibrium diversity values. For this reason, we might predict a lower extinction rate in temperate or boreal biomes with moderate losses in effective area, while the same amount of habitat destruction in the tropics should cause much higher extinction rates. Thus, because of their greater potential for higher extinction rates, areas in tropical biomes should be given the highest priority for conservation.

ACKNOWLEDGMENTS

We are grateful to R. Condit, E. Dinerstein, R. Foster, P. Grogan, C. Hawkins, H. Helmisaari, H. Hytteborne, J. LaFrankie, R. Leemans, S. Linder, W. Pruitt, A. Solomon, A. Svidenko, and W. Wettengel for responding to our queries relating to information on the world's biomes and tree diversity estimates. We thank S. Schnitzer, C. Dick, Z. Miller and one anonymous reviewer for helpful comments regarding the manuscript, and the Michigan Society of Fellows for support.

REFERENCES

Adams, J.M. and Woodward, F.I. (1989) Patterns in tree species richness as a test of the glacial extinction hypothesis. *Nature* 339, 699–701.

Baldwin, B.G. and Sanderson, M.J. (1998) Age and rate of diversification of the Hawaiian silversword alliance. *Proceedings of the National Academy of Sciences of the United States of America* 95, 9402–9406.

Bates, J.M., Hackett, S.J., and Goerck, J.M. (1999) High levels of mitochondrial DNA differentiation in two lineages of antbirds (*Drymophila* and *Hypocnemis*). *Auk* 116, 1093–1106.

Beerling, D.J. and Woodward, F.I. (2001) *Vegetation and the Terrestrial Carbon Cycle: Modeling the First 400 Million Years*. Cambridge University Press, Cambridge.

Brown, W.L. Jr. (1957) Centrifugal speciation. *Quarterly Review of Biology* 32, 247–277.

Bush, M.B. (1994) Amazonian speciation: a necessarily complex model. *Journal of Biogeography* 21, 5–17.

Chown, S.L. (1997) Speciation and rarity: separating cause from consequence. In W.E. Kunin and K.J. Gaston (eds), *The Biology of Rarity*. Chapman & Hall, London, pp. 91–109.

Chown, S.L. and Gaston, K.J. (2000) Areas, cradles, and museums: the latitudinal gradient in species richness. *Trends in Ecology and Evolution* 15, 311–315.

Crane, P.R. and Lidgard, S. (1990) Angiosperm diversification and paleolatitudinal gradients in Cretaceous floristic diversity. *Science* 246, 675–678.

Currie, D.J. and Paquin V. (1987) Large-scale biogeographical patterns of species richness of trees. *Nature* 329, 326–327.

Dowsett, H.J., Barron, J.A., Poore, R.Z. *et al.* (1999) Middle Pliocene paleoenvironmental reconstruction: PRISM2. US Geological Survey Open File Report 99–535.

Fine, P.V.A. (2001) An evaluation of the geographic area hypothesis using the latitudinal gradient in North American tree diversity. *Evolutionary Ecology Research* 3, 413–428.

Fine, P.V.A. and Ree, R.H. (2006) Evidence for a time-integrated species–area effect in the latitudinal gradient in tree diversity. *American Naturalist* 168, 796–804.

Francis, W.D. (1981) *Australian Rain Forest Trees, Including Notes on Some of the Tropical Rain Forests and Descriptions*, 4th Edition. Australian Government Publication Service, Canberra.

Gaston, K.J. (1996) Species-range size distributions: patterns, mechanisms and implications. *Trends in Ecology and Evolution* 11, 197–201.

Gaston, K.J. (2000) Global patterns in biodiversity. *Nature* 405, 220–227.

Gaston, K.J. and Blackburn, T.M. (2000) *Pattern and Process in Macroecology*. Blackwell Science, Oxford.

Gavrilets, S., Li, H., and Vose, M.D. (1974) Patterns of parapatric speciation. *Evolution* 54, 1126–1134.

Graham, A. (1999) *Late Cretaceous and Cenozoic History of North American Vegetation*. Oxford University Press, Oxford.

Hawkins, B.A. and Porter, E.E. (2001) Area and the latitudinal diversity gradient for terrestrial birds. *Ecology Letters* 4, 595–601.

Hoffmann, M.H. (1999) The phylogeny of *Actaea* (Ranunculaceae): a biogeographical approach. *Plant Systematics and Evolution* 216, 251–263.

Huston, M.A. (1994) *Biological Diversity*. Cambridge University Press, Cambridge.

Hytteborn, H., Maslov, A.A., Nazimova, D.I. and Rysin, L.P. (2005) Boreal forests of Eurasia. In F. Andersson (ed), *Coniferous Forests*, Vol. 6 *Ecosystems of the World* (ed D.W. Goodall). Elsevier, Amsterdam.

Jaramillo, C., Rueda, M., and Mora, G. (2006) Paleogene patterns of plant diversification in the Neotropics. *Science* 311, 1893–1896.

Judd, W.S., Sanders, R.W., and Donoghue, M.J. (1994) Angiosperm family pairs: preliminary phylogenetic analyses. *Harvard Papers in Botany* 5, 1–51.

Latham, R.E. and Ricklefs, R.E. (1993) Continental comparisons of temperate-zone tree species diversity. In R.E. Ricklefs and D. Schluter (eds), *Species Diversity in Ecological Communities: Historical and Geographical Perspectives*. University of Chicago Press, Chicago, IL, pp. 294–314.

Loehle, C. (1998) Height growth rate tradeoffs determine southern range limits for trees. *Journal of Biogeography* 25, 735–742.

MacArthur, R.M. (1972) *Geographical Ecology*. Harper & Row, New York.

McLachlan, J.S., Clark, J.S., and Manos, P.S. (2005) Molecular indicators of tree migration capacity under rapid climate change. *Ecology* 86, 2088–2098.

Magallón, S. and Sanderson, M.J. (2001) Absolute diversification rates in angiosperm clades. *Evolution* 55, 1762–1780.

Martin, P.R. and McKay, J.K. (2004) Latitudinal variation in genetic divergence of populations and the potential for future speciation. *Evolution* 58, 938–945.

Marzluff, J.M. and Dial, K.P. (1991) Life history correlates of taxonomic diversity. *Ecology* 72, 428–439.

Mayr, E. (1954) Change of genetic environment and evolution. In J.S. Huxley, A.C. Hardy, and E.B. Ford (eds), *Evolution as a Process*. Allen and Unwin, London, pp. 157–180.

Mayr, E. (1963) *Animal Species and Evolution*. Harvard University Press, Cambridge, MA.

McGlone, M.S. (1996) When history matters: scale, time, climate and tree diversity. *Global Ecology and Biogeography Letters* 5, 309–314.

Moritz, C., Patton, J.L., Schneider, C.J., and Smith, T.B. (2000) Diversification of rainforest faunas: an integrated molecular approach. *Annual Review of Ecology and Systematics* 31, 533–563.

Morley, R.J. (2000) *Origin and Evolution of Tropical Rain Forests*. John Wiley and Sons, Chichester.

Olson, D.M., Dinerstein, E., Wikramanayake, E.D. *et al.* (2001) Terrestrial ecoregions of the world: a new map of life on earth. *BioScience* 51, 933–938.

Palumbi, S.R. (1992) Marine speciation on a small planet. *Trends in Ecology and Evolution* 7, 114–118.

Parrish, J.T., Ziegler, A.M., and Scotese, C.R. (1982) Rainfall patterns and the distribution of coals and evaporites in the Mesozoic and Cenozoic. *Palaeogeography, Palaeoclimatology, Palaeoecology* 40, 67–101.

Pennington, R.T., Richardson, J.E., and Cronk, Q.C.B. (2004) Plant phylogeny and the origin of major biomes: introduction and synthesis. *Philosophical Transactions of the Royal Society of London Series B* 359, 1455–1465.

Petrides, G.A. (1998) *A Field Guide to the Eastern Trees: Eastern United States and Canada, Including the Midwest*. Houghton Mifflin, New York.

Petrides, G.A. and Petrides, O. (1992) *A Field Guide to Western Trees: Western United States and Canada*. Houghton Mifflin, New York.

Ree, R.H., Moore, B.M., Webb, C.O., and Donoghue, M.J. (2005) A likelihood approach to inferring the evolution of geographic range on phylogenetic trees. *Evolution* 59, 2299–2311.

Ricklefs, R.E. (1999) Global patterns of tree species richness in moist forests: distinguishing ecological influences and historical contingency. *Oikos* 86, 369–373.

Ricklefs, R.E. (2003) Global diversification rates of passerine birds. *Proceedings of Royal Society of London B* 270, 2285–2291.

Ricklefs, R.E. (2004) A comprehensive framework for global patterns in biodiversity. *Ecology Letters* 7, 1–15.

Ricklefs, R.E. and Schluter, D. (1993) Species diversity: regional and historical influences. In R.E. Ricklefs and D. Schluter (eds), *Species Diversity in Ecological Communities: Historical and Geographical Perspectives.* University of Chicago Press, Chicago, IL, pp. 350–364.

Rodriguez, R., Matthei O., and Quezada, M. (1983) *Flora arbórea de Chile.* Universidad de Concepción, Concepción, Chile.

Rohde, K. (1997) The larger area of the tropics does not explain latitudinal gradients in species diversity. *Oikos* 79, 169–172.

Rosenzweig, M.L. (1992) Species diversity gradients: we know more and less than we thought. *Journal of Mammalogy* 73, 715–730.

Rosenzweig, M.L. (1995) *Species Diversity in Space and Time.* Cambridge University Press, Cambridge.

Rosenzweig, M.L. and Sandlin, E. (1997) Species diversity and latitudes: listening to area's signal. *Oikos* 80, 172–176.

Sakai, A. and Weiser, C.J. (1973) Freezing resistance of trees in North America with reference to tree regions. *Ecology* 54, 118–126.

Scheen, A.C., Brochmann, C., Brysting, A.K. *et al.* (2004) Northern hemisphere biogeography of *Cerastium* (Caryophyllaceae): insights from phylogenetic analysis of noncoding plastid nucleotide sequences. *American Journal of Botany* 91, 943–952.

Schemske, D.W. (2002) Ecological and evolutionary perspectives on the origins of tropical diversity. In R. Chazdon and T. Whitmore (eds), *Foundations of Tropical Forest Biology: Classic Papers with Commentaries.* University of Chicago Press, Chicago, IL, pp. 163–173.

Scotese, C.R. (2004) Cenozoic and Mesozoic paleogeography: changing terrestrial biogeographic pathways. In M. Lomolino and L. Heaney (eds), *Frontiers of Biogeography: New Directions in the Geography of Nature.* Sinauer, Sunderland, MA, pp. 1–27.

Smith, A.G., Smith, D.G., and Funnell, B.M. (1994) *Atlas of Cenozoic and Mesozoic Coastlines.* Cambridge University Press, New York.

Stanley, S.M. (1979) *Macroevolution.* W.H. Freeman, San Francisco, CA.

Stebbins, G.L. (1974) *Flowering Plants.* Harvard University Press, Cambridge, MA.

Svenning, J-C. and Skov, F. (2004) Limited filling of the potential range in European tree species. *Ecology Letters* 7, 565–573.

ter Steege, H., Sabatier, D., and Castellanos, H. (2000) An analysis of floristic composition and diversity of Amazonian forests including those of the Guiana shield. *Journal of Tropical Ecology* 16, 801–828.

Terborgh, J. (1973) On the notion of favorableness in plant ecology. *American Naturalist* 107, 481–501.

Tiffney, B.H. and Niklas, K.J. (1990) Continental area, dispersion, latitudinal distribution, and topographic variety: a test of correlation with terrestrial plant diversity. In R.M. Ross and W.D. Almon (eds), *Causes of Evolution: A Paleontological Perspective.* University of Chicago Press, Chicago, IL, pp. 76–102.

Wallace, A.R. (1876) *The Geographic Distribution of Animals.* Macmillan, London.

Wiens, J.J. and Donoghue, M.J. (2004) Historical biogeography, ecology and species richness. *Trends in Ecology and Evolution* 19, 639–644.

Whittaker, R.J., Willis, K.J., and Field, R. (2001) Scale and species richness: towards a general hierarchical theory of species diversity. *Journal of Biogeography* 28, 453–470.

Willig, M.R., Kaufman, D.M., and Stevens, R.D. (2003) Latitudinal gradients of biodiversity: pattern, process, scale and synthesis. *Annual Review of Ecological Evolution and Systematics* 34, 273–309.

Willis, J.C. (1922) *Age and Area: A Study in Geographical Distribution and Origin in Species.* Cambridge University Press, Cambridge, MA.

Willis, K.J. and McElwain, J.C. (2002) *The Evolution of Plants.* Oxford University Press, Oxford.

Woodward, F.I. (1987) *Climate and Plant Distribution.* Cambridge University Press, Cambridge, MA.

Wright, D.H., Currie, D.J., and Maurer, B.A. (1993) Energy supply and patterns of species richness on local and regional scales. In R.E. Ricklefs and D. Schluter (eds), *Species Diversity in Ecological Communities: Historical and Geographical Perspectives.* University of Chicago Press, Chicago, IL, pp. 66–74.

Zink, R.M. (2004) The role of subspecies in obscuring avian biological diversity and misleading conservation policy. *Proceedings of Royal Society of London B* 271, 561–564.

Chapter 4

EXPLAINING GEOGRAPHIC RANGE SIZE BY SPECIES AGE: A TEST USING NEOTROPICAL Piper SPECIES

John R. Paul and Stephen J. Tonsor

OVERVIEW

Tropical plant species vary dramatically in their geographic range sizes. Theory predicts that narrowly endemic species may simply be young species that have not had sufficient time to expand their ranges. If two assumptions are met, namely that new species start with small range sizes and that the probability of extinction is inversely related to range size, then older species should, on average, have larger range sizes than younger species. This conjecture, originally formulated by John Willis as the age-and-area hypothesis, and recently predicted by models of neutral community dynamics, has not been adequately tested in tropical plant taxa. To test this hypothesis, we focused on neotropical species of the tropical understory shrub genus *Piper* (Piperaceae). We used published internal-transcribed spacer (ITS) sequences to infer species' divergence times using Bayesian relaxed-clock methods and herbarium records to estimate range sizes. We asked if there is a positive relationship between species age and range size. Using linear regression, we found that relative species age significantly explains a quarter of the variance in range size among species in this prominent tropical plant genus. This result confirms that species age can be a significant predictor of range size, and is notable in light of uncertainties in divergence time estimation using limited sequence data and incomplete sampling. We discuss the generality of our results with regard to other tropical plant taxa and briefly review the limited data on species-level age estimates from tropical plants. Furthermore, we discuss the potential limitations and difficulties of using divergence times as proxies for species ages, particularly when applied to analyses involving range and population sizes of new species. We suggest that the wealth of new genetic and biogeographic data on tropical plant species promises broader explorations of the impact of species age on species' range sizes in the near future.

INTRODUCTION

What accounts for rarity and endemism? Ecology, the study of distribution and abundance of species, remains without a coherent and consistent answer to this question. In tropical forest communities, the vast majority of species have few individuals and small geographic ranges (Wallace 1878, Dobzhansky 1950, Hubbell 2001a). Explaining how rare species differ from more common species, and elucidating the relative importance of various factors that regulate species' abundance and distribution, is a central goal of ecology. However, the complicating influence of both deterministic and stochastic forces acting at various levels of biological organization and temporal duration make this a difficult task. In this chapter, we concentrate on the role of evolutionary history in structuring the abundance and distribution of plant species in tropical forests. Specifically, we address how the age of species can help explain patterns of rarity and endemism.

The potential importance of species age as a predictor of range size was first championed by Willis (1922). His "age-and-area hypothesis" asserted that, on average, older species will have larger ranges than younger species. He drew much of his evidence from studies of the tropical flora of Ceylon (now Sri Lanka) where he observed that putatively ancestral species were more widely distributed than derived forms. Willis published a number of papers on the subject, and his ideas were subsequently debated and, in some cases, even ridiculed (e.g., Fernald 1924, Gleason 1924). In time, Willis's hypothesis failed to gain support (Stebbins and Major 1965) and his most lasting influence may actually have been in phylogenetics, via Yule's (1925) seminal paper that mathematically derived a model of a pure-birth speciation process, using Willis's ideas as the theoretical foundation.

Recently, the potential effect of historical processes on the distribution and abundance of organisms has received renewed attention (e.g., Ricklefs 2004, Wiens and Donoghue 2004). Much of this interest has been driven by two factors: the influx of molecular data on organisms that provide the potential to age the divergence dates of species, and the publication of Hubbell's *Neutral Theory of Biodiversity and Biogeography* (2001a), which incorporates the large-scale, long-term effects of speciation and extinction on the abundance and distribution of species. Hubbell's neutral theory also specifically predicts that most rare, endemic species will be young species, while most wide-ranging species will be old (Hubbell 2001a,b); in effect, Hubbell's model makes a prediction similar to Willis's hypothesis. This prediction can be viewed as a general expectation, rather than a prediction specific to Hubbell's model. A positive relationship between species age and range size can be expected if two assumptions are met: (1) species start with small population and range sizes; and (2) extinction risk is inversely proportional to population and/or range size. Under these assumptions, new (young) species will have small population and range sizes and will face a high probability of extinction, while species that do persist and increase in range size will

face a decreasing probability of extinction. As a result, on average, young species are expected to be narrowly endemic species, while wide-ranging species are expected to be old. Interestingly, some of the strongest criticism of Hubbell's neutral model has focused on the expected age of common species. Specifically, if common species reach high abundance via ecological drift, the expected age of these species is unrealistically old, because of the slow pace of drift (Leigh 1999, Ricklefs 2003, Nee 2005). In contrast, if fitness deviations are accepted in the model, species can reach high abundance or go extinct much more quickly (e.g., Yu *et al.* 1998, Fuentes 2004). As a result, a positive age and range size relationship may be expected to persist much longer in clades that have been primarily driven by neutral processes than in clades where selection has driven species with high relative fitness to occupy large ranges.

Of course, the relationship between species age and range size may take many forms, and Willis's age-and-area hypothesis (1922) is only one of several models of post-speciation range-size transformation. For example, Gaston and colleagues (Chown 1997, Gaston 1998, 2003, Chown and Gaston 2000) have summarized a series of models of post-speciation range-size transformations (e.g., cyclical, random, stasis, etc.) that could potentially better explain the age and area relationships of some species. For example, the age and area relationship may be explained by a hump-shaped curve, where species start with small range sizes, reach their maximum range size at an intermediate age, and then decline towards extinction when they are old. Such a pattern was found for the proportion of fossil assemblages occupied by Cenozoic mollusks (Foote *et al.* 2007). Because there are a variety of processes that can expand or reduce species' ranges, individual clades may have their own unique age–area relationships. Thus, the utility of species age as a broad explanatory variable remains to be seen. In this chapter, we briefly review the few empirical tests of age and area and present an analysis using a clade of tropical understory shrubs (*Piper*). We discuss how the species age and range size relationship can be viewed more broadly than the simple hypothesis presented by Willis (1922) and how this can lead to new hypotheses and understandings of the impact of historical processes on the current distribution and abundance of species.

EMPIRICAL TESTS OF AGE AND AREA

There have been few explicit tests of the age-and-area hypothesis. Two studies of marine fossil fauna have found evidence that indirectly supports a positive age and area relationship. Jablonski (1987) documented a positive relationship between age (species duration) and geographic range size in the beginning of fossil mollusk species' lifetimes, followed by long periods of stasis, but the focus of that study was on the possibility of species-level selection, rather than testing age and area *per se*. Similarly, Miller (1997) found that in Ordovician marine genera, older genera had larger ranges. Studying birds, Gaston and Blackburn (1997) found that for the entire New World avifauna, there was no relationship between mean range size of a clade and clade age, but there was a weak positive relationship between evolutionary age and total clade range size. In another study, Webb and Gaston (2000) examined six clades of birds and found various forms of the age and range size relationship. Overall, roughly 20–50% of the variance in range size could be accounted for by species age (inferred from standard mitochondrial DNA molecular clock divergence estimates of 2% divergence per million years, Ma), but only one clade showed a positive age and area relationship; three showed a negative relationship and two a hump-shaped relationship. A study on *Sylvia* warblers found a weakly significant positive relationship between breeding range size and species age, but in that study the relationship could be better explained by older species generally having better dispersal abilities than younger species (Böhning-Gaese *et al.* 2006). Finally, Jones *et al.* (2005) analyzed large molecular datasets of primates and carnivores and found evidence of a weakly negative age and area relationship (see that paper for a more detailed overview of Willis's age-and-area hypothesis and approaches to testing it).

Overall, a convincing positive age and area relationship predicted by Willis is not supported

by these empirical data. However, a careful look at the published data reveals two trends. First, analyses that use fossil samples and measures of species duration as a proxy for age tend to find some evidence for a significant age and area relationship (e.g., Jablonski 1987, Miller 1997). In contrast, studies that examine extant species using molecular divergence dates as a proxy for age generally tend to find either no significant relationship between species age and range size, or a mixture of positive and negative relationships (see table 7.1 in Jones *et al.* 2005). This discrepancy may be due, at least partly, to the different sampling methods. For example, a species' fossil record potentially allows sampling along the entire history of a species' range-size trajectory over time (Figure 4.1a). This is the ideal situation, in which the range size for a given species can be estimated at multiple ages. In contrast, molecular dating methods generally permit a single snapshot of a species' age and range size at a given point in time, and by looking at multiple species we can infer the general trend of the age and area relationship for a group of organisms. Having only snapshots of a species age and range size relationship can introduce considerable variance into the relationship, particularly if all species follow varying range-transformation trajectories over time (even if the general shape of the relationship is similar, e.g., hump-shaped; Figure 4.1b). However, it is likely that the majority of future age estimates for most taxa will be derived through molecular-based inference; thus, understanding how these measures can potentially bias relationships such as age and area is critical to robust interpretation of results.

In addition to the potential discrepancies introduced through fossil versus molecular analysis of age and area, studies on extant species suggest that the phylogenetic level of the analysis is important. In studies of large clades containing many well-defined and potentially divergent subgroups (e.g., mammals, carnivores, or birds), general analyses of age and area find no or weak relationships (Jones *et al.* 2005), while studies of individual clades within these broad groups often find significant, but inconsistent, relationships (e.g., the six clades of birds studied by Webb and Gaston 2000). This discrepancy suggests that the signal of an age and area relationship may be

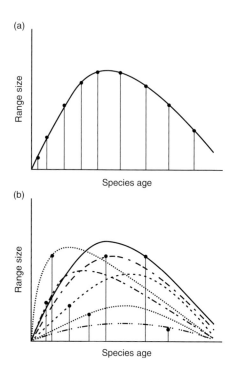

(a)

Range size

Species age

(b)

Range size

Species age

Figure 4.1 Graphical depictions of range-size trajectories of species over time. Black dots indicate sampling points in time. Ideally, fossil analyses can allow the range size of a species to be assessed at multiple time points (a), effectively sampling over the lifespan of a species. When using molecular estimates of ages, species can usually be sampled at only a single point in time (b). By sampling multiple species (different lines on the graph), a general relationship between species age and range size can be inferred. However, even if all species show roughly the same shape for an age and area relationship (e.g., hump-shaped), if they follow varying range-transformation trajectories, sampling single points over time will introduce considerable variation into the species age and range size relationship and make inferring general trends more difficult.

obscured when clades with distinct evolutionary histories are combined.

AN EMPIRICAL TEST USING A TROPICAL PLANT GENUS

Willis developed the age-and-area hypothesis thinking about tropical floras, and even his critics

acknowledged that the hypothesis might be more important in the tropics (Gleason 1924), which were seen as stable and relatively homogeneous. Despite this early attention to the tropics, to our knowledge there have been no explicit tests of the hypothesis using tropical plants. The immense diversity of tropical plant species is only beginning to receive a genetic treatment, and our estimates of species' range sizes are imperfect, but slowly improving (e.g., Pitman et al. 2001). Most of the molecular dating of tropical plants to date has been conducted at higher phylogenetic levels; typically these studies are concerned with the general age of families and genera, and inferring when and where these groups of species diversified (e.g., Davis et al. 2005, Zerega et al. 2005, Muellner et al. 2006). In contrast, analyses of age and area require species-level resolution to properly address the hypothesis.

Here we examine the relationship between relative species age and range size in the diverse shrub genus *Piper* (Piperaceae) using publicly available internal-transcribed spacer (ITS) sequences from GenBank (www.ncbi.nlm.nih.gov). Most of these sequences were originally published in Jaramillo and Manos (2001) and Jaramillo and Callejas (2004a,b). We chose *Piper* because its species are prominent and important members of many rainforest communities throughout the world (Jaramillo and Manos 2001, Marquis 2004), there is a reasonably large amount of species-level informative genetic data available, and this taxon is an ideal model system for the study of ecology and evolution (Dyer and Palmer 2004). We focused our analysis on neotropical species because many sequences were available for these species, the biogeography of neotropical species has been studied (Marquis 2004, Quijano-Abril et al. 2006), and the range sizes of many species could be estimated using data from the Missouri Botanical Garden's online database, W³Tropicos (http://mobot.mobot.org/W3T/Search/vast.html).

We used Bayesian inference to infer a phylogenetic tree, and then used this tree topology to estimate relative divergence dates among the species using the program BEAST v1.3 (Drummond and Rambaut 2007), which uses a Bayesian relaxed-clock approach to divergence time estimation (Drummond et al. 2006). For the phylogenetic inference, we aligned 113 sequences from 101 *Piper* (and *Macropiper*) species and five outgroup species using ClustalW (Thompson *et al.* 1994), followed by manual corrections. We used Modeltest (Posada and Crandall 1998) to evaluate the most appropriate model of molecular evolution for our analysis, which was determined by Akaike's information criterion (AIC) model selection to be the general time reversible model with gamma distributed rates and proportion of invariable sites (GTR+I+G). We ran our analysis in MrBayes 3.1.1 (Ronquist and Huelsenbeck 2003), using model specifications for the GTR+I+G model, with a Dirichlet prior on substitution rates and state frequencies, and an unconstrained, exponential prior distribution on branch lengths. All analyses with MrBayes used two concurrent runs, each with four Markov Chain Monte Carlo (MCMC) chains (one "cold" and three "heated" chains). We examined an initial run of 2 million generations of MCMC simulations to assess if the chain had reached a stable distribution. Although the log-likelihood values stabilized by approximately 200,000 generations, clade probabilities failed to stabilize until nearly 1.5 million generations (assessed using the program "Are We There Yet?," Wilgenbusch *et al.* 2004, Nylander *et al.* 2008). As a result, we ran a second analysis for 5 million generations, discarding the initial 2 million generations as burnin. This analysis effectively sampled from a stable distribution (with samples taken every 100 generations), resulting in a total of 60,000 trees after combining the two runs, from which a majority rule consensus tree was derived (Figure 4.2). This tree recovered the major clades described for *Piper* in previous work on ITS sequences (Jaramillo and Callejas 2004b).

We then used the topology of this phylogenetic tree as our input tree for the relative age analysis in BEAST. We held the topology of the tree constant for the analysis and fixed the mean substitution rate to one. BEAST uses MCMC sampling to assess branch lengths and divergence times by varying substitution parameters and the rate distribution based on a model of molecular evolution (we used the GTR+I+G). A preliminary analysis running for 2 million generations did not stabilize and the effective sample sizes of many parameters were low. The

analysis presented here ran for 10 million generations, with the first 4 million discarded as burnin. The resulting samples (taken every 100 generations) showed a stable log-likelihood distribution and good effective sample sizes for all parameters. We assessed the posterior probability densities of ages (divergence times of two species subtending these nodes) for 47 nodes on the phylogenetic tree (Figure 4.2). The mean divergence time values of these nodes were used to determine the relative ages of the neotropical *Piper* species for the age and area analysis (Table 4.1). Since BEAST analyses have a stochastic element, we also ran the same analysis two additional times. The results were nearly identical (e.g., correlation coefficients of node ages between runs were >0.99) so only the first run results are presented here.

To estimate range sizes, we counted the number of $1° \times 1°$ latitude–longitude squares occupied by geo-referenced herbarium records in W[3]Tropicos. This is effectively an area of occurrence measure (Gaston 1994). A few species for which we determined the age did not have records in W[3]Tropicos; most of these were species listed as endemic to Colombia in Trelease and Yuncker (1950). Therefore, we present our analysis excluding these species; however, we also provided generous range-size estimates for these species and ran the analyses including them – the results were nearly identical and thus are not included here. The distribution of range sizes we calculated for the species with W[3]Tropicos records is presented in Figure 4.3. The distribution is characterized by a few species with large range sizes and a long tail of species with small ranges (<10 of $1° \times 1°$ latitude–longitude squares).

To assess the relationship between relative species age and range size, we used linear least-squares regression using SAS 8.2 (SAS Institute 2001). We log-transformed both the mean species' ages and range sizes of the 58 neotropical *Piper* species for which we had data. We found a highly significant positive relationship ($y = 0.9399x + 2.6143, P < 0.001$) that explains 25% ($r^2 = 0.252$) of the variation in range size for these *Piper* species (Figure 4.4). Thus, our analysis supports the simple, positive relationship between species age and range size predicted by

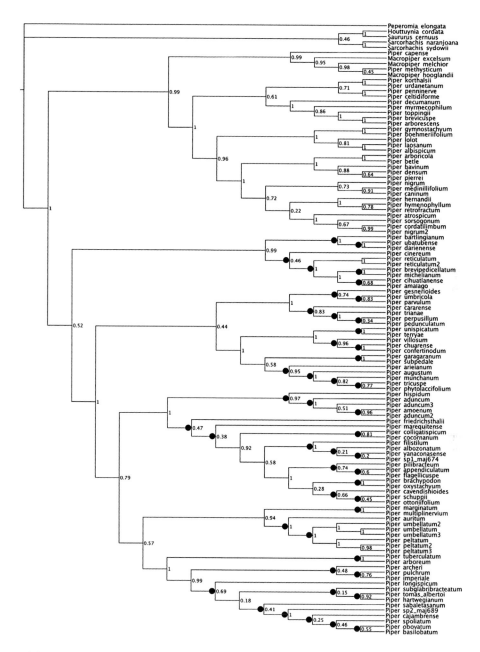

Figure 4.2 A phylogenetic hypothesis of *Piper* species relationships inferred by a Bayesian analysis of ITS sequences. Posterior probabilities of clades are shown at the nodes. Black dots depict the nodes for which relative ages were calculated in a separate Bayesian analysis in which this tree topology was used (see text for details).

Table 4.1 Relative ages of the *Piper* species estimated by Bayesian relaxed-clock analysis. The table shows the mean ages and standard deviations (SD), median ages, and highest posterior density (HPD) distributions.

Species	Mean age	SD	Median age	95% HPD (lower)	95% HPD (upper)
Piper aduncum	1.0E–02	2.3E–04	9.6E–03	3.2E–03	2.0E–02
Piper albozonatum	3.5E–03	6.6E–05	3.0E–03	2.3E–04	7.9E–03
Piper amalago	7.1E–03	1.5E–04	6.3E–03	1.4E–03	1.5E–02
Piper amoenum	1.0E–02	2.3E–04	9.6E–03	3.2E–03	2.0E–02
Piper appendiculatum	1.2E–02	1.6E–04	1.1E–02	4.3E–03	2.0E–02
Piper arboreum	3.2E–02	6.4E–04	3.1E–02	1.9E–02	4.7E–02
Piper archeri	2.3E–02	3.7E–04	2.2E–02	1.0E–02	3.6E–02
Piper arieianum	3.4E–02	4.7E–04	3.3E–02	2.2E–02	5.0E–02
Piper augustum	2.2E–02	3.5E–04	2.2E–02	1.2E–02	3.4E–02
Piper auritum	4.2E–02	5.9E–04	4.2E–02	2.6E–02	6.2E–02
Piper bartlingianum	4.8E–02	9.3E–04	4.7E–02	2.3E–02	7.5E–02
Piper basilobatum	3.9E–03	6.2E–05	3.5E–03	7.6E–04	7.6E–03
Piper brachypodon	1.3E–02	1.6E–04	1.2E–02	4.4E–03	2.1E–02
Piper brevipedicellatum	2.9E–03	5.4E–05	2.3E–03	4.5E–05	7.6E–03
Piper cajambrense	8.6E–03	1.8E–04	8.3E–03	4.1E–03	1.4E–02
Piper cararense	1.9E–02	3.6E–04	1.8E–02	7.2E–03	3.3E–02
Piper cavendishioides	1.8E–02	3.5E–04	1.8E–02	1.0E–02	2.6E–02
Piper chuarense	9.7E–03	2.3E–04	8.9E–03	3.4E–03	1.9E–02
Piper cihuatlanense	7.1E–03	1.5E–04	6.3E–03	1.4E–03	1.5E–02
Piper cinereum	7.6E–02	1.2E–03	7.5E–02	4.6E–02	1.1E–01
Piper cocornanum	1.8E–02	3.1E–04	1.8E–02	4.5E–03	3.3E–02
Piper colligatispicum	1.8E–02	3.1E–04	1.8E–02	4.5E–03	3.3E–02
Piper confertinodum	9.7E–03	2.3E–04	8.9E–03	3.4E–03	1.9E–02
Piper darienense	1.7E–02	3.5E–04	1.5E–02	4.3E–03	3.2E–02
Piper filistilum	7.1E–03	1.4E–04	6.4E–03	1.4E–03	1.4E–02
Piper flagellicuspe	1.2E–02	1.6E–04	1.1E–02	4.3E–03	2.0E–02
Piper friedrichsthalii	4.4E–02	5.4E–04	4.4E–02	3.1E–02	5.9E–02
Piper garagaranum	2.4E–02	4.7E–04	2.4E–02	7.9E–03	3.9E–02
Piper gesnerioides	1.9E–02	3.9E–04	1.8E–02	7.6E–03	3.3E–02
Piper hartwegianum	8.2E–03	1.2E–04	7.7E–03	2.0E–03	1.5E–02
Piper hispidum	3.7E–02	5.1E–04	3.6E–02	1.8E–02	5.6E–02
Piper imperiale	1.6E–02	3.0E–04	1.5E–02	5.6E–03	3.0E–02
Piper longispicum	1.9E–02	3.2E–04	1.8E–02	1.1E–02	2.9E–02
Piper marequitense	3.9E–02	4.8E–04	3.9E–02	2.6E–02	5.2E–02
Piper marginatum	2.9E–02	6.8E–04	2.8E–02	9.1E–03	5.0E–02
Piper michelianum	2.9E–03	5.4E–05	2.3E–03	4.5E–05	7.6E–03
Piper multiplinervium	2.9E–02	6.8E–04	2.8E–02	9.1E–03	5.0E–02
Piper munchanum	1.8E–02	3.2E–04	1.7E–02	8.5E–03	2.8E–02
Piper obovatum	3.9E–03	6.2E–05	3.5E–03	7.6E–04	7.6E–03
Piper ottoniifolium	1.6E–02	3.3E–04	1.6E–02	8.9E–03	2.5E–02
Piper oxystachyum	1.3E–02	1.6E–04	1.2E–02	4.4E–03	2.1E–02
Piper parvulum	1.2E–02	2.6E–04	1.1E–02	3.1E–03	2.2E–02
Piper pedunculatum	5.0E–03	9.2E–05	4.5E–03	7.8E–04	1.0E–02
Piper peltatum	2.2E–02	3.1E–04	2.1E–02	1.0E–02	3.4E–02
Piper perpusillum	5.0E–03	9.2E–05	4.5E–03	7.8E–04	1.0E–02

Continued

Table 4.1 Continued

Species	Mean age	SD	Median age	95% HPD (lower)	95% HPD (upper)
Piper phytolaccifolium	1.2E–02	2.4E–04	1.2E–02	5.0E–03	2.0E–02
Piper pilibracteum	1.8E–02	2.0E–04	1.7E–02	9.3E–03	2.7E–02
Piper pulchrum	1.6E–02	3.0E–04	1.5E–02	5.6E–03	3.0E–02
Piper reticulatum	4.4E–02	7.4E–04	4.3E–02	2.2E–02	6.9E–02
Piper sabaletasanum	1.4E–02	2.4E–04	1.4E–02	7.6E–03	2.2E–02
Piper schuppii	1.6E–02	3.3E–04	1.6E–02	8.9E–03	2.5E–02
Piper sp1maj674	1.8E–03	3.4E–05	1.5E–03	3.1E–05	4.5E–03
Piper sp2maj689	1.0E–02	2.0E–04	9.7E–03	4.7E–03	1.6E–02
Piper spoliatum	7.4E–03	1.6E–04	7.1E–03	3.3E–03	1.2E–02
Piper subglabribracteatum	1.3E–02	1.9E–04	1.3E–02	5.4E–03	2.2E–02
Piper subpedale	2.4E–02	4.7E–04	2.4E–02	7.9E–03	3.9E–02
Piper terryae	7.1E–03	1.3E–04	6.0E–03	7.1E–04	1.7E–02
Piper tomas–albertoi	8.2E–03	1.2E–04	7.7E–03	2.0E–03	1.5E–02
Piper trianae	7.5E–03	1.4E–04	6.9E–03	1.8E–03	1.4E–02
Piper tricuspe	1.2E–02	2.4E–04	1.2E–02	5.0E–03	2.0E–02
Piper tuberculatum	3.2E–02	6.4E–04	3.1E–02	1.9E–02	4.7E–02
Piper ubatubense	1.7E–02	3.5E–04	1.5E–02	4.3E–03	3.2E–02
Piper umbellatum	2.2E–02	3.1E–04	2.1E–02	1.0E–02	3.4E–02
Piper umbricola	1.2E–02	2.6E–04	1.1E–02	3.1E–03	2.2E–02
Piper unispicatum	7.1E–03	1.3E–04	6.0E–03	7.1E–04	1.7E–02
Piper villosum	2.1E–02	3.6E–04	2.0E–02	1.0E–02	3.4E–02
Piper yanaconasense	1.8E–03	3.4E–05	1.5E–03	3.1E–05	4.5E–03

the age-and-area hypothesis. The strength of this relationship is notable in light of the various factors that can potentially obscure a positive age and area relationship.

There are some important caveats to this initial analysis of age and area in a group of tropical plants. First, our ages were based on divergence times of *Piper* species. Our analysis represents only about 5–10% of the approximately 700 (Jaramillo and Manos 2001) to 1150 (Quijano-Abril *et al.* 2006) neotropical *Piper* species. Taxon sampling affects age estimates, because missing taxa would alter the estimated divergence times of species if they were included in the analysis (Linder *et al.* 2005). Missing taxa can lead to an overestimation of ages (Chown and Gaston 2000, Webb and Gaston 2000, Jones *et al.* 2005). However, given the strength of the positive age and area relationship that we found based on the *Piper* sequences available, and no reason to expect an inherent bias to the species that were selected to sequence or to the locations of missing taxa on

the tree, we suspect the positive age and area relationship found here will be borne out in future analyses of larger datasets.

WHAT DO OTHER TROPICAL PLANT CLADES TELL US?

Aside from *Piper*, there are very few molecular datasets available for specific clades of tropical plants that can be effectively used to assess age and area relationships. Considerable molecular data have amassed recently on tropical plant lineages and their divergence dates, but most of these data examine higher phylogenetic levels (e.g., families or higher; Renner *et al.* 2001, Davis *et al.* 2005, Lavin *et al.* 2005) and have focused on the origin and age of the clades and species that make up current tropical communities. These data tell an interesting story, but do not yet provide any clear expectations for the generality of the kind of age and area relationship found for *Piper*.

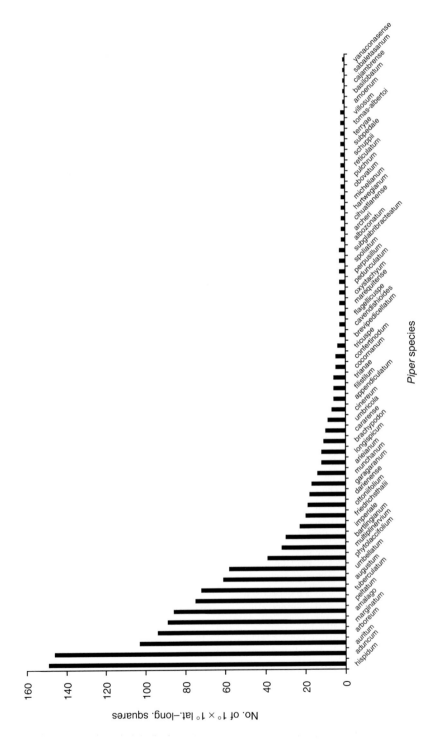

Figure 4.3 The distribution of range sizes of the neotropical *Piper* species used in the analysis of age and area.

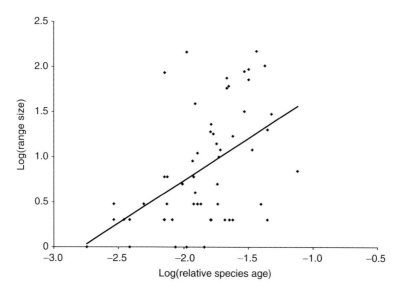

Figure 4.4 Linear regression of log-transformed relative species age and log-transformed range size; $y = 0.9399x + 2.6143$, $r^2 = 0.252$, $P < 0.001$.

Species-rich genera like *Piper* have a wide range of ages, based on the available evidence from molecular dating. *Piper* is a member of the basal angiosperms (Angiosperm Phylogeny Group 2003), and may be a rather old lineage (based on *Piper* and *Peperomia* divergence, ~40 Ma; Wikström *et al.* 2001). In contrast, analysis of the diverse legume genus *Inga* suggests that it is a young genus and many species originated on the scale of 2–10 Ma (Richardson *et al.* 2001). In light of evidence of the existence of rainforests from the late or mid-Cretaceous (~100 Ma; Morley 2000; Davis *et al.* 2005), *Inga* species must be considered quite young (Bermingham and Dick 2001). Despite its relatively recent origin, this clade has spread throughout the forests of South and Central America, and at many sites *Inga* species are important forest components in terms of both number and biomass (Richardson *et al.* 2001). In fact, legume clades in general may be remarkably young given their widespread distribution and numerical importance in tropical forests (~4–16 Ma; Lavin *et al.* 2004). Other speciose tropical clades are considerably older, such as those in the Annonaceae (e.g., *Xylopia*, *Annona*) which appear to be on the scale of

approximately 15–25 Ma (Richardson *et al.* 2004, Pirie *et al.* 2006). Like *Piper*, many of these clades have pantropical or even cosmopolitan distributions; in fact, one of the most widespread tropical plant species, *Symphonia globulifera* (Clusiaceae), also ages to the mid-Tertiary (~28 Ma; Dick *et al.* 2003). In Africa, the origin of the herbaceous begonias (Begoniaceae) is also on the scale of approximately 30 Ma, but many of the species in this group diverged relatively recently (from ~1 to 10 Ma; Plana *et al.* 2004).

In another widespread herbaceous genus, *Costus* (Costaceae), the neotropical species appear to have diversified rapidly and recently (Kay *et al.* 2005). In the case of very recent diversification of clades like *Inga* and *Costus*, widespread species within these genera provide evidence that common members of these clades are not particularly old. However, the relationships of age and area within these and other genera have not been assessed. In a rapidly diversifying genus, if more widespread species were found to be older, the expected slope of the age and area relationship would simply be very steep. However, finding young but common species would certainly not be surprising in light of recent evidence confirming a rare species advantage in many tropical forests, probably resulting from lower density-dependent or frequency-dependent mortality (e.g., Harms *et al.* 2000, Volkov *et al.* 2005, Wills *et al.* 2006). Rare species that have a fitness advantage are expected to increase in abundance much more rapidly than predicted under neutral drift, for example, resulting in younger species that have large range and population sizes. Thus, if new species do indeed start with small population and range sizes, some of these species may be expected to increase their population and range sizes rapidly. Overall, the generality of a positive age and area relationship in tropical plant species awaits future analyses, particularly of densely sampled, speciose clades.

Fortunately, there is considerable promise that in the near future we can gain a broader perspective on age and area relationships in tropical plants. For example, work on the diverse tropical herbaceous genus *Begonia* (Begoniaceae) has provided insight into the phylogenetics and timing of diversification in this pantropical genus (e.g., Forrest and Hollingsworth 2003, Plana *et al.* 2004). Likewise, phylogenetic work on the diverse pantropical genus *Psychotria* (Rubiaceae; Nepokroeff *et al.* 1999, J. Paul unpublished data) promises to provide evidence from a genus that in many ways mirrors *Piper* in its species' ecology, abundance, and distribution (e.g., high local and regional species richness, numerical abundance, understory and gap habitat, etc.), although it is phylogenetically distantly related. Interestingly, Hamilton (1989) suggested that within the Mesoamerican members of *Psychotria* subgenus *Psychotria*, species groups often contained one basal member with a large geographic range, and putatively derived members with narrow ranges.

AN AGE-AND-AREA HYPOTHESIS FOR MODERN TIMES

The strong positive age and area relationship found for neotropical *Piper* species warrants further investigation into the generality of this relationship in tropical plants. If, in general, many rare species are found to be young species, this information may be crucial to incorporate into our understanding of the variation in range size among species and, at the local scale, variation in abundance, which often shows a positive relationship with range size (Gaston 1994). In order to effectively integrate species age information derived from molecular inference (as most future data promise to be) into our understanding of tropical forest community structure, we need to recognize the potential sources of error in these data, as well as take a broader view on the simple age-and-area hypothesis proposed by Willis (1922).

First, one of the obvious shortcomings of the traditional age-and-area hypothesis (Willis 1922) is its failure to account for old species with small ranges. Empirical evidence suggests that in some cases, the age and area relationship may be a hump-shaped relationship (Webb and Gaston 2000), where both old and young species have small ranges, and intermediate age species have the largest ranges (or the greatest degree of ecological occupancy, e.g. Foote *et al.* 2007). Clearly, many old species must either go through range contraction as they age, or have their range sizes reduced through the process of speciation. As a result, a complete age-and-area hypothesis needs to account for these species, recognizing that a positive age and area relationship may be limited to the lower end of the temporal axis. For example, if the assumption that new species start with small ranges is accepted, then the general positive relationship between species age and range size can be expected to persist until some threshold, and then the relationship will become flat or negative, as older species lose range size. Almost all of the models of post-speciation range-size transformation presented in Gaston (1998), for example, have an initial phase in which there is a roughly linear positive relationship between species age and range size. The differences in these lines is the steepness of their slope and their temporal duration; some models, such as a cyclical and stasis models, predict a rapid increase in range size post-speciation, while the traditional age and area model is depicted as a gradual increase. However, depending on the total age of a clade of interest, and the rate at which transformations occur, all of these models are similar in their initial prediction of a positive species age and range size relationship. Thus, the more important question may be, when does a positive age and area relationship cease to exist, and why? Furthermore, analyses that examine clades of species and ask if on average rare species are younger than old species, rather than simply looking for a positive slope of an age and area relationship, may be more informative.

Second, the positive age and area expectations of most models of post-speciation transformation are primarily driven by the assumption that new species start with small population sizes. But do they? It has been asserted that much speciation in tropical woody plants arises through isolation of small local populations (e.g., Ehrendorfer 1982, Leigh *et al.* 2004), but strong empirical evidence to support this position is generally lacking. Since the population sizes of new species cannot practically be measured, inference must be used to estimate the sizes of ranges and populations. For example, fossil evidence supports African large-mammal populations starting as small, narrowly ranging populations (Vrba and DeGusta 2004). Unfortunately, the sparse fossil record for many taxa, particularly plants in the tropics, makes inference based on fossil evidence rare. The data presented here for *Piper* are certainly suggestive that newer species have small range sizes, as evidenced by the preponderance of young species with small range sizes and the lack of young species with large ones. Future analyses of age and area relationships in tropical plants may help to fill in the gaps of our knowledge of new species population and range sizes that are unlikely to be filled by fossil evidence.

Third, a practical difficulty arises from using divergence times of species as proxies for ages. When speciation is defined as a cladogenic (splitting) event, such as on a dichotomously branching phylogenetic tree, any speciation event yields at least two new species, both assigned the same age. These new species have range and population sizes defined by the boundaries of their newly isolated gene pools (or lineages). Thus, when speciation is viewed as a splitting process with a geographical component, new species will often have smaller range and population sizes than their direct ancestor, because the ancestral range

(and the distribution of individuals defining it) is subdivided. If the relative range and population sizes of sister species are markedly skewed, there will be considerable variance in the distribution of population sizes of the new species. For example, when a new species (B) is introduced via a point-mutation model of speciation (where one individual is assigned a new species status based on some new defining character, *sensu* Hubbell 2001a), its ancestor species (A) with population size N must also be deemed a new species (C), with a population size $N - 1$. Since species B and C are assigned the same age, the youngest species in the community are represented by species with both small (B) and large (C) population and range sizes. In other words, when a widespread species gives rise to a narrowly endemic sister species, but the widespread species persists essentially unchanged in its ecological and genetic attributes, both sister species are assigned the same age. This is potentially at odds with the meaning of species age in an evolutionary sense. It also clearly creates difficulty in analyzing age and area, as such a process will obscure any expectation of a positive relationship if such asymmetric range splits are commonplace in a clade. In light of this potential source of noise in the age and area relationship, it is all the more remarkable that a positive relationship explaining a good portion of the variance in range size was found in our analysis of *Piper* species.

Finally, molecular age estimates are potentially subject to many different kinds of errors and uncertainties (Arbogast *et al.* 2002). For example, the model of molecular evolution used, the degree of consensus between gene trees examined and true species trees (Nichols 2001), the reliability of any fossil ages used for calibration, and success of an analytical model dealing with rate heterogeneity can all introduce potential errors in estimates of ages (Sanderson *et al.* 2004, Renner 2005).

In summary, future studies on age and area relationships in tropical plants have the potential to provide insight into the role that the simple explanatory variable species age can play in explaining patterns of rarity and endemism. Of course, as Willis himself recognized, age by itself cannot be the mechanistic driver of these patterns we observe. Rather, age acts as a proxy for the playing-out of various ecological interactions at different spatial and temporal scales. If a positive age and area relationship is found for a group of taxa, this finding can point to valuable lines of research for future studies (Jones *et al.* 2005). For example, if such a relationship is found in a 20 Ma clade of plants, could this be an indication that the range-size transformations within this group are rather slow and potentially governed by the ecological drift? If only certain guilds of plants (e.g., understory shrubs) show a positive age and area relationship, could this be related to the potential dispersal limitations imposed on these plants through their canopy position and reliable seed dispersers? In addition, how do clades that have many old species with small ranges differ from those clades like *Piper* which apparently lack many old species with small ranges? An updated view of the age-and-area hypothesis thus allows researchers to inquire about much more than whether the age and area relationship in a given group of organisms is linear and positive. The shape of the relationship in a given clade can be used to infer the importance of various factors in the range transformation of species, and suggest if new species start with small range sizes.

CONCLUSIONS

An explanation for why many tropical forest species are rare and endemic may simply be the relatively young age of these species. We have reviewed the limited empirical work addressing age and area relationships, none of which came from strictly tropical taxa, and showed that support for the traditional age-and-area hypothesis is equivocal. Using neotropical *Piper* species as a case study, we conducted the first age and area analysis for a tropical plant clade, and found significant support for a positive age and area relationship that explains a quarter of the variance in range size among species. Speculation about the age and area relationships within other taxonomic groups is difficult, however, because species-level data on either ages or ranges are sparse. Although inferring species ages from molecular data and phylogenetic trees can introduce difficulties when interpreting results of age and area analyses, we predict that in the near future broader analyses of

age and area will be plausible with many clades of tropical plants.

ACKNOWLEDGMENTS

We thank Walter Carson and Stefan Schnitzer for inviting us to contribute this chapter. We thank four anonymous reviewers, Anthony Baumert, Anthony Bledsoe, James Cronin, Richard Gomulkiewicz, Susan Kalisz, Mark McPeek, April Randle, Scott Stark, John J. Wiens, and the Wiens lab for their comments on earlier versions of this chapter. We also thank the original authors of the *Piper* sequences for making these data publicly available for use by the scientific community.

REFERENCES

Angiosperm Phylogeny Group (2003) An update of the Angiosperm Phylogeny Group classification for the orders and families of flowering plants: APG II. *Botanical Journal of the Linnean Society* 141, 399–436.

Arbogast, B.S., Edwards, S.V., Wakeley, J. *et al.* (2002) Estimating divergence times from molecular data on phylogenetic and population genetic timescales. *Annual Review in Ecology and Systematics* 33, 707–740.

Bermingham, E. and Dick, C. (2001) The *Inga* – newcomer or museum antiquity? *Science* 293, 2214–2216.

Böhning-Gaese, K., Caprano, T., van Ewijk, K. *et al.* (2006) Range size: disentangling current traits and phylogenetic and biogeographic factors. *American Naturalist* 167, 555–567.

Chown, S.L. (1997) Speciation and rarity: separating cause from consequence. In W.E. Kunin and K.J. Gaston (eds), *The Biology of Rarity: Causes and Consequences of Rare and Common Differences.* Chapman & Hall, London, pp. 91–109.

Chown, S.L. and Gaston, K.J. (2000) Areas, cradles and museums: the latitudinal gradient in species diversity. *Trends in Ecology and Evolution* 15, 311–315.

Davis, C.C., Webb, C.O., Wurdack, K.J., Jaramillo, C.A., and Donoghue, M.J. (2005) Explosive radiation of Malpighiales supports a mid-Cretaceous origin of modern tropical rain forests. *American Naturalist* 165, E36–E65.

Dick, C.W., Abdul-Salim, K., and Bermingham, E. (2003) Molecular systematic analysis reveals cryptic tertiary diversification of a widespread tropical rain forest tree. *American Naturalist* 162, 691–703.

Dobzhansky, T. (1950) Evolution in the tropics. *American Scientist* 38, 209–221.

Drummond, A.J. and Rambaut, A. (2007) BEAST: Bayesian evolutionary analysis by sampling trees. *BMC Evolutionary Biology* 7, 214, http://beast.bio.ed.ac.uk/. Last accessed: August 2006.

Drummond, A.J., Ho, S.Y.W., Phillips, M.J. *et al.* (2006) Relaxed phylogenetics and dating with confidence. *PLoS Biology* 4, e88.

Dyer, L.A. and Palmer, A.D.N. (eds) (2004) *Piper. A Model Genus for Studies of Phytochemistry, Ecology, and Evolution.* Kluwer Academic, New York.

Ehrendorfer, F. (1982) Speciation patterns in woody angiosperms of tropical origin. In C. Barrigozzi (ed.), *Mechanisms of Speciation.* Liss, New York, pp. 479–509.

Fernald, M.L. (1924) Isolation and endemism in Northeastern America and their relation to the age-and-area hypothesis. *American Journal of Botany* 11, 558–572.

Foote, M., Crampton, J.S., Beu, A.G. *et al.* (2007) Rise and fall of species occupancy in Cenozoic fossil mollusks. *Science* 318, 1131–1134.

Forrest, L.L. and Hollingsworth, P.M. (2003) A recircumscription of *Begonia* based on nuclear ribosomal sequences. *Plant Systematics and Evolution* 241, 193–211.

Fuentes, M. (2004) Slight differences among individuals and the unified neutral theory of biodiversity. *Theoretical Population Biology* 66, 199–203.

Gaston, K.J. (1994) *Rarity.* Chapman & Hall, London.

Gaston, K.J. (1998) Species-range size distributions: products of speciation, extinction and transformation. *Philosophical Transactions of the Royal Society of London B* 353, 219–230.

Gaston, K.J. (2003) *The Structure and Dynamics of Geographic Ranges.* Oxford University Press, Oxford.

Gaston, K.J. and Blackburn, T.M. (1997) Age, area, and avian diversification. *Biological Journal of the Linnean Society* 62, 239–253.

Gleason, H.A. (1924) Age and area from the viewpoint of phytogeography. *American Journal of Botany* 11, 541–546.

Hamilton, C.W. (1989) A revision of Mesoamerican *Psychotria* subgenus *Psychotria* (Rubiaceae), Part I: Introduction and species 1–16. *Annals of the Missouri Botanical Garden* 76, 67–111.

Harms, K.E., Wright, S.J., Calderón, O. *et al.* (2000) Pervasive density-dependent recruitment enhances seedling diversity in a tropical forest. *Nature* 404, 493–495.

Hubbell, S.P. (2001a) *The Unified Neutral Theory of Biodiversity and Biogeography.* Princeton University Press, Princeton.

Hubbell, S.P. (2001b) The unified neutral theory of biodiversity and biogeography: a synopsis of the theory and some challenges ahead. In J. Silvertown and J. Antonovics (eds), *Integrating Ecology and Evolution in a Spatial Context.* Blackwell Science, Oxford, pp. 393–411.

Jablonski, D. (1987) Heritability at the species level: analysis of geographic ranges of Cretaceous molluscs. *Science* 238, 360–363.

Jaramillo, M.A. and Callejas, R. (2004a) A reappraisal of *Trianaeopiper* Trelease: convergence of dwarf habitat in some *Piper* species of the Chocó. *Taxon* 53, 269–278.

Jaramillo, M.A. and Callejas, R. (2004b). Current perspectives on the classification and phylogenetics of the genus *Piper* L. In L.A. Dyer and A.D.N. Palmer (eds), *Piper. A Model Genus for Studies of Phytochemistry, Ecology, and Evolution.* Kluwer Academic, New York, pp. 179–198.

Jaramillo, M.A. and Manos, P.S. (2001) Phylogeny and patterns of floral diversity in the genus *Piper* (Piperaceae). *American Journal of Botany* 88, 706–716.

Jones, K.E., Sechrest, W., and Gittleman, J.L. (2005) Age and area revisited: identifying global patterns and implications for conservation. In A. Purvis, J.L. Gittleman, and T. Brooks (eds), *Phylogeny and Conservation.* Cambridge University Press, Cambridge, pp. 142–165.

Kay, K.M., Reeves, P.A., Olmstead, R.G. *et al.* (2005) Rapid speciation and the evolution of hummingbird pollination in Neotropical *Costus* subgenus *Costus* (Costaceae): evidence from nrDNA ITS and ETS sequences. *American Journal of Botany* 92, 1899–1910.

Lavin, M., Herendeen, P.S., and Wojciechowski, M.F. (2005) Evolutionary rates analysis of Leguminosae implicates a rapid diversification of lineages during the Tertiary. *Systematic Biology* 54, 575–594.

Lavin, M., Schrire, B.P., Lewis, G.P. *et al.* (2004) Metacommunity process rather than continental tectonic history better explains geographically structured phylogenies in legumes. *Philosophical Transactions of the Royal Society of London Series B* 359, 1509–1522.

Leigh, E.G. Jr. (1999) *Tropical Forest Ecology, a View from Barro Colorado Island.* Oxford University Press, Oxford.

Leigh, E.G. Jr., Davidar, P., Dick, C.W. *et al.* (2004) Why do some tropical forests have so many species of trees? *Biotropica* 36, 447–473.

Linder, H.P., Hardy, C.R., and Rutschmann, F. (2005) Taxon sampling effects in molecular clock dating: an example from the African Restionaceae. *Molecular Phylogenetics and Evolution* 35, 569–582.

Marquis, R.J. (2004) Biogeography of Neotropical *Piper*. In L.A. Dyer and A.D.N. Palmer (eds), *Piper. A Model Genus for Studies of Phytochemistry, Ecology, and Evolution*. Kluwer Academic, New York, pp. 78–96.

Miller, A.I. (1997) A new look at age and area: the geographic and environmental expansion of genera during the Ordovician radiation. *Paleobiology* 23, 410–419.

Morley, R.J. (2000) *Origin and Evolution of Tropical Rainforests*. John Wiley, Chichester.

Muellner, A.N., Savolainen, V., Samuel, R. *et al.* (2006) The mahogany family "out-of-Africa": divergence time estimation, global biogeographic patterns inferred from plastid *rbc*L DNA sequences, extant, and fossil distribution of diversity. *Molecular Phylogenetics and Evolution* 40, 236–250.

Nee, S. (2005) The neutral theory of biodiversity: do the numbers add up? *Functional Ecology* 19, 173–176.

Nepokroeff, M., Bremer, B., and Sytsma, K.J. (1999) Reorganization of the genus *Psychotria* and Tribe Psychotrieae (Rubiaceae) inferred from ITS and *rbc*L sequence data. *Systematic Botany* 24, 5–27.

Nichols, R. (2001) Gene trees and species trees are not the same. *Trends in Ecology and Evolution* 16, 358–364.

Nylander, J.A.A., Wilgenbusch, J.C., Warren, D.L., and Swofford, D.L. (2008) AWTY (are we there yet?): A system for graphical exploration of MCMC convergence in Bayesian phylogenetics. *Bioinformatics* 24, 581–583.

Pirie, M.D., Chatrou, L.W., Mols, J.B. *et al.* (2006) "Andean-centered" genera in the short-branch clade of Annonaceae: testing biogeographical hypotheses using phylogeny reconstruction and molecular dating. *Journal of Biogeography* 33, 31–46.

Pitman, N.C.A., Terborgh, J.W., Silman, M.R. *et al.* (2001) Dominance and distribution of tree species in upper Amazonian terra firma forests. *Ecology* 82, 2101–2117.

Plana, V., Gascoigne, A., Forrest, L.L. *et al.* (2004) Pleistocene and pre-Pleistocene *Begonia* speciation in Africa. *Molecular Phylogenetics and Evolution* 31, 449–461.

Posada, D. and Crandall, K.A. (1998) Modeltest: testing the model of DNA substitution. *Bioinformatics* 14, 817–818.

Quijano-Abril, M.A., Callejas-Posada, R., and Miranda-Esquivel, D.R. (2006) Areas of endemism and distribution patterns for Neotropical *Piper* species (Piperaceae). *Journal of Biogeography* 33, 1266–1278.

Renner, S.S. (2005) Relaxed molecular clocks for dating historical plant dispersal events. *Trends in Plant Science* 10, 550–558.

Renner, S.S., Clausing, G., and Meyer, K. (2001) Historical biogeography of Melastomataceae: the roles of tertiary migration and long-distance dispersal. *American Journal of Botany* 88, 1290–1300.

Richardson, J.E., Pennington, R.T., Pennington, T.D. *et al.* (2001) Rapid diversification of a species-rich genus of Neotropical rain forest trees. *Science* 293, 2242–2245.

Richardson, J.E., Chatrou, L.W., Mols, J.B. *et al.* (2004) Historical biogeography of two cosmopolitan families of flowering plants: Annonaceae and Rhamnaceae. *Philosophical Transactions of the Royal Society of London Series B, Biological Sciences* 359, 1495–1508.

Ricklefs, R.E. (2003) A comment on Hubbell's zero-sum ecological drift model. *Oikos* 100, 185–192.

Ricklefs, R.E. (2004) A comprehensive framework for global patterns in biodiversity. *Ecology Letters* 7, 1–15.

Ronquist, F. and Huelsenbeck, J.P. (2003) MrBayes 3: Bayesian phylogenetic inference under mixed models. *Bioinformatics* 19, 1572–1574.

Sanderson, M.J., Thorne, J.L., Wikström, N. *et al.* (2004) Molecular evidence on plant divergence times. *American Journal of Botany* 91, 1656–1665.

SAS Institute Inc. (2001) SAS Software, Release 8.2. Cary, NC: SAS Institute Inc.

Stebbins, G.L. and Major, J. (1965) Endemism and speciation in the California flora. *Ecological Monographs* 35, 1–35.

Thompson, J.D., Higgins, D.G., and Gibson, T.J. (1994) ClustalW: improving the sensitivity of progressive multiple sequence alignment through sequence weighting, position-specific gap penalties and weight matrix choice. *Nucleic Acids Research* 22, 4673–4680.

Trelease, W. and Yuncker, T.G. (1950) *The Piperaceae of Northern South America*. University of Illinois Press, Urbana.

Volkov, I., Banavar, J.R., He, F. *et al.* (2005) Density dependence explains tree species abundance and diversity in tropical forests. *Nature* 438, 658–661.

Vrba, E.S. and DeGusta, D. (2004) Do species populations really start small? New perspectives from the Late Neogene fossil record of African mammals. *Philosophical Transactions of the Royal Society of London Series B* 359, 285–292.

Wallace, A.R. (1878) *Tropical Nature and Other Essays*. Macmillan, New York and London.

Webb, T.J. and Gaston, K.J. (2000) Geographic range size and evolutionary age in birds. *Proceedings of the Royal Society of London Series B* 267, 1843–1850.

Wiens, J.J. and Donoghue, M.J. (2004) Historical biogeography, ecology, and species richness. *Trends in Ecology and Evolution* 19, 639–644.

Wikström, N., Savolainen, V., and Chase, M.W. (2001) Evolution of the angiosperms: calibrating the family tree. *Proceedings of the Royal Society of London Series B* 268, 2211–2220.

Wilgenbusch, J.C., Warren, D.L., and Swofford, D.L. (2004) AWTY: A system for graphical exploration of MCMC convergence in Bayesian phylogenetic inference. http://ceb.csit.fsu.edu/awty. Last accessed: September 8, 2006.

Willis, J.C. (1922) *Age and Area.* Cambridge University Press, Cambridge.

Wills, C., Harms, K.E., Condit, R. *et al.* (2006) Nonrandom processes maintain diversity in tropical forests. *Science* 311, 527–531.

Yu, D.W., Terborgh, J.W., and Potts, M.D. (1998) Can high tree species richness be explained by Hubbell's null model? *Ecology Letters* 1, 193–199.

Yule, G.U. (1925) A mathematical theory of evolution, based on the conclusions of Dr J.C. Willis, F.R.S. *Philosophical Transactions of the Royal Society of London Series B, Biological Sciences* 213, 21–87.

Zerega, N.J.C., Clement, W.L., Datwyler, S.L. *et al.* (2005) Biogeography and divergence times in the mulberry family (Moraceae). *Molecular Phylogenetics and Evolution* 37, 402–416.

APPENDIX

GenBank accession numbers of the species used in this study.

Species	GenBank accession numbers
Houttuynia cordata	AF275211
Macropiper excelsum	AF275193
Macropiper hooglandii	AF275192
Macropiper melchior	AF275191
Peperomia elongata	AF275213
Piper aduncum	AF275159
Piper aduncum2	AF275158
Piper aduncum3	AF275157
Piper albispicum	AY572317
Piper albozonatum	AY326195
Piper amalago	AF275186
Piper amoenum	AF275160
Piper appendiculatum	AY326196
Piper arborescens	AF275202
Piper arboreum	AF275180
Piper arboricola	AY572319
Piper archeri	AF275178
Piper arieianum	AF275163

Species	GenBank accession numbers
Piper atrospicum	AY572318
Piper augustum	AF275165
Piper auritum	AF275175
Piper bartlingianum	AF275183
Piper basilobatum	AY326197
Piper bavinum	AF275199
Piper betle	AF275201
Piper boehmeriifolium	AF275204
Piper brachypodon	AY326198
Piper brevicuspe	AY572321
Piper brevipedicellatum	AF275189
Piper cajambrense	AY326199
Piper caninum	AF275195
Piper capense	AY326200
Piper cararense	AY326201
Piper cavendishioides	AF275153
Piper celtidiforme	AF275205
Piper chuarense	AY326202
Piper cihuatlanense	AF275187
Piper cinereum	AF275190
Piper cocornanum	AY326203
Piper colligatispicum	AY326204
Piper confertinodum	AF275166
Piper cordatilimbum	AY572323
Piper darienense	AF275181
Piper decumanum	AF275203
Piper densum	AY615963
Piper filistilum	AF275155
Piper flagellicuspe	AF275154
Piper friedrichsthalii	AY326205
Piper garagaranum	AF275162
Piper gesnerioides	AY326206
Piper gymnostachyum	AY572325
Piper hartwegianum	AY326207
Piper hernandii	AY572324
Piper hispidum	AF275156
Piper hymenophyllum	AY572327
Piper imperiale	AF275176
Piper korthalsii	AF275208
Piper laosanum	AY572326
Piper lolot	AY326208
Piper longispicum	AY326209
Piper marequitense	AY326210
Piper marginatum	AY326211
Piper medinillifolium	AY667455
Piper methysticum	AF275194
Piper michelianum	AF275188

Continued

Species	GenBank accession numbers
Piper multiplinervium	AF275168
Piper munchanum	AF275164
Piper myrmecophilum	AY572328
Piper nigrum	AF275198
Piper nigrum2	AF275197
Piper obovatum	AY326212
Piper ottoniifolium	AY326213
Piper oxystachyum	AF275152
Piper parvulum	AF275167
Piper pedunculatum	AY326214
Piper peltatum	AF275171
Piper peltatum2	AF275170
Piper peltatum3	AF275169
Piper penninerve	AF275206
Piper perpusillum	AY326215
Piper phytolaccifolium	AY326216
Piper pierrei	AF275200
Piper pilibracteum	AY768829
Piper pulchrum	AF275177
Piper reticulatum	AF275185
Piper reticulatum2	AF275184
Piper retrofractum	AF275196
Piper sabaletasanum	AY326217
Piper schuppii	AY326218

Species	GenBank accession numbers
Piper sorsogonum	AY572320
Piper sp1 maj674	AY326219
Piper sp2 maj689	AY326230
Piper spoliatum	AF275179
Piper subglabribracteatum	AY326220
Piper subpedale	AF275161
Piper terryae	AY326221
Piper tomas-albertoi	AY326222
Piper toppingii	AY572322
Piper trianae	AY326224
Piper tricuspe	AY326225
Piper tuberculatum	AY326223
Piper ubatubense	AF275182
Piper umbellatum	AF275174
Piper umbellatum2	AF275173
Piper umbellatum3	AF275172
Piper umbricola	AY326226
Piper unispicatum	AY326227
Piper urdanetanum	AF275207
Piper villosum	AY326228
Piper yanaconasense	AY326229
Sarcorhachis naranjoana	AF275210
Sarcorhachis sydowii	AF275209
Saururus cernuus	AF275212

Chapter 5

PATTERNS OF HERBIVORY AND DEFENSE IN TROPICAL DRY AND RAIN FORESTS

Rodolfo Dirzo and Karina Boege

OVERVIEW

Studies on tropical herbivory are largely based on information from tropical rain forests (TRFs) but we argue that our understanding of this and other phenomena of tropical community ecology and evolutionary biology will benefit from comparisons with tropical dry forests (TDFs). In particular we analyze the possible consequences of rainfall seasonality. We develop a line of reasoning as to how rainfall seasonality and phenology of TRF and TDF plants bring about fundamental differences in the availability of foliage for folivores and how these differences can in turn lead to predictable and contrasting patterns of herbivory and defense. We then compare available information, from the literature and from our own ongoing work, on herbivory and defense in plants from both forest types. We found strong evidence that higher constancy of foliage, implying greater risk and impact of herbivory in TRF plants in ecological time, may lead to a greater evolutionary history of herbivory, favoring greater selection for increased defense and lower herbivory. The predicted patterns were evident when we controlled for interspecific heterogeneity in herbivory and defense within both TDF (due to contrasts in life history and growth) and TRF (due to contrasts in phenology). In addition, the expected patterns were mirrored using more controlled intra-site (dry forest) comparisons looking at plants of contrasting phenologies. Moreover, preliminary evidence suggests that the observed patterns might hold independently of, or in addition to, phylogenetic influence.

INTRODUCTION

The term "tropical forest" conceals the existence of a complex and diverse variety of vegetation entities and plant associations with distinct physiognomic features and ecological characteristics, ranging from the savanna-like formations to the usual rain forests. Among this exuberant variety of tropical forest formations, two major categories of lowland (≤ 1000 m above sea level) tropical forest can be distinguished (Dirzo 2001): rain (including wet and moist, i.e., evergreen) and seasonally dry (i.e., deciduous) forests.

Although we recognize the existence of a seasonality gradient, for the purposes of the arguments we develop in this chapter we distinguish tropical dry forests (TDFs) from tropical rain forests (TRFs) as those in which the number of dry months (rainfall ≤ 100 mm) per year is five or more, and during which time the vegetation is almost entirely devoid of foliage. Such contrast in foliage availability has the potential to affect the evolutionary ecology of plant–herbivore interactions, as has been demonstrated by studies looking at the consequences of plant phenology on herbivory within a single site at both the community (Janzen 1981, Janzen and Waterman 1984, Filip *et al.* 1995) and population (Feeny 1970, Aide 1988, Forkner *et al.* 2004) levels.

Studies on the biology of tropical forests have a strong bias on TRFs, and herbivory is no exception. For example, a review of the literature on patterns of tropical herbivory shows that the number of studies (published since 1970) dealing with TRF plants is about three times greater than those on TDF plants (Table 5.1). This imbalance compromises our understanding of both forest types. In this study we attempt to broaden our current perspective on tropical plant–herbivore interactions by developing a comparative analysis of herbivory in TRF versus TDF.

Herbivory is a central process in tropical forest biology. Phytophagous insects and their food plants constitute a large proportion (at least 50%) of the number of known species on earth (Price 1997) and both groups are disproportionately represented in tropical forests (Dirzo and Raven 2003). On the other hand, although levels of herbivory for most tropical plants are low to moderate, on occasion, damage can approach complete defoliation both in TRF species (Dirzo and Mota 1997) and in TDF species (Janzen 1981). In addition, intense defoliation has been shown to affect several components of plant fitness both in rain forest species (e.g., *Piper arieianum*, Marquis 1984) and in dry forest species (e.g., *Erythroxylum havanense*, Dirzo and Domínguez 1995). Although the number of studies directly documenting the role of herbivores as important selective pressures for plants is very limited (see Marquis 1992), the importance of herbivores as selective agents of great preponderance in tropical forests has been inferred from the over-representation of putative defensive compounds such as alkaloids (Levin 1976) when compared with extra-tropical plants. Moreover, in a review specifically directed to assess patterns of herbivory and defense in tropical versus temperate forest plants, Coley and Aide (1991) conclude that herbivory and defense are significantly greater in tropical forests than in temperate forests.

While ecogeographic analyses of tropical herbivory have emphasized the tropical–temperate comparison (Coley and Aide 1991), comparative studies within the tropical realm are restricted to interspecific comparisons within a given site (Coley 1982, Dirzo 1984) or to limited citations of observed levels of herbivory in TDFs as compared with TRFs (Coley and Aide 1991, Dirzo and Domínguez 1995, Coley and Barone 1996, Dyer and Coley 2002). To our knowledge, no detailed comparison of herbivory and defense between TRF and TDF has been undertaken, and no specific theoretical frameworks for expected patterns have been developed.

THEORETICAL FRAMEWORKS AND PREDICTIONS

Two specific theoretical constructs have been developed to explain interspecific variation in herbivory (leading to differential defensive responses) within a site: the resource availability hypothesis (Coley *et al.* 1985, see also Janzen 1974) and the plant apparency hypothesis (Feeny 1976). To what extent can we use either or both of these constructs to attempt to explain patterns of variation in herbivory and defense between TRF and TDF plants? The first hypothesis posits that slow-growing species, adapted to live in habitats of low resource availability (e.g., shaded forest understory, poor-quality soils), in which the cost of loss of tissue to herbivores is very high, should have higher investments in defense than fast-growing species adapted to persist in habitats of high resource availability (i.e., forest gaps, more fertile soils). This hypothesis is not directly applicable for a comparison between tropical dry and wet forests, given that, in addition to the differences in the relevant limiting resources between both types of forest (light or soil nutrients in TRF and water in TDF), it does not consider a central difference between them: rainfall seasonality (see below). The second hypothesis argues that the probability of a plant being encountered by a herbivore (apparency) determines the risk of herbivory and the investment in defense: apparent plants (large, long lived, growing in dense populations, etc.) have a greater probability of being found and, therefore, selection should favor greater investment in defense than in less apparent plants (small, short lived, growing in sparse populations). This hypothesis could be challenged because of the subjectivity of the concept of apparency and the lack of a consistent fit with empirical data (Stamp 2003). Moreover, the argued apparency

Table 5.1 Sources of information used in this chapter.

Location	Precipitation (mm)	Temperature (°C)	Seasonality (months with precipitation ≤100 mm)	No. of species	Forest type	Herbivory (% LAE)	TP (% dry mass)	CT	Toughness (g mm^{-2})	Source
Chamela, Mexico (19°30'N, 105°03'W)	788	20–27	5–6	16	Dry	7.49	–	–	12.93	Filip et al. (1995)
Chamela, Mexico (19°30'N, 105°03'W)	788	20–27	5–6	1	Dry	15.81	8.1	4.3	8.7	Boege (2005)
Chamela, Mexico (19°30'N, 105°03'W)	788	20–27	5–6	3	Dry	13.46	6.23	3.85	8.7	Boege (2004)
Chamela, Mexico (19°30'N, 105°03'W)	788	20–27	5–6	30	Dry	5.48	–	–	–	Herrerías-Diego (1999)
Chamela, Mexico (19°30'N, 105°03'W)	788	20–27	5–6	5	Dry	9.2	–	–	–	Martínez (2000)
Chamela, Mexico (19°30'N, 105°03'W)	788	20–27	5–6	9	Dry	6.78	–	–	–	O. Sánchez (unpublished data)
Guanacaste, Costa Rica (10°45'N, 85°30'W)	1565	NA	5–6	23	Dry	–	–	–	8.7	Leffler and Enquist (2002)
Huautla, Mexico (18°25'N, 99°30'W)	1039	20–29	7–8	1	Dry	26.84	–	–	–	Sánchez-Montoya (2002)
Huautla, Mexico (18°25'N, 99°30'W)	1039	20–29	7–8	20	Dry	11.52	–	–	–	R. Dirzo (unpublished data)
Huautla, Mexico (18°25'N, 99°30'W)	1039	20–29	7–8	21	Dry	0.125/day[a]	–	–	21.65	Carrazco-Carballido (2002)
Mundumalai, India (11°30'N, 77°27'W)	1100	12–35	5–6	275	Dry	10.8	–	–	–	Murali and Sukumar (1993)
Palo Verde, Costa Rica (10°21'N, 85°21'W)	1500	30	5	18	Dry	10.69	–	–	–	R. Dirzo (unpublished data)
Palo Verde, Costa Rica (10°21'N, 85°21'W)	1500	30	5	NA	Dry	10.75	–	–	–	Dirzo and Domínguez (1995)
Santa Rosa, Costa Rica (10°21'N, 85°21'W)	1500	25	6–7	77	Dry	–	5.66	5.72	–	Janzen and Waterman (1984)
Barro Colorado Island, Panama (9°10'N, 79°51'W)	2623	32	3–4	6	Rain	–	7.6	–	–	Milton (1979)
Barro Colorado Island, Panama (9°10'N, 79°51'W)	2623	32	3–4	6	Rain	0.206/day[a]	–	–	–	Coley (1982)

Continued

Table 5.1 Continued

Location	Precipitation (mm)	Temperature (°C)	Seasonality (months with precipitation ≤100 mm)	No. of species	Forest type	Herbivory (% LAE)	TP (% dry mass)	CT	Toughness (g mm^{-2})	Source
Barro Colorado Island, Panama (9°10'N, 79°51'W)	2623	32	3–4	6	Rain	0.136/day[a]	6.15	5.68	66.37	Coley (1983)
Chajul, Mexico (14°04'N, 91°30'W)	3859	22	2	27	Rain	9.72	–	–	–	R. Dirzo (unpublished data)
Manaus, Campo 41, Brazil (2°30'N, 60°W)	2600	26	4–5	3	Rain	11.00[b]	–	–	–	Benitez-Malvido and Kossmann-Ferraz (1999)
Manaus, Campo 41, Brazil (2°30'N, 60°W)	2600	26	4–5	16	Rain	10.42	–	–	–	R. Dirzo (unpublished data)
Dorrigo, Australia (30°20'S, 153°E)	1527	21–25	3–4	3	Rain	15.12	–	–	–	Lowman (1992)
Douala-Edea, Cameroon (3°35'N, 9°54'E)	3750	23–32	1	16	Rain	–	7.52	6.36	–	McKey et al. (1978)
Douala-Edea, Cameroon (3°35'N, 9°54'E)	3750	23–32	1	16	Rain	–	6.80	5.84	–	Waterman et al. (1980)
Douala-Edea, Cameroon (3°35'N, 9°54'E)	3750	23–32	1	16	Rain	–	6.95	5.66	–	Gartlan et al. (1980)
El Verde, Puerto Rico (18°19'N, 65°45'W)	3460	21–25	1	13	Rain	7	–	–	–	Odum and Ruiz-Reyes (1970)
Kibale, Uganda (0°13'N, 30°19'E)	1671	12–25	2–4	14	Rain	–	3.47	2.57	–	McKey et al. (1978)
Kibale, Uganda (0°13'N, 30°19'E)	1671	12–25	2–4	14	Rain	–	3.35	2.40	–	Waterman et al. (1980)
Kibale, Uganda (0°13'N, 30°19'E)	1671	12–25	2–4	14	Rain	–	3.63	3.45	–	Gartlan et al. (1980)
Kuala Lumpur, Malaysia (3°43'N, 102°17'E)	2000	23–33	NA	1	Rain	–	4.12	4.93	–	Waterman and Ross (1988)
Les Nourages, French Guiana (4°05'N, 52°40'W)	NA	NA	NA	NA	Rain	5.3	–	–	–	Sterck et al. (1992)
La Selva, Costa Rica (10°26'N, 83°59'W)	3962	26	0–1	1	Rain	27.35	–	–	–	Marquis and Clark (1989)
Los Tuxtlas, Mexico (19°04'N, 18°35'W)	4725	17–29	2	61	Rain	8.9	–	–	–	Dirzo (1987)

Locality									Reference
Los Tuxtlas, Mexico (19°04'N, 18°35'W)	4725	17–29	2	18	Rain	11.9	–	–	R. Dirzo (unpublished data)
Los Tuxtlas, Mexico (19°04'N, 18°35'W)	4725	17–29	1–2	51	Rain	9.435[b]	–	–	de la Cruz and Dirzo (1987)
Los Tuxtlas, Mexico (19°04'N, 18°35'W)	4725	17–29	1–2	2	Rain	15.05[b]	–	–	Núñez-Farfán and Dirzo (1988)
Los Tuxtlas, Mexico (19°04'N, 18°35'W)	4725	17–29	1–2	1	Rain	20.70[b]	–	–	García-Guzman and Benítez-Malvido (2003)
Luquillo, Puerto Rico (18°20'N, 66°00'W)	3460	21–25	1	2	Rain	17.33	6.18	25.87	Myster (2002)
Petit Saut, French Guiana (5°04'N, 53°04'W)	3150	2–3	NA	NA	Rain	5.4	–	–	Sterck et al. (1992)
Papua New Guinea (North) (5°25'S, 147°16'E)	NA	NA	NA	7	Rain	14	–	–	Wint (1983)
Papua New Guinea (South) (5°25'S, 147°16'E)	NA	NA	NA	23	Rain	12	–	–	Wint (1983)
San Blas, Panama (8°48'N, 77°40'W)	2155	NA	4	23	Rain	12.5	–	–	Wint (1983)
Sepilok, Malaysia (5°54'N, 118°04'E)	3110	27	0–1	1	Rain	–	5.83	8.78	Waterman and Ross (1988)
Sierra Leone, Tiwai Island (7°33'N, 11°21'W)	3300	NA	3–4	4	Rain	–	4.77	10.06	Mole et al. (1988)
Malaysia	NA	NA	NA	2	Rain	5.74[b]	–	–	Howlett and Davidson (2001)
Queensland, Australia	NA	NA	NA	1	Rain	5.40[b]	–	–	Jackson and Bach (1999)
Australia	NA	NA	NA	1	Rain	4.50[b]	–	–	Iddles et al. (2003)
Singapore (1°20'N, 103°50'E)	1900	30	NA	25	Rain	–	6.56	3.15	Turner (1995)
Wongabel, Australia (17°19'S, 145°30'E)	1560	NA	NA	2	Rain	–	–	14.64	Hurley (2000)

Notes: Includes data on herbivory (leaf area eaten, LAE), total phenolics (TP), condensed tannins (CT), and leaf toughness (where available). Also provided is information on the location of study sites (locality, country), and their corresponding precipitation, temperature, and seasonality. NA, data not available.

[a] Indicates rates of herbivory.
[b] Data correspond to seedlings.

becomes irrelevant from the dry forest perspective, given that during the dry season, when the plants are not apparent, herbivores (the target of the apparency signal) are not active either, implying that abiotic control, rather than apparency, may be a more important determinant of the risk of herbivory. Although some elements of these hypotheses can be applicable to our TRF/TDF comparison, their application is only partial, as we discuss below.

We propose, instead, a scenario of contrasting responses of herbivory and defense based on rainfall seasonality as the starting point leading to a contrast in resource availability for herbivores between TDFs and TRFs (Figure 5.1). Precipitation, an important component in the physical environment of both TRFs and TDFs, operates as a fundamental driver of contrasting plant phenological responses, in particular foliage availability, the resource base for leaf-feeding herbivores. Foliage availability, as affected by rainfall seasonality, will in turn lead to differences in herbivore risk of attack/damage, which will determine differences in the potential impact of herbivores on plants. Under these conditions, differential selective pressures for the evolution of defense and, ultimately, patterns of herbivory, should be expected. The specific predictions regarding these variables for TRF and TDF are as follows (Figure 5.1):

1 *Foliage availability.* Under the conditions of strong seasonality in TDFs foliage availability to herbivores should be episodic, restricted to 50% of the time or less, while availability of foliage should be continuous throughout the year in the vast majority of plants in TRFs.

2 *Risk of attack.* Given the above contrast, and assuming no initially (i.e., back in evolutionary time) marked differences in the levels of defense between TDF and TRF plants, the risk of attack should be restricted to a few months of the year for TDF plants, and should be continuous (i.e., all months of the year) for TRF plants.

3 *Potential impact of herbivores.* In addition, contrasts in rainfall seasonality will promote marked differences in the lifespan of leaves in plants from both types of forest, with consequences for herbivory and plant fitness. The higher leaf turnover rate and shorter leaf longevity of dry forest plants (during the rainy season) as

compared with those of the slow-growing rain forest species (Coley and Aide 1991) should allow for a greater capacity to replace tissue loss, rendering the fitness costs of herbivory lower in the former.

4 *Selective pressure for defense.* The expected contrast in impact of herbivory should lead to differences in the selective pressure regime exerted by herbivores. The lower impact of herbivory in TDF plants as compared with TRF plants is expected to operate as a selective pressure of relatively lower intensity for the former and one of higher intensity for the latter.

5 *Investment in defense.* Accordingly, investment in defense is predicted to be comparatively lower in plants of seasonally dry tropical forests, as compared with plants of aseasonal/less seasonal rain forests.

6 *Herbivory.* Consequently, our final prediction is for the level of herbivory to be comparatively higher in plants of seasonally dry forests, whereas plants of tropical rain forests are predicted to sustain lower levels of herbivory.

The apparent contradiction of the fact that high levels of herbivory do not promote the evolution of high levels of defense in plants of TDF could be explained, as we argued above, by the deployment of attributes that confer some degree of tolerance to herbivory, in particular their higher leaf turnover rates and shorter leaf lifespans. In contrast, the low levels of herbivory in TRF plants do not favor the evolution of reduced defense given the continuous risk of attack/damage and the higher fitness costs of herbivory in this environment. These contrasts are compatible with the theoretical expectations of plants to evolve a variety of responses to herbivory, ranging from tolerance of varying degrees, to defense, or a given combination thereof (Strauss and Agrawal 1999, Stowe *et al.* 2000, Fornoni *et al.* 2003). Moreover, our arguments do not imply that plants of the TDF should be devoid of defenses; we argue, rather, that in comparison with TRF species, levels of defense are likely to be lower and levels of tolerance are predicted to be higher in TDF species. The abiotic control of herbivore populations (population reductions during the dry season) concomitant with foliage availability should enhance the effect of rainfall seasonality.

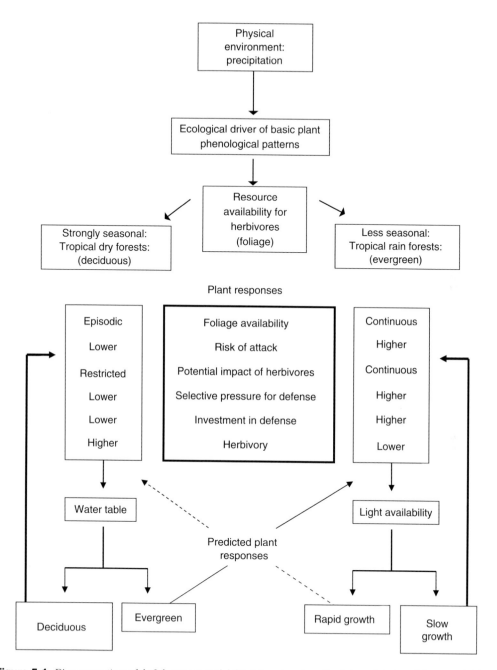

Figure 5.1 Diagrammatic model of the water availability/phenology hypothesis to explain how foliage availability leads to a series of contrasting responses between plants of seasonally dry tropical forest (TDF) and evergreen tropical rain forest (TRF), regarding herbivory and defense. A further subdivision of TDF plants into deciduous and evergreen, and TRF plants into rapid-growth and slow-growth species, shows relevant interspecific heterogeneity within each major forest type, and size of boxes are indicative of the relative representation of these four types of plants. Ascending arrows indicate the predicted responses for plants of each of the four groups. See text for details.

Testing these predictions is fraught with complications due to the effect of a host of uncontrolled variables, particularly the interspecific heterogeneity in the patterns of defense and herbivory of plants within each forest type. In particular, plants of the TRF can be classified into different relative positions along a shade tolerance gradient: in the extremes of the gradient, they will be shade tolerant and slow growing or shade intolerant and fast growing. According to the resource availability/growth hypothesis (Janzen 1974, Coley et al. 1985), slow-growth plants of the mature forest are expected to be better defended than rapid-growth plants typical of forest gaps, and empirical data support this expectation (Coley 1982). Evidence shows that the predominant growth strategy among plants of this type of forest is that of slow growth, shade tolerance, with relative proportions that range between 89% (Martínez-Ramos 1994) and 95% (Welden et al. 1991). Such a distinction is not applicable to, or is of lower importance, for TDF plants, where shading is not such a critical aspect of the environment (Mooney et al. 1985).

On the other hand, in TDFs, two distinct categories of phenological patterns can readily be distinguished due to the presence of riparian habitats. Given their higher water table and conservation of soil moisture, riparian habitats are populated by evergreen species (Dirzo and Domínguez 1995) which, in contrast to the deciduous species, retain their foliage throughout the year. Evergreen species of the TDF, although a minor fraction of the total flora (e.g., 1.1–9.7% in a variety of sites from Mesoamerica; Lott and Atkinson 2002), represent a significant contrast in phenology that is likely to lead to differences in patterns of herbivory and defense. For the comparisons we develop in this chapter we attempted to control for such sources of interspecific variation by looking at the levels of herbivory and defense separating plants according to these strategies (slow growth and fast growth in TRF; evergreen and deciduous in TDF), in addition to the overall, total-species comparisons (see Figure 5.1).

In addition, we cannot rule out the possibility that variables other than contrast in water availability and seasonality will differ between

TDF and TRF, including different herbivore communities, historical factors, etc. Ideally, the relative effect of these additional, potentially confounding, factors should be controlled. A limited, but useful, mimic of such control can be provided by the riparian habitats of TDFs. Riparian habitats intermingled in the predominant vegetation of TDF allow for an intra-site comparison under the same climatic regime, potential herbivore community, and several historical factors to test our water availability/phenology hypothesis. Such comparison may have the caveat that riparian communities represent relatively small habitat islands for insects. However, they can maintain resident populations of some insect species, and/or they can harbor insects of other (non-dormant) species that migrate to them during the dry season (see Janzen 1973). Therefore, an additional, convergent prediction is that the evergreen plant species of the riparian habitats in TDFs should have attributes that resemble those of the slow-growth plants that predominate in TRFs. It follows, therefore, that in order to test our predictions for TDF versus TRF plants, it is the deciduous species of TDFs that should be compared with the slow-growth species of TRF (bold arrows in Figure 5.1). Finally, the rapid-growth plant species of TRF, although not directly relevant for our water availability/phenology hypothesis, are predicted to resemble more the responses exhibited by the deciduous species of TDF (diagonal broken arrow in Figure 5.1).

THE INFORMATION BASE TO TEST THE PREDICTIONS

We carried out an extensive search of all published studies of herbivory or defense or both, in tropical systems between 1970 and 2004, using the Web of Science and BIOSIS. In addition, we included data from our own unpublished studies on tropical herbivory in the Neotropics, but particularly in Mexico, for both tropical rain and dry forests. Specific information on study site locations, their seasonality, measurements, and sources of data are given in Table 5.1. The information compiled in this table provides sources of data for the reader interested in this topic, but, for the purpose of the comparisons of this chapter, we filtered the information as follows. We selected only those studies from which specific data regarding plant phenology (i.e., evergreen or deciduous in TDF) and adaptation to growth in specific microhabitats (i.e., slow in shaded understory or rapid in forest gaps of TRF) could be obtained to separate species accordingly. In addition, regarding more practical aspects, given that the predominant information available in the literature consists of standing levels of herbivory, we restricted our analyses to studies reporting herbivory as a discrete measure of accumulated herbivore attack throughout the lifespan of leaves (see discussions in Lowman 1992 and Filip et al. 1995). This meant that unfortunately we were not able to use some important studies based on herbivory rates for comparisons of herbivory, although for some of those studies we used other data presented therein (e.g., plant defense). Regarding defensive characteristics (total phenolics, tannins, and toughness), our comparisons are restricted to data corresponding to mature leaves. The most detailed study available for this aspect (Coley 1983) clearly shows that interspecific variation can be better predicted by the analysis of mature leaves. An important insight regarding variation in herbivory is the heterogeneity associated with plant ontogeny (Boege and Marquis 2006) and it would have been important to explore to what extent the data based on juvenile/mature plants (as we have done here) are consistent or not with the data corresponding to seedlings. Unfortunately, data on seedling herbivory were available only for TRF plants, including a substantial number of species (de la Cruz and Dirzo 1987, see also Table 5.1), but we found no equivalent or even partial information for TDF species. This is an aspect that warrants further assessment. Finally, our comparisons are based on data corresponding to each individual species, instead of averaging values for a given study (i.e., site) when it involved several taxa. In those few cases in which a given species was studied several times in a single site, we averaged the values. Table 5.1 provides all sources of information used in this study; unpublished data reported therein are available from the authors upon request.

DOES EMPIRICAL EVIDENCE MATCH PREDICTIONS?

Herbivory

A first approach was to compare overall TDF species versus TRF species (Figure 5.2, upper panel). This comparison shows that the two groups of plants are statistically indistinguishable (Mann–Whitney's $Z = 1.69$, d.f. $= 149$, $P = 0.09$), not supporting our predicted outcome. However, when we separated the TRF and TDF data into evergreen and deciduous species of TDF on the one hand and slow-growth and rapid-growth species from TRF on the other (Figure 5.2, lower panel), the

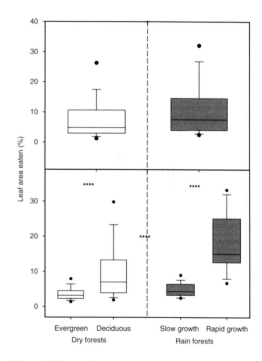

Figure 5.2 Levels of herbivory for TDF species (open boxes) and TRF species (shaded boxes), comparing the total conglomerate of species of each forest type (upper panels) and the species separated according to their phenology in TDFs (lower panel, left) and growth habit in TRFs (lower panel, right). Data represent medians, quartiles (25–75% and 5–95%), and extreme values. ****$P < 0.0001$.

contrast between deciduous versus slow-growth species is highly significant: the median damage of TDF deciduous species is about two-fold greater than that of slow-growth TRF species (Mann–Whitney's $Z = -3.88$, d.f. $= 96$, $P = 0.0001$). This result supports the expected pattern: TDF species, with episodic leaf availability to herbivores and presumably lower investment in defenses, are considerably more damaged than slow-growing species of TRF with continuous foliage availability (see bold arrows in Figure 5.1).

It is conceivable that the differences in herbivory between the plants of the two forest types could be associated with a host of environmental variables beyond those related to our hypothesis of water availability/phenology. A more controlled test of our prediction would be to compare levels of herbivory between deciduous and evergreen species of dry forests (see discussion above). Some of the data in Table 5.1 permit such comparison, particularly based on data from Mesoamerican TDF (Costa Rica and Mexico). Such comparison yielded highly significant differences (Figure 5.2, lower panel, left): TDF deciduous species had a median herbivory value 2.8 times as large as that of the evergreen species (Mann–Whitney's $Z = -4.19$, d.f. $= 79$, $P < 0.0001$). These results are consistent with, and slightly sharper than, those of the comparison between deciduous TDF species versus slow-growth TRF species. Thus, in general, the levels of herbivory of the deciduous species of the dry forest mirror the original expectation, while levels of herbivory of the evergreen species resemble those of the slow-growing plants of the less seasonal TRF (see ascending diagonal arrow for TDF in Figures 5.1 and 5.2).

Separating slow-growing versus fast-growing species in TRF provides results that are consistent with previous findings (Coley 1983, Dirzo 1987) and with the resource availability hypothesis (Janzen 1974, Coley et al. 1985): species of rapid growth sustained levels of herbivory 3.4 times as large as those found in the slow-growing species (Mann–Whitney's $Z = 6.62$, d.f. $= 68$, $P < 0.0001$). Given such contrast, rain forest species of rapid growth tend to resemble the response of deciduous TDF species (see ascending broken arrow in Figure 5.1), although their levels

of herbivory are significantly different (Mann–Whitney's $Z = 4.4$, d.f. $= 94$, $P < 0.0001$).

Plant defense

Data were available to compare three attributes related to plant defense: total phenolics, condensed tannins, and leaf toughness (Table 5.1, Figure 5.3). However, data for total phenolics and condensed tannins in the case of TDF plants, although available for a large number of species (64), correspond to a single though comprehensive study in a single locality (Janzen and Waterman 1984), while data for these two variables for TRF plants are derived from eight independent studies and a large number of species (152). (The measurement of phenolic compounds in these studies is largely based on the use of quebracho standards; see Appel et al. 2001 for a discussion on the limitations to evaluating phenolic compounds using such standards.) Likewise, data for leaf toughness are derived from three studies for each of the two forest types, involving 42 species for TDF and 69 species for TRF (see Table 5.1).

Overall, considering all types of plants within dry and rain forests, total phenolics concentration was only marginally greater in TRF plants ($P = 0.06$; Figure 5.3, upper panel, left). However, when we separate the species according to their phenology in the case of TDF plants and their growth rate strategy in the case of TRF plants, significant differences are readily observed (Kruskal–Wallis $\chi^2 = 12.00$, d.f. $= 3$, $P = 0.007$; Figure 5.3, lower panel, left). The comparisons show that slow-growing species of TRF have a two times greater concentration of total phenolics when compared with deciduous species from TDF (Mann–Whitney's $Z = -3.2$, d.f. $= 119$, $P = 0.001$). Again, an extended comparison between evergreen and deciduous species of the dry forest shows a consistent pattern: evergreen species had 60% greater concentrations of total phenolics, though differences are marginally significant ($P = 0.06$).

Condensed tannins had a similar concentration when considering the overall group of species from both forest types (Figure 5.3, upper panel, center). Furthermore, a separation of the

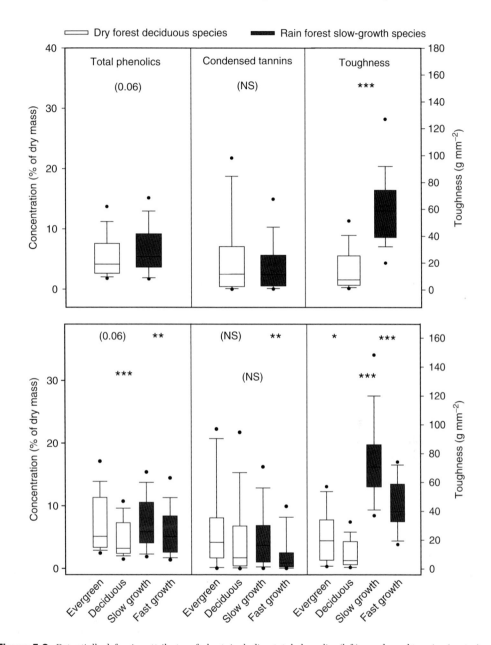

Figure 5.3 Potentially defensive attributes of plants including total phenolics (left), condensed tannins (center), and toughness (right) from TDF (open boxes) and TRF (shaded boxes), comparing the total conglomerate of species of each forest type (upper panels) and the species separated according to their phenology in TDFs and growth habit in TRFs (lower panels). Data represent medians, quartiles (25–75% and 5–95%), and extreme values. *P < 0.05, **P < 0.01, ***P < 0.001, NS, not significant.

data considering differences in phenology and growth rate in TDF and TRF plants, respectively (Figure 5.3, lower panel, center), detected a significant contrast only between slow-growing versus fast-growing species – an expected result, but of secondary interest to our hypothesis. Thus, while total phenolics supported the original predictions (see Figure 5.1), condensed tannins did not. However, in addition to the limitations of the dataset we used for these comparisons, it is necessary to bear in mind that these two groups of compounds, assayed on such a variety of species, may not necessarily reflect defensive responses, given the variety of secondary metabolites known to be important as anti-herbivore mechanisms (Rosenthal and Janzen 1979). Again, this is an aspect that warrants further investigation involving additional sites and other defensive compounds.

Leaf toughness, in contrast, exhibited a very consistent difference between TDF and TRF species. The overall difference was highly significant (Figure 5.3, upper panel, right): rain forest species were approximately six times tougher than dry forest species, and this difference arises despite the fact that plants from both forest types in this comparison included different phenologies and growth rates. Therefore, when data are disaggregated, among-group differences are highly significant (Kruskal–Wallis $\chi^2 = 69.9$, d.f. = 3, $P = 0.0001$) and the contrast between deciduous plants from TDF and slow-growing species from TRF is even more marked – a 10-fold difference (Mann–Whitney's $Z = -6.41$, d.f. = 57, $P < 0.0001$; Figure 5.3, lower panel, right). Furthermore, evergreen species from dry forest were about two-fold tougher than their deciduous counterparts. Regarding the comparison between slow-growing and fast-growing species from TRF (Figure 5.3, lower panel, right) we detected that slow-growth species were 1.75 times as tough as rapid-growth species.

In all three cases of comparison of defensive characteristics between the two groups of TRF plants (i.e., slow- and fast-growth species), differences were highly significant, indicating that defensive attributes were significantly greater in slow-growing species, as would be expected from previous studies and the resource availability

hypothesis (Janzen and Waterman 1984, Coley *et al.* 1985).

Despite our consistent findings, it is important to take into consideration that the three attributes we used to infer defense may not necessarily reflect ultimate evolutionary responses resulting from previous selective pressure exerted by herbivores on plants. The importance of the distinction between proximal and ultimate factors responsible for currently observed characteristics in plant–animal interactions has been discussed in other studies (Dirzo 1984, Farrell and Mitter 1993) and warrants further work in the context of our water availability/phenology hypothesis.

PHYLOGENETIC INERTIA

Our results have shown that plants from TRF sustain significantly lower levels of herbivory and possess attributes that confer, at least at the proximal level, greater defense. Moreover, results from the intra-site comparisons in the case of TDFs, looking at deciduous versus evergreen species, support such an argument. The latter comparison, in addition to highlighting the internal heterogeneity of TDF plants, provided a useful comparison by exerting some control of environmental, *extrinsic* factors. However, other *intrinsic* factors that might be responsible for the observed patterns may need to be taken into account. In particular, recent studies argue for the importance of controlling for phylogenetic inertia (see Armbruster 1992): observed ecological patterns may be, at least partly, the result of differences due to phylogeny. Unfortunately, our dataset is very limited in allowing assessment of the importance of phylogeny in our comparisons. Nevertheless, our dataset allows for five intra-generic comparisons of herbivory between related species from TDF (deciduous species) and TRF (slow-growth species) (*Lonchocarpus, Cordia, Erythroxylum, Trichilia*, and *Piper*) and three intra-family comparisons (two species of each of these families: Sapindaceae, Apocynaceae, Euphorbiaceae). If all eight pairs of taxa are used (Figure 5.4), we observe that in six of the comparisons herbivory is greater in plants from TDF, as compared with their TRF relatives, while in one

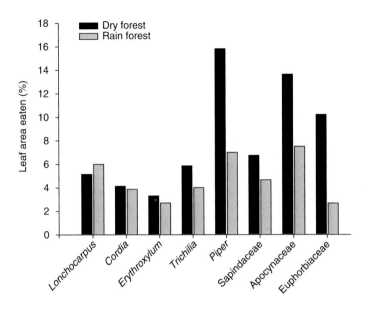

Figure 5.4 Levels of herbivory in pairs of taxa comparing TDF (black bars) and TRF (gray bars) species within the same genus or within the same family. Herbivory data are derived from the measurement of a bulk sample of 50 randomly collected leaves from each of three plants of each of the species.

of them the difference is negligible, and the comparison between the two species of *Lonchocarpus* is slightly higher in the rain forest species. An overall paired comparison indicated that differences are statistically significant (signed-rank test = 15, $P = 0.04$). A more controlled study and one with a larger number of comparisons, such as that performed by Fine *et al.* (2004), should be useful to assess the relevance of phylogenetic inertia in studies of water availability/phenology related to herbivory.

Due to their scarcity, the available data are far from satisfactory for teasing apart the relative importance of phylogeny in the patterns we uncovered. Nevertheless, it is of interest that the limited information points in the direction of water availability/phenology having an important role for our predicted patterns, independent of, or in addition to, phylogenetic inertia.

CONCLUSIONS

Some studies of ecogeographic variation in plant attributes have looked for patterns of variation

and then assessed the possible mechanisms responsible for such patterns. Our approach was to develop a line of reasoning as to why rainfall seasonality and phenology of TRF and TDF plants bring about fundamental differences in the availability of foliage for folivores and how these differences may in turn lead to predictable patterns of herbivory and defense. We then confronted the predictions with the available data and found a strong suggestion that higher constancy of foliage, implying greater risk and impact of herbivory in TRF plants in ecological time, may lead to a greater evolutionary history of herbivory, favoring greater selection for increased defense and lower herbivory. The predicted patterns were evident when we controlled for interspecific heterogeneity in herbivory and defense within both TDF and TRF. In addition, the expected patterns were mirrored using more controlled intra-site (dry forest) comparisons looking at plants of contrasting phenologies. Moreover, patterns seem to hold independently of, or in addition to, phylogenetic influence.

As is well known, local and ecogeographic patterns of herbivory are influenced not only by bottom-up factors but also by top-down controls (see Dyer and Coley 2002). Clearly, our emphasis on the former provides a limited panorama and further work is needed to assess how the observed patterns accommodate to the possible variations in influence from the third trophic level. This is an exciting field of research that warrants consideration.

Finally, this study highlights the importance of TDF to our general understanding of tropical biology, including a variety of aspects of ecological and evolutionary significance, such as herbivory. The value of the perspectives that comparisons with TDFs may offer, and the need to shift the balance of attention towards a more dry-centric interest, can hardly be overemphasized, given their great concentration of biodiversity and endemism (see Trejo and Dirzo 2002), their ecological services to society (Maass *et al.* 2005), and the fact that TDFs are considered the most threatened tropical ecosystems (Janzen 1988, Trejo and Dirzo 2000).

ACKNOWLEDGMENTS

We thank Paul Fine, Lissy Coley, Mike Donoghue, and Radika Bahskar for constructive comments. Logistical support to write this chapter was provided by Hal Mooney's lab at Stanford University. We are grateful to Walter Carson and Stefan Schnitzer for their invitation and support to write this chapter. This study was partly supported by a UNAM-DGAPA grant to R.D.

REFERENCES

Aide, T.M. (1988) Herbivory as a selective agent on the timing of leaf production in a tropical understory community. *Nature* 336, 574–575.

Appel, H.M., Govenor, H.L., D'Ascenzo, M., Siska, E., and Schultz, J.C. (2001) Limitations of foliar assays of foliar phenolics in ecological studies. *Journal of Chemical Ecology* 27, 765–778.

Armbruster, W.S. (1992) Phylogeny and the evolution of plant–animal interactions. *BioScience* 42, 12–20.

Benitez-Malvido, J. and Kossmann-Ferraz, I.D. (1999) Litter cover variability affects seedling performance and herbivory. *Biotropica* 31, 598–606.

Boege, K. (2004) Induced responses in three tropical dry forest plant species: direct and indirect effects on herbivory. *Oikos* 107, 541–548.

Boege, K. (2005) Herbivore attack in *Casearia nitida* influenced by plant ontogenetic variation in foliage quality and plant architecture. *Oecologia* 143, 117–125.

Boege, K. and Marquis, R.J. (2006) Plant quality and predation risk mediated by plant ontogeny: consequences for herbivores and plants. *Oikos* 115, 559–572.

Carrazco-Carballido, P.V. (2002) *Variación intraespecífica en la herbivoría en plantas de fenología contrastante en la selva baja de Huautla.* Tesis de Licenciatura, Universidad Nacional Autónoma de México, Mexico City.

Coley, P.D. (1982) Rates of herbivory on different tropical trees. In E.G.J. Leigh, A.S. Rand, and D.M. Windsor (eds), *Ecology of a Tropical Forest: Seasonal Rhythms and Long-term Changes.* Smithsonian Institution Press, Washington, DC, pp. 123–132.

Coley, P.D. (1983) Herbivory and defensive characteristics of tree species in a lowland tropical forest. *Ecological Monographs* 53, 209–233.

Coley, P.D. and Aide, M.T. (1991) Comparison of herbivory and plant defenses in temperate and tropical broad-leaved forests. In P.W. Price, M. Lewinsohn, G.W. Fernandes, and W.W. Bneson (eds), *Plant–Animal Interactions: Evolutionary Ecology in Tropical and Temperate Regions.* John Wiley & Sons.

Coley, P.D. and Barone, J.A. (1996) Herbivory and plant defenses in tropical forests. *Annual Review of Ecology and Systematics* 27, 305–335.

Coley, P.D., Bryant, J.P., and Chapin, F.S. (1985) Resource availability and plant herbivore defense. *Science* 230, 895–899.

de la Cruz, M. and Dirzo, R. (1987) A survey of the standing levels of herbivory in seedlings from a Mexican rain forest. *Biotropica* 19, 98–106.

Dirzo, R. (1984) Herbivory: a phytocentric overview. In R. Dirzo and J. Sarukhan (eds), *Perspectives on Plant Ecology.* Sinauer, Sunderland, MA, pp. 141–165.

Dirzo, R. (1987) Estudios sobre interacciones planta-herbívoro en "Los Tuxtlas," Veracruz *Revista de Biología Tropical* 35, 119–131.

Dirzo, R. (2001) Tropical forests. In F.S. Chapin, O.E. Sala, and E. Hubert-Shannwald (eds), *Global Biodiversity in a Changing Environment.* Springer, New York, pp. 251–276.

Dirzo, R. and Domínguez, C. (1995) Plant–herbivore interactions in Mesoamerican tropical dry forests. In S. Bullock, H.A. Mooney, and I.E. Medina (eds), *Tropical Dry Forests.* Cambridge University Press, New York, pp. 304–323.

Dirzo, R. and Mota, L.M. (1997) *Omphalea oleifera.* In E. Gonzalez-Soriano, R. Dirzo, and R.C. Vogt (eds),

Historia Natural de Los Tuxtlas. UNAM-CONABIO, Mexico City.

Dirzo, R. and Raven, P.H. (2003) Global state of biodiversity and loss. *Annual Review of Environment and Resources* 28, 137–167.

Dyer, L.A. and Coley, P.D. (2002) Tritrophic interactions in tropical versus temperate communities. In T. Tscharntke and A. Hawkins (eds), *Multitrophic Level Interactions*. Cambridge University Press, Cambridge, pp. 67–88.

Farrell, B.D. and Mitter, C. (1993) Phylogenetic determinants of insect/plant community diversity. In R.E. Ricklefs and D. Schluter (eds), *Species Diversity in Ecological Communities: Historical and Geographical Perspectives*. University of Chicago Press, Chicago, pp. 253–266.

Feeny, P. (1970) Seasonal changes in oak leaf tannins and nutrients as a cause of spring feeding by winter moth caterpillars. *Ecology* 51, 565–581.

Feeny, P. (1976) Plant apparency and chemical defense. *Recent Advances in Phytochemistry* 10, 1–40.

Filip, V., Dirzo, R., Maass, J.M., and Sarukhán, J. (1995) Within- and among-year variation in the levels of herbivory on the foliage from a Mexican tropical deciduous forest. *Biotropica* 27, 78–86.

Fine, P.V.A., Mesones, I., and Coley, P.D. (2004) Herbivores promote habitat specialization by trees in Amazonian forests. *Science* 305, 663–665.

Forkner, R.E., Marquis, R.J., and Lill, J.T. (2004) Feeny revisited: condensed tannins as anti-herbivore defences in leaf-chewing herbivore communities of *Quercus*. *Ecological Entomology* 29, 174–187.

Fornoni, J.S., Nuñez-Farfán, J., and Valverde, P.L. (2003) Evolutionary ecology of tolerance to herbivory advances and perspectives. *Comments on Theoretical Biology* 8, 643–663.

Garcia-Guzman, G. and Benitez-Malvido, J. (2003) Effect of litter on the incidence of leaf-fungal pathogens and herbivory in seedlings of the tropical tree *Nectandra ambigens*. *Journal of Tropical Ecology* 19, 171–177.

Gartlan, J.S., McKey, D.B., Waterman, P.G., Mbi, C.N., and Struhsaker, T.T. (1980) A comparative study of the phytochemistry of two African rain forests. *Biochemical Systematics and Ecology* 8, 401–422.

Herrerias-Diego, Y. (1999) *Variación en los niveles de daño causado por folivoros y su relación con características foliares*. Universidad Nacional Autónoma de Mexico, Mexico DF.

Howlett, B.E. and Davidson, D.W. (2001) Herbivory on planted dipterocarp seedlings in secondary logged forests and primary forests of Sabah, Malaysia. *Journal of Tropical Ecology* 17, 285–302.

Hurley, M. (2000) Growth dynamics and leaf quality of the stinging trees *Dendrocnide moroides*

and *Dendrocnide cordifolia* (Family Urticaceae) in Australian tropical rainforest: implications for herbivores. *Australian Journal of Botany* 48, 191–201.

Iddles, T.L., Read, J., and Sanson, G.D. (2003) The potential contribution of biomechanical properties to anti-herbivore defence in seedlings of six Australian rainforest trees. *Australian Journal of Botany* 51, 119–128.

Jackson, R.V. and Bach, C.E. (1999) Effects of herbivory on growth and survival of seedlings of a rainforest tree *Alphitonia whitei*. *Australian Journal of Ecology* 24, 178–267.

Janzen, D.H. (1973) Sweep samples of tropical foliage insects: effects of seasons, vegetation types, elevation, time of day and insularity. *Ecology* 54, 687–708.

Janzen, D.H. (1974) Tropical blackwater rivers, animals, and mast fruiting by the Dipterocarpaceae. *Biotropica* 6, 69–103.

Janzen, D.H. (1981) Patterns of herbivory in a tropical deciduous forests. *Biotropica* 13, 271–282.

Janzen, D.H. (1988) Tropical dry forests: the most endangered major tropical ecosystem. In E.O. Wilson (ed.), *Biodiversity*. National Academy Press, Washington, DC, pp. 130–137.

Janzen, D.H. and Waterman, P.G. (1984) A seasonal census of phenolics, fiber and alkaloids in foliage of forest trees in Costa Rica: some factors influencing their distribution and relation to host selection by Sphingidae and Saturniidae. *Biological Journal of the Linnean Society of London* 21, 439–454.

Leffler, A.J. and Enquist, B.J. (2002) Carbon isotope composition of tree leaves from Guanacaste, Costa Rica: comparison across tropical forests and tree life history. *Journal of Tropical Ecology* 18, 151–159.

Levin, D.A. (1976) Alkaloid-bearing plants: an eco-geographic perspective. *American Naturalist* 110, 261–284.

Lott, E.J. and Atkinson, T.H. (2002) Biodiversidad y fitogeografica de Chamela-Cuixmala Jalisco. In F. Noguera, J.H. Vega Rivera, A.N. García Aldrete, and M. Quesada (eds), *Historia natural de Chamela*. Instituto de Biologia UNAM, Mexico City, pp. 83–97.

Lowman, M.D. (1992) Leaf growth dynamics and herbivory in five species of Australian rainforest. *Journal of Ecology* 80, 433–447.

Maass, J., Balvanera, P., Castillo, A. *et al.* (2005) Ecosystem services of tropical dry forests: insights from long-term ecological and social research on the Pacific Coast of Mexico. *Ecology and Society* 10(1), 17. [online] URL: http://www.ecologyandsociety.org/vol10/iss1/art17/

Marquis, J.R. and Clark, D.B. (1989) Habitat and fertilization effects on leaf herbivory in *Hampea appendiculata* (Malvaceae) implications for tropical firewood systems. *Agriculture, Ecosystems and Environment* 25, 165–174.

Marquis, R.J. (1984) Leaf herbivores decrease fitness of a tropical plant. *Science* 226, 537–539.

Marquis, R.J. (1992) The selective impact of herbivores. In R.S. Fritz and E.L Simms (eds), *Plant Resistance to Herbivore and Pathogens. Ecology, Evolution and Genetics*. University of Chicago Press, Chicago, pp. 301–325.

Martínez, B.R.I. (2000) *Estudio comparativo del herbivorismo en dos especies de Croton en la selva baja caducifolia de Chamela, Jalisco México*. Universidad Nacional Autónoma de México, Mexico.

Martínez-Ramos, M. (1994) Regeneración natural y diversidad de especies arbóreas. *Boletín de la Sociedad Mexicana de Botánica* 54, 179–224.

McKey, D., Waterman, P.G., Mbi, C.N., Gartlan, J.S., and Struhsaker, T.T. (1978) Phenolic content of vegetation in two African rain forests: ecological implications. *Science* 202, 61–64.

Milton, K. (1979) Factors influencing leaf choice by howler monkeys: a test of some hypothesis of food selection by generalist herbivores. *American Naturalist* 114, 362–378.

Mooney, H.A., Bullock, S.H., and Medina, E. (1985) Introduction. In S.H. Bullock, H.A. Mooney, and E. Medina (eds), *Seasonally Dry Tropical Forests*. Cambridge University Press, Cambridge, pp. 1–8.

Mole, S., Ross, J.A.M., and Waterman, P.G. (1988) Light-induced variation in phenolic levels in foliage of rain-forest plants. *Journal of Chemical Ecology* 14, 1–21.

Murali, K.S. and Sukumar, R. (1993) Leaf flushing phenology and herbivory in a tropical dry deciduous forest, southern India. *Oecologia* 94, 114–119.

Myster, R.W. (2002) Foliar pathogen and insect herbivore effects on two landslide tree species in Puerto Rico. *Forest Ecology and Management* 169, 231–242.

Núñez-Farfán, J. and Dirzo, R. (1988) Within-gap spatial heterogeneity and seedling performance in a Mexican tropical forest. *Oikos* 51, 274–284.

Odum, H.T. and Ruiz-Reyes, H.T. (1970) Holes in the leaves and the grazing control mechanism. In H.T. Odum and R. Pigeon (eds), *A Tropical Rain Forest*. Division of Technical Information, U.S. Atomic Energy Commission, Tennessee, pp. 11–69.

Price, P.W. (1997) *Insect Ecology*. John Wiley & Sons, New York.

Rosenthal, G.A. and Janzen, D.H. (1979) *Herbivores: Their Interaction with Secondary Plant Metabolites*. Academic Press, New York.

Sánchez-Montoya, J.G.J. (2002) *Variación intraespecífica en la herbivoría y su impacto en algunos componentes del éxito reproductivo masculino y femenino de Ipomoea pauciflora en una selva baja caducifolia*. Universidad Nacional Autónoma de México, Mexico City.

Stamp, N. (2003) Out of the quagmire of plant defense hypothesis. *Quarterly Review of Biology* 78, 23–55.

Sterck, F., van der Meer, P., and Bongers, F. (1992) Herbivory in two rain forest canopies in French Guiana. *Biotropica* 24, 97–99.

Stowe, K.A., Marquis, R.J., Hochwender, C.G., and Simms, E.L. (2000) The evolutionary ecology of tolerance to consumer damage. *Annual Review of Ecology and Systematics* 31, 565–595.

Strauss, Y.S. and Agrawal, A.A. (1999) The ecology and evolution of plant tolerance to herbivory. *Trends in Ecology and Evolution* 14, 179–185.

Trejo, R.I. and Dirzo, R. (2000) Deforestation of seasonally dry forest towards its northern distribution: a national and local analysis in Mexico. *Biodiversity and Conservation* 94, 133–142.

Trejo, R.I. and Dirzo, R. (2002) Floristic diversity of Mexican seasonally dry tropical forests. *Biodiversity and Conservation* 11, 2063–2084.

Turner, I.M. (1995) Foliar defences and habitat adversity of three woody plant communities in Singapore. *Functional Ecology* 9, 279–284.

Waterman, P.G., Mbi, C.N., McKey, D.B., and Gartlan, J.S. (1980) African rainforest vegetation and rumen microbes: phenolic compounds and nutrients as correlates of digestibility. *Oecologia* 47, 22–33.

Waterman, P.G. and Ross, J.A.M. (1988) Comparison of the floristics and leaf chemistry of the tree flora in two Malaysian rain forests and the influence of leaf chemistry on populations of colombine monkeys in the Old World. *Biological Journal of the Linnean Society* 34, 1–32.

Welden, C.W., Hewett, S.W., Hubbell, S.P., and Foster, R.B. (1991) Saplings survival, growth and recruitment: relationship to canopy height in a neotropical forest. *Ecology* 72, 35–50.

Wint, G.R.W. (1983) Leaf damage in tropical rain forest canopies. In S.L. Sutton, T.C. Whitmore, and A.C. Chadwick (eds), *Tropical Rain Forest: Ecology and Management*. Blackwell Scientific, Oxford, pp. 229–240.

Chapter 6

ECOLOGICAL ORGANIZATION, BIOGEOGRAPHY, AND THE PHYLOGENETIC STRUCTURE OF TROPICAL FOREST TREE COMMUNITIES

Campbell O. Webb, Charles H. Cannon, and Stuart J. Davies

OVERVIEW

The assembly of local tropical forest tree communities is influenced by abiotic filters from a larger regional species pool (e.g., habitat differentiation, mass effects, dispersal limitation) and local biotic interactions (e.g., density dependence, resource competition; summarized in Figure 6.1). These assembly processes are mediated by the phenotypic similarities or differences of individuals, which are the outcome of evolutionary change in historical communities, and ultimately the composition of taxa in a regional species pool is the outcome of biogeographic processes. Given the great diversity of tropical tree species, we are unlikely ever to know enough about the ecologically important phenotypes or precise spatial ranges of species to be able to predict local community species composition based on detailed attributes of every species. However, we suggest in this chapter that because species similarity and difference are strongly influenced by common ancestry, as is the presence or absence of a taxonomic clade in a geographic area, a phylogenetic approach may be most effective for understanding and predicting local community composition.

In this chapter, we briefly review current understanding of abiotic and biotic controls of local species composition, and of evolutionary patterns in ecological characters. We then describe phylogenetic analyses that explore the outcomes of neighborhood interactions, habitat filtering, climatic gradients, and biogeographic history by analyzing the phylogenetic patterns of species composition at nested spatial scales (Figure 6.1). We test these methods with data from forests in Southeast Asia. Finally, we discuss the association of ecological and biogeographic characteristics with internal nodes of plant phylogenies and the creation of predictive models for the general taxonomic and ecological composition of communities.

ECOLOGICAL ORGANIZATION OF TROPICAL TREE COMMUNITIES

What is the relative importance of abiotic limitation, biotic interactions, and chance events on contemporary species abundance and distribution of tropical trees? The non-random distribution of taxa on abiotic gradients (i.e., differing realized niches; Hutchinson 1957) has been demonstrated repeatedly, for topography (Ashton 1964, Clark et al. 1999, Webb and Peart 2000, Valencia et al. 2004), soil nutrients (Potts et al. 2002, Hall et al. 2004, Palmiotto et al. 2004, Russo et al. 2005, Paoli et al. 2006), geology (Cannon and Leighton 2004), elevation (van Steenis 1972, Lieberman *et al.* 1985, Ashton 2003a), rainfall (Gentry 1982, Schnitzer 2005), understory light (Swaine and Whitmore 1988, Clark and Clark 1992, Davies *et al.* 1998), and architectural position (Kohyama 1993). However, many species at a local site also appear to share the same realized niche (Potts *et al.* 2004, Valencia *et al.* 2004). While species may sort into their appropriate habitats at a local scale, it is unlikely that they occupy all sites on a landscape where they might grow, because of the continual perturbation of climate oscillations and temporal variation in biotic interactions. Over short time scales (100–1000 years), the geographic distributions of some taxa will be expanding, and those of others will be shrinking (Bennett 1997). Over long time scales (10,000–1,000,000 years), biogeographic connections (e.g., land-bridges) and barriers (e.g., mountain ranges) change, and at even longer time scales (10–100 My [million years]), land areas and geologies will be appearing and disappearing, again changing the potential geographic distribution of taxa. Hence, the geographic distribution of most taxa will not be in equilibrium with the contemporary abiotic environment, but will represent a dynamic balance of large-scale climatic oscillations and gradients, location of species origin, rate of dispersal, and availability of dispersal routes. This disequilibrium is vital to keep in mind when fitting environmental niche envelopes (on axes of rainfall, elevation, temperature, etc.) using geographical information systems (GIS)-based interpolation (e.g., Austin 2002, Graham *et al.* 2004).

While species differ in their local realized distribution, it is less clear to what extent local biotic interactions, such as ubiquitous competition for light and physical space (Hubbell 2001, Kitajima and Poorter Chapter 10, this volume) or for pollinators and dispersers, modify growth and survival under these different abiotic conditions. For example, is the absence of "poor-soil" species on rich-soil sites due to some fundamental cost associated with ecological specialization that restricts their fundamental niche, or to their local exclusion from rich sites (included in their fundamental niche) by faster-growing but less stress-tolerant "rich-soil" species

(Fine *et al.* 2004)? Some seedling growth experiments (e.g., Hall *et al.* 2003, 2004, Palmiotto *et al.* 2004) suggest that optimal performance in the absence of competition is achieved in the soils on which a species is most abundant, implying that competition may not greatly shift the position of the peak of the realized niche away from that of the fundamental niche. However, the ability of many species to prosper under conditions in which they are not normally found, when potential competition is reduced (e.g., in botanical gardens), suggests that generalized competition may also play a large role in compressing the boundaries of species' fundamental niches.

Even if competition for space and/or light is experienced by all forest plants, does the negative effect of neighboring plants vary with the neighbor's identity, that is, whether a neighbor is a conspecific, a phylogenetically closely related species or a distantly related one? In a temperate forest, Canham *et al.* (2004, 2006) detected different effects on focal species of different neighbor species. These effects might also be mediated by competition for "mobile links," the pollinators and dispersers plants depend upon (Vamosi *et al.* 2006), or experienced as indirect competition resulting from pathogen or herbivore population dynamics. There is abundant evidence that plants do respond more negatively to increasing density of conspecifics than heterospecifics ("negative density dependence"; reviewed by Wright 2002). However, ecological exclusion and eventual character displacement in sibling species, and ecological speciation (Schluter 2001), depend upon the most closely related taxa experiencing the strongest negative interactions. The limited data for tropical trees support this relationship at some spatial scales (Uriarte *et al.* 2004, Webb *et al.* 2006). However, the ultimate outcome of this process, resulting in "checkerboard" patterns where certain combinations of species in the same habitat are never found due to strong competitive effects among species (e.g., Graves and Gotelli 1993), has not yet been reported. The only demonstrated example of over-dispersion of a character in tropical forest trees, of the kind generally thought to indicate biotic, competitive structuring of a community (Bowers and Brown 1982, Wilson 1999), is the segregation of flowering and fruiting times (*Miconia*, Snow 1966; *Shorea* section *Mutica*, Ashton *et al.* 1988).

On the other hand, the frequently reported association of species of particular taxonomic groups with different habitats (e.g., Gentry 1988, Webb 2000) suggests that while generalized competition for shared resources in forests is ubiquitous, and biotic interactions may negatively influence the performance of similar or phylogenetically close neighboring plants, species distributions at habitat-wide scales generally result from the "attractive" effect of abiotic conditions. We note that positive advantages of having related taxa nearby (e.g., higher pollinator and seed disperser availability, sharable ectomycorrhizae) may reinforce similarity in the distributions of related taxa (see Momose *et al.* 1998).

EVOLUTION OF ECOLOGICAL CHARACTERS IN TROPICAL FOREST TREES

If ecological processes affecting today's tropical forest trees are the same as in the past (i.e., ecological uniformitarianism on a time scale of millions of years), we expect the evolution of ecological characters to have been shaped by the ecological conditions discussed above, that is, competition among large numbers of similar taxa filtered into particular forest habitats (ridge-tops, gullies, swamps, gaps). The diversity of such communities (itself probably ancient, e.g., 64 My; Johnson and Ellis 2002) means that neighborhood encounters between any specified pair of species will be relatively infrequent, and divergent selection among phenotypically similar taxa may be very weak (Connell 1980, Stevens 1980, Ashton 1988). Additionally, the long generation time of many tropical trees (100 + years), in combination with climate changes over 1000–10,000 years, will further weaken such divergent selection. This is a very different community scenario to the species-poor, small island systems which have provided much of our empirical knowledge of ecological divergence (Darwin 1859). The animal and plant species on islands may frequently experience sustained pairwise competition (causing trait "push"; Silvertown 2004) and the frequent

opportunity to fill empty niches (trait "pull"), leading to rapid selective divergence in ecological characters (Givnish 1998, Schluter 2000, Silvertown 2004). Species of tropical forest trees clearly differ in autecology in many ways, but these differences are likely to be the result of drift within a species or subset of metapopulations, or weak selection on peripheral/founder populations in slightly different conditions. If the accumulation of ecological changes is essentially random and relatively slow compared with speciation, then closely related taxa will generally share the same character ("symplesiomorphy"), showing an overall conservative pattern of character evolution (Harvey and Pagel 1991, Ackerly and Donoghue 1998, Webb *et al.* 2002).

What evidence is there that a random accumulation of character changes best represents ecological evolution in tropical trees? The most rigorous assessments of the pattern of ecological character evolution use standard methods of phylogenetic ancestral state reconstruction (or divergent tendency; see Moles *et al.* 2005) for niche-related characters (see Linder and Hardy 2005 for a good example). There are now a handful of phylogenetically based studies of ecological character evolution in tropical and subtropical forest plant genera (Davies 1996, Givnish *et al.* 2000, Dubuisson *et al.* 2003, Bramley *et al.* 2004, Cavender-Bares *et al.* 2004, Fine *et al.* 2004, Fine *et al.* 2005, Plana *et al.* 2004), but none that we know of that has attempted to sample all the extant species in a lineage. There is no consensus view arising from these studies. Some studies report that closely related species are ecologically similar, others that related species differ in a way that may indicate adaptive divergence, but since the taxon sampling is usually sparse, and the characters examined so different, it is impossible yet to generalize. No study we are aware of has rigorously tested for significant conservatism in ecological characters in a densely sampled group of closely related tree species, as has been done for animals (e.g., Losos *et al.* 2003, Stephens and Wiens 2004).

Because standardized data on niche parameters for tropical trees are hard to collect, an alternative approach is to estimate niche parameters by modeling the distribution of species in multidimensional environment space, which has been successful for other taxa (e.g., Peterson and Holt 2003, Raxworthy *et al.* 2003). Tools such as GARP (Scachetti-Pereira 2002), BIOCLIM (http://biogeo.berkeley.edu/worldclim) and Why-Where (http://biodi.sdsc.edu) use the values of various spatially modeled factors (rainfall, temperature, elevation, etc.) at geographical points of known species occurrence to create a niche envelope in multidimensional factor space. This envelope can be re-projected onto the GIS landscape to predict the potential range of the species. Niche parameters derived in this way can be reconstructed on a phylogeny, as with other characters (Graham *et al.* 2004).

Supplemental information on ecological evolution comes from comparing the ecological character of congeners that systematists consider to be sister species (although usually no molecular evidence exists to confirm this impression). These supposed close relatives are often ecologically similar (Forman 1966), but have non-overlapping (allopatric) distributions (e.g., Stevens 1980 table 4, Prance and White 1988). This suggests that sibling taxa cannot co-occur until they have diverged sufficiently to overcome local competitive exclusion, and that rapid ecological divergence is common. A similar interpretation might apply to studies that find substantial variation in ecological character among locally co-occurring congeners (Grubb and Metcalfe 1996, Osunkoya 1996, Thomas 1996, Davies 1998, Smith-Ramirez *et al.* 1998), and cases of congeners occupying markedly different habitats within a site (Valencia *et al.* 2004).

But is local niche variation more than expected from models of slow character evolution? We suggest not. Because much speciation in tropical trees appears to be allopatric (e.g., Stevens 1980, Ashton 1988, Gentry 1989, although again this has seldom been confirmed with phylogenetic or population genetic data), the spatial segregation of ecologically similar, new sibling species is expected to be observed frequently. Over time niche parameters may then drift randomly and, independently, species ranges will shift such that sibling species become sympatric (Barraclough and Vogler 2000). Despite the great noise added to a system due to climate oscillations and temporal variation in biotic interactions, we expect in a system dominated by allopatric speciation that degree of range overlap will be positively associated with time since divergence (but see critique by Losos and Glor 2003). If variance in trait characters is also correlated with time since divergence, then we expect that the greater the range overlap of sibling taxa the more likely they are to differ in autecology. An appropriate null model is required to test whether niche segregation in congeners, implying either contemporary competitive exclusion or historical selection for niche divergence, is greater than expected by chance.

We must also note that in addition to cases of striking habitat variation there are also many cases of congeners sharing habitats (Webb and Peart 2000), and of general association of large taxonomic groups (e.g., families) with particular habitats (Ashton 1988, Gentry 1988, Davis *et al.* 2005). The existence of ecological conservatism (Wiens 2004) in plants has also been demonstrated by the ability to predict a species' distribution based on observations of related species on a different continent (Huntley *et al.* 1989, Ricklefs and Latham 1992). Morphological characters with clear ecological significance (e.g., pollen ultrastructure, pollination syndrome, seed dispersal mode, and tree architecture) are also usually strongly conserved. Overall, phylogenetically based information on the evolution of autecology is scanty, and we have no basis to reject the "null" hypothesis that ecological character changes have generally accumulated at random in tropical forest tree species, and thus display a pattern of significant phylogenetic conservatism in most species (accepting the existence of great variation in evolutionary history).

A final question, fundamental to interpreting patterns of species association at geographic scales is: How consistent in space are species ecological characters and species boundaries? Some species are morphologically and ecologically constant, and studies of their breeding systems show them to be strongly resistant to hybridization (e.g., cerrado species studied by Barros 1989). Other species may not be ecologically constant across their range, and attempts to detect community assembly rules (Gotelli and McCabe 2002) over large areas may fail because species behave differently in different places. Where species exhibit large genotypic variation, subspecies have often been recognized (van Steenis 1948), usually in disjunct populations – clinal (continuous) variation in morphology is extremely rare in tropical tree species. Variation is not problematic in a phylogenetic context if different morphs are monophyletic (i.e., all share the same most recent common ancestor). However, we are becoming increasingly aware of complex gene-flow reticulations in some tropical tree clades within and even among species (Cannon and Manos 2003; see also Mallet 2005). More than in perhaps any other functional group of organisms, individuals of tropical forest trees are long lived and rare, which has profound consequences for the evolutionary coherence of these species. The relatively few generations between major climatic events, and the danger of erosion of necessary genetic variability when rare, may select for individual-level characters that maintain openness to gene flow (van Valen 1976, Grant and Grant 1996). The existence of "swarms" of related species or subspecies with near-complete combinations of morphological traits ("ochlospecies"; White 1962, Cronk 1999) may indicate that the terminal phylogenetic structure of some tropical tree lineages may actually be highly reticulate (Veron 1995, Funk and Omland 2003). Analyzing variation in single genes with methods that always produce a bifurcating phylogeny will seriously mislead us about the true history of evolution in these groups. If genetic information controlling the ecological niche of a taxon can be easily exchanged, then we should really treat either the inclusive clade as the effective ecological entity, or just the single individual. Incorporating a more dynamic species concept into community ecology will be a major enterprise over the coming years.

THE PHYLOGENETIC STRUCTURE OF SPECIES ASSEMBLAGES

While clade-based studies of ecological character evolution in tropical forest trees are few, the number of studies of community composition for tropical forests is far greater. A phylogenetic approach to examining assemblages can reveal patterns of non-random species composition which are otherwise hidden, patterns resulting from contemporary ecological processes (Simberloff 1970, Enquist *et al.* 2002), biogeographic history, and the evolutionary history of ecological characters. The important components for analyzing the composition of assemblages are: (1) lists of species in a sample at some explicit scale, and an estimate of the pool of species from which sampled taxa are drawn; (2) a phylogenetic hypothesis for the species in the pool; and (3) means of quantifying phylogenetic structure. In this section, we provide an overview of each of these three components separately, and then bring them together to discover and interpret the phylogenetic structure of tree communities at three different spatial scales.

Species composition, pools, and samples. Defining meaningful spatial scales in ecology has always been a problem, partly because the scale of sampling must relate to the scale of phenomena that we want to measure (but commonly do not know), and partly because the scales of ecological and evolutionary processes can merge continuously (Chave Chapter 2, this volume). That said, we can still define and refer to explicit scales, and we can attempt to pick natural breaks in the continuum that correspond to the scale of biogeographic, climatic, lithologic, and topographic transitions. We will refer to six potential levels: global (the total extent of tropical forests), continental (1000–10,000 km, in which climatic and biogeographic gradients may occur, e.g., Borneo), regional (10–1000 km, fairly homogeneous in climate and biogeographic history, e.g., northwest Borneo), community or "local" (1–10 km, a scale of mixing of tree seeds in one or a few tree generations), habitat (10–1000 m, a lithologically or topographically defined patch, e.g., a sandstone ridge-top), and neighborhood (0–10 m, the scale of direct inter-plant interactions). The meaningful definition of these scales will vary by taxon and geography, but the key characteristics of nested scale from an evolutionary ecology point of view are (1) a pool of species influenced by biogeographic history (including climatic barriers), and (2) a sample influenced by contemporary ecological interactions (Figure 6.1).

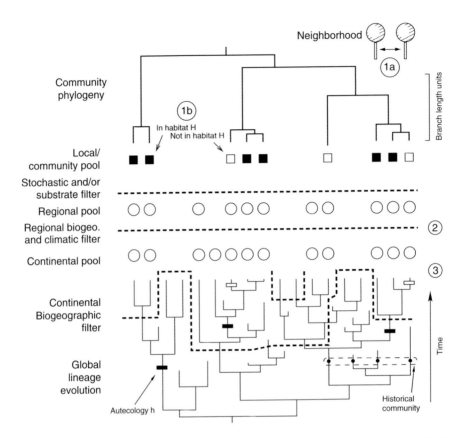

Figure 6.1 Illustrative example of the sampling of lineages through different scales, depending on ecological characters and biogeography. Negative (or positive) interactions may occur among individual neighbors (1a). Habitat filtering for traits that permit survival in abiotic habitat H may occur among taxa in a community pool, modified by competitive interactions (1b). The community pool is in turn an environment- and dispersal-dependent sampling of a regional pool (2), which is structured by the geographic history of lineage diversification (3). The particular physical size of each sample/pool pair, and the number of levels of nestedness, depend on biological and physical circumstances (see text). Note the intentional similarity to the "life-cycle filtering" model in Harper (1977).

Phylogenies. The great expansion in numbers of plant species sequenced and included in phylogenetic analyses offers an increasingly resolved picture of the relationships among angiosperms (Stevens 2001, Angiosperm Phylogeny Group 2003); for example, as of mid-2005, some 6500 taxa in GenBank had been sequenced for *rbc*L (A. Driskell personal communication). Although relatively few tropical forest tree species are included in these analyses, enough exist to draw a reasonable picture of relationships at the "generic" level, and it has become increasingly possible for

ecologists themselves to undertake the molecular work required to produce a phylogeny. There are even online tools that permit the rapid retrieval of a coarse phylogenetic hypothesis for any list of angiosperm species (Webb and Donoghue 2005).

A number of serious caveats are necessary concerning phylogenies for tropical trees. First, these phylogenies are hypotheses, and some areas of the angiosperm tree may still undergo serious re-organization, especially with the increasing evidence that "deep" hybridization events may have been frequent (Davis and Wurdack 2004,

Soltis *et al.* 2004, Mallet 2005). Second, the lack of resolution and comprehensive sampling among congeners restricts our understanding of the vital recent evolution of species' autecology (Malcomber 2002); data on the rates of divergence in ecological character among sibling species are perhaps the key missing information in our models of tropical tree speciation and ecological evolution (see character evolution section above). Third, even if resolution among species existed, we may need to use individuals as the terminal units for ecological analysis, given the great intraspecific variation that has been noted for some tropical forest trees (Cannon and Manos 2003, Dick *et al.* 2003). Despite these issues, when forest communities comprise many different genera, we do have sufficient power to detect non-random patterns in phylogenetic structure.

Metrics of phylogenetic structure. The distribution of a subset of taxa on a phylogeny can be summarized with single indices (Phylogenetic Diversity, Faith 1992; Net Relatedness Index [NRI], Nearest Taxon Index [NTI], Webb 2000; Webb *et al.* 2002, Cavender-Bares *et al.* 2004). The primary characteristic of the distribution captured by these indices is how phylogenetically concentrated or clustered the subset is. In this, they are functionally equivalent to metrics of trait conservatism (Consistency Index; QVI, Ackerly and Donoghue 1998): community structure can also be analyzed by treating the presence of a taxon in a sample as a binary trait (Chazdon *et al.* 2003). More detailed aspects of phylogenetic structure can be assessed by observing the ratio of whole-tree clustering (NRI) to "tip-clustering" (NTI), which indicates whether samples occur in a single cluster on a phylogeny (high NRI/NTI) or in several (low NRI/NTI). By comparing the number of sampled taxa subtending each node in a phylogeny with the number expected at each node under an appropriate null model of random phylogenetic structure, the over- and under-representation of sampled taxa in each clade can be determined (the "Nodesig" algorithm in Phylocom, Webb *et al.* 2004; Figure 6.2). This "node-loading" result permits overall measures of phylogenetic distribution to be interpreted in terms of bias in individual clades. Finally, the phylogenetic similarity of multiple samples can be compared by using mean phylogenetic distances between all pairs of taxa in each of every pair of samples to build a pairwise phylogenetic distance matrix for samples (the "Comdist" algorithm in Phylocom, Webb *et al.* 2004). This distance matrix can be visualized using standard clustering or ordination techniques, and rather than reflecting the shared presence or absence of taxa, it represents the "shared evolutionary heritage" among samples, and can show similarity in deep phylogenetic structure even when no taxa are shared (see below and Figure 6.3).

Bringing the above components together, we can think of the members of any sample as being distributed on the phylogeny of the larger pool of species. At each scale transition, different ecological and biogeographic processes determine which species will be "filtered" into the sample (Figure 6.1), influencing the phylogenetic distribution of the sample members, or their "phylogenetic structure." The ecological interpretation of phylogenetic structure is therefore wholly dependent on the particular scales involved. The following analyses move up in increasing geographic scale, from a single habitat with different topographic features, to a single watershed containing a number of habitats, and finally to the continental scale between landmasses.

Local processes: plant–plant interactions and habitat filtering from the community pool

The evolutionary distribution of ecological characters, either conserved or convergent, interacts with the ecological organizing processes in communities, which either draw phenotypically similar taxa together in habitats ("phenotypic attraction") or force them apart via local exclusion of ecologically similar individuals ("phenotypic repulsion"; Webb 2000, Webb *et al.* 2002, Cavender-Bares *et al.* 2004, Cavender-Bares *et al.* 2006). The balance between the abiotic filter (or "funnel") and biotic "spreader" will determine the gross phylogenetic structure of the assemblages on habitats, that is, whether closely related taxa co-occur, or whether the taxa in a sample are

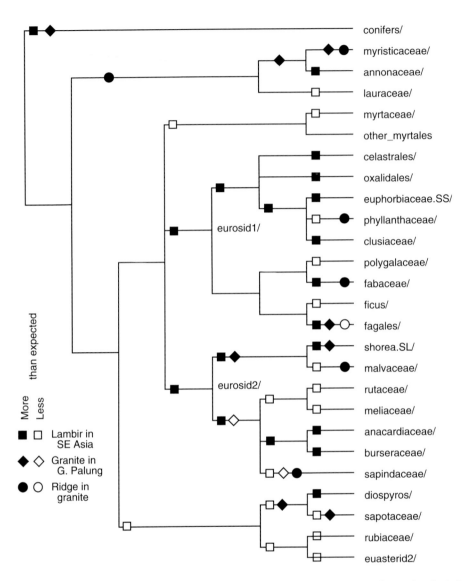

Figure 6.2 Results of significance tests on the association of clades of the angiosperms with samples of rainforest at different spatial scales. Black symbols indicate significantly more taxa in the subtended clade are in the "sample" than expected by chance on the larger-scale "pool," while white symbols indicate fewer. "Ridge in granite" refers to the distribution of species sampled in ridge-top 0.16 ha plots on the pool of all species in the lowland granite zone at Gunung Palung (Webb and Peart 2000). "Granite in G. Palung" refers to the distribution of species sampled in the granite zone (all habitats) within the list of all species at Gunung Palung (Cannon and Leighton 2004). Both of these represent "habitat within community" sampling processes (see text). "Lambir in SE Asia" refers to the distribution of species sampled in the Lambir 50 ha CTFS plot within the full species list of Lambir plus Pasoh plus Huai Kha Khaeng, and represents "region within continent" sampling.

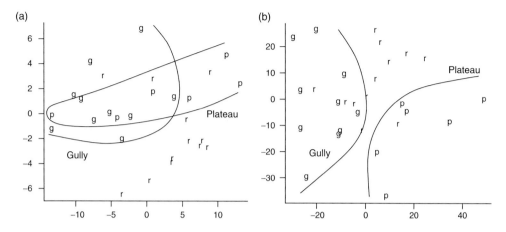

Figure 6.3 Comparison of an ordination (using multidimensional scaling, MDS, axes 1 and 2 shown for both (a) and (b)) of 28 plots in lowland granite forest at Gunung Palung based on (a) Euclidean presence/absence distance versus (b) mean phylogenetic distance between plots (nearest taxon method). Plot edaphic classes: r, ridge; g, gully; p, plateau. Note the greater separation of habitats in (b).

less related than expected by chance. Again, the scale of samples (patches) will determine both the processes acting and the assemblage outcome. For example, it is possible that competitive, antagonistic forces among similar taxa may act on very small, neighborhood scales, for example among seedlings, and could lead to local exclusion in sub-habitat sized patches ("1a" in Figure 6.1), but that at the integrated scale of a habitat patch (ridge-top, river-bank, etc.) abiotic niche filtering appears to draw similar taxa together (Webb *et al.* 2006). The processes that determine habitat-scale assemblages occur on a spatial scale at which sample locations (habitats) are linked via seed dispersal, and at which any taxon could possibly occur in any sample within a few reproductive cycles. The appropriate pool of taxa is thus the sum of all taxa in all habitats. In tropical rainforest, where seed dispersal by vertebrates is common, this community scale is ca. 1–10 km.

When phylogenetic structure of habitat-sized samples has been examined in complete forest communities, a variety of results has been found. Webb (2000) found the tree taxa in 0.16 ha plots in Borneo to be more closely related than expected by chance, across the whole angiosperm phylogeny, which can result only from overall conserved traits and phenotypic "attraction" (Webb

et al. 2002). H. Steers (personal communication) found a positive correlation between taxa co-occurrence in small plots in Mexican dry forest and their phylogenetic relatedness. Cavender-Bares *et al.* (2006) found phylogenetic clustering in plots in subtropical Florida woodland when all woody taxa were included in the analysis, but she found samples to contain taxa evenly distributed on a phylogeny when only the oaks were considered (Cavender-Bares *et al.* 2004). Kembel and Hubbell (2006) found that taxa in the 50 ha plot at Barro Colorado Island on some habitats (plateaus and secondary forest) were phylogenetically clustered, while in swamps the taxa were evenly distributed. Only Cavender-Bares *et al.* (2004) have simultaneously assessed phylogenetic structure and the distribution of ecological characters.

If we consider the dominant process structuring local assemblages to be abiotic niche filtering (above), then we can interpret these mixed patterns as implying that most niche traits are conserved phylogenetically, though a few are significantly convergent. This mixed pattern may correspond to "deep" and old fixedness of β-niches (habitats) and more lability in α-niches (intra-habitat niches; Pickett and Bazzaz 1978, Ackerly *et al.* 2006, Silvertown *et al.* 2006; see also Streelman and Danley 2003). Better phylogenetic resolution and trait data are needed: our lack of resolution at the tips of community phylogenies may be hiding more character convergence than currently detected, or may be preventing the detection of the local exclusion of the most closely related taxa on single habitat patches, taxa that share a phylogenetically conserved ecological niche.

Within the limits of our phylogenetic resolution, we can observe the overall distribution of clades associated with different habitats using the node-loading method described above. This can be thought of as an analysis of the association of habitat with species, genera, and families simultaneously. As an example, we see that myristicaceae/, phyllanthaceae/, malvaceae/, and sapindaceae/ clades are significantly associated with ridge-top plots in lowland hill forest on granite at Gunung Palung (we use "/" as a mark to indicate the name of a rank-free clade), while species in fagaceae/ occur less often than expected on this habitat (Figure 6.2). An ordination of mean phylogenetic distances among species in different plots, as opposed to their Euclidean presence/absence distance, indicates that taxa in plateau and gully habitats differ at a deeper phylogenetic level than either do with taxa in the ridge habitats (Figure 6.3). At a larger spatial and elevational scale, but still within the general model of "habitat-within-community," we note that myristicaeae/, fagales/, shorea/, and sapotaceae/ are over-represented in forest on granite, relative to an area incorporating five habitat types (data from Cannon and Leighton 2004; Figure 6.2).

Figures 6.2 and 6.3 offer a phylogenetic characterization of the observation that the family composition of forests on different habitats appears to be quite predictable (Ashton 1988, Gentry 1988). For example, one of the most striking patterns in taxonomic turnover is the repeated change in family dominance with increasing elevation and/or decreasing fertility, from Euphorbiaceae, Meliaceae through Lauraceae (and Fagaceae in the Old World) to Ericaceae (Gentry 1982, 1988, Lieberman *et al.* 1985, Ashton 1988). This predictable floristic turnover occurs because many of the species in each of these groups share ecological characters. This may be due to repeated homoplasy, but we

suggest it is more likely a consequence of one or more clades within the group being symplesiomorphic for a particular niche. We must of course be very careful to recognize the sampling bias inherent in making evolutionary conclusions based on taxa in an ecological sample. If most of the members of a clade present at a particular site (e.g., the euphorbiaceae/ at Gunung Palung) occur on rich, lowland soils, we cannot make deductions about the ecology of ancestral euphorbs, because the family (even in its modern definition) contains thousands of species, most of which are not rainforest inhabitants. It is not that the approach of inferring ancestral states from contemporary characters is flawed, but that our sampling must be a random subset of the clade (Ackerly 2000). The bias in a sample depends on the global biogeographic extent of the clade. Fortunately, we are in a better position to assess ancestral ecologies using tropical forest species than we would be in, say, a temperate grassland, because the ancestors of many of the lineages present in the former, unlike the latter, probably originated in environments and biotic habitats not too dissimilar from present day conditions (e.g., Davis *et al.* 2005, but see Schrire *et al.* 2005 for a seasonally dry, rainforest margin origin for Fabaceae).

Despite these caveats, studies that attempt to incorporate ecological data for the full extent of a clade can begin to make deductions about its ecological nature (e.g., Davis *et al.* 2005), an approach which we feel will be increasingly powerful when global ecological databases (e.g., the Center for Tropical Forest Science [CTFS] network; Ashton *et al.* 2004) are joined with the various emerging Tree of Life projects.

Assembly of regional pools from the continental pool

Regional species pools (10–1000 km) reflect a combination of the remnants of intra-continental speciation patterns and similar species responses of distribution to recent climatic changes. This is the "phylogeographic scale," at which population and species phylogenies and networks can be reconstructed using genetic markers (Cannon and

Manos 2003, Dick *et al.* 2003), and spatial history of range expansion and contraction and population movement can be inferred (Templeton *et al.* 1995). The analysis of multi-clade assemblages at these scales has taken two main approaches. The first approach is non-phylogenetic similarity analysis, which interprets variation among local areas as the result of permanent range restrictions in some species, and climatic responses in other species that move more freely on a landscape (e.g., van Balgooy 1987). Terborgh and Andresen (1998) used family abundances as units in an analysis of patterns of Amazonian tree distribution. They were unable to use species or genera because there was so little overlap in taxonomic composition using these ranks. They suggested that the strong differentiation they found among regions in the Amazon basin was a result of both historical biogeographic factors (the historical isolation of the Guianan shield) and the interaction between species autecology and contemporary conditions, where families tend to contain many species with similar autecology.

The second approach is explicitly phylogenetic, requiring phylogenies for all members of the assemblages included, and interprets conserved congruence among the "area cladograms" for different lineages to reflect the history of land splitting (vicariance), and convergence, or homoplasy, to reflect historical dispersal events (and/or extinction and sympatric speciation; e.g., Brooks Parsimony Analysis [BPA], Brooks and McLennan 2001; Turner *et al.* 2001, van Welzen *et al.* 2003). However, because of its assumption that vicariance is the primary driver of clade-area association, area-cladogram congruence is not suitable for spatial scales where the signal from migration (or dispersal) resulting from climate cycles may be stronger than residual vicariance events, a situation likely to happen within continents. "Event-based" biogeographic methods, however, can be parametrized to allow for frequent dispersal ("trees-within-trees"; Dispersal Vicariance Analysis [DIVA], Ronquist 1997; Page 2002). The related questions of how much mixing has occurred, how fast species can move on a landscape, and the balance between residual allopatric speciation signal and recent (<10,000 years) climate tracking are fundamental to understanding

tropical forests at regional scales. For example, are diversity gradients and taxonomic turnover patterns in Borneo due to a post-glacial expansion of species out of a refugium in the northwest (Ashton 2003b) or to contemporary west–east rainfall gradients (see Slik *et al.* 2003)? While north temperate forests appear to have re-established generic composition and diversity fairly rapidly after each recent glacial retreat (100,000 years to the present), perhaps soon reaching an equilibrium, is it likely that the species composition of high-diversity tropical forests would similarly rebound?

The analysis of phylogenetic structure (as described above) at this continental/regional scale, with the pool being the continental flora and the sample being a regional flora, may help answer some of these questions. Intra-continental diversification would be observed on the pool phylogeny as numerous small phylogenetic clusters of taxa in a regional-scale sample (high NTI). However, deeper association of clades with different regional samples would indicate that samples differ in their edaphic and climatic factors and taxa within the associated clades share suitable ecological characters. Using inter-plot phylogenetic distances to ordinate samples can extract ecological signal from the data that would be missed in ordinary presence/absence or abundance ordination, because across continental scales there may be sufficient species turnover that few species are shared by sample units (e.g., Terborgh and Andresen 1998). Finally, if the taxa in regional-scale samples are evenly distributed on the pool phylogeny (i.e., sibling taxa seldom occur together on the regional scale; low NTI), then this would indicate either extensive regional competitive exclusion or the persistent signal of allopatric speciation in all clades. The latter is unlikely, given the repeated mixing at continental scales caused by climate cycles.

As an example, we examined species turnover between three 50 ha Asian plots in the CTFS network: Lambir (on Borneo; Lee *et al.* 1999), Pasoh (Peninsular Malaysia; Manokaran *et al.* 1992), and Huai Kha Khaeng (HKK; Western Thailand; Davies *et al.* unpublished data). We treat the plots as samples of the regional floras, within the continental context of Southeast Asia; they share a number of taxa by historical mixing of the Southeast Asian flora, but there is also a strong seasonality gradient from north to south. We note first that the Euclidean distance (based on species presence/absence) between HKK and Pasoh is 0.83 times the distance between Pasoh and Lambir, and HKK–Lambir is 0.97 times Pasoh–Lambir, indicating relatively low *species* similarity between Pasoh and Lambir, and overlap of common species among all three plots. However, the mean nearest taxon distance (the phylogenetic distance, or age, between most closely related taxa; Webb *et al.* 2002) between HKK and Pasoh is 3.0 times the distance between Pasoh and Lambir, and HKK–Lambir is 4.4 times Pasoh–Lambir, indicating more phylogenetic and therefore ecological similarity between Pasoh and Lambir, which are both aseasonal rainforests. Lambir also showed a significant association with rainforest clades, such as annonaceae/, sapindaceae/, and anacardiaceae/ (Figure 6.2), and a significant under-representation of taxa more common in seasonal forests: phyllanthaceae/, myrtales/, and asteraceae/.

Global biogeography and assembly of continental-scale biota

Most of the species diversity within continents is generated by intra-continental speciation, and the phylogenetic structure of continental samples on a global pool will show clustering of taxa within separate clades, which does not necessarily reflect any ecological signal. The larger question is, are deeper angiosperm clades distributed more evenly around the globe than expected by chance? Clearly, some deep clades are restricted geographically (e.g., dipterocarpaceae/ in Southeast Asia). However, in an influential paper, Gentry (1988) noted that the family-level taxonomic structure of rainforest plots was very similar across the three tropical zones. This could occur by (1) a more similar than expected phylogenetic composition of continental pools or (2) continental-to-regional, regional-to-local, or local-to-habitat processes that selectively create habitat-scale plots with a globally similar deep phylogenetic structure. The tricky part of addressing this question

is deciding what the appropriate null models should be. An appropriate null model for the global assortment of plant lineages must take into account the history of land movements and climatic zones (e.g., Morley 2000), the emerging understanding of the biogeographic tracks of major lineages (e.g., Davis *et al.* 2002), and the possibility of extensive intercontinental dispersal (Pennington and Dick 2004). The time is nearing when a grand review of the movement of higher plant lineages in space–time will be possible, an update of Raven and Axelrod's (1974) landmark paper. However, we know of no processes that would cause an over-dispersed (or even) global distribution of major clades. The stems of clades that compose tropical rainforests are old enough that there has been time for members of most major clades to disperse globally, but we expect and observe significant variation among continents in the phylogenetic composition of each clade.

Any similarity in composition of forest plots around the world must therefore lie either in selective intra-continental diversification patterns, such that continental species pools have a more similar than expected clade composition, or in continental-to-local processes that may cause local plot composition to be globally more similar than expected. The former is a possibility, given the conservatism of reproductive characters and their potential influence on speciation rates in different clades. Alternatively, what is the possibility that the taxonomic components of a forest "fit together" – that, for example, a "rubiaceae/ and lauraceae/ and sapotaceae/" set is more stable in some way than a "rubiaceae/ and rubiaceae/ and rubiaceae/" set? This resembles the old question of whether there are hundreds of niches in a forest so that hundreds of species can coexist. Rather than hundreds of niches, there may be a smaller number of niches that do occur regularly on the scale of a forest sample plot (Valencia *et al.* 2004), which, combined with clade-wide ecological characters, stabilize the higher-level taxonomic and phylogenetic structure of a forest. For example, we might eventually attempt to deconstruct forest phylogenetic composition in terms of conserved functional characters (Wilson 1999), for example: (1) N-fixers (fabaceae/), (2) emergents (some clades of dipterocarpaceae/), (3) understory, bird dispersed (some clades of rubiaceae/), etc. Within these broad "niche–clade" associations there are many equivalent species, the abundance of which may be determined predominantly by biotic density dependence or even by chance (Carson *et al.* Chapter 13 this volume, Chave Chapter 2 this volume).

TOWARDS PREDICTING TROPICAL CLADE COMPOSITION

As climate models become more detailed and powerful and the rate of forest conversion (hopefully) stabilizes, predictive models of forest species composition could be used to estimate the vulnerability of various forest areas to species loss and invasion. However, at a first examination, predicting which species are likely to occur in an unvisited forest community appears impossible, because the composition of tropical forest, at the local scale we perceive as we walk through it, is influenced by the complex interplay of abiotic environment acting on ecological characters, biotic interactions, historical causes of species range, and chance (e.g., Connell 1980, Ricklefs and Schluter 1993, Hubbell 2001). With *l* species in a local community from which the *h* members of a plot on a particular habitat are drawn, we might appear to need to know every entry in an $l \times E$ matrix of E ecological attributes, an $h \times h$ matrix of biotic interactions. To predict the *l* community members from a regional species pool of size *r*, we would need to know the contents of a vector of dispersal probabilities of length *r*, and so on up to the global scale. Beyond being unrealistic because of the amount of ecological data necessary, this approach still would not allow the prediction of taxonomic structure in unvisited areas for which the species are unknown.

A phylogenetic approach to ecological prediction helps us because it reduces the dimensionality of the problem. For *n* taxa, a fully resolved phylogeny requires approximately $(n^2/2)$ bits of information, but the work of systematists is making this level of resolution a possibility. Once we have the phylogeny, the distribution of character states for the terminal taxa (e.g., species)

can be coded with as few as a single change somewhere in the tree, or as many as $(n/2)$ changes (for a highly homoplasious character), but never approaching the *n* states required if the taxa were all independent. Similarly, biogeographic distributions can be reconstructed on the tree, and there are good reasons to believe that interspecific biotic interactions should also have a phylogenetic signal (Webb *et al.* 2006). Overall, the dimensionality of the problem is greatly reduced. The phylogenetic approach provides the capacity to make reasonable predictions about the characters of unknown taxa, propagating information outwards on the tree of life (D. Ackerly personal communication).

For example, one potential algorithm for predicting the clade composition, or phylogenetic structure, of a habitat in tropical forest (but not the precise species composition) is:

1 Identify the ancestral ecological condition of *c* clades of tropical forest trees: ε_c.
2 Class the clades as having evolved ecologically in a particular fashion; for example, silverswords of Hawaii, with high divergence and homoplasy, or Rhizophoraceae with high conservatism: ϕ_c.
3 Modify each clade by the general pattern of ecological evolution to give an expected vector of ecological character for the extant members of the clade:

$$\vec{E}_c = f(\phi_c, \varepsilon_c).$$

4 Filter the species in all clades through a clade-specific biogeographic β_c and general climatic filter (κ) to give the clade composition of a local species pool:

$$\vec{L} = \Sigma f(\beta_c, \vec{E}_c | \kappa).$$

5 Identify the niche template of a habitat-scale site: $\vec{v}$.
6 Combine the above to give a potential species composition at a habitat-scale site:

$$\vec{H} = f(\vec{v}, \Sigma \vec{E}, \vec{L}).$$

7 Add the modifying roles of chance (θ), and of phylogenetically correlated biotic interactions (ς), in influencing abundance within an ecologically equivalent group (and therefore presence/absence in small plots), to give a final vector of species composition:

$$\vec{h} = f(\vec{H}, \theta, \varsigma).$$

The species in vector $\vec{h}$ should have the same phylogenetic structure as the observed community, although more detailed information on the evolution of ecological character in all clades would be required to predict the precise species composition.

CONCLUSIONS

The characters of taxa that determine the outcome of local ecological interactions can be examined in a phylogenetic context. The nature of diverse tropical forest communities suggests that these characters may have generally evolved slowly (over millions of years) and essentially randomly, probably leading to a phylogenetically conservative pattern of autecology for most species, although ecologists and systematists need to work together to better document the precise pattern of character evolution. Local ecological biotic interactions are probably diffuse and lead to the co-occurrence of many related taxa in particular habitats, but the exclusion of similar taxa at smaller scales. Preliminary analysis of the phylogenetic structure of communities at various spatial scales reveals patterns of relatedness within habitats, regions, and continents, and supports the general theory of the evolutionary ecology of tropical forest trees outlined here. Phylogenetic models of ecological characters and biotic interactions may soon permit us to predict taxonomic composition of tropical forest communities. Such models will become increasingly important as the effects of both local and global human activities become more profound.

ACKNOWLEDGMENTS

We thank Peter Ashton, David Ackerly, Michael Donoghue, Mark Leighton, Gary Paoli, David Peart, and Miles Silman for discussions that helped form this chapter. Stefan Schnitzer and two anonymous reviewers offered very helpful comments. Sarayudh Bunyavejchewin, Somboon Kiratiprayoon, Md. Nur Supardi Noor, Abdul Rahman Kassim, and Sylvester Tan kindly provided access to the Thai and Malaysian CTFS plot data. The first author was supported by an NSF grant (DEB-0212873).

REFERENCES

Ackerly, D.D. (2000) Taxon sampling, correlated evolution, and independent contrasts. *Evolution* 54, 1480–1492.

Ackerly, D.D. and Donoghue, M. (1998) Leaf size, sapling allometry, and Corner's rules: phylogeny and correlated evolution in Maples (*Acer*). *American Naturalist* 152, 767–791.

Ackerly, D.D., Schwilk, D.W., and Webb, C.O. (2006) Niche evolution and adaptive radiation: testing the order of trait divergence. *Ecology* 87, S50–S61.

Angiosperm Phylogeny Group (2003) An update of the Angiosperm Phylogeny Group classification for the orders and families of flowering plants: APG II. *Botanical Journal of the Linnean Society* 141, 399–436.

Ashton, P.S. (1964) Ecological studies in the mixed dipterocarp forests of Brunei State. *Oxford Forestry Memoirs* 25.

Ashton, P.S. (1988) Dipterocarp biology as a window to the understanding of tropical forest structure. *Annual Review of Ecology and Systematics* 19, 347–370.

Ashton, P.S. (2003a) Floristic zonation of tree communities on wet tropical mountains revisited. *Perspectives in Plant Ecology, Evolution and Systematics* 6, 87–104.

Ashton, P.S. (2003b) Introduction. In P.S. Ashton, A.S. Kamariah, and I.M. Said, (eds), *A Field Guide to the Forest Trees of Brunei Darusssalam and the Northwest Borneo Hotspot, Volume 1*. Universiti Brunei Darussalam, Brunei.

Ashton, P.S. and CTFS Working Group (2004) Floristics and vegetation of the forest dynamics plots. In E.C. Losos and E.G. Leigh (eds), *Tropical Forest Diversity and Dynamism*. University of Chicago Press, Chicago, pp. 90–106.

Austin, M.P. (2002) Spatial prediction of species distribution: an interface between ecological theory and statistical modelling . *Ecological Modelling* 157, 101–118.

Barraclough, T.G. and Vogler, A.P. (2000) Detecting the geographical pattern of speciation from species-level phylogenies. *American Naturalist* 155, 419–434.

Barros, M.A.G. (1989) *Studies on the pollination biology and breeding systems of some genera, with sympatric species in the Brazilian cerrados.* Doctoral dissertation, University of St. Andrews, Scotland.

Bennett, K.D. (1997) *Ecology and Evolution.* Cambridge University Press, Cambridge.

Bowers, M.A., and Brown, J.H. (1982) Body size and coexistence in desert rodents: chance or community structure? *Ecology* 63, 391–400.

Bramley, G.L.C., Pennington, R.T., Zakaria, R., Tjitrosoedirdjo, S.S., and Cronk, Q.C.B. (2004) Assembly of tropical plant diversity on a local scale: *Cyrtandra* (Gesneriaceae) on Mount Kerinci, Sumatra. *Biological Journal of the Linnean Society* 81, 49–62.

Brooks, D.R. and McLennan, D.A. (2001) A comparison of a discovery-based and an event-based method of historical biogeography. *Journal of Biogeography* 28, 757–767.

Canham, C.D., LePage, P.T., and Coates, K.D. (2004) A neighborhood analysis of canopy tree competition: effects of shading versus crowding. *Canadian Journal of Forest Research* 34, 778–787.

Canham, C.D., Papaik, M.J., Uriarte, M., McWilliams, W.H., Jenkins, J.C., and Twery, M.J. (2006) Neighborhood analyses of canopy tree competition along environmental gradients in New England forests. *Ecological Applications* 16, 540–554.

Cannon, C.H. and Leighton, M. (2004) Tree species distributions across five habitats in a Bornean rain forest. *Journal of Vegetation Science* 15, 257–266.

Cannon, C.H. and Manos, P.S. (2003) Phylogeography of the Southeast Asian stone oaks (Lithocarpus). *Journal of Biogeography* 30, 211–226.

Cavender-Bares, J., Ackerly, D.D., Baum, D.A., and Bazzaz, F.A. (2004) Phylogenetic overdispersion in Floridian oak communities. *American Naturalist* 163, 823–843.

Cavender-Bares, J., Keen, A., and Miles, B. (2006) Phylogenetic structure of Floridian plant communities depends on taxonomic and spatial scale. *Ecology* 87, S109–S122.

Chazdon, R.L., Careaga, S., Webb, C.O., and Vargas, O. (2003) Community and phylogenetic structure of reproductive traits of woody species in wet tropical forests. *Ecological Monographs* 73, 331–348.

Clark, D.A. and Clark, D.B. (1992) Life history diversity of canopy and emergent trees in a neotropical rain forest. *Ecological Monographs* 62, 315–344.

Clark, D.B., Palmer, M.W., and Clark, D.A. (1999) Edaphic factors and the landscape-scale distributions of tropical rain forest trees. *Ecology* 80, 2662–2675.

Connell, J.H. (1980) Diversity and the coevolution of competitors, or the ghost of competition past. *Oikos* 35, 131–138.

Cronk, Q.C.B. (1999) The ochlospecies concept. In C.R. Huxley, J.M. Lock, and D.F. Cutler (eds), *Chorology, Taxonomy and Ecology of the Floras of Africa and Madagascar.* Royal Botanic Gardens, Kew, pp. 155–170.

Darwin, C. (1859) *The Origin of Species by Means of Natural Selection.* Murray, London.

Davies, S.J. (1996) *The comparative ecology of Macaranga (Euphorbiaceae).* Doctoral dissertation, Harvard University, MA.

Davies, S.J. (1998) Photosynthesis of nine pioneer *Macaranga* species from Borneo in relation to life-history. *Ecology* 79, 2292–2308.

Davies, S.J., Palmiotto, P.A., Ashton, P.S., Lee, H.S., and LaFrankie, J.V. (1998) Comparative ecology of 11 sympatric species of *Macaranga* in Borneo: tree distribution in relation to horizontal and vertical resource heterogeneity. *Journal of Ecology* 86, 662–673.

Davis, C.C., Bell, C.D., Mathews, S., and Donoghue, M.J. (2002) Laurasian migration explains Gondwanan disjunctions: evidence from Malpighiaceae. *Proceedings of the National Academy of Sciences of the United States of America* 99, 6833–6837.

Davis, C.C., Webb, C.O., Wurdack, K.J., Jaramillo, C.A., and Donoghue, M.J. (2005) Explosive radiation of Malpighiales supports a mid-Cretaceous origin of modern tropical rain forests. *American Naturalist* 165, E36–E65.

Davis, C.C., and Wurdack, K.J. (2004) Host-to-parasite gene transfer in flowering plants: phylogenetic evidence from Malpighiales. *Science* 305, 676–678.

Dick, C.W., Abdul-Salim, K., and Bermingham, E. (2003) Molecular systematic analysis reveals cryptic Tertiary diversification of a widespread tropical rain forest tree. *American Naturalist* 162, 691–703.

Dubuisson, J.Y., Hennequin, S., Rakotondrainibe, F., and Schneider, H. (2003) Ecological diversity and adaptive tendencies in the tropical fern *Trichomanes* L. (Hymenophyllaceae) with special reference to climbing and epiphytic habits. *Botanical Journal of the Linnean Society* 142, 41–63.

Enquist, B.J., Haskell, J.P., and Tiffney, B.H. (2002) General patterns of taxonomic and biomass partitioning in extant and fossil plant communities. *Nature* 419, 610–613.

Faith, D.P. (1992) Conservation evaluation and phylogenetic diversity. *Biological Conservation* 61, 1–10.

Fine, P.V.A., Daly, D.C., Villa Muñoz, G., Mesones, I., and Cameron, K.M. (2005) The contribution of edaphic heterogeneity to the evolution and diversity of Burseraceae trees in the western Amazon. *Evolution* 59, 1464–1478.

Fine, P.V.A., Mesones, I., and Coley, P.D. (2004) Herbivores promote habitat specialization by trees in Amazonian forests. *Science* 305, 663–665.

Forman, L.L. (1966) Generic delimitation in the Castaneoideae. *Kew Bulletin* 18, 421–426.

Funk, D.J. and Omland, K.E. (2003) Species-level paraphyly and polyphyly: frequency, causes and consequences, with insights from animal mitochondrial DNA. *Annual Review of Ecology, Evolution and Systematics* 34, 397–423.

Gentry, A.H. (1982) Patterns of Neotropical plant species diversity. *Evolutionary Biology* 15, 1–84.

Gentry, A.H. (1988) Changes in plant community diversity and floristic composition on environmental and geographical gradients. *Annals of the Missouri Botanical Garden* 75, 1–34.

Gentry, A.H. (1989) Speciation in tropical forests. In L.B. Holm-Nielsen, I.C. Nielsen, and H. Balslev (eds), *Tropical Forests: Botanical Dynamics, Speciation and Diversity*. Academic Press, London, pp. 113–134.

Givnish, T.J. (1998) Adaptive radiation of plants on oceanic islands: classical patterns, molecular data, new insights. In P. Grant (ed.), *Evolution on Islands*. Oxford University Press, Oxford, pp. 281–304.

Givnish, T.J., Evans, T.M., Zjhra, M.L., Patterson, T.B., Berry, P.E., and Sytsma, K.J. (2000) Molecular evolution, adaptive radiation, and geographic diversification in the amphiatlantic family Rapateaceae: evidence from *ndh*F sequences and morphology. *Evolution* 54, 1915–1937.

Gotelli, N.J. and McCabe, D.J. (2002) Species co-occurrence: a meta-analysis of J.M. Diamond's assembly rules model. *Ecology* 83, 2091–2096.

Graham, C.H., Ron, S.R., Santos, J.C., Schneider, C.J., and Moritz, C. (2004) Integrating phylogenetics and environmental niche models to explore speciation mechanisms in dendrobatid frogs. *Evolution* 58, 1781–1793.

Grant, P.R. and Grant, B.R. (1996) Speciation and hybridization in island birds. *Philosophical Transactions of the Royal Society of London, Series B: Biological Sciences* 351, 765–772.

Graves, G.R. and Gotelli, N.J. (1993) Assembly of avian mixed-species flocks in Amazonia. *Proceedings of the National Academy of Sciences of the United States of America* 90, 1388–1391.

Grubb, P.J. and Metcalfe, D.J. (1996) Adaptation and inertia in the Australian tropical lowland rain-forest flora: Contradictory trends in intergeneric and intrageneric comparisons of seed size in relation to light demand. *Functional Ecology* 10, 512–520.

Hall, J.S., Ashton, P.M.S., and Berlyn, G.P. (2003) Seedling performance of four sympatric *Entandrophragma* species (Meliaceae) under simulated fertility and moisture regimes in a Central African rain forest. *Journal of Tropical Ecology* 19, 55–66.

Hall, J.S., McKenna, J.J., Ashton, P.M.S., and Gregoire, T.G. (2004) Habitat characterizations underestimate the role of edaphic factors controlling the distribution of *Entandrophragma*. *Ecology* 85, 2171–2183.

Harper, J.L. (1977) *Population Biology of Plants*. Academic Press, London.

Harvey, P.H. and Pagel, M.D. (1991) *The Comparative Method in Evolutionary Biology*. Oxford University Press, Oxford.

Hubbell, S.P. (2001) *The Unified Neutral Theory of Biodiversity and Biogeography*. Princeton University Press, Princeton.

Huntley, B., Bartlein, P.J., and Prentice, I.C. (1989) Climatic control of distribution and abundance of beech (*Fagus* L.) in Europe and North America. *Journal of Biogeography* 16, 551–60.

Hutchinson, G.E. (1957) Concluding remarks. Population studies: animal ecology and demography. *Cold Springs Harbor Symposia on Quantitative Biology* 22, 415–427.

Johnson, K.R. and Ellis, B. (2002) A tropical rainforest in Colorado 1.4 million years after the Cretaceous-Tertiary boundary. *Science* 296, 2379–2383.

Kembel, S.W. and Hubbell, S.P. (2006) The phylogenetic structure of a neotropical forest tree community. *Ecology* 87, S86–S99.

Kohyama, T. (1993) Size-structured tree populations in gap-dynamic forest: the forest architecture hypothesis for the stable coexistence of species. *Journal of Ecology* 81, 131–143.

Lee, H.S., LaFrankie, J.V., Tan, S. *et al.* (1999) *The 52-ha Forest Research Plot at Lambir Hills National Park Sarawak, Malaysia*. Volume 2: Maps and diameter tables. Sarawak Forest Department, Kuching, Sarawak, Malaysia.

Lieberman, M., Lieberman, D., Hartshorne, G.S., and Peralta, R. (1985) Small-scale altitudinal variation in lowland wet tropical forest vegetation. *Journal of Ecology* 73, 505–516.

Linder, H.P. and Hardy, C.R. (2005) Species richness in the Cape flora: macroevolutionary and macroecological perspective. In F.T. Bakker, L.W. Chatrou, B. Gravendeel, and P.B. Pelser (eds), *Plant Species-Level Systematics: New Perspectives on Pattern and Process*. Lubrecht and Cramer.

Losos, J.B. and Glor, R.E. (2003) Phylogenetic comparative methods and the geography of speciation. *Trends in Ecology and Evolution* 18, 220–227.

Losos, J.B., Leal, M., Glor, R.E. *et al.* (2003) Niche lability in the evolution of a caribbean lizard community. *Nature* 424, 542–545.

Malcomber, S.T. (2002) Phylogeny of *Gaertnera* Lam. (Rubiaceae) based on multiple DNA markers: evidence of a rapid radiation in a widespread, morphologically diverse genus. *Evolution* 56, 42–57.

Mallet, J. (2005) Hybridization as an invasion of the genome. *Trends in Evolution and Ecology* 20, 229–237.

Manokaran, N., LaFrankie, J.V., Kochummen, K.M. *et al.* (1992) *Stand Table and Distribution of Species in the Fifty Hectare Research Plot at Pasoh Forest Reserve*. Forest Research Institute of Malaysia, Research Data, No. 1, Kuala Lumpur.

Moles, A.T., Ackerly, D.D., Webb, C.O. *et al.* (2005) A brief history of seed size. *Science* 307, 576–580.

Momose, K., Ishii, R., Sakai, S., and Inoue, T. (1998) Plant reproductive intervals and pollinators in the aseasonal tropics: a new model. *Proceedings of the Royal Society of London, Series B: Biological Sciences* 265, 2333–2339.

Morley, R.J. (2000) *Origin and Evolution of Tropical Rain Forests*. John Wiley, New York.

Osunkoya, O.O. (1996) Light requirements for regeneration in tropical forest plants: taxon-level and ecological attribute effects. *Australian Journal of Ecology* 21, 429–441.

Page, R.D.M. (ed.) (2002) *Tangled Trees: Phylogeny, Cospeciation and Coevolution*. University of Chicago Press, Chicago.

Palmiotto, P.A., Davies, S.J., Vogt, K.A. *et al.* (2004) Soil-related habitat specialization in dipterocarp rain forest tree species in Borneo. *Journal of Ecology* 92, 609–623.

Paoli, G.D., Curran, L.M., and Zak, D.R. (2006) Soil nutrients and beta diversity in the Bornean Dipterocarpaceae: evidence for niche partitioning by tropical rain forest trees. *Journal of Ecology* 94, 157–170.

Pennington, R.T., and Dick, C.W. (2004) The role of immigrants in the assembly of the South American rainforest tree flora. *Philosophical Transactions of The Royal Society of London Series B – Biological Sciences* 359, 1611–1622.

Peterson, A.T. and Holt, R.D. (2003) Niche differentiation in Mexican birds: using point occurrences to detect ecological innovation. *Ecology Letters* 6, 774–782.

Pickett, S.T.A. and Bazzaz, F.A. (1978) Organization of an assemblage of early successional species on a soil moisture gradient. *Ecology* 59, 1248–1255.

Plana, V., Gascoigne, A., Forrest, L.L., Harris, D., and Pennington R.T. (2004) Pleistocene and pre-Pleistocene *Begonia* speciation in Africa. *Molecular Phylogenetics and Evolution* 31, 449–461.

Potts, M.D., Ashton, P.S., Kaufman, L.S., and Plotkin, J.B. (2002) Habitat patterns in tropical rain forests: A comparison of 105 plots in Northwest Borneo. *Ecology* 83, 2782–2797.

Potts, M.D., Davies, S.J., Bossert, W.H., Tan, S., and Nur Supardi, M.N. (2004) Habitat heterogeneity and niche structure of trees in two tropical rain forests. *Oecologia* 139, 446–453.

Prance, G.T. and White, F. (1988) The genera of Chrysobalanaceae: a study in practical and theoretical taxonomy and its relevance to evolutionary biology. *Philosophical Transactions of the Royal Society of London: Series B* 320, 1–184.

Raven, P.H. and Axelrod, D.I. (1974) Angiosperm biogeography and past continental movements. *Annals of the Missouri Botanical Garden* 61, 539–673.

Raxworthy, C.J., Martinez-Wells, M.L., Horning, N. *et al.* (2003) Predicting distributions of known and unknown reptile species in Madagascar. *Nature* 426, 837–841.

Ricklefs, R.E. and Latham, R.E. (1992) Intercontinental correlation of geographical ranges suggests stasis in ecological traits of relict genera of temperate perennial herbs. *American Naturalist* 139, 1305–1321.

Ricklefs, R.E. and Schluter, D. (1993) Species diversity: regional and historical influences. In R.E. Ricklefs and D. Schluter (eds), *Species Diversity in Ecological Communities: Historical and Geographical Perspectives*. University of Chicago Press, Chicago, pp. 350–363.

Ronquist, F. (1997) Dispersal-vicariance analysis: a new approach to the quantification of historical biogeography. *Systematic Biology* 46, 195–203.

Russo, S.E., Davies, S.J., King, D.A., and Tan, S. (2005) Soil-related performance variation and distributions of tree species in a Bornean rain forest. *Journal of Ecology* 93, 879–889.

Scachetti-Pereira, R. (2002) *Desktop GARP*. University of Kansas Natural History Museum and Biodiversity Research Center.

Schluter, D. (2000) *The Ecology of Adaptive Radiation*. Oxford University Press, Oxford.

Schluter, D. (2001) Ecology and the origin of species. *Trends in Ecology and Evolution* 16, 372–380.

Schnitzer, S.A. (2005) A mechanistic explanation for global patterns of liana abundance and distribution. *American Naturalist* 166, 262–276.

Schrire, B.D., Lavin, M., and Lewis, G.P. (2005) Global distribution patterns of the Leguminosae: insights from recent phylogenies. *Biologiske Skrifter* 55, 375–422.

Silvertown, J. (2004) The ghost of competition past in the phylogeny of island endemic plants. *Journal of Ecology* 92, 168–173.

Silvertown, J., Dodd, M., Gowing, D., and Lawson, C. (2006) Phylogeny and the hierarchical organization of plant diversity. *Ecology* 87, S39–S49.

Simberloff, D.S. (1970) Taxonomic diversity of island biotas. *Evolution* 24, 23–47.

Slik, J.W.F., Poulsen, A.D., Ashton, P.S. *et al.* (2003) A floristic analysis of the lowland dipterocarp forests of Borneo. *Journal of Biogeography* 30, 1517–1531.

Smith-Ramirez, C., Armesto, J.J., and Figueroa, J. (1998) Flowering, fruiting and seed germination in Chilean rain forest Myrtaceae: ecological and phylogenetic constraints. *Plant Ecology* 136, 119–131.

Snow, D.W. (1966) A possible selective factor in evolution of fruiting seasons in tropical forest. *Oikos* 15, 274–281.

Soltis, D.E., Albert, V.A., Savolainen, V. *et al.* (2004) Genome-scale data, angiosperm relationships, and "ending incongruence": a cautionary tale in phylogenetics. *Trends in Plant Science* 9, 477–483.

Stephens, P.R. and Wiens, J.J. (2004) Convergence, divergence, and homogenization in the ecological structure of emydid turtle communities: the effects of phylogeny and dispersal. *American Naturalist* 164, 244–254.

Stevens, P.F. (1980) A revision of the Old World species of *Calophyllum* (Guttiferae). *Journal of the Arnold Arboretum* 61, 117–699.

Stevens, P.F. (2001) *Angiosperm Phylogeny Website (v4.0)*. http://www.mobot.org/MOBOT/research/APweb.

Streelman, J.T. and Danley, P.D. (2003) The stages of vertebrate evolutionary radiation. *Trends in Ecology and Evolution* 18, 126–131.

Swaine, M.D. and Whitmore, T.C. (1988) On the definition of ecological species groups in tropical rain forests. *Vegetatio* 75, 81–86.

Templeton, A.R., Routman, E., and Phillips, C.A. (1995) Separating population structure from population history: a cladistic analysis of the geographical distribution of mitochondrial DNA haplotypes in the tiger salamander, *Ambystoma tigrinum*. *Genetics* 140, 767–782.

Terborgh, J. and Andresen, E. (1998) The composition of Amazonian forests: patterns at the local and regional scales. *Journal of Tropical Ecology* 14, 645–664.

Thomas, S.C. (1996) Asymptotic height as a predictor of growth and allometric characteristics in Malaysian rain forest trees. *American Journal of Botany* 83, 556–566.

Turner, H., Hovenkamp, P., and van Welzen, P.C. (2001) Biogeography of Southeast Asia and the West Pacific. *Journal of Biogeography* 28, 217–230.

Uriarte, M., Condit, R., Canham, C.D., and Hubbell, S.P. (2004) A spatially explicit model of sapling growth in a tropical forest: does the identity of neighbours matter? *Journal of Ecology* 92, 348–360.

Vamosi, J.C., Knight, T.M., Steets, J.A., Mazer, S.J., Burd, M., and Ashman, T.-L. (2006) Pollination decays in biodiversity hotspots. *PNAS* 103, 956–961.

Valencia, R., Foster, R.B., Villa, G. *et al.* (2004) Tree species distributions and local habitat variation in the Amazon: large forest plot in eastern Ecuador. *Journal of Ecology* 92, 214–229.

van Balgooy, M.M.J. (1987) A plant geographical analysis of Sulawesi. In T.C. Whitmore (ed.), *Biogeographical Evolution of the Malay Archipelago*. Clarendon Press, Oxford, pp. 94–102.

van Steenis, C.G.G.J. (1948) General considerations. *Flora Malesiana Series I* 4, xiii–lxix.

van Steenis, C.G.G.J. (1972) *Mountain Flora of Java*. Brill, Leiden.

van Valen, L. (1976) Ecological species, multispecies, and oaks. *Taxon* 25, 233–239.

van Welzen, P.C., Turner, H., and Hovenkamp, P. (2003) Historical biogeography of Southeast Asia and the West Pacific, or the generality of unrooted area networks as historical biogeographic hypotheses. *Journal of Biogeography* 30, 181–192.

Veron, J.E.N. (1995) *Corals in Space and Time: The Biogeography and Evolution of the Scleractinia*. Comstock, Ithaca, NY.

Webb, C.O. (2000) Exploring the phylogenetic structure of ecological communities: an example for rain forest trees. *American Naturalist* 156, 145–155.

Webb, C.O., Ackerly, D.D., and Kembel, S. (2004) *Phylocom: software for the phylogenetic analysis of communities and characters; version 3.22*. http://www.phylodiversity.net/phylocom.

Webb, C.O., Ackerly, D.D., Mcpeek, M.A., and Donoghue, M.J. (2002) Phylogenies and community ecology. *Annual Review of Ecology and Systematics* 33, 475–505.

Webb, C.O., and Donoghue, M.J. (2005) Phylomatic: tree assembly for applied phylogenetics. *Molecular Ecology Notes* 5, 181–183.

Webb, C.O., Gilbert, G.S., and Donoghue, M.J. (2006) Phylodiversity dependent seedling mortality, size structure, and disease. *Ecology* 87, S123–S131.

Webb, C.O., and Peart, D.R. (2000) Habitat associations of trees and seedlings in a Bornean rain forest. *Journal of Ecology* 88, 464–478.

White, F. (1962) Geographic variation and speciation in Africa with particular reference to *Diospyros*. In D. Nichols (ed.), *Taxonomy and Geography*. Systematics Association, London, pp. 71–103.

Wiens, J.J. (2004) Speciation and ecology revisited: phylogenetic niche conservatism and the origin of species. *Evolution* 58, 193–197.

Wilson, J.B. (1999) Assembly rules in plant communities. In E. Weiher and P. Keddy (eds), *Ecological Assembly Rules: Perspectives, Advances, Retreats.* Cambridge University Press, Cambridge, pp. 97–113.

Wright, S.J. (2002) Plant diversity in tropical forests: a review of mechanisms of species coexistence. *Oecologia* 130, 1–14.

Chapter 7

LARGE TROPICAL FOREST DYNAMICS PLOTS: TESTING EXPLANATIONS FOR THE MAINTENANCE OF SPECIES DIVERSITY

Jess K. Zimmerman, Jill Thompson, and Nicholas Brokaw

OVERVIEW

Large tropical forest dynamics plots (FDPs) play a significant role in developing theory and testing hypotheses that may explain the species diversity of tropical forests. Here we summarize the contribution of FDPs belonging to the network coordinated by the Center for Tropical Forest Science (CTFS). In CTFS FDPs all trees and shrubs with diameter at breast height of 1 cm or more are identified and measured for diameter, and their locations are mapped in plots that range in size from 2 to 52 ha (most are 16–50 ha). By virtue of their large size, the FDPs present a comprehensive picture of relative species abundances and species distribution in tropical forests, including the contribution of rare species. The plots have demonstrated that the shape of the species abundance curves is similar in a wide geographical and structural range of tropical forests. Mapped locations of trees and repeated censuses (usually at 5-year intervals) provide data on spatial and temporal dynamics critical to testing theoretical explanations for high species diversity in tropical forests. In this chapter we concentrate on three potential explanations for the high level of species diversity observed in tropical forests that have received particular attention using data from the FDPs: (1) neutral theory, which is a null model for community dynamics; (2) negative density dependence (NDD); and (3) gap specialization and dynamics. The differences among these explanations reflect, among other things, rare species advantage and the relative importance of the life-history characteristics of each species, and how they may determine the community dynamics. The value of neutral theory as a general explanation for the high species diversity of tropical forests is weakened by the abundant phenomenological evidence from FDPs that demonstrates that NDD operates in most tropical forest. Although a small number of "pioneer" tree species and non-tree species are specialized for growing in canopy gaps, information from large FDPs does not support a strong role for gap specialization as an explanation for the high diversity of tropical forest trees.

INTRODUCTION

How to explain the great species diversity of most tropical forests is a fundamental issue in tropical ecology (Connell 1978, Leigh *et al.* 2004), and one that also demonstrates the great difficulty of testing ecological theory in the tropics. In most tropical forest communities, only a few species are common, and many species are rare. In such diverse communities it is difficult to obtain a sufficient sample of trees to provide an estimate of community diversity and structure, or determine the demographic parameters of individual species, in order to test ecological theory. One solution is to census large areas (i.e., tens of hectares) of forest, thereby assuring that the sample comprises a reasonable number of individuals of rare species in addition to common species. Observing the changes in the tree spatial arrangement over time sheds light on the factors that affect a species' population growth and survival and the outcome of species interactions (Condit 1995). In order to study the diversity and ecology of tropical forests around the world, some researchers have adopted the approach of using large tropical forest dynamics plots (FDPs), repeatedly censused over the long term. By comparing the spatio-temporal dynamics of populations in forest communities observed in the FDPs with the patterns predicted by different explanations for community diversity, we may be able to understand the mechanisms that drive the community patterns. The research in many tropical FDPs is coordinated by the Center for Tropical Forest Science (CTFS) of the Smithsonian Institution (Losos and Leigh 2004). Most of the FDPs are 50 ha and are usually re-censused every 5 years. This chapter describes the contribution of these large tropical FDPs to testing explanations for the diversity, structure, and composition of tropical forest tree communities.

ECOLOGICAL EXPLANATIONS FOR SPECIES DIVERSITY

In this chapter, we focus on three explanations regarding the interpretation of species diversity, structure, and community dynamics in tropical forest that have dominated the literature, and that have been tested using large FDPs (Chesson 2000, Wright 2002, Leigh *et al.* 2004). First, we address neutral theory (Hubbell 2001), an extension of the theory of island biogeography (MacArthur and Wilson 1967) that incorporates speciation, and uses individuals, not species, as the basic units of propagation. Neutral theory is controversial (Chave 2004, Missa 2005) because, unlike other theories, it assumes that a tree's prospects of death or reproduction are not affected by what species it is, or what species are in its neighborhood, and thus establishes that differences among tree species are irrelevant to maintaining tree species diversity. Even if neutral theory is not an adequate explanation on its own, it does, however, constitute a null model for community

structure and dynamics, against which other potential explanations for high species diversity in ecological communities can be assessed (Chave 2004). We consider two departures from the neutral theory null model: the first is negative density dependence (NDD), a tenet of the Janzen–Connell hypothesis (Janzen 1970, Connell 1971) and other biological effects that provide for a rare species advantage. With NDD, unlike in neutral theory, an individual's reproductive prospects differ according to whether it belongs to a species that is common or rare. The second departure from neutral theory we consider is the role of canopy disturbance and gap specialization, and the degree to which variation in light availability allows light-demanding and shade-tolerant species to coexist via niche partitioning (Connell 1978).

DEVELOPMENT OF THE CTFS NETWORK OF LARGE FDPs

CTFS plots are located in different biogeographic regions, and were placed to encompass the extremes and means of tropical forest environments (Ashton 1998, Condit 1998). Relevant environment considerations included total annual rainfall, its seasonal distribution, soils and topography, and natural disturbance regimes (frequency of fires, hurricanes, etc.; Losos 2004). Most CTFS plots are located at altitudes between 0 and 500 m; they have between 0 and 6 dry months, and have yellow–red zonal soils ranging from basic to acid. The specific locations of CTFS plots were dictated by the need to balance accessibility and available research resources, and freedom from potential human disturbance. The CTFS network in Asia has a larger program because, so far, it has been better funded than those in Africa and Latin America (Table 7.1). There are currently 18 FDPs in the CTFS network, in which scientists are monitoring approximately three million trees representing 6000 species. Most plots have completed at least two censuses (Table 7.1) to allow the study of the static distribution of trees and also the short-term forest dynamics. The FDP on Barro Colorado Island (BCI) in Panama has been censused more times than

Table 7.1 The locations and site descriptions for the large tropical forest dynamics plots belonging to the Center for Tropical Forest Science (CTFS) network.

Center for Tropical Forest Science program	Collaborating institutions	Year initiated and censuses completed	Plot size	Altitude	Annual rainfall, and duration of dry season	No. of species and trees
Latin America						
Barro Colorado Island Nature Monument, Panama	CTFS/Smithsonian Tropical Research Institute (STRI), University of Georgia (USA)	1980; 5th census: 2000 6th census: ongoing	50 ha	120–160 m	2551 mm 3 months	300 spp. 213,800 trees
Luquillo Experimental Forest, Puerto Rico	University of Puerto Rico (USA), US Forest Service, STRI	1990; 3rd census: 2001	16 ha	333–428 m	3548 mm 0 months	138 spp. 67,100 trees
Yasuní National Park, Ecuador	Pontificia Universidad Católica del Ecuador, University of Aarhus (Denmark), STRI	1995; 2nd census: 2003	50 ha	215–245 m	3081 mm 0 months	1104 spp. 152,400 trees (25 ha)
La Planada Nature Reserve, Colombia	Instituto de Investigación de Recursos Biológicos "Alexander von Humboldt", STRI	1997; 2nd census: 2002	25 ha	1796– 1891 m	4087 mm 0 months	228 spp. 115,100 trees
Biological Dynamics of Forest Fragments, Manaus, Brazil	Brazilian Institute for Research in the Amazon (INPA)	Future site; census: ongoing	50 ha	40–80 m	2600 mm 5 months	–
Asia						
Pasoh Forest Reserve, Peninsular Malaysia	Forest Research Institute Malaysia, National Institute of Environmental Studies (Japan), Harvard University (USA), STRI	1986; 4th census: 2001 5th census: ongoing	50 ha	70–90 m	1788 mm 1 month	814 spp. 335,400 trees
Mudumalai Wildlife Sanctuary, India	Center for Ecological Sciences of the Indian Institute of Science, STRI	1988; 5th census: 2004	50 ha	980–1120 m	1250 mm 6 months	71 spp. 18,000 trees

Site	Institutions	Census	Area	Elevation	Rainfall / dry months	Species / trees
Huai Kha Khaeng Wildlife Sanctuary, Thailand	Royal Thai Forest Department, Kasetsart and Mahidol universities (Thailand), Harvard University (USA), STRI	1992; 3rd census: 2004	50 ha	549–638 m	1476 mm 6 months	251 spp. 72,500 trees
Sinharaja World Heritage Site, Sri Lanka	University of Peradeniya (Sri Lanka), Sri Lanka Department of Forestry, Harvard and Yale universities (USA), STRI	1993; 2nd census: 2002	25 ha	424–575 m	5016 mm 0 months	204 spp. 193,400 trees
Lambir Hills National Park, Sarawak, Malaysia	Sarawak Forest Department, Harvard University (USA), Osaka City, Ehime, and Kyoto universities (Japan), STRI	1990; 3rd census: 2003	52 ha	104–244 m	2664 mm 0 months	1182 spp. 359,600 trees
Bukit Timah Nature Reserve, Singapore	National Institute of Education/Nanyang Technological University (Singapore), Singapore National Parks Board, STRI	1993; 4th census: 2003	4 ha	150 m	2473 mm 0 months	329 spp. 11,900 trees (2 ha)
Palanan Wilderness Area, Philippines	Isabela State University College of Forestry (Philippines), Conservation International, PLAN International, Harvard University (USA), STRI	1994; 2nd census: 2004	16 ha	85–140 m	3379 mm 4 months	335 spp. 66,000 trees
Doi Inthanon National Park, Thailand	Royal Thai Forest Department, Osaka City, Kyoto, Maejo, and Utsunomiya universities (Japan), Kasetsart University (Thailand), Chiba Natural History Museum and Institute (Japan), STRI	1997; 1st census: 2000	15 ha	1660–1740 m	1908 mm 6 months	162 spp. 73,700 trees

Continued

Table 7.1 Continued

Center for Tropical Forest Science program	Collaborating institutions	Year initiated and censuses completed	Plot size	Altitude	Annual rainfall, and duration of dry season	No. of species and trees
Fushan Nature Reserve, Taiwan	Taiwan Forestry Research Institute, Tunghai University (Taiwan), STRI	2002; census: ongoing	25 ha	650–733 m	4067 mm 0 months	–
Nanjenshan Nature Reserve, Taiwan	Tunghai University (Taiwan), STRI	1989; 3rd census: 2004	6 ha	300–340 m	3582 mm 0 months	125 spp. 36,400 trees (3 ha)
Khao Chong Wildlife Refuge, Thailand	Royal Thai Forest Department, National Institute of Environmental Studies (Japan), Harvard University (USA), STRI	1998; 1st census: 2003	24 ha	50–300 m	2700 mm 2–3 months	612 spp. 100,800 trees
Africa						
Ituri Forest/Okapi Wildlife Reserve, Democratic Republic of Congo	Centre de Formation et de Recherche en Conservation Forestiere (Democratic Republic of Congo), Wildlife Conservation Society (USA), STRI	1994; 2nd census: 2000	40 ha	700–850 m	1730 mm 3–4 months	420 spp. 299,000 trees
Korup National Park, Cameroon	BioResources Development and Conservation Programme (Cameroon), Oregon State University (USA), International Cooperative Biodiversity Group, STRI	1997; 1st census: 1999	50 ha	150–240 m	5272 mm 3 months	494 spp. 329,000 trees

Source: Center for Tropical Forest Science.

any other; its sixth census in 2005 marked a total of 25 years of investigation.

Historical perspective

Although the BCI FDP, established in 1980, is often considered to be the first large FDP (Table 7.1; Losos and Leigh 2004), the first large area (13.4 ha) of tropical forest was censused by Hubbell (1979) in tropical dry forest in Costa Rica. Originally intended for an investigation into the diet and leaf selection of leaf-cutting ants, the tree map of this 13.4 ha plot was used by Hubbell (1979) to test the hypothesis that adult tropical trees have relatively uniform dispersion as predicted by the Janzen–Connell hypothesis. Hubbell (1979) also introduced a neutral model to explain the relative abundances of tree species in this Costa Rican plot. From this experience Hubbell recognized the power of large plots to investigate the patterns of relative abundance and diversity in tropical forests (Hubbell 2004), and with his collaborator Robin Foster sought a site for a permanent plot that could be resampled at regular intervals. They chose BCI because of the logistical support available, its well-known flora (Croat 1978), the relatively flat topography of the island summit (formerly a hill, that was isolated by the waters of Lake Gatun during the development of the Panama Canal), and the low number of poisonous snakes (S.P. Hubbell personal communication). It was difficult to obtain funding to establish the first large FDP (Hubbell 2004), in large part because many biologists, including those familiar with the characteristics of tropical forest, were skeptical of the usefulness of such a large plot. As the value of the rich detail of data on species' relative abundance, diversity, and forest dynamics provided by the BCI FDP was recognized (Hubbell and Foster 1986a,b, 1988), the need for comparative data elsewhere in the tropics became the motivating factor for the development of other plots, which began with the Pasoh plot in Malaysia (Table 7.1). As more plots were established, the CTFS network was founded to maintain consistent methodologies and foster collaboration among researchers from the participating plots.

Tropical forest community structure revealed by FDPs

The first census of the FDP on BCI, Panama revealed 305 species among 235,000 trees and saplings ≥ 1 cm diameter at breast height (dbh) (169 species per ha; Hubbell and Foster 1986a). Tree diversity in the BCI plot is relatively low when compared with some of the other FDPs (Table 7.1). The first census of the Pasoh plot in Peninsular Malaysia (50 ha, completed in 1988) found 814 species ≥ 1 cm dbh (495 species per ha; Manokaran et al. 2004), while the Lambir plot, in Sarawak, Malaysia (established in 1990), had even more species, with 1182 in 52 ha (618 species per ha; Lee et al. 2004). A relatively small plot of only 25 ha in Yasuní, Ecuador (established in 1995) had 1104 species (655 per ha; Valencia et al. 2004a): this is the most diverse in the CTFS FDP network.

In addition to total species richness, dominance–diversity curves (Figure 7.1) are another way to compare tropical diversity among FDPs. Dominance–diversity curves plot the proportional abundance (on a $\log_{10}$ scale) versus the rank order abundance of an FDP's tree species, and the curve nearly always has a long tail. Dominance–diversity curves demonstrate that most FDPs contain few common and many rare tree species. For example, the first census of the 50 ha BCI plot revealed that 22 species were represented by a single individual (Hubbell and Foster 1986a), while just under 50% of species were represented by 99 or fewer individuals – in a total sample of 235,000 trees! In the Pasoh FDP, the equivalent ratio is 367 of 815 species (45%). If one considers that there are only about 1000 species of tree north of the Mexican border in North America (Hubbell and Foster 1986a), the magnitude of tropical forest diversity becomes evident. The high degree of species rarity across large areas of tropical forests, and the need to determine the spatial distribution and sufficient samples of these rare species, is the single greatest justification for such large plots.

In tropical forests growing on islands (Luquillo and Sinharaja), where diversity is relatively low, on former mainlands (BCI), and on true mainlands (Pasoh), the dominance–diversity curves

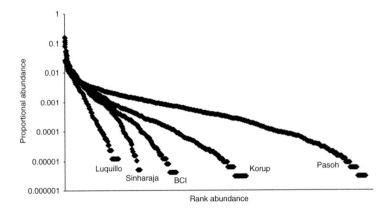

Figure 7.1 Dominance–diversity curves for large FDPs from Luquillo, Puerto Rico (16 ha; J. Thompson *et al.* unpublished data), Sinharaja, Sri Lanka (25 ha; Gunatilleke *et al.* 2004), Barro Colorado Island, Panama (BCI; 50 ha; Condit *et al.* 1996), Korup, Cameroon (50 ha; Thomas *et al.* 2003), and Pasoh, Malaysia (50 ha; Manokaran *et al.* 1992).

have the same structure and show that all the forests have a few common species and relatively large numbers of rare species. The extent of each curve depends, in part, on the forest diversity and the size of the plot (Figure 7.1). These patterns suggested to Hubbell (2001) that similar factors underlie the structuring of the forest communities. In the following section we discuss three potential explanations for the patterns of species diversity and community structure of different tropical forests that have been tested using large FDPs.

LARGE FDPs AND ECOLOGICAL THEORY

Neutral theory – a null model for communities

In developing neutral theory, Hubbell (2001, Chapter 9, this volume) united ideas on island biogeography (MacArthur and Wilson 1967) and relative species abundances, incorporating assumptions regarding speciation, immigration, and extinction at local (community) and global (metacommunity) scales. Borrowing ideas originally and independently discovered by population geneticists (Hubbell 2001, Chave 2004), neutral

theory can generate expected distributions of relative species abundances in local communities very similar to the pattern of curves observed for natural forest communities (Figure 7.1). Another important aspect of neutral theory is the distinction between the local community and the metacommunity (or regional species pool) and the degree to which dispersal limitation allows the local extinction of species (i.e., by preventing a species' presence in a local community to be "rescued" by dispersal from outside the local community). One resulting contribution of neutral theory is that it makes clear that the logseries distribution (Fisher *et al.* 1943), which in contrast to neutral theory predicts a greater number of rare species relative to common species in the community, actually applies to the metacommunity, not the measured, local community. Further, neutral theory results in the replacement of Preston's (1948, 1962) symmetrical lognormal distribution with a type of multinomial distribution, thereby explaining the asymmetrical shape of the distribution of relative species abundances in local communities (Figure 7.2). Forest dynamics plots are relatively large and sample a large portion of the "local" community, and as a result are able to reveal the true shape of the distribution of relative species abundances. The FDP results appeared to confirm one of the key predictions of neutral

(a)

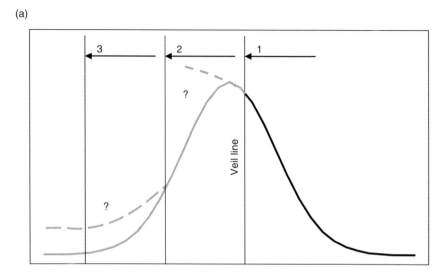

(b)

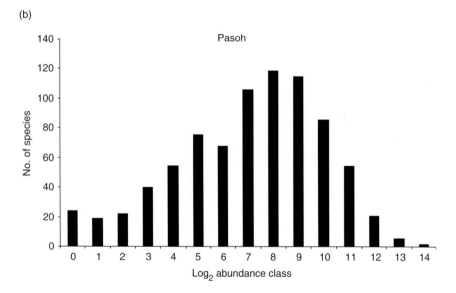

Figure 7.2 (a) Disagreements over the shape of the distribution of relative species abundances in communities resulted, in part, from the inability to sample the rarest species. As sample size increases, the "veil line" moves to the left, and questions ("?") are eliminated regarding the shape of the distributions of relative species abundances. Going from (1) to (2) the logseries distribution (Fisher *et al.* 1943) is distinguished from a lognormal distribution (Preston 1948, 1962), by including a mode. Then, finally, including the left-hand tail, (3) indicates that the distribution is asymmetrical, as suggested by the zero-sum multinomial distribution of Hubbell (2001). (b) Example of distribution of species relative abundances from the plot at Pasoh (Manokaran *et al.* 1992).

theory, as seen in the example for the Pasoh plot (Figure 7.2b). The ability of neutral theory to resolve some long-standing issues in community ecology and to generate realistic relative species abundance curves made the theory expounded in Hubbell's (2001) book appear compelling to many scientists. However, because of some of the assumptions used in Hubbell's (2001) treatment of neutral theory, and because it promoted the idea that biological identity is unimportant in community dynamics, it has drawn much controversy and criticism (summarized by Chave 2004 and Missa 2005).

The key problem with Hubbell's approach to testing neutral theory is that, however compelling it is at explaining the relative species abundances among FDPs (Figure 7.1), other explanations, including niche-based ones (Chave 2004, Purves and Pacala 2005), can also generate the same patterns. A recent modification of a mathematical formulation of neutral theory (Volkov *et al.* 2003) by Hubbell and his colleagues, employing density dependence (rare species advantage), makes this precise point (Volkov *et al.* 2005). The pattern of relative species abundances reveals nothing about the processes that generate them, and one must delve further into the dynamics of tropical forests to distinguish neutral theory from other potential explanations.

How can we use data from large FDPs to help resolve this issue? Taking the viewpoint that neutral theory constitutes a null model for community dynamics, we consider two departures from the neutral theory as potential explanations for local species diversity: NDD and gap specialization. Where appropriate in the following discussion, we make the distinction between equalizing versus stabilizing effects, and their contribution to the maintenance of species diversity (Chesson 2000). Equalizing effects minimize the impact of the differences in species fitness that lead to competitive exclusion, and thereby promote the community dynamics envisioned by neutral theory. Stabilizing effects, on the other hand, actively contribute to the maintenance of diversity by increasing the impact of negative interactions between conspecifics relative to interactions among species. Discovery of stabilizing effects and the implication that species' identity

and their particular life-history attributes contribute to community structure would, therefore, refute neutral theory.

Negative density dependence

One important assumption of neutral theory, as formulated by Hubbell and colleagues (Hubbell 2001, Volkov *et al.* 2003) is that species are ecologically equivalent and competitively neutral, that is, competition between neighboring conspecifics is no different than that between heterospecific neighbors. Negative density dependence, on the other hand, allows for species-specific differences and the likelihood that an individual occurring in a high density of conspecifics (that compete for the same resources or share pathogens/herbivores) is less likely to survive and reproduce than the same individual occurring among heterospecifics at the same density (which suffer less competition for resources or slower pathogen/herbivore transmission). The effect is often termed "negative density dependence" although it really refers to the importance of conspecific versus heterospecific density. A number of studies have utilized data from large tropical FDPs to address this issue (Hubbell *et al.* 1990, 2001, Condit *et al.* 1992, 1994, Gilbert *et al.* 1994, Wills *et al.* 1997, 2006, Wills and Condit 1999, Peters 2003, Losos and Leigh 2004, Uriarte *et al.* 2004a,b, 2005a,b), often couching their studies as tests of the Janzen–Connell hypothesis. In simple terms, the Janzen–Connell hypothesis considers density- or distance-dependent recruitment that is driven by specialized seed or seedling predators or pathogens, which make the areas around parents inhospitable for the establishment of conspecifics (Janzen 1970, Connell 1971). Taking the case of seedlings, for example, the Janzen–Connell hypothesis predicts that seedlings of a species that establish away from their parent and near to adults of a different species gain a growth and survival advantage, because these seedlings are less likely to be attacked by the species-specific pathogens and predators that are associated with an adult of another species. In addition, dispersal limitation will initially cause higher densities of seedlings near to the parent trees, while a relatively small number of seedlings will escape the parental neighborhood with a high density of conspecifics. As a consequence intraspecific competition between siblings dispersed further away will be less. This mechanism, then, promotes the survival of a variety of species in any location, and therefore forest diversity is increased. Because it is a stabilizing effect (Chesson 2000), it offers a direct contrast to neutral theory.

The development of the Janzen–Connell hypothesis, and the various ways it has been studied in tropical forests, is treated in more detail by Adler and Muller-Landau (2005) and Carson *et al.* (Chapter 13, this volume; also Leigh Chapter 8, this volume). Many studies of individual species have found evidence for the Janzen–Connell hypothesis, often with direct information on the pests or pathogens causing the effect (Hammond and Brown 1998, Carson *et al.* Chapter 13, this volume). Nonetheless, a species by species approach makes it difficult to determine how widespread this effect is in structuring tropical forest communities. Negative density dependence is implicitly spatially dependent, and therefore the data from the FDPs on tree size and mapped locations makes them ideally suited to investigate this explanation. It is important to understand, however, that studies of demography and community structure can only reveal the existence of NDD, not the mechanism causing NDD.

Many of the first tests of NDD in an FDP were conducted using the BCI and Pasoh plots by looking for spatial patterns in the abundance of recruits or mortality, with respect to distance between individuals, or density of conspecifics within subplots of the FDP (Hubbell *et al.* 1990, 2001, Wills *et al.* 1997, Wills and Condit 1999). The impression from these studies was that only the most common species, encompassing about 10% of the community, exhibited NDD (see also Condit *et al.* 1992). However, using subplots of an FDP to test density dependence can be criticized because trees that are close to the edge of a subplot present little information about the tree composition of the real neighborhood around each individual for analysis. Using an individual-based analysis, where neighborhoods of varying radii are constructed around each individual tree, Peters (2003) detected density-dependent mortality in a pattern consistent with the Janzen–Connell hypothesis, in more than 80% of species he investigated in the BCI and Pasoh plots. Peters' (2003) methods, however, did not account for spatial autocorrelation in the data (see Hubbell *et al.* 2001 as a contrast), among other things, and the study likely overstates the number of species exhibiting NDD. Using a different statistical approach to that of Peters (2003), for individual-based neighborhood analyses, Uriarte *et al.* (2004a, 2005a) analyzed patterns of growth on the BCI plot and found that 26 of 60 species exhibited negative conspecific interactions, while 16 of 50 species exhibited negative conspecific interactions for survival. In the hurricane-driven Luquillo FDP, Uriarte *et al.* (2004b) found negative conspecific growth effects in all of the 11 species tested and, for mortality, in 7 of 12 species tested. Taken together, these results (Uriarte *et al.* 2004a,b, 2005b) suggest that NDD may be common in tropical forests. The individual-based approaches are limited by the sample sizes needed to detect an effect of density on plant performance, and they may not reveal the full extent of density dependence in tropical forests (Uriarte *et al.* 2005a). Returning to the subplot approach, Wills *et al.* (2006) asked how the diversity of subplots changed as a result of the survival and recruitment of trees in different dbh size classes in repeated censuses, and consistently found that in most of the forests tested diversity increased with time, consistent with the predictions of NDD and two other potential explanations (see below; Wills *et al.* 2006). This study encompassed seven of the CTFS plots and the results were robust; only one comparison in one FDP failed to show the expected pattern.

The key difference between neutral theory (Hubbell 2001, Volkov *et al.* 2003) and NDD is that they make fundamentally different predictions about the status of rare species (Volkov *et al.* 2005). Neutral theory implies that rare species, compared with species of more modest abundance, are on the verge of "winking out" of the community, while the rare species advantage (as a result of NDD) specifies that they are on their way to becoming more common (Volkov *et al.* 2005).

The recent work of Wills *et al.* (2006) makes it clear that a rare species advantage may operate in many tropical forests, strongly suggesting that neutral theory (Hubbell 2001, Volkov *et al.* 2003) may not explain the underlying patterns of relative species abundance.

We reiterate that the evidence provided by FDPs is phenomenological, and cannot determine whether pathogens, predators, or competition for resources are responsible for the observed patterns, so these analyses do not directly test the Janzen–Connell hypothesis. In fact, the results are also consistent with greater intra- versus interspecific competition and facilitation, or the "species herd" effect (positive density dependence; Peters 2003, Wills *et al.* 2006). Peters (2003) showed a "species herd" effect at Pasoh (but not at BCI), a pattern in which overall survival *increased* with increasing numbers of heterospecifics. Overall, the results described in this section provide strong evidence that species differences are important for maintaining species diversity.

Most FDPs have not yet satisfactorily investigated the critical seed-to-seedling and seedling-to-sapling stages of forest dynamics. However, information on the seed-to-seedling stage was provided by Harms *et al.* (2000; see also Wright *et al.* 2005) in the BCI FDP when they compared seed arrival with the adjacent abundance of seedlings and found evidence of density-dependent seedling recruitment for all 53 species studied. In the Luquillo FDP, Uriarte *et al.* (2005b) studied density dependence in seedling survival relative to the location of conspecific adults using seedling census data recorded shortly after Hurricane Georges and 2 years later. They used different individual-based models that included terms for dependent mortality of seedlings and compared these with models that did not. They found that including the density-dependence term made the model a better fit to the data and was significant for all nine species tested, very dramatically so for some species. Incorporating the density-dependence term in the model increased the distance away from the parent tree for peak seedling survival, and the apparent clumping (high local density) of seedlings was less pronounced (solid lines in Figure 7.3) compared with when it was not included. We note that the hurricane may have enhanced the degree of NDD due to the large numbers of seedlings recruiting into open areas created by the disturbance.

Gap specialization

Gap specialization and colonization–competition trade-offs are two related explanations for tropical species diversity, sometimes lumped together under the intermediate disturbance hypothesis (Connell 1978, Hubbell *et al.* 1999). Gap specialization and colonization–competition trade-offs focus on the reliance of some species on canopy gaps for their establishment and growth, emphasizing (1) specialization of pioneers for gaps (i.e., niche differentiation), and (2) the ability of species to arrive at and colonize a gap before more effective competitors usurp the space – the competition–colonization trade-off. These two processes are theoretically distinct (Pacala and Rees 1998), though they are often treated as related issues. Both are stabilizing effects (Chesson 2000) that, when operating, counter neutral theory.

The competition–colonization trade-off is reviewed in this volume by Muller-Landau (Chapter 11) with regard to trade-offs in seed size and various competitive traits in tropical forests. Here we focus on gap specialization because competition–colonization trade-offs are expected to be less important for maintaining species diversity in forested ecosystems with small disturbances (Pacala and Rees 1998). Exploring patterns of growth and mortality in saplings and adult trees and shrubs in large FDPs, we consider the evidence for (1) partitioning of gap and non-gap forest areas, and (2) life-history trade-offs in community dynamics and demographic variability.

The role of gap dependence in community dynamics has been a consistent theme in forest ecology for many years (Grubb 1977, Hartshorn 1978, Whitmore 1978, Brokaw 1987, Denslow 1987, Pacala *et al.* 1996, Hubbell *et al.* 1999). One approach has been to directly assess the degree of gap dependence by comparing the growth, mortality, or diversity of species in gap and non-gap areas. Although there is evidence

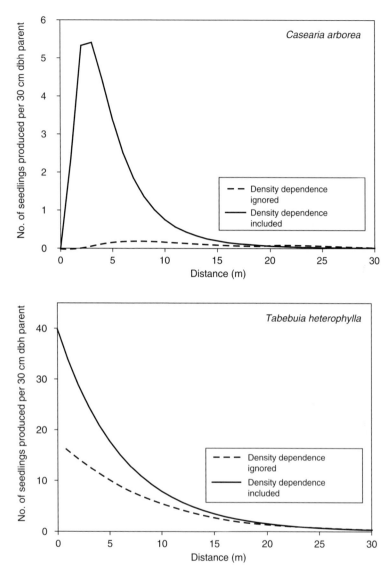

Figure 7.3 The shape of seedling dispersal curves for two species in the Luquillo plot (Uriarte *et al.* 2005b). The continuous line represents the model that includes density dependence prior to any effects of density on seedling mortality (i.e., as if no seedling had died from density-dependent effects). The dotted line represents a model that does not include density-dependent effects (i.e., as if the effect were ignored).

that some tree species grow best in different-sized gaps or under different light levels (Brokaw 1987, Agyeman *et al.* 1999), it is not clear the degree to which this influences forest-wide community composition, and the degree of gap

specialization may depend on the species' functional group (Schnitzer and Carson 2001). Several studies have addressed gap dependence in woody species in the BCI FDP. Welden *et al.* (1991) analyzed the recruitment, growth, and survival of individuals 1–4 cm dbh for 108 tree and shrub species, with respect to low (i.e., gaps) versus high canopy areas, and concluded that few species were either gap or understory specialists – most species (~80%) did equally well in gap and non-gap areas. Similarly, Hubbell *et al.* (1999) found that tree and shrub species diversity of gaps versus non-gaps was not different in the BCI FDP but they failed to account for differences in tree stem density that was higher in gaps and examined only tree species diversity (Chazdon *et al.* 1999, Tilman 1999). After accounting for density via rarefaction of both pioneer trees and lianas, Schnitzer and Carson (2001) did find that gaps in the BCI FDP had higher species richness. Pioneer and liana species (with lianas predominating) account for more than 30% of all woody species on BCI. For trees and shrubs, Wright *et al.* (2003) have taken something of a middle ground showing that, on the BCI plot, there is a continuum of response, and emphasizing that there are species that are either extremely

shade tolerant or light demanding (Agyeman *et al.* 1999).

Gap specialization has often been thought of as representing a fundamental life-history axis, where there are trade-offs in life-history characteristics and demographic patterns among species. Large FDPs allow us to ask (Swaine and Whitmore 1988, Alvarez-Buylla *et al.* 1992, Agyeman *et al.* 1999, Kyereh *et al.* 1999): (1) are there trade-offs in life-history characteristics related to growth and mortality patterns, and (2) are life histories discontinuous (e.g., "pioneer" versus "non-pioneer" [otherwise known as "shade-intolerant" versus "shade-tolerant" or "mature"] species forming bimodal groups) or are they continuous (i.e., unimodal or having uniform variation over each life-history characteristic)? Hubbell and Foster (1992) demonstrated a negative relationship between mortality of species in the shade and their rate of growth in gaps (Hubbell and Foster 1992; Figure 7.4a), suggesting a life-history trade-off, but the pattern

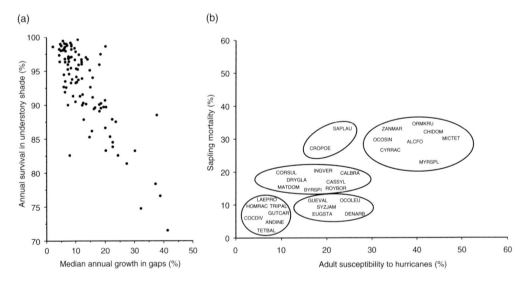

Figure 7.4 Evidence from FDPs for trade-offs in demographic characteristics among tropical forest trees and the presence of distinct species groups (e.g., "pioneers"). (a) The BCI FDP, showing variation among species in mortality in the shade and growth in gaps (Hubbell and Foster 1992). (b) The Luquillo FDP (Uriarte *et al.* 2004a), where hurricane susceptibility was measured as the proportion of stems greater than 10 cm dbh damaged in Hurricane Hugo in 1989 (see Uriarte *et al.* 2004a for species names) and compared with post-hurricane mortality. The groupings shown are arbitrary and were chosen to include rare species in the analyses of neighborhood competition.

does not suggest two distinct groups of species, conforming to "pioneer" or "mature" species. In fact, the large number of species with low mortality in the shade and relatively low growth rates in high light was viewed as evidence that most tree and shrub species are generalists, and pioneer-like species are infrequent in the community (Figure 7.4a; Hubbell and Foster 1992). Uriarte et al. (2004a) demonstrated life-history variation in the hurricane-disturbed Luquillo FDP by showing hurricane susceptibility (likelihood of stem breakage or tip-up during a hurricane) is positively correlated with post-hurricane sapling mortality (a measure of shade tolerance; Figure 7.4b). These two studies directly address the issue of modality in life-history characteristics, showing that there are no modes of variation corresponding to "pioneer" and "non-pioneer" species typologies (Swaine and Whitmore 1988). Rather, variation in life histories is uniform or unimodal, with species at the extremes representing pioneer or mature species (Alvarez-Buylla et al. 1992, Zimmerman et al. 1994, Agyeman et al. 1999, Wright et al. 2003), what Pacala and his colleagues call the "life history manifold" (Moorcroft et al. 2001, Purves and Pacala 2005).

Focusing on the variability in species' demography alone, efforts have been made to show that the more diverse plots have greater demographic variability among species (indicative of greater variation in life-history types) than less diverse plots (Condit et al. 1999, 2006). Because so many factors lead to the evolution of a species' demographic characteristics, not just gap dependence, this is only an indirect test of that particular aspect of a species' life history. However, if it were to be shown that more diverse plots have greater demographic variability, it would be consistent with the idea that demographic variability facilitates high species diversity. Condit et al. (1999) found that the Pasoh plot contains fewer species with extremely high growth rates (one of many life-history characteristics of "pioneers") than BCI (Condit et al. 1999) and gaps are smaller and less frequent than those on BCI (Putz and Appanah 1987, Leigh et al. 2004). Yet, species diversity is much higher on Pasoh, nearly three times that of BCI, contrary to the prediction that species diversity

is promoted by greater numbers of demographic niches. Condit et al. (2006) recently conducted a much larger comparison of the tree mortality and relative growth rate among 10 of the CTFS plots. Again, contrary to the expectation that high demographic variability among species in a plot should be related to high species diversity, the most diverse plots had the least demographic variation. Those plots with high demographic variability included the American plots of BCI (Figure 7.4a), Yasuní and La Planada, where species belonging to the genus Cecropia had some of the highest demographic rates (Condit et al. 2006), so some degree of species diversity may be explained by demographic variability. Condit et al. (2006) thus conclude that the results do not completely eliminate a role of demographic variability in explaining species diversity of trees and shrubs, but suggest this role is very limited.

From the perspective of theory, it is important to note that the existence of life-history trade-offs does not provide for the stabilizing effects needed to disprove neutral theory. These trade-offs, in fact, constitute equalizing effects (Chesson 2000) that will not contribute to the maintenance of species diversity in a way that refutes neutral theory, as noted by Hubbell (2001). Only distinct differentiation of species' gap response or any form of niche differentiation, as described in a theoretical context by Pacala and Rees (1998), can provide for the stabilizing effects that maintain species diversity. Yet, there is no convincing evidence, within or between FDPs, to support this explanation for the high tree and shrub species diversity of tropical forests. Schnitzer and Carson (2001) argued that gaps were likely to be critical for the maintenance of pioneer trees and lianas in addition to herbs, shrubs, and herbaceous vines that together account for more than 60% of the plant diversity on BCI. This is undoubtedly true for the many species trapped in the understory. Similarly, Wright (2002) suggested gaps were necessary for the small-statured tree species that otherwise might fail to reproduce if permanently shaded in the understory. However, for taller-statured tree species, the species addressed in the FDPs, gap specialization appears to make a limited contribution to explaining species diversity. This is in contrast to the long-standing belief that gap dependence was a key factor in the maintenance of diversity of trees and shrubs in tropical forests (e.g., Grubb 1977, Denslow 1987).

Habitat specialization beyond gap dependence could provide another explanation of high species diversity. In the relatively flat plot at BCI, which occupies the top of a hill, the scope of habitat specialization appeared limited (Hubbell and Foster 1986, Harms et al. 2000) although a handful of species occurred only in a swampy area and along slopes. Valencia et al. (2004b) recently investigated topographic habitat specialization in the more topographically varied Yasuní plot in Ecuador. There was some evidence of niche partitioning among valley, mid-slope, and upper ridge areas. While about 25% of species had large abundance differences among topographic positions, another 25% were complete generalists. Thus, partitioning among topographic niches at Yasuní provided no explanation of the co-occurrence of generalists, or the hundreds of species sharing the ridge-tops. In the case of the Yasuní FDP, Valencia et al. (2004b) concluded that this type of habitat specialization makes only a minor contribution to local species diversity. Studies of habitat specialization in more FDPs will be needed before we fully understand the precise role of habitat specialization in explaining the species diversity of tropical forests.

Overall, the CTFS network of large FDPs has done a great deal to distinguish among three prominent explanations for the diversity of tropical forests. Analyses of data from the CTFS network of large FDPs make clear that there is some rare species advantage common to tropical forests and also some evidence for habitat preference that promotes species diversity. The evidence from the network of plots does not support a strong role for gap specialization in explaining patterns of tree and shrub species diversity, and more information is needed on other potential habitat niches that may influence diversity. The results obtained so far in the FDPs suggest that on balance species diversity in tropical forests is not consistent with the null model of species interactions provided by the neutral theory of Hubbell (2001). In addition, the phenomenological data collected in the FDPs must be substantiated by direct experimental tests of the mechanisms involved.

LIMITATIONS

The large-plot approach has some limitations with regard to testing ecological theory and here we briefly touch on these. Some ecologists object to the use of one large plot because individual contiguous plots within one large forested area are "pseudoreplicated" because there are no randomly and independently selected replicates (Hurlbert 1984, Scheiner 2001). Thus, the results may not be generalized to any forested area away from the plot. The only way to correct this would be to have replicated large plots, perhaps of a smaller size, randomly located throughout the forest. This presents obvious logistical difficulties and places limits on the types of questions that could be addressed (e.g., seed dispersal distance, or neighborhood effects) in several smaller plots. Small plots suffer from the edge effect, insufficient numbers of complete tree neighborhoods, and an absence of data from local tree neighborhoods that extend outside of the plot edges. Sampling smaller plots away from an FDP may be effectively used to place the FDP in a larger, regional context of forest variability (e.g., Condit et al. 2002). Statistical techniques that take into account spatial autocorrelation (i.e., the lack of independence) of trees or subplots can also be employed to address this problem (Robertson 1987, Rossi et al. 1992, Hubbell et al. 2001). Thus, while not insurmountable, it is a real problem that each FDP is a single sample, albeit a very large one. The strength in the approach, however, lies in the network of plots throughout the world. While not true replicates in a statistical sense, robust results from a variety of forest types, such as those provided by Wills et al. (2006) and Condit et al. (2006), support the overall approach.

Another limitation of using FDPs to test hypotheses explaining the high diversity of their forest communities is that the results are phenomenological (Hubbell 2004). They thus typically do not provide evidence that a particular mechanism is operating (e.g., the Janzen–Connell hypothesis), and the observed patterns could be the result of other factors not considered. For example, Wills et al. (2006) cite three factors that may explain increasing plot diversity with time. Nonetheless, the consistent lack of evidence for a particular mechanism, even phenomenological, will eventually allow researchers to discount its importance. Once detected phenomenologically, however, experimental studies are needed to identify causative factors.

A further limitation of FDPs is a result of the census method, in which only stems ≥ 1 cm dbh are censused throughout the whole plot. This size limit omits the most dynamic size class, the seedlings, in which the majority of mortality occurs. The huge number of seedlings precludes the monitoring of the total seedling population, but several FDPs now conduct repeated censuses in numerous small plots to sample seedling dynamics (e.g., Harms *et al.* 2000, Uriarte *et al.* 2005b). Most FDPs exclude lianas from their samples (one exception is the Korup plot, in Cameroon), because it is almost impossible to reliably recognize an individual, and most of the liana biomass is supported up in the tree canopies. We recognize that lianas may significantly affect gap dynamics in some tropical forests (Schnitzer *et al.* 2000, Schnitzer and Bongers 2002).

Finally, there is the "bigfoot effect," a corollary of the Heisenberg uncertainty principle (Malakoff 2004): the trampling caused by field workers may obscure the real forest dynamics (Phillips *et al.* 2002, Wright 2005). Most FDPs are re-censused at 5-year intervals or longer, hopefully minimizing any severe impact. Our observations at Luquillo, however, suggest that the overall research activity at FDPs causes significant human disturbance, more than suggested by a once-every-5-years visit. However, it is reassuring that Goldsmith *et al.* (2006) found no significant differences in seedling density and dispersion, height-class distributions, species richness, evenness, and overall composition between plots inside and outside the FDP at BCI. Because the number of plots sampled was large, the comparison had high statistical power.

FUTURE RESEARCH IN FDPs

Research from large tropical FDPs in the CTFS network, particularly large, multi-plot comparisons (Condit *et al.* 2006, Wills *et al.* 2006), is beginning to distinguish among explanations

for the diversity of tropical forest shrubs and trees. What role do large tropical FDPs have in future research? Wills *et al.* (2006) clearly show that diversity enhancement through time is a common feature in the FDPs, but cannot distinguish between the Janzen–Connell hypothesis, niche complementarity, or the species herd effect (facilitation). Understanding the precise mechanisms underlying diversity enhancement should be a focus of future research. Second, some theoretical explanations remain untested in large FDPs (e.g., the storage effect; Chesson 2000). Detailed studies of monodominant forests in the tropics (Connell and Lowman 1989), such as that provided by the Ituri plots (Makana *et al.* 2004), will provide an important perspective on low-diversity forests, as will complementary studies of temperate forests (e.g., HilleRisLambers *et al.* 2002). As a large network of forest plots measured consistently over time and around the globe, the plots will be essential in evaluating the impacts of climate change on tropical forests (Condit *et al.* 1996, Phillips *et al.* 2002, Wright 2005). Finally, human disturbance has fragmented many tropical forests, and many others are now in secondary growth (e.g., Grau *et al.* 2003, Wright 2005). Studies are needed to determine if there are different mechanisms that promote species diversity in old-growth forest communities versus second-growth forests (Thompson *et al.* 2002, Thomas 2004) or if it is only the relative importance of the different mechanisms that has changed with disturbance. Such investigations will be important in developing management options for human-disturbed forests. In sum, large FDPs will continue to be important in refining our understanding of tropical forest diversity, be it high versus low or under the influence of natural or anthropogenically modified conditions.

ACKNOWLEDGMENTS

We acknowledge and express our appreciation to the many people, too numerous to mention, who have undertaken the huge task of censusing these forest dynamics plots. We thank Elizabeth Losos, Kyle Harms, Richard Condit, and two anonymous reviewers for their comments on the manuscript. Helene Muller-Landau, Egbert Leigh, and Walter Carson deserve special thanks for their help with this manuscript. Preparation of this chapter was supported by the Luquillo Long-Term Ecological Research Program, which is funded by the National Science Foundation, the University of Puerto Rico, and the USDA Forest Service's International Institute of Tropical Forestry. The CTFS kindly provided Table 7.1. Jess Zimmerman developed this chapter while serving at the National Science Foundation. Any opinions, findings, and conclusions or recommendations expressed in this chapter are those of the authors and do not necessarily reflect the views of the National Science Foundation. Jill Thompson worked on this chapter while in receipt of a Bullard Fellowship at Harvard Forest.

REFERENCES

Adler, F.R. and Muller-Landau, H.C. (2005) When do localized natural enemies increase species richness? *Ecology Letters* 8, 438–447.

Agyeman, V.K., Swaine, M.D., and Thompson, J. (1999) Responses of tropical forest tree seedlings to irradiance and the derivation of a light response index. *Journal of Ecology* 87, 815–827.

Alvarez-Buylla, E.R. and Martínez-Ramos, M. (1992) Demography and allometry of *Cecropia obtusifolia*, a neotropical pioneer tree – an evaluation of the climax-pioneer paradigm for tropical rain forests. *Journal of Ecology* 80, 275–290.

Ashton, P.S. (1998) A global network of plots for understanding tree species diversity in tropical forests. In F. Dallmeier and J.A. Comiskey (eds), *Forest Biodiversity Research, Monitoring and Modeling: Conceptual Background and Old World Case Studies*. Man and the biosphere series, Volume 20. UNESCO, Paris and Parthenon Publishing Group, New York, pp. 47–62.

Brokaw, N.V.L. (1987) Gap-phase regeneration of three pioneer tree species in a tropical forest. *Journal of Ecology* 75, 9–19.

Chazdon, R.L., Colwell, R.K., and Denslow, J.S. (1999) Tropical tree richness and resource-based niches. *Science* 285, 1459a.

Chave, J. (2004) Neutral theory and community ecology. *Ecology Letters* 7, 241–253.

Chesson, P. (2000) Mechanisms of maintenance of species diversity. *Annual Review of Ecology and Systematics* 31, 343–346.

Condit, R. (1995) Research in large, long-term tropical forest plots. *Trends in Ecology and Evolution* 10, 18–22.

Condit, R. (1998) *Tropical Forest Census Plots*. Springer-Verlag, Berlin, and R. G. Landes Company, Georgetown, Texas.

Condit, R., Hubbell, S.P., and Foster, R.B. (1992) Recruitment near conspecific adults and the maintenance of tree and shrub diversity in a neotropical forest. *American Naturalist* 140, 261–286.

Condit, R., Hubbell, S.P., and Foster, R.B. (1994) Density dependence in two understory tree species in a neotropical forest. *Ecology* 75, 671–705.

Condit, R., Hubbell, S.P., and Foster, R.B. (1996) Changes in tree species abundance in a neotropical forest: impact of climate change. *Journal of Tropical Ecology* 12, 231–256.

Condit, R., Pitman, N., Leigh, E.G. *et al.* (2002) Beta-diversity in tropical forest trees. *Science* 295, 666–669.

Condit, R., Ashton, P.S., Manokaran, N., LaFrankie, J.V., Hubbell, S.P., and Foster, R.B. (1999) Dynamics of the forest communities at Pasoh and Barro Colorado: comparing two 50-ha plots. *Philosophical Transactions of the Royal Society of London Series B* 354, 1739–1748.

Condit, R., Ashton, P.S., Bunyavejchewin, H.S. *et al.* (2006) The importance of demographic niches to tree diversity. *Science* 313, 98–101.

Connell, J.H. (1971) On the role of natural enemies in preventing competitive exclusion in some marine animals and in rain forest trees. In P.J. den Boer and G.R. Gradwell (eds), *Dynamics of Populations*. Center for Agricultural Publishing and Documentation, Wageningen, The Netherlands, pp. 298–310.

Connell, J.H. (1978) Diversity in tropical rain forests and coral reefs. *Science* 199, 1302–1310.

Connell, J.H. and Lowman, M.D. (1989) Low-diversity tropical rain forests – some possible mechanisms for their existence. *American Naturalist* 134, 88–119.

Croat, T. (1978) *Flora of Barro Colorado Island*. Stanford University Press, California.

Denslow, J.S. (1987) Tropical rainforest gaps and tree species diversity. *Annual Review of Ecology and Systematics* 18, 431–451.

Fisher, R.A., Corbet, A.S., and Williams, C.B. (1943) The relation between the number of species and the number of individuals in a random sample of an animal population. *Journal of Animal Ecology* 12, 42–58.

Gilbert, G.S., Hubbell, S.P., and Foster, R.B. (1994) Density and distance-to-adult effects of a canker disease of trees in a moist tropical forest. *Oecologia* 98, 100–108.

Goldsmith, G.R., Comita, L.S., Morefield, L.L., Condit, R., and Hubbell, S.P. (2006) Long-term research impacts on seedling community structure and composition in a permanent forest plot. *Forest Ecology and Management* 234, 34–39.

Grau, H.R., Aide, T.M., Zimmerman, J.K. *et al.* (2003) The ecological consequences of socioeconomic and land use changes in postagriculture Puerto Rico. *BioScience* 53, 1159–1168.

Grubb, P.J. (1977) The maintenance of species-richness in plant communities: the importance of the regeneration niche. *Biological Reviews* 52, 107–145.

Gunatilleke, C.V.S., Gunatilleke, I.A.U.N., Ethugala, A.U.K., and Esufali, S. (2004) *Ecology of Sinharaja Rain Forest and the Forest Dynamics Plot in Sri Lanka's Natural World Heritage site*. Colombo, WHT Publications.

Hammond, D.S. and Brown, V.K. (1998) Disturbance, phenology, and life-history characteristics: factors influencing distance/density-dependent attack on tropical seeds and seedlings. In D.M. Newberry, H.H.T. Prins, and N.D. Brown (eds), *Dynamics of Tropical Forest Communities*. Blackwell Science, Oxford, pp. 51–78.

Harms, K.E., Condit, R., Hubbell, S.P., and Foster, R.B. (2001) Habitat associations of trees and shrubs in a 50-ha neotropical forest plot. *Journal of Ecology* 89, 947–959.

Harms, K.E., Wright, S.J., Calderón, O., Hernández, A., and Herre, E.A. (2000) Pervasive density-dependent recruitment enhances seedling diversity in a tropical forest. *Nature* 404, 493–495.

Hartshorn, G.S. (1978) Treefalls and tropical forest dynamics. In P.B. Tomlinson and M.H. Zimmerman (eds), *Tropical Trees as Living Systems*. Cambridge University Press, Cambridge, pp. 617–638.

HilleRisLambers, J., Clark, J.S., and Beckage, B. (2002) Density-dependent mortality and the latitudinal gradient in species diversity. *Nature* 417, 732–735.

Hubbell, S.P. (1979) Tree dispersion, abundance, and diversity in a tropical dry forest. *Science* 203, 1299–1309.

Hubbell, S.P. (1997) A unified theory of biogeography and relative species abundance and its application to tropical rain forests and coral reefs. *Coral Reefs* 16, S9–S21.

Hubbell, S.P. (2001) *The Unified Neutral Theory of Biodiversity and Biogeography*. Princeton University Press, Princeton and Oxford.

Hubbell, S.P. (2004) Two decades of research on the BCI Forest Dynamics plot: where we have been and where we are going. In E. Losos and E.G. Leigh (eds), *Tropical Forest Diversity and Dynamism: Findings from a Large-scale Plot Network*. University of Chicago Press, pp. 8–30.

Hubbell, S.P., Ahumada, J.A., Condit, R., and Foster, R.B. (2001) Local neighborhood effects on long-term survival of individual trees in a neotropical forest. *Ecological Research* 16, 859–875.

Hubbell, S.P., Condit, R., and Foster, R.B. (1990) Presence and absence of density dependence in a neotropical tree community. *Philosophical Transactions of the Royal Society of London Series B* 330, 269–282.

Hubbell, S.P. and Foster, R.B. (1986a) Commonness and rarity in a neotropical forest: implications for tropical tree conservation. In M.E. Soulé (ed.), *Conservation Biology: The Science of Scarcity and Diversity*. Sinauer Associates, Sunderland, MA, pp. 205–231.

Hubbell, S.P. and Foster, R.B. (1986b) Biology, chance, and history and the structure of tropical rain forest tree communities. In J. Diamond and T.J. Case (eds), *Community Ecology*. Harper and Rowe, New York, pp. 314–329.

Hubbell, S.P. and Foster, R.B. (1988) Diversity of canopy trees in a neotropical forest and implications for conservation. In S.L. Sutton, T.C. Whitmore, and A.C. Chadwick (eds), *Tropical Rain Forest: Ecology and Management*. Blackwell Scientific, Oxford, pp. 25–41.

Hubbell, S.P. and Foster, R.B. (1992) Short-term dynamics of a neotropical forest: why ecological research matters to tropical conservation and management. *Oikos* 63, 48–61.

Hubbell, S.P., Foster, R.B., and O'Brien, S.T. (1999) Light-gap disturbances, recruitment limitation, and tree diversity in a neotropical forest. *Science* 283, 554–557.

Hurlbert, S.H. (1984) Pseudoreplication and the design of ecological field experiments. *Ecological Monographs* 54, 187–211.

Janzen, D.H. (1970) Herbivores and the number of tree species in tropical forests. *American Naturalist* 104, 501–528.

Kyereh, B., Swaine, M.D., and Thompson, J. (1999) Effect of light on the germination of forest trees in Ghana. *Journal of Ecology* 87, 772–783.

Lee, H.S., Tan, S., Davies, S.J. *et al.* (2004) Lambir Forest Dynamics Plot, Sarawak, Malaysia. In E. Losos and E.G. Leigh (eds), *Tropical Forest Diversity and Dynamism: Findings from a Large-scale Plot Network*. University of Chicago Press, pp. 527–539.

Leigh, E.G. Jr., Davidar, P., and Dick, C.W. (2004) Why do some tropical forests have so many species of trees? *Biotropica* 36, 447–473.

Losos, E. (2004) The whole is greater than the sum of the plots. In E. Losos and E.G. Leigh (eds), *Tropical Forest Diversity and Dynamism: Findings from a Large-Scale Plot Network*. University of Chicago Press, pp. 31–36.

Losos, E. and Leigh, E.G. (eds) (2004) *Tropical Forest Diversity and Dynamism: Findings from a Large-Scale Plot Network*. University of Chicago Press.

MacArthur, R.R. and Wilson, E.O. (1967) *The Theory of Island Biogeography*. Princeton University Press, Princeton.

Malakoff, D. (2004) Measuring the significance of a scientist's touch. *Science* 306, 801.

Makana, J., Hart, T.B., Hibbs, D.E., and Condit, R. (2004) Stand structure and species diversity in the Ituri Forest Dynamics Plot: a comparison of monodominant and mixed forest stands. In E. Losos and E.G. Leigh (eds), *Tropical Forest Diversity and Dynamism: Findings from a Large-scale Plot Network*. University of Chicago Press, pp. 159–174.

Manokaran, M., LaFrankie, J.V., Kochummen, K., Quah, E., Ashton, P.S., and Hubbell, S.P. (1992) *Stand Table and Distribution of Species in the 50-ha Research Plot at Pasoh Forest Reserve*. FRIM Research Data. Forest Research Institute Malaysia, Kepong, Malaysia.

Manokaran, M., Eng Seng, Q., Ashton, P.S. *et al.* (2004) Pasoh Forest Dynamics Plot, Peninsular Malaysia. In E. Losos and E.G. Leigh (eds), *Tropical Forest Diversity and Dynamism: Findings from a Large-scale Plot Network*. University of Chicago Press, pp. 585–598.

Missa, O. (2005) The unified theory of biodiversity and biogeography: alive and kicking. *Bulletin of the British Ecological Society* 36, 12–17.

Moorcroft, P.R., Hurtt, G.C., and Pacala, S.W. (2005) A method for scaling vegetation dynamics: the ecosystem demography model (ED). *Ecological Monographs* 71, 557–585.

Pacala, S.W. and Rees, M. (1998) Models suggesting field experiments to test two hypotheses explaining successional diversity. *American Naturalist* 152, 729–737.

Pacala, S.W., Canham, C.D., Saponara, J., Silander, J.A., Kobe, R.K., and Ribbens, E. (1996) Forest models defined by field measurements. II. Estimation, error analysis, and dynamics. *Ecological Monographs* 66, 1–44.

Peters, H.A. (2003) Neighborhood regulated mortality: the influence of positive and negative density dependence on tree populations in species-rich tropical forests. *Ecology Letters* 6, 757–765.

Phillips, O.L., Malhi, Y., Vinceti, B. *et al.* (2002) Changes in growth of tropical forests: evaluating potential biases. *Ecological Applications* 12, 576–587.

Preston, F.W. (1948) The commonness, and rarity, of species. *Ecology* 29, 254–283.

Preston, F.W. (1962) The canonical distribution of commonness and rarity. *Ecology* 43, 185–215, 410–432.

Purves, D.W. and Pacala, S.W. (2005) Ecological drift in niche-structured communities: neutral pattern doesn't imply neutral process. In D.F.R.P. Burslem, M. Pinard, and S. Hartley (eds), *Biotic Interactions in the Tropics*. Cambridge University Press, pp. 107–140.

Robertson, G.P. (1987) Geostatistics in ecology: interpolating with known variance. *Ecology* 68, 744–748.

Rossi, R.E., Mulla, D.J., Journel, A.G., and Franz, E.H. (1992) Geostatistical tools for modelling and interpreting ecological spatial dependence. *Ecological Monographs* 62, 277–314.

Scheiner, S.M. (2001) Theories, hypotheses, and statistics. In S.M. Scheiner and J. Gurevitch (eds), *Design and Analysis of Ecological Experiments*, 2nd Edition. Oxford University Press, pp. 3–13.

Schnitzer, S.A. and Bongers, F. (2002) The ecology of lianas and their role in forest. *Trends in Ecology and Evolution* 17, 223–230.

Schnitzer, S.A. and Carson, W.P. (2001) Treefall gaps and the maintenance of species diversity in a tropical forest. *Ecology* 82, 913–919.

Schnitzer, S.A., Dalling, J.W., and Carson, W.P. (2000) The impact of lianas on tree regeneration in tropical canopy gaps: evidence for an alternative pathway in gap-phase regeneration. *Journal of Ecology* 88, 655–666.

Schupp, E.W., Milleron, T., and Russo, S.E. (2002) Dissemination limitation and the origin and maintenance of species rich tropical forests. In D.J. Levey, W.R. Silva, and M. Galetti (eds), *Seed Dispersal and Frugivory: Ecology, Evolution, and Conservation*. CABI Publishing, Wallingford, pp. 19–33.

Swaine, M.D. and Whitmore, T.C. (1988) On the definition of ecological species groups in tropical rain forests. *Vegetatio* 75, 81–86.

Thomas, D., Kenfack, D., Chuyong, G.B. *et al.* (2003) *Tree Species of Southwester Cameroon: Tree Distribution Maps, Diameter Table, and Species Documentation of the 50-hectare Korup Forest Dynamics Plot*. Center for Tropical Forest Science of the Smithsonian Tropical Research Institute and Conservation Programme-Cameroon, Washington, DC.

Thomas, S.C. (2004) Ecological correlates of tree species persistence in tropical forest fragments. In E. Losos and E.G. Leigh (eds), *Tropical Forest Diversity and Dynamism: Findings from a Large-scale Plot Network*. University of Chicago Press, pp. 279–314.

Thompson, J., Brokaw, N., Zimmerman, J.K. *et al.* (2002) Land use history, environment, and tree composition in a tropical forest. *Ecological Applications* 12, 1344–1363.

Tilman, D. (1999) Diversity by default. *Science* 283, 495–496.

Uriarte, M., Canham, C.D., Thompson, J., and Zimmerman, J.K. (2004a) A maximum-likelihood, neighborhood analysis of tree growth and survival in a tropical forest. *Ecological Monographs* 74, 591–614.

Uriarte, M., Condit, R., Canham, C.D., and Hubbell, S.P. (2004b) A spatially explicit model of sapling growth in a tropical forest: does the identity of neighbors matter? *Journal of Ecology* 92, 348–360.

Uriarte, M., Hubbell, S.P., John, R., Condit, R., and Canham, C.D. (2005a) Neighbourhood effects on sapling growth and survival in a neotropical forest and the ecological equivalence hypothesis. In D.F.R.P. Burslem, M.A. Pinard, and S.E. Hartley (eds), *Biotic Interactions in the Tropics: Their Role in the Maintenance of Species Diversity*. Cambridge University Press, Cambridge, pp. 89–106.

Uriarte, M., Canham, C.D., Thompson, J., Zimmerman, J.K., and Brokaw, N. (2005b) Seedling recruitment in a hurricane-driven forest: light limitation, density-dependence, and the spatial distribution of parent trees. *Journal of Ecology* 93, 291–304.

Valencia, R., Condit, R., Foster, R.B. *et al.* (2004a) Yasuní Forest Dynamics Plot, Ecuador. In E. Losos and E.G. Leigh (eds), *Tropical Forest Diversity and Dynamism: Findings from a Large-Scale Plot Network*. University of Chicago Press, pp. 609–620.

Valencia, R., Foster, R.B., Villa, G. *et al.* (2004b) Tree species distributions and local habitat variation in the Amazon: large forest plot in eastern Ecuador. *Journal of Ecology* 92, 214–229.

Volkov, I., Banavar, J.R., He, F., Hubbell, S.P., and Maritan, A. (2005) Density dependence explains tree species abundance and diversity in tropical forests. *Nature* 438, 658–661.

Volkov, I., Banavar, J.R., Hubbell, S.P., and Maritan, A. (2003) Neutral theory and relative species abundance in ecology. *Nature* 424, 1035–1037.

Welden, C.W., Hewett, S.W., Hubbell, S.P., and Foster, R.B. (1991) Sapling survival, growth and recruitment: relationship to canopy height in a neotropical forest. *Ecology* 72, 35–50.

Whitmore, T.C. (1978) Gaps in the forest canopy. In P.B. Tomlinson and M.H. Zimmerman (eds), *Tropical Trees as Living Systems*. Cambridge University Press, Cambridge, pp. 639–655.

Wills, C. and Condit, R. (1999) Similar non-random processes maintain diversity in two tropical rainforests. *Proceedings of the Royal Society of London* 266, 1445–1452.

Wills, C., Condit, R., Foster, R.B., and Hubbell, S.P. (1997) Strong density- and diversity-related effects help to maintain tree species diversity in a neotropical forest. *Proceedings of the National Academy of Sciences of the United States of America* 94, 1252–1257.

Wills, C., Harms, K.E., Condit, R. *et al.* (2006) Nonrandom processes maintain diversity in tropical forests. *Science* 311, 527–531.

Wright, S.J. (2002) Plant diversity in tropical forests: a review of mechanisms of species coexistence. *Oecologia* 130, 1–14.

Wright, S.J. (2005) Tropical forest in a changing environment. *Trends in Ecology and Evolution* 10, 553–560.

Wright, S.J., Muller-Landau, H.C., Calderón, O., and Hernández, A. (2005) Annual and spatial variation in seedfall and seedling recruitment in a neotropical forest. *Ecology* 86, 848–860.

Wright, S.J., Muller-Landau, H.C., Condit, R., and Hubbell, S.P. (2003) Gap-dependent recruitment, realized vital rates, and size distributions in tropical trees. *Ecology* 84, 3174–3185.

Zimmerman, J.K., Everham III, E.M., Waide, R.B., Lodge, D.J., Taylor, C.M., and Brokaw, N.V.L. (1994) Responses of tree species to hurricane winds in subtropical wet forest in Puerto Rico: implications for tropical tree life histories. *Journal of Ecology* 82, 911–922.

TESTING THEORIES OF FOREST REGENERATION AND THE MAINTENANCE OF SPECIES DIVERSITY

Chapter 8

TROPICAL FOREST ECOLOGY: STERILE OR VIRGIN FOR THEORETICIANS?

Egbert G. Leigh, Jr

OVERVIEW

Trees interact primarily with near neighbors, so mathematical theory of forest ecology should be an interactive dynamics of spatial arrangements of trees of different species. No such dynamics yet exists. Nonetheless, crude theory of forest structure suggests:

1 Trees' competition for light causes wildly unequal distribution of light among leaves, greatly reducing forest productivity.
2 Competition for nutrients favors fine-root investment far beyond that which maximizes forest production.
3 Over a wide range of soil quality, total productivity of lowland tropical forest changes far less than above- versus below-ground allocation.
4 Trees adapted to poorer soils live longer, have denser wood, and have longer-lived leaves.
5 Long-lasting leaves avoid being eaten by being tough, and avoid drying out by limiting stomatal conductance, thus reducing photosynthetic capacity.

Testing Hubbell's neutral theory prediction of how fast initially rare species can spread shows that two tree clades invading South America 20 million years ago spread non-randomly quickly. They did not replace other clades, implying that differences between tropical tree species allow them to coexist.

Finally, to understand tropical forest one must consider animals and pathogens. Lotka–Volterra theory predicts:

1 Trees can reduce herbivory by more effective defense (which reduces growth) or by being rare.
2 Many tree species coexist if specialist pests keep each rare enough, as appears true in most tropical forests.
3 In more productive forests, predators limit herbivory.

Therefore, employing animals as pollinators and seed dispersers allowed diverse, productive flowering forest to replace slower-growing, better-defended gymnosperm forest.

THE QUESTIONS: WHAT MUST WE UNDERSTAND ABOUT TROPICAL FORESTS?

Understanding tropical forest ecology requires answers to several questions.

1 *What controls forest productivity?* First, why is the annual gross production of lowland moist or wet tropical forests near 3 $kg\,C\,m^{-2}$ (Leigh 1999, Loescher *et al.* 2003)? Second, what governs allocation to above- versus below-ground activities, such as stem- versus root-making, and to earlier reproduction versus longer life? Third, why is average annual mortality among a lowland tropical forest's tree's nearly the same for all sizes between 10 and 70 cm diameter at breast height (dbh) (Leigh 1999, p. 122)? Fourth, how do soil quality and herbivore pressure influence a tree's allocation between long life and early reproduction, and how do trees' life-history allocation and anti-herbivore defenses influence their soil?

2 *What governs forest structure?* First, what limits tree height? Why are tropical trees far shorter, and far shorter-lived, than redwoods (the world's tallest trees)? Second, what limits tree density and basal area? Why do most tropical forests have a basal area of 30 ± 10 m^{-2} ha^{-1}, regardless of soil quality (Lewis *et al.* 2004, pp. 429–430; Losos *et al.* 2004, p. 71)? Third, the ratio of tree to liana biomass and production is similar in tropical forests the world around (Schnitzer 2005): why? Fourth, what principles govern tree shape? Why are there so many different tree architectures (Hallé and Oldeman 1970)? What are the relative costs and benefits of leaves at different heights (Givnish 1984, 1987)? Finally, how does forest structure (canopy roughness, tree crown size and shape, tree density, liana abundance) vary with climate and soil quality?

3 *How do herbivores shape the characteristics of tropical forests?* Herbivores and pathogens help drive the trade-off between growing fast in high light versus surviving in shade (King 1994, Kitajima 1994) and the trade-off between growing fast on good soil versus surviving on poor soil (Fine *et al.* 2004). How do tropical plants allocate resources between enabling young to escape their parents' pests and pathogens (Janzen 1970) versus investing in defense (Regal 1977)? How does this allocation affect soil fertility, forest productivity, and the abundance and diversity of animal consumers (Corner 1964)? Finally, how does a forest's productivity affect the role of animals in controlling its herbivores (Oksanen *et al.* 1981)?

4 *Why are there so many kinds of tropical tree?* How do different allocations between early reproduction versus long life, growing in bright light versus surviving in shade, and tolerating versus evading drought contribute to tree diversity (Tyree 2003, Tyree *et al.* 2003, Wright *et al.* 2003)? Do specialized pests play an essential role in maintaining tree diversity? How are different strategies of anti-herbivore defense related to the great differences among tree species in the lifetime, toughness, and photosynthetic capacity of their leaves (Wright *et al.* 2004)?

FRAMING A MATHEMATICAL THEORY OF FOREST ECOLOGY

This chapter considers how theoretical concepts and their mathematical formulation can help answer the questions of forest ecology. Forest ecology lacks a coherent frame of deductive mathematical theory analogous to those of the genetical theory of natural selection (Fisher 1930), population genetics (Crow and Kimura 1970), or animal ecology (MacArthur 1972). Because animals move about, "averaging" their environments, theorists like Volterra (1931) could derive useful predictions by assuming that population densities are uniform in space. In contrast, the number of trees in each species does not suffice to predict a forest's dynamics because:

1 A tree competes for resources with a few near neighbors (Schaffer and Leigh 1976), while its successful young usually compete with the young of trees far beyond the competitive reach of its seed-parent's neighbors, a circumstance that favors less destructive or spiteful forms of competition (Wilson 1980, Leigh 1994). Therefore theorists must consider the dynamics of the spatial arrangements of trees of different species (Schaffer and Leigh 1976). This is a difficult proposition, even in the neutral case (Bramson *et al.* 1996, 1998, Chave and Leigh 2002).

2 A tree's competitive impact on its neighbors depends on its crown's height, size, shape and total leaf area, the density and distribution of its roots, and so forth. The theory of tree shape must account for the costs and benefits of leaves at different heights (Givnish 1987). No current mathematical theory is adequate to resolve these questions. It appears that only computer models like SORTIE (Pacala *et al.* 1996) can be modified to handle both the dynamics of tree arrangement and the changes in sizes and shapes of the trees involved, but simulations are a poor substitute for analytic theory.

Two general theories have recently been proposed for forest ecology. Hubbell's (2001) neutral theory of forest dynamics and tree diversity assumes that each tree has identical prospects of death and reproduction, regardless of its species or those of its neighbors. Hubbell's focus was the distribution of tree species abundances in plots ≤ 50 ha and biogeographic patterns of species distributions. West et al. (1997) and Enquist (2002) framed a theory of forest structure, production, and dynamics. They assumed that (1) only terminal twigs carry leaves; (2) each terminal twig has the same diameter and leaf area, regardless of its tree's species; (3) trunks and non-leafy branches all fork into n successors, each shorter by a factor $n^{-1/3}$ and narrower by a factor $n^{-1/2}$ than its predecessor; (4) a tree's height H is proportional to the total path length from its root collar to the tip of any leafy twig; (5) a tree's mass M is proportional to its total above-ground wood volume, which is proportional to $D^2 H$, where D is its trunk diameter; and (6) a tree's dry matter production is proportional to its leaf area LA. Thus a tree's height is proportional to $D^{2/3}$, its mass is proportional to $D^2 H$, and therefore to $D^{8/3}$, and its leaf area LA and dry matter production are proportional to D^2, which is proportional to $M^{3/4}$. Many other "laws" of forest structure and production have been derived from these relations (Enquist et al. 1998, 1999, Enquist and Niklas 2001, Niklas and Enquist 2001).

Both theories appear to fit masses of data. Their explanatory power, however, is limited. The neutral theory ignores differences between species that are crucial to understanding tree diversity (Leigh et al. 2004, Wills et al. 2006). Moreover, like Volterra, most neutral theorists track numbers of individuals in different species rather than tackling spatial arrangements, as Bramson et al. (1996, 1998) and Chave and Leigh (2002) began to do. Enquist's (2002) theory assumes that a given leaf area is equally productive in the canopy or at the forest floor, which is nonsense (Muller-Landau et al. 2006a). Moreover, it assumes that a tree's height H is related to its diameter D as $H = cD^{2/3}$, when in fact this relation varies according to the heights of its neighbors (King 1986, 1996). Indeed, the relation $1/H = 1/H_{max} + 1/aD^b$, where a and b are fitted positive constants, matches data for the trees of a mature forest species much better (Kato et al. 1978, Thomas 1996a). Enquist's theory fails to predict the productivity/biomass

allometry in seedlings and saplings (Reich et al. 2006), or the relation between the diameter, height, total photosynthesis and diameter growth rate among a forest's trees (Muller-Landau et al. 2006a).

WHAT MATHEMATICAL THEORY HAS DONE

Despite the obstacles, mathematical theory has contributed substantially to forest ecology. I now review some of these accomplishments, showing what theory has illumined, how it can mislead, and what still needs doing.

Limits on gross production

Gross production is governed by the area of leaves a forest deploys per unit area of ground (its leaf area index, or LAI), the light these leaves receive, and the photosynthesis this light supports. We now consider the first two of these items. Most tropical forests have a leaf area index between 6 and 8 (Leigh 1999). At Pasoh Reserve, Malaysia, where $LAI = 8$ (Kato et al. 1978), 0.3% of the light above the canopy reaches the ground. The relation of the vertical distributions of leaf area and light abundance from the canopy downward suggests that each unit of LAI halves the light passing through it (Yoda 1974, 1978). In forests where this is so, if $LAI = 7$, only $1/2^7$ (1/128) of the above-canopy light reaches the forest floor. Most tropical forests let about 1% of the light above the canopy reach the ground (Leigh 1999). In the shaded understory, coverage by seedlings and ground-herbs is low enough to suggest that they receive barely enough light to survive (Leigh 1975, Givnish 1988). In tropical forest, the few data available suggest that LAI rarely exceeds 8, presumably because extra leaves are a losing proposition, and LAI is seldom less than 6, for each unit decrease in LAI doubles the light reaching the forest floor. Accurate information on LAI is rare, but forests on soils of very different quality apparently support a similar dry weight of leaves (Malhi et al. 2004, p. 575), about 8 tons ha^{-1} (Leigh 1999).

How much photosynthesis does this leaf area carry out? Let a leaf receiving $I\,\mu\mathrm{E\,s}^{-1}$ of photosynthetic photons m^{-2} leaf surface photosynthesize at the rate $A(I)$, where

$$A(I) = mI/(1 + mI/A_{\max})\ \mu\mathrm{mol}\,\mathrm{C\,s}^{-1}\,\mathrm{m}^{-2}\ \text{leaf}$$

Here $m = 0.05$ and $A_{\max}$ is the leaf's maximum rate of photosynthesis. In canopy leaves of tropical forest, taking all species together, $A_{\max}$ usually averages $2\,\mathrm{g\,C\,hour}^{-1}\,\mathrm{m}^{-2}$ leaf, or $12.5\,\mu\mathrm{mol}\,\mathrm{C\,s}^{-1}\,\mathrm{m}^{-2}$ leaf (Kira 1978, p. 571, Zotz and Winter 1993). If the leaf area index above our leaf is L, set $I = Qe^{-kL}$, where Q is the above-canopy light level and $k = 0.7$, which ensures that $e^{-kL} = Q/2$ when $L = 1$. L need not be an integer: if half the sky overhead is covered by non-overlapping horizontal leaves, then $L = 1/2$ and $Qe^{-k/2} = Q/\sqrt{2}$.

To calculate the total photosynthetic rate P_T of a forest with total leaf area index LAI, receiving $Q\mu\mathrm{E\,s}^{-1}$ of light per square meter of ground, following Leigh (1999), set

$$P_T(Q) = \int_0^{LAI} \frac{mQe^{-kL}dL}{1 + mQe^{-kL}/A_{\max}}$$

$$= \frac{A_{\max}}{k}\ln\left[\frac{1 + mQ/A_{\max}}{1 + mQe^{-kLAI}/A_{\max}}\right]$$

To do the integral, set $u = 1 + mQe^{-kL}/A_{\max}$, $du = -kmQe^{-kL}/A_{\max}$. If $mQe^{-kLAI} \ll A_{\max}$, then we may approximate P_T by $(A_{\max}/k)\ln(1 + mQ/A_{\max})$.

To estimate this forest's total daily photosynthesis P_{daily}, let $Q = 0$ from 6 p.m. to 6 a.m., rise linearly to $4Q^*$ between 6 a.m. and noon, and decline linearly back to zero at 6 p.m. Here, Q^* is the long-term average light level above the canopy. Then P_{daily} is

$$\int_{6\ \mathrm{a.m.}}^{6\ \mathrm{p.m.}} P_T[Q(t)]dt = \frac{0.0432A_{\max}^2}{4kmQ^*}$$

$$\times\left[\left(1 + \frac{4mQ^*}{A_{\max}}\right)\ln\left(1 + \frac{4mQ^*}{A_{\max}}\right) - \frac{4mQ^*}{A_{\max}}\right]$$

If $Q^* = 390\,\mu\mathrm{E\,s}^{-1}\,\mathrm{m}^{-2}$ leaf (close to the pantropical forest average) and if $k = 0.7$, $m = 0.05$, and $A_{\max} = 12.5\,\mu\mathrm{mol}\,\mathrm{C\,s}^{-1}\,\mathrm{m}^{-2}$ leaf, then $P_{\mathrm{yearly}} = 365\,P_{\mathrm{daily}} = 4.4\,\mathrm{kg\,C\,m}^{-2}$. The real value is about $3\,\mathrm{kg\,m}^{-2}$ (Loescher et al. 2003). This theory gives a "ball-park" estimate of gross production, but it contains too many "givens." It does not explain why each layer of leaves should take up half the remaining light, even though canopy leaves are more steeply inclined than understory ones, or why $A_{\max}$ of canopy sun leaves should average about $12.5\,\mu\mathrm{mol}\,\mathrm{C\,s}^{-1}\,\mathrm{m}^{-2}$ leaf.

If foliage is equally productive in different forests, gross production should depend little on soil quality unless the soil is extremely poor. Fertilizing a Hawaiian *Eucalyptus* plantation in a rainforest climate at $20°\mathrm{N}$ increased its LAI from 4.7 to 6.5 and its gross production from 3 to $4\,\mathrm{kg\,C\,m}^{-2}$ (Giardina et al. 2003). On the other hand, annual gross production on poor soil in central Amazonia and on much richer soil in Costa Rica are both near $3\,\mathrm{kg\,C\,ha}^{-1}$ (Table 8.1), as if, in the long term, gross production were independent of soil quality.

Competition for light, in which trees grow tall trunks to shade their neighbors, creates majestic forests of great beauty. Yet the competition among trees to shade each other is a "tragedy of the commons" (Hardin 1968), which reduces the forest's productivity (Iwasa et al. 1984, King 1990, Falster and Westoby 2003). If light were distributed more evenly among a forest's leaves, its productivity would be much higher. Sea palms, *Postelsia palmaeformis*, annual intertidal kelps of the northeastern Pacific, can maintain over $14\,\mathrm{m}^2$ fronds m^{-2} substrate, and produce up to $7\,\mathrm{kg}$ dry matter m^{-2} substrate in 6 months (Leigh et al. 1987, Holbrook et al. 1991), far higher than a rainforest's annual dry matter production. This productivity is possible because these kelps are restricted to the most wave-beaten shores (Paine 1979, 1988), where the waves keep them short (Denny 1999), and continually stir their narrow, light-weight fronds, assuring them far more nearly equal access to light than a forest's leaves receive. Indeed, a forest's productivity declines sharply when its canopy closes and the access of different trees to light becomes progressively less equal (Binkley 2004).

Table 8.1 Gross production, soil respiration, litterfall, and below-ground respiration, in tons C ha^{-1} year^{-1}, in two lowland tropical rainforests.

Site	Gross production	Soil respiration	Litterfall	Below-ground respiration
La Selva, Costa Rica	29.5	12.5	4.4	8.1
Cuieiras, central Amazonia, Brazil	30.4	No data	No data	13.7

Sources: La Selva: Gross production from table 6 of Loescher *et al.* (2003); other data are averages of residual and old alluvium from figure 6.1 of Schwendenmann (2002). Cuieiras: All data from table 6 of Malhi *et al.* (1999).

Tree height, tree shape, and forest structure

Tree height

To understand forest structure, we must first learn what limits forest height. Early models assumed that forests grew until the costs of maintaining unproductive woody biomass left no resources for further growth (Bossel and Krieger 1991). This view is no longer credible. In the rainforest at La Selva, Costa Rica, above-ground tree-trunk respiration is only 7–12% of gross production (Ryan *et al.* 1994). This proportion does not increase quickly enough with tree height to limit the height of lodgepole pine forest (Ryan and Waring 1992).

The height of coast redwoods appears to be limited by the difficulty of lifting water to their crowns (Koch *et al.* 2004). Why are most trees shorter than redwoods? Since diameters of most canopy trees keep growing long after their height growth stops (King 1990, Ryan and Yoder 1997), resources are not limiting. King (1990, p. 809) therefore concluded that "adult tree height reflects an evolutionary balance between the costs and benefits of stature." To learn what limits forest height, King (1990) considered a tree's height as its strategy in a game played against its neighbors. What is the appropriate growth strategy of a tree in a forest of identical neighbors, each with height H_0 and leaf area LA? Set the tree's stemwood production $dw/dt = cLA(1 - H_0/A)$, where cLA is the wood production the crown would support if height imposed no extra costs, and H_0/A denotes the proportion by which height-associated costs reduce wood production. Now suppose that one

tree has the same leaf area LA and crown width w_0 as each of its neighbors, but a height $H \neq H_0$; let the light it receives, and its wood production, be $[1 + (H - H_0)z/w_0]$ times that of each neighbor, when each neighbor has height H_0. For trees at 45°N with conical crowns twice as tall as they are wide, $z = 0.76$ (King 1990). If crown width w is proportional to tree height H, that is to say, if $w = bH$, then the height H_c that maximizes each tree's wood production when they all have the same height is given by $\partial(dw/dt)/\partial H = 0$. This optimum height is $H_c = A/(1 + zb)$: a tree's wood production declines if it grows beyond this height. Using stand tables to estimate the stand height A at which wood production of surviving trees declines to zero, this prediction approximates observation for many, but not all, even-aged stands (King 1990).

King recognized that a tree's height growth depends on its neighbors', and assumes that in a mature forest, all canopy trees are equally tall. Trees of the genus *Tachigali* (Leguminosae) are monocarpic (flowering and fruiting only once before dying), so they grow much faster than their iterocarpic neighbors without sacrificing strength (as measured by wood density), but they grow no taller than their canopy neighbors (Poorter *et al.* 2005). On the other hand, King assumed that height-associated costs reduce wood production in linear proportion to tree height, and did not predict the constant of proportionality, that is, the height where canopy trees' wood production stops. These costs, however, may increase non-linearly with tree height. For example, the probability that an Amazonian canopy tree carries large lianas, which slow its growth and triple its annual probability of dying, increases disproportionately with the tree's diameter (Phillips *et al.* 2005). Finally, King ignored reproduction, the central purpose of tree life. Reproduction is costly: diameter of *Tachigali* trees increases four times faster than those of iterocarpic canopy neighbors (Poorter *et al.* 2005). Malaysian rainforest trees do not reproduce until they are well enough lit to achieve a substantial proportion of their annual reproductive potential (Thomas 1996a,b,c), in accord with theoretical predictions of Iwasa and Cohen (1989). A proper theory of tree height must incorporate the trade-off between growth, reproduction, and survival.

Trunk taper and tree shape

A tree's shape reflects the trade-off between the advantage of better-lit leaves and the costs of supporting and supplying a taller crown (Givnish 1988). The first step to understanding this trade-off is to learn what factors govern the design of tree-trunks.

One criterion proposed for designing a canopy tree's trunk is that, when wind blows upon its crown, the proportional stretch (the strain) on the most stressed fiber is the same for all distances above the ground. If so, the cube $D^3(y)$ of this trunk's diameter y meters below the crown's center of mass is proportional to y (Dean and Long 1986, West *et al.* 1989). To show this, let a force F on the crown's center of mass exert a torque yF about a point on the trunk y meters below. Then the strain on the most stressed fiber at level y will be proportional to $yFD(y)/K(y)$, which we assume equal to a constant c independent of y. Here, $K(y)$ is the countertorque excited in the trunk by this stress. This countertorque is exerted by the joint action of the pull of the fibers on the stretched side of the trunk and their push on the compressed side. To calculate this countertorque, consider paired fibers a distance x to either side of the unstretched neutral plane that splits the trunk longitudinally into stretched and compressed halves. The countertorque from this pair of fibers is proportional to the restoring force kx resulting from each fiber's strain, times their lever arm x from the neutral plane. For the "average" pair, this countertorque is proportional to $D^2(y)$.

The total countertorque is proportional to $D^4(y)$ – the number of fiber-pairs involved, which is proportional to the trunk's cross-sectional area, and thus to $D^2(y)$, times the average torque per fiber-pair, also proportional to $D^2(y)$ (see Leigh 1999, pp. 90, 113). If $yFD(y)/K(y)$ is proportional to $yFD(y)/D^4(y) = c$, y is proportional to $D^3(y)$. The cube of trunk diameter increases linearly with distance below the crown's center of mass until one reaches the butt swell for 45-year-old Douglas firs, *Pseudotsuga menziesii*, in western Washington (Long *et al.* 1981), mature lodgepole pine, *Pinus contorta*, in northern Utah (Dean and Long 1986), and mountain-ash, *Eucalyptus regnans*, planted in Tasmania (West *et al.* 1989).

A rival criterion of tree-trunk design is that they have a fixed safety factor against buckling under their own weight (McMahon 1973, McMahon and Kronauer 1976). If so, 20% less wood is needed to support the crown with a given safety factor if $D^2(y)$, the square of a trunk's diameter y meters below its crown's center of mass, rather than $D(y)$ or $D^4(y)$ (King and Loucks 1978, p. 149). $D^2(y)$ declines linearly with distance up the trunk for aspens, *Populus tremuloides*, in Wisconsin (King and Loucks 1978, p. 155), and, starting 3 m above the ground, for *Schefflera morototoni* in Panama (table 5.5, p. 92 in Leigh 1999).

Nonetheless, our failure to progress beyond these crude models of support costs is a major obstacle to understanding tree shape. Tropical forest has a great variety of "tree architectures" (Hallé and Oldeman 1970, Hallé *et al.* 1978, Leigh 1999). Our inability to predict their support costs is a primary reason why we usually cannot detect what advantage, if any, is peculiar to a given model.

Forest structure

Most natural forests have trees of all ages, diameters, and heights up to the maximum. What governs a forest's distribution of tree diameters and tree heights? Kohyama *et al.* (2003) showed how to relate a forest's distribution of tree diameters to tree death and growth rates. Let the number $N(D)$ of trees per hectare with diameters between D and $D+1$ be constant for all integers $D \geq 10$ cm.

During the time interval dt, let $mN(D)dt$ of these trees die, let $G(D-1)N(D-1)dt$ trees grow in from the next lower diameter class, and let $G(D)N(D)dt$ trees grow out to the next diameter class. Finally, let the tree death rate m be independent of D (Leigh 1999). Since ingrowth must balance outgrowth and mortality,

$$G(D-1)N(D-1) = G(D)N(D) + mN(D).$$

Set

$$G(D)N(D) - G(D-1)N(D-1)$$
$$\approx d[G(D)N(D)]/dD = -mN(D),$$
$$\{1/[G(D)N(D)]\}d[G(D)N(D)]/dD$$
$$= d\{\ln[G(D)N(D)]\}/dD = -m/G(D)$$
$$N(D) = [R/G(D)]\exp - m\int_{10\,cm}^{D} dx/G(x)$$

where $R = N(9\ cm\ dbh)G(9\ cm\ dbh)$ is the recruitment rate into the 10 cm diameter class. If we know the death rate m, the average height $H(D)$, and average rate of diameter increase $G(D)$, of trees of all diameters D, we can calculate forest structure (Kohyama et al. 2003, Muller-Landau et al. 2006b).

This, however, is just book-keeping, which does not predict the death rate, diameter growth, or average height of trees with diameter D. In fact, a tropical forest has many tree species, which have adapted to different levels of the forest by dealing with the trade-off between survival, growth, and reproduction in different ways. So far, few have tried to predict the vertical distribution of a forest's leaves, its "foliage height profile." Although Iwasa et al. (1984) made a start, theorists have not adequately related a forest's foliage height profile to the heights of its tallest trees, the light leaves at different heights need to pay for their construction, support, and supply (Givnish 1988) and the amount of light leaves at each height must let pass below.

Soil quality and forest structure

Soil and above- versus below-ground allocation

A forest's leaf biomass, and its gross production, depends little on soil fertility, although leaf fall (and therefore leaf production) is lower on poorer soil. Trees derive all their energy from leaves, yet the returns from each successive unit increase of LAI are half those from its predecessor, while the costs of making these leaves decline far more slowly.

The returns from additional fine roots decline even more slowly. Let a forest take up U kg nitrogen (N) ha^{-1} year^{-1}. U is roughly $U_0/(1 + R_v/R)$ (King 1993), where U_0 is the average rate of supply of "available" nutrients to the soil from litterfall and external sources such as rainfall and weathering bedrock, R_v is the dry mass of fine roots needed to take up nutrients at half the rate U_0 of supply – in good soil, $R_v \approx 100$ kg ha^{-1} (King 1993) – and R kg ha^{-1} is the forest's actual dry mass of fine roots. To take up 99% of the supply, R must be $99R_v$, whereas taking up 99% of the incoming light requires only seven times the leaf area needed to take up half of it.

A further problem is that, just as competition to avoid being shaded by neighbors reduces a forest's total photosynthesis, competing with neighbors for nutrients represents a tragedy of the commons that reduces a forest's wood production, not to mention its reproductive investment (King 1993, Gersan et al. 2001). Consider a forest of otherwise identical trees, each with a root biomass that yields optimum wood production per tree. A "selfish" tree can increase its nutrient uptake by increasing its root biomass and extending its roots under neighbors. These neighbors must do likewise, to compensate their losses to the original selfish tree. In the end, trees wind up extending their roots under an average of six neighbors (King 1993). The inability of trees to exclude neighbors' roots from under their own crowns, or mutually enforce a cooperative optimum in root production, sharply limits their wood production.

King (1993) derived equations showing how soil nitrogen supply governs allocation to leaves, wood, and roots, assuming that nitrogen, N, is the limiting nutrient. To summarize his argument, let there be x gNg^{-1} dry weight of leaf, and let a forest's dry matter production DP be $2(x/y - x^2/2y^2)DP_{max}$ when x is less than the maximum useful foliar nitrogen concentration y, and DP_{max} when $x \geq y$. Let p_f and p_r be the proportions of

DP devoted to making leaves and fine roots, and m_f and m_r their loss rates. Let foliage dry mass F and root dry mass R, in tons ha^{-1}, be in steady state, so $p_f DP = m_f F$ and $p_r DP = m_r R$. Finally, let wood production dw/dt (including branches, bark, and coarse roots) use the resources left over from making leaves and fine roots, so $dw/dt = (1 - p_f - p_r)DP$ tons ha^{-1} $year^{-1}$.

The rate U of the forest's N uptake is the soil N used in making new leaves, fine roots, and wood. Let the N concentration be x in leaves, $a_r x$ in roots, and $a_w x$ in new wood; let trees reabsorb fractions z_f and z_r of the N in dying leaves and roots, so that fractions $1 - z_f$ and $1 - z_r$ of the N in new leaves and roots must be drawn from the soil, and let plants translocate a proportion z_w of the nitrogen in new wood from older wood. Then

$$U = x p_f (1 - z_f)DP + a_r x p_r (1 - z_r)DP$$
$$+ a_w x (1 - z_w)dw/dt$$

Since $p_f DP = m_f F + p_r DP = m_r R$,

$$U = x(1 - z_f)m_f F + a_r x(1 - z_r)m_r R$$
$$+ a_w x(1 - z_w)dw/dt$$

As DP is a function of U_0, the average rate of supply of nutrients from litterfall, atmosphere, and bedrock, we can calculate DP from U_0 and the forest's allocation to wood, leaves, and roots. King (1993) assumed that root biomass was in a competitive equilibrium where no tree could benefit by increasing its root biomass, and he used these equations, appropriately parametrized, to calculate numerically what allocation gives highest DP for a given U_0. He found that foliage allocation stayed constant, but that as U_0 declined, the proportion of DP devoted to fine roots increased at the expense of wood production.

As predicted, a higher proportion of forest production and biomass is below ground on poorer soil. In Venezuelan Amazonia, the ratio of above- to below-ground biomass is lower on poorer soil (Medina and Cuevas 1989). Despite its better soil, the rainforest at La Selva, Costa Rica, has slightly lower gross production than central Amazonian rainforest at Cuieiras, but the Cuieiras

Table 8.2 Total annual dry matter production (DP_{tot}), above-ground production (DP_{above}), and below-ground production (DP_{below}), in metric tons ha^{-1}, in two nearby 40-year-old Douglas fir stands on soils of very different quality.

Soil	DP_{tot}	DP_{above}	DP_{below}
Good soil	17.8	13.7	4.1
Poor soil	15.4	7.3	8.1

Source: Keyes and Grier (1981).

forest devotes a proportion of its gross production to below-ground activities 7/4 times that at La Selva (Table 8.1). In two nearby 40-year-old stands of Douglas fir on soils of contrasting quality in western Washington, total dry matter production was 16% higher on good soil, but above-ground production was nearly twice as high on good soil, while the reverse was true for below-ground production (Table 8.2, from Keyes and Grier 1981).

Nutrient conservation

What factors influence the availability of nutrients in leaf and soil? How can trees affect this availability? The preceding section's last equation can be solved for leaf nitrogen concentration x:

$$x = U/\{[(1 - z_f)m_f F + a_r(1 - z_r)m_r R$$
$$+ a_w(1 - zw)]dw/dt\}$$

For fixed U, x can be increased by decreasing the turnover rates m_f of leaves and m_r of roots, increasing the proportions of nutrients z_f and z_r recovered from dying leaves and roots, increasing the translocation z_w from old to new wood, or decreasing wood production dw/dt.

To learn what factors influence the rate U_0 of supply of nitrogen per unit area of soil, King (1993) assumes that a fraction k of the unused supply $U_0 - U$ leaches away. When a tree falls, its roots die, and its root biomass takes a few years to recover: King (1993) assumes that, each year, the whole of U_0 is lost in a fraction of the forest's area equal to twice the proportion m of trees dying per year. Finally, let K be the rate at which rainfall, dust from the air, and weathered bedrock supply nitrogen per unit area of soil. Then gain balances loss of soil nitrogen when

$$k(U_0 - U) + 2mU_0 = K$$

Since the forest takes up nutrient at the rate $U = U_0/(1 + R_v/R) = U_0R/(R + R_v)$,

$$U_0 - U = [U_0(R + R_v) - U_0R]/(R_v + R)$$
$$= U_0R_v/(R_v + R);$$
$$K = 2mU_0 + k(U_0 - U)$$
$$= 2mU_0 + kU_0R_v/(R_v + R);$$
$$U_0 = K(R_v + R)/[kR_v + 2m(R_v + R)]$$
$$U = U_0R/(R + R_v)$$
$$= KR/[kR_v + 2m(R_v + R)]$$

Nutrient uptake is increased by increasing fine root biomass, especially if $2mR \ll (k + m)R_v$, and by decreasing tree mortality, especially when $2m(R_v + R) \gg kR_v$.

Poor soils favor reducing nutrient losses. Therefore, King's theory predicts that on poorer soil, trees, branches, and leaves should be longer lived, wood production lower, and nutrients more efficiently translocated from dying plant parts and old wood to their new counterparts.

A fertility gradient in Amazonia supports some of these predictions. Soils are more fertile at Amazonia's western edge, near the Andes, than in

central and eastern Amazonia (Malhi *et al.* 2004). As predicted, wood production is lower on poorer soil. Baker *et al.* (2004a, p. 360) found that wood production, the annual increase in above-ground woody biomass ΔAGB, in tons ha^{-1} year^{-1}, of a plot's trees ≥ 10 cm dbh surviving from one census to the next, was related to their annual basal area increase ΔBA, in m^2 ha^{-1} year^{-1}, by the regression

$$\Delta AGB = 9.57(\Delta BA) + 0.12(r^2 = 0.89)$$

Between 1985 and 1992, annual basal area increase averaged 0.64 m^2 ha^{-1} in western Amazonia and 0.4 m^2 ha^{-1} in less fertile central and eastern Amazonia (Lewis *et al.* 2004); accordingly wood production averaged 6.3 tons ha^{-1} year^{-1} in western Amazonia and 4 tons ha^{-1} year^{-1} further east.

Also as predicted, tree mortality is lower on poorer soils. Between 1985 and 1992, annual mortality rate for trees ≥ 10 cm dbh averaged 2% in western Amazonia and only 1% further east, whether measured as the proportion of trees dying per year, or the proportion of basal area comprised by these dying stems (Table 3 in Lewis *et al.* 2004). Trees in central and eastern Amazonia are indeed built to last: their wood density (g dry weight cm^{-3} fresh volume), a good measure of wood strength (Putz *et al.* 1983), averages 0.684, compared with 0.571 in western Amazonia (Baker *et al.* 2004b). Poor soil so promotes longevity over reproduction that, in nutrient-starved heath forest of Borneo,

Table 8.3 Number of free-standing woody plants ≥ 10 cm dbh (N), number of species among them (S), and Fisher's α, in relation to annual rainfall (P) and rainfall during the driest quarter (P_3) (both in mm) in 25 ha subplots of selected continental forest dynamics plots of the Center for Tropical Forest Science.

Site	N	S	α	P	P_3
Yasuni, Ecuadorian Amazonia	17,546	820	178	3081	594
Lambir, Sarawak, Malaysia	15,916	851	193	2664	498
Pasoh Reserve, Malaysia	13,276	604	130	1788	318
Korup, Cameroon	12,296	261	47	5272	172
Barro Colorado Island, Panama	10,728	206	36	2551	131
Huai Kha Khaeng, Thailand	10,938	185	32	1476	46

Note: Fisher's α is defined by the relation $S = \alpha \ln(1 + N/\alpha)$ (Condit *et al.* 2004, p. 79).
Sources: Climate data from table 4.3 of Leigh (2004); N, S, and L from table 7.1 of Condit *et al.* (2004).

regeneration from a clearing consists primarily of stump sprouts, whereas in nearby dipterocarp forest on better soil, regeneration is driven by seed fall and seedling growth (Riswan and Kartawinata 1991).

Finally, evidence from other sources suggests that, as predicted, leaf turnover is lower on poorer soil. In montane forest of Jamaica, leaf turnover (dry biomass of leaves divided by dry biomass of annual leaf fall) is 20% lower on a mor ridge than on a nearby, more fertile, mull ridge (Tanner 1980a,b). In Venezuelan Amazonia, leaves are longer lived on poorer soils (Reich et al. 2004, pp. 18–19). In sum, poor soil promotes nutrient conservation.

How conserving nutrients affects forest characteristics

The primary impacts on a forest of conserving nutrients arise from making longer-lived leaves. To live long, a leaf must avoid being eaten. To deter herbivores, leaves must be thick and tough, with low N concentration (Coley 1983, Waterman et al. 1988). In Venezuelan Amazonia, longer-lived leaves are tougher, and have lower N concentration (Reich et al. 1991), as this argument predicts. The relation between soil quality and leaf anti-herbivore defenses can become a vicious circle: defensive compounds in long-lived leaves slow their decomposition when they fall, further diminishing soil quality (Bruening 1996, p. 23).

To live long, a leaf must also avoid drying out. To do so, leaves must limit transpiration, and therefore stomatal conductance g_s (rate of water loss per m^2 leaf area per kPa vapor pressure deficit outside the leaf, $kg\,m^{-2}\,s^{-1}\,kPa^{-1}$) (Givnish 1984). Because leaves must release water vapor to let in CO_2, trees whose leaves have lower maximum stomatal conductance $g_{s\,max}$ must have lower A_{max}, lower photosynthetic capacity per unit area (Reich et al. 1991, Tyree 2003, Santiago et al. 2004). In turn, trees with less conductive leaves can make do with lower hydraulic conductance in their wood. As predicted, trees with lower A_{max} have leaf-bearing branches with lower leaf-specific hydraulic conductance k_L (kg water moving through the xylem per second per m^2 leaf in response to a unit change

in water potential, MPa, per meter of stem), in both the Old and New World tropics (Brodribb and Feild 2000, Brodribb et al. 2002, Santiago et al. 2004). Katul et al. (2003) derived mathematical theory predicting such relationships among A_{max}, $g_{s\,max}$, and k_L.

To keep their long-lived leaves from drying out, trees on poor soil also design leaves and crowns to restrict transpiration by limiting both leaf temperature and turbulent airflow around leaves (Givnish 1984). In nutrient-poor forests of heath, peat swamp, and white sand in Borneo and Venezuelan Amazonia, canopy leaves avoid overheating by being smaller, more reflective, and more nearly vertical than their counterparts on better soil (Brunig 1970, 1983, Medina et al. 1990). On these poor soils, the canopy is far smoother than on nearby oxisols (Brunig 1983), reducing turbulent airflow (Bruenig 1996), and allowing thick, transpiration-limiting boundary layers to build up around leaves and whole tree crowns (Meinzer et al. 1993). These characteristics were considered adaptations to occasional water shortage in sandy, easily drained soils (Brunig 1983). Perhaps because of these features, heath forests and Amazon caatinga are no more sensitive to water shortage than adjacent forests on oxisols (Coomes and Grubb 1998).

Similarly, if a tree is to live long, its wood should be strong and dense (Putz et al. 1983), but trees with denser wood tend to have narrower vessels and lower hydraulic conductance, which limits their leaves' A_{max} (Santiago et al. 2004). In denser-wooded trees, however, lower water potentials are needed to make leaves wilt (Gartner and Meinzer 2005) or to cause embolism in the xylem (Santiago et al. 2004, Hacke et al. 2005). Thus trees with denser wood can photosynthesize under drier conditions (Santiago et al. 2004). Furthermore, in leaves with lower maximum stomatal conductance and lower A_{max}, stomatal conductance is less sensitive to increasing vapor pressure deficit (Oren et al. 1999). Therefore, average transpiration is much the same in a mature Bornean heath forest as in a nearby mature forest on better soil (Becker 1996): the heath forest's more sustained transpiration and photosynthesis makes up for its neighbor's episodes of high photosynthesis.

Diversity

In some everwet equatorial forests, a 25 ha square contains over 800 species of tree ≥ 10 cm dbh; a similar square of seasonal tropical forest has about 200 species, far more than in a similar area of temperate-zone forest (Table 8.3). Why are there so many kinds of tropical tree?

Testing Hubbell's neutral theory

Do all these species of tropical tree differ in ways that allow them to coexist? To answer, first consider how fast a tree species can spread in a neutral world (Hubbell 2001) where it makes no difference what species a tree belongs to. Let time be measured in tree generations: if trees have annual death rate m, then t years corresponds to $T = mt$ tree generations. Consider a neutral species that begins with n mature trees at time 0. Let each tree alive at time t have probability mdt of dying by time $t + dt$ and equal probability of producing a seed by then that instantly becomes a mature tree, independently of the fates of all other trees of its species. Using the methods of branching processes, this neutral theory predicts that if the species still survives $T \gg n$ tree generations later, the probability that this species has over kT reproductive trees then is e^{-k} (Leigh 2007; see also Fisher 1930, p. 80).

Some tree species have spread much faster than chance would allow. About 20 million years ago, the tree species *Symphonia globulifera* (Guttiferae) first appeared in the Neotropics, after dispersing across the Atlantic from Africa (Dick *et al.* 2003). If the death rate of this tree species is 2% year^{-1}, its descendants crossed 400,000 tree generations ago. Now this species averages about two trees ≥ 10 cm dbh ha^{-1} all through Amazonia, and it has spread into Central America. This species must have over 10 million reproductive adults in the Neotropics (Leigh *et al.* 2004). An initially rare neutral species lucky enough to survive so long would have probability e^{-25}, less than 10^{-10}, of having so many reproductive trees after 400,000 tree generations. Similarly, the genus *Ocotea* (Lauraceae) first appeared in South America about 20 million years ago, after dispersing across the sea from North America

(Chanderbali *et al.* 2001, p. 139). Now *Ocotea* averages about five trees ≥ 10 cm dbh ha^{-1} all through Amazonia, representing hundreds of species: this clade, too, has spread far faster than chance would allow. Many other tree species have multiplied extensively in tropical forests after dispersing across oceans (Pennington and Dick 2004).

Although a decisive advantage spread *Symphonia* and *Ocotea* throughout Amazonia, they form only a small minority of Amazonia's trees. Other tree species must differ enough from these invaders to avoid competitive displacement (Leigh *et al.* 2004). These species coexist because no one species can do all things well (MacArthur 1961). Trees, like other organisms, face trade-offs: enhancing one ability usually entails sacrifices in others.

Another way to show that tree diversity reflects differences that allow different species to coexist is to show that natural selection driven by trade-offs causes tree speciation. Tree speciation is usually allopatric (Coyne and Orr 2004, Leigh *et al.* 2004). Nonetheless, if a tree species is divided into two completely isolated populations, millions of years may elapse before they become mutually intersterile (Ehrendorfer 1982). Closely related species isolated by having different pollinators or flowering times are often completely fertile when crossed artificially (Gentry 1989, Kay and Schemske 2003). In one sympatric pair of large herbs, *Costus*, plants of each species will not accept pollen from a nearby plant of the other, but can be fertilized by pollen from plants of the other 400 km away (Kay 2002), as if reproductive isolation were favored by selection (presumably driven by reduced fitness of hybrids for either parent's way of life). Selection against hybrids arises if a peripheral population occupies a habitat requiring adaptations unsuitable for the parental habitat (Stebbins 1982). Gillett (1962) suggested that rainforest trees had congeners in savanna, a habitat demanding a very different physiology (Hoffmann *et al.* 2004), because novel anti-herbivore defenses allow invasions of other, already occupied, habitats. Tree speciation associated with transitions from wet forest to dry forest or savanna (when dry habitats were expanding at the beginning of glacial cycles) and vice versa (when wet forest was expanding at the beginning of interglacials) was frequent in the Neotropics during the Pleistocene (Pennington et al. 2004). Further research on how tree species originate is urgently needed.

Dangers of theory: analysis of a trade-off

What trade-offs enable pioneer tree species to coexist with superior competitors? Answering this question shows how simple theory can clarify thinking, and how it can mislead. Two tree species can coexist if each species can invade a forest consisting only of the other. To invade a forest of superior competitors, a pioneer species that colonizes treefall gaps needs a steady supply of colonizable gaps. Pioneers will spread from gap to gap only if a pioneer in a new gap grows fast enough to set seed before superior competitors overtop it and crowd it out, and if, on the average, more than one of its seeds repeats this feat in other gaps. Similarly, the superior competitor can invade a forest of pioneers because its seeds can germinate in the pioneers' shade and its saplings grow up to overtop them. It is a question for theorists how big and how frequent treefall gaps must be, how fast pioneers must grow, and how fast and how densely they must scatter their seeds, to allow a pioneer species to spread from gap to gap.

Using mathematical techniques of Horn and MacArthur (1972), Tilman (1994) developed a schematic model of how pioneers coexist with superior competitors. If pioneers are absent, let a proportion $C(t)$ of the forest's space be occupied by superior competitors in year t. Following Tilman, let mortality empty a proportion m of this occupied space each year, and let superior competitors take over a proportion $rC(t)$ of the empty space. Then,

$$C(t+1) = C(t) + rC(t)[1 - C(t)] - mC(t)$$

At equilibrium, when $C(t+1) = C(t) = C$, $1-C = m/r$. Now consider a pioneer species that does not slow the superior competitor's recruitment. When can it invade? Let $P(t)$ be the proportion of the forest's space occupied in year t by the pioneer. Each year, let the pioneer take over a proportion

$krP(t)(1 - rC)$ of the empty space, where $k > 1$, and let superior competitors replace a proportion rC of the pioneers. Then, if all pioneers die from replacement by superior competitors,

$$P(t+1) = P(t) + krP(t)(1 - rC)$$
$$\times [1 - C - P(t)] - rCP(t)$$

The pioneer invades if $P(t+1) > P(t)$ when $P(t) \approx 0$. This happens when

$$kr(1 - rC)(1 - C) > rC, k > C/[(1 - C)(1 - rC)]$$
$$= (r - m)/[m(1 - r + m)]$$

In this model, pioneers can invade if they colonize empty space over $(r - m)/[m(1 - r + m)]$ times more rapidly than superior competitors.

This theory misled ecologists in two ways. First, extrapolating this theory led Tilman (1994) to conclude that the trade-off between high k and competitive superiority would allow an indefinite number of tree species to coexist, each species j being competitively superior but having lower k than all species $i < j$. This conclusion only holds, however, if a species j replaces any competitively inferior species $i < j$ as rapidly as if the competitive inferior's space were empty. If the competitive advantage of a species j over a species $i < j$ decreases with $j - i$, the trade-off between k and competitive ability allows only a few species to coexist (Adler and Mosquera 2000). In fact, gap-creating disturbance cannot explain diversity gradients in tropical forests (Condit et al. 2006). Pasoh Reserve, Malaysia, is much more diverse than Barro Colorado, with 815 species among the 335,000 stems ≥ 1 cm dbh on 50 ha, compared with Barro Colorado's 305 species among 235,000 such stems on 50 ha (Condit et al. 1999). At Pasoh, however, gaps are much smaller, less frequent, and less varied in size (Putz and Appanah 1987). Accordingly, the Pasoh plot has only three pioneer species with sapling diameter increase averaging more than 4 mm year^{-1} among its 422 species of canopy tree, compared with 16 of 141 on Barro Colorado's plot (Condit et al. 1999).

Tilman's (1994) theory also led him to conclude that coexistence between pioneers and mature forest tree species was driven by a trade-off between colonizing ability and competitive superiority. True, to survive, pioneer species must convey seeds to gaps soon after they open, or have them waiting in the soil. Most gaps, however, are already occupied by seedlings and saplings of superior competitors: pioneers must outgrow them and reproduce before being crowded out. Therefore, the principal factor allowing pioneers to coexist with superior competitors is not a competition–colonization trade-off, but the trade-off between growing fast in bright light and surviving in shade (Brokaw 1987, King 1994, Kitajima 1994). This trade-off affects all aspects of tree life. More light-demanding species tend to have more of their seedlings in gaps, higher mortality rates in their seedlings and saplings, lower sapling wood density, higher growth rates, and fewer saplings per reproductive adult than more shade-tolerant counterparts (Wright *et al.* 2003). Sun leaves have high photosynthetic capacity, which inflates respiratory costs but permits abundant photosynthesis in bright light; shade leaves with little opportunity for rapid photosynthesis reduce respiratory costs by maintaining low photosynthetic capacity. Trees with multilayer crowns designed to spread light over as much leaf surface as possible grow faster in bright light, whereas "monolayer" crowns designed to concentrate as much light as possible on a single layer of leaves allows herbs and saplings to survive better in shade (Horn 1971). Lower investment in antiherbivore defense allows pioneers, *Cecropia*, to maintain a dry matter production per unit leaf area double that of shade-tolerant competitors, but they need more light than shade-tolerants to be able to replace their short-lived leaves before herbivores eat them (King 1994). Sacrificing durability by making wood with density only 25% of the forest-wide average translates *Cecropia*'s dry matter production into height growth no shade-tolerant competitor can match (King 1994).

Obstacles to theory: pest pressure and tree diversity

Effects on tree diversity of the trade-offs plants face in resisting different pests and pathogens (and the corresponding trade-offs pests and pathogens face in attacking different plants) are of central interest to theorists. If "the jack of all trades is master of none" (MacArthur 1961), specialist pests and pathogens should inflict the most damage. Even in the tropics, generalist caterpillars of some species, such as *Hylesia lineata* (Saturniidae) can defoliate whole trees (Janzen 1984). Nonetheless, tropical plants suffer most from species- or genus-specific pests (Janzen 1988, Novotny *et al.* 2002). At all latitudes, diverse forests and tree plantations suffer less from herbivores, especially relatively specialized herbivores, than single-species stands. Moreover, a particular species is less damaged by specialist pests in plots where it is rarer (Jactel and Brockerhoff 2007). Pest pressure is more intense in the tropics, where no winter knocks back pest populations (Janzen 1970), and where caterpillars, the most damaging insect pests, are more specialized (Scriber 1973, Dyer *et al.* 2007). Young tropical leaves are therefore far more rapidly eaten, despite being far more poisonous, than temperate-zone dicot counterparts. Gillett (1962), Janzen (1970) and Connell (1971) therefore proposed that there are so many kinds of tropical plants because specialized pests keep each species rare enough to make room for many others. This idea unifies a great variety of data. But is it true? This question still arouses vigorous argument (Leigh *et al.* 2004, Leigh 2007). Can mathematical theory help resolve it?

A mathematical theory of pest pressure and tree diversity must be based on two propositions:

1 If a tree species is rare enough, the abundance of consumers specialist upon it declines, whereas if these consumers are abundant enough, the density of their host trees declines.

2 A forest's trees are also limited by light and by suitable space in which to grow.

To see how pest pressure might influence tree diversity, consider a community of n tree species, each with its own species of specialized pest. Let the biomass per unit area at time t of tree species i and its specialist consumer be $N_i(t)$ and $C_i(t)$, respectively. Let the total density of trees $N(t)$ be $N_1(t) + N_2(t) + N_3(t) + \cdots + N_n(t)$, where $1 \leq i \leq n$. Let r_i be the per capita rate of increase of tree species i when all trees are rare, let consumer species i diminish the population growth rate dN_i/dt of tree species i by the amount $a_iN_iC_i$, let total tree biomass N diminish per capita growth $d \ln N_i/dt$ of every species i by an amount bN, let m be the per capita mortality for each consumer species, and let $\lambda_ia_iN_i$ be the per capita birth rate of consumer species i. Then,

$$dN_i/dt = r_iN_i - a_iN_iC_i - bN_iN \qquad (8.1a)$$

$$dC_i/dt = -mC_i + \lambda_ia_iN_iC_i$$

$$= \lambda_ia_iC_i(N_i - m/\lambda_ia_i) \qquad (8.1b)$$

As postulated, consumer species i declines if its host tree's biomass falls below m/λ_ia_i and the consumers no longer encounter their hosts often enough to maintain themselves, whereas if $a_iC_i > r - bN$, overabundant consumers cause the population of tree species i to decline. Experiments with microorganisms show that a consumer population declines when the abundance of prey is below a critical threshold, while the abundance of prey declines when their consumers are too abundant (Maly 1969, 1978).

At equilibrium, when $dC_i/dt = dN_i/dt = 0$, $N_i = m/\lambda_ia_i$, the tree's population is sufficient to support its consumer. Here, lower λ_ia_i – improved defense in tree species i – increases N_i and reduces the proportion of these trees killed by consumers. To see how pest pressure influences tree diversity, set $r_i = r$, $a_i = a$, and $\lambda_i = \lambda$ for all tree species and their consumers. Then $C_i > 0$ if $r - bN = r - bnm/\lambda a > 0$ for all consumer species i. The more intense the pressure λa on each tree species from specialized pests, the more tree species can coexist.

In fact, pests are far smaller than their host trees. The population of tree species i should be modeled as a set of islands in a sea of trees of other species, with satellite islets representing young appearing and disappearing among larger islands representing adults. Consumers finding a tree of their host species colonize it, and their descendants disperse in search of other members of this species, young and adult. Three factors influence a pest's pressure on its hosts (Webb and Peart 1999). The proportion of infested adults is lower where a species is rarer; pests in an infested adult are more likely to find and damage nearby young of this species; and a seedling's

pests spread more readily to conspecific seedlings if they are nearer. Modeling these processes, however, requires a dynamics of spatial arrangement, which we have not yet got. Therefore, we cannot predict the precise relationship between the strength of these influences and the tree diversity they support.

The pest pressure hypothesis has been tested primarily by asking whether trees recruit or grow more slowly, or die faster, when closer to conspecifics (Hubbell *et al.* 2001, Peters 2003, Wills *et al.* 2006), and whether seeds germinate less frequently, and seedlings die faster, where they have more conspecific neighbors (Harms *et al.* 2000). Pest pressure, however, is not the only explanation for these patterns (Wills *et al.* 2006).

Theory, pest pressure, and the ecology and evolution of tropical forest

Pests and pathogens exert a pervasive influence on tropical forest. Herbivores influence the timing of leaf flush and fruit fall. In south India, where the dry season lasts 6 months, canopy trees flush leaves before the rains come, while insects are still rare, reducing herbivore damage (Murali and Sukumar 1993). Many tree species in the great dipterocarp forests of Southeast Asia fruit in synchrony every few years: the long intervening periods of fruit scarcity depress populations of seed predators, allowing many of the seeds produced during fruiting peaks to escape being eaten (Janzen 1974, Chan 1980).

Pests and pathogens also drive the trade-offs trees face between growing fast on clay soil versus surviving on white sand (Fine *et al.* 2004), and growing fast in bright light versus surviving in shade (Kitajima 1994). Slow-growing plants must be well defended, whereas if rapid growth is possible, plants can "outrun" herbivores rather than deploy costly defenses (Coley *et al.* 1985).

The central importance of herbivores is revealed by their role in driving the evolution and spread of flowering plants. Unlike grasslands, which depend on herbivores to exclude woody competitors, trees are adapted to reduce herbivory (McNaughton 1985). The last section's theory suggests that effective anti-herbivore defense and rarity are the two ways a tree species can reduce consumption by, and mortality from, pests and pathogens. Jurassic forests were dominated by conifers and other wind-pollinated plants (Corner 1964). Wind pollination works only for plants close to conspecifics (Regal 1977, Davis *et al.* 2004), so wind-pollinated trees must invest heavily in anti-herbivore defense. Modern wind-pollinated conifers have tough, long-lived leaves whose toxins poison the soil when the leaves fall (Northup *et al.* 1995, Reich *et al.* 1995, pp. 28–29). Rare trees can survive only by attracting animals that convey their pollen to distant conspecifics. Angiosperms began in the tropics (Wing and Boucher 1998), where pest pressure is most intense (Coley and Barone 1996). Lineages with flowers attracting pollinators willing to seek floral rewards from distant conspecifics diversified extensively (Crepet 1984). The ability to enlist animals as faithful pollinators (Crepet 1984) and as dispersers of large seeds (Wing *et al.* 1993, Tiffney and Mazer 1995) enabled a diverse set of rare, fast-growing flowering trees to replace a less diverse array of common, better-defended, slower-growing gymnosperms (Regal 1977). Opting for animal pollination and seed dispersal triggered the evolution of flowering rainforest whose diversity reduced the depredations of pests with less loss of productivity.

The theory just presented predicts that the evolution of a diverse forest of rare trees would enhance forest productivity. Extending this theory suggests that where a tropical forest's productivity is higher, predators contribute more to its anti-herbivore defense (Oksanen *et al.* 1981). To see this, let the biomass per hectare at time t of plants, consumers (herbivores), and predators be $N(t)$, $C(t)$, and $P(t)$, and set

$$\frac{d \ln N}{dt} = r - aC - bN = b\left(\frac{r}{b} - \frac{aC}{b} - N\right)$$

$$\frac{d \ln C}{dt} = -m + \lambda aN - a'P$$

$$\frac{d \ln P}{dt} = -m' + \lambda'a'C = \lambda'a'\left(C - \frac{m'}{\lambda'a'}\right)$$

Here r, a, b, m, and λ have the meanings of r_i, a_i, b, m, and λ in Equations 8.1, $a'P$ is the

decrease from predation of the per capita increase of herbivores, and m' and $\lambda'a'C$ are the predators' per capita death and birth rates. Predators can invade only if the consumer abundance C exceeds the density $m'/\lambda'a'$ needed to maintain predator numbers. If $m/\lambda a < r/b$, then, at the equilibrium with predators absent ($P = 0$), $N = m/\lambda a$ and $C = (r - bm/\lambda a)/a = (b/a)(r/b - m/\lambda a)$. More productive plant populations have higher r. If r is so high that $(b/a)(r/b - m/\lambda a) > m'/\lambda'a'$, predators can invade. Then the equilibrium becomes

$$C = \frac{m'}{\lambda'a'} < \frac{b}{a}\left(\frac{r}{b} - \frac{m}{\lambda a}\right), N = \left(\frac{r}{b} - \frac{a'm}{\lambda'a'b}\right),$$

$$P = \left[\lambda a\left(\frac{r}{b} - \frac{a'm}{\lambda'a'b}\right) - m\right]\bigg/a'$$

This theory predicts that if predators are removed, consumer abundance C increases more when r, and therefore plant productivity, is higher. Moreover, increased r increases plant biomass N and predator abundance P, leaving consumer abundance C unchanged. Wootton and Power (1993) tested the latter prediction in river-bottom enclosures. Here, algae grew on the rocky bottom, snails, mayfly nymphs and the like grazed the algae, and sticklebacks, *Gasterosteus* and dragonfly nymphs ate the grazers. Different levels of shade were imposed on these enclosures to create differences in algal productivity. As predicted, increased light increased algal and predator, but not herbivore, biomass (Wootton and Power 1993).

In dry forest canopy in Panama, excluding birds during the productive season of leaf flush increases insect populations and leaf damage, as this theory predicts. Excluding birds from canopy branches in wetter, more evergreen forest, or from understory plants in either forest, where leaves flush at a low rate all year long, does not increase insect abundance or leaf damage, perhaps because arthropods in the exclusion zone "take up the slack" (van Bael and Brawn 2005). Finally, in wet and dry forest, insects are equally common in canopy accessible to birds, even though insect-eating birds are more common in dry forest canopy, especially in the season of leaf flush (van Bael and Brawn 2005), in accord with the prediction that, where predators are present, insect abundance should not increase with plant productivity.

How much do tropical forests depend on predators for protection from herbivores? Birds help defend other tropical forests from insects. On Barro Colorado Island, over a third of the foliage eaten by insects that birds eat, and this appears true at other sites as well (Leigh 1999, pp. 167–168). More direct evidence comes from fragmentation of forest by reservoirs (whose impact is not confounded with the effects of fire, cows, or invading pioneer trees), which offers forest ecologists their nearest equivalent to the marine ecologist's exclusion experiments (Leigh *et al.* 2002). Fragmentation causes extinctions, especially of predators, releasing prey populations, with effects that cascade through the fragments' communities (Terborgh *et al.* 2001). When the Guri reservoir fragmented dry forest in Venezuela, leaf-cutter ants exploded on newly isolated islets (Terborgh *et al.* 1997). Islets less than 1 ha carried up to six mature leaf-cutter colonies ha^{-1}, whereas islands greater than 80 ha, with a full complement of mammals, carried two colonies per 3 ha (Rao 2000). As leaf-cutters on large islands have access to more preferred trees (Terborgh *et al.* 2006), they can hardly be limited by seasonal shortage of food: they must be controlled by some predator or pathogen (Terborgh *et al.* 2001). Army ants and armadillos, now extinct from small islets, limit recruitment (Rao 2000) and perhaps survival (Swartz 1998), of leaf-cutter ants on large islands and the mainland. The explosion of leaf-cutters on small islets severely reduced recruitment of trees, especially canopy trees, and favored well-defended species (Terborgh *et al.* 2001, 2006, Rao *et al.* 2002).

In sum, this crude theory suggests that employing animals as pollinators and seed dispersers, which allows plants to escape specialist pests through being rare, could enable a more diverse and productive flowering forest to replace a less productive, less diverse, better-defended, wind-pollinated gymnosperm forest. Extending this theory shows how the higher productivity of flowering forest enables predators to assume a greater role in its anti-herbivore defense. This theory brings home the lesson that we cannot understand tropical forest if we ignore animals.

CONCLUSIONS

In sum, mathematical theory related to forest structure and production has accomplished various useful "odd jobs." Simple theory suggests how soil quality affects forest dynamics, the hydraulic architecture and wood density of its trees, characteristics of their leaves, and the apportionment of resources between above- and below-ground activities, without altering forest production very much. Very crude theory shows how a tree species's physiology and life history are related to its response to the trade-off between growing fast in bright light and surviving in shade. Theory also shows how forest productivity is reduced by how trees compete for light and nutrients. Despite the efforts of Enquist (2002), however, a comprehensive theory of tree shapes and forest structure and production seems far off.

Mathematical theory in ecology has largely been concerned with the maintenance of species diversity (MacArthur 1972). How species arise has attracted much less interest, even though the process of speciation might reveal much about how species coexist.

Testing a null hypothesis derived from Hubbell's neutral theory of forest ecology suggests that we cannot understand why there are so many kinds of tropical trees without knowing how the differences among tree species allow them to coexist.

Are trees so much more diverse in the tropics, where no winter depresses populations of pests and pathogens, because each species there is kept rare enough by specialist pests and pathogens to make room for many others (Janzen 1970, Connell 1971)? This hypothesis is plausible. Crude theory, and some evidence, suggests that the more diverse a forest, the less its species suffer from specialist pests and the faster they can grow. Testing the pest pressure hypothesis of tree diversity is crucial to understanding how tree diversity affects tropical forest productivity, but it has been difficult to put to a decisive test. Because a tree's fate is governed mostly by its relationships with near neighbors, a mathematical theory of tree diversity based on a dynamics of the spatial arrangements of trees is needed to predict useful relationships between tree diversity and the pattern of pest pressure.

ACKNOWLEDGMENTS

I am most grateful to a variety of seminar audiences for encouragement in this enterprise, to David King for reading this chapter and warning me away from various traps for the unwary, to Walter Carson for working so hard to make this chapter readable, to the librarians of the Smithsonian Tropical Research Institute, especially Vielka Chang-Yau and Angel Aguirre, for procuring me a host of needed documents, and to the plants and animals of Barro Colorado Island for reminding me of what mathematical theory should try to explain.

REFERENCES

Adler, F.R. and Mosquera, J. (2000) Is space necessary? Interference competition and limits to biodiversity. *Ecology* 81, 3226–3232.

Baker, T.R., Phillips, O.L., Malhi, Y. *et al.* (2004a) Increasing biomass in Amazon forest plots. *Philosophical Transactions of the Royal Society of London B* 359, 353–365.

Baker, T.R., Phillips, O.L., Malhi, Y. *et al.* (2004b) Variation in wood density determines spatial patterns in Amazonian forest biomass. *Global Change Biology* 10, 545–562.

Becker, P. (1996) Sap flow in Bornean heath and dipterocarp forest trees during wet and dry periods. *Tree Physiology* 16, 295–299.

Binkley, D. (2004) A hypothesis about the interaction of tree dominance and stand production through stand development. *Forest Ecology and Management* 190, 265–271.

Bossel, H. and Krieger, H. (1991) Simulation model of natural tropical forest dynamics. *Ecological Modelling* 59, 37–71.

Bramson, M., Cox, J.T., and Durrett, R. (1996) Spatial models for species area curves. *Annals of Probability* 24, 1727–1751.

Bramson, M., Cox, J.T., and Durrett, R. (1998) A spatial model for the abundance of species. *Annals of Probability* 26, 658–709.

Brodribb, T.J. and Field, T.S. (2000) Stem hydraulic supply is linked to leaf photosynthetic capacity: evidence from New Caledonian and Tasmanian rainforests. *Plant, Cell and Environment* 23, 1381–1388.

Brodribb, T.J., Holbrook, N.M., and Gutiérrez, M.V. (2002) Hydraulic and photosynthetic co-ordination in seasonally dry tropical forest trees. *Plant, Cell and Environment* 25, 1435–1444.

Brokaw, N.V.L. (1987) Gap-phase regeneration of three pioneer tree species in a tropical forest. *Journal of Ecology* 75, 9–19.

Bruenig, E.F. (1996) *Conservation and Management of Tropical Rainforests*. CAB International, Wallingford.

Brunig, E.F. (1970) Stand structure, physiognomy and environmental factors in some lowland forests in Sarawak. *Tropical Ecology* 11, 26–43.

Brunig, E.F. (1983) Vegetation structure and growth. In F.B. Golley (ed.), *Tropical Rain Forest Ecosystems*. Elsevier, Amsterdam, pp. 49–75.

Chan, H.T. (1980) Reproductive biology of some Malaysian dipterocarps II. Fruiting biology and seedling studies. *Malaysian Forester* 43, 438–450.

Chanderbali, A.S., van der Werff, H., and Renner, S.S. (2001) Phylogeny and historical biogeography of Lauraceae: evidence from the chloroplast and nuclear genomes. *Annals of the Missouri Botanical Garden* 88, 104–134.

Chave, J. and Leigh, E.G. Jr. (2002) A spatially explicit neutral model of β-diversity in tropical forests. *Theoretical Population Biology* 62, 153–168.

Coley, P.D. (1983) Herbivory and defensive characteristics of tree species in a lowland tropical forest. *Ecological Monographs* 53, 209–233.

Coley, P.D. and Barone, J.A. (1996) Herbivory and plant defenses in tropical forests. *Annual Review of Ecology and Systematics* 27, 305–335.

Coley, P.D., Bryant, J.P., and Chapin, III, F.S. (1985) Resource availability and plant anti-herbivore defense. *Science* 230, 895–899.

Condit, R., Ashton, P., Bunyavejchewin, S. *et al.* (2006) The importance of demographic niches to tree diversity. *Science* 313, 98–101.

Condit, R., Ashton, P.S., Manokaran, N. *et al.* (1999) Dynamics of the forest communities at Pasoh and Barro Colorado: comparing two 50-ha plots. *Philosophical Transactions of the Royal Society of London B* 354, 1739–1748.

Condit, R., Leigh, E.G. Jr., Loo de Lao, S. *et al.* (2004) Species–area relationships and diversity measures in the forest dynamics plots. In E.C. Losos and E.G. Leigh, Jr. (eds), *Tropical Forest Diversity and Dynamism: Findings from a Large-Scale Plot Network*. University of Chicago Press, Chicago, pp. 79–89.

Connell, J.H. (1971) On the role of natural enemies in preventing competitive exclusion in some marine animals and in rain forest trees. In P.J. den Boer and G.R. Gradwell (eds), *Dynamics of Populations*. Centre for Agricultural Publication and Documentation, Wageningen, The Netherlands, pp. 298–312.

Coomes, D.A. and Grubb, P.J. (1998) Responses of juvenile trees to above- and belowground competition in nutrient-starved Amazonian rain forest. *Ecology* 79, 768–782.

Corner, E.J.H. (1964) *The Life of Plants*. World Publishing, Cleveland, OH.

Coyne, J.A. and Orr, H.A. (2004) *Speciation*. Sinauer Associates, Sunderland, MA.

Crepet, W.L. (1984) Advanced (Constant) insect pollination mechanisms: patterns of evolution and implications vis-à-vis angiosperm diversity. *Annals of the Missouri Botanical Garden* 71, 607–630.

Crow, J.F. and Kimura, M. (1970) *An Introduction to Population Genetics Theory*. Harper & Row, New York.

Davis, H.G., Taylor, C.M., Lambrinos, J.G. *et al.* (2004) Pollen limitation causes an Allee effect in a wind-pollinated invasive grass (*Spartina alterniflora*). *Proceedings of the National Academy of Sciences of the United States of America* 101, 13804–13807.

Dean, T.J. and Long, J.N. (1986) Validity of constant-stress and elastic-instability principles of stem formation in *Pinus contorta* and *Trifolium pratense*. *Annals of Botany* 58, 833–840.

Denny, M.W. (1999) Are there mechanical limits to size in wave-swept organisms? *Journal of Experimental Biology* 202, 3463–3467.

Dick, C.W., Abdul-Salim, K., and Bermingham, E. (2003) Molecular systematic analysis reveals cryptic Tertiary diversification of a widespread tropical rain forest tree. *American Naturalist* 162, 691–703.

Dyer, L.A., Singer, M.S., Lill, J.T. *et al.* (2007) Host specificity of Lepidoptera in tropical and temperate forests. *Nature* 448, 696–699.

Ehrendorfer, F. (1982) Speciation patterns in woody angiosperms of tropical origin. In C. Barrigozzi (ed.), *Mechanisms of Speciation*. Liss, New York, pp. 479–509.

Enquist, B.J. (2002) Universal scaling in tree and vascular plant allometry: toward a general quantitative theory linking plant form and function from cells to ecosystems. *Tree Physiology* 22, 1045–1064.

Enquist, B.J., Brown, J.H., and West, G.B. (1998) Allometric scaling of plant energetics and population density. *Nature* 395, 163–165.

Enquist, B.J. and Niklas, K.J. (2001) Invariant scaling relations across tree-dominated communities. *Nature* 410, 655–660.

Enquist, B.J., West, G.B., Charnov, E.L. *et al.* (1999) Allometric scaling of production and life-history variation in vascular plants. *Nature* 401, 907–911.

Falster, D.S. and Westoby, M. (2003) Plant height and evolutionary games. *Trends in Ecology and Evolution* 18, 337–343.

Fine, P.V.A., Mesones, I., and Coley, P.D. (2004) Herbivores promote habitat specialization by trees in Amazonian forests. *Science* 305, 663–665.

Fisher, R.A. (1930) *The Genetical Theory of Natural Selection*. Clarendon Press, Oxford.

Gartner, B.L. and Meinzer, F.C. (2005) Structure-function relationships in sapwood water transport and storage. In N.M. Holbrook and M.A. Zwieniecki (eds), *Vascular Transport in Plants*. Elsevier, Amsterdam, pp. 307–331.

Gentry, A.H. (1989) Speciation in tropical forests. In L.B. Holm-Nielsen, I.C. Nielsen, and H. Balslev (eds), *Tropical Forests: Botanical Dynamics, Speciation and Diversity*. Academic Press, London, pp. 113–134.

Gersani, M., Brown, J.S., O'Brien, E.E., Maina, G.M., and Abramsky, Z. (2001) Tragedy of the commons as a result of root competition. *Journal of Ecology* 89, 660–669.

Giardina, C.P., Ryan, M.G., Binkley, D. *et al.* (2003) Primary production and carbon allocation in relation to nutrient supply in a tropical experimental forest. *Global Change Biology* 9, 1438–1450.

Gillett, J.B. (1962) Pest pressure, an underestimated factor in evolution. *Systematics Association Publication Number* 4, 37–46.

Givnish, T.J. (1984) Leaf and canopy adaptations in tropical forests. In E. Medina, H.A. Mooney, and C. Vasquez-Yanes (eds), *Physiological Ecology of Plants of the Wet Tropics*. W. Junk, the Hague, the Netherlands, pp. 51–84.

Givnish, T.J. (1987) Comparative studies of leaf form: assessing the relative roles of selective pressures and phylogenetic constraints. *New Phytologist* 106(Suppl.), 131–160.

Givnish, T.J. (1988) Adaptation to sun and shade: a whole-plant perspective. *Australian Journal of Plant Physiology* 15, 63–92.

Hacke, U.G., Sperry, J.S., and Pittermann, J. (2005) Efficiency versus safety tradeoffs for water conduction in angiosperm vessels vs gymnosperm tracheids. In N.M. Holbrook and M.A. Zwieniecki (eds), *Vascular Transport in Plants*. Elsevier, Amsterdam, pp. 333–353.

Hallé, F. and Oldeman, R.A.A. (1970) *Essai sur l'Architecture et la Dynamique de Croissance des Arbres Tropicaux*. Masson et Cie, Paris.

Hallé, F., Oldeman, R.A.A., and Tomlinson, P.B. (1978) *Tropical Trees and Forests: An Architectural Analysis.* Springer-Verlag, Berlin.

Hardin, G. (1968) The tragedy of the commons. *Science* 162, 1243–1248.

Harms, K.E., Wright, S.J., Calderón, O. *et al.* (2000) Pervasive density-dependent recruitment enhances seedling diversity in a tropical forest. *Nature* 404, 493–495.

Hoffmann, W.A., Orthen, B., and Franco, A.C. (2004) Constraints to seedling success of savanna and forest trees across the savanna-forest boundary. *Oecologia* 140, 252–260.

Holbrook, N.M., Denny, M.W., and Koehl, M.A.R. (1991) Intertidal "trees": consequences of aggregation on the mechanical and photosynthetic properties of sea-palms *Postelsia palmaeformis* Ruprecht. *Journal of Experimental Marine Biology and Ecology* 146, 39–67.

Horn, H.S. (1971) *The Adaptive Geometry of Trees.* Princeton University Press, Princeton.

Horn, H.S. and MacArthur, R.H. (1972) Competition among fugitive species in a harlequin environment. *Ecology* 53, 749–752.

Hubbell, S.P. (2001) *The Unified Neutral Theory of Biodiversity and Biogeography.* Princeton University Press, Princeton.

Hubbell, S.P., Ahumada, J.A., Condit, R. *et al.* (2001) Local neighborhood effects on long-term survival of individual trees in a neotropical forest. *Ecological Research* 16, 859–875.

Iwasa, Y. and Cohen, D. (1989) Optimal growth schedule of a perennial plant. *American Naturalist* 133, 480–505.

Iwasa, Y., Cohen, D., and Leon, J.A. (1984) Tree height and crown shape, as results of competitive games. *Journal of Theoretical Biology* 112, 279–297.

Jactel, H. and Brockerhoff, E.G. (2007) Tree diversity reduces herbivory by forest insects. *Ecology Letters* 10, 835–848.

Janzen, D.H. (1970) Herbivores and the number of tree species in tropical forests. *American Naturalist* 104, 501–528.

Janzen, D.H. (1974) Tropical blackwater rivers, animals, and mast fruiting by the Dipterocarpaceae. *Biotropica* 6, 69–103.

Janzen, D.H. (1984) Natural history of *Hylesia lineata* (Saturniidae: Hemileucinae) in Santa Rosa National Park, Costa Rica, *Journal of the Kansas Entomological Society* 57, 490–514.

Janzen, D.H. (1988) Ecological characterization of a Costa Rican dry forest caterpillar fauna. *Biotropica* 20, 120–135.

Kato, R., Tadaki, Y., and Ogawa, H. (1978) Plant biomass and growth increment studies in Pasoh Forest. *Malayan Nature Journal* 30, 211–224.

Katul, G., Leuning, R., and Oren, R. (2003) Relationship between plant hydraulic and biochemical properties derived from a steady-state coupled water and carbon transport model. *Plant, Cell and Environment* 26, 339–350.

Kay, K. (2002) Evidence for reinforcement of speciation in pollinator-sharing Neotropical *Costus*. In *Tropical Forests: Past, Present, Future (Programs and Abstracts).* Association for Tropical Biology Annual Meeting, Panama, p. 54.

Kay, K.M. and Schemske, D.W. (2003) Pollinator assemblages and visitation rates for 11 species of Neotropical *Costus* (Costaceae). *Biotropica* 35, 198–207.

Keyes, M.R. and Grier, C.C. (1981) Above- and belowground net production in 40-year-old Douglas-fir stands on low and high productivity sites. *Canadian Journal of Forest Research* 11, 599–605.

King, D.A. (1986) Tree form, height growth, and susceptibility to wind damage in *Acer saccharum*. *Ecology* 67, 980–990.

King, D.A. (1990) The adaptive significance of tree height. *American Naturalist* 135, 809–828.

King, D.A. (1993) A model analysis of the influence of root and foliage allocation on forest production and competition between trees. *Tree Physiology* 12, 119–135.

King, D.A. (1994) Influence of light level on the growth and morphology of saplings in a Panamanian forest. *American Journal of Botany* 81, 948–957.

King, D.A. (1996) Allometry and life history of tropical trees. *Journal of Tropical Ecology* 12, 25–44.

King, D.A. and Loucks, O.L. (1978) The theory of tree bole and branch form. *Radiation and Environmental Biophysics* 15, 141–165.

Kira, T. (1978) Community architecture and organic matter dynamics in tropical lowland rain forests of Southeast Asia with special reference to Pasoh Forest, West Malaysia. In P.B. Tomlinson and M.H. Zimmermann (eds), *Tropical Trees as Living Systems.* Cambridge University Press, New York, pp. 561–590.

Kitajima, K. (1994) Relative importance of photosynthetic traits and allocation patterns as correlates of seedling shade tolerance of 13 tropical trees. *Oecologia* 98, 419–428.

Koch, G.W., Sillett, S.C., Jennings, G.M. *et al.* (2004) The limits to tree height. *Nature* 428, 851–854.

Kohyama, T., Suzuki, E., Partomihardjo, T. *et al.* (2003) Tree species differentiation in growth, recruitment and allometry in relation to maximum height in a Bornean mixed dipterocarp forest. *Journal of Ecology* 91, 797–806.

Leigh, E.G. Jr. (1975) Structure and climate in tropical rain forest. *Annual Review of Ecology and Systematics* 6, 67–86.

Leigh, E.G. Jr. (1994) Do insect pests promote mutualism among tropical trees? *Journal of Ecology* 82, 677–680.

Leigh, E.G. Jr. (1999) *Tropical Forest Ecology*. Oxford University Press, New York.

Leigh, E.G. Jr. (2004) How wet are the wet tropics? In E.C. Losos and E.G. Leigh, Jr. (eds), *Tropical Forest Diversity and Dynamism: Findings from a Large-Scale Plot Network*. University of Chicago Press, Chicago, pp. 43–55.

Leigh, E.G. Jr. (2007) Neutral theory: a historical perspective. *Journal of Evolutionary Biology* 20, 2075–2091.

Leigh, E.G. Jr., Cosson, J.-F., Pons, J.-M. *et al.* (2002) En quoi l'étude des îlots forestiers permet-elle de mieux connaître le fonctionnement de la forêt tropicale? *Revue d'Écologie (La Terre et la Vie)* 57, 181–194.

Leigh, E.G. Jr., Davidar, P., Dick, C.W. *et al.* (2004) Why do some tropical forests have so many species of trees? *Biotropica* 36, 447–473.

Leigh, E.G. Jr., Paine, R.T., Quinn, J.F. *et al.* (1987) Wave energy and intertidal productivity. *Proceedings of the National Academy of Sciences of the United States of America* 84, 1314–1318.

Lewis, S.L., Phillips, O.L., Baker, T.R. *et al.* (2004) Concerted changes in tropical forest structure and dynamics: evidence from 50 South American long-term plots. *Philosophical Transactions of the Royal Society of London B* 359, 421–436.

Loescher, H.W., Oberbauer, S.F., Gholz, H.L. *et al.* (2003) Environmental controls on net ecosystem-level carbon exchange and productivity in a Central American tropical forest. *Global Change Biology* 9, 396–412.

Long, J.N., Smith, F.W., and Scott, D.R.M. (1981) The role of Douglas-fir stem sapwood and heartwood in the mechanical and physiological support of crowns and development of stem form. *Canadian Journal of Forest Research* 11, 459–464.

Losos, E.C. and the CTFS working group. (2004) The structure of tropical forests. In E. Losos and E.G. Leigh, Jr. (eds), *Tropical Forest Diversity and Dynamism*. University of Chicago Press, Chicago, pp. 69–78.

MacArthur, R.H. (1961) Population effects of natural selection. *American Naturalist* 95, 195–199.

MacArthur, R.H. (1972) *Geographical Ecology*. Harper & Row, New York.

Malhi, Y., Baker, T.R., Phillips, O.L. *et al.* (2004) The above-ground coarse wood productivity of 104 Neotropical forest plots. *Global Change Biology* 10, 563–591.

Malhi, Y., Baldocchi, D.D., and Jarvis, P.G. (1999) The carbon balance of tropical, temperate and boreal forests. *Plant, Cell and Environment* 22, 715–740.

Maly, E.J. (1969) A laboratory study of the interaction between the predatory rotifer *Asplanchna* and *Paramecium*. *Ecology* 50, 59–73.

Maly, E.J. (1978) Stability of the interaction between *Didinium* and *Paramecium*: effects of dispersal and predator time lag. *Ecology* 59, 733–741.

McMahon, T.A. (1973) Size and shape in biology. *Science* 179, 1201–1204.

McMahon, T.A. and Kronauer, R.E. (1976) Tree structures: deducing the principles of mechanical design. *Journal of Theoretical Biology* 59, 443–466.

McNaughton, S.J. (1985) Ecology of a grazing ecosystem: the Serengeti. *Ecological Monographs* 55, 259–294.

Medina, E. and Cuevas, E. (1989) Patterns of nutrient accumulation and release in Amazonian forests of the upper Rio Negro basin. In J. Proctor (ed.), *Mineral Nutrients in Tropical Forest and Savanna Ecosystems*. Blackwell Scientific, Oxford, pp. 217–240.

Medina, E., Garcia, V., and Cuevas, E. (1990) Sclerophylly and oligotrophic environments: relationships between leaf structure, mineral nutrient content, and drought resistance in tropical rain forests of the upper Rio Negro region. *Biotropica* 22, 51–64.

Meinzer, F.C., Goldstein, G., Holbrook, N.M. *et al.* (1993) Stomatal and environmental control of transpiration in a lowland tropical forest tree. *Plant, Cell and Environment* 16, 429–436.

Muller-Landau, H.C., Condit, R.S., Chave, J. *et al.* (2006a) Testing metabolic ecology theory for allometric scaling of tree size, growth and mortality in tropical forests. *Ecology Letters* 9, 575–588.

Muller-Landau, H.C., Condit, R.S., Harms, K.E. *et al.* (2006b) Comparing tropical forest tree size distributions with the predictions of metabolic ecology and equilibrium models. *Ecology Letters* 9, 589–602.

Murali, K.S. and Sukumar, R. (1993) Leaf flushing phenology and herbivory in a tropical dry deciduous forest, southern India. *Oecologia* 94, 114–119.

Niklas, K.J. and Enquist, B.J. (2001) Invariant scaling relationships for interspecific plant biomass production rates and body size. *Proceedings of the National*

Academy of Sciences of the United States of America 98, 2922–2927.

Northup, R.R., Yu, Z., Dahlgren, R.A. *et al.* (1995) Polyphenol control of nitrogen release from pine litter. *Nature* 377, 227–230.

Novotny, V., Miller, S.E., Basset, Y. *et al.* (2002) Predictably simple: assemblages of caterpillars (Lepidoptera) feeding on rainforest trees in Papua New Guinea. *Proceedings of the Royal Society of London B* 269, 2337–2344.

Oksanen, L., Fretwell, S.D., Arruda, J. *et al.* (1981) Exploitation ecosystems in gradients of primary productivity. *American Naturalist* 118, 240–261.

Oren, R., Sperry, J.S., Katul, G.G. *et al.* (1999) Survey and synthesis of intra- and interspecific variation in stomatal sensitivity to vapor pressure deficit. *Plant, Cell and Environment* 22, 1515–1526.

Pacala, S.W., Canham, C.D., Saponara, J. *et al.* (1996) Forest models defined by field measurements: estimation, error analysis and dynamics. *Ecological Monographs* 66, 1–44.

Paine, R.T. (1979) Disaster, catastrophe, and local persistence of the sea palm *Postelsia palmaeformis*. *Science* 205, 685–687.

Paine, R.T. (1988) Habitat suitability and local population persistence of the sea palm *Postelsia palmaeformis*. *Ecology* 69, 1787–1794.

Pennington, R.T. and Dick, C.W. (2004) The role of immigrants in the assembly of the South American rainforest tree flora. *Philosophical Transactions of the Royal Society of London B* 359, 1611–1622.

Pennington, R.T., Lavin, M., Prado, D.E. *et al.* (2004) Historical climate change and speciation: neotropically seasonally dry forest plants show patterns of both Tertiary and Quaternary diversification. *Philosophical Transactions of the Royal Society of London B* 359, 515–538.

Peters, H.A. (2003) Neighbour-related mortality: the influence of positive and negative density dependence on tree populations in species-rich tropical forests. *Ecology Letters* 6, 757–765.

Phillips, O.L., Vásquez-Martínez, R., Monteagudo Mendoza, A. *et al.* (2005) Large lianas as hyperdynamic elements of the tropical forest canopy. *Ecology* 86, 1250–1258.

Poorter, L., Zuidema, P.A., Peña-Claros, M. *et al.* (2005) A monocarpic tree species in a polycarpic world: how can *Tachigali vasquezii* maintain itself so successfully in a tropical rain forest community? *Journal of Ecology* 93, 268–278.

Putz, F.E. and Appanah, S. (1987) Buried seeds, newly dispersed seeds, and the dynamics of a lowland forest in Malaysia. *Biotropica* 19, 326–333.

Putz, F.E., Coley, P.D., Lu, K. *et al.* (1983) Uprooting and snapping of trees: structural determinants and ecological consequences. *Canadian Journal of Forest Research* 13, 1011–1020.

Rao, M. (2000) Variation in leaf-cutter ant (*Atta* sp.) densities in forest isolates: the potential role of predation. *Journal of Tropical Ecology* 16, 209–225.

Rao, M., Terborgh, J., and Nuñez, P. (2002) Increased herbivory in forest isolates: implications for plant community structure and composition. *Conservation Biology* 15, 624–633.

Regal, P.J. (1977) Ecology and evolution of flowering plant dominance. *Science* 196, 622–629.

Reich, P.B., Kloeppel, B.D., Ellsworth, D.S. *et al.* (1995) Different photosynthesis-nitrogen relations in deciduous hardwood and evergreen coniferous tree species. *Oecologia* 104, 24–30.

Reich, P.B., Tjoelker, M.G., Machado, J.-L. *et al.* (2006) Universal scaling of respiratory metabolism, size and nitrogen in plants. *Nature* 439, 457–461.

Reich, P.B., Uhl, C., Walters, M.B. *et al.* (1991) Leaf lifespan as a determinant of leaf structure and function among 23 amazonian tree species in amazonian forest communities. *Oecologia* 86, 16–24.

Reich, P.B., Uhl, C., Walters, M.B. *et al.* (2004) Leaf demography and phenology in Amazonian rain forest: a census of 40,000 leaves of 23 tree species. *Ecological Monographs* 74, 3–23.

Riswan, S. and Kartawinata, K. (1991) Species strategy in early stage of secondary succession associated with soil properties status in a lowland mixed dipterocarp forest and kerangas forest in East Kalimantan. *Tropics* 1, 13–34.

Ryan, M.G., Hubbard, R.M., Clark, D.A. *et al.* (1994) Woody-tissue respiration for *Simarouba amara* and *Minquartia guianensis*, two tropical wet forest trees with different growth habits. *Oecologia* 100, 213–220.

Ryan, M.G. and Waring, R.H. (1992) Maintenance respiration and stand development in a subalpine lodgepole pine forest. *Ecology* 73, 2100–2108.

Ryan, M.G. and Yoder, B.J. (1997) Hydraulic limits to tree height and tree growth. *BioScience* 47, 235–242.

Santiago, L.S., Goldstein, G., Meinzer, F.C. *et al.* (2004) Leaf photosynthetic traits scale with hydraulic conductivity and wood density in Panamanian forest canopy trees. *Oecologia* 140, 543–550.

Schaffer, W.M. and Leigh, E.G. Jr. (1976) The prospective role of mathematical theory in plant ecology. *Systematic Botany* 1, 209–232.

Schnitzer, S.A. (2005) A mechanistic explanation for global patterns of liana abundance and distribution. *American Naturalist* 166, 262–276.

Schwendenmann, L.C. (2002) *Below-ground Carbon Dynamics as a Function of Climate Variability in Undisturbed Soils of a Neotropical Rain Forest.* Forschungszentrum Waldökosysteme der Universität Göttingen, Göttingen, Germany.

Scriber, J.M. (1973) Latitudinal gradients in larval feeding specialization of the world Papilionidae (Lepidoptera). *Psyche* 80, 355–373.

Stebbins, G.L. (1982) Plant speciation. In C. Barrigozzi (ed.), *Mechanisms of Speciation.* Liss, New York, pp. 21–39.

Swartz, M.B. (1998) Predation on an *Atta cephalotes* colony by an army ant, *Nomamyrmex esenbeckii.* *Biotropica* 30, 682–684.

Tanner, E.V.J. (1980a) Studies on the biomass and productivity in a series of montane rainforests in Jamaica. *Journal of Ecology* 68, 573–588.

Tanner, E.V.J. (1980b) Litterfall in montane rain forests of Jamaica and its relation to climate. *Journal of Ecology* 68, 833–848.

Terborgh, J., Feeley, K., Silman, M. *et al.* (2006) Vegetation dynamics of predator-free land-bridge islands. *Journal of Ecology* 94, 253–263.

Terborgh, J., Lopez, L., Nuñez, V.P. *et al.* (2001) Ecological meltdown in predator-free forest fragments. *Science* 294, 1923–1926.

Terborgh, J., Lopez, L., Tello, J. *et al.* (1997) Transitory states in relaxing ecosystems of land bridge islands. In W.F. Laurance and R.O. Bierregaard, Jr. (eds), *Tropical Forest Remnants.* University of Chicago Press, Chicago, pp. 256–274.

Thomas, S.C. (1996a) Asymptotic height as a predictor of growth and allometric characteristics in Malaysian rain forest trees. *American Journal of Botany* 83, 556–566.

Thomas, S.C. (1996b) Reproductive allometry in Malaysian rain forest trees: biomechanics vs. optimal allocation. *Evolutionary Ecology* 10, 517–530.

Thomas, S.C. (1996c) Relative size at onset of maturity in rain forest trees: a comparative analysis of 37 Malaysian species. *Oikos* 76, 145–154.

Tiffney, B.H. and Mazer, S.J. (1995) Angiosperm growth habit, dispersal and diversification reconsidered. *Evolutionary Ecology* 9, 93–117.

Tilman, D. (1994) Competition and biodiversity in spatially structured habitats. *Ecology* 75, 2–16.

Tyree, M.T. (2003) Hydraulic limits on tree performance: transpiration, carbon gain and growth of trees. *Trees* 17, 95–100.

Tyree, M.T., Engelbrecht, B.M.J., Vargas, G. *et al.* (2003) Desiccation tolerance of five tropical seedlings in Panama. Relationship to a field assessment of drought performance. *Plant Physiology* 132, 1439–1447.

van Bael, S. and Brawn, J.D. (2005) The direct and indirect effects of insectivory by birds in two contrasting Neotropical forests. *Oecologia* 143, 106–116.

Volterra, V. (1931) *Leçons sur la Théorie Mathématique de la Lutte pour la Vie.* Gauthier-Villars, Paris.

Waterman, P.G., Ross, J.A.M., Bennett, E.L. *et al.* (1988) A comparison of the floristics and leaf chemistry of the tree flora in two Malaysian rain forests and the influence of leaf chemistry on populations of colobine monkeys in the Old World. *Biological Journal of the Linnean Society* 34, 1–32.

Webb, C.O. and Peart, D.R. (1999) Seedling density dependence promotes coexistence of Bornean rain forest trees. *Ecology* 80, 2006–2017.

West, G.B., Brown, J.H., and Enquist, B.J. (1997) A general model for the origin of allometric scaling laws in biology. *Science* 276, 122–126.

West, P.W., Jackett, D.R., and Sykes, S.J. (1989) Stresses in, and the shape of, tree stems in forest monoculture. *Journal of Theoretical Biology* 140, 327–343.

Wills, C.R., Harms, K.E., Condit, R. *et al.* (2006) Nonrandom processes maintain diversity in tropical forests. *Science* 311, 527–531.

Wilson, D.S. (1980) *The Natural Selection of Populations and Communities.* Benjamin/Cummings, Menlo Park, CA.

Wing, S.L. and Boucher, L.D. (1998) Ecological aspects of the Cretaceous flowering plant radiation. *Annual Review of Earth and Planetary Sciences* 26, 379–421.

Wing, S.L., Hickey, L.J., and Swisher, C.C. (1993) Implications of an exceptional fossil flora for late Cretaceous vegetation. *Nature* 363, 342–344.

Wootton, J.T. and Power, M.E. (1993) Productivity, consumers, and the structure of a river food chain. *Proceedings of the National Academy of Sciences, USA* 90, 1384–1387.

Wright, I.J., Reich, P.B., Westoby, M. *et al.* (2004) The worldwide leaf economics spectrum. *Nature* 428, 821–827.

Wright, S.J., Muller-Landau, H.C., Condit, R. *et al.* (2003) Gap-dependent recruitment, realized vital rates, and size distributions of tropical trees. *Ecology* 84, 3174–3185.

Yoda, K. (1974) Three-dimensional distribution of light intensity in a tropical rain forest of west Malaysia. *Japanese Journal of Ecology* 24, 247–254.

Yoda, K. (1978) Three-dimensional distribution of light intensity in a tropical rain forest of west Malaysian. *Malayan Nature Journal* 30, 161–177.

Zotz, G. and Winter, K. (1993) Short-term photosynthesis measurements predict leaf carbon balance in tropical rain-forest canopy plants. *Planta* 191, 409–412.

Chapter 9

APPROACHING ECOLOGICAL COMPLEXITY FROM THE PERSPECTIVE OF SYMMETRIC NEUTRAL THEORY

Stephen P. Hubbell

OVERVIEW

I argue, seemingly paradoxically, that the most rapid path to understanding ecological systems, especially complex systems such as species-rich tropical tree communities, is through an interaction of empirical science, guided by strong inference, with theories that start very simply, with few free parameters and assumptions, and add complexity reluctantly, kicking and screaming, only when absolutely necessary to obtain some desired level of fit to the data. Neutral theory is one such starting point. Although it is only a first approximation, neutral theory is a remarkably good approximation to many of the patterns of relative tree species abundance we observe in tropical forests worldwide. In this chapter, I briefly review some of the major developments in neutral theory since publication of my book in 2001, and try to clear up several persistent misconceptions about neutral theory. One common misconception is that a finding of density dependence falsifies neutrality, which it does not, provided that all species exhibit approximate symmetry in their density dependence. I conclude with some new findings about the dynamics of the tropical tree community on Barro Colorado Island, Panama, over the past quarter century that are more consistent with neutrality and drift than they are with stable population fluctuations around fixed carrying capacities, the expectation of classical niche-assembly theory.

INTRODUCTION

Soon after coming to the United States during World War II, Enrico Fermi, the great Italian physicist, was told by some US flag officers that so-and-so was a great general. What is the definition of a great general, Fermi asked? After some thought, they agreed that winning five major battles made a great general. And how many generals are great? After some more back and forth, they replied only about 3%. Well, Fermi replied, suppose armies in battles are equally matched, and the probabilities of winning or losing are equal and random. Then you would find by chance that about 3% of generals win five battles!

The point of relating this story, paraphrased from Sagan and Druyan (1997), is not to discuss whether *military intelligence* is an oxymoron, but whether one always needs complex theory to explain apparently complex ecological phenomena. In my view, much of the complexity of contemporary theory in ecology is probably unnecessary and actually impedes the advancement of the ecological sciences. Nearly half a century ago, Platt (1964) urged molecular biologists to pursue a program of strong-inference driven science, and now we need a comparable program in theoretical and empirical ecology. Using this approach, profound discoveries revealing the nearly identical molecular machinery of all life have been made in evolutionary developmental biology and molecular phylogeny. These discoveries show a remarkable conservatism and simplicity in the fundamental regulatory control mechanisms underlying the vast phenotypic diversity in eukaryotic organisms in the world today (Carroll *et al.* 2001, Cracraft and Donoghue 2004, Carroll 2005, Donoghue 2005). Could it be that a similar conservatism and simplicity underlie ecological phenomena?

Ecologists may have a hard time answering this question unless we fundamentally reinvent our way of doing science. We need to collectively and routinely ask, what elements of our theories are absolutely essential to explain this or that phenomenon? What is the simplest set of assumptions that is sufficient? What assumptions are necessary? Ecology still lingers in the narrative stage of its development, the stage of collecting case studies of ecological phenomena. Some ecologists take the almost post-modernist view that a collection of unique narratives is all that ecology can ever achieve and that the search for generality is pointless and quixotic. Others who believe generality exists do meta-analyses on the collected case studies to look for it. However, there are serious problems with most current meta-analyses in ecology and evolutionary biology (Travis 2006). A big problem is that many of our studies are confirmatory and are not designed to reject our favorite hypotheses, but to support them. A few years ago, I attended a symposium in Japan on indirect effects in communities (e.g., indirect competition sensu Holt 1977). After listening to many papers confirming the importance of indirect effects in this or that system, and several meta-analysis talks asserting the nearly universal prevalence of indirect effects, I asked the question: So, how many studies were explicitly done on systems in which indirect effects were not expected to be important? I would describe the reaction as stunned silence. All one can honestly say is – of systems picked for study because researchers expected indirect effects to be important in them, in a majority the investigator's hunch proved to be correct. The only negative

results would occur in surprise cases when the investigator's prior hypothesis was not correct. Confirmatory results get an added boost from the reluctance of investigators and journals to publish negative results. The good-faith effort to falsify hypotheses thus faces a persistent and systematic bias, a triple-jeopardy handicap at all stages of our science, from conception, to execution, to publication. So it is very difficult to assess the significance of meta-analyses which almost never discuss the selection criteria used to pick which systems to study (e.g., McGill *et al.* 2006). We are all guilty of these biases. Do we ever counsel our students to pick an ecological system in which we think the process they want to study does not occur?

Some of our most cherished hypotheses have become sacred cows that are virtually impermissible to seriously challenge. I have personal experience of the difficulties of going against the grain of conventional wisdom in community ecology. Nearly 30 years ago, I asked the question, what would the patterns of relative tree species abundance in closed-canopy forests be like if they were determined purely by demographic stochasticity in birth, death, and dispersal rates (Hubbell 1979)? This paper, my first foray into neutral theory, was a study of a tropical dry forest in Costa Rica. It unfortunately appeared during the height of the wars over null community assembly rules (Strong *et al.* 1984), ideas that were a logical outgrowth of the theory of island biogeography (MacArthur and Wilson 1967). The sacred cow in question was then, as now, the hypothesis that ecological communities are "niche-assembled," that is, limited-membership, equilibrium assemblies of niche-differentiated species, each the best competitor in its own niche, coexisting with the other species in competitive equipoise. These wars were sufficiently off-putting that they delayed any further serious discussion of neutral theory in ecology for nearly 20 years. In my own case, I revisited neutral theory only in the mid-1990s, when a student in my Princeton biogeography class asked, why doesn't the theory of island biogeography include a process of speciation, and what would happen if it did? I did not know, and I set about finding out.

Neutral theory, as it appeared in my book (Hubbell 2001) and in two earlier papers, published obscurely (Hubbell 1995, 1997), was the result, also discovered independently by Graham Bell (2000, 2001). The obscurity of my first two papers was not by design. I tried to publish my second paper, which ultimately appeared in *Coral Reefs* (Hubbell 1997), in many prominent places, including *Nature, Science, Proceedings of the National Academy of Sciences of the United States of America, American Naturalist*, and *Ecology*, all of which rejected the manuscript, three without review ("not of sufficient interest"), and two with very brief and very chilly reviews. Excerpts: "Ecology is not ready for yet another null model of community assembly. Let sleeping dogs lie." "Any theory [of community assembly] based on such an obviously false assumption does not merit publication here, or anywhere else." "If this paper is published, I will never review for this journal again." "Has Hubbell lost his mind? I don't understand the math, but his conclusions are anti-intuitive and must be wrong." It is perhaps ironic that these and many other leading journals, collectively, have published a very large number of papers on neutral theory in ecology since 2001, including special issues in some journals fully devoted to the subject. I do not think the editors of *Coral Reefs* wanted to publish the paper either, but I was one of the keynote speakers for the Eighth International Coral Reef Symposium in Panama in 1996, and the editors had agreed in advance to publish the keynote addresses. The take-home message of this experience: if you can't get your iconoclastic paper accepted in a peer-reviewed journal, expand its scope and write a book.

In my synopsis of some of the high points of neutral theory here, I could proceed historically with how the ideas originally developed, but I have chosen instead to develop the theory conceptually in steps of increasing complexity, starting from the simplest possible model. This approach draws on some major advances in the theory that occurred after my book was published (Hubbell 2001). I was fortunate that my book attracted the attention and interest of a number of brilliant statistical physicists who have considerably improved and generalized the theory I originally presented. Because this chapter is not a review of all the developments in neutral theory

since 2001, I apologize to the many people with papers on the subject to which I do not refer.

THEORETICAL RECIPE: START SIMPLY, ADD AS FEW FREE PARAMETERS AS POSSIBLE, STIR VIGOROUSLY

A *free parameter* is a number in a theory that cannot be derived from the theory itself. In my view, the best theories are those which make the largest number of testable predictions per free parameter, a qualitative judgment in the spirit of the Akaike information criterion (AIC), which penalizes models the more free parameters they have. Neutral theory is attractive because it has very few free parameters, all of which have ready biological interpretations. Yet from its small set of free parameters, neutral theory leverages a large number of predictions about diverse phenomena in community ecology and biogeography. Here I limit discussion to its predictions for the static and dynamic patterns of relative species abundance, the subject receiving the most attention in recent years, and I will confront these predictions with data on tropical tree communities. Before doing so, however, there is a rather long theoretical preamble, which I feel is necessary but for which I apologize to readers more interested in the biological punch line.

The simplest possible neutral model of relative species abundance is to imagine a large, self-contained and homogeneous, biogeographic region in which the only processes at work are speciation and extinction. In neutral theory, this is called the *metacommunity*. The metacommunity is the evolutionary–biogeographic unit in which most member species live out their entire evolutionary lifespans. In our bare-bones model, let us for the moment ignore almost all of our favorite ecological processes such as density- and frequency dependence, niche differences, dispersal limitation, and so on. Let us further imagine that the metacommunity is in an *interregnum* period of steady-state species richness between evolutionary punctuational events, so that the speciation and extinction rates are in balance. What patterns of relative species abundance do we expect? This is clearly a non-equilibrium community in the taxonomic sense because any given species experiences a finite lifespan with a "birth" and a "death." But there is nevertheless a non-trivial stochastic steady-state distribution of relative species abundance in the metacommunity among the slowly turning over species. Neutral theory proves that this distribution is Fisher's logseries (Hubbell 2001, Volkov *et al.* 2003).

Fisher's logseries emerged from one of the two most celebrated papers ever written on relative species abundance, one by Fisher *et al.* (1943), and the other by Preston (1948), which sparked a theoretical controversy – about which more will be said in a moment – a controversy that persists to the present day (McGill 2003, Volkov *et al.* 2003, 2007, Dornelas *et al.* 2006). Fisher found an excellent fit with the logseries to relative abundance data on Lepidoptera from Britain and Malaysia. Under the logseries, the expected number of species with *n* individuals $\langle \phi_n \rangle$ is given by

$$\langle \phi_n \rangle = \alpha \frac{x^n}{n} \tag{9.1}$$

where parameter x is a positive number less than (but very close to) unity, and α is a diversity parameter known as Fisher's α. One of the remarkable properties of Fisher's α is that it is relatively stable in the face of increasing sample size. This stability makes Fisher's α one of the preferred measures of species diversity, but why is it so stable? Until the development of neutral theory, Fisher's logseries was simply a phenomenological statistical distribution fit to relative abundance data. There was no clear biological explanation for either Fisher's α or parameter x in the logseries that could be derived from population biology.

One of the most remarkable results of neutral theory is proof that the celebrated diversity parameter, Fisher's α (θ in neutral theory), is proportional to the product of the speciation rate and the size of the metacommunity. This offers an explanation of the stability of α: these are two very stable numbers, one the average per capita speciation rate in the entire metacommunity, a very small number, and the other the size of the

metacommunity – the sum of the population sizes of all species in the metacommunity – a very large number (Hubbell 2001). Fisher's α (or θ) is a biodiversity number that crops up all over neutral theory, so in a real sense it is a fundamental number in the theory. But what is the biological meaning of parameter x?

To explain parameter x, we need to introduce the so-called master equation of neutral theory, which describes the stochastic population dynamics of species in the metacommunity (Volkov *et al.* 2003). Let $b_{n,k}$ and $d_{n,k}$ be the probabilities of birth and death of an arbitrary species k at abundance n. Let $p_{n,k}(t)$ be the probability that species k is at abundance n at time t. Then, the rate of change of this probability is given by

$$\frac{dp_{n,k}(t)}{dt} = p_{n+1,k}(t)d_{n+1} + p_{n-1,k}(t)b_{n-1}$$
$$- p_{n,k}(t)(b_{n,k} + d_{n,k}) \tag{9.2}$$

This equation is not hard to understand. The first term on the right represents the transition from abundance $n + 1$ to n, due to a death. The second term is the transition from abundance $n - 1$ to n due to a birth. The last two terms are losses to $p_{n,k}(t)$ because they are transitions away from abundance n to either $n + 1$ or $n - 1$ through a birth or death, respectively. When one first sees Equation (9.2), it appears to be little more than a book-keeping exercise, but it is actually much more. Note that it is a recursive function of abundance, so we can use it to find an equilibrium solution for species of arbitrary abundance n. If we set derivatives at all abundances equal to zero, then each abundance transition is in equilibrium. Let $P_{n,k}$ denote this equilibrium. Then $P_{n,k} = P_{n-1,k} \cdot [b_{n-1,k}/d_{n,k}]$, and more generally, this corresponds to an equilibrium solution for the metacommunity:

$$P_{n,k} = P_{0,k} \prod_{i=0}^{n-1} \frac{b_{i,k}}{d_{i+1,k}} \tag{9.3}$$

Note that the probability of being at abundance n is a function of the product of the birth rate to death rate ratios over all abundances up to $n - 1$. Because the $P_{n,k}$'s are probabilities and must sum to unity, we can find the value of $P_{0,k}$ from this sum, and all other terms as well.

Actually, the master equation (Equation (9.2)) applies much more generally than to neutral theory alone. It can also describe the dynamics of non-neutral communities if we let species have species-specific birth, death, and speciation rates, and it is completely general in regard to what factors may control these rates. So, for example, the birth and death rates could be density dependent, they could depend on competition or predation, and so on. But for now, hewing to the philosophy of adding complexity in small, considered steps, consider a symmetric neutral community of S species that are all alike on a per capita demographic basis, such that they all have the same per capita birth and death rates, that is, $b_{n,k} \equiv b_n$ and $d_{n,k} \equiv d_n$ (i.e., the species identifier k does not matter). We can introduce speciation by recognizing a special "birth rate" in this general metacommunity solution, that is, $b_0 = v$, the speciation rate. The mean number of species with n individuals, $\langle \phi_n \rangle$, in a community of S identical species is simply proportional to P_n:

$$\langle \phi_n \rangle = SP_0 \prod_{i=0}^{n-1} \frac{b_i}{d_{i+1}} \qquad (9.4)$$

What does all this have to do with Fisher's logseries? It turns out that Equation (9.4) is Fisher's logseries if we make birth and death rates density independent. Herein lies one of the most profound insights to come from neutral theory: obtaining Fisher's logseries *necessarily* implies *density independence* in population growth on metacommunity spatial scales. If one's relative species abundance data fit Fisher's logseries on large biogeographic scales, one can definitively conclude that the population dynamics of species on large scales behave in a stochastically density-independent manner. This theoretical result has potentially paradigm-shifting implications in ecology for the scale dependence of population regulation, for the structure and dynamics of communities on large spatial scales, for conservation biology, and for the evolution of biotas and phylogeography (Hubbell 2008a).

We can easily derive this result. What does it mean to have density-independent population growth? It means that the per capita birth and death rates remain constant as population density varies. Mathematically, we write that the absolute birth rate of a species of current abundance n is simply n times the birth rate of a species with abundance 1, that is, $b_n = nb_1$, the definition of density independence. Similarly, suppose that the death rates are density independent, $d_n = nd_1$. Substituting these expressions into Equation (9.4), we immediately obtain Fisher's logseries:

$$\langle \phi_n \rangle_M = S_M P_0 \frac{b_0 b_1 \cdots b_{n-1}}{d_1 d_2 \cdots d_n} = \theta \frac{x^n}{n} \qquad (9.5)$$

where the subscript M refers to the metacommunity, $x = b_n/d_n = b_1/d_1 = b/d$, $b_0 = v$, and $\theta = S_M P_0 v/b = \alpha$ of Fisher's logseries.

The derivation of Equation (9.5) reveals that the mysterious parameter x of the logseries is, in fact, biologically interpretable as the ratio of the average per capita birth rate to the average per capita death rate in the metacommunity. Note that when one introduces speciation, parameter x must be slightly less than 1 to maintain a finite metacommunity size. At very large spatial scales, the total birth and death rates have to be nearly in mass balance, resulting in a metacommunity b/d ratio only infinitesimally less than unity. The very slight deficit in birth rates versus death rates at the metacommunity biodiversity equilibrium is made up by the very slow input of new species.

NEXT STEP: ADD A BIT MORE COMPLEXITY – LOCAL COMMUNITIES AND DISPERSAL LIMITATION

A few years after Fisher and company's paper, Preston (1948) published a critique of Fisher's logseries. The logseries predicts that the rarest abundance category of singletons – species sampled only once – should have the most species; and indeed, the curve described by Equation (9.1) is almost hyperbolic in shape (i.e., $\langle\phi_n\rangle \approx \theta/n$ when x is very close to unity). Preston concluded that the shape of Fisher's logseries was a sampling artifact. He argued that if sample sizes were increased, and if relative species abundances were log-transformed, then they would be nearly normally distributed with the most species occurring at intermediate log abundance classes. This meant that the log-transformed distribution would display a bell shape with an "interior" mode, not a mode in the abundance class of singleton species. There were now two competing statistical hypotheses for the distribution of relative species abundance: Fisher's logseries and Preston's lognormal. Over the half century since this debate was enjoined, Preston's lognormal has probably been fit to relative species abundance data more often than Fisher's logseries, but Preston's hypothesis could not explain the stability of Fisher's α, and Fisher's hypothesis could not explain distributions with an interior mode. Can neutral theory reconcile these two disparate explanations?

So far, the theory we have developed above only gives rise to Fisher's logseries for the relative species abundance distribution at steady-state between speciation and extinction in the metacommunity. This distribution is appropriate on macroecological scales of space and time, but what patterns of relative species abundance do we expect on small spatial and temporal scales in local communities? In the metacommunity model, we ignored dispersal and the spatial substructure of the metacommunity, assuming that births and deaths, and species, were randomly distributed over the metacommunity landscape. What happens if we put substructure and dispersal into the theory, a bit of added complexity? Suppose we subdivide the metacommunity of size J_M individuals into local communities of size J individuals, and allow dispersal between the local communities with probability rate m per birth in a given local community. We can now describe the patterns of relative abundance in local communities as semi-isolated samples of the metacommunity. What do we get?

It turns out that under dispersal limitation, when species cannot move with impunity anywhere in the metacommunity in an infinitesimal time step, the distribution of relative species abundance in local communities will not be Fisher's logseries. The new insight from the slightly more complex version of neutral theory is that the shape of the local distribution of relative species abundance depends on the dispersal probability, m. If m is large and near unity, then the distribution approaches Fisher's logseries (i.e., no dispersal limitation), with no interior mode at intermediate abundances. However, if m is small, such that local communities are very isolated and do not often receive immigrants from the surrounding metacommunity, then the distribution becomes more lognormal-like, with an interior mode.

Although the distributions of local relative species abundance are not lognormals, they can be closely approximated by (and confused with) lognormals when the parameters are within certain ranges. Volkov et al. (2003) derive the analytical expression for relative species abundance in a local community undergoing immigration from a much larger metacommunity, analogous to the classical island–mainland problem in the theory of island biogeography. Again, let $\langle\phi_n\rangle$ be the mean number of species with n individuals. Then

$$\langle\phi_n\rangle = \theta \frac{J!}{n!(J-n)!} \frac{\Gamma(\gamma)}{\Gamma(J+\gamma)}$$
$$\times \int_0^\gamma \frac{\Gamma(n+y)}{\Gamma(1+y)} \frac{\Gamma(J-n+\gamma-y)}{\Gamma(\gamma-y)}$$
$$\times \exp\left(\frac{-y\theta}{\gamma}\right) dy \qquad (9.6)$$

where $\Gamma(z) = \int_0^\infty t^{z-1}e^{-t}dt$ which is equal to $(z-1)!$ for integer z, and $\gamma = m(J-1)/(1-m)$. Composite parameter γ is another fundamental number in neutral theory: it is the scale-independent fundamental dispersal parameter that takes out the effect of local community size J. Equation (9.6) can be solved numerically quite accurately. As the immigration rate m decreases, the distribution of relative species abundance in the local community given by Equation (9.6) becomes progressively more skewed, confirming the simulation-based results in my book (Hubbell 2001). Thus, as islands or local communities become more isolated, rare species become rarer, and common species become commoner. This is because, under low immigration rates, when rare species go locally extinct, they take longer to re-immigrate, so that the steady-state number of rare species locally is lower than would be expected if local communities were a random sample of the metacommunity, that is, if dispersal were unlimited.

Herein lies a potential mechanistic explanation for both Fisher's and Preston's hypotheses: the first applies to macroecological scales – or for pooled, multiple scattered samples from a landscape that overcomes dispersal limitation – and the second to local scales. Not everyone agrees with this inter-pretation (e.g., McGill 2003), but we will return to this question when we confront the theory with data on tropical tree species abundances.

NOW, ADD YET MORE COMPLEXITY: SYMMETRIC DENSITY- AND FREQUENCY DEPENDENCE

Density- and frequency dependence are often regarded as the classical signatures of niche-assembled communities because they imply the regulation of species abundances by their real-ized carrying capacities set either by the limiting resources available to the species in its real-ized niche, or by top-down predator control, as by Janzen–Connell effects (Janzen 1970, Connell 1971). However, if these effects are symmetric, meaning that every species experiences the same per capita density dependence when it is of equiv-alent abundance, then neutral theory can be generalized to accommodate density dependence (Volkov et al. 2005). There are any number of ways to put density dependence into the master equation, but the simplest is to make the ratios of per capita birth and death rates in Equation (9.4) be functions of abundance n. So, define a density-dependent birth: death ratio at abundance n as $\hat{r} = (b_n/d_{n+1})f(n)$. We can choose any appro-priate function $f(n)$ that has the property that $f(n) > 1$ when n is small and $f(n) \to 1$ when n is large. Note that as $n \to \infty, \hat{r} \to x$, where x is one of the two parameters of Fisher's logseries (recall, $x = b_n/d_{n+1} = b/d$). This means that

when n is small, $\hat{r}$ will be greater than unity, and births outnumber deaths (rare-species advan-tage). In Fisher's logseries, b/d does not change with density (i.e., density-independent growth). In the spirit of choosing simple over complex, we set $f(n) = n/(n+c)$, which requires only one additional free parameter, and the same for all species (symmetry). This function in the mas-ter equation results in per capita death rates that are independent of n, but per capita birth rates that increase as n gets smaller. This for-mulation results in a very simple and elegant modification of Fisher's logseries distribution for the mean number of species $\langle \phi_n \rangle$ having abun-dance n:

$$\langle \phi_n \rangle = \theta \frac{x^n}{n+c} \qquad (9.7)$$

where c is the density dependence parameter, with units of individuals, that measures the mean strength of density dependence in the community. In neutral theory, we call this dis-tribution the hyperlogseries because it has sta-tistical properties similar to the hypergeometric. How parameter c influences the b/d ratio is shown in Figure 9.1. When c is small, only very small populations enjoy a birth rate larger than the death rate. However, as c becomes bigger, larger and larger populations enjoy a frequency-dependent advantage. Note how the function for b/d crosses the line of population replacement ($b/d = 1$) at higher and higher values of n as c increases (Figure 9.1). In the symmetric case described by Equation (9.7), since all species enjoy the same advantage when rare, this func-tion describes both intraspecific density depen-dence and interspecific frequency dependence. Note that these stochastic population dynam-ics never lead to population equilibrium. Even though species on average are near replacement at large n, they do not exhibit central tenden-cies to a dynamical attractor at a fixed carrying capacity. The fate of every species in the the-ory is extinction, although time to extinction increases approximately at a rate proportional to $n \cdot \ln(n)$ with increasing abundance (Hubbell 2001). This fact will become important in the later discussion.

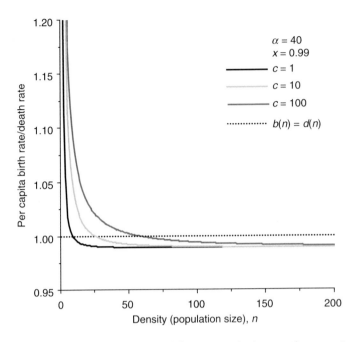

Figure 9.1 Theoretically expected curves of the ratio of the per capita birth rate to the per capita death rate in the density dependence version of neutral theory. The dotted line is the line of population replacement, when birth and death rates are equal. Parameter c controls the strength of density dependence, and the size of the population that experiences a rare-species advantage increases with the value of c. The two other parameters of the theory are the biodiversity number θ (Fisher's α) and x.

CONFRONTING THE THEORY WITH RELATIVE ABUNDANCE DATA ON TROPICAL TREE COMMUNITIES

Since my book was published, an old controversy has re-emerged over which distribution fits relative abundance data better, Fisher's logseries, Preston's lognormal, or now, the distribution predicted by neutral theory. McGill (2003) and McGill et al. (2006) contend that Preston's lognormal fits most available data on relative abundance better than the distributions from neutral theory. I have many problems with their analysis in addition to my earlier-stated objections to meta-analyses. The first problem is based on the number of free parameters. Under the dispersal limitation version of neutral theory (Hubbell 2001), the distribution has two free parameters (only θ and m – the number of species is a prediction), whereas the lognormal has three (mean, variance, and

the modal number of species), so from an AIC perspective one needs to devalue the lognormal hypothesis relative to neutral theory for having more parameters. Second, even without this devaluation, the fit of the lognormal to the Barro Colorado Island (BCI) data is actually slightly worse than neutral theory in the case of BCI (Volkov et al. 2003) which was the basis of McGill's original assertion (McGill 2003). In a subsequent paper, we showed that both versions of neutral theory – the dispersal limitation version and the newer version with symmetric density dependence (Volkov et al. 2005) – fit the static relative tree species abundance data from six tropical forests very well (Figure 9.2). The forests in question have very different evolutionary histories and ecology, but despite this, they are all fit quite well by the same neutral model with different values of the free parameters (Table 9.1).

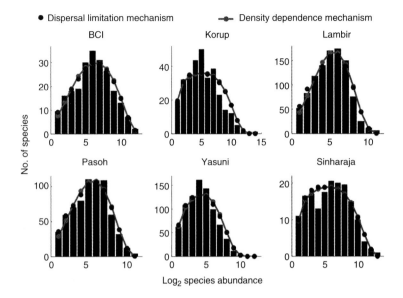

Figure 9.2 Preston-style relative tree species abundance distributions for six 50 ha plots across the New and Old World tropics, showing the equally good fits of two neutral models that have very different mechanisms. The black bars are the data. The *x* axes of each graph are abundance categories binning species abundances into classes of log to the base 2 individuals per species. The *y* axis is the number of species in each abundance class. The circles are the fit of the dispersal limitation hypothesis of Hubbell (2001) and island biogeography theory. The line is the fit of the symmetric density dependence hypothesis (Volkov *et al.* 2005). The sites are as follows: "BCI" is Barro Colorado Island, Panama; "Korup" is Korup National Park, Cameroon, West Africa; "Lambir" is Lambir Hills National Park, Sarawak, Malaysian Borneo; "Pasoh" is Pasoh Research Forest, Peninsular Malaysia; "Yasuni" is Yasuni National Park, Amazonian Ecuador; "Sinharaja" is Sinharaja Forest Reserve, Sri Lanka.

Table 9.1 Parameter values for the two versions of neutral theory, dispersal limitation and density dependence, for the six large plots of tropical forest whose distributions of relative tree species abundance are illustrated in Figure 9.2.

Forest	From data		b/d x	Dispersal limitation		Density dependence	
	S	J		θ_1	m	θ_2	c
Barro Colorado Island, Panama	225	21,457	0.9978	48.1	0.09	47.5	1.80
Yasuni, Ecuador	825	17,546	0.9883	204.2	0.43	213.2	0.51
Pasoh, Malaysia	678	26,554	0.9932	192.5	0.09	189.5	1.95
Korup, Cameroon	308	24,591	0.9979	52.9	0.54	53.0	0.24
Lambir, Sarawak	1004	33,175	0.9915	288.2	0.11	301.0	2.02
Sinharaja, Sri Lanka	167	16,936	0.9982	27.3	0.55	28.3	0.38

Notes: These data are for trees >10 cm diameter at breast height. Parameters *S* (total number of species) and *J* (total number of individuals) are from the data and do not need to be estimated. θ_1 and θ_2 are the biodiversity numbers estimated under the dispersal limitation and density dependence models, respectively. Parameter *m* is the probability of immigration per birth under the dispersal limitation model, and *c* is the density dependence parameter in the density dependence model (after Volkov *et al.* 2005).

Third, fitting issues aside, I have more fundamental problems with the lognormal hypothesis. First of all, the lognormal is essentially a generic statistical distribution whose parameters are not derived from population biological processes, whereas every parameter of neutral theory has a straightforward biological interpretation. Although there have been several attempts to construct theories of community organization that result in lognormal-like distributions, such as the sequential broken stick model (Sugihara 1980) or the nested niche hierarchy model (Sugihara *et al.* 2003), the evidence marshaled to support these models is not ecologically or evolutionarily compelling, in my opinion, and once again, the parameters do not have straightforward connections to population biology. Proponents of the lognormal have not adequately addressed two really serious problems with the distribution. The first problem is the assumption of a fixed variance, or spread, between the commonest and rarest species in a community, the so-called canonical hypothesis (Preston 1962). The canonical hypothesis means that, for example, with a doubling in sample size, each species should increase in logarithmic proportion, so that the variance in log abundances remains constant even as mean species abundance doubles. The only way this can happen is if the abundances of the rarest species also increase in logarithmic lockstep with the common species. But this is never observed in real samples. In reality the rarest species in a sample are almost always singletons, regardless of sample size, so as the common species increase in abundance, the variance in relative abundances also increases. There is no such canonical assumption in neutral theory. The second problem is that the lognormal fails as a dynamical model of communities; it can be fit only to static relative abundance data (Hubbell and Borda de Água 2004). This limitation does not apply to neutral theory, as we will see below.

Another critique of neutral theory was made by Dornelas *et al.* (2006), who analyzed relative abundance patterns in a geographically large collection of reef communities (the "reef metacommunity"). They found that the metacommunity had a relative abundance distribution with an interior mode, resembling a lognormal distribution, whereas local reefs had a Fisher logseries-like distribution, just the opposite to the prediction in my book (Hubbell 2001). They asserted that this observation "refutes" neutral theory. Of course they did nothing of the sort – any more than one can "refute" the Hardy–Weinberg equilibrium – because both are mathematical theorems. What one can legitimately say, however, is this case does not fit neutral theory as presented in my book. In doing good science, there should be regular feedback between empirical work and theory development. When a theory fails in a particular case, the next thing to do is to figure out why it happened. We began a quest for possible causal factors that might invert the pattern of relative species abundance. After considering the biogeographic differences between the coral reef and rainforest systems, we realized that two contrasting theoretical scenarios of metacommunity structure would lead to inverted patterns of relative abundance on local and regional scales (Volkov *et al.* 2007). On the one hand, one could have relatively small, partially isolated local communities surrounded by a very large metacommunity acting as a source of immigrants, which is the structure I envisioned in my book for tropical rainforests (Hubbell 2001). On the other hand, one could have spatially very isolated island communities whose assemblage in aggregate acts as the metacommunity, as in coral reefs. In the tropical forest scenario, the time for species turnover in the metacommunity is extremely long relative to turnover in local communities, so that the relative abundance distribution in the metacommunity is essentially fixed or "frozen" as a source of immigrants compared with the fast dynamics of the local community. This is very different from the coral reef scenario. In the latter case, each local community receives immigration from all the surrounding, isolated island communities, in each of which relative species abundances are not frozen. This means that the effect of dispersal limitation emerges only on metacommunity scales, not on local scales. As a consequence, one can prove that logseries-like distributions will be found in local reef communities, but not on regional scales (Volkov *et al.* 2007). So the new insight is that the relative abundance pattern you predict depends on whether the metacommunity is large and continuous, or is an archipelago of very isolated islands.

But what about the structure of tropical rainforest tree communities on large spatial scales? If the assumption of a large, continuous forest is correct, then neutral theory asserts that Fisher's logseries should be the distribution of relative species abundance in the metacommunity, and not the lognormal. Is this, in fact, correct? Over the last two decades, a dataset consisting of 288,973 individual tree records has been assembled from a large number of small plots all over the Amazon basin (ter Steege *et al.* 2006). The plots extend throughout the Brazilian Amazon into Amazonian Colombia, Ecuador, and Peru to the west, and into the Guianan shield to the northeast. According to neutral theory, samples aggregated from many small scattered samples collected over a large area reduce the impact of dispersal limitation and will better reflect the distribution of relative abundances for the entire metacommunity. Although many taxonomic problems remain at the species level with these data, the generic-level determinations are much more reliable. This is fortunate because we can test the fit of the logseries and the lognormal to the abundances of Amazonian genera. Neutral theory asserts that generic- and familial-level clades should also obey

the same metacommunity dynamics as species, the only difference being that they should have lower rates of origination and extinction than species do.

Neutral theory predicts that the abundances of Amazonian tree genera and families should be distributed according to Fisher's logseries, and not Preston's lognormal, and this prediction is, in fact, correct (Figure 9.3; Hubbell *et al.* 2008). The figure shows a tight fit of the logseries with a value of θ (Fisher's α) of approximately 71. The inset graph shows a Preston-type plot of species binned into doubling abundance classes. The flat top of the Preston curve over many doubling abundance classes for rare species is predicted by the logseries in species-rich assemblages on large spatial scales, but does not agree with the pattern predicted by Preston's lognormal. Given this result, it is highly unlikely that the species distribution will be a Preston canonical lognormal.

Perhaps the most challenging test for neutral theory in community ecology is to predict community dynamics, a much more stringent test than simply showing that it can fit snapshot static data on relative species abundance (e.g., Volkov *et al.* 2003). Only one or perhaps two of the large-plot studies of the Center for Tropical Forest Science have time series that are

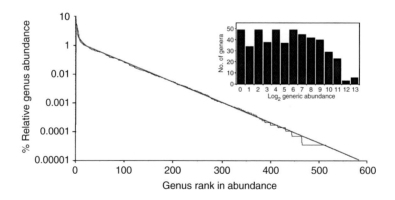

Figure 9.3 Fit of Fisher's logseries to the relative abundance data of Amazonian tree genera (data provided by ter Steege). Fisher's α (θ) is about 71. The rank abundance curve is shown in the full figure, and the Preston-style histogram of species binned into doubling classes of abundance is shown as the inset graph (after Hubbell *et al.* 2008). The flatness of the Preston curve for seven doubling classes of abundance at the rare-species end of the curve is not predicted by the canonical lognormal of Preston (1962). (After Hubbell *et al.* 2008).

Table 9.2 Changes in abundance in Barro Colorado Island tree species in the 50 ha plot over a 23-year interval from 1982 to 2005.

% absolute change in abundance, 1982–2005	No. of species	Mean 1982 species abundance	Standard deviation in 1982 species abundance
0–25	128	1379.6	4380.7
25–50	66	568.0	1038.7
50–75	36	306.4	622.8
75–100	24	317.5	654.1
>100	31	83.1	154.9

Notes: Only the 285 species with two or more individuals in 1982 are included. Percentage change was calculated as the absolute value of the difference in abundance of a species in 2005 and 1982 times 100 divided by the abundance of the species in 1982. The mean percent change ±1 standard deviation over all species was 47.6 ± 67.0%.

sufficiently long to perform such a test. One of these is the plot on Barro Colorado Island, for which we have a quarter-century record of forest change and turnover (Hubbell 2008c). The BCI forest has exhibited remarkable dynamism over the last 25 years (Table 9.2). Of the 285 species sufficiently abundant to test, nearly a third (31.9%) changed by more than 50% in total abundance, and these changes were not limited to rare species. These changes are not purely successional because two-thirds of the species (63.9%) do not exhibit monotonic directional changes in abundance. In and of itself, this level of dynamism would be challenging to most equilibrium theories of community organization. However, we need to test the possibility that these changes are simply Gaussian stochastic fluctuations centrally tending around fixed carrying capacities in an equilibrium, niche-assembled community. Alternatively, these fluctuations may be more accurately described by drift, with no central tendency, the prediction of neutral theory.

A simple measure of community change over time is the decay in the coefficient of determination, R^2, of community composition. In order to normalize changes on a per capita basis, one should evaluate changes in abundance on a log scale. At a time lag of zero, no change can yet have occurred, and the auto-regression of species abundances on themselves therefore yields an R^2 of unity. But as the time between censuses increases, we expect a decay in the value of R^2 of the regression of log species abundances at time $t + \tau$ on the log abundances of the same species at previous time t. We can test the predictions of niche assembly versus neutrality for community dynamics because these theories make very different predictions of the expected patterns of decay in R^2 over time. The prediction from neutral theory, when the metacommunity is very species-rich, is nearly perfectly linear decay in similarity (as measured by R^2 of the time-lagged regression of log abundances). In contrast, under Gaussian stochastic fluctuations around a stable equilibrium of a niche-assembled community, one expects a relatively fast, curvilinear decay in R^2, to an asymptotic R^2 value, reflecting the underlying community stability and the tendency of species to approach their niche-determined carrying capacities.

To test the niche-assembly predictions, we randomly sampled the distribution of observed intrinsic rates of increase of the BCI species over the past quarter century, which are approximately normally distributed with a mean of zero (Figure 9.4). We then applied the randomly sampled intrinsic rates, normalized to 5-year census intervals, to each species and computed the decay in R^2 of log species abundances over 25 years and six censuses. We repeated this procedure for an ensemble of 1000 stochastic runs, and averaged the results.

To test the predictions of neutral theory and a drifting community, we first had to consider whether to use the dispersal limitation version of the theory, or the version with symmetric density dependence. Recall that on the basis of the fit to static relative abundance data, these two versions of the theory cannot be distinguished. However, on the basis of the dynamics data, there is a clear winner. Recall from Figure 9.1 that under symmetric density dependence, we expect that the ratio of per capita birth rate to per capita decay rate should exceed unity at small population sizes, conferring a growth rate advantage on species when rare. However, there is no evidence

of the necessary rare-species advantage at the whole plot level (Figure 9.5). Rare and common species alike have mean values of b/d that do not differ significantly from unity. Thus, we modeled a drifting neutral community under the original dispersal limitation mechanism (Hubbell 2001) and island biogeography theory (MacArthur and Wilson 1967).

Dispersal limitation and symmetric density dependence are not mutually exclusive mechanisms, so in principle both can operate simultaneously. In fact, we know that density dependence is pervasive and strong in the BCI tree community (Hubbell 2008b,c). Harms et al. (2000) demonstrated very strong density dependence in the seed-to-seedling transition, as measured by the difference between the number of seeds collected in a network of traps and the number of seedlings that germinate in seedling plots adjacent to the traps. There are also negative conspecific density effects on sapling growth and survival (Hubbell et al. 2001, Ahumada et al. 2004, Uriarte et al. 2005). How do we reconcile these observations with the results in Figure 9.5? The answer is that all of these density effects weaken to background within short distances (most <20 m, to 30–40 m in a few species; Hubbell et al. 2001, Hubbell 2008b) and, as a result, they do not regulate the populations of BCI tree species at the scale of 50 ha.

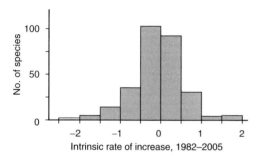

Figure 9.4 Distribution of intrinsic rates of increase among BCI tree species over a 23-year period from 1982 to 2005.

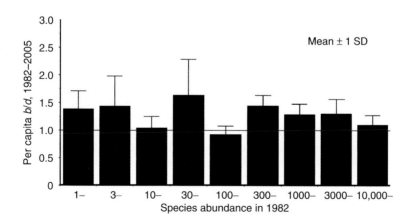

Figure 9.5 Lack of evidence of density- and frequency dependence in the BCI tree community over the entire 23 years of the study. The mean per capita birth rate/death rate (b/d) ratio ±1 standard deviation for species binned into half log base 10 intervals of abundance. In all abundance categories the confidence limits for b/d bracket unity.

Janzen (1970) and Connell (1971) independently suggested that an interaction between seed dispersal and host-specific seed and seedling herbivores and predators could limit the local population density of tropical tree species, but their focus was not on population regulation but on explaining the high local tree species richness of tropical forests by preventing any one species from becoming monodominant. One can show theoretically that Janzen–Connell effects do maintain more species locally at equilibrium (Chave et al. 2002, Hubbell and Lake 2003, Adler and Muller-Landau 2005). However, Janzen–Connell effects impose only a very weak dynamical constraint on the abundance of any given species in species-rich communities. Imagine a "perfect" Janzen–Connell effect that completely prevents species i from replacing itself in the same place in the forest. The individual of species j that replaces a given tree of species i can be any one of $S − 1$ other species in the forest, each of which is not constrained by the same enemy. If we turn this argument around, species i has complete freedom to replace any tree of the $S − 1$ other species in the forest, which occupy all sites not occupied by species i. So long as species i does not approach monodominance, the dynamical constraint on the population growth of species i is very weak. The more species-rich is the tree community, the weaker the dynamical constraint becomes on any given species.

Returning to the question of whether drift or niche-assembly hypotheses better describe the dynamics of the BCI tree community over the past quarter century, the answer is once again clear (Figure 9.6). There is an almost perfectly linear decay in community similarity over time, with no short-term evidence of a plateau of community similarity – as measured by the coefficient of determination of lagged species composition. In contrast, the simulation of community dynamics under the stochastic equilibrium community under the hypothesis of niche assembly clearly shows decelerating curvilinearity and an approach to an asymptotic R^2 value. It should be noted that ultimately the neutral model decay curve also approaches an asymptote – but much later and at a much lower positive R^2 value than under niche assembly – set by

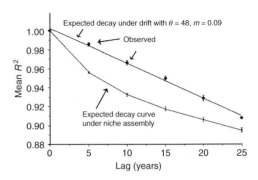

Figure 9.6 Expected decay in community similarity, as measured by the decay in the coefficient of determination (R^2) of log species abundances at time $t + \tau$ auto-regressed on the log abundances of the same species at time t. The black circles are the observed R^2 values, which fall along a straight line itself with an R^2 of 0.997. However, the straight line through the observed values is not a regression, but is the prediction from neutral theory derived from fitting the static relative abundance data from the first census in 1982, for which $\theta = 48$ and $m = 0.09$. The error bars are ± 1 standard deviation of all possible 5-, 10-, 15-, and 20-year time lags between censuses. Lags involving 1982 were pooled with the nearest 5-year lag. In contrast, the curve for the niche-assembly hypothesis, assuming Gaussian stochastic variation carrying capacities (lower cure), the niche-assembly hypothesis, is not what is observed in the BCI tree community. Error bars on the niche-assembly curve are ± 1 standard error of the mean based on an ensemble of 100 simulations.

the immigration–extinction steady-state with the metacommunity.

The drift prediction is actually much more robust than simply demonstrating linear decay in the coefficient of determination of community similarity. We can predict quite precisely the community dynamics from parameters measured only on the static relative abundance data (Figure 9.6). The two key parameters from neutral theory are the biodiversity number θ (Fisher's α) and the immigration probability, m. These values were estimated from the static first-census data for 1982 as $\theta = 48$ and $m = 0.09$ (Table 9.1; Volkov et al. 2005). The fit is essentially exact. This is a powerful test of neutral theory in the case of the BCI tropical forest, and it passed with flying colors.

CONCLUSIONS

Neutral theory should be viewed and used as a powerful tool in advancing the ecological sciences using strong inference in combination with continual feedback between observation and theory improvement. Neutral theory is a first-approximation theory that asks, what are the expected properties of model ecological communities if all species are demographically alike on a per capita basis? This is a very important question because we have no idea what differences among species are critical to explaining properties of ecological communities until we ask the appropriate question. Conventional theory in ecology starts from the premise that species differences are essential to understanding the assembly rules of ecological communities, but does this apply to all differences equally? Of course not. But if not, which differences are critical for explaining the patterns and processes in communities, and which ones can be safely ignored and yet still achieve some predetermined level of precision?

I have three modest recommendations for the advancement of ecology. The first is to value approximate theories. We have little serious discussion in ecology of the standards of precision and generality to which we hold our theories, and we tend to be harder on competing theories than our own. We need more widespread appreciation that all theories are approximations and have value. Neutral theory is an approximation. However, Lotka–Volterra competition theory, island biogeography theory, and Newtonian mechanics are also approximations. My physicist colleagues are taken aback at the intolerance for approximation in ecology, when all of their theories are approximations. The goal of theory should be to teach and to provide answers to problems with a predetermined standard of accuracy. How much can one explain with a minimum set of assumptions and free parameters? In the context of valuing approximate theory, we desperately need to move beyond the mindset of "*t*-test" rejection, following the advice of Hilborn and Mangel (1997). Because virtually all theories are approximations, all theories are, or should be, rejectable. This is not to devalue rejection as a cornerstone of the program of strong inference, but in too many cases there is no follow-up examination of why the theory failed, much less an attempt to make corrective changes to the theory to improve it.

The second recommendation is always – always – to start with minimalist theory, and add complexity only when absolutely necessary to explain some phenomenon of interest to some predetermined level of accuracy. This means that theory should evolve and, we hope, improve with time. Some critics have actually complained that neutral theory is a "moving target," but this is a curious, anti-scientific objection. Science is not (or should not be) static, but continuously evolving through an intimate and continual feedback between empirical and theoretical research. For example, the failure of pre-existing neutral theory to agree with the data on coral reefs stimulated us to explore new theory about metacommunity structure and its consequences, at a cost of a minimal amount of added complexity.

The third recommendation is for honesty, not advocacy in our science. Advocacy is dishonest and the hand-maiden of confirmatory science. Honest theories are those that provide imbedded, explicit, quantitative tools for their own rejection. By this standard, how many theories in ecology are honest? Neutral theory provides such tools, and perhaps this is one reason for the more frequent claims of its rejection than other theories in ecology. For example, the test of neutral theory regarding community dynamics above is an honest and stringent test. I suspect that the spate of rejection is partly because neutral theory makes many ecologists who are heavily vested in our current narratives uncomfortable. Is a paper rejecting neutral theory more likely to be published than one failing to reject neutral theory? If we are ever to go beyond the narrative phase of our science, we must be prepared to challenge each and every one of our comfortable stories in an open, honest, and deep way. As I discussed in the introduction, I am no fan of confirmatory science whether we are talking about niche-assembly theory or neutral theory. I am a fan of honest science.

These are exciting times in ecology, and given the press of global change, the ecological sciences have never been more critical to the future of humanity and non-human life on earth. To successfully meet this daunting challenge, we will need to significantly overhaul the way we do science.

ACKNOWLEDGMENTS

I thank Walter Carson and Stefan Schnitzer for inviting me to contribute to their book, and for being so patient with me for delivery of my chapter. I thank the National Science Foundation, the John D. and Catherine T. MacArthur Foundation, the A. K. Mellon Foundation, the Pew Charitable Trusts, the Guggenheim Foundation, the Celara Foundation, the Smithsonian Tropical Research Institute, and the more than 100 collaborators and field assistants who have supported or worked on the BCI Forest Dynamics project. I particularly thank Robin Foster, Rick Condit, Joe Wright, Rolando Perez, and Salomon Aguilar for their long-term collaboration on the BCI project. I thank Ira Rubinoff and Peter Ashton for major efforts in launching and financing the Center for Tropical Forest Science (CTFS), and Elizabeth Losos and Stuart Davies for their direction of CTFS.

REFERENCES

Adler, F.R. and Muller-Landau, H.C. (2005) When do localized natural enemies increase species richness? *Ecology Letters* 8, 438–447.

Ahumada, J.A., Hubbell, S.P., Condit, R., and Foster, R.B. (2004) Long-term tree survival in a neotropical forest: the influence of local biotic neighborhood. In E. Losos and E.G. Leigh, Jr. (eds), *Forest Diversity and Dynamism: Findings from a Network of Large-Scale Tropical Forest Plots*. University of Chicago Press, Chicago, pp. 408–432.

Bell, G. (2000) The distribution of abundance in neutral communities. *American Naturalist* 155, 606–617.

Bell, G. (2001) Neutral macroecology. *Science* 293, 2413–2418.

Carroll, S.B. (2005) *Endless Forms Most Beautiful: The New Science of Evo-devo and the Making of the Animal Kingdom*. W.W. Norton & Co., New York.

Carroll, S.J. Grenier, J.K., and Weatherbee, S. (2001) *From DNA to Diversity: Molecular Genetics and the Evolution of Animal Design*. Blackwell Publishing, Oxford.

Chave, J.H., Muller-Landau, C., and Levin, S.A. (2002) Comparing classical community models: theoretical consequences for patterns of diversity. *American Naturalist* 159, 1–23.

Connell, J.H. (1971) On the role of natural enemies in preventing competitive exclusion in some marine animals and in rain forest trees. In P.J. den Boer and G.R. Gradwell (eds), *Dynamics of Populations*. Centre for Agricultural Publishing and Documentation, Wageningen, The Netherlands, pp. 298–312.

Cracraft, J. and Donoghue, M.J. (2004) *Assembling the Tree of Life*. Oxford University Press, New York.

Donoghue, M.J. (2005) Key innovations, convergence, and success: macroevolutionary lessons from plant phylogeny. *Paleobiology* 21, 77–93.

Dornelas, M., Connolly, S.R., and Hughes, T.P. (2006) Coral reef diversity refutes the neutral theory of biodiversity. *Nature* 440, 80–82.

Fisher R.A., Corbet, A.S., and Williams, C. (1943) The relation between the number of species and the number of individuals in a random sample of an animal population. *Journal of Animal Ecology* 12, 42–58.

Hilborn, R. and Mangel, M. (1997) *The Ecological Detective*. Princeton University Press, Princeton, NJ.

Holt, R.D. (1977) Predation, apparent competition, and the structure of prey communities. *Theoretical Population Biology* 12, 197–229.

Hubbell, S.P. (1979) Tree dispersion, abundance and diversity in a tropical dry forest. *Science* 203, 1299–1309.

Hubbell, S.P. (1995) Towards a theory of biodiversity and biogeography on continuous landscapes. In G.R. Carmichael, G.E. Folk, and J.L. Schnoor *et al.* (eds), *Preparing for Global Change: A Midwestern Perspective*. Academic Publishing, Amsterdam, The Netherlands, pp. 173–201.

Hubbell, S.P. (1997) A unified theory of biogeography and relative species abundance and its application to tropical rain forests and coral reefs. *Coral Reefs* 16 (Suppl.), S9–S21.

Hubbell, S.P. (2001) *The Unified Neutral Theory of Biodiversity and Biogeography*. Princeton University Press, Princeton.

Hubbell, S.P. (2008a) Neutral theory and the scaling of process in population biology, community ecology, and evolution. *American Naturalist* (submitted).

Hubbell, S.P. (2008b) To know a tropical forest: what factors maintain high tree diversity on Barro Colorado Island, Panama? In I. Bilick and M. Price (eds), *The Ecology of Place*. University of Chicago Press, Chicago.

Hubbell, S.P. (2008c) Neutral theory and island biogeography theory: perspectives from a twenty-five year study of the tropical forest on Barro Colorado Island,

Panama. In J. Losos and R.E. Ricklefs (eds) *The Theory of Island Biogeography at 40: Impacts and Prospects.* Princeton University Press, Princeton.

Hubbell, S.P. and Borda de Água, L. (2004) The unified neutral theory of biodiversity and biogeography: reply. *Ecology* 85, 3175–3178.

Hubbell, S.P., He, F.-L., Condit, R., Borda de Água, L., Kellner, J., and ter Steege, H. (2008) How many tree species are there in the Amazon, and how many of them will go extinct? *Proceedings of the National Academy of Sciences of the United States of America* (in press).

Hubbell, S.P. and Lake, J. (2003) The neutral theory of biogeography and biodiversity: and beyond. In T. Blackburn and K. Gaston (eds) *Macroecology: Concepts and Consequences.* Blackwell Publishing, Oxford, pp. 45–63.

Janzen, D. (1970) Herbivores and the number of tree species in tropical forests. *American Naturalist* 104, 501–528.

MacArthur, R.H. and Wilson, E.O. (1967) *The Theory of Island Biogeography.* Princeton University Press, Princeton.

McGill, B.J. (2003) A test of the unified neutral theory of biodiversity. *Nature* 422, 881–885.

McGill, B.J., Maurer, B.A., and Weiser, M.J. (2006) Empirical evaluation of the neutral theory. *Ecology* 87, 1411–1423.

Platt, J.R. (1964) Strong inference. *Science* 146, 3642–3645.

Preston, F.W. (1948) The commonness, and rarity, of species. *Ecology* 29, 254–283.

Preston, F.W. (1962) The canonical distribution of commonness and rarity. *Ecology* 43, 185–215, 410–432.

Sagan, C. and Druyan, A. (1997) *The Demon-Haunted World: Science as a Candle in the Dark.* Ballantine Books, New York.

Strong, D.R., Simberloff, D.S., Abele, L.G., and Thistle, A.B. (eds) (1984) *Ecological Communities: Conceptual Issues and the Evidence.* Princeton University Press, Princeton.

Sugihara, G. (1980) Minimal community structure: an explanation of species abundance patterns. *American Naturalist* 116, 770–787.

Sugihara, G., Bersier, L.F., Pimm, S.L., Southwood, T.R., and May, R.M. (2003) A correspondence between two classical notions of community structure. *Proceedings of the National Academy of Sciences of the United States of America* 100, 5246–5251.

ter Steege, H., Pitman, N.C.A., Phillips, O. *et al.* (2006) Continental-scale patterns of canopy tree composition and function across Amazonia. *Nature* 443, 444–446.

Travis, J. (2006) Is it what we know or who we know? Choice of organism and robustness of inference in ecology and evolutionary biology. *American Naturalist* 167, 303–314.

Uriarte, M., Hubbell, S.P., John, R., Condit, R., and Canham, C.D. (2005) Neighbourhood effects on sapling growth and survival in a Neotropical forest and the ecological equivalence hypothesis. In D.F. Burslem, M.A. Pinard, and S.E. Hartley (eds), *Biotic Interactions in Tropical Forests: Their Role in the Maintenance of Species Diversity.* Cambridge University Press, Cambridge, pp. 89–106.

Volkov, I., Banavar, J.R., He, F.-L., Hubbell, S.P., and Maritan, A. (2005) Density dependence explains tree species abundance and diversity in tropical forests. *Nature* 438, 658–661.

Volkov, I., Banavar, J.R., Hubbell, S.P., and Maritan, A. (2003) Neutral theory and the relative abundance of species in ecology. *Nature* 424, 1035–1037.

Volkov, I., Banavar, J.R., Hubbell, S.P., and Maritan, A. (2007) Patterns of relative species abundance in rain forests and coral reefs. *Nature* 450, 45–49.

Chapter 10

FUNCTIONAL BASIS FOR RESOURCE NICHE PARTITIONING BY TROPICAL TREES

Kaoru Kitajima and Lourens Poorter

OVERVIEW

The resource niche, that is, specialization along resource availability gradients, is one of the frequently hypothesized mechanisms for coexistence of tropical tree species. Here, we evaluate physiological mechanisms that may lead to partitioning of resource gradients by tropical trees, with particular attention to light as the key limiting resource. The functional basis for light competitiveness is the extent to which individual tree crowns reduce light. Pioneer trees cannot invade the shaded space occupied by shade-tolerant tree crowns that maintain multiple layers of terminal shoots and leaves. Within the lowest stratum of a forest, light niche specialization by seedlings is better explained by the growth–survival trade-off, rather than by performance rank reversal between low and high light environments. In other words, seedlings of early successional species that specialize for treefall gaps tend to grow faster in both high and low light but suffer higher mortality than late successional species that regenerate in the shaded understory. As individuals grow beyond the seedling stage, ontogenetic shifts in light niche may contribute significantly to tree species coexistence in forests. Analysis of crown exposure index in relation to individual height demonstrates that nearly all species experience higher light levels when they grow towards the canopy, with frequent ontogenetic shifts of light niches. Adult size is also an important species-specific trait that influences fundamental light niches of trees. Compared with tall species, small species that have to invest less into support have sufficient carbon surplus to reproduce in shade of canopy dominants. Future studies need to address whether similar mechanisms underlie niche specialization along other resource gradients, how interaction of multiple resource axes and ontogenetic shifts may expand niche hyperspace, and how biotic factors modify the realized niche space along resource gradients.

NICHE – DEFINITIONS AND IMPLICATIONS

The ecological niche can be defined as a set of environmental factors that a species requires in order to persist in a community, as an integrative result of the collective impacts of environmental factors on the focal species as well as the focal species' impacts on its environment (see Chase and Leibold 2003 for variations from this definition). The niche of a species may be viewed as its position in the ecological multivariate space, often referred to as niche hyperspace (or hypervolume) defined by the relevant multiple environmental factors. In particular, specialization to a limited range of resource availability has been considered central to the niche theory that attempts to explain species coexistence (e.g., MacArthur and Levins 1967, Ricklefs 1977, Denslow 1980, Chesson 2000, Chase 2005). For plants, resources that potentially constrain population growth rates include light, water, and soil mineral nutrients. The competitive exclusion principle, historically attributed to Gause (1936), states that only one species can persist as the competitive winner when two species consume a single resource. Two or more types of resources are required to allow coexistence in theoretical models that consider only resource competition. However, often a single resource constrains growth rates, as stated by Liebig's law of the minimum (Tilman 1982). Thus, it is more likely that the fundamental niche of a species along a resource axis is defined by its tolerance of high and low levels of the most limiting resource, while interactions with competing neighbors and other biotic factors, including natural enemies, narrow each species distribution down to its realized niche where it is superior to its competitors. In theoretical models, the influence of natural enemies, such as predation rates, can be treated as a second niche axis (Chase and Leibold 2003).

Light is generally the most limiting resource in closed-canopy forests, and light competition is the major driver of successional change where soil resources are not limiting (Grime 1979, Tilman 1988). The total daily light varies more than 1000-fold within a forest from the forest floor to the canopy top, as well as from the shaded understory to an open clearing (Yoda 1974, Chazdon and Fetcher 1984). Yet, the light preferences of trees may be difficult to distinguish beyond a few broadly defined guilds (Hubbell and Foster 1986, Brown and Jenning 1998, Hubbell et al. 1999, Brokaw and Busing 2000), prompting development of a radically different view of species coexistence based solely on stochastic processes, known as the neutral theory (Hubbell 2001, 2005). However, the neutral theory fails to explain observations that unrelated species occupying similar habitats converge in physiological and life-history traits (ter Steege and Hammond 2001, Cavender-Bares et al. 2004a,b, Zanne et al. 2005). Furthermore, widely separated communities sharing a regional species pool often converge on similar species composition (Tuomisto and Poulsen 2000, Clark and McLachlan 2003). These observations support a strong role of niche-based mechanisms for distribution and abundance of species at local and regional scales (Condit et al. 2002). The real world perhaps falls somewhere between these two extreme theoretical views, such that stochastic processes interact with niche preferences of species (Svenning et al. 2004). An example of such compromise is a stochastic-niche

model of Tilman (2004), which assumes that a new species can invade the space already occupied by an established species only if the former can survive stochastic mortality while growing to maturity consuming the resources left unconsumed by the established competitor. Indeed, contemporary views on ecological niches recognize demographic and environmental stochasticity and consider spatial and temporal dimensions in biotic and abiotic factors (Chase 2005).

Hereafter, we evaluate functional mechanisms underlying resource niche specialization by trees and evidence for resource niche partitioning in the three-dimensional space within tropical forests. Which resource constitutes a significant niche axis depends on the temporary and spatial scale of investigation. Soil nutrient availability and water regime determined by bedrock and topography are relatively stable throughout the lifetime of a tree, creating a coarse matrix of niches for tropical trees and contributing to species turnover (= beta-diversity) at regional scales (Schulz 1960, Clark et al. 1998, Harms et al. 2001, Condit et al. 2002, Svenning et al. 2004). In contrast, the light niche is important as a potential mechanism to promote local (alpha) diversity, because light creates much finer and complex environmental heterogeneity in three dimensions within a forest. Light also exhibits unpredictable changes, such that a variety of ontogenetic trajectories for light niche preference could be potentially successful at a spot where a seed may arrive (Figure 10.1a). Thus, we focus primarily on light as the key limiting resource that shapes species-specific traits underlying trade-offs essential for niche partitioning. However, parallels can be drawn where and when nutrients or water are the limiting resource of species distribution and abundance.

This review consists of four parts. First, we consider the functional basis for light competition. Second, we review contrasting types of trade-offs that may contribute to light niche partitioning within a horizontal plane of the forest. Third, we review how light niche may be partitioned vertically. Lastly, we briefly review how more niches may be created through interaction of multiple resource axes, as well as how pests and mutualists may influence the realized niche breadth of each species.

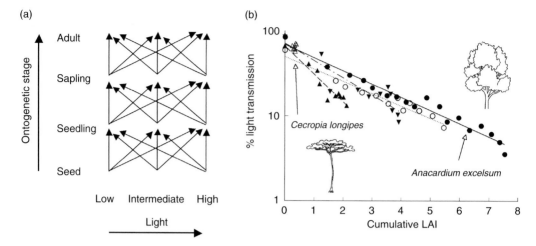

Figure 10.1 (a) Possible ontogenetic trajectories of light niches for tree species inferred by a crown-exposure index (modified from Poorter *et al.* 2005). Within a given horizontal plane, light availability is roughly classified into low, intermediate, and high based on crown exposure. The preferred light environment of a species may remain constant from seed to adult stage (vertical arrows) or may shift from one stage to the next (diagonal arrows). However, crown position alone cannot infer the average leaf light environment, which is also a function of leaf display patterns. (b) Species differences in % light transmission (on log scale) as a function of cumulative leaf area index (LAI) through individual crowns of five canopy tree species in a seasonal dry forest in Panama (modified from Kitajima *et al.* 2005). A pioneer, *Cecropia longipes* (open triangle at upper left) exhibits steep light extinction within a shallow crown of total LAI less than 1. In contrast, long-lived dominant *Anacardium excelsum* (closed circle) creates less steep light extinction through its crown, but its high total LAI casts much deeper shade underneath. Three other species exhibit intermediate characteristics in relation to architecture and successional status.

FUNCTIONAL MECHANISM FOR LIGHT COMPETITION

At the heart of niche theory is competitive asymmetry. An established plant cannot be competitively displaced by another plant, unless the latter is sufficiently superior in competition for the most limiting resource in that particular location. Does the functional basis for resource competitiveness lie in the rate of biomass accumulation per unit of the limiting resource consumed, or in the thoroughness with which the limiting resource is consumed? The latter is supported by the theoretical and empirical studies of nitrogen competition (Tilman 1982, Wedin and Tilman 1993). Growth rates of individuals and populations decline as competing individuals use up the resource in the shortest supply relative to demand. The resource level at which the net growth rate is zero is known as R^*, and theory predicts that the lower

the R^*, the greater the competitiveness for that resource (Tilman 1982, Chase and Leibold 2003). R^* varies among species and is quantified as the minimum level to which a monospecific stand of this species eventually drives down resource availability (Wedin and Tilman 1993). However, does this idea apply to the competitive interaction of trees for light in a humid tropical forest?

Observations of light utilization by adult tree crowns indeed suggest differences in R^* for light between early and late successional tree species. The light intensity received by the horizontal plane above the forest canopy is conventionally called "full sun," that is, the reference relative to which light intensity on an inclined surface or light transmitted through the canopy is expressed. Both in temperate (Canham *et al.* 1994) and tropical forests (Kabakoff and Chazdon 1996, Kitajima *et al.* 2005), the percentage of light transmitted below adult crowns of late successional trees is much lower than the percentage of light transmitted through early successional tree crowns (Figure 10.1b). The ability to cast deep shade requires maintenance of multiple leaf layers, as well as the tolerance of self-shading experienced by leaves positioned low within the crown. Adults of later successional canopy trees have more steeply inclined leaves, such that even the leaves at the uppermost layer of the canopy experience <20% of full sun. At the same time, more steeply inclined leaves (and also terminal shoots) result in a shallower regression slope of log (% of full sun transmitted) plotted against the cumulative leaf area index (Figure 10.1b). In other words, a leaf should receive more light and a greater net carbon gain when leaves above it are displayed at steeper inclinations. Hence, low light extinction coefficients allow a deep crown consisting of multiple leaf layers that collectively absorb more light, leaving less light to a small neighbor in its vicinity (Kitajima *et al.* 2005).

Greater leaf lifespan of late successional trees also provides an important physiological basis for maintenance of multiple leaf layers and competitive ability to cast deep shade. The most shaded leaves within the crown of late successional species experience light availability similar to that in the understory (e.g., much less than 5% of full sun). In contrast, individual leaves of the pioneer *Cecropia longipes* experience much higher average light levels per unit leaf area, and none of its leaves persist in light below 10% of full sun (Figure 10.1b). Thus, *Cecropia* trees cannot invade the space already occupied by deep crowns of long-lived trees. More studies are needed to reveal light competition strategies and coexistence of trees in the uppermost strata of the forest. It will also be interesting to examine leaf display and light extinction by liana crowns that may competitively suppress growth and reproduction of canopy trees (Avalos *et al.* 2007). In summary, the light competitiveness of the upper canopy trees and lianas may be predictable from their functional leaf traits, even though their competitive dominance cannot be inferred merely by their positions at a given time.

Smaller trees, including saplings and seedlings, experience strongly asymmetric light competition imposed by canopy trees, even though competition among tree seedlings is probably rare in the understory (Svenning *et al.* 2008) because they occur at low densities (e.g., 1–6 seedlings per m^2 in a neotropical rainforest; Harms *et al.* 2004). Under the closed canopy of humid tropical forests, only 0.5–3% of full sun reaches seedlings and saplings (Chazdon and Fetcher 1984, Montgomery and Chazdon 2002). The degree to which they can tolerate shade and maintain a positive net carbon balance must be an important determinant of juvenile distribution and abundance. Even slight increases in shade cast by an understory neighbor may have large consequences for seedling carbon balance (Montgomery and Chazdon 2002, Montgomery 2004).

In contrast, juveniles in treefall gaps compete with each other to pre-empt the higher strata and to cast shade upon their competitors. Hence, casual observers may predict that the fastest-growing individual will be the competitive winner in a given gap. But is this true? In reality, new treefall gaps are simultaneously colonized and occupied by seedlings of early successional and late successional species, depending on dispersal limitation and other chance events. Thus, species composition in a gap cannot be predicted simply from gap size or age (Popma *et al.* 1988, Hammond and Brown 1998, Schnitzer and Carson 2001, Dalling *et al.* 2004). Shade-tolerant tree juveniles may persist in a newly created gap for many years after being surpassed by a nearby pioneer tree. However, the former may eventually grow to displace the latter, possibly after the latter matures and senesces. Which one should be called the winner of light competition in this gap? Do gaps represent a niche position along the light gradient (Denslow 1980), a successional niche (Pacala and Rees 1998), or merely a phase in transient dynamics (Tilman 1988)? The answers to these questions differ depending on the spatio-temporal scale at which demographic dynamics are examined in gaps.

Furthermore, gaps of different sizes do not create discrete niches. Instead, light availability varies continuously across the gap–shade continuum in relation to heterogeneity of the overstory canopy and position within each gap (Brown 1993). Seedling light requirements also vary among species in a continuous manner (Augspurger 1984a). As a result, the abundance and establishment probabilities of seedlings exhibit different, yet overlapping, distributions among species in relation to light availability (Montgomery and Chazdon 2002, Poorter and Arets 2003). In other words, the competitive edge of one species over another at any particular light availability appears to be a matter of probability, which may be very subtle. What types of trade-offs lead to such continuous variations in preferred light environment of seedlings?

TRADE-OFFS PROMOTING SPECIES RICHNESS WITHIN A HORIZONTAL PLANE

Species-specific traits associated with size and biomass allocation patterns are thought to underlie various trade-offs that contribute to species sorting along a niche axis. Niche theory posits that coexistence of species A and B is possible when species A outperforms species B in one environment, but species B outperforms species A in a second environment. Individual fitness components, such as growth rates, survival rates, or fecundity of individuals, can be used to evaluate performance of potentially competing species in contrasting environments. However, these individual performance measures may not be positively correlated with each other, nor equally important in their relative contribution to overall fitness at different positions along a niche axis. Thus, two types of trade-off must be distinguished.

In the first type of trade-off, adaptations to one type of environment preclude optimal trait combinations in another environment, leading to a rank reversal in a fitness component between the two environments (Latham 1992, McPeek 1996, Chesson 2000). The second type of trade-off occurs between two fitness components, such as growth rates and survival (Brokaw 1987, Kitajima 1994, Poorter and Bongers 2006). If the relative importance of these fitness components shifts between two environments, it can lead to an overall performance rank reversal. A hybrid of

these two types of trade-off is often reported in the literature as a strong empirical pattern in relation to light environment, that is, high light growth versus low light survival trade-off (Kobe *et al.* 1995, Wright 2002, Baraloto *et al.* 2005, Hubbell 2005). But does this trade-off exist because species that grow fast in high light are somehow prevented from growing fast in the shade (the first type of trade-off), or because growth and survival, two fitness components, exhibit negative cross-species correlation regardless of the environment (the second type)? These two alternative hypotheses predict contrasting cross-species correlations when growth or survival is compared between two contrasting environments (Box 10.1).

For quantitative evaluations of these two alternative hypotheses, how survival and growth respond to light gradients must be quantified for multiple species, as shown in Figure 10.2. These types of response curves are known in population ecology as phenotypic reaction norms, and a rank reversal (i.e., crossing of reaction norms) represents a strong case of genotype × environment interaction (Schlichting 1986). Many species increase growth rates as light availability increases from deep shade (0.5–2% full sun) up to the light levels found in treefall gaps (e.g., 10–40% full sun), followed by a plateau and possibly a decline when light levels exceed the optimum (Figure 10.2b). Survival also tends to increase at higher light availability associated with larger treefall gaps, but shows no response or even a decline with gap size in some species (Figure 10.2a). In general, species differences in growth rates tend to be larger at higher light availability, while species differences in survival tend to be greater at low light than at high light.

How strong is the evidence for rank reversals of growth or survival rate along a light gradient when these two key performance traits are examined separately? For objective evaluation of the frequency of rank reversals, either parametric or non-parametric statistics may be used (Box 10.1). All statistical tests require that a sufficient number of species are compared between two contrasting environments that represent the two purported niches (e.g., gap versus shaded understory), or when possible, along the entire light gradient observed in the community.

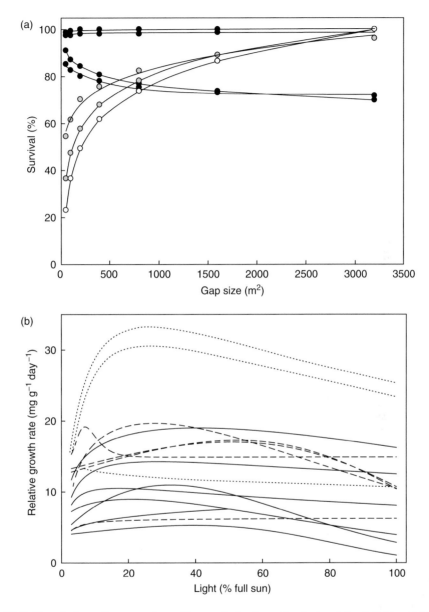

Figure 10.2 Phenotypic reaction norms for seedling survival and growth rates in relation to light availability. (a) Seedling survival of seven Guyanan tree species along a gradient of gap size (from Rose 2000). The open symbols represent gap-dependent species, while closed symbols represent shade-tolerant species. (b) Seedling relative growth rates of 15 Bolivian tree species in relation to experimental shading expressed as % full sun (data from Poorter 1999). The trend for each species is shown by best fitting polynomial regression.

Box 10.1 How do we test whether performance ranks of multiple species reverse between two environments?

A null expectation is that two randomly chosen species reverse their ranks at a chance of 50%. Given this, how many rank reversals are necessary in a community containing n species to conclude that rank reversals occur more frequently than expected by random chances? The answer depends on the total number of species in the test. In a three species system, three pair-wise comparisons are possible, and the null probability of finding 0, 1, 2, and 3 rank reversals, is 1/8, 3/8, 3/8, and 1/8. Thus, even when all possible pairs exhibit rank reversals in a 3 species system, the

null hypothesis of random rank reversal cannot be rejected at the conventional significance level of 0.05. In a four species system, only when all six possible pairwise comparisons exhibit rank reversals, the null hypothesis can be rejected at $P = 0.023$. Clearly, it is advisable to compare many more species in analysis of performance rank reversals. More generally, the maximum number of rank reversals for n species is $C_{n,2} = n(n-1)/2$, and the probability of finding equal to or more than k rank-reversals in a n-species system is given as:

$$\frac{\sum_{i=k}^{[n(n-1)/2]} C_{[n(n-1)/2],i}}{\sum_{i=1}^{[n(n-1)/2]} C_{[n(n-1)/2],i}}$$

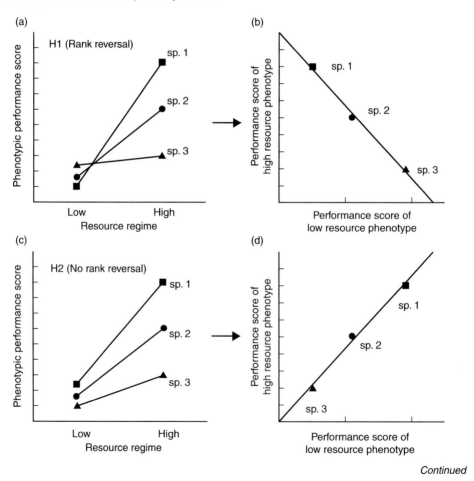

(a) H1 (Rank reversal) — Phenotypic performance score vs Resource regime (Low, High); sp. 1, sp. 2, sp. 3. (b) Performance score of high resource phenotype vs Performance score of low resource phenotype; sp. 1, sp. 2, sp. 3.

(c) H2 (No rank reversal) — Phenotypic performance score vs Resource regime (Low, High); sp. 1, sp. 2, sp. 3. (d) Performance score of high resource phenotype vs Performance score of low resource phenotype; sp. 1, sp. 2, sp. 3.

Continued

Box 10.1 Continued

This is mathematically equivalent of testing with a Kendall's rank concordance (B. Bolker, personal communication). It can be also tested with a Monte-Carlo simulation in which species ranks are randomly shuffled between two environments.

These non-parametric approaches are statistically conservative and blind to the magnitudes of differences among species within and across environments. However, real data of species performance often show greater difference among species in either high or resource regime (e.g., Figure 10.2, among species, survival rates vary more at the low light end, while growth rates differ at the high light end). Thus, parametric tests may provide not only statistically powerful approach, but also biologically meaningful insight, as graphically illustrated below. A community consisting of three species may exhibit two extreme patterns: H1) performance rank reversals in all three pair-wise combinations (a), and H2) no rank reversal in any pair of species (c). When these reaction norms are plotted as correlation plots, the rank-reversal case (a) is seen as a negative correlation (b), while the case of rank concordance (c) is seen as a positive correlation (d). More generally, whether species ranks are concordant or reverse among n species can be tested by Pearson's correlation

between performance measured between the high and low resource regimes as shown in the panels (b) and (d).

There is one caveat for both parametric and non-parametric tests described above. For example, most rank reversals of growth rates occur at very low light levels where species differences are often very small. Extrapolation of the reaction norms in the panel (c) suggests that these reactions norms might cross if a sufficiently lower light regime was used for the test. Indeed, some published nursery studies used light regimes of 3–5% full sun as the low light regime, which is higher than the typical level in the understory of mature tropical forests (0.5–3% of full sun). A parametric method proposed by Sack and Grubb (2001) circumvents this problem by calculating light at which the two lines cross, which is essentially an extrapolation of reaction norms (see Kitajima and Bolker 2003; Sack and Grubb 2003 for details of statistical pros and cons associated with this as well as other methods). Given pros and cons associated with all statistical methods, it is essential that the experiments to be conducted with sufficient number of species, using realistic contrasts to represent the resource regimes observed in the field.

The seedling survival data of Rose (2000; Figure 10.2a) across a large gradient of gap size show that rank reversals are observed only above gap size greater than 1000 m^2, below which species exhibit a perfect rank concordance; the four shade-tolerant species survive better than the three light-demanding species. Two of the four shade-tolerant species maintain almost 100% survival across the entire gap-size gradient, while the other two exhibit lower survival at larger gaps. Three light-demanding species, in contrast, respond strongly and positively to increasing gap size. Yet, survival probabilities of three light-demanding species do not surpass that of two shade-tolerant species even in the largest gaps. Similarly, Kobe (1999) found that seedling survival of four species responded positively to higher light availability, except survival of one of the two shade-tolerant species decreased above 20% full sun (i.e., light level typically found in treefall gaps).

The fact that survival rank reversals are observed at light levels higher than typical for treefall gaps speaks against the relevance of rank reversals for light partitioning between shade-tolerant and gap-dependent species.

Seedling survivorship of nine species studied on Barro Colorado Island (BCI) exhibits only five out of the maximum possible 36 rank reversals between shaded understory and gaps (Figure 10.3a, Augspurger 1984a). This frequency is lower than expected by a statistical null model (rejected at $P = 0.009$ with a Monte Carlo analysis with 1000 shuffles, assuming 50% as a chance of rank reversal between any randomly selected pair of species; see Box 10.1). Hence, species that survive well relative to other species in shade also survive well in gaps (Figure 10.3b; Pearson's correlation coefficient $r = 0.76$, $P < 0.02$, and Kendall's $\tau = 0.70$, $P < 0.009$). A positive correlation of survival rate was also found

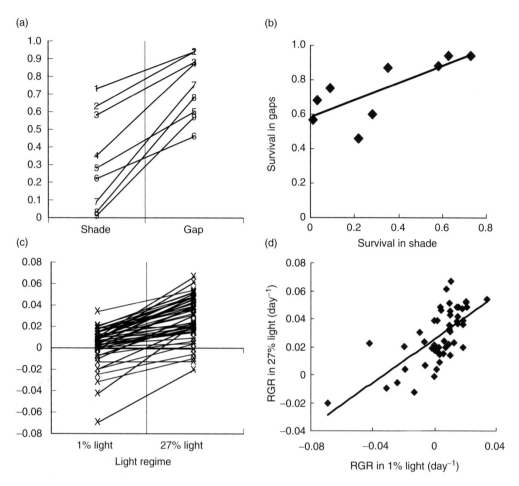

Figure 10.3 Survival and growth responses of woody seedlings on Barro Colorado Island (BCI), Panama, to sun and shade conditions, presented as phenotypic reaction norms (a,c; two points connected by a line represent a species) and as correlation plots between the two light environments (b,d; each point is a species mean). These figures are shown in a manner to correspond to the analysis shown in Box 10.1. (a,b) Proportion of seedlings of nine canopy tree species surviving from germination to 2 months in the shaded understory and treefall gaps (Augspurger 1984a). (c,d) Seedling relative growth rate (RGR) of 50 woody species on BCI determined under 1% and 27% of full sun in a screened enclosure from time of the expansion of first true leaves until cotyledons were lost or 10 weeks later (data from Kitajima 1992).

across 18 species grown under high and low light in a nursery (Augspurger 1984b).

Do species ranks of growth rates reverse between two light environments? Relative growth rate of seedling biomass (RGR, rate of size increment per unit biomass per unit time) is ideal for such analysis as it standardizes for size

differences among species. Relative growth rate of seedlings exhibits an overall rank concordance between high and low light environments among 15 Bolivian tree species along a wide range of light (Figure 10.2b) and among 50 woody species on BCI grown under 1% and 27% full sun (Figure 10.3c,d). Non-parametric statistics applied to the latter dataset shows that the frequency of rank reversals was significantly less than the null expectation (313 out of 1225 maximum possible rank reversals, $P < 0.001$ by a Monte Carlo simulation of 1000 shuffles; Kendall's $\tau = 0.49$, $P < 0.0001$; Pearson's $r = 0.52$, $P < 0.0001$). Significant concordance of RGR between sun and shade is also demonstrated in many other studies (Ellison *et al.* 1993, Kitajima 1994, Osunkoya *et al.* 1994, Valladares *et al.* 2000, Bloor and Grubb 2003, Dalling *et al.* 2004, Baraloto *et al.* 2005), with few studies showing a lack of a relationship (Popma and Bongers 1988) or the opposite pattern (Agyeman *et al.* 1999). Thus, species switch their growth rate ranks between the understory and gaps less frequently than expected according to null models.

Parametric analysis demonstrates that most rank reversals in growth rates are expected to occur at extremely low light availability (Sack and Grubb 2001). However, differences in growth rates are so small in deep shade that carefully replicated experiments should be used to detect rank reversals (Kitajima and Bolker 2003). Because most rank reversals for growth rates occur between species that are similar to each other, growth rank reversals cannot explain the habitat difference between pioneers and shade-tolerant species (Kitajima 1994), and not even between pioneers that prefer large versus small gaps (Dalling *et al.* 2004). Thus, performance rank reversals in either growth rate or survival alone do not provide a general mechanism underlying light preference of seedling distribution.

However, overall performance of species must be evaluated in an integrative manner taking into account both growth rates and survival, because of the trade-off between these two performance measures (Kitajima 1994, Kobe 1999). Among the six species overlapping between the two datasets shown in Figure 10.4, growth rates are negatively correlated with survival probability in low light ($r = -0.97$, $P = 0.007$; $\tau = -0.80$, $P = 0.05$), as well as in high light ($r = -0.57$, $P = 0.23$; $\tau = -0.60$, $P = 0.15$). Such negative cross-species correlation can be explained by allocation-based trade-offs; allocation patterns that enhance growth

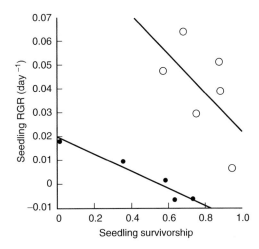

Figure 10.4 Growth–survival trade-offs for seedlings grown under high and low light regimes (open and closed symbols, respectively) among six and five species that overlap in each light regime between the two datasets shown in Figure 10.3.

rates may come at the cost of reduced biomass allocation to survival-enhancing functions, such as structural and chemical defenses and storage (Kitajima 1996). More importantly, species' positions along the growth–survival trade-off line are associated with their light niches in Figure 10.4; along each regression line, gap species occupy the "fast-growth, low-survival" end, whereas shade-tolerant species occupy the "slow-growth, high-survival" end. A similar association between light preference and the position along the growth–survival trade-off line has been found for naturally recruited seedlings of 22 liana and 31 tree species in Panama (Gilbert *et al.* 2006), as well as for saplings of 53 rainforest tree species in Bolivia (Poorter and Bongers 2006).

Would species performance ranks reverse more often between low and high light when growth and survival, which are negatively correlated with each other, are integrated into one performance measure? In general, growth–survival trade-offs should have an equalizing effect, that is, yielding more similar reaction norms across species for a performance measure integrating both growth and survival, compared with the reaction norms for either growth or survival alone. Most likely, such an equalizing effect means greater importance of stochasticity over niches. Yet, this simple prediction does not explain the observed match of a species' position along the trade-off line with its habitat preference (Poorter and Arets 2003, Wright *et al.* 2003, Poorter and Bongers 2006). This puzzle may be solved if the relative fitness value of growth and survival change non-linearly along the light gradient. Fitness value of ability to grow fast should be high only under high light availability, as fast-growing species can achieve their full growth potential only when there is sufficient light. In contrast, fitness value of traits that enhance survival is greater in lower light, as it takes a long time to recover from damage incurred by disease, herbivores, and falling debris in shade. How to integrate growth and survival in order to evaluate overall performance of species is an important challenge left to future studies.

Differences in individual functional traits clearly underlie species differences in growth rates and survival of seedlings. Relative to more light-demanding species, shade-tolerant trees tend to have large seeds, large initial seedling size, storage cotyledons, dense stem and leaf tissue, low specific leaf area (SLA, leaf area divided by leaf mass), low leaf area ratio (LAR, leaf area divided by whole-plant mass), and high root:shoot ratio (Kitajima 1994, 1996, Osunkoya *et al.* 1994, Cornelissen *et al.* 1996, Veneklaas and Poorter 1998, Poorter 1999, Zanne *et al.* 2005). Higher stem wood density enhances survival of seedlings (Augspurger 1984b), probably because higher tissue density increases biomechanical strength for protection against physical disturbance (Clark and Clark 1985), herbivory, and disease (Alvarez-Clare and Kitajima 2007). Large seedling size not only helps seedlings to emerge from the litter layer (Molofsky and Augspurger 1992), but also enhances seedling survival in the understory through a larger pool of non-structural carbohydrate reserves (which is the product of carbohydrate concentration and biomass; Myers and Kitajima 2007, Poorter and Kitajima 2007). Indeed, carbohydrate pool size, rather than seed mass, seedling mass, or cotyledon mass, predicts survival and tolerance of seedlings following experimental shading

(0.08% of full sun for 2 months) and defoliation (Myers and Kitajima 2007). These multiple trait associations together provide the mechanistic basis for the slower growth as well as higher survival for seedlings of shade-tolerant species relative to light-demanding species (Kitajima 1996, Veneklaas and Poorter 1998).

VERTICAL LIGHT GRADIENTS AND ONTOGENETIC SHIFTS

Ontogenetic shifts, that is, switching of growth and survival ranks of species between size classes, may represent a trade-off that contributes to species coexistence (Baraloto *et al.* 2005). The null model for this idea is ontogenetic concordance, which is expected when relative differences among species in functional traits, such as allocation patterns and leaf traits, are maintained regardless of plant size (Poorter 2007). Do pioneers always outgrow shade-tolerant species (null hypothesis), or do ontogenetic shifts in performance occur (alternative hypothesis) for physiological reasons? Physically well-defended leaves (Coley 1988, Poorter *et al.* 2004) and stems (Guariguata 1998, Van Gelder *et al.* 2006), as well as carbohydrate reserves (Kobe 1997, Poorter and Kitajima 2007), are important for survival not only at the seedling stage, but also at the sapling stage. Across Panamanian tree species, species that grow fast as a sapling in the understory also grow fast in gaps (Pearson's $r = 0.61$, $P < 0.001$, $n = 115$ species; data from Welden *et al.* 1991), but those that grow fast in the understory are also likely to die fast ($r = 0.52$, $P < 0.001$, $n = 108$). Hence, the growth–survival trade-off is as important in the sapling stage as in the seedling stage. More importantly, species' relative positions along the growth–survival trade-off lines are generally concordant for saplings and seedlings among tree species (Gilbert *et al.* 2006).

Yet, there are functional reasons to suspect ontogenetic shifts in growth and survival rates among species. Light-demanding species are thought to have a growth advantage over shade-tolerant species because of their high LAR. Many light-demanding species are small seeded, and they deploy their leaf area rapidly to become autotrophic (Kitajima 2002). Interspecific differences in LAR are especially marked at the seedling stage, but tend to decrease over time when plants increase in size (Poorter and Rose 2005). Larger plants with larger leaves require greater support, causing ontogenetic declines in SLA, leaf mass ratio (= ratio of leaf mass to the total biomass), and LAR (Boot 1996, Veneklaas and Poorter 1998, Delagrange et al. 2004). These changes, in turn, cause a reduction in RGR. The size-dependent decline in LAR has important consequences for the whole-plant light compensation point. As the ratio of photosynthesizing to respiratory tissue decreases, the plant needs to encounter brighter light conditions to support the greater respiratory mass (Givnish 1988). All species exhibit size-dependent shifts in allocation patterns that should result in slower growth rates under exactly the same light availability when they are larger. Yet, juveniles of all six species examined by Clark and Clark (1992) exhibit increasing survival and growth rates as taller individuals, partly because they receive more light at higher strata of the forest. The critical question is whether such ontogenetic declines in LAR are predictably faster for pioneers than for shade-tolerant species, so as to result in ontogenetic shifts of species ranking.

Lusk (2004) analyzed ontogenetic changes in LAR for four temperate rainforest species growing in the understory. Pioneers started out with a higher LAR, because of a high SLA. Their LAR declined rapidly with height due to their fast leaf turnover, whereas shade-tolerant species maintained their LAR because of their long leaf retention times. Consequently, shade-tolerant species have a consistently higher LAR than light-demanding species at the sapling stage (Lusk 2002). Interestingly, the higher carbon gain rate per mass expected from the higher LAR did not lead to differences in growth rates, suggesting possible interspecific differences in allocation to storage and defense. Pioneer species continue to produce new leaves and to extend shoots, in an attempt to receive more light available at greater height. Such a strategy may pay off in dense gap vegetation that creates a steep increase in light with height (Denslow 1995). It may fail, however, in the understory, where the vertical light gradient is less steep (Montgomery 2004). Pioneer species may be unable to sustain such a rapid leaf turnover in the light-limited understory and literally may grow themselves to death when their LAR falls below a critical threshold level.

Differences in size at maturity may also contribute to partitioning of the vertical height gradient and species coexistence (Richards 1952, Terborgh 1985). Compared with the gap–shade paradigm, the small–large paradigm has received considerably less attention, despite its importance for many aspects of the life cycle of a tree (Kohyama 1993, Westoby 1998, Thomas and Bazzaz 1999, Turner 2001, Falster and Westoby 2005, King et al. 2006). Yet, niche specialization of adults along the vertical height gradient may be evolutionarily more important than light niche preference of juveniles. Phylogenetic constraints may make related species to occupy similar height niches. Different families prefer different canopy positions (e.g., species in Annonaceae and Rubiaceae tend to be small understory specialists, and those in Dipterocarpaceae and Fabaceae tend to mature as canopy dominants). At the same time, niche diversification may be important for related taxa with similar ecological requirements to avoid competition. Indeed, species-rich genera in Malaysia can show remarkable size variation among sympatric species (Thomas 1996a).

A trade-off between maximization of current versus future light interception is one of the functional mechanisms leading to vertical light niche partitioning by adult trees. Species that mature in the forest understory often differ from canopy species in their architecture, light utilization strategy, and shade tolerance. Understory species maximize current light interception by making wide crowns, whereas canopy species maximize future light interception by making narrow crowns, which allows them to grow quickly to the canopy (King 1990, Kohyama and Hotta 1990). Understory species have relatively thick stems to support the wide crowns and resist dynamic loading due to falling debris (King 1986, van Gelder et al. 2006), whereas canopy species have slender stems, to rapidly attain the canopy at low costs for construction and support (Kohyama et al. 2003, Poorter *et al.* 2003). Indeed, species with slender stems and narrow crowns show a faster height-related increase in crown exposure (Poorter *et al.* 2005). For these species the extension function of architecture is more important than the light interception function; they gamble upon reaping the benefits of a better and brighter future in the canopy. Accordingly, Kohyama (1987) referred to these species as "optimists," whereas the understory species were referred to as "pessimists." A theoretical model by Kohyama (1993) predicts that small species are able to coexist with tall species only if the former have a higher recruitment rate. Such a relationship was indeed observed for 27 tree species that co-occurred in a Bornean dipterocarp forest (Kohyama *et al.* 2003). Sapling recruitment rate per adult basal area was negatively correlated with adult height (cf. King *et al.* 2006). Similarly, in 45 Costa Rican wet forest species, the per capita recruitment rate was negatively correlated with tree lifespan (Lieberman *et al.* 1985), which is closely associated with the maximal size of the species.

The second mechanism underlying the trade-off between small versus tall adult size is the cost of reproduction. Tree height increases steeply with diameter at breast height (dbh), and levels off when species start to reproduce (Thomas 1996a). Small species start to reproduce at a smaller dbh than large species (Lieberman *et al.* 1985, Thomas 1996b, van Ulft 2004, Wright *et al.* 2005). Carbon allocation to reproduction cannot be invested in height growth, and small species are thus left behind in the race for the canopy (Turner 2001). Large species often delay their reproduction until they are in the canopy, and can expand their tree crown. Greater photosynthetic productivity of large and well-exposed crowns enables these species to produce large seeds (Hammond and Brown 1995, Metcalfe and Grubb 1995) and/or a large seed crop (van Rheenen 2005). Annual seed production is therefore positively correlated with the adult stature of the species (Davies and Ashton 1999). It might well be that the high seed production balances the delayed reproduction, leading to a similar lifetime seed production for small and large species (Moles *et al.* 2004). Yet, good comparative data to support this hypothesis are still lacking for tropical rainforest trees.

The two mechanisms described above, that is, the trade-off between current versus future light interception, and the trade-off between early versus late reproduction, can explain vertical niche segregation of short versus tall species. While light availability is positively correlated with height in general (Yoda 1974), the exact future light environment is unpredictable for a given seedling because of unpredictability associated with overstory canopy characteristics and dynamics. Thus, temporal unpredictability of light availability may equalize fitness associated with a variety of ontogenetic trajectories of light preference and contributes to species coexistence. How do tree species vary in ontogenetic trajectories for light environment within each adult stature class (e.g., among canopy tree species), as well as between adult stature classes (e.g., between subcanopy versus canopy species)?

There are many different ways for trees to grow and mature (Figure 10.1a). Poorter *et al.* (2005) evaluated the height–light trajectories of 53 co-occurring Liberian wet forest tree species, using a crown-exposure index (Dawkins and Field 1978). Nine different height–light trajectories were distinguished based on the light environments of juveniles and adults, compared with the average vertical light profile in the forest canopy. The majority of the species simply followed the vertical light profile in the forest canopy (Figure 10.5a). Only one species occurred consistently at higher light levels than the average light profile (whole-life light demander), and one species occurred at consistently lower light levels (whole-life shade tolerant). One species (*Syzygium gardneri*) experienced decreasing light when growing in height. This species germinates in the high light environment of gaps, but becomes quickly overshaded by faster-growing neighbors. It therefore switches from a light demander as a seedling to a shade tolerant as a sapling. A similar strategy has been observed for *Alseis blackiana* in Panama (Dalling *et al.* 2001). Species with such behavior are also known as "gamblers" (Oldeman and van Dijk 1991) or "cryptic pioneers" (Hawthorne 1995). Some species exhibit more complicated trajectories, appearing to be relatively shade tolerant in the middle stage (Clark and Clark 1992).

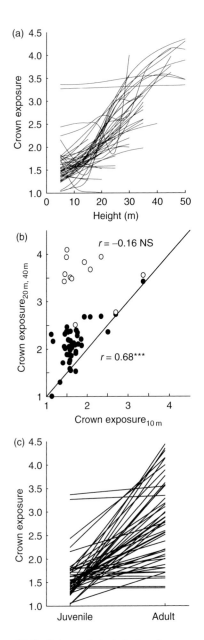

Figure 10.5 Ontogenetic trajectories of crown-exposure index for 53 Liberian rainforest trees. (a) Height-dependent change in crown exposure. (b) Correlation between species' crown exposure at 10 m height and at 20 m (closed symbols) and 40 m (open symbols). (c) Changes in crown exposure between the juvenile and adult phases. Each species is presented by a line (a,c) or a symbol (b). From Poorter *et al.* (2005).

However, a species' distribution in relation to light environment by itself is only a snapshot that merely suggests the light environment in which its growth and survival are optimal. For example, saplings may be shrinking in size, even when they persist in shade by relying on carbohydrate reserves accumulated when light availability is temporarily high. Thus, for more rigorous evaluation, growth and survival patterns need to be examined in relation to light, as well as relative to patterns exhibited by other species. Indeed, *Alseis blackiana* does not shift relative position along the interspecific growth–survival trade-off relationship between seedling and sapling stages (Gilbert *et al.* 2006). Such a shift was observed in only one of the 30 tree species for which growth–survival trade-offs were analyzed for seedlings and saplings; *Inga marginata* is a shade-tolerant species with slow growth as seedlings, but a fast-growing species with higher mortality as saplings (Gilbert *et al.* 2006).

Species differ in crown exposure when compared at a similar size (Figure 10.5a). But to what extent is this species ranking in crown exposure maintained when they increase in height? The crown exposure is positively correlated when trees of 10 and 20 m height are compared (Figure 10.5b); species that have a high crown exposure when small also have a high crown exposure when tall. However, this consistency in ranking disappears when species approach the canopy and attain similar full light levels. Furthermore, rank reversals in crown exposure are common between juveniles and adults (Figure 10.5c). Tall species have higher population-level crown exposures, higher adult crown exposures, and make larger switches in their crown exposure from juvenile to adults compared with small species. Adult stature is therefore an important life-history trait for which many species differentiate (Thomas and Bazzaz 1999, Turner 2001). Tall species should be very plastic in their traits, given the large switch they make from low light as seedling to high light as a canopy tree (e.g., Rijkers *et al.* 2000), but if there is a limit to the acclimation potential of species, then it follows that tall canopy species should be more light demanding than shade-tolerant species (Thomas and Bazzaz 1999). Indeed, a positive correlation has been found between adult stature and juvenile light requirements (e.g., Poorter *et al.* 2003, 2006, Sheil *et al.* 2006), but the relationship is not tight. Furthermore, functional traits of adult leaves can be explained better by regeneration niches rather than adult niches, suggesting long-lasting selective importance of seedling regeneration stage (Poorter 2007).

In summary, high heterogeneity of light in time and space creates many niche opportunities through various types of trade-off, including trade-offs between fast growth versus high survival, current versus future light interception, and early maturity versus large fecundity. Along each type of trade-off, multiple suites of traits are associated in a convergent manner. The relative position along the growth–survival trade-off is indicative of the species' preferred light environment at a given size class. This position is generally concordant through ontogenetic stages even though notable exceptions exist (Gilbert *et al.* 2006). Still, light is but one ecological factor that impacts growth and survival. Do we see similar growth–survival trade-offs in relation to the species' position along niche axes defined by other resources? Will there be more niche opportunities when other ecological factors interact with light?

NICHE HYPERSPACE

The total volume of niche hyperspace that allows coexistence of similar life-forms may be expanded in a multiplicative manner if niche axes are orthogonal to each other (i.e., varying independently). If two or more resources limit plant performance, then the total number of performance rank reversals may be greater than the number of reversals that occur along a single resource gradient (Latham 1992, Burslem *et al.* 1996, Walters and Reich 1996). This requires orthogonality of not only resource gradients, but also functional traits; adaptations in relation to one resource gradient may be independent of adaptations in relation to another resource gradient. Orthogonality of niche axes may be suggested in the multivariate space defined by functional traits of potentially competing species. Principal components analysis and other multivariate

statistics reduce the dimensionality of multiple trait spaces, often to just two dimensions defined by the first and second principal components that are orthogonal to each other. The traits associated with growth–survival trade-offs, such as SLA, LAR, leaf lifespan, photosynthetic rates per unit leaf mass, and nitrogen per unit mass, form the first principal component axis, while photosynthetic water use efficiency and nitrogen per unit area form the second principal component axis (Poorter and Bongers 2006). Thus, functional traits correlated with light gradients form the first principal component axis, while traits associated with use of soil resources form the second principal component axis. Orthogonality of adaptations to two different resource axes can also be shown experimentally; shade and drought tolerance were uncorrelated, and therefore orthogonal to each other, among temperate shrubs (Sack *et al.* 2003).

Yet, availabilities of light, nutrients, and water may not vary independently of each other in the field. Nutrient and water availabilities are strongly associated with each other in relation to topography, such that moist sites tend be more fertile (Svenning *et al.* 2004). Likewise, the rainfall gradient strongly influences nutrient regimes in tropical forests (Schuur and Matson 2001, Santiago *et al.* 2004), such that very wet forests tend to be infertile due to greater degrees of leaching. Thus, niche differentiation due to moisture and nutrient availabilities may be difficult to distinguish, as they change together in relation to topography, soil texture, and rainfall. Light gradient is also not totally independent of soil resource availability, as wetter forests tend to support denser vegetation leading to darker understories. Certain combinations, such as "fertile and very open forest" or "infertile and very dark forest," are very rare.

Distribution of species in relation to topography may not reflect adaptations to contrasting nutrient or water availability, but rather adaptations to other ecological factors that change with topography. The difference in distribution of two *Mora* species in relation to topography in Guyana could be explained not by seedling drought tolerance, but by differences in flood tolerance of seeds (ter Steege 1994). In dry forest and savanna biomes, fire-related adaptations explain species distribution in relation to topography better than adaptations to nutrient and water availability (Hoffmann et al. 2004). Topography at small spatial scale also creates rare niches, modulating resource availability regimes in time and space (Grubb 1977, 1996). Steep ravines, for example, offer moist and shady microhabitat, but its unstable substrate leads to frequent disturbance that benefits light demanders and resprouters. Abiotic microsite characteristics, such as tip-up mounds in gaps (Putz 1983) and litter-free slopes in shade (Metcalfe and Grubb 1995), benefit small-seeded species that would not be able to tolerate burial by litterfall otherwise.

Biotic factors are as important as abiotic factors in structuring niche hyperspace (Colwell and Fuente 1975, McPeek 1996, Chase and Leibold 2003). Growth–survival trade-offs involving herbivore defense are important in explaining niche position not only along light gradients (Coley 1988, Kitajima 1994) but also along soil fertility gradients (Fine et al. 2004). Positive cross-species correlations for growth between high and low fertilities among 34 species in Mexico (Huante et al. 1995) and nine species in French Guyana (Baraloto et al. 2005) suggest that specialization to rich versus poor soil cannot be explained by rank reversal of growth rates across the soil gradient. Indeed, species specialized to fertile sites tend to grow faster across soil gradients in both temperate (Schreeg et al. 2005) and tropical forests (Huante et al. 1995, Fine et al. 2004). Species that specialize in infertile white sand allocate more to defense and grow slowly regardless of soil type, but they survive significantly better than clay specialists when seedlings are grown in the white sand and exposed to herbivores (Fine et al. 2004).

Natural enemies are important for species coexistence not only via density-dependent predator control according to the Janzen–Connell mechanism (see Carson and Schnitzer Chapter 13), but also in changing the realized niche position and breadth of a species from its fundamental niche (Gilbert et al. 2001, Ahumada et al. 2004). Cases of biological invasions, where species are introduced to environments free of natural enemies, can offer excellent natural experiments to test this idea. *Clidemia hirta* invades the understory

of rainforest in Hawaii, where it persists in much shadier environment than in its native habitats in Costa Rica (DeWalt et al. 2004). Herbivory rates, as well as plant response to experimental fumigation, are lower in Hawaii than in Costa Rica. These results demonstrate that the absence of natural enemies enhances the carbon balance of the plant, reducing the minimum light requirement and expanding the realized light niche (DeWalt et al. 2004). Such truncating effects of natural enemies can be treated in a theoretical model as a second niche axis (Chase and Leibold 2003). The presence of mutualists would have an opposite effect to those of natural enemies and expand the realized niche breadth. We hope to see many more experimental studies to reveal the interactions between biotic and abiotic factors in shaping niche hyperspace in species-rich tropical forests.

CONCLUDING REMARKS

The role of niches in explaining species coexistence has been debated as long as the history of ecology as a scientific discipline (Chase and Leibold 2003). The classical and strictly deterministic view of niche is no longer favored by many ecologists. The niche theory is most strongly objected by the neutral theory proposed by Hubbell (2001) who uses tropical tree communities for empirical support of his theory. Yet, species are not equal as postulated by the neutral theory, but exhibit wide variations in multiple functional traits and life-history strategies in relation to their preferred light environment. Hence, the contemporary views of ecological niches incorporate a variety of ecological factors in relation to life-history trade-offs (Chase and Leibold 2003, Leigh et al. 2004). At a first glance, growth–survival trade-offs associated with contrasting allocation patterns may simply provide an equalizing force that contributes to ecological similarity of the species as postulated by the neutral theory, instead of a stabilizing force to allow each species to persist via its competitive superiority in its own niche in the community (Chesson 2000). Yet, there is an empirical link between growth–survival trade-offs and apparent light niches of tree species. Species whose allocation patterns place a high priority on fast growth thrive in gaps, but such allocation patterns are accompanied by high susceptibility to disease, herbivores, and physical damage that makes it impossible to survive in shaded understory. A contrasting allocation pattern is to place priority in persistence through allocation to defense and storage, which results in inherently slow growth. Thus, we propose that a key to resolve this apparent paradox lies in the role of natural enemies in modulating the relative importance of growth and survival along the resource availability gradient. For proper evaluation of niches as a mechanism to promote species coexistence, future studies need to address how adult trees exert asymmetric competition for light and other resources, as well as demographic integration of growth–survival trade-offs through multiple ontogenetic stages.

ACKNOWLEDGMENTS

We gratefully acknowledge constructive comments at various stages of preparation of this chapter from D. Tilman, S. Pacala, M. Walters, S. Mulkey, C. Augspurger, E. Herre, F. Valladares, B. Bolker, E. Leigh, L. Sack, F. Putz, S.J. Wright, H. Muller-Landau, F. Bongers, R. Kobe, and an anonymous reviewer. K.K. was supported by NSF-EEP0093303. L.P. was supported by Veni grant 863.02.007 from the Netherlands Organization of Scientific Research (NWO).

REFERENCES

Agyeman, V.K., Swaine, M.D., and Thompson, J. (1999) Responses of tropical forest tree seedlings to irradiance and the derivation of a light response index. *Journal of Ecology* 87, 815–827.

Ahumada, J.A., Hubbell, S.P., Condit, R., and Foster, R.B. (2004) Long-term tree survival in a Neotropical forest: the influence of local biotic neighborhood. In E. Losos and E.G. Leigh, Jr. (eds), *Tropical Forest Diversity and Dynamism.* University of Chicago Press, Chicago, IL, pp. 408–432.

Alvarez-Clare, S. and Kitajima, K. (2007) Physical defense traits enhance seedling survival of neotropical tree species. *Functional Ecology* 21, 1044–1054.

Augspurger, C.K. (1984a) Seedling survival of tropical tree species; interactions of dispersal distance, light gaps, and pathogens. *Ecology* 65, 1705–1712.

Augspurger, C.K. (1984b) Light requirements of neotropical tree seedlings: a comparative study of growth and survival. *Journal of Ecology* 72, 777–795.

Avalos, G., Mulkey, S.S., Kitajima, K., and Wright, S.J. (2007) Canopy colonization strategies of two liana species in a tropical dry forest. *Biotropica* 39, 393–399.

Baraloto, C., Goldberg, D.E., and Bonal, D. (2005) Performance trade-offs among tropical tree seedlings in contrasting microhabitats. *Ecology* 86, 2461–2472.

Bloor, J.M.G. and Grubb, P.J. (2003) Growth and mortality in high and low light: trends among 15 shade-tolerant tropical rain forest tree species. *Journal of Ecology* 91, 77–85.

Boot, R.G.A. (1996) The significance of seedling size and growth rate of tropical rain forest tree seedlings for regeneration in canopy openings. In M.D. Swaine (ed.), *The Ecology of Tropical Forest Tree Seedlings,* UNESCO, Paris, pp. 267–284.

Brokaw, N. and Busing, R.T. (2000) Niche versus chance and tree diversity in forest gaps. *Trends in Ecology and Evolution* 15, 184–188.

Brokaw, N.V.L. (1987) Gap-phase regeneration of three pioneer tree species in a tropical forest. *Journal of Ecology* 75, 9–19.

Brown, N. (1993) The implications of climate and gap microclimate for seedling growth conditions in a Bornean lowland rain forest. *Journal of Tropical Ecology* 9, 153–168.

Brown, N.D. and Jennings, S. (1998) Gap-size niche differentiation by tropical rainforest trees: a testable hypothesis or a broken-down bandwagon. In D.M. Newbery, H.H.T. Prins, and N.D. Brown (eds), *Dynamics of Tropical Communities.* Blackwell Science, Oxford, pp. 79–94.

Burslem, D.F.R.P., Grubb, P.J., and Turner, I.M. (1996) Responses to simulated drought and elevated nutrient supply among shade-tolerant tree seedlings of lowland tropical forest in Singapore. *Biotropica* 28, 636–648.

Canham, C.D., Finzi, A.D., Pacala, S.W., and Burbank, D.H. (1994) Causes and consequences of resource heterogeneity in forests: interspecific variation in light transmission by canopy trees. *Canadian Journal of Forest Research* 24, 337–340.

Cavender-Bares, J., Ackerly, D.D., Baum, D.A., and Bazzaz, F.A. (2004a) Phylogenetic overdispersion in Floridian oak communities. *American Naturalist* 163, 823–843.

Cavender-Bares, J., Kitajima, K., and Bazzaz, F.A. (2004b) Multiple trait associations in relation to habitat differentiation among 17 Floridian oak species. *Ecological Monographs* 74, 635–662.

Chase, J.M. (2005) Towards a really unified theory for metacommunities. *Functional Ecology* 19, 182–186.

Chase, J.M. and Leibold, M.A. (2003) *Ecological Niches: Linking Classical and Contemporary Approaches*. University of Chicago Press, Chicago, p. 212.

Chazdon, R.L. and Fetcher, N. (1984) Photosynthetic light environments in a lowland tropical rain forest in Costa Rica. *Journal of Ecology* 72, 553–564.

Chesson, P.L. (2000) Mechanisms of maintenance of species diversity. *Annual Review of Ecology and Systematics* 31, 343–366.

Clark, D.A. and Clark, D.B. (1992) Life history diversity of canopy and emergent trees in a neotropical rain forest. *Ecological Monographs* 62, 315–344.

Clark, D.B. and Clark, D.A. (1985) Seedling dynamics of a tropical tree: impact of herbivory and meristem damage. *Ecology* 66, 1884–1892.

Clark D.B., Clark, D.A., and Read, J.M. (1998) Edaphic variation and the mesoscale distribution of tree species in a neotropical rain forest. *Journal of Ecology* 86, 101–112.

Clark, J.S. and McLachlan, J.S. (2003) Stability of forest biodiversity. *Nature* 423, 635–638.

Coley, P.D. (1988) Effects of plant growth rate and leaf lifetime on the amount and type of anti-herbivore defense. *Oecologia* 74, 531–536.

Colwell, R.K. and Fuentes, E.R.(1975) Experimental studies of the niche. *Annual Review of Ecology and Systematics* 6, 281–310.

Condit, R., Pitman, N., Leigh, E.G. *et al.* (2002) Beta-diversity in tropical forest trees. *Science* 295, 666–669.

Cornelissen, J.H.C., Diez, P.C., and Hunt, R. (1996) Seedling growth, allocation and leaf attributes in a wide range of woody plant species and types. *Journal of Ecology* 84, 755–765.

Dalling, J.W., Winter, K., and Hubbell, S.P. (2004) Variation in growth responses of neotropical pioneers to simulated forest gaps. *Functional Ecology* 18, 725–736.

Dalling, J.W., Winter, K., Nason, J.D., Hubbell, S.P., Murawski, D.A., and Hamrick, J.L. (2001) The unusual life history of *Alseis blackiana*: a shade-persistent pioneer tree? *Ecology* 82, 933–945.

Davies, S.J. and Ashton, P.S. (1999) Phenology and fecundity in 11 sympatric pioneer species of *Macaranga* (Euphorbiaceae) in Borneo. *American Journal of Botany* 86, 1786–1795.

Dawkins, H.C. and Field, D.R.B. (1978) *A long-term surveillance system for British woodland vegetation*. Occasional Papers No. 1, Department of Forestry, Oxford University.

Delagrange, S., Messier, C., Lechowicz, M.J., and Dizengremel, P. (2004) Physiological, morphological and allocational plasticity in understory deciduous trees: importance of plant size and light availability. *Tree Physiology* 24, 775–784.

Denslow, J.S. (1980) Gap partitioning among tropical rainforest trees. *Tropical Succession* 12, 47–55.

Denslow, J.S. (1995) Disturbance and diversity in tropical rain-forests – the density effect. *Ecological Applications* 5, 962–968.

DeWalt, S.J., Denslow, J.S., and Ickes, K. (2004) Natural-enemy release facilitates habitat expansion of an invasive shrub, *Clidemia hirta*. *Ecology* 85, 471–483.

Ellison, A.M., Denslow, J.S., Loiselle, B.A., and Brenes, M.D. (1993) Seed and seedling ecology of neotropical Melastomataceae. *Ecology* 74, 1733–1749.

Falster, D.S. and Westoby, M. (2005) Alternative height strategies among 45 dicot rainforest species from tropical Queensland, Australia. *Journal of Ecology* 93, 521–535.

Fine, P.V.A., Mesones, I., and Coley, P.D. (2004) Herbivores promote habitat specialization in Amazonian Forests. *Science* 305, 663–665.

Gause, G.F. (1936) *The Struggle for Existence*. Williams and Wilkins, Baltimore.

Gilbert, G.S., Harms, K.E., Hamill, D.N., and Hubbell, S.P. (2001) Effects of seedling size, El Niño drought, seedling density, and distance to nearest conspecific adult on 6-year survival of *Ocotea whitei* seedlings in Panama. *Oecologia* 127, 509–516.

Gilbert, B., Wright, S.J., Muller-Landau, H.C., Kitajima, K., and Hernandéz, A. (2006) Life history trade-offs in tropical trees and lianas. *Ecology* 87, 1281–1288.

Givnish, T.J. (1988) Adaptation to sun and shade: a whole-plant perspective. *Australian Journal of Plant Physiology* 15, 63–92.

Grime, J.P. (1979) *Plant Strategies and Vegetation Processes*. John Wiley, New York.

Grubb, P.J. (1977) The maintenance of species-richness in plant communities: the importance of the regeneration niche. *Biological Review* 52, 107–145.

Grubb, P.J. (1996) Rainforest dynamics: the need for new paradigms. In D.S. Edwards, W.E. Booth, and S.C. Choy (eds), *Tropical Rainforest Research – Current Issues*. Kluwer Academic Publishers, Dordrecht, pp. 215–233.

Guariguata, M.R. (1998) Response of forest tree saplings to experimental mechanical damage in lowland Panama. *Forest Ecology and Management* 102, 103–111.

Hammond, D.S. and Brown, V.K. (1995) Seed size of woody plants in relation to disturbance, dispersal, soil type in wet Neotropical forests. *Ecology* 76, 2544–2561.

Hammond, D.S. and Brown, V.K. (1998) Disturbance, phenology and life-history characteristics: factors influencing distance/density-dependent attack on tropical seeds and seedlings. In D.M. Newbery, H.H.T. Prins, and N.D. Brown (eds), *Dynamics of Tropical Communities*. Blackwell Science, Oxford, pp. 51–78.

Harms, K.E., Condit, R., Hubbell, S.P., and Foster, R.B. (2001) Habitat associations of trees and shrubs in a 50-ha neotropical forest plot. *Journal of Ecology* 89, 947–959.

Harms, K.E., Powers, J.S., and Montgomery, R.A. (2004) Variation in small sapling density, understory cover, and resource availability in four Neotropical forests. *Biotropica* 36, 40–51.

Hawthorne, W.D. (1995) *Ecological profiles of Ghanaian forest trees*. Tropical Forestry Papers 29. Oxford Forestry Institute, Oxford.

Hoffmann, W.A., Orthen, B., and Franco, A.C. (2004) Constraints to seedling success of savanna and forest trees across the savanna-forest boundary. *Oecologia* 140, 252–260.

Huante, P., Rincon, E., and Acosta, I. (1995) Nutrient availability and growth of 34 woody species from a tropical deciduous forest in Mexico. *Functional Ecology* 9, 849–858.

Hubbell, S.P. (2001) *The Unified Neutral Theory of Biodiversity and Biogeography*. Princeton University Press, Princeton.

Hubbell, S.P. (2005) Neutral theory in community ecology and the hypothesis of functional equivalence. *Functional Ecology* 19, 166–172.

Hubbell, S.P. and Foster, R.B. (1986) Commonness and rarity in a Neotropical forest: implications for tropical tree conservation. In M.E. Soulé (ed.), *Conservation Biology: The Science of Scarcity and Diversity*. Sinauer Associates, Sunderland, MA, pp. 205–231.

Hubbell, S.P., Foster, R.B., O'Brien, S.T. *et al.* (1999) Light-gap disturbances, recruitment limitation, and tree diversity in a neotropical forest. *Science* 283, 554–557.

Kabakoff, R.P. and Chazdon R.L. (1996) Effects of canopy species dominance on understory light availability in low-elevation secondary forest stand in Costa Rica. *Journal of Tropical Ecology* 12, 779–788.

King, D.A. (1986) Load bearing capacity of understory treelets of a tropical wet forest. *Bulletin of the Torrey Botanical Club* 114, 419–428.

King, D.A. (1990) Allometry of saplings and understorey trees in a Panamanian forest. *Functional Ecology* 5, 85–492.

King, D.A. (1994) Influence of light level on the growth and morphology of saplings in a Panamanian forest. *American Journal of Botany* 81, 948–957.

King, D.A., Wright, S.J., and Connell, J.H. (2006) The contribution of interspecific variation in maximum tree height to tropical and temperate variation in diversity. *Journal of Tropical Ecology* 22, 11–24.

Kitajima, K. (1992) *The importance of cotyledon functional morphology and patterns of seed reserve utilization for the physiological ecology of neotropical tree seedlings*. Doctoral dissertation, University of Illinois, Urbana.

Kitajima, K. (1994) Relative importance of photosynthetic traits and allocation patterns as correlates of seedling shade tolerance of 13 tropical trees. *Oecologia* 98, 419–428.

Kitajima, K. (1996) Ecophysiology of tropical tree seedlings. In S.S. Mulkey, R.L. Chazdon, R.L., and A.P. Smith (eds), *Tropical Forest Plant Ecophysiology*. Chapman and Hall, New York, pp. 559–596.

Kitajima, K. (2002) Do shade-tolerant tropical tree seedlings depend longer on seed reserves? Functional growth analysis of three Bignoniaceae species. *Functional Ecology* 16, 433–444.

Kitajima, K. and Bolker, B.M. (2003) Testing performance rank reversals among coexisting species: crossover point irradiance analysis by Sack and Grubb (2001) and alternatives. *Functional Ecology* 17, 276–281.

Kitajima, K., Mulkey, S.S., and Wright, S.J. (2005) Variation in crown light utilization characteristics among tropical canopy trees. *Annals of Botany* 95, 535–547.

Kobe, R.K. (1997) Carbohydrate allocation to storage as a basis of interspecific variation in sapling survivorship and growth. *Oikos* 80, 226–233.

Kobe R.K. (1999) Light gradient partitioning among tropical tree species through differential seedling mortality and growth. *Ecology* 80, 187–201.

Kobe, R.K., Pacala, S.W., Silander, J.A.J., and Canham, C.D. (1995) Juvenile tree survivorship as a component of shade tolerance. *Ecological Applications* 5, 517–532.

Kohyama, K. (1987) Significance of architecture and allometry in saplings. *Functional Ecology* 1, 399–404.

Kohyama, T. (1993) Size-structured tree populations in gap-dynamic forest - the forest architecture hypothesis for the stable coexistence of species. *Journal of Ecology* 81, 131–143.

Kohyama, T. and Hotta, M. (1990) Significance of allometry in tropical saplings. *Functional Ecology* 4, 515–521.

Kohyama, T., Suzuki, E., Partomihardjo, T., Yamada, T., and Kubo, T. (2003) Tree species differentiation in growth, recruitment and allometry in relation to maximum height in a Bornean mixed dipterocarp forest. *Journal of Ecology* 91, 797–806.

Latham, R.E. (1992) Co-occurring tree species change rank in seedling performance with resources varied experimentally. *Ecology* 73, 2129–2144.

Leigh, E.G., Davidar, P., Dick, C.W. *et al.* (2004) Why do some tropical forests have so many species of trees? *Biotropica* 36, 447–473.

Lieberman, D., Lieberman, M., Hartshorn, G., and Peralta, R. (1985) Growth rates and age–size relationships of tropical wet forest trees in Costa Rica. *Journal of Tropical Ecology* 1, 97–109.

Lusk, C.H. (2002) Leaf area accumulation helps juvenile evergreen trees tolerate shade in a temperate rainforest. *Oecologia* 132, 188–196.

Lusk, C.H. (2004) Leaf area and growth of juvenile temperate evergreens in low light: species of contrasting shade tolerance change rank during ontogeny. *Functional Ecology* 18, 820–828.

MacArthur, R.H. and Levins, R. (1967) The limiting similarity, convergence, and divergence of coexisting species. *American Naturalist* 101, 377–385.

McPeek, M.A. (1996) Trade-offs, food web structure, and the coexistence of habitat specialists and generalists. *American Naturalist* 148(Supplement), 124–138.

Metcalfe, D.J. and Grubb, P.J. (1995) Seed mass and light requirements for regeneration of Southeast Asian rain forest. *Canadian Journal of Botany* 73, 817–826.

Moles, A.T., Falster, D.S., Leishman, M.R., and Westoby, M. (2004) Small-seeded species produce more seeds per square metre of canopy per year, but not per individual per lifetime. *Journal of Ecology* 92, 384–396.

Molofsky, J. and Augspurger, C.K. (1992) The effect of leaf litter on early seedling establishment in a tropical forest. *Ecology* 73, 68–77.

Montgomery, R.A. (2004) Effects of understory foliage on patterns of light attenuation near the forest floor. *Biotropica* 36, 33–39.

Montgomery R.A. and Chazdon R.L. (2002) Light gradient partitioning by tropical tree seedlings in the absence of canopy gaps. *Oecologia* 131, 165–174.

Myers, J.A. and Kitajima, K. (2007) Carbohydrate storage enhances seedling shade and stress tolerance in a neotropical forest. *Journal of Ecology* 95, 383–395.

Oldeman, R.A.A. and van Dijk, J. (1991) Diagnosis of the temperament of tropical rain forest trees. In A. Gómez-Pompa, T.C. Whitmore, and M. Hadley (eds), *Rain Forest Regeneration and Management*, Man and the Biosphere Series 6. UNESCO, Paris, pp. 21–65.

Osunkoya, O.O., Ash, J.E., Hopkins, M.S., and Graham, A.W. (1994) Influence of seed size and seedling ecological attributes on shade-tolerance of rain forest tree species in Northern Queensland. *Journal of Ecology* 82, 149–163.

Pacala, S.W. and Rees, M. (1998) Models suggesting field experiments to test 2 hypotheses explaining successional diversity. *American Naturalist* 152, 729–737.

Poorter, L. (1999) Growth response of 15 rain forest tree species to a light gradient: the relative importance of morphological and physiological traits. *Functional Ecology* 13, 396–410.

Poorter, L. (2007) Are species adapted to their regeneration niche, adult niche, or both? *American Naturalist* 169, 433–442.

Poorter, L. and Arets, E. (2003) Light environment and tree strategies in a Bolivian tropical moist forest: an evaluation of the light partitioning hypothesis. *Plant Ecology* 166, 295–306.

Poorter, L. and Bongers, F. (2006) Leaf traits are good predictors of plant performance across 53 rain forest species. *Ecology* 87, 1733–1743.

Poorter, L., Bongers, F., Sterck, F.J., and Woll, H. (2003) Architecture of 53 rain forest tree species differing in adult stature and shade tolerance. *Ecology* 84, 602–608.

Poorter, L., Bongers, F., Sterck, F.J., and Woll, H. (2005) Beyond the regeneration phase: differentiation of height–light trajectories among tropical tree species. *Journal of Ecology* 93, 256–267.

Poorter, L., Bongers, L., and Bongers, F. (2006) Architecture of 54 moist forest species: traits, trade-offs, and functional groups. *Ecology* 87, 1289–1301.

Poorter, L. and Kitajima, K. (2007) Carbohydrate storage and light requirements of tropical moist and dry forest tree species. *Ecology* 88, 1000–1011.

Poorter, L. and Rose, S.A. (2005) Light-dependent changes in the relationship between seed mass and seedling traits: a meta-analysis for rain forest tree species. *Oecologia* 142, 378–387.

Poorter, L., van de Plassche, M.V., Willems, S., and Boot, R.G.A. (2004) Leaf traits and herbivory rates of tropical tree species differing in successional status. *Plant Biology* 6, 746–754.

Popma, J. and Bongers, F. (1988) The effect of canopy gaps on growth and morphology of seedlings of rain forest species. *Oecologia* 75, 625–632.

Popma, J., Bongers, F., Martínez-Ramos, M., and Veneklaas, J. (1988) Pioneer species distribution in treefall gaps in Neotropical rain forest; a gap definition and its consequences. *Journal of Tropical Ecology* 4, 77–88.

Putz, F.E. (1983) Treefall pits and mounds, buried seeds, and the importance of soil disturbance to pioneer trees on Barro Colorado Island, Panama. *Ecology* 64, 1069–1074.

Richards, P.W. (1952) *The Tropical Rain Forest*. Cambridge University Press, London.

Ricklefs, R.E. (1977) Environmental heterogeneity and plant species diversity: a hypothesis. *American Naturalist* 111, 376–381.

Rijkers, T., Pons, T.L., and Bongers, F. (2000) The effect of tree height and light availability on photosynthetic leaf traits of four neotropical species differing in shade tolerance. *Functional Ecology* 14, 77–86.

Rose, S.A. (2000) Seeds, seedlings and gaps – size matters. *A study in the tropical rain forest of Guyana*. Doctoral dissertation, Utrecht University, Tropenbos-Guyana series 9. Ipskamp, Enschede.

Sack, L., Grubb, P.J., and Maranon, T. (2003) The functional morphology of juvenile plants tolerant of strong summer drought in shaded forest understories in southern Spain. *Plant Ecology* 168, 139–163.

Sack, L. and Grubb, P.J. (2001) Why do species of woody seedlings change rank in relative growth rate between low and high irradiance? *Functional Ecology* 15, 145–154.

Sack, L. and Grubb, P.J. (2003) Crossovers in seedling relative growth rates between low and high irradiance: analyses and ecological potential (reply to Kitajima and Bolker 2003). *Functional Ecology* 17, 281–287.

Santiago, L.S., Kitajima, K., Wright, S.J., and Mulkey, S.S. (2004) Coordinated changes in photosynthesis, water relations and leaf nutritional traits of canopy trees along a precipitation gradient in lowland tropical forest. *Oecologia* 139, 495–502.

Schlichting, C.D. (1986) The evolution of phenotypic plasticity in plants. *Annual Review of Ecology and Systematics* 17, 667–693.

Schnitzer, S.A. and Carson, W.P. (2001) Treefall gaps and the maintenance of species diversity in a tropical forest. *Ecology* 82, 913–919.

Schreeg, L.A., Kobe, R.K., and Walters, M.B. (2005) Tree seedling growth, survival and morphology in response to landscape-level variation in soil resource availability in northern Michigan. *Canadian Journal of Forest Research* 35, 263–273.

Schulz, J.P. (1960) *Ecological studies on rain forest in northern Suriname*. Mededelingen van het Botanisch Museum en Herbarium van de Rijsuniversiteit te Utrecht No. 163. N.V. Noord-Hollandsche Uitgevers Maatschappij, Amsterdam.

Schuur, E.A.G. and Matson, P.A. (2001) Net primary productivity and nutrient cycling across a mesic to wet precipitation gradient in Hawaiian montane forest. *Oecologia* 128, 431–442.

Sheil, D., Salim, A., Chave, J., Vanclay, J.D., and Hawthorne, W.D. (2006) Illumination-size relationships of coexisting tropical trees. *Journal of Ecology* 94, 494–507.

Svenning, J.-C., Fabbro, T. and Wright, S.J. (2008) Seedling interactions in a tropical forest in Panama. *Oecologia* 155, 143–150.

Svenning, J.-C., Kinner, D.A., Stallard, R.F., Engelbrecht, B.M.J. and Wright, S.J. (2004) Ecological determinism in plant community structure across a tropical forest landscape. *Ecology* 85, 2526–2538.

ter Steege, H. (1994) Flooding and drought tolerance in seeds and seedlings of two *Mora* species segregated along a soil hydrological gradient in the tropical rain forest of Guyana. *Oecologia* 100, 356–367.

ter Steege, H. and Hammond, D.S. (2001) Character convergence, diversity, and disturbance in tropical rain forest in Guyana. *Ecology* 82, 3197–3212.

Terborgh, J. (1985) The vertical component of plant species diversity in temperate and tropical forests. *American Naturalist* 126, 760–776.

Thomas, S.C. (1996a) Relative size at onset of maturity in rain forest trees: a comparative analysis of 37 Malaysian species. *Oikos* 76, 145–154.

Thomas, S.C. (1996b) Reproductive allometry in Malaysian rain forest trees: biomechanics versus optimal allocation. *Evolutionary Ecology* 10, 517–530.

Thomas, S.C. and Bazzaz, F.A. (1999) Asymptotic height as a predictor of photosynthetic characteristics in Malaysian rain forest trees. *Ecology* 80, 1607–1622.

Tilman, D. (1982) *Resource Competition and Community Structure*. Princeton University Press, Princeton.

Tilman, D. (1988) *Dynamics and Structure of Plant Communities*. Princeton University Press, Princeton.

Tilman, D. (2004) Niche tradeoffs, neutrality, and community structure: a stochastic theory of resource competition, invasion, and community assembly. *Proceedings of the National Academy of Sciences of the United States of America* 101, 10854–10861.

Tuomisto, H. and Poulsen, A.D. (2000) Pteridophyte diversity and species composition in four Amazonian rain forests. *Journal of Vegetation Science* 11, 383–396.

Turner, I.M. (2001) *The Ecology of Trees in the Tropical Rain Forest*. Cambridge University Press, Cambridge.

Valladares, F., Wright, S.J., Lasso, E., Kitajima, K., and Pearcy, R.W. (2000) Plastic phenotypic response to light of 16 congeneric shrubs from a Panamanian rain forest. *Ecology* 81, 1925–1936.

Van Rheenen, H.M.P.J.B. (2005) *The Role of Seed Trees and Seedling Regeneration for Species Maintenance in Logged-over Forest.* PROMAB Scientific Series 9. PROMAB, Riberalta, Bolivia.

van Ulft, L.H. (2004) The effect of seed mass and gap size on seed fate of tropical rain forest tree species in Guyana. *Plant Biology* 6, 214–221.

van Gelder, A., Poorter, L., and Sterck, F.J. (2006) Wood mechanics, allometry, and life-history variation in a tropical rain forest tree community. *New Phytologist* 171, 367–378.

Veneklaas, E.J. and Poorter, L. (1998) Growth and carbon partitioning of tropical tree seedlings in contrasting light environments. In H. Lambers, H. Poorter, and M.M.I. Van Vuuren (eds), *Inherent Variation in Plant Growth. Physiological Mechanisms and Ecological Consequences.* Bunkhuys Publishers, Leiden, pp. 337–361.

Walters, M.B. and Reich, P.B. (1996) Are shade tolerance, survival, and growth linked? Low light and nitrogen effects on hardwood seedlings. *Ecology* 77, 841–853.

Wedin, D. and Tilman, D. (1993) Competition among grasses along a nitrogen gradient: initial conditions and mechanisms of competition. *Ecological Monographs* 63, 199–229.

Welden, C.W., Hewett, S.W., Hubbell, S.P., and Foster, R.B. (1991) Sapling survival, growth, and recruitment: relationship to canopy height in a neotropical forest. *Ecology,* 72, 35–50.

Westoby, M. (1998) A leaf-height-seed (LHS) plant ecology strategy scheme. *Plant and Soil* 199, 213–227.

Wright, S.J. (2002) Plant diversity in tropical forests: a review of mechanisms of species coexistence. *Oecologia* 130, 1–14.

Wright, S.J., Jaramillo, M.A., Pavon, J., Condit, R., Hubbell, S.P., and Foster, R.B. (2005) Reproductive size thresholds in tropical trees: variation among individuals, species and forests. *Journal of Tropical Ecology* 21, 307–315.

Wright, S.J., Muller-Landau, H.C., Condit, R., and Hubbell, S.P. (2003) Gap-dependent recruitment, realized vital rates, and size distributions of tropical trees. *Ecology* 84, 3174–3185.

Yoda, K. (1974) Three-dimensional distribution of light intensity in a tropical rain forest of west Malaysia. *Japanese Journal of Ecology* 24, 247–254.

Zanne, A.E., Chapman, C.A., and Kitajima, K. (2005) Evolutionary and ecological correlates of early seedling morphology in East African trees and shrubs. *American Journal of Botany* 92, 972–978.

Chapter 11

COLONIZATION-RELATED TRADE-OFFS IN TROPICAL FORESTS AND THEIR ROLE IN THE MAINTENANCE OF PLANT SPECIES DIVERSITY

Helene C. Muller-Landau

OVERVIEW

Interspecific trade-offs involving colonization ability can contribute strongly to the maintenance of plant species diversity, and are often cited as a potential mechanism underlying high tropical forest diversity. The well-known competition–colonization trade-off, between the ability to win a regeneration site after arrival and the ability to arrive, can in theory maintain very high species diversity, but only if there is strong competitive asymmetry among species, such that the best competitor present is highly disproportionately likely to win. Other, less-studied trade-offs involving colonization ability can contribute to diversity maintenance given appropriate habitat heterogeneity, by facilitating habitat niche partitioning. Specifically, a trade-off between fecundity and stress tolerance combined with corresponding variation in stress among regeneration sites can lead to coexistence between more tolerant species able to win high stress sites and more fecund species that are numerically more likely to win low stress sites. A trade-off between fecundity and dispersal can similarly contribute to coexistence given spatial variation in the density of suitable regeneration sites. Empirical studies of species trait relationships, current understanding of the asymmetry of competitive interactions among seedlings, and results of a seed addition experiment all suggest that the classical competition–colonization trade-off is not present among tropical trees, and thus does not contribute to their coexistence. In contrast, trait relationships do provide evidence for the presence of a tolerance–fecundity trade-off mediated by seed size, with small-seeded species having higher fecundity and lower stress tolerance than large-seeded species. Evidence concerning the existence of a dispersal–fecundity trade-off is mixed and inconclusive. To further elucidate the roles of these colonization-related trade-offs, and specifically to assess their contributions, if any, to species coexistence in tropical forests, we need additional studies of how spatio-temporal variation in environmental conditions and seed arrival contribute to regeneration success, in natural systems, field experiments, and/or models.

INTRODUCTION

Interspecific trade-offs involving species' abilities to reach or "colonize" regeneration sites can play multiple roles in niche partitioning and diversity maintenance among tropical forest tree species and in other communities. The best-known examples are competition–colonization trade-offs between a species' ability to reach sites with its recruits and the per recruit ability to win sites at which recruits arrive (Tilman and Pacala 1993). There is a long history of theoretical work on this mechanism of coexistence, which shows that these trade-offs can potentially make a strong contribution to species diversity maintenance in homogeneous environments (Skellam 1951, Levins and Culver 1971, Horn and MacArthur 1972, Armstrong 1976, Hastings 1980, Tilman 1994). However, the conditions for competition–colonization trade-offs to maintain diversity are stringent and likely to be uncommon in nature, and in the absence of these conditions these trade-offs need not be a powerful or even significant force enhancing coexistence (Geritz *et al.* 1999, Levine and Rees 2002, Kisdi and Geritz 2003a).

Other colonization-related trade-offs can contribute to habitat partitioning among species, and thereby to coexistence in heterogeneous environments. A trade-off between fecundity and the ability to tolerate low resource conditions or harsh habitats can mediate coexistence when there is spatial variation in resource availability or the harshness of conditions among local regeneration sites (Levine and Rees 2002). A trade-off between fecundity and dispersal distance can mediate coexistence in the presence of spatial variation in the density of sites suitable for regeneration (Yu and Wilson 2001).

While competition–colonization trade-offs have been the subject of extensive theoretical investigation and of empirical research in a number of plant communities (Turnbull *et al.* 1999, Coomes and Grubb 2003), other colonization-related trade-offs have rarely been studied. In tropical forests, there has been little research even on competition–colonization trade-offs. The limited consideration of such trade-offs in tropical forests has focused mainly on whether they might contribute to the coexistence of shade-tolerant and gap-dependent species (Connell 1978, Leigh *et al.* 2004). Nonetheless, there are a number of relevant empirical studies in tropical forests whose results shed light on the potential for colonization-related trade-offs to contribute to the maintenance of species richness in these diverse plant communities.

In this chapter, I review the theory and evidence regarding the contribution of colonization-related trade-offs to diversity maintenance in tropical forests. Throughout, I devote the most space to the competition–colonization trade-off, because of its premier position in the literature on colonization-related trade-offs and thus the abundance of relevant theory and empirical studies. I begin by briefly reviewing the relevant theory, identifying the key assumptions and predictions of models in which these trade-offs contribute to diversity maintenance. I then consider the methods that can be used to document these trade-offs and their roles in real communities, and evaluate the relevant empirical evidence from tropical forests in particular. I end with recommendations for future research and a summary of what we can conclude thus far.

THEORY ON COLONIZATION-RELATED TRADE-OFFS AND DIVERSITY MAINTENANCE

Equalizing versus stabilizing influences

Colonization-related trade-offs have the potential to exert equalizing and/or stabilizing effects on diversity maintenance, *sensu* Chesson (2000). Equalizing influences minimize fitness differences among species that would otherwise lead to competitive exclusion (Chesson 2000), making dynamics less exclusionary and more neutral (*sensu* Hubbell 2001). If there is merely partial equalization so that species remain less than perfectly equal, dynamics are near-neutral, and the weaker species are deterministically excluded (Zhang and Lin 1997, Yu *et al.* 1998), albeit at a slower rate than they would be without the trade-off (Figure 11.1a–d). If there is perfect or complete equalization, species become equal in competitive ability, and thus are subject to neutral drift (Hubbell 2001) (Figure 11.1e,f). In contrast, stabilizing influences actively contribute to diversity maintenance by increasing negative intraspecific interactions relative to negative interspecific interactions (Chesson 2000). This ensures that each species is relatively advantaged when rare and disadvantaged when common, which tends to keep species from extinction or monodominance (Figure 11.1g,h). Stabilizing influences make dynamics less neutral, but in a

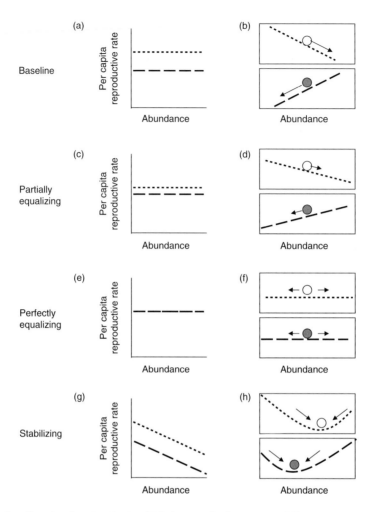

Figure 11.1 Consider a baseline situation in which there are fixed competitive differences among species in their per capita reproductive rates (a) – differences that would deterministically lead to the competitive exclusion of the species with the lower reproductive rate (dashed line) by the species with the higher reproductive rate (dotted line). In this case, we can think of each species as a ball precariously located on a steep slope (b), down which it will inevitably roll, with the weaker species moving towards zero abundance and the common one towards dominance. If we add a partially equalizing influence, the reproductive rates of the two species become more similar (c), but because one is still superior, the weaker species will still inevitably be lost, albeit at a slower rate (d). In the extreme case of perfectly equalizing influences, the reproductive rates of the two species become identical (e). This case is analogous to one in which both species are balls on a flat tabletop (f): there is no slope tending to make them increase or decrease in abundance, but both are subject to random drift which could result in their abundance going to zero or to dominance. If instead we add a stabilizing influence, then each species' reproductive rate decreases as it becomes more abundant, and increases as it becomes more rare (g); here, there are pairs of abundances at which the species have equal reproductive rates and can stably coexist. In this case, it is as though each species is a ball sitting in a bowl (h): any perturbation of its abundance to higher or lower levels will induce negative feedbacks that will return it to its stable equilibrium position. For example, if its abundance is depressed, its reproductive rate will increase, and thus it will return to its equilibrium abundance.

way that tends to maintain diversity rather than lead to competitive exclusion (Figure 11.1g,h).

The most well-known theoretical model of the competition–colonization trade-off is stabilizing and thus has tremendous diversity-maintaining potential (Skellam 1951, Tilman 1994). Many documented competition–colonization trade-offs, however, consist of trait relationships that in themselves are only equalizing. For example, a trade-off between seed production and seed survival alone can at best perfectly equalize species' competitive abilities by ensuring that all species have the same number of seedlings per adult. Similarly, while habitat partitioning mechanisms are invariably stabilizing when the theoretical conditions under which they are defined are met, at the exact boundary of those conditions they are merely perfectly equalizing, and on the other side of the conditions they operate as partially equalizing. Thus, before we can evaluate the role of the colonization-related trade-offs in real communities, we need to take a close look at which model assumptions are critical to determining the existence and magnitude of stabilizing influences on diversity.

Competition–colonization trade-offs in homogeneous environments

The simple competition–colonization trade-off model first introduced by Skellam (1951) encapsulates the inherent potential of such trade-offs to contribute to diversity maintenance in homogeneous environments in a stabilizing manner. Its dynamics have been fruitfully explored in many subsequent papers, most notably Hastings (1980) and Tilman (1994). In this model, space is divided into discrete sites each occupied by a single adult. Adults produce seeds that are distributed randomly among all sites, and die at a fixed rate. Species have strict competitive rankings that are the exact inverse of their rankings in seed production. When a seed arrives at a site occupied by an adult of an inferior competitor, it immediately displaces the occupant and becomes the new adult at the site. Under these conditions, a potentially infinite number of species differing in competition and colonization abilities can stably

coexist (Tilman 1994). While this model usefully illustrates the potential strength of the trade-off, its assumptions of perfectly asymmetric competition (the better competitor always wins even if only a tiny bit better) and immediate displacement are highly unrealistic for plant communities, and its behavior is also a poor match to real community dynamics. For example, species with higher competitive abilities are more abundant, and simultaneously more vulnerable to habitat loss (Tilman et al. 1997) – which contradicts abundant evidence that rare species are most endangered (Wilcove et al. 1998). Further, this model is evolutionarily unstable: if species traits are allowed to evolve, each species evolves to higher and higher competitive ability and lower fecundity, and thus eventual extinction (Jansen and Mulder 1999).

Alternative models of competition–colonization trade-offs encapsulating a range of more realistic assumptions show that a crucial requirement for stable coexistence under this mechanism is strong competitive asymmetry (Rees and Westoby 1997, Geritz et al. 1999, Adler and Mosquera 2000, Levine and Rees 2002, Kisdi and Geritz 2003a). The classical model described above encapsulates perfect competitive asymmetry – the better competitor always wins the site. In contrast, if competition is purely symmetric such that competitive differences are merely density independent (e.g., if there is interspecific variation in density-independent seed survival and all surviving propagules are equally likely to win a site), then stable coexistence via this mechanism alone is impossible (Comins and Noble 1985). The quantitative importance of asymmetry is elegantly demonstrated by Geritz et al. (1999) in their model of annual plants, in which seed size mediates a trade-off between seed production and competitive ability. Competitive ability is encapsulated by both a density-independent survival term (an equalizing force) as well as the probability of winning in the face of competition (a stabilizing influence). The per capita probability of winning is characterized as an exponential function of seed mass that includes a parameter for the degree of competitive asymmetry: as this asymmetry parameter increases, the species with the highest seed mass becomes ever more likely to win the site. Geritz *et al.* (1999) consider the evolutionary as well as ecological dynamics of this model, and show that as the degree of competitive asymmetry increases, the number of types (species) that evolve and stably coexist increases. Adler and Mosquera (2000) analytically derive the conditions under which one, two, and infinite numbers of species can coexist via the competition–colonization trade-off, specifically showing that infinite coexistence is possible only under perfect asymmetry. Kisdi and Geritz (2003a) demonstrate similar effects of varying asymmetry in models of perennial plants. The role of asymmetry in these models is consistent with the results of Tilman (1994) and Kisdi and Geritz (2003b), who find coexistence of infinite numbers of species in models with perfectly asymmetric competition, and of Levine and Rees (2002) who find limited coexistence under low asymmetry.

Thus, the classical competition–colonization trade-off can be a strong stabilizing force for diversity maintenance, but only if there is sufficient competitive asymmetry. Perfect asymmetry, which is unrealistic for real communities (Adler and Mosquera 2000), is required for the effectively infinite coexistence attained in the original theoretical models (Tilman 1994). In contrast, if competition is perfectly symmetric, then the contribution of this trade-off alone to diversity maintenance can be only equalizing. The perfectly equalizing case is essentially infinitely unlikely to occur; even small deviations that make the trade-off partially rather than perfectly equalizing are sufficient to make some species superior competitors and shorten coexistence (Zhang and Lin 1997, Yu *et al.* 1998). Finally, if competition is partially asymmetric, the most likely case in real communities, then the trade-off may be able to contribute to stable coexistence of a few species, or it may be merely a partially equalizing force.

COLONIZATION-RELATED TRADE-OFFS AND HABITAT PARTITIONING

While only certain competition–colonization trade-offs can be a stabilizing influence on diversity maintenance in homogeneous environments, a wider range of colonization-related trade-offs can have stabilizing influences given appropriate spatial or temporal environmental heterogeneity. Specifically, these trade-offs can contribute to diversity maintenance if the combination of each species' colonization and competitive abilities on the different habitats is such that each species has the highest population growth rate in some time or place (Chesson and Warner 1981, Comins and Noble 1985, Yu and Wilson 2001). Because both habitat heterogeneity and variation in species performance on different habitats are ubiquitous in real ecosystems, these trade-offs have the potential to play important roles in diversity maintenance. Here I consider two specific examples – tolerance–fecundity trade-offs and dispersal–fecundity trade-offs.

A tolerance–fecundity trade-off can mediate coexistence when there is spatial variation in resource availability and thus in the level of recruit provisioning needed to tolerate local conditions and have a chance at winning the regeneration site. In this case, a trade-off between recruit provisioning (e.g., seed mass) and fecundity (e.g., seed production) can mediate coexistence by allowing the more fecund species to succeed disproportionately often in sites where little provisioning is needed, and thus make up for the consistent success of the better-provisioned species on sites where resource availability is low (Levine and Rees 2002). In principle, many species can coexist given sufficient variation in habitat quality among sites, and appropriate consistency in the trade-off between habitat tolerance and fecundity. Specifically, such a trade-off will be stabilizing if the fecundity of each less tolerant species exceeds that of the next more tolerant species by a particular multiple, with that multiple depending on their relative habitat tolerances, and seed survival, if relevant. If the fecundity of the less tolerant species is less than (or equal to) this multiple of the fecundity of the more tolerant species, then the trade-off will be partially (or perfectly) equalizing.

A dispersal–fecundity trade-off can allow two competitors to coexist given spatial variation in the density of potential regeneration sites (Yu and Wilson 2001). The more fecund species is more successful in areas of high site density and the better disperser is more successful in areas of low site density, enabling coexistence. In principle, many species could thus coexist given sufficient spatial variation in the density of regeneration sites (Yu and Wilson 2001). Again, there is a specific quantitative condition for the relationships of species' fecundities and dispersal abilities beyond which the trade-off is stabilizing, at which it is perfectly equalizing, and below which it is partially equalizing.

Tolerance–fecundity and dispersal–fecundity trade-offs are but two examples of colonization-related trade-offs that can contribute to diversity maintenance given habitat heterogeneity. Both of these mechanisms partition spatial heterogeneity. Trade-offs involving dormancy or dispersal in time more generally can play a role in partitioning temporal heterogeneity, and thus in stabilizing coexistence in temporally varying environments (Chesson and Warner 1981). There is an extensive literature on species coexistence via habitat partitioning; however, the focus has mainly been on species differences in competitive ability in the different habitats (Amarasekare 2003). Similarly, the focus of research on colonization-related trade-offs has been on coexistence due to these trade-offs alone in homogeneous environments (Amarasekare 2003). Additional theoretical work is needed to explore how colonization differences among species can interact with habitat heterogeneity to contribute to species coexistence.

METHODS FOR EVALUATING THE PRESENCE AND ROLE OF COLONIZATION-RELATED TRADE-OFFS

There are multiple possible approaches to investigating colonization-related trade-offs in real communities. The most common approach is to simply measure particular species traits and analyze correlations among these traits to test for the presence of a particular trade-off among species. This provides useful information on the *presence* of the trade-off, but in and of itself says little about the *role* of the trade-off in species coexistence; measurements of other key features of

the community or individual interactions within it (such as competitive asymmetry in the case of the competition–colonization trade-off) are generally necessary to evaluate theoretical conditions for coexistence. An alternative approach examines spatio-temporal variation in recruitment success in the field and tests the degree to which it can be explained by model predictions. The role of the trade-offs can also be assessed through community-level field experiments, which again can test either general model predictions or specific predictions based on additional information. In principle, any of the above efforts could be used to parametrize models of the hypothesized mechanisms, and thereby to enable further theoretical tests of whether conditions for stabilizing coexistence are met, either analytically or through simulations.

Clearly, a colonization-related trade-off can play a role in community dynamics only if it is present. Thus, a first question is whether species traits trade off in the hypothesized manner. This question is generally addressed through correlation or regression analyses of species traits. In part because most available data are collected for other (or at least broader) purposes, they often concern not the most relevant measures for the trade-offs, but rather some component contributing trait. In addition, the most useful integrative traits (e.g., "colonization ability," "competitive ability") are often particularly difficult to measure or even define (Clark *et al.* 2005). As a consequence, the resulting correlation analyses rarely provide definitive answers regarding even the presence of the overall trade-off.

Studies relating spatio-temporal variation in environmental conditions, seed arrival, and successful recruitment have long been used to assess the relative importance of seed arrival and habitat suitability to population-level recruitment patterns (e.g., LePage *et al.* 2000, HilleRisLambers and Clark 2003). Analyses of spatial patterns of environment and species distributions have also been used to examine the importance of habitat partitioning at the community level (e.g., Plotkin *et al.* 2000). Similar approaches could be used to specifically evaluate the predictions of the competition–colonization, tolerance–fecundity, and dispersal–fecundity trade-off models for patterns of recruitment success given patterns of seed arrival and environmental conditions. Such studies could simply test for the general patterns expected under different models – for example, for the competition–colonization trade-off, the expectation would be that some species win whenever they arrive, and others only when these dominant competitors fail to arrive. Alternatively, independent information on species traits (e.g., competitive rankings) could be used to predict specific patterns of which species are expected to win where.

Community-level seed addition experiments are the most powerful way to investigate the competition–colonization trade-off, providing the means to assess whether the trade-off is stabilizing or equalizing, and to what degree – even if the competitive rankings of species are not known. If the trade-off is stabilizing, then species that are poorer colonists and better competitors should increasingly exclude those that are better colonists and poorer competitors as more seeds are added, with stronger compositional shifts indicating a more powerful stabilizing effect. If instead the trade-off is merely equalizing, or if habitat-mediated tolerance–fecundity trade-offs alone are stabilizing, then increasing seed rain of all species by the same multiplicative factor should have no impact on species composition. Further insight into the relative importance of a competition–colonization trade-off specifically to the coexistence of early successional pioneers and late successional shade tolerants can be obtained by combining seed addition experiments with early successional removal experiments (Pacala and Rees 1998). As outlined by Pacala and Rees (1998), the latter involves removing early successional seedlings from sites at which late successional seedlings have also arrived. The combination of this experiment with a community-level seed addition experiment makes it possible to quantify to what degree successional diversity is maintained by a successional niche – that is, by some species being better competitors in early successional (high light) sites and others better competitors in late successional (low light) sites – and to what degree it is maintained by a competition–colonization trade-off (Pacala and Rees 1998).

Information gleaned from measurements of species traits, field studies of determinants of spatio-temporal variation in recruitment success, and/or field experiments can potentially be used to parametrize models that allow for further investigation of the roles of colonization-related trade-offs. Explicit consideration of model requirements and estimation of key traits can make it possible to quantitatively evaluate whether analytical conditions for stable coexistence of particular species are met (Geritz et al. 1999, Kisdi and Geritz 2003a). Further, the parametrization and application of individual-based community models offers the possibility of running virtual experiments that would be impractical in the real world (e.g., Pacala et al. 1996). Such simulation experiments could include all the experiments described above, which could of course be run for much longer time periods and larger spatial scales in models than they could feasibly be executed in the field.

EMPIRICAL EVIDENCE IN TROPICAL FORESTS

While there are in principle many ways to investigate colonization-related trade-offs in tropical forests, available evidence at this time is largely limited to trait relationships. As in temperate systems, research on colonization-related trade-offs has focused on the potential for seed-size mediated trade-offs (Westoby et al. 1996, Leishman et al. 2000). Specifically, the hypothesis is that species may be good colonists producing many small seeds of low competitive ability and/or low stress tolerance, or they may be good competitors and/or stress tolerators, producing few large seeds of high competitive ability and/or high stress tolerance. There is widespread empirical support for these relationships in extra-tropical systems (Westoby et al. 1996, Leishman et al. 2000, Moles et al. 2004, Moles and Westoby 2004), and it has been hypothesized that the advantages of large seeds should be even stronger in tropical forests (Foster 1986). Here, I first consider the relationship of seed mass to fecundity, which underlies all three trade-offs examined in this chapter, and then evaluate the other types of evidence for each trade-off in turn.

Seed mass and fecundity appear to be negatively related in tropical forests, just as in other plant communities, although relatively little data are available. In their global meta-analysis of seed size and seed production, Moles *et al.* (2004) have reproductive data for only five tropical forest species, among which there is no significant relationship. Dalling and Hubbell (2002) demonstrate a negative correlation between seed mass and seed density in the soil seed bank for 15 pioneer species in a Panamanian wet tropical forest; this is consistent with a negative relationship between seed mass and seed production, although seed density also includes the effects of adult abundance (Figure 11.2a). In the same forest, Muller-Landau *et al.* (2008) find a strong negative relationship between seed mass and per basal area seed production among 40 tree species of varying life history strategy, a relationship well-fit by a power function.

Competition–colonization trade-offs

No tropical studies have specifically examined the degree to which seed mass predicts total competitive ability – the outcome of competition among seedlings. Seed mass appears to be positively related with some traits expected to provide a competitive advantage, but not all. Seed mass is positively related to seedling size at germination and in the first 2 years (Rose and Poorter 2002, Green and Juniper 2004a, Svenning and Wright 2005), but because small-seeded species have higher relative growth rates (Poorter and Rose 2005), this advantage decays as seedlings age (Rose and Poorter 2002). Seed mass is also positively correlated with the probability that a seed will become an established seedling (Muller-Landau 2001, Dalling and Hubbell 2002, Svenning and Wright 2005), a transition probability that encompasses seed survival, germination probability, and early seedling survival. The evidence regarding the relationship of seed mass with later seedling survival is mixed – some studies have found a positive relationship, while others have found no relationship (Augspurger 1984,

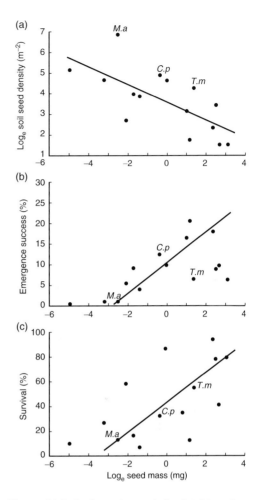

Figure 11.2 Seed mass is negatively related to seed density in the soil (a) and positively related to establishment probability (b) and seedling survival (c) among 15 gap-dependent tree species in a wet tropical forest in Panama. Three common pioneer species are identified: *Cecropia peltata* (C.p), *Miconia argentea* (M.a), and *Trema micrantha* (T.m). Reprinted from Dalling and Hubbell (2002) with permission of Blackwell Publishing.

Rose and Poorter 2002, Svenning and Wright 2005). Separate consideration of studies conducted under different light levels reveals that seed mass is positively related to seedling survival in the shade, but unrelated to seedling survival in high light (Rose and Poorter 2002, Poorter and Rose 2005).

While interspecific trait relationships appear to show some evidence consistent with a competition–colonization trade-off when habitat-independent competitive traits are measured, habitat-specific analyses suggest that the trade-off is not of the type encapsulated in the classical model of this mechanism (Tilman 1994). Seed mass is negatively related to fecundity (the classical measure of colonization ability), but is not uniformly positively related to competitive traits in all environments. For larger-seeded species to be better competitors as envisioned under the standard competition–colonization trade-off, they should consistently outcompete small-seeded species in all environments whenever they are present, and regardless of the habitat conditions. Evidence instead shows that smaller-seeded species are as good or better at winning high light sites, with no seedling survival disadvantage and higher growth rates that rapidly make up for their smaller initial size (Rose and Poorter 2002). This suggests smaller-seeded species do not merely win by default when larger-seeded (and presumably more competitive) species are absent – they can win in high light environments even when those competitors are present, whether because of their numerical dominance (fecundity–tolerance trade-off) or their specific adaptations for these environments (successional niche, *sensu* Pacala and Rees 1998). This clearly contradicts the predictions of the competition–colonization trade-off model with perfect asymmetry. Further, the evidence that large-seeded species do not even have a lasting per capita advantage in seedling survival and size in high light sites is not consistent with even partial asymmetry. As Leigh *et al.* (2004) argue, we can at this point set aside the competition–colonization model as a possible explanation for the coexistence of pioneer and shade-tolerant species.

Nonetheless, it could be argued that competition–colonization trade-offs could still play a role in species coexistence within understory sites, where large seeds do seem to enjoy consistent advantages. A series of population-level seed addition experiments by Svenning and Wright (2005) provide evidence that even in the shaded forest understory, any seed-size mediated

competition–colonization trade-off can be neither stabilizing nor perfectly equalizing. Svenning and Wright (2005) added seeds of 32 shade-tolerant species whose seed masses spanned three orders of magnitude to understory sites. If the trade-off is merely equalizing and is based on seed mass, then increasing seed rain of all species by the same absolute total mass of seed added should result in the same absolute increase in seedling abundance. If the trade-off is stabilizing, then the same increase in seed mass arriving should result in greater increases in seedling abundance in species that are good competitors and poor colonists than in those that are poor competitors and good colonists. Svenning and Wright (2005) found that the probability that an added seed would germinate, establish, and survive to 1 or 2 years did not differ significantly with seed size. This suggests that the addition of similar biomass of seeds of all species would result in a disproportionate increase in seedling numbers of small-seeded species – contrary to the competition–colonization hypothesis based on seed size which predicts that large-seeded species should benefit most. The advantage of small-seeded species decreased from year 1 to year 2, though (Svenning and Wright 2005), so it is possible that a long enough seed addition experiment would eventually find an equal or greater effect of seed addition in large-seeded species, consistent with a perfectly equalizing or stabilizing effect of a competition–colonization trade-off. At this point, however, the evidence from the first 2 years of the experiment can at best be interpreted as showing a very weakly equalizing effect. Altogether, the results further discount the possibility that a competition–colonization trade-off in the classical sense contributes to diversity maintenance in tropical forests.

Tolerance–fecundity trade-offs

Empirical studies have examined the relationship of seed mass to tolerance among tropical tree species. The results show large-seeded species have higher survival rates in the face of some hazards and in low light sites. Species with larger seeds tend to retain larger reserves in storage cotyledons

(Green and Juniper 2004a) and thus are more likely to be able to resprout after severe seedling herbivory or damage (Harms and Dalling 1997, Green and Juniper 2004b). However, large seed mass does not appear to be associated with better survival of either pre-dispersal seed predation or post-dispersal seed removal in a meta-analysis of data for tropical species (Moles et al. 2003). Seed mass is positively related to seedling survival in the shade, even though it is unrelated to seedling survival in high light (reviewed in Rose and Poorter 2002, Poorter and Rose 2005). Because light is a key limiting resource in tropical forests and especially for seedlings (Chazdon 1988, Montgomery and Chazdon 2002), this suggests that large seed mass specifically conveys an advantage in tolerating low resource conditions.

The accumulating evidence of a negative relationship of seed size with fecundity and a positive relationship of seed mass with tolerance of low light, herbivory, and damage increasingly suggests the presence of a seed-size mediated fecundity–tolerance trade-off in tropical forests. Such a trade-off could contribute to habitat partitioning of regeneration sites among species based on the resource and stress levels of the sites and the stress tolerance of the species. Specifically, the data are consistent with the idea that large-seeded species win sites that are too low in resources or high in stress for small-seeded species to tolerate, and small-seeded species disproportionately win in high resource, low stress sites where their numerical dominance in seed arrival becomes a dominance in seedling recruits. The many studies on spatial variation in understory light levels (Becker and Smith 1990, Nicotra et al. 1999), and on the stochasticity of physical damage (Clark and Clark 1989), further provide evidence for abundant relevant heterogeneity to be partitioned. However, it is important to note that the numerical success of small-seeded species in high resource sites is probably also due in part to their specific adaptations for these environments, and thus at least in part to the successional niche mechanism (Pacala and Rees 1998) rather than to a fecundity–tolerance trade-off. Additional research is needed to quantify the relative importance of these two mechanisms.

Dispersal–fecundity trade-offs

If seed mass and fecundity are strongly negatively related among species, then dispersal and fecundity can be strongly negatively related (i.e., trade off) only if seed mass and dispersal are themselves positively related. Such a relationship has been hypothesized for animal-dispersed species, but the opposite relationship is expected which constitute 70–100% of plants in wet tropical forests (Willson et al. 1989), among wind-dispersed species. Among animal-dispersed species, it is hypothesized that larger-seeded fruits tend to be eaten by animal species with larger body sizes (Kalko et al. 1996, Grubb 1998, Peres and van Roosmalen 2002), and that these animal species in turn tend to have slower gut passage time and larger home ranges (Brown 1995, Kalko et al. 1996), which together should produce longer dispersal distances (Murray 1988). Further, among scatter-hoarding rodents, dispersal distances are expected to increase with seed size because larger seeds offer more reward for the effort of caching (Jansen et al. 2002). Among wind-dispersed species, in contrast, larger-seeded species are expected to have higher terminal velocities and thus shorter dispersal distances, a prediction supported by empirical studies (Augspurger 1986, Muller-Landau 2001). It is important to note that a dispersal–fecundity trade-off, like any coexistence mechanism, could play a role in the coexistence of one group of species (e.g., those dispersed by a particular type of animal) even if it were not present in all.

There are relatively few data on the relationship of seed mass with seed dispersal among animal-dispersed tropical species at this point. Holbrook and Smith (2000) showed that among nine taxa dispersed by hornbills, gut passage times and thus estimated dispersal distances were longer in larger-seeded taxa, while Levey (1986) found gut passage times among nine species of birds were shorter for larger seeds. Westcott and Graham (2000) show that there is a positive, almost linear, relationship between disperser body mass and median dispersal distance among eight tropical bird species, which would imply a positive relationship between seed size and dispersal distance if disperser body size is positively related with seed size – but there is little evidence regarding this hypothesized relationship. Further, because seeds of any given tropical plant species are typically dispersed by many different animal species (Muller-Landau and Hardesty 2005), total dispersal by all agents must be examined in order to estimate the total pattern of seed dispersal. An inverse modeling study by Muller-Landau et al. (2008) found that seed mass was negatively related to estimated mean dispersal distances (by all animal species combined) among 31 animal-dispersed species in Panama; however, the data and methods used are inadequate to quantify long-distance dispersal.

At this point, the limited evidence suggests that dispersal–fecundity trade-offs are not generally present across all tropical species, although they may be present within some groups. Specifically, there is some empirical support for their presence among bird-dispersed species, but evidence for the opposite pattern (a positive dispersal–fecundity correlation) among wind-dispersed species and among animal-dispersed species in general. It remains unclear whether there is sufficient spatial heterogeneity in the density of suitable regeneration sites to facilitate coexistence via a dispersal–fecundity trade-off if one is present. Thus, further research is necessary to assess the role and even presence of dispersal–fecundity trade-offs among tropical trees.

CONCLUSIONS AND FUTURE DIRECTIONS

Theory demonstrates that colonization-related trade-offs can contribute to diversity maintenance in weak (equalizing) and/or strong (stabilizing) ways. Theoretical and empirical attention has long focused on the competition–colonization trade-off, which was early demonstrated to have the potential to contribute strongly to diversity maintenance of many species. However, conditions for these contributions are stringent, and current evidence suggests that this trade-off is not present in tropical forests in its classical form, and thus is not contributing to diversity maintenance in this ecosystem. More recent research has identified two other colonization-related trade-offs – between fecundity and tolerance of low

resources or high stress, and between fecundity and dispersal – as potentially important diversity-maintaining mechanisms. Further, it appears that tolerance–fecundity trade-offs are present in tropical forests, with small-seeded species having higher fecundity and lower ability to tolerate low resource or high stress habitats than large-seeded species. Theory and data remain insufficient to evaluate the potential of the tolerance–fecundity trade-off to exert stabilizing or equalizing influences in tropical forests. The limited data on dispersal–fecundity trade-offs suggest they are not generally present – but cannot exclude the possibility of a role within some groups.

Further theoretical and empirical research, and novel integration of the two, is needed to investigate the potential and actual role of tolerance–fecundity and dispersal–fecundity trade-offs in tropical forests. Theoretical work on the tolerance–fecundity trade-off is necessary to determine the conditions under which this mechanism is stabilizing for various scenarios of community dynamics – specifically, how must the fecundity and tolerance of two species be related in order for them to stably coexist. Theory should also consider how different mechanisms might interact, and how their influences can be disentangled – in this context, a particularly important issue is the relative role of fecundity–tolerance trade-offs versus species trade-offs in performance among habitats in contributing to habitat niche partitioning. Further measurement of habitat tolerances, fecundities, and dispersal abilities of species would make it possible to better characterize the trade-offs, to determine what environmental axes are involved in the tolerance mechanism, and to discover if/where the fecundity–dispersal trade-off is present. Empirical assessments of dispersal should include not only dispersal distance but also differential dispersal to particular habitats (including directed dispersal), clumping, and more complex phenomena influencing arrival rates, and should examine not only correlations with seed size but also other possible trade-offs. The distribution of the relevant habitat types being partitioned – environmental conditions, or densities of favorable sites – must also be measured. In combination with information on species traits and theory, this should make possible some calculations of the role of colonization-related trade-offs.

Finally, research on these trade-offs should go beyond simply trait measurements and theory, to other and stronger tests of the roles of the trade-offs in real communities. Community-level field studies of spatio-temporal variation in environmental conditions, seed arrival, and successful recruitment should be conducted to determine the degree to which different mechanisms can explain recruitment patterns. Field experiments involving manipulations of seed arrival, seedling recruitment, and environmental conditions could provide even stronger tests of the mechanisms. For time scales and spatial scales over which such field studies and experiments are infeasible, models parametrized from field data can provide a useful tool to explore long-term, large-scale dynamical implications of documented processes. With this combination of tools, we should be able to achieve a much better understanding of colonization-related trade-offs in tropical forests in the future.

ACKNOWLEDGMENTS

I thank Joe Wright, Steve Pacala, and two anonymous reviewers for helpful comments on this chapter.

REFERENCES

Adler, F.R. and Mosquera, J. (2000) Is space necessary? Interference competition and limits to biodiversity. *Ecology* 81, 3226–3232.

Amarasekare, P. (2003) Competitive coexistence in spatially structured environments: a synthesis. *Ecology Letters* 6, 1109–1122.

Armstrong, R.A. (1976) Fugitive species: experiments with fungi and some theoretical considerations. *Ecology* 57, 953–963.

Augspurger, C.K. (1984) Seedling survival of tropical tree species: interactions of dispersal distance, light-gaps, and pathogens. *Ecology* 65, 1705–1712.

Augspurger, C.K. (1986) Morphology and dispersal potential of wind-dispersed diaspores of neotropical trees. *American Journal of Botany* 73, 353–363.

Becker, P.F. and Smith, A.P. (1990) Spatial autocorrelation of solar radiation in a tropical moist forest understory. *Agricultural and Forest Meteorology* 52, 373–379.

Brown, J.H. (1995) *Macroecology.* University of Chicago Press, Chicago, IL.

Chazdon, R.L. (1988) Sunflecks and their importance to forest understorey plants. *Advances in Ecological Research* 18, 1–63.

Chesson, P. (2000) Mechanisms of maintenance of species diversity. *Annual Reviews of Ecology and Systematics* 31, 343–366.

Chesson, P.L. and Warner, R.R. (1981) Environmental variability promotes coexistence in lottery competitive systems. *American Naturalist* 117, 923–943.

Clark, D.B. and Clark, D.A. (1989) The role of physical damage in the seedling mortality regime of a neotropical forest. *Oikos* 55, 225–230.

Clark, J.S., LaDeau, S., and Ibanez, I. (2005) Fecundity of trees and the colonization-competition hypothesis. *Ecological Monographs* 74, 415–442.

Comins, H.N. and Noble, I.R. (1985) Dispersal, variability, and transient niches: species coexistence in a uniformly variable environment. *American Naturalist* 126, 706–723.

Connell, J.H. (1978) Diversity in tropical rain forests and coral reefs. *Science* 199, 1302–1310.

Coomes, D.A. and Grubb, P.J. (2003) Colonization, tolerance, competition and seed-size variation within functional groups. *Trends in Ecology and Evolution* 18, 283–291.

Dalling, J.W. and Hubbell, S.P. (2002) Seed size, growth rate and gap microsite conditions as determinants of recruitment success for pioneer species. *Journal of Ecology* 90, 557–568.

Foster, S.A. (1986) On the adaptive value of large seeds for tropical moist forest trees: a review and synthesis. *The Botanical Review* 52, 260–299.

Geritz, S.A.H., van der Meijden, E., and Metz, J.A.J. (1999) Evolutionary dynamics of seed size and seedling competitive ability. *Theoretical Population Biology* 55, 324–343.

Green, P.T. and Juniper, P.A. (2004a) Seed–seedling allometry in tropical rain forest trees: seed mass-related patterns of resource allocation and the "reserve effect." *Journal of Ecology* 92, 397–408.

Green, P.T. and Juniper, P.A. (2004b) Seed mass, seedling herbivory and the reserve effect in tropical rain forest seedlings. *Functional Ecology* 18, 539–547.

Grubb, P.J. (1998) Seeds and fruits of tropical rainforest plants: interpretation of the range in seed size, degree of defence and flesh/seed quotients. In D.M. Newbery, H.H.T. Prins, and N.D. Brown (eds), *Dynamics of Tropical Communities.* Blackwell Scientific, Oxford, pp. 1–24.

Harms, K.E. and Dalling, J.W. (1997) Damage and herbivory tolerance through resprouting as an advantage of large seed size in tropical trees and lianas. *Journal of Tropical Ecology* 13, 617–621.

Hastings, A. (1980) Disturbance, coexistence, history, and competition for space. *Theoretical Population Biology* 18, 363–373.

HilleRisLambers, J. and Clark, J.S. (2003) Effects of dispersal, shrubs, and density-dependent mortality on seed and seedling distributions in temperate forests. *Canadian Journal of Forest Research-Revue Canadienne de Recherche Forestiere* 33, 783–795.

Holbrook, K.M. and Smith, T.B. (2000) Seed dispersal and movement patterns in two species of *Ceratogymna* hornbills in a West African tropical lowland forest. *Oecologia* 125, 249–257.

Horn, H.S. and MacArthur, R.H. (1972) Competition among fugitive species in a harlequin environment. *Ecology* 53, 749–752.

Hubbell, S.P. (2001) *The Unified Neutral Theory of Biodiversity and Biogeography*. Princeton University Press, Princeton.

Jansen, P.A., Bartholomeus, M., Bongers, F. *et al.* (2002) The role of seed size in dispersal by a scatter-hoarding rodent. In D.J. Levey, W.R. Silva, and M. Galetti (eds), *Seed Dispersal and Frugivory: Ecology, Evolution and Conservation*. CAB International, Wallingford, Oxfordshire, pp. 209–225.

Jansen, V.A.A. and Mulder, G. (1999) Evolving biodiversity. *Ecology Letters* 2, 379–386.

Kalko, E.K.V., Herre, E.A., and Handley, C.O. (1996) Relation of fig fruit characteristics to fruit-eating bats in the New and Old World tropics. *Journal of Biogeography* 23, 565–576.

Kisdi, E. and Geritz, S.A.H. (2003a) Competition–colonization trade-off between perennial plants: exclusion of the rare species, hysteresis effects and the robustness of co-existence under replacement competition. *Evolutionary Ecology Research* 5, 529–548.

Kisdi, E. and Geritz, S.A.H. (2003b) On the coexistence of perennial plants by the competition–colonization trade-off. *American Naturalist* 161, 350–354.

Leigh, Egbert G., Davidar, P. *et al.* (2004) Why do some tropical forests have so many species of trees? *Biotropica* 36, 447–473.

Leishman, M.R., Wright, I.J., Moles, A.T. *et al.* (2000) The evolutionary ecology of seed size. In M. Fenner (ed.), *Seeds: The Ecology of Regeneration in Plant Communities*. CAB International, Wallingford, Oxfordshire, pp. 31–57.

LePage, P.T., Canham, C.D., Coates, K.D. *et al.* (2000) Seed abundance versus substrate limitation of seedling recruitment in northern temperate forests of

British Columbia. *Canadian Journal of Forest Research* 30, 415–427.

Levey, D.J. (1986) Methods of seed processing by birds and seed deposition patterns. In A. Estrada and T.H. Felming (eds), *Frugivores and Seed Dispersal*. Dr W. Junk Publishers, Dordrecht, pp. 147–158.

Levine, J.M. and Rees, M. (2002) Coexistence and relative abundance in annual plant assemblages: the roles of competition and colonization. *American Naturalist* 160, 452–467.

Levins, R. and Culver, D. (1971) Regional coexistence of species and competition between rare species. *Proceedings of the National Academy of Sciences of the United States of America* 68, 1246–1248.

Moles, A.T., Falster, D.S., Leishman, M.R. *et al.* (2004) Small-seeded species produce more seeds per square metre of canopy per year, but not per individual per lifetime. *Journal of Ecology* 92, 384–396.

Moles, A.T., Warton, D.I., and Westoby, M. (2003) Do small-seeded species have higher survival through seed predation than large-seeded species? *Ecology* 84, 3148–3161.

Moles, A.T. and Westoby, M. (2004) Seedling survival and seed size: a synthesis of the literature. *Journal of Ecology* 92, 372–383.

Montgomery, R.A. and Chazdon, R.L. (2002) Light gradient partitioning by tropical tree seedlings in the absence of canopy gaps. *Oecologia* 131, 165–174.

Muller-Landau, H.C. (2001) *Seed dispersal in a tropical forest: empirical patterns, their origins and their consequences for forest dynamics*. PhD dissertation, Princeton University, Princeton.

Muller-Landau, H.C. and Hardesty, B.D. (2005) Seed dispersal of woody plants in tropical forests: concepts, examples, and future directions. In D. Burslem, M. Pinard, and S. Hartley (eds), *Biotic Interactions in the Tropics: Their Role in the Maintenance of Species Diversity*. Cambridge University Press, Cambridge, pp. 267–309.

Muller-Landau, H.C., Wright, S.J., Calderón, O. *et al.* (2008) Interspecific variation in primary seed dispersal in a tropical forest. *Journal of Ecology* in press.

Murray, K.G. (1988) Avian seed dispersal of three neotropical gap-dependent plants. *Ecological Monographs* 58, 271–298.

Nicotra, A.B., Chazdon, R.L., and Iriarte, S.V.B. (1999) Spatial heterogeneity of light and woody seedling regeneration in tropical wet forests. *Ecology* 80, 1908–1926.

Pacala, S.W., Canham, C.D., Saponara, J. *et al.* (1996) Forest models defined by field measurements: estimation, error analysis and dynamics. *Ecological Monographs* 66, 1–43.

Pacala, S.W. and Rees, M. (1998) Models suggesting field experiments to test two hypotheses explaining successional diversity. *American Naturalist* 152, 729–737.

Peres, C.A. and van Roosmalen, M. (2002) Primate frugivory in two species-rich neotropical forests: implications for the demography of large-seeded plants in overhunted areas. In D.J. Levey, W.R. Silva, and M. Galetti (eds), *Seed Dispersal and Frugivory: Ecology, Evolution and Conservation*. CAB International, Wallingford, Oxfordshire, UK, pp. 407–421.

Plotkin, J.B., Potts, M.D., Leslie, N. *et al.* (2000) Species–area curves, spatial aggregation, and habitat specialization in tropical forests. *Journal of Theoretical Biology* 207, 81–99.

Poorter, L. and Rose, S. (2005) Light-dependent changes in the relationship between seed mass and seedling traits: a meta-analysis for rain forest tree species. *Oecologia* 142, 378–387.

Rees, M. and Westoby, M. (1997) Game-theoretical evolution of seed mass in multi-species ecological models. *Oikos* 78, 116–126.

Rose, S. and Poorter, L. (2002) The importance of seed mass for early regeneration in tropical forest: a review. In H. Ter Steege (ed.), *Long Term Changes in Composition and Diversity: Case Studies from the Guyana Shield, Africa, Borneo and Melanesia*. Tropenbos Foundation, Wageningen, The Netherlands, pp. 5–11.

Skellam, J.G. (1951) Random dispersal in theoretical populations. *Biometrika* 38, 196–218.

Svenning, J.-C. and Wright, S.J. (2005) Seed limitation in a Panamanian forest. *Journal of Ecology* 93, 853–862.

Tilman, D. (1994) Competition and biodiversity in spatially structured habitats. *Ecology* 75, 2–16.

Tilman, D., Lehman, C.L., and Yin, C. (1997) Habitat destruction, dispersal and deterministic extinction in competitive communities. *American Naturalist* 149, 407–435.

Tilman, D. and Pacala, S. (1993) The maintenance of species richness in plant communities. In R.E. Ricklefs and D. Schluter (eds), *Species Diversity in Ecological Communities: Historic and Geographical Perspectives*. University of Chicago Press, Chicago, IL, pp. 13–25.

Turnbull, L.A., Rees, M., and Crawley, M.J. (1999) Seed mass and the competition/colonization trade-off: a sowing experiment. *Journal of Ecology* 87, 899–912.

Westcott, D.A. and Graham, D.L. (2000) Patterns of movement and seed dispersal of a tropical frugivore. *Oecologia* 122, 249–257.

Westoby, M., Leishman, M., and Lord, J. (1996). Comparative ecology of seed size and dispersal. *Philosophical Transactions of the Royal Society of London B* 351, 1309–1318.

Wilcove, D.S., Rothstein, D., Dubow, J. *et al.* (1998) Quantifying threats to imperiled species in the United States. *BioScience* 48, 607–615.

Willson, M.F., Irvine, A.K., and Walsh, N.G. (1989) Vertebrate Dispersal syndromes in some Australian and New-Zealand plant-communities, with geographic comparisons. *Biotropica* 21, 133–147.

Yu, D.W., Terborgh, J.W., and Potts, M.D. (1998) Can high tree species richness be explained by Hubbell's null model? *Ecology Letters* 1, 193–199.

Yu, D.W. and Wilson, H.B. (2001) The competition–colonization trade-off is dead; long live the competition–colonization trade-off. *American Naturalist* 158, 49–63.

Zhang, D.-Y. and Lin, K. (1997) The effects of competitive asymmetry on the rate of competitive displacement: how robust is Hubbell's community drift model? *Journal of Theoretical Biology* 188, 361–367.

Chapter 12

TREEFALL GAPS AND THE MAINTENANCE OF PLANT SPECIES DIVERSITY IN TROPICAL FORESTS

Stefan A. Schnitzer, Joseph Mascaro, and Walter P. Carson

OVERVIEW

Treefall gaps, one of the key forms of disturbance in tropical forests, are hypothesized to maintain species diversity via three main and non-mutually exclusive ways. First, they create high light habitats, providing a regeneration niche for early successional shade-intolerant and intermediate-tolerant species to reach reproductive maturity, and thus prevent their competitive exclusion by more shade-tolerant species. Second, species may specialize on and partition resources along resource gradients that vary strongly from the gap center to forest interior, thus permitting species coexistence. Third, species may specialize along a gradient of gap sizes, with some species regenerating in small gaps and others in large gaps, which would also permit stable species coexistence. Support for the gap hypothesis is mixed, but evidence suggests that some plant groups may benefit from gaps more than others.

Pioneer tree species and at least some species of lianas appear to require or capitalize on gaps for successful regeneration or to reach reproductive maturity. This may also be true for shrubs and herbaceous species, but these growth forms are rarely considered in studies of gap dynamics. Gaps provide not only an essential regeneration niche for some growth forms, but they also may provide the resources necessary for reproduction; this latter aspect of gap dynamics has been largely ignored. In contrast, shade-tolerant tree diversity does not appear to be maintained by gaps, possibly due to a combination of seed, dispersal, and recruitment limitation, the last possibly due to competition with other plants, particularly lianas. Nevertheless, treefall gaps maintain the diversity of some plant groups, which, in many tropical forests, may comprise a large proportion of the vascular plant community.

INTRODUCTION

The formation of treefall gaps and their influence on forest regeneration and dynamics has a long history in ecology. Whitmore (1989) suggested that "gaps, openings in the forest canopy, drive the forest cycle," and that "the gap phase is thus the most important part of the growth cycle for the determination of floristic composition."

The genesis of the gap hypothesis began with the studies of Watt (1925), Aubréville (1938), and Jones (1945), who described the patchy nature of mature forest communities (cited in Swaine and Hall 1988, Peet 1991). Watt (1947) extended these ideas by describing "the gap phase" as a general phenomenon whereby succession occurred within small patches of relatively stable plant communities, regardless of the specific ecosystem. In the 1950s, treefall gaps became recognized as foci for regeneration and succession in both temperate and tropical forests (e.g., Richards 1952, Beard 1955, Bray 1956, Webb 1959). Following the development of non-equilibrium explanations for the maintenance of diversity (e.g., Connell 1978, Huston 1979), a number of authors more fully developed the concept of gap-phase regeneration into an important theory to explain the maintenance of species diversity in tropical forests (e.g., Ricklefs 1977, Whitmore 1978, Denslow 1980, 1987, Hartshorn 1980, Brokaw 1982, 1985a, 1985b, Orians 1982, Hubbell and Foster 1986). The variety of ideas on the role of gaps in the maintenance of diversity can be synthesized into a single "gap hypothesis." The formation of canopy gaps by the death of one to a few canopy trees creates sufficient resource heterogeneity to allow for resource partitioning and niche differentiation (*sensu* Grinnell 1917), or releases sufficient resources (e.g., light and nutrients) to permit the establishment or reproduction of plant species that would otherwise be excluded from the forest in the absence of gaps.

While much of the contemporary theory for the gap hypothesis was formulated in the late 1970s and early 1980s, community-level empirical tests of this hypothesis remained uncommon until the late 1990s (e.g., Hubbell *et al.* 1999, Kobe 1999, Schnitzer and Carson 2001). Given the historical importance of the gap hypothesis, it is surprising that few studies explicitly tested this hypothesis in tropical forests. Previous tests have focused on the capacity of plants to partition resources or to respond rapidly to high resource availability. Collectively, these tests suggested that the diversity of some plant growth forms were maintained by gaps, whereas that of other growth forms was not. Consequently, evidence for the gap hypothesis as a general mechanism to maintain plant species diversity is likely to depend upon the functional group or growth form under study.

The goal of this chapter is to summarize new and previously published data to provide a coherent picture of the role of treefall gaps in tropical forest regeneration and dynamics, and to determine the degree to which there is empirical support for the gap hypothesis. Specifically, we will describe and examine the following key topics: (1) the main processes and pathways of

gap-phase regeneration; (2) the models and mechanisms by which treefall gaps are proposed to maintain diversity; (3) the empirical evidence for the gap hypothesis as applied to major plant growth forms or functional groups; (4) variation in the impact of gaps across broad environmental gradients; and (5) the relationship between the gap hypothesis and the intermediate disturbance hypothesis. We restrict our definition of the gap hypothesis to canopy gaps that are formed by the standing death or toppling of one to a few trees or a significant limb-fall. Larger-scale stand replacement disturbances are outside the scope of the gap hypothesis. We acknowledge that there is a continuum of disturbance events that create small gaps to events that create very large clearings (Lieberman *et al.* 1989, Vandermeer *et al.* 1996); nonetheless, single- to several-tree canopy gaps are the most frequent type of disturbance in many tropical forests (e.g., Hubbell and Foster 1986).

PROCESSES AND PATHWAYS OF GAP-PHASE REGENERATION

The process of gap colonization can be divided into four discrete categories, only some of which are mutually exclusive.

1 *From seed.* Plants recruit from seeds that were present in the soil seed bank prior to gap formation (Dalling *et al.* 1998) or are dispersed into the gap soon thereafter by wind or animals (Schemske and Brokaw 1981, Levey 1988, Wunderle *et al.* 2006). Seeds of shade-intolerant pioneer species typically require the high light and temperature conditions of gaps for germination (Pearson *et al.* 2003a). Once established, these pioneers can partially fill gaps via extremely rapid growth rates (Brokaw 1985a). Shade-tolerant species may also recruit from seed immediately following gap formation (Kitajima and Poorter Chapter 10, this volume).

2 *From advance regeneration.* Shade-tolerant seedlings and saplings present in the understory prior to gap formation grow rapidly to fill the gap (Uhl *et al.* 1988).

3 *From vegetative reproduction.* Trees or shrubs within the gap or lianas pulled into the gap by fallen trees produce numerous clonal shoots

(Putz 1984a, Vandermeer *et al.* 1996, 2000, Bond and Midgly 2001, Schnitzer *et al.* 2004).

4 *From spreading laterally into the gap from the surrounding intact forest.* Some growth forms, typically lianas and some herbaceous vines, can recruit into and fill treefall gaps by growing laterally from the adjacent intact forest (Peñulosa 1984, Schnitzer *et al.* 2000).

The first two regeneration pathways and occasionally the third are common to trees, whereas lianas may exploit all four pathways due to their ability to disperse their seeds throughout the forest (Wright *et al.* 2007), persist in the shaded understory (Gilbert *et al.* 2006), rapidly produce clonal shoots from their fallen stems (Putz 1984a, Schnitzer *et al.* 2000, 2004), and maintain positive growth rates in the understory (Schnitzer 2005). Lateral growth of tree crowns may also fill small gaps from above and, while crown growth is not a true recruitment pathway in the sense of those mentioned above, it is likely responsible for the partial or complete closure of small gaps originating from the death of a single small tree or a large limb (Hubbell and Foster 1986).

Following gap formation, plant recruitment and growth are exceptionally high for the first few years. Thinning, or the reduction in total stem number, typically begins within 3–6 years of gap formation, as individuals increase in size and close the canopy, thereby decreasing light reaching the understory and increasing mortality (Brokaw 1985a, Hubbell and Foster 1986, Fraver *et al.* 1998). Gap-phase regeneration is complete when one or a few large trees attain the height of the intact canopy, thus effectively closing the gap. Gap-phase regeneration can be a rapid process in relatively productive tropical forests, with the height of the canopy increasing by as much as 5–7 m per year when certain pioneer trees are present (e.g., *Trema micrantha*, *Cecropia insignis*, and *Zanthoxylum* spp.; Putz 1983, Brokaw 1985a, 1987), and by several meters per year for non-pioneer trees (e.g., *Simarouba amara*, *Virola sebifera*, and *Protium panamensis*; Brokaw 1985a).

Gap-phase regeneration, however, can also become stalled or arrested at a low canopy height for many years (Figure 12.1). This alternative regeneration pathway can occur when lianas are in high abundance initially and the falling trees drag even more lianas into the gap. Most lianas survive the initial treefall (Putz 1984a), after which they copiously produce new stems in the high resource environment, forming a dense tangle of vegetation (Babweteera *et al.* 2000, Schnitzer *et al.* 2000, 2004). Lianas colonizing the gap from seed or from advance regeneration, or growing into gaps from the intact forest all contribute to these liana tangles (Appanah and Putz 1984, Peñulosa 1984, Putz 1984a, Putz and Chai 1987), which can continue to expand in size and density if lianas fail to find trellises (Peñulosa 1984). Once a liana tangle forms, it will block and delay gap-phase regeneration of trees by some combination of below-ground competition, light pre-emption, and mechanical interference (Schnitzer *et al.* 2000, 2004, 2005, Tabanez and Viana 2000). Gap-phase regeneration may be stalled at a low canopy height by lianas for at least 13 years, and probably much longer (Schnitzer *et al.* 2000). Eventually, some trees (usually pioneers; Putz 1984a, Schnitzer *et al.* 2000) may ultimately escape vertically through the liana tangle and begin to close the canopy. The legacy of these formerly arrested gaps is often an impenetrable thicket of liana stems that remain in the understory (Figure 12.2).

Palms may cause a completely different successional trajectory of gap-phase regeneration, when they are abundant in a newly formed gap (Figure 12.1). While the gap dynamics of palms has not been explicitly studied, understory palms suppress seedling regeneration by casting deep shade and dropping large fronds that decay slowly and smother seedlings (Denslow *et al.* 1991). Consequently, there is often a depauperate seedling layer beneath palms in intact forests prior to gap formation (Farris-Lopez *et al.* 2004, Peters *et al.* 2004, Wang and Augspurger 2006). Palms may also act as a filter to regeneration, whereby only very shade-tolerant species that are also resistant to mechanical damage (*sensu* Clark and Clark 1991) are available to fill the newly opened gap. Thus, the suite of species found in palm-dominated gaps may be very different than that in either liana-dominated or tree-dominated gaps, and the pathway of gap-phase regeneration in a given forest will likely depend on the relative abundance of the different growth forms.

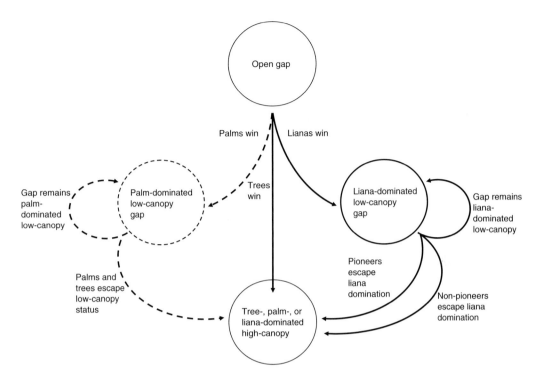

Figure 12.1 Model of possible pathways of gap-phase regeneration. Following a treefall, gaps are assumed to go from a low-canopy, open gap through a series of progressive stages, culminating at a high-canopy, intact forest condition. However, alternative pathways of gap-phase regeneration also may occur. Modified from Schnitzer *et al.* (2000).

HYPOTHESES AND MECHANISMS BY WHICH TREEFALL GAPS ARE PROPOSED TO MAINTAIN DIVERSITY

Theories that invoke disturbance as a mechanism to maintain diversity often assume that communities never reach an equilibrium state because disturbance prevents competitively dominant species from excluding competitively inferior, early successional species (e.g., Connell 1978). The gap hypothesis may be considered a non-equilibrium mechanism for the maintenance of diversity because the death of a canopy tree and subsequent formation of a treefall gap initiate a successional sequence that begins with pioneer species and eventually transitions to dominance by shade-tolerant species. Thus, treefall gaps provide a regeneration niche for shade-intolerant

pioneer species or intermediate shade-tolerant species to establish and regenerate, preventing their competitive exclusion from the community (e.g., Swaine and Whitmore 1988, Whitmore 1989, Dalling *et al.* 2001). This is the simplest form of the gap hypothesis, which, at the landscape scale, permits the coexistence of both early successional species occupying gaps and late successional species occupying the surrounding matrix of intact forest (Whitmore 1978, Connell 1979).

The gap hypothesis also provides an equilibrium and stabilizing (*sensu* Chesson 2000) niche-based explanation for the maintenance of diversity. Resources (e.g., light, soil moisture, and soil nutrients), which vary strongly from the edge to the interior of gaps (Ricklefs 1977, Denslow 1980, 1987, Chazdon and Fetcher 1984, Becker *et al.* 1988), may be partitioned by species with different regeneration requirements. If each

Figure 12.2 Lianas in the understory on Barro Colorado Island, Panama. Photograph by S. Schnitzer (2005).

species competes optimally at a particular combination of resources, then within-gap resource gradients may allow species to coexist in equilibrium over the landscape, provided that species are not seed limited and can disperse their seeds to newly formed gaps (Dalling *et al.* 1998, Hubbell *et al.* 1999, Muller-Landau Chapter 11, this volume). Species may also specialize along a gradient of gap sizes, with some species specializing on small gaps and others on large gaps (Denslow 1980, Orians 1982, Brokaw 1987). Under this latter scenario, gaps promote species' coexistence by providing a heterogeneous environment at the landscape scale, with gaps of different sizes providing various levels of resources on which different species specialize. In both cases, the resource niche view (reviewed by Chase and Leibold 2003, Kitajima and Poorter Chapter 10, this volume) requires that gaps create sufficiently large resource gradients, either within the gap or among gaps, to allow species to stably coexist at equilibrium (Ricklefs 1977).

These three proposed mechanisms by which gaps are proposed to maintain diversity provide the following testable predictions: (1) resources will be measurably more heterogeneous within gaps (or among gaps of different sizes and characteristics) than in a comparable area within the intact forest; (2) some species will require the increase in resource quantity or heterogeneity from canopy gaps to establish and survive

(e.g., Dalling *et al.* 2001); (3) some species will require the enhanced resources available in a gap to initiate reproduction; for example, understory trees, shrubs, and herbs that may be able to establish in the absence of a gap but fail to reproduce in the shaded understory; and (4) individuals within gaps (or near the gap edge) will have substantially higher fecundity or a larger proportion of them will reach reproductive maturity on a per area or per stem basis than those individuals in the intact forest. Thus, even if thinning reduces diversity as gaps close, gaps may maintain diversity if they are sites of high fecundity for many species (Schnitzer 2001). These second two predictions of the gap hypothesis remain little studied.

EMPIRICAL TESTS: IS THERE EVIDENCE TO SUPPORT THE GAP HYPOTHESIS?

There is evidence that gaps maintain a significant level of tropical plant species diversity, particularly for some plant groups. For example, pioneer tree species require gaps for colonization and regeneration and are almost always absent in the intact shaded understory (e.g., Brokaw 1985a, 1987, Clark *et al.* 1993, Whitmore and Brown 1996, Schnitzer and Carson 2000, 2001). Thus, the cyclical and predictable disturbance from treefall gaps is necessary for pioneer trees to remain in the community. Under the idealized successional pathway in tropical forests, pioneer trees recruit into gaps soon after gap formation and are later replaced by shade-tolerant species. Pioneer tree diversity may be maintained by resource partitioning if species are uniquely adapted to resources in different zones of a single gap or among gaps of different sizes. For example, in a rainforest at Los Tuxtlas in Mexico, Popma *et al.* (1988) reported that many pioneer species had clear preferences for regeneration in either the gap center or edge, but not both. In a Panamanian moist forest, Brokaw (1987) reported that *Trema micrantha*, *Miconia argentea*, and *Cecropia insignis* all specialized in gaps of different sizes (see also Barton 1989, Van der Meer *et al.* 1998, Pearson *et al.* 2003b,c). While gap-size partitioning may occur when gaps differ greatly in size (and thus microclimate), Brown (1993) argued that gap-size partitioning is likely to be uncommon because the relationship between gap size and microclimate is unpredictable due to large spatial and temporal variation in microclimate. For example, in a Malaysian tropical forest, Brown and Whitmore (1992) and Whitmore and Brown (1996) found no evidence to support the hypothesis that pioneer tree species specialized on gaps of different sizes. Nevertheless, there is no disagreement about the general necessity of gaps to maintain light-demanding pioneer trees in tropical forests.

Is shade-tolerant tree diversity maintained by treefall gaps via within- or among-gap resource partitioning? Plants require light at a level that compensates for a minimum level of metabolism (i.e., light compensation point), and empirical studies demonstrate that many tree species vary in their growth rate or have a trade-off between growth and survival that is dependent on their light compensation point (Swaine and Hall 1988, Barik et al. 1992, Kitajima 1994, Rao et al. 1997, Kobe 1999, Poorter 1999, Kitajima and Poorter Chapter 10, this volume). While pioneer and shade-tolerant species differ substantially in their light compensation requirements, lesser differences among shade-tolerant species may also be sufficient to allow their coexistence along light gradients, which are known to exist within and among gaps, as well as in the intact forest (Chazdon and Fetcher 1984, Becker et al. 1988, Lieberman et al. 1989, 1995, Montgomery and Chazdon 2002). Thus, diversity of shade-tolerant species could possibly be maintained by resource partitioning; however, in situ demonstrations of niche partitioning leading to species coexistence are necessary to substantiate this claim.

Currently, there is only limited evidence that shade-tolerant tree species partition resources within or among gaps (Brandani et al. 1988, Clark and Clark 1992, Clark et al. 1993, Zanne and Chapman 2005). For example, Brandani et al. (1988) reported that several shade-tolerant tree species were non-randomly associated with different areas within gaps (e.g., the root, bole, and crown zones; but see Brown and Whitmore 1992, Brown 1996, Whitmore and Brown 1996). Collectively, however, there is little conclusive

evidence that gaps maintain more than a few of the thousands of shade-tolerant tree species in tropical forests via within- or among-gap resource partitioning (see reviews by Brown and Jennings 1998, Brokaw and Busing 2000). Consequently, the gap hypothesis as a mechanism to maintain tree species diversity has received considerable criticism because it apparently fails to explain the maintenance of diversity for a predominant group of species: the shade-tolerant tree species (Welden et al. 1991, Whitmore and Brown 1996, Brown and Jennings 1998, Hubbell et al. 1999, Brokaw and Busing 2000; but see Chazdon et al. 1999, Kobe 1999).

For example, Hubbell et al. (1999) conducted a census of saplings in over 1200 canopy gap and non-gap sites in a permanent 50 ha old-growth forest plot on Barro Colorado Island (BCI) in central Panama. After correcting for density between gap and non-gap sites, they concluded that gaps played a "relatively neutral role in maintaining [tree] species richness." Similarly, Brown and Jennings (1998) questioned whether a gradient in light availability was a viable axis for niche differentiation for the majority of tropical trees. These authors further argued that the "excessive emphasis" on treefall gaps has deflected attention away from other important processes that more likely determine community composition. Consequently, there remain sharply contrasting views regarding the viability of the gap hypothesis as an explanation for the maintenance of species diversity in tropical forests.

Have we missed the forest for the trees?

Most studies of gaps have failed to consider important plant groups other than trees (Brokaw and Busing 2000, Schnitzer and Carson 2000), and have failed to consider the impact of gaps on plant reproduction. For example, although lianas are rarely considered in gap studies, they are an important component of many tropical forests in terms of high stem density and leaf area, are highly diverse, and directly and uniquely impact gap-phase regeneration (Gentry 1991, Schnitzer and Bongers 2002). In many tropical forests,

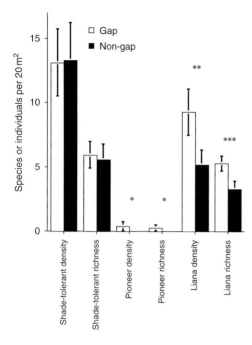

Figure 12.3 Mean density and species richness in gap versus non-gap sites ($n = 17$) for shade-tolerant trees, pioneer trees, and lianas. Asterisks represent significant differences (* = $P < 0.05$, ** = $P < 0.01$, *** = $P < 0.001$). Modified from Schnitzer and Carson (2001).

lianas commonly compose approximately 25% of the woody stems and nearly 30% of the woody species in intact forest (Gentry 1991). In a test of the gap hypothesis in the same 50 ha area of forest on BCI where Hubbell *et al.* (1999) found no difference in tree diversity between gap and non-gap sites, Schnitzer and Carson (2001) reported that both liana and pioneer tree abundance (density) and richness were significantly higher in both 5- and 10-year-old gaps than in non-gap sites on both a per area and per stem basis (Figure 12.3; Schnitzer and Carson 2001). Lianas and pioneer trees combined can represent more than 40% of the woody species diversity in many tropical forests, demonstrating that gaps are important for maintaining woody species diversity, even if they fail to maintain shade-tolerant tree diversity (Schnitzer and Carson 2000, 2001).

Lianas may be particularly abundant and diverse in gaps because they can colonize gaps in four ways, whereas trees typically use only two. Lianas colonize gaps both from seed and via advance regeneration, as do trees. Lianas can compose from 18 to 32% of the advance regeneration (<2 m tall) under the intact canopy in tropical forests (Putz 1984a, Putz and Chai 1987, Schnitzer and Carson unpublished). Lianas can also colonize gaps as adults, which is less common in other growth forms (but see Bond and Midgley 2001). Putz (1984a) reported that approximately 90% of the lianas that were pulled into a gap during the treefall survived and regenerated in the gap. Lianas can also colonize gaps from the intact forest by growing along the forest floor (Peñulosa 1984). Upon arrival in a gap, lianas vigorously grow and produce new stems at a rapid rate (Putz 1984a, Schnitzer *et al.* 2000, 2004), which may promote greater survivorship. Even though many lianas may be shade tolerant (Gilbert *et al.* 2006), because they can arrive in high numbers and survive in gaps for a long period of time, gaps may be integral to maintaining their diversity.

Liana species richness may be maintained by gaps via two main mechanisms. First, lianas may be able to partition the abundant and heterogeneous light resource in gaps (*sensu* Ricklefs 1977, Denslow 1980). While this mechanism lacks strong empirical support, liana diversity was higher in gaps than in comparable-sized areas of intact forest, even after correcting for density (Schnitzer and Carson 2001), suggesting that resource partitioning in gaps is possible. Second, although lianas are sometimes considered to be gap-dependent pioneers (Peñulosa 1984, DeWalt *et al.* 2000), they actually appear to have attributes of both pioneer and shade-tolerant species due to their tolerance of low light (Gilbert *et al.* 2006) and exceptionally rapid growth rates in high light (Schnitzer *et al.* 2004). This rapid growth rate is probably related to their reputedly high ratio of photosynthetic to structural tissue (Gartner 1991, Schnitzer *et al.* 2008), which allows them to fix more carbon per unit biomass compared with other growth forms. Consequently, treefalls may increase liana diversity forest-wide by creating ideal habitats for these species. Only detailed studies of the life history of numerous liana species will determine the percentage that require gaps to persist in tropical forests.

In addition, most studies have failed to consider the per capita impact of gaps on size- or age-specific rates of reproduction of species. For example, if gaps increase light and allow shade-tolerant trees to become reproductive or produce more seeds while still in the understory, then they may promote diversity by increasing fecundity, even if per capita species diversity is not higher than in non-gap sites (Schnitzer 2001). This aspect of the gap hypothesis has been neglected and may be particularly relevant to herbs, herbaceous vines, shrubs, and mid-sized trees – groups that do not typically reach the canopy, but may depend on treefall gaps to initiate reproduction (Gentry and Dodson 1987, Levey 1988, Denslow 1990, Dirzo *et al.* 1992, Goldblum 1997, Schnitzer and Carson 2000). For example, the fecundity of forest herbs and shrubs may be substantially higher in gaps than in nearby intact forest (Levey 1988, Denslow 1990, Dirzo *et al.* 1992, Goldblum 1997). On BCI, these groups of understory plants constitute around one third of the vascular plant flora (Figure 12.4); when combined with lianas and pioneer tree species, they represent 65% of all plant species on BCI and the majority of the flora in tropical forests worldwide (Gentry and Dodson 1987). Thus, gaps may maintain the majority of

the flora in many tropical forests when both reproduction and diversity are examined. To adequately determine the role of treefall gaps in maintaining species diversity, both the growth and survival of species as well as their reproductive output must be considered.

Shade-tolerant trees, lianas, and treefall gaps

Although there are still only a handful of relevant studies, gaps do not appear to have a strong influence on the per area or per capita diversity of shade-tolerant tree species. Typically, shade-tolerant trees establish prior to gap formation and are present as advance regeneration; thus, processes that occur prior to gap formation probably determine the composition and abundance of species that are available to take advantage of a newly formed gap (Uhl *et al.* 1988, Brown and Whitmore 1992). Additionally, many shade-tolerant tree species are limited by low seed production or poor dispersal, and thus they cannot distribute sufficient propagules into newly formed treefall gaps to take advantage of these ephemeral, high resource environments (see Dalling *et al.* 1998, Hubbell *et al.* 1999, Brokaw and Busing 2000, Muller-Landau Chapter 11, this volume). Finally, shade-tolerant tree abundance and diversity may be reduced in treefall gaps if gap-phase regeneration is co-opted by lianas or palms (Schnitzer *et al.* 2000).

Shade-tolerant trees may have structural characteristics that make them particularly susceptible to competition from lianas. Shade-tolerant tree species grow slowly and are adapted to maximize light interception by producing many branches, which can act as trellises for lianas, allowing them to climb and sometimes smother shade-tolerant trees under a blanket of foliage. Conversely, pioneer trees and palms have characteristics that may allow them to shed or avoid lianas, such as rapid growth, smooth or peeling bark, and an unbranched, monopodial trunk (Putz 1984b). While the severity of liana competition may vary with tree species identity, with lianas affecting some tree species or guilds more than others (e.g., Putz 1984a, Peréz-Salicrup and

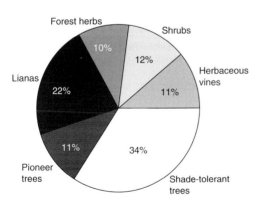

Figure 12.4 Percentage of species in different vascular plant groups on Barro Colorado Island, Panama. Data originally from Croat (1978) and summarized in Schnitzer and Carson (2000).

Barker 2000, Schnitzer *et al.* 2000, Schnitzer and Bongers 2002), overall, lianas may create conditions in gaps that are inimical to shade-tolerant species.

To test this liana competition hypothesis, we monitored recruitment, growth, and survivorship of all woody species for 8 years in experimental liana removal and control gaps in central Panama (Schnitzer and Carson unpublished). We found that the presence of lianas reduced recruitment of shade-tolerant trees (≥ 1.3 m tall) in all five sampling years, and by a total of 51% after 8 years ($n = 9$ removal, $n = 8$ control gaps). Lianas also substantially reduced relative growth rates of shade-tolerant trees (Schnitzer and Carson unpublished). Mortality of shade-tolerant trees, however, was not significantly altered by lianas, suggesting that lianas reduce shade-tolerant tree density by limiting recruitment and growth rather than increasing their mortality. Nonetheless, shade-tolerant tree mortality was slightly higher where lianas were present, and thus the total accumulation of shade-tolerant trees (recruitment minus mortality) after 8 years was 70% higher in gaps where we had removed lianas. Thus, gaps may fail to maintain shade-tolerant tree diversity because lianas substantially restrict shade-tolerant tree recruitment into gaps. Previously, the lack of shade-tolerant tree recruitment in gaps has been interpreted as evidence for seed limitation or dispersal limitation (reviewed by Brokaw and Busing 2000). Our data, however, demonstrate that lianas also play a role in limiting shade-tolerant tree recruitment in gaps. To fully explain why gaps fail to maintain shade-tolerant tree diversity, further research is necessary to determine the relative importance of plant competition versus seed and dispersal limitation.

VARIATION IN THE IMPACT OF GAPS ACROSS BROAD ENVIRONMENTAL GRADIENTS

The impact of gaps on the maintenance of species diversity and forest regeneration likely varies over large-scale environmental gradients. Theoretically, the influence of gaps should be greatest in forests where treefalls create steep resource gradients from the gap center to the intact understory because these steep gradients provide the highest potential for resource partitioning (Ricklefs 1977). Gaps should be most important in aseasonal tropical wet forests, which tend to have dark understories (Asner *et al.* 2003) and relatively poor soils (Denslow and Hartshorn 1994). These dark understories are caused by a combination of high year-round cloud cover, few deciduous trees, and multiple understory layers that efficiently intercept light before it reaches the forest floor. In contrast, gradients in light levels from a gap center to the intact forest are likely to be much lower in seasonally moist and dry forests, which tend to have lower cloud cover and a much higher proportion of deciduous trees (Condit *et al.* 2000), and thus allow far more light to penetrate the intact canopy into the understory, especially during the dry season. Dry forests, in particular, tend to have shorter stature, less complex structure, and lower leaf area, resulting in greater year-round light penetration.

The steepness of nutrient gradients in treefall gaps often parallels that of light gradients: increasing with greater rainfall. Upon tree death, the nutrients in the phytomass move into the soil where they are quickly assimilated by resident vegetation or leached from the soil. The availability of nutrients in a treefall gap may provide a steep albeit ephemeral gradient, especially beneath the fallen crown where leaves release a pulse of nutrients into the soil (Brokaw 1985a, Vitousek and Denslow 1986, Ostertag 1998). Nutrient gradients should be steepest in aseasonal wet forests, which may have lower nutrient levels because heavy year-round precipitation leaches nutrients out of the soil, although exceptions to this general rule certainly exist (Denslow and Hartshorn 1994).

Treefall gaps may also be more common in aseasonal forests than in seasonally dry forests because the year-round prevalence of unstable, waterlogged soils tends to increase treefall rates (Hartshorn 1978, Brandani *et al.* 1988). If true, the greater frequency of gaps may somewhat ameliorate dispersal limitation in wet forests by reducing the distance between gaps and propagule sources, thus providing more opportunities for resource partitioning across the landscape. Additionally, the negative impact of lianas on tree recruitment and growth of shade-tolerant trees may be far lower in aseasonal wet forests because liana abundance is much lower in these forests (Schnitzer 2005). Overall, we predict that gaps will have the strongest impact on the maintenance of diversity and forest regeneration in aseasonal wet forests because light in the understory is most limiting, nutrient gradients are more dramatic, and gaps are likely more frequent. Comparative studies of gaps along precipitation gradients are required to more thoroughly address this hypothesis.

TREEFALL GAPS PROVIDE A POOR MODEL TO TEST THE INTERMEDIATE DISTURBANCE HYPOTHESIS

The intermediate disturbance hypothesis (IDH) states that diversity will be highest at intermediate levels of disturbance size and frequency, and time since the last disturbance (Connell 1978). According to the IDH, large and frequent disturbances reduce diversity by physically removing individuals and thus extirpating species. When disturbances are too infrequent, competitive exclusion occurs, which also leads to lower species diversity. Only at some intermediate level of disturbance will diversity peak (see Connell 1978).

If treefall gaps represent an intermediate level of disturbance in tropical forests, then the gap hypothesis might be considered within the framework of the IDH (e.g., Hubbell 1999, Molino and Sabatier 2002). The fundamental difficulty with this approach, however, is that it is doubtful that treefall gaps are sufficiently large enough or range in size enough to rigorously test the IDH. Even if gaps spanned a large enough range of disturbance so that diversity varied with gap size, testing the IDH with treefall gaps is still problematic because the absence of a unimodal response does not reject the IDH. For instance, a linear increase in diversity with gap size may indicate that the gap disturbance is on the low disturbance side of the unimodal IDH curve (Sheil and Burslem 2003). However, a positive relationship between gap size and diversity provides no information on whether the true curve will eventually become unimodal with decreasing disturbance. For example, Hubbell (1999) refuted the IDH with species-individual curves from BCI, showing that species accumulation (diversity) increased as gap size decreased, with the non-gap forest having the highest diversity accumulation. The IDH may still hold, however, if diversity decreases when levels of disturbance become lower than the background (non-gap) disturbance regime on BCI (Sheil and Burslem 2003). Even the old-growth forest on BCI has a history of disturbance, possibly indicating that it is still undergoing succession, and thus diversity could still decrease over time in the non-gap forest, which would be consistent with the IDH (Sheil and Burslem 2003). In addition, with the exception of pioneer trees, Hubbell et al. (1999) omitted growth forms most likely to be disturbance dependent in light-limited forests (e.g., lianas, shrubs, and herbs; Schnitzer and Carson 2000). A rigorous test of the IDH requires the consideration of key relevant growth forms along a disturbance gradient ranging from minimal to catastrophic. These considerations make rigorous tests of the IDH extremely challenging.

CONCLUSIONS

Treefall gaps provide both equilibrium and non-equilibrium explanations for the maintenance of species diversity in tropical forests. Although the gap hypothesis is one of the major hypotheses proposed to explain the maintenance of species diversity in tropical forests, considerable work remains to be done to test the full range of predictions that stem from this hypothesis. Currently, the degree to which gaps maintain diversity likely depends on the growth form and life-history characteristics of the species examined. Liana and pioneer tree diversity, and possibly that of shrubs and herbaceous plants, appears to be maintained by treefall gaps to a significant degree. Little evidence suggests that gaps maintain shade-tolerant tree diversity, apparently because of dispersal and recruitment limitation, which may be exacerbated by competition with lianas. A full and rigorous test of the predictions of the gap hypothesis requires an examination of the impact of gaps on the diversity of all dominant plant functional groups (e.g., trees, lianas, shrubs, and herbs), as well as examination of whether gaps permit some species to remain in the community by enhancing size-, age-, or growth-form-specific rates of reproduction.

REFERENCES

Appanah, S. and Putz, F.E. (1984) Climber abundance in virgin dipterocarp forest and the effect of pre-felling climber cutting on logging damage. *The Malaysian Forester* 47, 335–342.

Asner, G.P., Scurlock, J.M.O., and Hicke, J.A. (2003) Global synthesis of leaf area index observations: implications for ecological and remote sensing studies. *Global Ecology and Biogeography* 12,191–205.

Aubréville, A. (1938) La forêt coloniale. *Annales Académie des Sciences Coloniales (Paris)* 9, 1–245.

Babweteera, F., Plumptre, A., and Obua, J. (2000) Effect of gap size and age on climber abundance and diversity in Budongo Forest Reserve, Uganda. *African Journal of Ecology* 38, 230–237.

Barik, S.K., Pandey, H.N., Tripathi, R.S., and Rao, P. (1992) Microenvironmental variability and species-diversity in treefall gaps in a subtropical broadleaved forest. *Plant Ecology* 103, 31–40.

Barton, A.M., Fetcher, N., and Redhead, S. (1989) The relationship between treefall gap size and light flux in a Neotropical rain forest in Costa Rica. *Journal of Tropical Ecology* 5, 437–439.

Beard, J.S. (1955) The classification of tropical American vegetation-types. *Ecology* 36, 89–100.

Becker, P., Rabenold, P.E., Idol, J.R., and Smith, A.P. (1988) Water potential gradients for gaps and slopes in a Panamanian tropical moist forests dry season. *Journal of Tropical Ecology* 4, 173–184.

Bond, W.J. and Midgley, J.J. (2001) Ecology of sprouting in woody plants: the persistence niche. *Trends in Ecology and Evolution* 16, 45–51.

Brandani, A., Hartshorn, G.S., and Orians, G.H. (1988) Internal heterogeneity of gaps and species richness in Costa Rican tropical wet forest. *Journal of Tropical Ecology* 4, 99–119.

Bray, R.J. (1956) Gap phase replacement in a Maple-Basswood forest. *Ecology* 37, 598–600.

Brokaw, N.V.L. (1982) The definition of treefall gap and its effect on measures of forest dynamics. *Biotropica* 14, 158–160.

Brokaw, N.V.L. (1985a) Gap-phase regeneration in a tropical forest. *Ecology* 66, 682–687.

Brokaw, N.V.L. (1985b) Treefalls, regrowth, and community structure in tropical forests. In S.T.A. Picket and P.S. White (eds), *The Ecology of Natural Disturbance and Patch Dynamics.* Academic Press, New York, pp. 53–69.

Brokaw, N.V.L. (1987) Gap-phase regeneration of three pioneer tree species in a tropical forest. *Journal of Ecology* 75, 9–20.

Brokaw, N.V.L. and Busing, R.T. (2000) Niche versus chance and tree diversity in forest gaps. *Trends in Ecology and Evolution* 15, 183–188.

Brown, N.D. (1993) The implications of climate and gap microclimate for seedling growth conditions in a Bornean lowland rain forest. *Journal of Tropical Ecology* 9, 153–168.

Brown, N.D. (1996) A gradient of seedling growth from the centre of a tropical rain forest canopy gap. *Forest Ecology and Management* 82, 239–244.

Brown, N.D. and Jennings, S.B. (1998) Gap-size niche differentiation by tropical rainforest trees: a testable hypothesis or a broken-down bandwagon? In D.M. Newberry, H.H.T. Prins, and N.D. Brown (eds), *Dynamics of Tropical Communities.* Blackwell Science, Oxford, pp. 79–94.

Brown, N.D. and Whitmore, T.C. (1992) Do dipterocarp seedlings really partition tropical rain forest gaps? *Philosophical Transactions of the Royal Society of London (Series B)* 335, 369–378.

Chase, J.M. and Leibold, M.A. (2003) *Ecological Niches: Linking Classical and Contemporary Approaches.* University of Chicago Press, Chicago.

Chazdon, R.L., Colwell, R.L., and Denslow, J.S. (1999) Tropical tree richness and resource-based niches. *Science* 285, 1459a.

Chazdon, R.L. and Fetcher, N. (1984) Photosynthetic light environments in a lowland tropical rainforest in Costa Rica. *Journal of Ecology* 72, 553–564.

Chesson, P. (2000) Mechanisms of maintenance of species diversity. *Annual Review of Ecology and Systematics* 31, 343–366.

Clark, D.A. and Clark, D.B. (1991) The impact of physical damage on canopy tree regeneration in tropical forest. *Journal of Ecology* 79, 447–457.

Clark, D.A. and Clark, D.B. (1992) Life history diversity of canopy and emergent trees in a neotropical rain forest. *Ecological Monographs* 62, 315–344.

Clark, D.B., Clark, D.A., and Rich, P.M. (1993) Comparative analysis of microhabitat utilization by saplings of nine tree species in neotropical rain forest. *Biotropica* 25, 397–407.

Condit, R., Ashton, P.S., Baker, P. *et al.* (2000) Spatial patterns in the distribution of tropical tree species. *Science* 288, 1414–1418.

Connell, J.H. (1978) Diversity in tropical rainforests and coral reefs. *Science* 199, 1302–1310.

Connell, J.H. (1979) Tropical rain forests and coral reefs as open nonequilibrium systems. In R.M. Anderson, L.R. Taylor, and B.D. Turner (eds), *Population Dynamics*. Blackwell, Oxford, pp. 141–163.

Croat, T.B. (1978) *Flora of Barro Colorado Island*. Stanford University Press, Stanford.

Dalling, J.W., Hubbell, S.P., and Silvera, K. (1998) Seed dispersal, seedling establishment and gap partitioning among tropical pioneer trees. *Journal of Ecology* 86, 674–689.

Dalling, J.W., Winter, K., Nason, J.D., Hubbell, S.P., Murawski, D.A., and Hamrick, J.L. (2001) The unusual case of *Alseis blackiana*: a shade-persistent pioneer tree? *Ecology* 82, 933–945.

Denslow, J.S. (1980) Gap partitioning among tropical rainforest trees. *Biotropica (supplement)* 12, 47–55.

Denslow, J.S. (1987) Tropical rainforest gaps and tree species diversity. *Annual Review of Ecology and Systematics* 18, 431–451.

Denslow, J.S. (1990) Disturbance and diversity in tropical rainforests: the density effect. *Ecological Applications* 5, 962–968.

Denslow, J.S. and Hartshorn, G.S. (1994) Tree-fall gap environments and forest dynamic processes. In L.A. McDade, K.S. Bawa, H.A. Hespenhide, and G.S. Hartshorn (eds), *La Selva: Ecology and Natural History of Neotropical Rain Forest*. University of Chicago Press, Chicago, pp. 120–127.

Denslow, J.S., Newell, E., and Ellison, A.M. (1991) The effect of understory palms and cyclanths on the growth and survival of *Inga* seedlings. *Biotropica* 23, 225–234.

DeWalt, S.J., Schnitzer, S.A., and Denslow, J.S. (2000) Density and diversity of lianas along a chronosequence in a central Panamanian lowland forest. *Journal of Tropical Ecology* 16, 1–19.

Dirzo, R., Horvitz, C.C., Quevedo, H., and Lopez, M.A. (1992) The effects of gap size and age on the understory herb community of a tropical Mexican rain forest. *Journal of Ecology* 80, 809–822.

Farris-Lopez, K., Denslow, J.S., Moser, B., and Passmore, H. (2004) Influence of a common palm, *Oenocarpus mapora*, on seedling establishment in a tropical moist forest in Panama. *Journal of Tropical Ecology* 20, 429–438.

Fraver, S., Brokaw, N.V.L., and Smith, A.P. (1998) Delimiting the gap phase in the growth cycle of a Panamanian forest. *Journal of Tropical Ecology* 14, 673–681.

Gartner, B.L. (1991) Structural stability and architecture of vines vs shrubs of poison oak, *Toxicodendron diversilobum*. *Ecology* 72, 2005–2015.

Gentry, A.H. (1991) The distribution and evolution of climbing plants. In F.E. Putz and H.A. Mooney (eds), *The Biology of Vines*. Cambridge University Press, Cambridge, pp. 3–49.

Gentry, A.H. and Dodson, C. (1987) Contribution of nontrees to species richness of a tropical rain forest. *Biotropica* 19, 149–156.

Gilbert, B., Wright, S.J., Muller-Landau, H.C., Kitajima, K., and Hernandez, A. (2006) Life history trade-offs in tropical trees and lianas. *Ecology* 87, 1281–1288.

Goldblum, D. (1997) The effects of treefall gaps on understory vegetation in New York State. *Journal of Vegetation Science* 8, 125–132.

Grinnell, J. (1917) Field tests and theories concerning distributional control. *American Naturalist* 51, 115–128.

Hartshorn, G.S. (1978) Tree falls and tropical forest dynamics. In P.B. Tomlinson and M.H. Zimmermann (eds), *Tropical Trees as Living Systems*. Cambridge University Press, Cambridge, pp. 617–638.

Hartshorn, G.S. (1980) Neotropical forest dynamics. *Biotropica (Supplement)* 12, 23–30.

Hubbell, S.P. (1999) Tropical tree richness and resource-based niches. *Science* 285, 1495a.

Hubbell, S.P. and Foster, R.B. (1986) Canopy gaps and the dynamics of a neotropical forest. In M.J. Crawley (ed.), *Plant Ecology*. Blackwell Scientific Publications, Oxford, pp. 77–96.

Hubbell, S.P., Foster, R.B., O'Brien, S.T. *et al.* (1999) Light-gap disturbances, recruitment limitation, and tree diversity in a neotropical forest. *Science* 283, 554–557.

Huston, M.A. (1979) A general hypothesis of species diversity. *American Naturalist* 113, 81–101.

Inouye, R.S., Huntly, N., and Wasley, G.A. (1997) Effects of pocket gophers (*Geomys bursarius*) on microtopographic variation. *Journal of Mammalogy* 78, 1144–1148.

Jones, E.W. (1945) The structure and reproduction of the virgin forests of the north temperate zone. *New Phytologist* 44, 130–148.

Kitajima, K. (1994) Relative importance of photosynthetic traits and allocation patterns as correlates of seedling shade tolerance of 13 tropical trees. *Oecologia* 98, 419–428.

Kobe, R.K. (1999) Light gradient partitioning among tropical tree species through differential seedling mortality and growth. *Ecology* 80, 187–201.

Levey, D.J. (1988) Spatial and temporal variation in Costa Rican fruit and fruit-eating bird abundance. *Ecological Monographs* 58, 251–269.

Lieberman, M., Lieberman, D., and Peralta, R. (1989) Forests are not just swiss cheese: canopy stereogeometry of nap-gaps in tropical forests. *Ecology* 70, 550–552.

Lieberman, M., Lieberman, D., Peralta, R., and Hartshorn, G.S. (1995) Canopy closure and the distribution of tropical forest tree species at La Selva, Costa Rica. *Journal of Tropical Ecology* 11, 161–178.

Molino, J-F. and Sabatier, D. (2002) Humped pattern of diversity: fact or artifact? *Science* 297, 1763a.

Montgomery, R. and Chazdon, R. (2002) Light gradient partitioning by tropical tree seedlings in the absence of canopy gaps. *Oecologia* 131, 165–174.

Orians, G.H. (1982) The influence of tree-falls in tropical forests on tree species richness. *Tropical Ecology* 23, 255–279.

Ostertag, R. (1998) Belowground effects of canopy gaps in a tropical wet forest. *Ecology* 79, 1294–1304.

Pearson, T.R.H., Burslem, D.F.R.P., Goeriz, R.E., and Dalling, J.W. (2003b) Interactions of gap size and herbivory on establishment, growth and survival of three species of neotropical pioneer trees. *Journal of Ecology* 91, 785–796.

Pearson, T.R.H., Burslem, D.F.R.P., Goeriz, R.E., and Dalling, J.W. (2003c) Regeneration niche partitioning in neotropical pioneers: effects of gap size, seasonal drought and herbivory on growth and survival. *Oecologia* 137, 456–465.

Pearson, T.R.H., Burslem, D.F.R.P., Mullins, C.E., and Dalling, J.W. (2003a) Functional significance of photoblastic germination in neotropical pioneer trees: a seed's eye view. *Functional Ecology* 17, 394–402.

Peet, R.K. (1991) Lessons from nature. In L.A. Real and J.H. Brown (eds), *Foundations of Ecology*. The University of Chicago Press, Chicago, pp. 605–615.

Peñulosa, J. (1984) Basal branching and vegetative spread in two tropical rain forest lianas. *Biotropica* 16, 1–9.

Peréz-Salicrup, D.R. and Barker, M.G. (2000) Effect of liana cutting on water potential and growth of adult *Senna multijuga* (Caesalpinioideae) trees in a Bolivian tropical forest. *Oecologia* 124, 469–475.

Peters, H.A., Pauw, A., Silman, M.R., and Terborgh, J.W. (2004) *Falling palm fronds structure Amazonian rainforest sapling communities*. Proceedings of the Royal Society of London B (Suppl.) 271, S367–S369.

Popma, J., Bongers, F., and Martinez-Ramos, M. (1988) Pioneer species distributions in treefall gaps in Neotropical rain forest: a gap definition and its consequences. *Journal of Tropical Ecology* 4, 77–88.

Poorter, L. (1999) Growth responses of 15 rain-forest tree species to a light gradient: the relative importance of morphological and physiological traits. *Functional Ecology* 13, 396.

Putz, F.E. (1983) Liana biomass and leaf area of a "tierra firme" forest in the Rio Negro Basin, Venezuela. *Biotropica* 15, 185–189.

Putz, F.E. (1984a) The natural history of lianas on Barro Colorado Island, Panama. *Ecology* 65, 1713–1724.

Putz, F.E. (1984b) How trees avoid and shed lianas. *Biotropica* 16, 19–23.

Putz, F.E. and Chai, P. (1987) Ecological studies of lianas in Lambir National Park, Sarawak, Malaysia. *Journal of Ecology* 75, 523–531.

Rao, P., Barik, S.K., Pandey, H.N., and Tripathi, R.S. (1997) Tree seed germination and seedling establishment in treefall gaps and understorey in a subtropical forest of northeast India. *Australian Journal of Ecology* 22, 136–145.

Richards, P.W. (1952) *The Tropical Rainforest*. Cambridge University Press, Cambridge.

Ricklefs, R.E. (1977) Environmental heterogeneity and plant species diversity, a hypothesis. *American Naturalist* 111, 376–381.

Schemske, D.W. and Brokaw, N. (1981) Treefalls and the distribution of understory birds in a tropical forest. *Ecology* 62, 938–945.

Schnitzer, S.A. (2001) *Treefall gaps and the maintenance of species diversity: redefining and expanding the gap hypothesis*. Doctoral dissertation, University of Pittsburgh, Pittsburgh.

Schnitzer, S.A. (2005) A mechanistic explanation for global patterns of liana abundance and distribution. *American Naturalist* 166, 262–276.

Schnitzer, S.A. and Bongers, F. (2002) The ecology of lianas and their role in forests. *Trends in Ecology and Evolution* 117, 223–230.

Schnitzer, S.A. and Carson, W.P. (2000) Have we forgotten the forest because of the trees? *Trends in Ecology and Evolution* 15, 375–376.

Schnitzer, S.A. and Carson, W.P. (2001) Treefall gaps and the maintenance of species diversity in a tropical forest. *Ecology* 82, 913–919.

Schnitzer, S.A., Dalling, J.W., and Carson, W.P. (2000) The impact of lianas on tree regeneration in tropical forest canopy gaps: evidence for an alternative pathway of gap-phase regeneration. *Journal of Ecology* 88, 655–666.

Schnitzer, S.A., Kuzee, M., and Bongers, F. (2005) Disentangling above-and below-ground competition between lianas and trees in a tropical forest. *Journal of Ecology* 93, 1115–1125.

Schnitzer, S.A., Londré, R.A., Klironomos, J., and Reich, P.B. (2008) Biomass and toxicity responses of poison ivy (*Toxicodendron radicans*) to elevated atmospheric CO_2: Comment. *Ecology* 89, 581–585.

Schnitzer, S.A., Parren, M.P.E., and Bongers, F. (2004) Recruitment of lianas into logging gaps and the effects of pre-harvest climber cutting in a lowland forest in Cameroon. *Forest Ecology and Management* 190, 87–98.

Sheil, D. and Burslem, D.F.R.P. (2003) Disturbing hypotheses in tropical forests. *Trends in Ecology and Evolution* 18, 18–26.

Swaine, M.D. and Hall, J.B. (1988) The mosaic theory of forest regeneration and the determination of forest composition in Ghana. *Journal of Tropical Ecology* 4, 253–269.

Swaine, M.D. and Whitmore, T.C. (1988) On the definition of ecological species groups in tropical rain forests. *Vegetation* 75, 81–86.

Tabanez, A.A.L. and Viana, V.M. (2000) Patch structure within Brazilian Atlantic forest fragments and implications for conservation. *Biotropica* 32, 925–933.

Uhl, C., Clark, K., Dezzeo, N., and Maquino, P. (1988) Vegetation dynamics in Amazonian treefall gaps. *Ecology* 69, 751–763.

Van der Meer, P.J., Sterk, F.J., and Bongers, A. (1998) Tree seedling performance in canopy gaps in a tropical rain forest at Nouragues, French Guiana. *Journal of Tropical Ecology* 14, 119–137.

Vandermeer, J., Boucher, D., Perfecto, I., and de la Cerda, I.G. (1996) A theory of disturbance and species diversity: evidence from Nicaragua after Hurricane Joan. *Biotropica* 28, 600–613.

Vandermeer, J., de la Cerda, I.G., Boucher, D., Perfecto, I., and Ruiz, J. (2000) Hurricane disturbance and tropical tree species diversity. *Science* 290, 788–791.

Vitousek, P.M. and Denslow, J.S. (1986) Nitrogen and phosphorous availability in treefall gaps of a lowland tropical rainforest. *Journal of Ecology* 74, 1167–1178.

Wang, Y.H. and Augspurger, C. (2006) Comparison of seedling recruitment under arborescent palms in two Neotropical forests. *Oecologia* 147, 533–545.

Watt, A.S. (1925) On the ecology of British beechwoods with special reference to their regeneration. Part II: The development and structure of beech communities on the Sussex Downs. *Journal of Ecology* 13, 27–73.

Watt, A.S. (1947) Pattern and process in the plant community. *Journal of Ecology* 35, 1–22.

Webb, L.J. (1959) Environmental relationships of the structural types of Australian rain forest vegetation. *Ecology* 49, 296–311.

Welden, C.W., Hewett, S.W., Hubbell, S.P., and Foster, R.B. (1991) Sapling survival, growth, and recruitment: relationship to canopy height in a neotropical forest. *Ecology* 72, 35–50.

Whitmore, T.C. (1978) Gaps in the forest canopy. In P.B. Tomlinson and M.H. Zimmermann (eds), *Tropical Trees as Living Systems*. Cambridge University Press, Cambridge, pp. 639–655.

Whitmore, T.C. (1989) Canopy gaps and the two major groups of forest trees. *Ecology* 70, 536–538.

Whitmore, T.C. and Brown, N.D. (1996) Dipterocarp seedling growth in rain forest canopy gaps during six and a half years. *Philosophical Transactions of the Royal Society of London (Series B)* 351, 1195–1203.

Wright, S.J., Hernandez, A., and Condit, R. (2007) The bushmeat harvest alters seedling banks by favoring lianas, large seeds and seeds dispersed by bats, birds and wind. *Biotropica* 39, 363–371.

Wunderle, J.M., Henriques, L.M.P., and Willig, M.R. (2006) Short-term responses of birds to forest gaps and understory: an assessment of reduced-impact logging in a lowland Amazon forest. *Biotropica* 38, 235–255.

Zanne, A.E. and Chapman, C.A. (2005) Diversity of woody species in forest, treefall gaps, and edge in Kibale National Park, Uganda. *Plant Ecology* 178, 121–139.

Chapter 13

CHALLENGES ASSOCIATED WITH TESTING AND FALSIFYING THE JANZEN–CONNELL HYPOTHESIS: A REVIEW AND CRITIQUE

Walter P. Carson, Jill T. Anderson, Egbert G. Leigh, Jr, and Stefan A. Schnitzer

OVERVIEW

The Janzen–Connell hypothesis proposes that density- and distance-dependent natural enemies regulate plant populations, thereby enhancing alpha-diversity and potentially contributing to the latitudinal gradient in species richness. There have been over 50 studies designed to test predictions of this hypothesis, and our review shows that many tree species exhibit patterns consistent with Janzen–Connell effects. Here, we review studies that were designed to test the Janzen–Connell hypothesis and raise a number of general issues and challenges with regard to testing it.

First, the Janzen–Connell hypothesis is fundamentally a community-level hypothesis that predicts that enemies cause higher alpha-diversity; this key prediction remains poorly tested at the appropriate scale. Second, the Janzen–Connell hypothesis in its most general context is a special case of keystone predation, where specialist enemies keep species that are superior competitors in check. It remains unknown if the removal of enemies for any woody species will subsequently cause a reduction in alpha-diversity. Overall, the Janzen–Connell hypothesis is difficult to falsify because it may promote diversity if enemies act as keystone species by keeping only a relatively very small proportion of superior competitors in low abundance. Rare species that have shade-tolerant juveniles and produce large seeds may be the ones most likely to show Janzen–Connell effects yet least likely to be included in studies due to low population densities of adults. Third, complex trade-offs underlie Janzen–Connell effects, particularly a trade-off between competitive or establishment ability and vulnerability to enemies. Many tests of the Janzen–Connell hypothesis assume implicitly that traits that confer high survivorship in the shade are correlated with traits that enhance survivorship under prolonged pest pressure in the understory. This correlation does not hold for all shade-tolerant tree species and the tightness of this relationship needs to be directly tested. Consequently an often overlooked but important trade-off for plant species coexistence may be allocation to those physiological and morphological traits that confer survivorship at low light versus traits that confer survivorship under varying degrees of pest pressure. Fourth, diversity may be maintained, at least in part, by episodic outbreaks of specialist pests, which may reduce the survivorship, growth, and fecundity of adults whenever adults are particularly aggregated. This impact of enemies on adults, although originally emphasized by Janzen (1970), has received far less attention than the effect of enemies on juveniles even though it is well known to be important outside of the tropics. Challenges notwithstanding, Janzen–Connell effects are common in tropical systems and thus a likely key mechanism maintaining high plant diversity.

INTRODUCTION

The Janzen–Connell hypothesis states that specialist pests and pathogens keep key plant species rare enough or reduce their competitive ability enough so as to make space available for many other species (Janzen 1970, Connell 1971). This idea has a long history in ecology. Ridley (1930, p. xvi), nearly 80 years ago, remarked: "Where too many plants of one species are grown together, they are very apt to be attacked by some pest, insect or fungus …. It is largely due to this also, in Nature, that one-plant associations are prevented and nullified by better means of dispersal for the seeds." Later, Gillett (1962) presented a non-equilibrium form of this hypothesis. Finally, Janzen (1970) and Connell (1971) established this hypothesis as one of the commonplaces of tropical biology. They both presented evidence that the seedlings and saplings of trees exhibit repelled recruitment patterns around adults thereby potentially creating space for numerous plant species. MacArthur (1972, 191 ff.) accepted this general idea as true. Nonetheless, a meta-analysis by Hyatt *et al.* (2003) found little support for the distance-dependent prediction of the hypothesis and concluded that "further testing to explore this hypothesis as a diversity-maintaining mechanism is unnecessary." Although Leigh *et al.* (2004) provided much indirect evidence in favor of the role of pest pressure in maintaining tropical tree diversity, the evidence they provided could be said to allure, rather than extort. In the end, they were not able to establish the truth of this hypothesis beyond reasonable doubt. Below we review studies that tested different aspects of the Janzen–Connell hypothesis and address a number of issues and challenges associated with this hypothesis.

A REVIEW OF STUDIES TESTING FOR JANZEN–CONNELL EFFECTS

We searched Web of Science to locate articles published between 1970 and September 2006 that cited Janzen (1970) and explicitly addressed the Janzen–Connell hypothesis. We identified

53 appropriate studies (Tables 13.1 and 13.2). We excluded a relatively small number of studies (less than five) that assessed only static distribution patterns of one life-history stage as a function of distance from the nearest adult conspecific or a function of conspecific juvenile density. We felt that these studies were less informative than those that sampled focal plants through time to address how distance- and density-dependent effects influence performance (survivorship and growth rate), or compared two or more life-history stages to assess changes in distribution patterns due to density- or distance-dependent factors.

The majority of the studies (58%) focused on a single species, 21% studied between two and nine species, and 21% studied 10 or more species (Table 13.1). Most studies were purely observational (51%) whereas 34% were purely experimental, and 15% used both experimental and observational approaches. About 75% of the studies were restricted to seeds and seedlings and 17% focused on saplings. We found only one study that focused on adults (>10 cm dbh; Stoll and Newberry 2005) and only two studies considered all life-history stages (Connell *et al.* 1984, Silva Matos *et al.* 1999). Nearly half of the studies (47%) failed to mention the seed size of the species under study, a trait originally thought to be important by both Janzen and Connell. Fifty of the 53 studies found evidence consistent with either density or distance dependency but of these, half provided no evidence for the mechanism underlying the pattern. Where a putative mechanism was tentatively identified, there was a near even split among vertebrates (eight studies), invertebrates (10 studies), and pathogens (seven studies). Several studies speculated that intraspecific competition could underlie Janzen–Connell patterns (Connell *et al.* 1984, Silva Matos *et al.* 1999, Stoll and Newberry 2005). Host specificity, another trait thought to be important by both Janzen and Connell, was reported in only one third of the studies, probably because it was unknown. For the nine studies where it was evaluated, five reported high host specificity, three reported low specificity, and one reported the occurrence of both specialists and generalists.

Mean study duration was 3.5 years (±4 SD) and ranged from 18 years (Connell *et al.* 1984)

Table 13.1 A review of studies that tested for Janzen–Connell effects (either distance dependence, density dependence or both) for different developmental stages of plant species in various habitats, countries, and field stations around the world. Evidence was categorized as either experimental, observational or both.

Habitat	Country	Field site	Tested for density-, distance-dependency or both	# species studied	% of species that showed distance-dependency consistent with the J–C hypothesis	% of species that showed density-dependency consistent with the J–C hypothesis	Type of evidence	Developmental stage	Citation
Lowland seasonal	Mexico	Los Tuxtlas	Density	1	N/A	Survivorship: small plants: 0%; larger plants: 100%. Fecundity: 100%	Observational	Woody plants >0.3 cm dbh	Alvarez-Buylla (1994)
Terra firme forest	Panama	BCI	Both	8	100% (1 of 1)	>75% (24)	Experimental	Seeds and seedlings	Augspurger and Kelly (1984)
Terra firme forest	Panama	BCI	Both	1	100% (18)	100% (23)	Experimental	Seeds and seedlings	Augspurger and Kitajima (1992)
Terra firme forest	Panama	BCI	Both	1	100%	N/A	Observational	Seedlings	Augspurger (1983)
Terra firme forest	Panama	BCI	Distance	9	89%	N/A	Observational	Seedlings	Augspurger (1984)
Terra firme forest	Belize	Chiquibul Forest reserve	Density	1	N/A	100%	Experimental and observational	Seedlings	Bell et al. (2006)
Terra firme forest	Indonesia	Dipterocarp forest	Distance	4	50%	N/A	Observational	Seedlings	Blundell and Peart (1998)
Terra firme forest	Belize	Bladen Nature Reserve	Distance	1	100% for unburied, 0% for buried seeds	N/A	Experimental	Seeds and seedlings	Brewer and Webb (2001)
Lowland seasonal (2)	Mexico	Los Tuxtlas	Both	1	100% (20)	0%	Experimental	Seeds	Burkey (1994)
Terra firme forest	Peru	Cocha Cashu	Both	2	Seeds: 50%; seedlings: 50% (19)	Seeds: 50%; Seedlings: 0%	Experimental and observational	Seeds and seedlings	Cintra (1997)
Terra firme forest	Costa Rica	La Selva	Both	1	100%	100%	Observational	Seedlings	Clark and Clark (1984)
Terra firme forest	Panama	BCI	Distance	80	19% (16)	N/A	Observational	Woody plants >1 cm dbh	Condit et al. (1992)

Terra firme forest	Panama	BCI	Density	2	N/A	50%	Observational	Woody plants >1 cm dbh	Condit et al. (1994)
Evergreen moist forest (6)	Australia	Queensland	Both	>100 (15)	~10%		Experimental and observational	All	Connell et al. (1984)
Terra firme forest	Panama	BCI	Distance	1	Seeds: 0%; seedlings:100%	N/A	Experimental	Seeds and seedlings	De Steven and Putz (1984)
Terra firme forest	Panama	BCI	Density	3	N/A	33.3%	Observational	Seeds and seedlings	De Steven and Wright (2002)
Terra firme forest	Roraima, Brazil	Maraca Island Ecological Reserve	Distance	1	100%	N/A	Experimental	Seeds	Fragoso et al. (2003), Fragoso (1997)
Terra firme forest	Panama	BCI	Both	1	100%	100%	Observational	Woody plants >1 cm dbh	Gilbert et al. (1994)
Terra firme forest	Panama	BCI	Both	1	0%	100%	Observational	Seedlings	Gilbert et al. (2001)
Terra firme forest	Guyana	Mabura Hill	Distance	1	Final proportion of seeds consumed: 0%; time to seed consumption: 100%; germination success: 100%	N/A	Experimental	Seeds	Hammond et al. (1999)
Terra firme forest (1)	Panama	BCI (9)	Density	53	N/A	100%	Observational	Seeds and seedlings	Harms et al. (2000)
Temperate deciduous forest	North Carolina, USA	Coweeta Hydrological Laboratory	Both	7	57%	86%	Observational	Seeds and seedlings	HilleRisLambers et al. (2002)
Terra firme forest	Ghana	Neung South Forest Reserve	Distance	1	100%	N/A	Experimental	Seedlings	Hood et al. (2004)
Cool temperate forest (7)	Québec, Canada	Tantaré Ecological Reserve	Density	1	N/A	0%	Observational	Seeds and seedlings	Houle (1998)
Terra firme forest	Panama	BCI	Both	81 (10)	Survivorship: 18%, growth: 55%, recruitment: 33% (17)	50% (1 of 2)	Observational	Woody plants >1 cm dbh	Hubbell et al. (1990)

Continued

Table 13.1 Continued

Habitat	Country	Field site	Tested for density-, distance-dependency or both	# species studied	% of species that showed distance-dependency consistent with the J–C hypothesis	% of species that showed density-dependency consistent with the J–C hypothesis	Type of evidence	Developmental stage	Citation
Montane forest (8)	Spain	Anadusian highlands	Density	3	N/A	0%	Experimental	Seeds	Hulme (1997)
Tropical dry forest	India	Mudumalai	Density	16 (14)	N/A	Mortality: 25%; recruitment: 100%	Observational	Saplings >1 cm dbh	John et al. (2002)
Terra firme forest	Malaysia	Danum Valley	Density	1	N/A	100%	Experimental	Seedlings	Massey et al. (2006)
Temperate arid forests / shrublands	Spain		Both	1	0% (22)	100%, seed and seedling levels	Experimental and observational	Seeds and seedlings	Montesinos et al. (2006)
Terra firme forest	Para, Brazil	Kayapo Centre for Ecological Studies	Both	1	Seed predation: 0%; seedling recruitment: 100%	100% (26)	Experimental	Seeds and seedlings	Norghauer et al. (2006)
Terra firme forest	Northeastern Peru	Jenaro Herrera	Both	1	0%	0%	Experimental	Seeds	Notman et al. (1996)
Temperate deciduous forest	Bloomington Indiana, USA	Griffy Lake Nature Preserve	Both	1	100%	100%	Experimental and observational	Seeds and seedlings	Packer and Clay (2000)
Terra firme forest	Para, Brazil		Distance	1	100%	N/A	Experimental	Seeds	Peres et al. (1997)
Terra firme forest	Malaysia and Panama	Pasoh and BCI	Density	732 (12)	N/A	80% at both sites	Observational	Saplings, treelets, and trees	Peters (2003)
Temperate deciduous forest	Eastern USA		Distance	1	100%	N/A	Experimental	Seedlings	Reinhart et al. (2005)
Terra firme forest	Peru	Cocha Cashu	Density	1	N/A	0%	Observational	Seeds	Romo et al. (2004)
Floodplain forest (4)	Peru	Cocha Cashu	Both	1	100%	100%	Experimental and observational	Seeds	Russo and Augspurger (2004)
Terra firme forest	Panama	BCI	Density	1	Seeds: 100%; seedlings: 0%	N/A	Experimental	Seeds and seedlings	Schupp (1988)

Terra firme forest	Panama	BCI	Both	1	100%	0%	Experimental	Seeds and seedlings	Schupp (1992)
Temperate deciduous forest	Central Japan	Ogawa Forest Reserve	Both	4	50%	50%	Observational	Seeds and seedlings	Shibata and Nakashizuka (1995)
Subtropical moist forest (5)	São Paulo, Brazil	Municipal Reserve of Santa Genebra	Both	1	100% (21)	100% – smallest size/age class only	Observational	Seedling to adult (7 size categories)	Silva Matos et al. (1999)
Terra firme forest	Indonesia	Dipterocarp forest	Density	10	N/A	50%	Observational	Trees 10–100 cm dbh	Stoll and Newberry (2005)
Tropical dry forest	Costa Rica	Santa Rosa	Both	1	0%	100%	Observational	Saplings 0.5–10 m tall	Sullivan (2003)
Terra firme forest	Peru	Manu	Distance	5	20%	N/A	Experimental	Seeds	Terborgh et al. (1993)
Temperate deciduous forest	Northern Japan	Mt. Kurikoma	Both	1	100%	100%	Observational	Seeds	Tomita et al. (2002)
Terra firme forest	Panama	BCI	Density	60	N/A	~9%	Observational	Saplings >1 cm dbh, but <4 cm dbh	Uriarte et al. (2004)
Terra firme forest	Indonesia	Gunung Palung	Density	149–181 (11)	N/A	27% (25)	Experimental and observational	Seedlings	Webb and Peart (1999)
Montane forest (3)	Costa Rica	Monteverde	Distance	1	0% seeds consumed: 100%; seedling survivorship	N/A	Experimental and observational	Seeds and seedlings	Wenny (2000)
Terra firme forest	Eastern USA	Pasoh and BCI	Density	200 (13)	N/A	(27)	Observational	Woody plants >1 cm dbh	Wills and Condit (1999)
Terra firme forest	Panama	BCI	Density	84	N/A	80% recruitment, 64% intrinsic growth rate	Observational	Woody plants >1 cm dbh	Wills et al. (1997)

Continued

Table 13.1 Continued

Habitat	Country	Field site	Tested for density-, distance-dependency or both	# species studied	% of species that showed distance-dependency consistent with the J–C hypothesis	% of species that showed density-dependency consistent with the J–C hypothesis	Type of evidence	Developmental stage	Citation
Terra firme forest	Panama	BCI	Distance	1	100%	N/A	Observational	Seeds and seedlings	Wright and Duber (2001)
Terra firme forest	Panama	BCI	Distance	1	100%	N/A	Experimental	Seeds	Wright (1983)
Floodplain forest	Peru	Manu	Distance	2	100%	N/A	Observational	Seeds and seedlings	Wyatt and Silman (2004)

N/A = not applicable.

Notes:

1 – Lowland *terra firme* tropical forest.
2 – Lowland seasonal tropical forest.
3 – Montane tropical forest.
4 – Lowland floodplain tropical forest.
5 – Swampy area of subtropical moist forest.
6 – Subtropical and tropical evergreen moist forest.
7 – Cool temperate deciduous forest.
8 – Limestone outcrop with montane climate.
9 – Barro Colorado Island, Panama.
10 – Only 2 species are studied in depth.
11 – Species studied in different analyses.
12 – 544 in Pasoh,188 in BCI.
13 – 100 species from each plot.
14 – But only 11 species discussed in depth.
15 – Species number is unlisted.
16 – A higher percentage of large trees (35%) showed repelled recruitment than medium (17%) or small trees (11%) or shrubs (7%).
17 – Survivorship: 18% (2 of 11 species). Growth: 55% (6 of 11 species). Recruitment: 33% (27 of 81 total species).
18 – Results inconsistent. Seedling survivorship increased with distance at local scales, but was significantly lower in the extended tail distribution (1.8 km from the parental tree).
19 – But only for one of the two years of the study.
20 – But only at the closest distance to the adult.
21 – For youngest age class and closest distance category only.
22 – The probability of seedling survivorship increased with proximity to adult females.
23 – But only early in the seed to seedling transition (seedlings <2 months old) and only in the highest density plantings.
24 – Field experiment: 100% (1 of 1), Shadehouse experiment: 75% (6 of 8).
25 – 27% for species-level analyses (4 of 15 species). In community-level analysis, seedling mortality was directly related to species abundance (149 species).
26 – Tested against adult density.
27 – Analyses were conducted at the community-level. Conspecific density does not affect mortality in these species. Recruitment, however, is consistent with Janzen–Connell density-dependency.

to 0 years (Wright and Dubor 2001 where the distribution of seeds that showed evidence of vertebrate and invertebrate predation was evaluated). Most studies did not repeatedly sample individuals through time but only initially and at the end. Ninety-two percent of the studies were conducted at a single site and 40% of studies were conducted on or very near Barro Colorado Island (BCI). The bulk of the studies occurred in lowland and moist tropical forests (72%), with only 13% occurring in other tropical forests (e.g., floodplain, dry, etc.) and 15% in temperate regions (Table 13.3). The majority of studies (58%) either explicitly or implicitly considered a number of factors contributing to plant performance in addition to density or distance (e.g., light level, drought, heterospecific abundance).

Only seven studies investigated the impact of Janzen–Connell effects on species diversity (Connell *et al.* 1984, Condit *et al.* 1992, Wills *et al.* 1997, Webb and Peart 1999, Wills and Condit 1999, Harms *et al.* 2000, HilleRisLambers *et al.* 2002, see also Wills *et al.* 2006). Of these studies, three used data only from BCI and two from BCI and Pasoh (Table 13.1). Thus, the bulk of our generalizations regarding the Janzen–Connell hypothesis come from a single forest plot (BCI).

In Table 13.2, we summarize Janzen–Connell results, on a per species basis, for 173 species in 49 of the 53 studies from Table 13.1. The four remaining studies did not list their focal species (Connell *et al.* 1984, Wills *et al.* 1997, Harms *et al.* 2000, Peters 2003). Three of these studies occurred on BCI or at Pasoh (Wills *et al.* 1997, Harms *et al.* 2000, Peters 2003) and thus there would be some overlap with the species from these studies and those listed in Table 13.2. Nonetheless, we acknowledge that Table 13.2 underrepresents the species for which this hypothesis has been tested, especially since Peters (2003) found that approximately 80% of 732 species at BCI and Pasoh showed patterns consistent with the Janzen–Connell hypothesis.

The majority of species studied were canopy and understory trees (81%) whereas few lianas, palms, and shrubs were considered (Table 13.4). Density dependence was evaluated for 125 species (Table 13.2). A species was considered to exhibit density dependence in Table 13.2 if density

dependence occurred in any part of its life history. Negative density dependence occurred for 40% of the species whereas 57% showed no density dependence or exhibited positive density dependence and 3% had results that varied among studies. Distance dependence was evaluated for 129 species and 36% had patterns consistent with Janzen–Connell. Sixty percent had either no density dependence or survivorship decreased with distance from adult conspecifics and 4% had results that varied among studies.

Most species studied (79.8%) were from lowland tropical forests, some of which had pronounced dry seasons. Only 7% of species were from dry tropical forests and less than 2% were from mature tropical floodplain sites or swampy tropical habitats. Temperate forests accounted for 10% of the species studied while there was only one study in temperate arid forests and one in montane tropical forests. A surprising 63% of the species studied were studied on or near BCI.

Only 18% of the studies reported seed dry weights ($X = 1.8$ g $\pm$ 3.9 SD). The abundance of adults was not reported for 27 species and most species were simply classified as common (100 species), moderately abundant (one species), or uncommon (two species). The mean abundance of species where reported was 45 adults per ha (± 70.6 SD).

This brief review of studies leads us to the following conclusions. There have now been ample studies of the distance- and density-dependent predictions of the Janzen–Connell hypothesis and many species show Janzen–Connell effects (Tables 13.1 and 13.2). That said, however, there have been few studies outside of the lowland tropics, and Dirzo and Boege (Chapter 5, this volume) predict that pest pressure will be reduced when resource availability is more seasonal and episodic (e.g., dry forest). Additionally, there have been too few studies of life-forms other than trees (Tables 13.3 and 13.4) and too few studies of species at locations other than on or near BCI.

Far greater attention needs to be centered on the causes of Janzen–Connell effects and the degree to which they occur in later life-history stages (post small-sapling stages). Uncommon or rare species have been largely ignored yet may

Table 13.2 A review of studies that tested for Janzen–Connell effects (either distance dependence, density dependence or both) on a per species basis for different life-forms, plant families, habitats, locations, and countries around the world. We also included seed size or weight and information on the local abundance of the focal species if these data were reported.

Habitat	Location	Species name	Family	Distance-dependency?	Density-dependency?	Life form	Citation	Seed size (g)	Abundance of focal species (adults/ha)
Lowland terra firme	BCI (see Table 13.1)	Acalypha diversifolia	Euphorbiaceae	No	N/A	Shrub	Condit et al. (1992)		
Temperate deciduous	Coweeta (see Table 13.1)	Acer pennsyl- vanicum	Aceraceae	Yes	Yes	Canopy tree	HilleRisLambers et al. (2002)		
Temperate deciduous	Coweeta	Acer rubrum	Aceraceae	Yes	Yes	Canopy tree	HilleRisLambers et al. (2002)		
Lowland terra firme	BCI	Alibertia edulis	Rubiaceae	No	N/A	Understory tree	Condit et al. 1992		
Lowland terra firme	BCI	Alseis blackiana	Rubiaceae	Yes (Condit et al. 1992)	Yes (Uriarte et al. 2004)	Canopy tree	Condit et al. (1992), Uriarte et al. (2004)		
Lowland terra firme	BCI	Anaxagorea panamensis	Annonaceae	No	N/A	Shrub	Condit et al. (1992)		
Lowland terra firme	BCI	Annona acuminata	Annonaceae	No	N/A	Shrub	Condit et al. (1992)		
Tropical dry deciduous	Mudumalai (see Table 13.1)	Anogeissus latifolia	Combretaceae	N/A	Yes	Woody plants >1 cm dbh	John et al. (2002)	14.5	45.6
Lowland terra firme	BCI	Aspidosperma cruenta	Apocynaceae	Yes (Augspurger 1984)	No (Augspurger and Kelly 1984, Uriarte et al. 2004)	Canopy tree	Augspurger (1984), Augspurger and Kelly (1984), Uriarte et al. (2004)	Large	
Lowland terra firme	Manu, Peru	Astrocaryum macrocalyx	Arecaceae	Yes	N/A	Canopy tree	Terborgh et al. (1993)	0.0056	30
Lowland terra firme	Bladen, Belize	Astrocaryum mexicanum	Arecaceae	Yes	N/A	Understory palm	Brewer and Webb (2001)		
Lowland terra firme	Cocha Cashu, Peru	Astrocaryum murumuru	Arecaceae	Yes (both studies)	Yes (Cintra 1997)	Canopy tree	Cintra (1997), Wyatt and Silman (2004)	0.335	

Lowland terra firme	BCI	*Attalea butyraceae* (was *Scheelea zonensis*)	Arecaceae	Yes	N/A	Understory palm	Wright (1983), Wright and Duber (2001)	Large: 4 cm long	
Lowland terra firme	BCI	*Bactris major*	Arecaceae	No	N/A	Understory tree	Condit et al. (1992)		
Lowland terra firme	BCI	*Beilschmiedia pendula*	Lauraceae	Yes (Condit et al. 1992)	Yes (Uriarte et al. 2004)	Canopy tree	Condit et al. (1992), Uriarte et al. (2004)		
Lowland terra firme	Manu, Peru (Terborgh et al. 1993), Para, Brazil (Peres et al. 1997)	*Bertholletia excelsa*	Lecythidaceae	No (Terborgh et al. 1993), yes (Peres et al. 1997)	N/A	Canopy tree	Terborgh et al. (1993)		
Cool temperate deciduous	Tantaré, Canada (see Table 13.1)	*Betula alleghaniensis*	Betulaceae	N/A	No	Canopy tree	Houle (1998)	0.0188	280
Temperate deciduous	Coweeta	*Betula* sp.	Betulaceae	Yes	Yes	Canopy tree	HilleRisLambers et al. (2002)		
Lowland terra firme; lowland seasonal	BCI (Condit et al. 1992, Uriarte et al. 2004), Los Tuxtlas, Mexico (Burkey 1994)	*Brosimum alicastrum*	Moraceae	No (Condit et al. 1992, yes (Burkey 1994)	No (Uriarte et al. 2004, Burkey 1994)	Canopy tree	Condit et al. (1992), Uriarte et al. 2004, Burkey (1994)	Large seeded: 2 cm × 0.5 cm seeds	
Lowland terra firme	Manu, Peru	*Calatola venezuelana*	Icacinaceae	No	N/A	Canopy tree	Terborgh et al. (1993)		
Lowland terra firme	BCI	*Calophyllum longifolium*	Clusiaceae	No (Condit et al. 1992)	Yes in confamilial analysis (Uriarte et al. 2004)	Canopy tree	Condit et al. (1992), Uriarte et al. (2004)		
Lowland terra firme	BCI	*Capparis frondosa*	Capparidaceae	No	N/A	Shrub	Condit et al. (1992)		
Temperate deciduous	Ogawa, Japan	*Carpinus cordata*	Betulaceae	Yes	Yes	Tree	Shibata and Nakashizuka (1995)	4.0 × 3.5 mm	85
Temperate deciduous	Ogawa, Japan	*Carpinus japonica*	Betulaceae	Yes	No	Tree	Shibata and Nakashizuka (1995)	0.045	11.33

Continued

Table 13.2 Continued

Habitat	Location	Species name	Family	Distance-dependency?	Density-dependency?	Life form	Citation	Seed size (g)	Abundance of focal species (adults/ha)
Temperate deciduous	Ogawa, Japan	*Carpinus laxiflora*	Betulaceae	No	No	Tree	Shibata and Nakashizuka (1995)	5–8 cm long	88.67
Temperate deciduous	Ogawa, Japan	*Carpinus tschonoskii*	Betulaceae	No	Yes	Tree	Shibata and Nakashizuka (1995)	0.4	15.3
Lowland *terra firme*	BCI	*Casearia aculeata*	Flacourtiaceae	No (Condit et al. 1992)	No (Uriarte et al. 2004)	Understory tree	Condit et al. (1992)		
Tropical dry deciduous	Mudumalai, India	*Cassia fistula*	Fabaceae	N/A	Yes	Woody plants >1 cm dbh	John et al. (2002)	10	37.7
Lowland *terra firme*	BCI	*Cassipourea elliptica*	Rhizophoraceae	No (Condit et al. 1992)	No (Uriarte et al. 2004)	Midstory tree	Condit et al. (1992)		
Lowland *terra firme*	BCI	*Cavanillesia platanifolia*	Bombacaceae	Yes	N/A	Canopy tree	Augspurger (1984)		
Lowland *terra firme*	BCI	*Cecropia insignis*	Cecropiaceae	No	N/A	Canopy tree	Condit et al. (1992)		
Lowland seasonal	Los Tuxtlas, Mexico	*Cecropia obtusifolia*	Cecropiaceae	N/A	Yes	Pioneer tree	Alvarez-Buylla (1994)		
Lowland *terra firme*	BCI	*Ceiba pentandra*	Bombacaceae	No (Augspurger 1984)	No (Augspurger and Kelly 1984)	Canopy tree	Augspurger (1984), Augspurger and Kelly (1984)		
Lowland *terra firme*	Mabura Hill, Guyana	*Chlorocardium rodiei*	Lauraceae	Yes	N/A	Canopy tree	Hammond et al. (1999)	0.0795	12.2
Lowland *terra firme*	BCI	*Chrysophyllum argenteum*	Sapotaceae	N/A	No	Canopy tree	Uriarte et al. (2004)	Small	
Lowland *terra firme*	BCI	*Chrysophyllum panamense*	Sapotaceae	No	N/A	Canopy tree	Condit et al. (1992)		
Lowland *terra firme*	BCI	*Cochlospermum vitifolium*	Cochlospermaceae	N/A	No	Understory to canopy tree	Augspurger and Kelly (1984)		
Lowland *terra firme*	BCI	*Cordia alliodora*	Boraginaceae	Yes (Augspurger 1984)	No (Augspurger and Kelly 1984)	Understory to canopy tree	Augspurger (1984), Augspurger and Kelly (1984)		

Continued

Forest type	Location	Species	Family			Habit	Reference		
Lowland terra firme	BCI	*Cordia bicolor*	Boraginaceae	No (Condit et al. 1992)	No (Uriarte et al. 2004)	Midstory tree	Condit et al. (1992), Uriarte et al. (2004)		
Lowland terra firme	BCI	*Cordia lasiocalyx*	Boraginaceae	No (Condit et al. 1992)	No (Uriarte et al. 2004)	Midstory tree	Condit et al. (1992), Uriarte et al. (2004)		
Tropical dry deciduous	Mudumalai, India	*Cordia obliqua*	Boraginaceae	N/A	Yes	Woody plants >1 cm dbh	John et al. (2002)	9.9	3.8
Lowland terra firme	BCI	*Coussarea curvigemmia*	Rubiaceae	No (Condit et al. 1992)	Yes (Uriarte et al. 2004)	Understory tree	Condit et al. (1992), Uriarte et al. (2004)		
Montane forest	Anadusian highlands, Spain	*Crataegus monogyna*	Rosaceae	N/A	No	Unreported, probably canopy tree	Hulme (1997)		
Lowland terra firme	BCI	*Croton bilbergianus*	Euphorbiaceae	No (Condit et al. 1992)	No (Uriarte et al. 2004)	Understory tree	Condit et al. (1992), Uriarte et al. (2004)		
Lowland terra firme	BCI	*Cupania sylvatica*	Sapindaceae	No (Condit et al. 1992)	No (Uriarte et al. 2004)	Understory tree	Condit et al. (1992), Uriarte et al. (2004)		
Lowland terra firme	BCI	*Desmopsis panamensis*	Annonaceae	Yes (Condit et al. 1992)	No (Condit et al. 1994, Uriarte et al. 2004)	Understory tree	Condit et al. (1992), Condit et al. (1994), Uriarte et al. (2004)	Fresh mass: 1.4 ± 0.5 g, length: 17.0 ± 1.8 mm (mean ± SD)	
Tropical dry deciduous	Mudumalai, India	*Diospyros montana*	Ebenaceae	N/A	No	Woody plants >1 cm dbh	John et al. (2002)		
Lowland terra firme	Dipterocarp forests, Indonesia	*Dipterocarpus kerrii*	Dipterocarpaceae	N/A	No	Canopy trees	Stoll and Newberry (2005)		8.1
Lowland terra firme	Cocha Cashu, Peru (Cintra 1997), (Romo et al. 2004), Manu, Peru (Terborgh et al. 1993)	*Dipteryx micrantha*	Fabaceae	Yes (Cintra 1997), no (Cintra 1997, Terborgh et al. 1993)	No (Cintra 1997, Romo et al. 2004)	Canopy tree	Cintra (1997), Romo et al. (2004), Terborgh et al. (1993)	0.00126	6

Table 13.2 Continued

Habitat	Location	Species name	Family	Distance-dependency?	Density-dependency?	Life form	Citation	Seed size (g)	Abundance of focal species (adults/ha)
Lowland *terra firme*	BCI (De Steven and Putz 1984), La Selva, Costa Rica (Clark and Clark 1984)	*Dipteryx panamensis*	Fabaceae	Yes (both studies)	Yes (Clark and Clark 1984)	Canopy tree	De Steven and Putz (1984), Clark and Clark (1984)	8–10 mm in diameter, fresh mass = 0.29 ± 0.07 g (mean ± SD)	2
Lowland *terra firme*	BCI	*Drypetes standleyi*	Euphorbiaceae	No (Condit et al. 1992)	No (Uriarte et al. 2004)	Canopy tree	Condit et al. (1992), Uriarte et al. (2004)		
Tropical dry deciduous	Mudumalai, India	*Emblica officinalis*	Euphorbiaceae	N/A	Yes	Woody plants >1 cm dbh	John et al. (2002)	0.051	11.5
Tropical dry deciduous	Mudumalai, India	*Eriolaena quinquelocularis*	Sterculiaceae	N/A	No	Woody plants >1 cm dbh	John et al. (2002)		
Lowland *terra firme*	BCI	*Eugenia coloradensis*	Myrtaceae	Yes (Condit et al. 1992)	No (Uriarte et al. 2004)	Canopy tree	Condit et al. (1992), Uriarte et al. (2004)		
Lowland *terra firme*	BCI	*Eugenia galalonensis*	Myrtaceae	Yes (Condit et al. 1992)	No (Uriarte et al. 2004)	Understory tree	Condit et al. (1992), Uriarte et al. (2004)		
Lowland *terra firme*	BCI	*Eugenia nesiotica*	Myrtaceae	No (Condit et al. 1992)	No (Uriarte et al. 2004)	Midstory tree	Condit et al. (1992), Uriarte et al. (2004)		
Lowland *terra firme*	BCI	*Eugenia oerstedeanna*	Myrtaceae	Yes (Condit et al. 1992)	No (Uriarte et al. 2004)	Midstory tree	Condit et al. (1992), Uriarte et al. (2004)		
Subtropical moist forest	Santa Genebra, Brazil (see Table 13.1)	*Euterpe edulis*	Arecaceae	Yes	Yes	Subcanopy palm	Silva Matos et al. (1999)	2	284

Forest type	Location	Species	Family			Canopy position	References		
Temperate deciduous	Mt. Kurikoma, Japan	*Fagus crenata*	Fagaceae	Yes	Yes	Canopy tree	Tomita et al. (2002)		213.8
Lowland terra firme	BCI	*Faramea occidentalis*	Rubiaceae	Yes (Schupp 1988, 1992), no (Condit et al. 1992)	Yes (Condit et al. 1994), no (Schupp 1992, Uriarte et al. 2004)	Subcanopy tree	Schupp (1992), Condit et al. (1992), Condit et al. (1994), Uriarte et al. (2004), Schupp (1988)	0.75	
Temperate deciduous	Coweeta	*Fraxinus americana*	Oleaceae	No	Yes	Canopy tree	HilleRisLambers et al. (2002)		
Lowland terra firme	BCI	*Garcinia acuminata*	Clusiaceae	No	N/A	Subcanopy tree	Condit et al. (1992)		
Lowland terra firme	BCI	*Garcinia edulis*	Clusiaceae	No	N/A	Subcanopy tree	Condit et al. (1992)		
Lowland terra firme	BCI	*Garcinia intermedia*	Clusiaceae	N/A	No	Midstory tree	Uriarte et al. (2004)	Small	
Lowland terra firme	BCI	*Guarea guidonia*	Meliaceae	No Condit et al. (1992)	No (Uriarte et al. 2004)	Subcanopy tree	Condit et al. (1992), Uriarte et al. (2004)		
Lowland terra firme	BCI	*Guarea sp. nov.*	Meliaceae	No	N/A	Subcanopy tree	Condit et al. (1992)		
Lowland terra firme	BCI	*Guarea* unknown ("fuzzy")	Meliaceae	N/A	Yes in confamilial analysis	Midstory tree	Uriarte et al. (2004)		
Lowland terra firme	BCI	*Guatteria dumetorum*	Annonaceae	Yes (Condit et al. 1992)	No (Uriarte et al. 2004)	Canopy tree	Condit et al. (1992), Uriarte et al. (2004)		
Lowland terra firme	BCI	*Heisteria concinna*	Olacaceae	No (Condit et al. 1992)	Yes (Uriarte et al. 2004)	Understory tree	Condit et al. (1992), Uriarte et al. (2004)		
Lowland terra firme	BCI	*Herrania purpurea*	Sterculiaceae	No (Condit et al. 1992)	No (Uriarte et al. 2004)	Understory tree	Condit et al. (1992), Uriarte et al. (2004)		
Lowland terra firme	BCI	*Hirtella triandra*	Chrysobalanaceae	Yes (Condit et al. 1992)	No (Uriarte et al. 2004)	Understory tree	Condit et al. (1992), Uriarte et al. (2004)		
Lowland terra firme	Dipterocarp forests, Indonesia	*Hopea nervosa*	Dipterocarpaceae	N/A	Yes	Canopy trees	Stoll and Newberry (2005)	0.1237	12.4

Continued

Table 13.2 Continued

Habitat	Location	Species name	Family	Distance-dependency?	Density-dependency?	Life form	Citation	Seed size (g)	Abundance of focal species (adults/ha)
Lowland terra firme	BCI	Hybanthus prunifolius	Violaceae	No	N/A	Shrub	Condit et al. (1992)		
Lowland terra firme	Manu, Peru	Hymenaea courbaril	Fabaceae	No	N/A	Canopy tree	Terborgh et al. (1993)	0.003	1
Lowland terra firme	BCI	Inga marginata	Fabaceae	Yes (Condit et al. 1992)	No (Uriarte et al. 2004)	Canopy tree	Condit et al. (1992); Uriarte et al. (2004)		
Lowland terra firme	BCI	Inga nobilis	Fabaceae	N/A	No	Midstory tree	Uriarte et al. (2004)		
Lowland terra firme	BCI	Inga quaternata	Fabaceae	No	N/A	Subcanopy tree	Condit et al. (1992)		
Lowland terra firme	BCI	Inga sp. nov.	Fabaceae	No	N/A	Understory tree	Condit et al. (1992)		
Lowland terra firme	BCI	Inga umbellifera	Fabaceae	N/A	No	Midstory tree	Uriarte et al. (2004)		
Floodplain forest	Manu, Peru	Iriartea deltoidea	Arecaceae	Yes	N/A	Canopy tree	Wyatt and Silman (2004)	0.0024	45
Tropical dry deciduous	Mudumalai, India	Kydia calycina	Malvaceae	N/A	Yes	Woody plants >1 cm dbh	John et al. (2002)	0.044	103.5
Lowland terra firme	BCI	Lacistema aggregatum	Lacistemaceae	No (Condit et al. 1992)	No (Uriarte et al. 2004)	Understory tree	Condit et al. (1992); Uriarte et al. (2004)		
Lowland terra firme	BCI	Laetia thamnia	Flacourtiaceae	No (Condit et al. 1992)	Yes (Uriarte et al. 2004)	Understory tree	Condit et al. (1992); Uriarte et al. (2004)		
Lowland terra firme	BCI	Lafoensia punicifolia	Lythraceae	N/A	Yes	Understory to canopy trees	Augspurger and Kelly (1984)		
Tropical dry deciduous	Mudumalai, India	Lagerstroemia microcarpa	Lythraceae	N/A	Yes	Woody plants >1 cm dbh	John et al. (2002)	10–14 mm in diameter	79.6
Temperate deciduous	Coweeta	Liriodendron tulipfera	Magnoliaceae	Yes	Yes	Canopy tree	HilleRisLambers et al. (2002)		

Habitat	Location	Species	Family			Life form	Reference		
Lowland terra firme	BCI	*Lonchocarpus latifolia*	Fabaceae	No (Condit et al. 1992)	No (Uriarte et al. 2004)	Canopy tree	Condit et al. (1992), Uriarte et al. (2004)		
Lowland terra firme	BCI	*Lonchocarpus pentaphyllus*	Fabaceae	Yes	N/A	Canopy trees	Augspurger (1984)		
Lowland terra firme	BCI	*Luehea seemannii*	Tiliaceae	N/A	Yes	Understory to canopy trees	Augspurger and Kelly (1984)		
Lowland terra firme	Jenaro Herrera, Peru (see Table 13.1)	*Macoubea guianensis*	Apocynaceae	No	No	Canopy tree	Notman et al. (1996)	0.0699	1.5
Lowland terra firme	BCI	*Maquira costaricana*	Moraceae	No (Condit et al. 1992)	No (Uriarte et al. 2004)	Subcanopy tree	Condit et al. (1992), Uriarte et al. (2004)		
Lowland terra firme	Maraca Island, Brazil (see Table 13.1)	*Maximiliana maripa*	Arecaceae	Yes	N/A	Canopy tree	Fragoso et al. (2003), Fragoso (1997)	0.149	128
Lowland terra firme	BCI	*Miconia affinis*	Melastomataceae	No (Condit et al. 1992)	No (Uriarte et al. 2004)	Understory tree	Condit et al. (1992), Uriarte et al. (2004)		
Lowland terra firme	BCI	*Miconia argentea*	Melastomataceae	Yes (Condit et al. 1992)	No (Uriarte et al. 2004)	Subcanopy tree	Condit et al. (1992), Uriarte et al. (2004)		
Lowland terra firme	BCI	*Miconia nervosa*	Melastomataceae	No	N/A	Shrub	Condit et al. (1992)		
Lowland terra firme	Neung South, Ghana	*Milicia regia*	Moraceae	Yes	N/A	Canopy tree	Hood et al. (2004)		
Lowland terra firme	BCI	*Mouriri myrtylloides*	Melastomataceae	No	N/A	Shrub	Condit et al. (1992)		
Lowland terra firme	BCI	*Ochroma pyramidale*	Bombacaceae	N/A	No	Understory to canopy trees	Augspurger and Kelly (1984)		
Lowland terra firme	BCI	*Ocotea cernua*	Lauraceae	No	N/A	Subcanopy tree	Condit et al. (1992)		
Montane tropical forest	Monteverde, Costa Rica	*Ocotea endresiana*	Lauraceae	Yes	N/A	Canopy tree	Wenny (2000)		Large seeded: 1 cm × 3 cm seeds
Lowland terra firme	BCI	*Ocotea skutchii*	Lauraceae	No	N/A	Canopy tree	Condit et al. (1992)		

Continued

Table 13.2 Continued

Habitat	Location	Species name	Family	Distance-dependency?	Density-dependency?	Life form	Citation	Seed size (g)	Abundance of focal species (adults/ha)
Lowland terra firme	BCI	Ocotea whitei	Lauraceae	Yes (Gilbert et al. 1994), no (Gilbert et al. 2001)	Yes (both studies)	Canopy tree	Gilbert et al. (1994), (2001)		
Lowland terra firme	BCI	Ouratea lucens	Ochnaceae	Yes	N/A	Shrub	Condit et al. (1992)		
Lowland terra firme	BCI	Palicourea guianensis	Rubiaceae	No	N/A	Shrub	Condit et al. (1992)		
Lowland terra firme	Dipterocarp forests, Indonesia	Parashorea malaanonan	Dipterocarpaceae	N/A	No	Canopy trees	Stoll and Newberry (2005)	0.444	18.6
Lowland terra firme	BCI	Pentagonia macrophylla	Rubiaceae	N/A	No	Understory tree	Uriarte et al. (2004)		
Lowland terra firme	BCI	Picramnia latifolia	Picramniaceae	No (Condit et al. 1992)	No (Uriarte et al. 2004)	Understory tree	Condit et al. (1992), Uriarte et al. (2004)		
Lowland terra firme	BCI	Piper cordulatum	Piperaceae	No	N/A	Shrub	Condit et al. (1992)		
Lowland terra firme	BCI	Platypodium elegans	Fabaceae	Yes (all studies)	Yes (Augspurger and Kelly 1984)	Canopy tree	Augspurger (1983), (1984), Augspurger and Kelly (1984)		
Lowland terra firme	BCI	Poulsenia armata	Moraceae	Yes	N/A	Canopy tree	Condit et al. (1992)		
Lowland terra firme	BCI	Pouteria reticulata	Sapotaceae	N/A	No	Canopy tree	Uriarte et al. (2004)		
Lowland terra firme	Jenaro Herrera, Peru	Pouteria sp.	Sapotaceae	No	No	Canopy tree	Notman et al. (1996)	0.057	1.5
Lowland terra firme	BCI	Pouteria unilocularis	Sapotaceae	No	N/A	Canopy tree	Condit et al. (1992)		
Lowland terra firme	BCI	Prioria copaifera	Fabaceae	No (Condit et al. 1992)	No (Uriarte et al. 2004)	Canopy tree	Condit et al. (1992), Uriarte et al. (2004)		

Habitat	Site	Species	Family			Life form	References
Lowland *terra firme*	BCI	*Protium costaricense*	Burseraceae	No (Condit et al. 1992)	Yes in confamilial analysis (Uriarte et al. 2004)	Subcanopy tree	Condit et al. (1992), Uriarte et al. (2004)
Lowland *terra firme*	BCI	*Protium panamense*	Burseraceae	No (Condit et al. 1992)	Yes in confamilial analysis (Uriarte et al. 2004)	Subcanopy tree	Condit et al. (1992), Uriarte et al. (2004)
Lowland *terra firme*	BCI	*Protium tenuifolium*	Burseraceae	No (Condit et al. 1992)	Yes in confamilial analysis (Uriarte et al. 2004)	Subcanopy tree	Condit et al. (1992), Uriarte et al. (2004)
Montane forest	Anadusian highlands, Spain	*Prunus mahaleb*	Rosaceae	N/A	No	Unreported, but probably canopy tree	Hulme (1997)
Temperate deciduous	Griffy Lake, Indiana, USA (Packer and Clay 2000), Eastern USA (Reinhart et al. 2005)	*Prunus serotina*	Rosaceae	Yes (both studies)	Yes (Packer and Clay 2000)	Canopy tree	Packer and Clay (2000), Reinhart et al. (2005)
Lowland *terra firme*	BCI	*Pseudobombax septenatum*	Bombacaceae	N/A	Yes	Understory to canopy trees	Augspurger and Kelly (1984)
Lowland *terra firme*	BCI	*Psidium anglo-hondurese*	Myrtaceae	No	N/A	Understory tree	Condit et al. (1992)
Lowland *terra firme*	BCI	*Psychotria horizontalis*	Rubiaceae	No	N/A	Shrub	Condit et al. (1992)
Lowland *terra firme*	BCI	*Psychotria marginata*	Rubiaceae	No	N/A	Shrub	Condit et al. (1992)
Lowland *terra firme*	BCI	*Pterocarpus rohrii*	Fabaceae	No (Condit et al. 1992)	No (Uriarte et al. 2004)	Canopy tree	Condit et al. (1992), Uriarte et al. (2004)
Lowland *terra firme*	BCI	*Quararibea asterolepis*	Bombacaceae	No (Hubbell et al. 1990, Condit et al. 1992)	No (Hubbell et al. 1990, De Steven and Wright 2002, Uriarte et al. 2004)	Canopy tree	Hubbell et al. (1990), Condit et al. (1992), De Steven and Wright (2002), Uriarte et al. (2004)

Continued

Table 13.2 Continued

Habitat	Location	Species name	Family	Distance-dependency?	Density-dependency?	Life form	Citation	Seed size (g)	Abundance of focal species (adults/ha)
Temperate deciduous	Coweeta	*Quercus rubra*	Fagaceae	No	Yes	Canopy tree	HilleRisLambers et al. (2002)		
Lowland *terra firme*	BCI	*Randia armata*	Rubiaceae	No (Condit et al. 1992)	No (Uriarte et al. 2004)	Understory tree	Condit et al. (1992), Uriarte et al. (2004)		
Lowland *terra firme*	BCI	*Rinorea sylvatica*	Violaceae	No	N/A	Shrub	Condit et al. (1992)		
Lowland *terra firme*	Chiquibul, Belize	*Sebastiana longicuspis*	Euphorbiaceae	N/A	Yes	Tree	Bell et al. (2006)	Small	
Lowland *terra firme*	Dipterocarp forests, Indonesia	*Shorea argentifolia*	Dipterocarpaceae	N/A	Yes	Canopy tree	Stoll and Newberry (2005)	65.5 ± 22.3 g fresh weight (mean $\pm$ SD)	9.9
Lowland *terra firme*	Dipterocarp forests, Indonesia	*Shorea fallax*	Dipterocarpaceae	N/A	Yes	Canopy tree	Stoll and Newberry (2005)	0.273650324	45.8
Lowland *terra firme*	Dipterocarp forests, Indonesia	*Shorea hopeifolia*	Dipterocarpaceae	Yes	N/A	Canopy tree	Blundell and Peart (1998)	5	3.5
Lowland *terra firme*	Dipterocarp forests, Indonesia	*Shorea johorensis*	Dipterocarpaceae	N/A	Yes	Canopy tree	Stoll and Newberry (2005)	0.8	24.6
Lowland *terra firme*	Danum Valley, Malaysia	*Shorea leprosula*	Dipterocarpaceae	N/A	Yes	Canopy tree	Massey et al. (2006)	Small	
Lowland *terra firme*	Dipterocarp forests, Indonesia	*Shorea longisperma*	Dipterocarpaceae	No	N/A	Canopy tree	Blundell and Peart (1998)	0.0202	0.48
Lowland *terra firme*	Dipterocarp forests, Indonesia	*Shorea parvifolia*	Dipterocarpaceae	No (Blundell and Peart 1998)	No (Stoll and Newberry 2005)	Canopy tree	Blundell and Peart (1998), Stoll and Newberry (2005)	0.544	
Lowland *terra firme*	Dipterocarp forests, Indonesia	*Shorea pauciflora*	Dipterocarpaceae	N/A	No	Canopy tree	Stoll and Newberry (2005)	0.041	10.9

Lowland terra firme	Dipterocarp forests, Indonesia	*Shorea pilosa*	Dipterocarpaceae	N/A	Yes	Canopy tree	Stoll and Newberry (2005)	0.399	14.5
Lowland terra firme	Dipterocarp forests, Indonesia	*Shorea pinanga*	Dipterocarpaceae	Yes	N/A	Canopy tree	Blundell and Peart (1998)	0.032	1.8
Temperate arid	Spain	*Silene diclinis*	Caryophyllaceae	No	Yes	Long-lived herb	Montesinos et al. (2006)		
Lowland terra firme	BCI	*Simarouba amara*	Simaroubaceae	No (Condit et al. 1992)	No (Uriarte et al. 2004)	Canopy tree	Condit et al. (1992), Uriarte et al. (2004)		
Lowland terra firme	BCI	*Sloanea terniflora*	Elaeocarpaceae	N/A	No	Canopy tree	Uriarte et al. (2004)		
Lowland terra firme	BCI	*Socratea exorrhiza*	Arecaceae	No	N/A	Subcanopy tree	Condit et al. (1992)		
Lowland terra firme	BCI	*Soracea affinis*	Moraceae	No	N/A	Shrub	Condit et al. (1992)		
Lowland terra firme	BCI	*Stylogyne standleyi*	Myrsinaceae	No	N/A	Shrub	Condit et al. (1992)		
Lowland terra firme	BCI	*Swartzia simplex* var. *grandiflora*	Fabaceae	No (Condit et al. 1992)	No (Uriarte et al. 2004)	Understory tree	Condit et al. (1992), Uriarte et al. (2004)		
Lowland terra firme	BCI	*Swartzia simplex* var. *ochnacea*	Fabaceae	No (Condit et al. 1992)	No (Uriarte et al. 2004)	Understory tree	Condit et al. (1992), Uriarte et al. (2004)		
Lowland terra firme	Kayapo Centre, Para, Brazil (see Table 13.1)	*Swietenia macrophylla*	Meliaceae	Yes	Yes	Emergent tree	Norghauer et al. (2006)	0.0019	
Tropical dry forest	Santa Rosa, Costa Rica	*Tabebuia ochracea*	Bignoniaceae	No	Yes	Canopy tree	Sullivan (2003)		
Lowland terra firme	BCI	*Tabebuia rosea*	Bignoniaceae	Yes (Augspurger 1984)	Yes (Augspurger and Kelly 1984)	Canopy tree	Augspurger (1984) and Augspurger and Kelly (1984)		
Lowland terra firme	BCI	*Tabernaemontana arborea*	Apocynaceae	No (Condit et al. 1992)	No (Uriarte et al. 2004)	Canopy tree	Condit et al. (1992), Uriarte et al. (2004)		

Continued

Table 13.2 Continued

Habitat	Location	Species name	Family	Distance-dependency?	Density-dependency?	Life form	Citation	Seed size (g)	Abundance of focal species (adults/ha)
Lowland terra firme	BCI	Tachigalia versicolor	Fabaceae	Yes (Augspurger and Kitajima 1992), (Condit et al. 1992)	Yes (Augspurger and Kitajima 1992), no (Uriarte et al. 2004)	Canopy tree	Augspurger and Kitajima (1992), Condit et al. (1992), Uriarte et al. (2004)	Large	
Lowland terra firme	BCI	Talisia nervosa	Sapindaceae	N/A	No	Understory tree	Uriarte et al. (2004)		
Lowland terra firme	BCI	Talisia princeps	Sapindaceae	N/A	No	Midstory tree	Uriarte et al. (2004)		
Montane forest	Anadusian highlands, Spain	Taxus baccata	Taxaceae	N/A	No	Unreported, but probably canopy tree	Hulme (1997)		
Tropical dry deciduous	Mudumalai, India	Tectona grandis	Verbenaceae	N/A	Yes	Woody plants >1 cm dbh	John et al. (2002)	12.1	42.8
Lowland terra firme	BCI	Terminalia amazonica	Combretaceae	N/A	No	Understory to canopy trees	Augspurger and Kelly (1984)		
Tropical dry deciduous	Mudumalai, India	Terminalia crenulata	Combretaceae	N/A	No	Woody plants >1 cm dbh	John et al. (2002)	1.3–2.6 cm long, 0.8–1.3 cm wide	55.4
Lowland terra firme	BCI	Terminalia oblonga	Combretaceae	Yes Augspurger (1984)	No (Augspurger and Kelly 1984)	Canopy tree	Augspurger (1984), Augspurger and Kelly (1984)		
Lowland terra firme	BCI	Tetragastris panamensis	Burseraceae	Yes Condit et al. (1992)	Yes (Uriarte et al. 2004), no (De Steven and Wright 2002)	Canopy tree	Condit et al. (1992), De Steven and Wright (2002), Uriarte et al. (2004)	Large	

Lowland terra firme	BCI	*Trichilia pallida*	Meliaceae	No Condit et al. (1992)	No (Uriarte et al. 2004)	Subcanopy tree	Condit et al. (1992), Uriarte et al. (2004)		
Lowland terra firme	BCI	*Trichilia tuberculata*	Meliaceae	Yes (Hubbell et al. 1990, Condit et al. 1992)	Yes (Hubbell et al. 1990, De Steven and Wright 2002), no (Uriarte et al. 2004)	Canopy tree	Hubbell et al. (1990), Condit et al. (1992), De Steven and Wright (2002), Uriarte et al. (2004)		
Lowland terra firme	BCI	*Triplaris cumingiana*	Polygonaceae	Yes (Augspurger 1984)	Yes (Augspurger and Kelly 1984)	Canopy tree	Augspurger (1984), Augspurger and Kelly (1984)		
Lowland terra firme	BCI	*Unonopsis pittieri*	Annonaceae	Yes (Condit et al. 1992)	No (Uriarte et al. 2004)	Midstory tree	Condit et al. (1992), Uriarte et al. (2004)		
Lowland terra firme	Dipterocarp forests, Indonesia	*Vatica dulitensis*	Dipterocarpaceae	N/A	No	Canopy tree	Stoll and Newberry (2005)	3–5 cm long	8.1
Floodplain forest (4)	Cocha Cashu, Peru	*Virola calophylla*	Myristicaceae	Yes	Yes	Canopy tree	Russo and Augspurger (2004)	0.55	2.9
Lowland terra firme	BCI	*Virola sebifera*	Myristicaceae	No (Condit et al. 1992)	No (Uriarte et al. 2004)	Midstory tree	Condit et al. (1992), Uriarte et al. (2004)		
Temperate deciduous	Coweeta	*Vitis* sp.	Vitaceae	No	No	Vine	HilleRisLambers et al. (2002)		
Tropical dry deciduous	Mudumalai, India	*Xeromphis spinosa*	Rubiaceae	N/A	Yes	Woody plants >1 cm dbh	John et al. (2002)	0.412	15.4
Lowland terra firme	BCI	*Xylopia macarantha*	Annonaceae	No (Condit et al. 1992)	No (Uriarte et al. 2004)	Midstory tree	Condit et al. (1992), Uriarte et al. (2004)		

N/A = not applicable.

Table 13.3 Regions and forest types for studies that evaluated the Janzen–Connell hypothesis.

Habitat	No. of studies	%
Lowland tropical forest	38	71.7
Tropical forests in mature floodplain or swampy areas	2	3.8
Montane tropical forest	1	1.9
Tropical dry forest	2	3.8
Subtropical forest	2	3.8
Temperate deciduous forest	7	13.2
Temperate arid forest	1	1.9

Table 13.4 Number and percent of species in different life-history classes that were investigated for Janzen–Connell effects.

Life-form	No. of species	%
Herbaceous plant	1	0.6
Liana (woody vine)	1	0.6
Emergent tree	2	1.2
Canopy tree	81	47.7
Pioneer tree	2	1.2
Shrub	15	8.8
Species listed as tree (no indication of canopy position)	8	4.7
Understory palm	3	1.8
Understory tree	57	33.5
Total	170	

suffer strong pressure from pests (see below). Furthermore, if enemies are the cause of these effects then the degree that these enemies are specialists or facultative specialists needs to be quantified. In addition, far greater attention needs to be given to quantifying the relative abundance of focal species and more importantly to aspects of plant species life history, particularly seed size, dispersal mode, degree of shade tolerance, and overall habitat breadth. It is important to determine whether species that are kept in check by

their enemies are superior competitors, that is they have life-history traits that allow them to establish, form dense stands, and persist for long periods in the understory. Most importantly, we argue that the focus needs to be shifted away from whether these effects occur and towards the very difficult task of evaluating their impact on local and regional patterns of diversity (see below).

CHALLENGES AND ISSUES ASSOCIATED WITH TESTING, EVALUATING, AND FALSIFYING THE JANZEN–CONNELL HYPOTHESIS

The Janzen–Connell hypothesis is ultimately a community-level hypothesis

As Janzen (1970) concluded, host-specific or facultatively host-specific seed and seedling predators will decrease tree population density of a given tree species and/or increase distances between new adults. Either of these consequences of predation will lead "to more space in the habitat for other species of trees and therefore higher total number of tree species." Thus, the most unequivocal tests of Janzen–Connell, as with Paine's original test of keystone predation (Paine 1966), will come from studies that experimentally remove enemies or subsets of enemies over long periods of time and quantify the change in species diversity. This task, of course, is not a trivial undertaking. Nonetheless, other major challenges have been overcome in studies of tropical forests (e.g., establishing and maintaining 50 ha plots).

Janzen (1970, p. 517) proposed five field experiments or observational studies that would test predictions of his model. However, none of these experiments focused on the key prediction, that the exclusion of host-specific predators would cause a decrease in diversity as tree species with greater establishment or competitive ability formed low-diversity seedling and sapling communities where dominance was concentrated in a few species. Connell (1971) did propose such an experiment: "if all enemies of trees were removed from an entire forest, each species would probably form small groves and the more rapidly growing species would gradually spread over the habitat . . . . The final result would be a lower pattern diversity and as a consequence fewer species in any local area of forest." We suggest that new studies should now be designed to test the diversity prediction and thus build upon species-specific studies that have demonstrated patterns consistent with the Janzen–Connell model.

There have been a small number of community-level evaluations of Janzen–Connell effects in the tropics (Connell *et al.* 1984, Condit *et al.* 1992, Harms *et al.* 2000, Hubbell *et al.* 2001, Peters 2003, Wills *et al.* 2006). However, these studies did not *directly* test the diversity prediction, were not experimental, and did not determine the causes (e.g., pest pressure versus intraspecific competition) for patterns found to be consistent with Janzen–Connell effects. As Wright (2002) pointed out, "field measurements only demonstrate that niche differences, Janzen–Connell effects, and negative density dependence occur. Implications for species coexistence and plant diversity remain conjectural."

Is the Janzen–Connell hypothesis a special case of keystone predation?

We suggest that the Janzen–Connell hypothesis is a type of, or special case of, keystone predation. To some degree Janzen acknowledged this in his original paper (Janzen 1970, pp. 502, 522). Janzen wondered how you pack so many species in a tropical forest. His answer was that his research was an extension of Paine's (1966) findings that "local animal species diversity is related to the number of predators in the system and their efficiency in preventing single species from monopolizing" space or resources. Thus, we suggest the Janzen–Connell hypothesis can be viewed in this very general context (see also Connell 1971, p. 307). Under the keystone species concept, natural enemies limit the abundance of superior competitors that would otherwise displace subordinate species, thereby enhancing alpha-diversity. Thus, the suppression of superior competitors always has the capacity to maintain diversity.

The number of keystone predator species influencing the abundance of potentially dominant prey species (woody species that are superior competitors) does not change the nature of keystone predation though we acknowledge that classic ideas about keystone predation did not focus on rare species advantage. Regardless, what remains unresolved is: How many plant species would increase in abundance and depress diversity if their natural enemies were eliminated?

Complex trade-offs underlie the Janzen–Connell hypothesis: to what degree are tolerance to pest pressure and tolerance to low light correlated?

Implicit in the Janzen–Connell hypothesis is a trade-off between establishment or competitive ability and vulnerability to seed and seedling predation. Janzen (1970, p. 512) pointed out this trade-off (Janzen 1970, pp. 509, 516, 521) where large-seeded species are typically more vulnerable to seed predators or less likely to be produced in sufficient quantity to satiate predators but have a greater likelihood of establishing relative to small-seeded species, particularly in deeply shaded microsites. Janzen (1970) concluded that "a tree may persist in the face of very heavy predation if the occasional surviving seedling is a very superior competitor, and a tree with very light predation may be a very poor competitor yet survive by repeated trials at establishment." Connell (1971) suggested that the trade-off was between vulnerability to predation and rapid growth. Regardless, both suggested that trade-offs likely play a central role in how Janzen–Connell effects operate in tropical forests, an idea that has been recognized by others (Wright 2002, Leigh *et al.* 2004).

In both temperate and tropical forests, there appears to be a continuum of species from pioneers that have rapidly growing saplings in high light to mature forest species that have saplings that persist for years in the shaded understory (e.g., Wright 2002, Pacala *et al.* 2003, Wright *et al.* 2003, Leigh *et al.* 2004). Typically, species that are classified as shade tolerant have seedlings and saplings with a suite of correlated traits (establishment within shaded understories, dense wood, well-defended leaves, low photosynthetic capacity, slower growth, higher survivorship, and low responsiveness to increased light), which confer an advantage both in the shade and under conditions of prolonged exposure to pathogens and herbivores in the understory (Coley 1983, Coley *et al.* 1985, Sagers and Coley 1995, Kursar & Coley 1999, Wright 2002, Leigh *et al.* 2004). It remains unclear the *degree* to which traits that confer high survivorship under low light conditions are positively correlated with traits that confer high survivorship under prolonged pest pressure. In the literature, shade tolerance has come to mean the ability to survive in the understory for long periods of time at relatively small stature. This trait or strategy could be due to varying combinations of the ability to survive at low light and the ability to survive (tolerate, defend, or avoid) prolonged periods of browsing or herbivore damage prior to reaching a size refuge (e.g., from browsers) or reaching the canopy. In the extreme case, a species may be able to persist in the shaded understory only in years or locations where pest pressure is extremely low; thus juveniles might only rarely be encountered. This might lead to the erroneous conclusion that the species is shade intolerant.

Here we provide two examples from temperate deciduous forests in North America. Eastern hemlock (*Tsuga canadensis*) and American beech (*Fagus grandifolia*) can survive prolonged periods under deep shade. Beech is also highly browse tolerant while hemlock is not. Consequently, when browsers are abundant, hemlock may fail to regenerate and is restricted to refugia, whereas beech becomes extremely abundant (Horsley *et al.* 2003, Banta *et al.* 2005). We suggest that the degree of shade tolerance for any species will vary and likely decline with an increase in herbivore damage (Long *et al.* 2007).

For instance, in the presence of browsers in the understory of a temperate forest, we found that saplings of sugar maple (*Acer saccharum*), a putatively highly shade-tolerant species, had patterns of growth and mortality similar to saplings of black cherry (*Prunus serotina*), a shade-intolerant species (Long *et al.* 2007). In the absence of browsers, sugar maple had patterns of growth and mortality consistent with its classification as highly shade tolerant. These findings emphasize the need to critically evaluate the relationship between tolerance to low light and tolerance to herbivore damage and that these attributes will not always be highly correlated among coexisting species. Thus we propose that there may be another important trade-off among some coexisting species in forest understories, namely allocation to physiological and morphological traits that confer survivorship at low light versus traits that confer survivorship under varying degrees of pest pressure. There are hundreds of shade-tolerant species in tropical forests and they will vary in the degree to which they are tolerant to herbivores – and as herbivore damage increases, the degree of their shade tolerance relative to each other may change substantially. Testing for the existence of this trade-off or rigorously evaluating the relationship between low light survival and tolerance to damage by enemies will require studies of growth and survivorship along a continuum of light levels and simultaneously a continuum of herbivore or pathogen damage. If hierarchies of shade tolerance among species shift as pest damage increases, then models of forest dynamics will make different predictions of future canopy composition depending upon pest pressure (Royo and Carson 2006, Long *et al.* 2007).

The impact of enemies on aggregated adults and of outbreaks has been neglected

The main focus of tests of the Janzen–Connell model is how enemies create repelled patterns of juvenile recruitment around adults due to density- and distance-dependent predation. Studies have paid far less attention to the effect of natural enemies on adult plants; however, adults in dense aggregations could, in addition to their juveniles, be vulnerable to higher per capita rates of pest attack and damage. In a classic paper, Root (1973) formalized this concept and proposed the resource concentration hypothesis: "herbivores are more likely to find and remain on hosts that are growing in dense or nearly pure stands." Much evidence supports this hypothesis for plant populations (e.g., Andow 1991), yet its importance for the general maintenance of diversity in plant communities is underappreciated. Thus, insect herbivores and pathogens may act as keystone species by reducing the vigor, abundance, and fecundity of aggregated stands of adult conspecifics, thereby increasing the diversity of coexisting species (Carson and Root 2000, Long *et al.* 2003, Carson *et al.* 2004). In many instances these taxa are some of the most abundant woody species in the community (Carson *et al.* 2004). Thus, aggregations of conspecifics are in double jeopardy because both their adults and their juveniles are likely to be more vulnerable to enemies and suffer greater per capita rates of attack or damage.

In tropical forests far less attention has been given to how aggregations of adults make them more vulnerable to enemies. Janzen (1970) recognized that pre-dispersal seed predation would have a strong effect on the "intensity and patterns of seed shadows cast by parent trees" and considered this a key part of his hypothesis. Indeed, Janzen (1970) presented a graphical model whereby intense pre-dispersal seed predation would lead to peaks in seedling abundance that would be smaller and *closer* to the parent tree compared with light pre-dispersal seed predation. Thus, enemies are likely to have a greater impact on both juveniles and adults of any given plant species whenever adults are aggregated. Furthermore, these aggregations may lead to periods when insects become abundant or to episodic outbreaks of specialist enemies that function as keystone species that defoliate and sometimes kill adults over large areas, thereby increasing plant species diversity in the habitat (Carson and Root 2000, Carson *et al.* 2004).

In the tropics these outbreaks or periods of high insect abundance may be frequent when viewed from a phytocentric perspective. An outbreak of a specialist that occurs only once every 50 years means that it occurs multiple times in the life of a long-lived tree species. There are now a number of examples from tropical forests where outbreaks of specialists defoliate common or abundant woody species (Wolda and Foster 1978, Janzen 1981, Wong *et al.* 1990, Torres 1992, Nascimento and Proctor 1994). In Indonesian forests, Nair (2000, 2001) concluded that outbreaks often occurred

when tree species grew in aggregation and that low-diversity stands precipitated such outbreaks. It is likely that outbreaks are more common than previously thought because they are spatially and temporally patchy and occur high in the canopy (Wolda and Foster 1978, Lowman 1987). In addition, if relatively infrequent periods of high herbivore abundance regulate populations, then short-term studies of insect abundance and damage or incidence of specialists versus generalists will have little relevance for the Janzen–Connell hypothesis. Research is needed that links how insect abundance and damage on adults varies with host abundance over long time periods, and how this damage affects lifetime fecundity and juvenile mortality. The fact that insects and insect outbreaks have a strong top-down impact on plant communities is well recognized outside of the tropics (but see Janzen 1981 for tropical examples). Although conventional wisdom suggests otherwise, there is little solid empirical evidence at the appropriate temporal and spatial scales to suggest that outbreaks are either less common or less profound in their impact in tropical forests relative to temperate forests.

There are few studies of Janzen–Connell effects both among and within latitudes

Janzen–Connell effects may be stronger in tropical than temperate forests because of the higher abundances of natural enemies, a greater degree of specialization in aseasonal tropical habitats, and greater rates of damage even when leaves are better defended (Coley and Barone 1996, Novotny *et al.* 2002, Dyer 2007). Both Janzen (1970) and Connell (1971) proposed that these differences may help explain the latitudinal gradient in species richness. Nonetheless, very little effort has been expended in comparing Janzen–Connell effects across latitude. We found that the great majority of studies that have investigated the Janzen–Connell hypothesis have been conducted in tropical habitats (81% of 53 studies; Table 13.1). The few temperate studies that exist, however, have found evidence for density- and distance-dependent mortality in temperate forests.

For example, HilleRisLambers *et al.* (2002) concluded that density-dependent mortality is just as prevalent in temperate as in tropical forests, although they acknowledged that the strength of Janzen–Connell effects could be stronger in the tropics. In addition, the logic that implicates the Janzen–Connell hypothesis in the latitudinal gradient in species richness should also apply to habitats with substantively different levels of diversity within the same general latitude (Dirzo and Boege Chapter 5, this volume). Tropical forests with prolonged flooded or dry seasons generally have depressed diversity relative to aseasonal tropical forests (Ferreira 2000, Fajardo *et al.* 2005). This seasonality could also decrease the abundance and impact of natural enemies, therefore decreasing the strength of Janzen–Connell effects (Dirzo and Boege Chapter 5). Nevertheless, very few studies have investigated the Janzen–Connell hypothesis in abiotically stressful tropical habitats (but see John *et al.* 2002), and none have attempted to compare results from abiotically stressful and abiotically more benign forests. Overall, studies are required that assess the occurrence and strength of Janzen–Connell effects and their consequences both among and within latitudes or wherever there are steep stress gradients and sharply contrasting patterns of species diversity (e.g., length of dry season).

The Janzen–Connell hypothesis is difficult to test and falsify and may be most prevalent for uncommon or rare species

As Janzen (1970) pointed out (p. 521) heavy predation may keep some species rare and widely spaced and these species may also be the best competitors. If so, the species most likely to form dense aggregations and reduce diversity in local stands may be the ones least likely to be studied by ecologists either because they are so uncommon or it is reasoned that such rare species are unlikely to be regulated by density dependence (Wright 2002). This means that if only a fraction of woody species in diverse tropical forest are actually kept in check by their predators, the Janzen–Connell model still

holds if these species are the superior competitors in the community or can establish and grow rapidly throughout the habitat, or both. Therefore the failure to find Janzen–Connell effects for what could be hundreds of tree species does not reject the hypothesis (though it does reduce the importance of the hypothesis for explaining coexistence of all species in the community). For example, Condit *et al.* (1992) studied patterns of recruitment around reproductive adults of 80 species. They found repelled recruitment syndromes for just 15 species and concluded that Janzen–Connell effects occurred over short distances and for few species (note that recent work suggests they underestimated Janzen–Connell effects; see Leigh 1999). If, however, only a few of these tree species were excellent competitors and pests actually caused these patterns, then these pests would be keystone species and Janzen–Connell would be operating. Hyatt *et al.* (2003) did not recognize this possibility when they concluded that there was "no general support for the distance dependent prediction of the hypothesis and ... further testing to explore this hypothesis [Janzen–Connell] as a diversity-maintaining mechanism is unnecessary." Yet Hyatt *et al.* (2003) did find that there were "individual cases of conformity to the hypothesis," which is all that is needed for the hypothesis to work if the specific cases represent tree species that are excellent competitors, highly shade tolerant, or habitat generalists, or some combination of these traits (e.g., Silman *et al.* 2003). Testing whether repelled recruitment syndromes are strong for uncommon tree species, particularly shade-tolerant species with large seeds that are putatively attractive to seed predators or enemies, may give insight into this problem of testing the Janzen–Connell hypothesis. Additionally, exclusion experiments (using exclosures or insecticides) nested beneath these trees may lead to dense and depauperate stands of juvenile conspecifics in less than a decade.

As discussed above, the Janzen–Connell model is at its core a community-level model where the key prediction is that predation ultimately leads indirectly to the maintenance of high woody species diversity in tropical forests. To identify predation as the indirect cause of woody species diversity will require long-term experiments that preclude other explanations originally pointed out by Connell (1971, e.g., allelopathy and intraspecific competition; see also Wright 2002). Additionally, Connell (1971) argued that predation was more critical during the seedling and early sapling stage and not during the seed stage (but see Harms *et al.* 2000). Because there is little evidence for competition among seedlings of tropical trees even at high densities (Paine *et al.* in review), experiments may need to be conducted for many years up to and through the sapling stage as large and dense understory layers begin to thin (Chazdon Chapter 23, this volume). Furthermore, a Janzen–Connell effect may be driven by periodic outbreaks of specialist insects in the understory or overstory (Wong *et al.* 1990, Carson *et al.* 2004). If these outbreaks drive the Janzen–Connell effect, then studies will need to run through a typical outbreak cycle, which for the vast majority of species will almost certainly be lengthy.

There are additional serious logistical and conceptual challenges that impede testing the Janzen–Connell hypothesis. These difficulties include: (1) identifying the key enemies to exclude in experimental tests; (2) the ability to remove or substantively reduce entire or even partial enemy trophic levels for long periods of time (pathogens are a particular hurdle); (3) directly linking distance and/or density effects to their putative causes (i.e., enemies); and (4) demonstrating the entire chain of events necessary, specifically that enemies responding in a distance- or density-dependent fashion reduce the abundance of putative superior competitors and thereby cause an increase in alpha-diversity.

Overall, rejecting the Janzen–Connell model is extremely difficult. The key experiments testing the prediction of higher diversity remain to be done and will require large-scale, logistically difficult, long-term studies. An important but more modest step for testing Janzen–Connell is to link repelled recruitment syndromes to life-history strategies or traits that lead to high survivorship in low light in the absence of enemies. Clearly, theoretical approaches that incorporate results from field studies will also have to play a major role in evaluating the relative role of various diversity promoting mechanisms including the Janzen–Connell hypothesis.

For the Janzen–Connell hypothesis, if only a relatively small number of tree species are kept in check by their enemies (5–20%), then this may explain why these tree species do not spread to exclude others, but it still may not explain the coexistence of many additional species in these species-rich communities (see, e.g., Hubbell 1980, Penfold and Lamb 1999). Thus the Janzen–Connell hypothesis would be a necessary but insufficient explanation of tree diversity. Still, Janzen–Connell may readily promote coexistence in combination with other processes necessary to explain hyper-diverse communities (e.g., Wright 2002, Barot and Gignoux 2004, Leigh *et al.* 2004). We agree completely with Barot and Gignoux (2004) who concluded the critical task is to "assess the respective influence of each mechanism [of coexistence] for different communities … and determine the main processes that shape their biodiversity."

ACKNOWLEDGMENTS

We thank Tim Nuttle and Liza Comita for comments on earlier drafts. This work was supported by NSF grants to Walter Carson.

REFERENCES

Alvarez-Buylla, E.R. (1994) Density dependence and patch dynamics in tropical rain forests: matrix models and applications to a tree species. *American Naturalist* 143, 155–191.

Andow, D.A. (1991) Vegetational diversity and arthropod population response. *Annual Review of Entomology* 36, 561–586.

Augspurger, C.K. (1983) Seed dispersal of the tropical tree, *Platypodium elegans*, and the escape of its seedlings from fungal pathogens. *Journal of Ecology* 71, 759–771.

Augspurger, C.K. (1984) Seedling survival of tropical tree species – interactions of dispersal distance, light-gaps, and pathogens. *Ecology* 65, 1705–1712.

Augspurger, C.K. and Kelly, C.K. (1984) Pathogen mortality of tropical tree seedlings – experimental studies of the effects of dispersal distance, seedling density, and light conditions. *Oecologia* 61, 211–217.

Augspurger, C.K. and Kitajima, K. (1992) Experimental studies of seedling recruitment from contrasting seed distributions. *Ecology* 73, 1270–1284.

Banta, J.A., Royo, A.A., Kirschbaum, C. *et al.* (2005) Plant communities growing on boulders in the Allegheny National Forest: evidence for boulders as refugia from deer and as a bioassay of overbrowsing. *Natural Areas Journal* 25, 10–18.

Barot, S. and Gignoux, J. (2004) Mechanisms promoting plant coexistence: can all the proposed processes be reconciled. *Oikos* 106, 185–192.

Bell, T., Freckleton, R.P., and Lewis, O.T. (2006) Plant pathogens drive density-dependent seedling mortality in a tropical tree. *Ecology Letters* 9, 569–574.

Blundell, A.G. and Peart, D.R. (1998) Distance-dependence in herbivory and foliar condition for juvenile *Shorea* trees in Bornean dipterocarp rain forest. *Oecologia* 117, 151–160.

Brewer, S.W. and Webb, M.A.H. (2001) Ignorant seed predators and factors affecting the seed survival of a tropical palm. *Oikos* 93, 32–41.

Burkey, T.V. (1994) Tropical tree species-diversity – a test of the Janzen–Connell model. *Oecologia* 97, 533–540.

Carson, W.P., Cronin, J.P., and Long, Z.T. (2004) A general rule for predicting when insects will have strong top-down effects on plant communities: on the relationship between insect outbreaks and host concentration. In W.W. Weisser and E. Siemann (eds), *Insects and Ecosystem Function*. Springer-Verlag, Berlin, pp. 193–211.

Carson, W.P. and Root, R.B. (2000) Herbivory and plant species coexistence: community regulation by an outbreaking phytophagous insect. *Ecological Monographs* 70, 73–99.

Cintra, R. (1997) A test of the Janzen–Connell model with two common tree species in Amazonian forest. *Journal of Tropical Ecology* 13, 641–658.

Clark, D.A. and Clark, D.B. (1984) Spacing dynamics of a tropical rain-forest tree – evaluation of the Janzen–Connell model. *American Naturalist* 124, 769–788.

Coley, P.D. (1983) Herbivory and defensive characteristics of tree species in a lowland tropical forest. *Ecological Monographs* 53, 209–233.

Coley, P.D. and Barone, J.A. (1996) Herbivory and plant defenses in tropical forests. *Annual Review of Ecology and Systematics* 27, 305–335.

Coley, P.D., Bryant, J.P., and Chapin III, F.S. (1985) Resource availability and plant antiherbivore defense. *Science* 230, 895–899.

Condit, R., Hubbell, S.P., and Foster, R.B. (1992) Recruitment near conspecific adults and the maintenance of tree and shrub diversity in a Neotropical forest. *American Naturalist* 140, 261–286.

Condit, R., Hubbell, S.P., and Foster, R.B. (1994) Density-dependence in two understory tree species in a Neotropical forest. *Ecology* 75, 671–680.

Connell, J.H. (1971) On the role of natural enemies in preventing competitive exclusion in some marine mammals and in rain forest trees. In P.J. Boer and G. Gradwell (eds), *Dynamics of Populations*. PUDOC, The Netherlands, pp. 298–310.

Connell, J.H., Tracey, J.G., and Webb, L.J. (1984) Compensatory recruitment, growth, and mortality as factors maintaining rain forest tree diversity. *Ecological Monographs* 54, 141–164.

De Steven, D. and Putz, F. (1984) Impact of mammals on early recruitment of a tropical canopy tree, *Dipteryx panamensis*, in Panama. *Oikos* 43, 207–216.

De Steven, D. and Wright, S.J. (2002) Consequences of variable reproduction for seedling recruitment in three neotropical tree species. *Ecology* 83, 2315–2327.

Dyer, L.A., Singer, M.S., Lill, J.T. *et al.* (2007) Host specificity of Lepidoptera in tropical and temperate forests. *Nature* 448, 696–699.

Fajardo, L., Gonzalez, V., and Nassar, J.M. (2005) Tropical dry forests of Venezuela: characterization and current conservation status. *Biotropica* 37, 531–546.

Ferreira, L.V. (1997) Effects of the duration of flooding on species richness and floristic composition in three hectares in the Jau National Park in floodplain forests in central Amazonia. *Biodiversity and Conservation* 6, 1353–1363.

Ferreira, L.V. (2000) Effects of flooding duration on species richness, floristic composition and forest structure in river margin habitat in Amazonian blackwater floodplain forests: implications for future design of protected areas. *Biodiversity and Conservation* 9, 1–14.

Fragoso, J.M.V. (1997) Tapir-generated seed shadows: scale-dependent patchiness in the Amazon rain forest. *Journal of Ecology* 85, 519–529.

Fragoso, J.M.V., Silvius, K.M., and Correa, J.A. (2003) Long-distance seed dispersal by tapirs increases seed survival and aggregates tropical trees. *Ecology* 84, 1998–2006.

Gilbert, G.S., Harms, K.E., Hamill, D.N. *et al.* (2001) Effects of seedling size, El Nino drought, seedling density, and distance to nearest conspecific adult on 6-year survival of *Ocotea whitei* seedlings in Panama. *Oecologia* 127, 509–516.

Gilbert, G.S., Hubbell, S.P., and Foster, R.B. (1994) Density and distance-to-adult effects of a canker disease of trees in a moist tropical forest. *Oecologia* 98, 100–108.

Gillett, J.B. (1962) Pest pressure, an underestimated factor in evolution. *Systematics Association Publication* 4, 37–46.

Hammond, D.S., Brown, V.K., and Zagt, R. (1999) Spatial and temporal patterns of seed attack and germination. *Oecologia* 119, 208–218.

Harms, K.E., Wright, S.J., Calderon, O. *et al.* (2000) Pervasive density-dependent recruitment enhances seedling diversity in a tropical forest. *Nature* 404, 493–495.

HilleRisLambers, J., Clark, J.S., and Beckage, B. (2002) Density-dependent mortality and the latitudinal gradient in species diversity. *Nature* 417, 732–735.

Hood, L.A., Swaine, M.D., and Mason, P.A. (2004) The influence of spatial patterns of damping-off disease and arbuscular mycorrhizal colonization on tree seedling establishment in Ghanaian tropical forest soil. *Journal of Ecology* 92, 816–823.

Horsley, S.B., Stout, S.L., and de Calesta, D.S. (2003) White-tailed deer impact on the vegetation dynamics of a northern hardwood forest. *Ecological Applications* 13, 98–118.

Houle, G. (1998) Seed dispersal and seedling recruitment of *Betula alleghaniensis*: spatial inconsistency in time. *Ecology* 79, 807–818.

Hubbell, S.P. (1980) Seed predation and the coexistence of tree species in tropical forests. *Oikos* 35, 214–229.

Hubbell, S.P., Ahumada, J.A., Condit, R. *et al.* (2001) Local neighborhood effects on long-term survival of individual trees in a Neotropical forest. *Ecological Research* 16, 859–875.

Hubbell, S.P., Condit, R., Foster, R.B. *et al.* (1990) Presence and absence of density dependence in a Neotropical tree community [and discussion]. *Philosophical Transactions: Biological Sciences* 330, 269–281.

Hulme, P.E. (1997) Post-dispersal seed predation and the establishment of vertebrate dispersed plants in Mediterranean scrublands. *Oecologia* 111, 91–98.

Hyatt, L.A., Rosenberg, M.S., Howard, T.G. *et al.* (2003) The distance dependence prediction of the Janzen–Connell hypothesis: a meta-analysis. *Oikos* 103, 590–602.

Janzen, D. (1970) Herbivores and the number of tree species in tropical forests. *American Naturalist* 104, 501–528.

Janzen, D.H. (1981) Patterns of herbivory in a tropical deciduous forest. *Biotropica* 13, 271–282.

John, R., Dattaraja, H.S., Suresh, H.S. (2002) Density-dependence in common tree species in a tropical dry forest in Mudumalai, southern India. *Journal of Vegetation Science* 13, 45–56.

Kursar, T.A. and Coley, P.D. (1999) Contrasting modes of light acclimation in two species of the rainforest understory. *Oecologia* 121, 489–498.

Leigh, E.G. Jr (1999) *Tropical Forest Ecology: a View from Barro Colorado Island.* Oxford University Press, New York.

Leigh, E.G. Jr, Davidar, P., Dick, C.W. *et al.* (2004) Why do some tropical forests have so many species of trees? *Biotropica* 36, 447–473.

Long, Z.T., Mohler, C.L., and Carson, W.P. (2003) Extending the resource concentration hypothesis to plant communities: effects of litter and herbivores. *Ecology* 84(3), 652–665.

Long, Z.T., Pendergast, T.H., and Carson, W.P. (2007) The impact of deer on relationships between tree growth and mortality in an old-growth beech-maple forest. *Forest Ecology Management* 252, 230–238.

Lowman, M.D. (1997) Herbivory in forests – from centimeters to megameters. In A.D. Watt, N.E. Stork, and M.D. Hunter (eds), *Forests and Insects.* Chapman and Hall, New York, pp. 135–149.

MacArthur, R.H. (1972) *Geographical Ecology.* Harper & Row, New York.

Massey, F.P., Massey, K., Press, M.C. *et al.* (2006) Neighbourhood composition determines growth, architecture and herbivory in tropical rain forest tree seedlings. *Journal of Ecology* 94, 646–655.

Montesinos, D., Garcia-Fayos, P., and Mateu, I. (2006) Conflicting selective forces underlying seed dispersal in the endangered plant *Silene diclinis*. *International Journal of Plant Sciences* 167, 103–110.

Nair, K.S.S. (2000) *Insect Pests and Diseases in Indonesian Forests: An Assessment of the Major Threats, Research Efforts and Literature.* Center for International Forestry Research, Bogor, Indonesia.

Nair, K.S.S. (2001) *Pest Outbreaks in Tropical Forest Plantations: Is There a Greater Risk for Exotic Tree Species?* Center for International Forestry Research, Bogor, Indonesia.

Nascimento, M.T. and Proctor, J. (1994) Insect defoliation of a monodominant Amazonian rainforest. *Journal of Tropical Ecology* 10, 633–636.

Norghauer, J.M., Malcolm, J.R., Zimmerman, B.L. *et al.* (2006) An experimental test of density – and distant-dependent recruitment of mahogany (*Swietenia macrophylla*) in southeastern Amazonia. *Oecologia* 148, 437–446.

Novotny, V., Miller, S.E., Basset, Y. *et al.* (2002) Predictably simple: assemblages of caterpillars (Lepidoptera) feeding on rainforest trees in Papua New Guinea. *Proceeding of the Royal Society of London Series B* 269, 2377–2344.

Packer, A. and Clay, K. (2000) Soil pathogens and spatial patterns of seedling mortality in a temperate tree. *Nature* 404, 278–281.

Paine, R.T. (1966) Food web complexity and species diversity. *American Naturalist* 100, 65–75.

Paine, C.E.T., Harms, K.E., Schnitzer, S.A. et al. (2008) Weak competition among tropical tree seedlings: implications for species coexistence. *Biotropica*.

Penfold, G.C. and Lamb, D. (1999) Species co-existence in an Australian subtropical rain forest: evidence for compensatory mortality. *Journal of Ecology* 87, 316–329.

Peres, C.A., Schiesari, L.C., and DiasLeme, C.L. (1997) Vertebrate predation of Brazil-nuts (*Bertholletia excelsa*, Lecythidaceae), an agouti-dispersed Amazonian seed crop: a test of the escape hypothesis. *Journal of Tropical Ecology* 13, 69–79.

Peters, H.A. (2003) Neighbour-regulated mortality: the influence of positive and negative density dependence on tree populations in species-rich tropical forests. *Ecology Letters* 6, 757–765.

Reinhart, K.O., Royo, A.A., Van der Putten, W.H. et al. (2005) Soil feedback and pathogen activity in *Prunus serotina* throughout its native range. *Journal of Ecology* 93, 890–898.

Ridley, H.N. (1930) *The Dispersal of Plants Throughout the World*. Lovell, Reeve, Ashford, Kent.

Romo, M., Tuomisto, H., and Loiselle, B.A. (2004) On the density-dependence of seed predation in *Dipteryx micrantha*, a bat-dispersed rain forest tree. *Oecologia* 140, 76–85.

Root, R.B. (1973) Organization of a plant–arthropod association in simple and diverse habitats: the fauna of collards (*Brassica oleracea*). *Ecological Monographs* 43, 95–124.

Royo, A. and Carson, W.P. (2006) On the formation of dense understory layers in forests worldwide: consequences and implications for forest dynamics, biodiversity, and succession. *Canadian Journal of Forest Research* 36, 1345–1362.

Russo, S.E. and Augspurger, C.K. (2004) Aggregated seed dispersal by spider monkeys limits recruitment to clumped patterns in *Virola calophylla*. *Ecology Letters* 7, 1058–1067.

Sagers, C. and Coley, P.D. (1995) Benefits and costs of defense in a neotropical shrub. *Ecology* 76, 1835–1843.

Schnitzer, S.A. (2005) A mechanistic explanation for the global patterns of liana abundance and distribution. *American Naturalist* 166, 262–276.

Schupp, E.W. (1988) Seed and early seedling predation in the forest understory and in treefall gaps. *Oikos* 51, 71–78.

Schupp, E.W. (1992) The Janzen–Connell model for tropical tree diversity – population implications and the importance of spatial scale. *American Naturalist* 140, 526–530.

Shibata, M. and Nakashizuka, T. (1995) Seed and seedling demography of four co-occurring *Carpinus* species in a temperate deciduous forest. *Ecology* 76, 1099–1108.

Silman, M.R., Terborgh, J.W., and Kiltie, R.A. (2003) Population regulation of a dominant rain forest tree by a major seed predator. *Ecology* 84, 431–438.

Silva Matos, D.M., Freckleton, R.P., and Watkinson, A.R. (1999) The role of density dependence in the population dynamics of a tropical palm. *Ecology* 80, 2635–2650.

Stoll, P. and Newbery, D.M. (2005) Evidence of species-specific neighborhood effects in the Dipterocarpaceae of a Bornean rain forest. *Ecology* 86, 3048–3062.

Sullivan, J.J. (2003) Density-dependent shoot-borer herbivory increases the age of first reproduction and mortality of neotropical tree saplings. *Oecologia* 136, 96–106.

Terborgh, J., Losos, E., Riley, M.P. et al. (1993) Predation by vertebrates and invertebrates on the seeds of five canopy tree species of an Amazonian forest. *Vegetation* 108, 375–386.

Tomita, M., Hirabuki, Y., and Seiwa, K. (2002) Post-dispersal changes in the spatial distribution of *Fagus crenata* seeds. *Ecology* 83, 1560–1565.

Torres, J.A. (1992) Lepidoptera outbreaks in response to successional changes after the passage of hurricane Hugo in Puerto Rico. *Journal of Tropical Ecology* 8, 285–298.

Uriarte, M., Condit, R., Canham, C.D. et al. (2004) A spatially explicit model of sapling growth in a tropical forest: does the identity of neighbours matter? *Journal of Ecology* 92, 348–360.

Webb, C.O. and Peart, D.R. (1999) Seedling density dependence promotes coexistence of Bornean rain forest trees. *Ecology* 80, 2006–2017.

Wenny, D.G. (2000) Seed dispersal, seed predation, and seedling recruitment of a neotropical montane tree. *Ecological Monographs* 70, 331–351.

Wills, C. and Condit, R. (1999) Similar non-random processes maintain diversity in two tropical rainforests. *Proceedings of the Royal Society of London Series B-Biological Sciences* 266, 1445–1452.

Wills, C., Condit, R., Foster, R.B. et al. (1997) Strong density- and diversity-related effects help to maintain tree species diversity in a neotropical forest. *Proceedings of the National Academy of Sciences of the United States of America* 94, 1252–1257.

Wills, C., Harms, K.E., Condit, R. et al. (2006) Nonrandom processes maintain diversity in tropical forests. *Science* 311, 527–531.

Wolda, H. and Foster, R. (1978) *Zunacetha annulata* (Lepidoptera; Dioptidae), an outbreak insect in a neotropical forest. *Geo-Eco-Trop* 2, 443–454.

Wong, M., Wright, S.J., Hubbell, S.P. *et al.* (1990) The spatial pattern and reproductive consequences of outbreak defoliation in *Quararibea asterolepis*, a tropical tree. *Journal of Ecology* 78, 579–588.

Wright, S.J. (1983) The dispersion of eggs by a bruchid beetle among *Scheelea* palm seeds and the effect of distance to the parent palm. *Ecology* 64, 1016–1021.

Wright, S.J. (2002) Plant diversity in tropical forests: a review of mechanisms of species coexistence. *Oecologia* 130, 1–14.

Wright, S.J. and Duber, H.C. (2001) Poachers and forest fragmentation alter seed dispersal, seed survival, and seedling recruitment in the palm *Attalea butyraceae*, with implications for tropical tree diversity. *Biotropica* 33, 583–595.

Wright, S.J., Muller-Landau, H.C., Condit, R. *et al.* (2003) Gap-dependent recruitment, realized vital rates, and size distributions of tropical trees. *Ecology* 84, 3174–3185.

Wyatt, J.L. and Silman, M.R. (2004) Distance-dependence in two Amazonian palms: effects of spatial and temporal variation in seed predator communities. *Oecologia* 140, 26–35.

Chapter 14

SEED LIMITATION AND THE COEXISTENCE OF PIONEER TREE SPECIES

James W. Dalling and Robert John

OVERVIEW

Seed limitation, defined as the failure of seeds to arrive at sites favorable for recruitment, may be a critical force structuring plant communities. When seed limitation is strong, interspecific competition is reduced, and competitive exclusion may be slowed to the extent that diversity can be maintained through speciation and migration. In mature tropical forests, seed limitation may be especially important in determining the recruitment patterns of pioneer tree species. These species depend on dispersal to infrequent and ephemeral treefall gaps for successful seedling establishment. Despite this requirement, pioneers show wide variation in the life-history traits that affect dispersal ability. Here we use seed trap data for pioneers from the 50 ha forest dynamics plot on Barro Colorado Island (BCI), Panama, to show that seed limitation has a significant effect on seedling recruitment patterns. We then assess whether the effects of limited dispersal in space can be offset by prolonged dispersal in time through the accumulation of a persistent soil seed bank. Using a simulation model we show that variation in dispersal in space may have surprisingly little effect on overall seedling recruitment rates. This is because there is a trade-off between the number of gaps colonized and recruit density per gap. While long-term seed persistence increases the fraction of gaps colonized, it cannot fully compensate for limited dispersal in space and carries a substantial fitness cost resulting from increased generation time.

INTRODUCTION

Most mechanisms thought to contribute to the maintenance of species diversity (e.g., niche differentiation, competition, and density dependence) are predicated on the recruitment of individuals into the community. The first step in the recruitment process is the arrival of a viable seed at a site suitable for seedling establishment. The probability of successful arrival is fundamentally constrained by the resources available to a plant's reproduction, and is further influenced by a suite of adaptive compromises that determine the size and number of seeds produced, and the resources allocated to ensure seed dispersal (Muller-Landau Chapter 11, this volume). The consequence of these constraints on recruitment is "seed limitation." This has been defined as the failure of seeds to arrive at sites favorable for recruitment as a consequence of either limited seed production or limited dispersal of the seeds produced (Nathan and Muller-Landau 2000). At the population level, seed limitation potentially restricts rates of population spread and opportunities for the colonization of patches of new suitable habitat, and influences population genetic structure (Wright 1969, Hanski 2001, Muller-Landau *et al.* 2003).

At the community level, theoretical work has shown that strong seed limitation can promote species coexistence by greatly slowing competitive exclusion (Tilman 1994, Hurtt and Pacala

1995). This is because when plants are seed limited, recruitment sites are frequently won not by the strongest competitor in the community (Kitajima and Poorter Chapter 10, this volume) but by the best competitor among the restricted set of species that arrives at that site. If competitive exclusion can be slowed sufficiently then diversity might be maintained as species loss is balanced through speciation and migration into the local community (Hubbell 2001). Most evidence for seed limitation comes from analyses of seed captures in temperate forests. These studies indicate that even in stands containing high conspecific adult densities much of the soil surface fails to receive seeds of any one species. While this limitation was due primarily to limited seed dispersal, variance in the reproductive output of individual trees and temporal variation in seed production also contribute to the observed seed limitation (Clark *et al.* 1998, 1999, 2004, McEuen and Curran 2004).

Here we present evidence for seed limitation in tropical pioneer species. Pioneers face particular challenges to maintaining populations in mature forest. The traits that allow these species to achieve high growth rates in the juvenile phase also restrict their initial recruitment to light gaps (Schnitzer *et al.* Chapter 12, this volume). In most forests, these sites are formed predominantly by treefalls and landslides, and typically occur at low densities (<2% of the landscape per year; Uhl and Murphy 1981, Brokaw 1982, Lieberman *et al.* 1990). The rarity of these disturbances and the short duration for which these sites are available for colonization therefore suggests that recruitment of pioneer species could be strongly limited by their ability to disperse.

If the need to regenerate in gaps is a strong selective force shaping pioneer life histories then we might expect that traits influencing dispersal would differ between shade-tolerant and pioneer species. Although pioneers are noted for their small seed size and high fecundity (Swaine and Whitmore 1988), seed mass and reproductive output still varies over four orders of magnitude among coexisting pioneer species (Dalling *et al.* 2002, Dalling and Burslem 2005). As a consequence, pioneer species may vary in the degree of seed limitation that they experience, or may

have developed other mechanisms that offset the effects of seed limitation to increase the probability of colonization success. Two potential mechanisms are (1) non-random (directed) dispersal to gaps, and (2) long-term persistence of seeds in soil seed banks. Evidence for directed dispersal to gaps in tropical forests is limited (Wenny 2001), but may be important for wind-dispersed species (Loiselle *et al.* 1996). In contrast, seed persistence is common among pioneers (e.g., Hopkins and Graham 1987, Dalling *et al.* 1997), although the contribution these seeds make to gap colonization and net reproductive output has rarely been quantified (Murray 1988).

In this chapter we use the pioneer tree species of Barro Colorado Island (BCI), Panama, as a case study. We use seed trap data to estimate components of seed limitation, and to generate parameters for models of seed dispersal. We provide evidence that spatial variation in annual seed rain can help explain patterns of seedling recruitment for some pioneers but not for others. We explore whether long-term persistence of seeds in the soil seed bank can compensate for limited dispersal in space, and finally, we briefly review evidence for the operation of other mechanisms that can promote coexistence among these species.

ARE TROPICAL PIONEERS SEED LIMITED?

The best evidence for seed limitation (the failure of seeds to arrive at a site over time) comes from long-term studies of seed capture rates. On BCI, an ongoing study with two hundred $0.5 \, m^2$ seed traps placed in the 50 ha forest dynamics plot has shown that, on average, 88% of pioneer and shade-tolerant tree species fail to disperse a single seed *to any given trap* over 10 years, and that *no seeds at all* were collected in any of the 200 traps for more than 50 species during the same period (Hubbell *et al.* 1999, Harms *et al.* 2000). These analyses also show that all pioneer species on BCI are also seed limited based on annual trap data (Table 14.1) and would therefore fail to disperse seeds to all parts of every gap that forms each year. Seed limitation for some pioneers

Table 14.1 Estimated median dispersal distances, fecundity, and seed, source, and dispersal limitation for pioneer species varying in dispersal mode, seed mass, and abundance on the BCI 50 ha forest dynamics plot.

	Seed mass (mg)	n	Median dispersal (m)	Fecundity (seeds cm^{-2})	Seed bank (years)	Seed limitation		Source limitation		Dispersal limitation	
						1 year	2 years	1 year	2 years	1 year	2 years
Wind dispersal											
Alseis blackiana	0.1	784	3.2	907.2	<2	0.68	0.47	<0.01	<0.01	0.68	0.47
Cordia alliodora	12.5	54	7.0	18.1	0	0.90	0.86	0.41	0.16	0.82	0.83
Jacaranda copaia	4.7	193	21.2	67.3	<2	0.39	0.03	0.22	<0.01	0.25	0.03
Luehea seemannii	1.9	64	8.2	273.7	<2	0.24	0.06	<0.01	<0.01	0.26	0.06
Terminalia amazonia	3.8	20	83.1	61.3	N/A	0.29	0.10	0.10	0.01	0.21	0.09
Animal dispersal											
Alchornea costaricensis	38.5	135	1.3	33.5	N/A	0.96	0.93	0.01	<0.01	0.96	0.93
Apeiba aspera	14.2	141	3.2	7.8	>2	0.96	0.94	0.19	0.04	0.95	0.94
Cecropia insignis	0.5	182	0.8	225.6	<2	0.93	0.89	0.23	0.17	0.77	0.73
Palicourea guianensis	14.3	1055	5.6	50.5	N/A	0.95	0.91	0.74	0.55	0.52	0.67
Zanthoxylum spp.	11–36	108	0.8	22.7	>30	0.92	0.86	0.50	0.25	0.70	0.81
Ballistic dispersal											
Croton billbergianus	24.0	367	2.2	2.6	>40	0.99	0.99	0.94	0.89	0.41	0.61

Notes: For each species, *n* is the number of reproductive-sized individuals in the plot. Seed, source, and dispersal limitation are defined in the text, and are mean captures to two hundred 0.5 m^2 traps calculated over 12 years. Mean values for limitation metrics are presented for 1 or 2 consecutive years. Seed bank data provide the longest reported period of seed persistence in the soil. Data are presented for 11 of 24 pioneer taxa commonly encountered in treefall gaps on BCI (Dalling *et al.* 1998), and for which sufficient seed captures to traps allow dispersal parameters to be calculated. Data from Dalling *et al.* (1997, 2002) and Dalling and Brown (unpublished data). N/A = data not available.

remains remarkably high over much longer periods, suggesting their seed rain reaches only a small fraction of new gap sites (Dalling *et al.* 2002).

Seed limitation can arise because an insufficient number of seeds are produced, defined as "source limitation," or because seeds are nonrandomly dispersed across the landscape, defined as "dispersal limitation" (Clark *et al.* 1998). When seed trap data are available, the degree to which a tree population is source limited can be evaluated by randomly "redistributing" the total seed count among all the traps used in the study. Source limitation is then defined as the proportion of traps that are still expected to fail to capture a single seed (Table 14.1). Differences in source limitation among species are the consequence of variation in adult population density, adult size at reproduction, and mean seed mass. Once source limitation has been calculated, dispersal limitation can be determined as the measure of how the proportion of traps receiving seeds is further reduced above and beyond constraints due to source limitation. Dispersal limitation is calculated as 1 − (proportion of traps receiving seeds)/(1 − source limitation). Dispersal limitation can be expected to be high for species with high seed production and short dispersal distances or with highly clumped dispersal.

Among the BCI pioneers, three species (*Alchornea*, *Alseis*, and *Luehea*) effectively escape source limitation in a given year, with sufficient seed production at the population level that seeds could theoretically reach ≥99% of sites (Table 14.1). In contrast, none of the species escape dispersal limitation and thus seed limitation, such that seeds of even the best-dispersed species, *Luehea*, reached only 76% of traps. Differences between species with wind-versus animal-vectored seed dispersal are clear. While seeds of the five wind-dispersed species reached between 10 and 76% of traps, seeds of animal-dispersed species only managed to reach from 4 to 8% of traps. This difference reflects the more aggregated pattern of animal seed dispersal in which seeds are often defecated together in clumps at dining roosts, sleeping roosts, and latrine sites (e.g., Schupp *et al.* 2002, Wehncke *et al.* 2003). The species exhibiting the strongest seed limitation, however, was *Croton*

billbergianus, a subcanopy tree with ballistic dispersal (Table 14.1). This is one of the most abundant pioneers on BCI, and illustrates how pioneers can apparently recruit successfully despite extreme seed and source limitation.

DO PIONEER RECRUITMENT PATTERNS REFLECT SEED LIMITATION?

Measures of seed limitation, based on captures of single seeds to traps, represent minimum dispersal rates from which recruitment could theoretically occur. However, probabilities of seed survival to germination, and of seedling survival to emergence and establishment, can be very low, even when conditions for recruitment are favorable (Harms *et al.* 2000). Furthermore, seedling emergence and establishment probabilities are strongly seed-size dependent (Dalling and Hubbell 2002), and are affected by leaf litter density and other microsite conditions within gaps (Brandani *et al.* 1988, Vázquez-Yanes *et al.* 1990, Molofsky and Augspurger 1992). Initially high seed densities on the soil surface may also be greatly reduced by a variety of animals (Levings and Franks 1982, Kaspari 1993, Carson *et al.* Chapter 13, this volume), while fungal pathogens may prevent seeds accumulating in the soil (Alvarez-Buylla and Martínez-Ramos 1990, Dalling *et al.* 1998a). As a consequence, seedling recruitment may be largely uncoupled from seed abundance, or at least reflect an interaction between seed abundance and substrate favorability, as has been found in a north temperate forest community (LePage *et al.* 2000).

To determine the relationship between seed abundance and seedling recruitment, Dalling *et al.* (2002) compared predicted seed rain densities to observed seedling recruitment patterns in natural tree fall gaps. Seed rain densities in gaps were predicted using data on seed captures to traps in conjunction with information on the size and location of potential seed sources to parameterize a seed dispersal model (for more information on this approach see Ribbens *et al.* 1994, Clark *et al.* 1998, Nathan and Muller-Landau 2000). We have confidence in the seed rain predictions for wind and ballistically dispersed species, as fits of predicted against actual seed counts in traps were good ($r^2 = 0.49$–0.87; $n = 6$ species). However, predictions for animal dispersed species may be poor because model fits were weak for these species ($r^2 = 0.11$–0.44; $n = 7$ species; Dalling *et al.* 2002). The low predictive power of dispersal models for animal-dispersed species using seed trap data is consistent with observations that large birds and mammals frequently carry seeds several hundred meters and that seeds are often secondarily dispersed from initial aggregations (e.g. Clark *et al.* 1998, Wehncke *et al.* 2003).

Next, Dalling *et al.* (2002) compared the ability of models with and without parameters for estimated seed rain to predict observed seedling abundance in the gaps. In the first (null) model, the number of seedling recruits per species in a gap was assumed to be proportional to the area potentially colonizable to seedlings. The expected seedling number per gap in this model was calculated by dividing the total seedling number per species in all gaps by the total area of all gaps. In subsequent models, seedling abundance was fitted as either a linear or non-linear (i.e., density-dependent) function of the predicted seed rain to the gap. Models were compared using the Akaike information criterion (for more details see Dalling *et al.* 2002).

Comparison of the models showed that the abundance of seed rain did affect the probability of seedling recruitment, at least for some pioneers. Overall, models incorporating seed rain improved predictions of seedling recruitment over the null model for eight of 14 pioneer species. Variation in how well recruitment models fit the seedling abundance data in part reflected the fit of dispersal functions, but also reflected the commonness of adult trees in the plot, and the proximity of seed sources to gaps (Figure 14.1). Large *Jacaranda* trees are common in the plot, and most gaps contain at least a few seedlings of this species. In contrast, the fit for a rarer species, *Cordia*, reflects the presence of a single gap with high estimated seed rain. For *Croton*, another common pioneer species, the recruitment model fit surprisingly poorly despite a high confidence in the dispersal function. *Croton* seeds are ballistically dispersed, land close to the plant, and may be secondarily dispersed a few meters more by ants. For three gaps that lacked

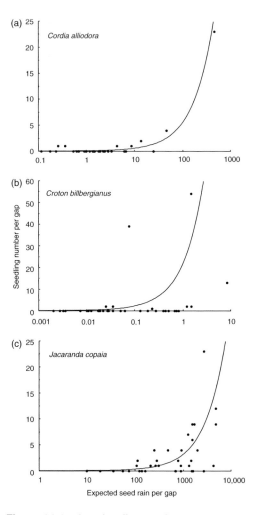

Figure 14.1 Plots of seedling number per gap against the expected seed rain to each gap (log scale) for three pioneer species, *Cordia alliodora*, *Croton billbergianus*, and *Jacaranda copaia*. Seedling data are from complete censuses of 36 treefall gaps on the BCI 50 ha plot (Dalling *et al.* 1998b). Seed rain to gap was estimated using a seed dispersal model (Dalling *et al.* 2002). The fitted curve represents the density-independent expectation for seedling number per gap, where seedling number is proportional to the expected seed rain × the seed–seedling transition probability (calculated from all 36 gaps combined). Redrawn from Dalling *et al.* (2002).

adult *Croton* trees within 30 m, seedling recruit abundance was an order of magnitude higher than estimated annual seed rain (Figure 14.1). For this species, we suspect that seedling recruitment may reflect many years of accumulation of viable seeds in the seed bank.

CAN LONG-TERM SEED PERSISTENCE COMPENSATE FOR LIMITED DISPERSAL IN SPACE?

In addition to annual seed rain, recruitment patterns may also reflect contributions from seeds that persist for many years in the soil. Although most pioneers are known to form soil seed banks (e.g., Guevara Sada and Gómez-Pompa 1972, Putz and Appanah 1987), little is known about the time scale of seed persistence for tropical pioneers, or the relative contributions of buried seeds versus seed rain for recruitment in gaps. Direct studies of seed persistence in the soil have used mesh bags to bury seeds several centimeters below the soil surface. These studies show that the majority of pioneer species retain some seed viability over 2 years (e.g., Perez-Nasser and Vázquez-Yanes 1986, Hopkins and Graham 1987, Dalling *et al.* 1997). Comparisons of annual seed rain inputs with soil seed bank densities in Costa Rican cloud forest also suggests that seed persistence over 5 years or more is common (Murray and Garcia 2002). On BCI, direct measurements of seed age using ^{14}C dating of seeds sieved from the soil have shown that viable seeds of three of the larger-seeded pioneer species with thick seed coats (*Trema micrantha*, *Zanthoxylum eckmannii*, and *Croton billbergianus*) buried at depths of less than 3 cm below the soil surface can be more than 30 years old (Dalling and Brown unpublished data).

In situ studies of seed persistence in the soil, however, may overestimate the contribution of the seed bank to seedling recruitment. This is because probabilities of successfully "entering" and "leaving" the seed bank are quite low (Williams-Linera 1990, Kennedy and Swaine 1992, Dalling and Hubbell 2002). Seeds dispersed onto the soil surface are especially susceptible to seed predation. Rates of seed removal by ants (and rodents for larger seeds) are very high in lowland tropical forests (e.g., Horvitz and Schemske 1986, Alvarez-Buylla and Martínez-Ramos 1990, Kaspari 1993, Fornara and Dalling 2005), with most seeds likely to be consumed (Levey and Byrne 1993). These high initial predation rates may explain the large discrepancy between estimated seed rain and soil seed bank densities. For two small-seeded pioneer species on BCI, *Cecropia insignis* and *Miconia argentea*, only 2% and 23% respectively of annual seed rain became incorporated into the seed bank (Dalling *et al.* 1998a).

Evidently, a direct evaluation of the contribution of persistent seeds to pioneer recruitment success would be difficult because this would require long-term data on seed survivorship and fate. We therefore built a spatially explicit simulation model to examine the potential consequences of seed persistence for recruitment success, and the interactions between persistence and other life-history traits. The model allows us to simulate seed dispersal, gap formation, and recruitment for a 1000 m × 500 m area. We used the model to explore the impact on recruitment success and population growth rate of species-specific parameters for fecundity, dispersal, seed persistence, and germination rates in gaps. Although seed burial experiments and ^{14}C dating studies have provided estimates of seed longevity, the exact survivorship curves for buried seeds are not adequately known. In the simulations described here we derived hypothetical seed survivorship curves whose functional form was based on a model used to describe the loss of viability of seeds stored in constant conditions (Ellis and Roberts 1980, see also Lonsdale 1988). This model assumes that seed mortality is normally distributed in time, and yields a type I survivorship curve, in which survivorship rate decreases with seed age. Here we use the model to explore how traits for dispersal and persistence interact to affect recruitment by comparing three hypothetical species with contrasting dispersal characteristics and by varying seed persistence from less than 1 year to about 20 years. Adult densities, fecundities, adult mortality rates, probabilities of incorporation into seed banks, and rates of germination in gaps also affect recruitment rates and were therefore kept constant in these simulations.

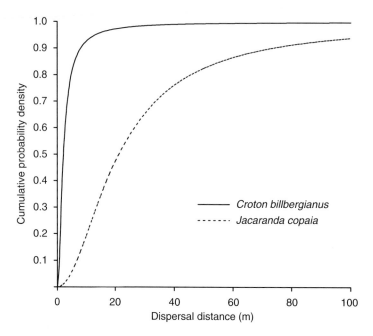

Figure 14.2 Cumulative probability density curves for the two-parameter 2Dt dispersal function for *Croton billbergianus* ($p = 0.73$, $u = 2.9$) and *Jacaranda copaia* ($p = 0.81$, $u = 328.8$). In *Croton* 99.9% of the seeds fall within 194 m of the parent tree, while that distance for *Jacaranda* is over 1000 m. Dispersal parameter estimates from Dalling *et al.* (2002).

Dispersal functions were chosen to match those of two pioneer species on BCI – *Croton billbergianus* (Euphorbiaceae), with ballistically dispersed seeds and an aggregated dispersal kernel (median dispersal distance = 2.2 m), and *Jacaranda copaia* (Bignoniaceae), with wind-dispersed seeds and relatively widespread dispersal (median = 21.2 m) (Figure 14.2). To investigate gap colonization under these contrasting dispersal scenarios, we simulated gap formation and closure based on empirical data from the BCI 50 ha plot, maintaining about 5% of the forest area under gaps at all times. Gap sizes varied from 25 m² through 625 m², with the size distribution of gaps declining as a power law of gap size (Hubbell *et al.* 1999, Schnitzer *et al.* 2000).

We found that gap colonization rates were substantially lower under aggregated dispersal (such as in *Croton*), compared with widespread dispersal (such as in *Jacaranda*). Under widespread dispersal, increasing seed persistence resulted in

a rapid increase in gap colonization, but it tended to reach an asymptote at longer seed persistence. Under highly aggregated dispersal as in *Croton*, the functional form of the relationship between seed persistence and gap colonization success was similar to that of *Jacaranda* but the initial increase was shallower and it did not saturate over the time scale of our simulations (Figure 14.3). Gap colonization success was, however, far lower under aggregated dispersal and seed persistence alone could not compensate for limited dispersal (Figure 14.3). Although aggregated seed dispersal results in lower gap colonization rates, overall seedling recruitment rates can still match those for more widely dispersing species if high-density clumps of seeds encounter gaps at a sufficient rate. We examined recruitment success by computing total lifetime reproduction of individuals under the two dispersal scenarios for different levels of seed persistence. Since adult densities and fecundities were considered equal for the two

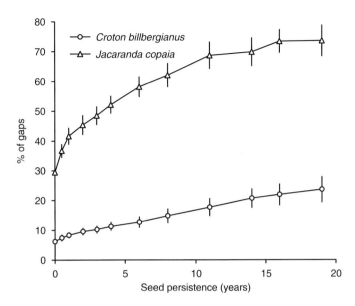

Figure 14.3 The percentage (± 1 standard deviation) of gaps that receive at least 1 seed m^{-2} of gap area (mean density). Under aggregated dispersal gap colonization rates remain poor even if seeds are able to persist for a long time. Simulations were carried out on a 1000 m × 500 m grid area with 5 m × 5 m quadrats. Seeds were dispersed from 50 randomly distributed adult trees, each dispersing about 8600 seeds per year. The total number of canopy gaps simulated was 419, more than 60% of which were ≤ 50 m^2 in area, while three gaps were about 650 m^2 in area.

species, the same numbers of seeds were dispersed each year by each species in our model, and the same per-seed recruitment probabilities were applied in gaps. Results from our simulations show that long-term mean recruitment rates increased with an increase in seed persistence but, surprisingly, were statistically indistinguishable between the two dispersal scenarios at all levels of seed persistence considered (Figure 14.4). Increasing seed persistence in the soil seed bank thus led to a general increase in long-term mean recruitment rates independent of dispersal.

Recruitment rates were, however, much more variable from year to year under aggregated dispersal compared with widespread dispersal (Figure 14.4). This was due to the differences in spatial variation in seed densities between the two dispersal scenarios. Distributions of seed densities in quadrats were much more skewed for aggregated dispersal compared with widespread dispersal, so although recruitment rates were often low with aggregated dispersal, pulses of

high recruitment were observed when light gaps occurred in quadrats with high densities of seeds. Increased spatial variation in seed densities therefore led to greater inter-annual fluctuations in recruitment, but as our simulations also show, long-term mean recruitment rates were similar for the two dispersal scenarios. This illustrates one potential way in which recruitment success can be equalized for species with different life histories, but also underlines the importance of spatial and temporal scales in understanding coexistence among pioneer species.

SEED LIMITATION IN CONTEXT

We have drawn attention to how the frequency and nature of gap disturbances could drive the evolution of life histories of pioneer species and to the potential importance of seed limitation for their coexistence. However, while steady-state seed limitation can help maintain diversity, species may

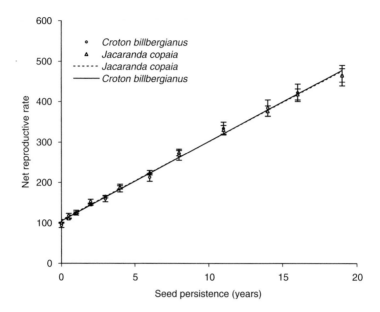

Figure 14.4 Long-term, plot-wide mean (± standard error) net reproductive rate at different levels of seed persistence for two different dispersal scenarios. The values are the number of germinating seedlings per adult tree during its lifetime. Adult density and seed production parameters are the same as those used in Figure 14.3. The per-seed probability of germination and establishment (seedling recruitment) was set to about 0.03 for both dispersal types. Mean seedling recruitment rates increase with seed persistence, but are evidently not different between the two dispersal scenarios. Inter-annual fluctuations are, however, greater under aggregated dispersal.

occasionally escape seed limitation when individuals become sufficiently abundant to saturate recruitment sites. For pioneers, which generally occur at low adult population densities, escaping seed limitation is probably rare except perhaps when windstorms or landslides open large areas that favor the recruitment of one or a few species. Shade-tolerant trees with seedlings that persist for years in the forest understory may more frequently overcome recruitment limitation. Wright (2002) describes the case of *Trichilia tuberculata*, a relatively large-seeded strongly shade-tolerant tree species that, in years of heavy seed set, recruits seedlings into the majority of seedling plots monitored. Density-dependent mortality is important in regulating population growth in these species, in both temperate and tropical forest (Harms *et al.* 2000, HilleRisLambers *et al.* 2002, Carson *et al.* Chapter 13, this volume).

Initial gap colonization patterns are also unlikely to be the sole determinant of adult distribution patterns. Variation in gap characteristics, coupled with constraints imposed on the ability of species to simultaneously disperse and establish at recruitment sites, also provides opportunities for species to coexist through niche differentiation (Kitajima and Poorter Chapter 10, this volume). The substantial variation in seed mass among pioneers reflects this constraint (Muller-Landau Chapter 11, this volume). Larger-seeded pioneers are able to establish at a wider range of microsites than small-seeded species, while small-seeded species reach more of the rare microsites they need because of their greater fecundity (Dalling *et al.* 1998b, Dalling and Hubbell 2002). A similar colonization–establishment trade-off also operates for temperate grassland pioneer communities (Turnbull *et al.* 2005).

A second axis of niche differentiation is important for pioneer species once seedlings outgrow their seed reserves. Growth rates of seedlings a few months old are uncorrelated with seed mass but strongly positively correlated with mortality rate. This correlation reflects two general trade-offs: first, between investment in growth versus defense against herbivores (Kitajima 1994, Dalling *et al.* 1998b, Dalling and Hubbell 2002), and second, between growth and susceptibility to drought-related mortality during the dry season (Pearson *et al.* 2003). Fast-growing species are therefore less likely to survive in gaps but more likely to reach reproductive size before a gap closes and may potentially shade out slower-growing competitors. Fast-growing species are also known to require larger gap sizes (Brokaw 1987), which may reflect greater opportunities to escape herbivory when growth rates are high.

CONCLUSIONS

We proposed that seed limitation should be particularly strong for pioneer species given their low population densities and the infrequency with which their recruitment sites become available. Evidence from seed trap data from BCI shows that seed limitation is strong for most pioneers, with only a few wind-dispersed species producing sufficient seeds that are widely enough dispersed to reach more than 90% of seed traps over two consecutive years of seed production. Long-term persistence of viable seeds in the soil seed bank can help compensate for strong seed limitation, but as our simulations show, gap colonization rates are still poor with aggregated dispersal. Further, theoretical studies seem to indicate that long-term seed persistence is unlikely to be selected for in perennial species due to the fitness cost incurred by extended generation time (Rees 1994). Nevertheless, there is direct evidence that seeds of some tropical pioneer species remain viable for decades in the soil seed bank. It is therefore unclear whether long-term seed persistence is a significant axis of niche differentiation among tropical pioneer species. Species with strong seed limitation are nevertheless successful in the BCI forest. The most seed-limited species, *Croton billbergianus*, with low fecundity and short-distance dispersal, has among the highest population density in this functional group. The ability of larger-seeded pioneers to maintain

populations in this forest indicates that post-dispersal processes must be important in determining recruitment patterns. We have shown that seedling distribution patterns reflect seed dispersal patterns for some species but have not examined the legacy left by dispersal on adult distribution patterns. Future work now awaits the development of a complete recruitment model for pioneers that includes dispersal, seed persistence, and the growth and mortality of seedlings to adulthood. This will provide the framework now needed to explore how variation in seed production, dispersal, and persistence affect species coexistence.

REFERENCES

Alvarez-Buylla, E. and Martínez-Ramos, R. (1990) Seed bank versus seed rain in the regeneration of a tropical pioneer tree. *Oecologia* 84, 314–325.

Brandani, A., Hartshorn, G.S., and Orians, G.H. (1988) Internal heterogeneity of gaps and species richness in Costa Rican tropical wet forest. *Journal of Tropical Ecology* 4, 99–199.

Brokaw, N.V.L. (1982) Treefalls: frequency, timing, and consequences. In E.G. Leigh, Jr., A.S. Rand, and D.M. Windsor (eds), *The Ecology of a Tropical Forest: Seasonal Rhythms and Long-Term Changes.* Smithsonian Institution Press, Washington, DC, pp. 101–108.

Brokaw, N.V.L. (1987) Gap-phase regeneration of three pioneer tree species in a tropical forest. *Journal of Ecology* 75, 9–19.

Clark, J.S., Ladeau, S., and Ibanez, I. (2004) Fecundity of trees and the colonization-competition hypothesis. *Ecological Monographs* 74, 415–442.

Clark, J.S., Macklin, E., and Wood, L. (1998) Stages and spatial scales of recruitment limitation in southern Appalachian forests. *Ecological Monographs* 68, 213–235.

Clark, J.S., Silman, M.R., Kern, R., Macklin, E., and HilleRisLambers, J. (1999) Seed dispersal near and far: generalized patterns across temperate and tropical forests. *Ecology* 80, 1475–1494.

Dalling, J.W. and Burslem D.F.R.P. (2005) Role of life-history and performance trade-offs in the equalization and differentiation of tropical tree species. In D. Burslem, M. Pinard, and S. Hartley (eds), *Biotic Interactions in the Tropics.* Cambridge University Press, Cambridge, pp. 65–88.

Dalling, J.W. and Hubbell, S.P. (2002) Seed size, growth rate and gap microsite conditions as determinants of recruitment success for pioneer species. *Journal of Ecology* 90, 557–568.

Dalling J.W., Hubbell, S.P., and Silvera, K. (1998b) Seed dispersal, seedling emergence and gap partitioning in gap-dependent tropical tree species. *Journal of Ecology* 86, 674–689.

Dalling, J.W., Muller-Landau, H.C., Wright, S.J., and Hubbell, S.P. (2002) Role of dispersal in the recruitment limitation of neotropical pioneer species. *Journal of Ecology* 90, 714–727.

Dalling, J.W., Swaine, M.D., and Garwood, N.C. (1997) Soil seed bank community dynamics in seasonally moist lowland forest, Panama. *Journal of Tropical Ecology* 13, 659–680.

Dalling, J.W., Swaine M.D., and Garwood, N.C. (1998a) Dispersal patterns and seed bank dynamics of pioneer trees in moist tropical forest. *Ecology* 79, 564–578.

Ellis, R.H. and Roberts, E.H. (1980) Improved equations for the prediction of seed longevity. *Annals of Botany* 45, 13–30.

Fornara, D.A. and Dalling, J.W. (2005) Post-dispersal removal of seeds of pioneer species from five Panamanian forests. *Journal of Tropical Ecology* 21, 1–6.

Guevara Sada, S. and Gómez-Pompa, A. (1972) Seeds from surface soils in a tropical region of Veracruz, Mexico. *Journal of the Arnold Arboretum* 53, 312–335.

Hanski, I (2001) Population dynamic consequences of dispersal in local populations and in metapopulations. In J. Clobert, E. Denchin, A.A. Dhondt, and J.D. Nichols (eds), *Dispersal*. Oxford University Press, Oxford, pp. 283–298.

Harms, K.E., Wright, S.J., Calderón, O., Hernandez, A., and Herre, E.A. (2000) Pervasive density-dependent recruitment enhances seedling diversity in a tropical forest. *Nature* 404, 493–495.

HilleRisLambers, J., Clark, J.S., and Beckage, B. (2002) Density-dependent mortality and the latitudinal gradient in species diversity. *Nature* 417, 732–735.

Hopkins, M.S. and Graham, A.W. (1987) The viability of seeds of rain forest species after experimental soil burials under tropical wet lowland forest in north-eastern Australia. *Australian Journal of Ecology* 12, 97–108.

Horvitz, C.C. and Schemske, D.W. (1986) Seed removal and environmental heterogeneity in a neotropical myrmecochore: variation in removal rates and dispersal distance. *Biotropica* 18, 319–323.

Hubbell, S.P. (2001) *The Unified Neutral Theory of Biodiversity and Biogeography*. Princeton University Press, Princeton.

Hubbell, S.P., Foster, R.B., O'Brien, S.T. *et al.* (1999) Light gap disturbances, recruitment limitation, and tree diversity in a neotropical forest. *Science* 283, 554–557.

Hurtt, G.C. and Pacala, S.W. (1995) The consequences of recruitment limitation: reconciling chance, history and competitive differences among plants. *Journal of Theoretical Biology* 176, 1–12.

Kaspari, M. (1993) Removal of seeds from neotropical frugivore droppings: ant responses to seed number. *Oecologia* 95, 81–88.

Kennedy, D.N. and Swaine, M.D. (1992) Germination and growth of colonizing species in artificial gaps of different sizes in dipterocarp rain forest. *Philosophical Transactions of the Royal Society London B* 335, 357–366.

Kitajima, K. (1994) Relative importance of photosynthetic traits and allocation patterns as correlates of seedling shade tolerance of thirteen tropical trees. *Oecologia* 98, 419–428.

LePage, P.T., Canham, C.D., Coates, K.D., and Bartemucci, P. (2000) Seed abundance versus substrate limitation of seedling recruitment in northern temperate forests of British Columbia. *Canadian Journal of Forest Research* 30, 415–427.

Levey, D.J. and Byrne, M.M. (1993) Complex ant–plant interactions: rain forest ants as secondary dispersers and post-dispersal seed predators. *Ecology* 74, 1802–1812.

Levings, S.C. and Franks, N.R. (1982) Patterns of nest dispersion in a tropical ground ant community. *Ecology* 63, 338–344.

Lieberman, D., Hartshorn, G.S., Lieberman, M., and Peralta, R. (1990) Forest dynamics at La Selva Biological Station, 1969–1985. In A.H. Gentry (ed.), *Four Neotropical Forests*. CABI Publishing, New York, pp. 509–521.

Loiselle, B.A., Ribbens, E., and Vargas, O. (1996) Spatial and temporal variation in seed rain in a tropical lowland wet forest. *Biotropica* 28, 82–95.

Lonsdale, W.M. (1988) Interpreting seed survivorship curves. *Oikos* 52: 361–364.

McEuen, A.B. and Curran, L.M. (2004) Seed dispersal and recruitment limitation across spatial scales in temperate forest fragments. *Ecology* 85, 507–518.

Molofsky, J. and Augspurger, C.L. (1992) The effect of leaf litter on early seedling establishment in a tropical forest. *Ecology* 73, 68–77.

Muller-Landau, H.C., Levin, S.A., and Keymer, J.E. (2003) Theoretical perspectives on evolution of long-distance dispersal and the example of specialized pests. *Ecology* 84, 1957–1967.

Murray, K.G. (1988) Avian seed dispersal of three neotropical gap-dependent plants. *Ecological Monographs* 58, 271–298.

Murray, K.G. and Garcia, M. (2002) Contributions of seed dispersal to recruitment limitation in a Costa

Rican cloud forest. In: D.J. Levey, R. Silva, and M. Galetti (eds), *Seed Dispersal and Frugivory: Ecology, Evolution and Conservation*. CAB International, Wallingford, pp. 323–338.

Nathan, R. and Muller-Landau, H.C. (2000) Spatial patterns of seed dispersal, their determinants and consequences for recruitment. *Trends in Ecology and Evolution* 15, 278–285.

Pearson, T.R.H., Burslem, D.F.R.P., Goeriz, R.E., and Dalling, J.W. (2003) Regeneration niche partitioning in neotropical pioneers: effects of gap size, seasonal drought, and herbivory on growth and survival. *Oecologia* 137, 456–465.

Perez-Nasser, N. and Vásquez-Yanes, C. (1986) Longevity of buried seeds from some tropical rain forest trees and shrubs of Veracruz, Mexico. *Malayan Forester* 49, 352–356.

Putz, F.E. and Appanah, B. (1987) Buried seeds, newly dispersed seeds, and the dynamics of a lowland forest in Malaysia. *Biotropica* 19, 326–339.

Rees, M. (1994) Delayed germination in seeds: a look at the effects of adult longevity, the timing of reproduction, and population age/stage structure. *American Naturalist* 144, 43–64.

Ribbens, E., Silander, J.A., and Pacala, S.W. (1994) Seedling recruitment in forests: calibrating models to predict patterns of tree seedling dispersion. *Ecology* 75, 1794–1806.

Schnitzer, S.A., Dalling, J.W., and Carson, W.P. (2000) The impact of lianas on tree regeneration in tropical forest canopy gaps: evidence for an alternative pathway of gap-phase regeneration. *Journal of Ecology* 88, 655–666.

Schupp, E.W., Milleron, T., and Russo, S.E. (2002) Dissemination limitation and the origin and maintenance of species-rich tropical forests. In D.J. Levey, W.R. Silva, and M. Galetti (eds), *Seed Dispersal and Frugivory: Ecology, Evolution, and Conservation*. CABI Publishing, New York, pp. 19–34.

Swaine, M.D. and Whitmore, T.C. (1988) On the definition of ecological species groups in tropical rain forests. *Vegetatio* 75, 81–86.

Tilman, D. (1994) Competition and biodiversity in spatially structured habitats. *Ecology* 75, 2–16.

Turnbull, L.A., Manley, L., and Rees, M. (2005) Niches, rather than neutrality, structure a grassland pioneer guild. *Proceedings of the Royal Society Series B* 272, 1357–1364.

Uhl, C. and Murphy, P.G. (1981) Composition, structure, and regeneration of a tierra firme forest in the Amazon Basin of Venezuela. *Tropical Ecology* 22, 219–237.

Vázquez-Yanes, C., Orozco-Segovia, A., Rincon, E. *et al.* (1990) Light beneath the litter in a tropical forest: effect on seed germination. *Ecology* 71, 1952–1958.

Wehncke, E.V., Hubbell, S.P., Foster, R.B., and Dalling, J.W. (2003) Seed dispersal patterns produced by white-faced monkeys: implications for the dispersal limitation of neotropical tree species. *Journal of Ecology* 91, 677–685.

Wenny, D.G. (2001) Advantages of seed dispersal: a re-evaluation of directed dispersal. *Evolutionary Ecology Research* 3, 51–74.

Williams-Linera, G. (1990) Origin and early development of forest edge vegetation in Panama. *Biotropica* 22, 235–241.

Wright, S. (1969) *Evolution and the Genetics of Populations. Volume 2. The Theory of Gene Frequencies.* University of Chicago Press, Chicago.

Wright, S.J. (2002) Plant diversity in tropical forests: a review of mechanisms of species coexistence. *Oecologia* 130, 1–14.

Chapter 15

ENDOPHYTIC FUNGI: HIDDEN COMPONENTS OF TROPICAL COMMUNITY ECOLOGY

A. Elizabeth Arnold

OVERVIEW

While the ecological importance of plant pathogenic fungi, decay fungi, and root symbionts is becoming well established in tropical biology, the contributions of one major group of ubiquitous symbionts – fungal endophytes of foliage – have yet to be explored. Fungal endophytes – fungi that live within plant tissues such as leaves without causing disease – are found in every lineage of plants and are especially common and diverse in tropical forests, where individual leaves may harbor dozens of species without any obvious indication of infection. Their ecological roles are only starting to be elucidated, but early evidence suggests that fungal endophytes play pervasive, if almost entirely overlooked, roles in tropical forest ecology. This chapter synthesizes current knowledge regarding the natural history of foliar endophytes in tropical forests, examines the varied evidence regarding their ecological roles, and highlights a series of tractable questions for future research. The overarching goal of this chapter is to encourage multidisciplinary research into the ecology of these little-known but omnipresent symbionts of tropical plants.

INTRODUCTION

Plant pathogenic and parasitic fungi play important roles in shaping tropical tree communities (Augspurger 1983, Dobson and Crawley 1994, Wills *et al.* 1997, Gilbert 2002, Gallery *et al.* 2007). Similarly, the nutrient cycling carried out by highly diverse saprophytic fungi is intrinsic to tropical ecosystem processes (reviewed by Hyde 1997). Less obvious to ecologists are the roles played by endosymbiotic fungi of living plants, which live within plant tissues without causing obvious detriment or visible symptoms (Arnold and Lutzoni 2007). The subset of these fungi that occur in the rhizosphere (mycorrhizal fungi) are increasingly recognized for their impact on tropical forest communities (e.g., Kiers *et al.* 2000, Husband *et al.* 2002, Mangan *et al.* 2004, Herre *et al.* 2005a). However, tropical plants also

harbor fungi in above-ground tissues such as leaves and stems. Broadly defined, these are fungal endophytes: fungi that colonize the interior of healthy plant tissues without causing disease (Petrini 1991). Endophytes are present in the photosynthetic tissues of every tropical plant studied to date, and their diversity is remarkable: individual leaves typically harbor over a dozen species, and the number of taxa associated with individual trees likely numbers in the thousands (Lodge *et al.* 1996, Fröhlich and Hyde 1999, Arnold *et al.* 2000, 2003, Arnold and Lutzoni 2007). Together, these poorly known fungi represent a trove of unexplored biodiversity, and a frequently overlooked component of tropical ecology.

The first study quantifying the richness and species composition of endophytes associated with a tropical dicotyledonous host was published only a decade ago (Lodge *et al.* 1996). In the intervening years, the few studies on tropical endophytes have been primarily descriptive, with some attention to the impact of tropical endophytes on estimates of global fungal diversity (e.g., Fröhlich and Hyde 1999, Arnold *et al.* 2000). Most of this work has focused on endophytes of leaves (foliar endophytes), which are especially diverse and abundant (Arnold *et al.* 2000). Four recent studies have provided the first evidence for ecologically relevant roles of tropical foliar endophytes, including increased host resistance to pathogens (Arnold *et al.* 2003) and physiological costs in terms of water relations and photosynthesis (Pinto *et al.* 2000, Herre *et al.* 2005b, Arnold and Engelbrecht 2007). These studies represent the tip of a very large iceberg: in as much as the mycota of all ecosystems are understudied, the endophytic fungi in any tropical forest remain extremely poorly known.

Due to a growing interest from ecologists, bioprospectors, and mycologists, an expansion of research infrastructure in the tropics, and the development of new methods, the study of tropical endophytes is more accessible now than ever

before. While alpha taxonomic studies are still sorely needed, the stage is also set for experimental manipulations of endophyte abundance and diversity, and for addressing ecological questions. In this context, the purposes of this chapter are three-fold: (1) to synthesize current knowledge regarding the natural history of foliar endophytes in tropical forests; (2) to examine current evidence regarding their ecological roles; and (3) to highlight a series of tractable questions for future research. The overarching goal of this chapter is to encourage multidisciplinary research into the ecology of these little-known but ubiquitous and potentially important symbionts of tropical plants.

ENDOPHYTE TRANSMISSION IN TROPICAL FORESTS

The vast majority of fungal endophytes associated with leaves (hereafter, endophytes) are Ascomycota, including all major lineages of non-lichenized, filamentous ascomycetes (Boxes 15.1

Box 15.1   What is an endophyte?

Endophytes are organisms that "at some time in their life ... colonize internal plant tissues without causing apparent harm" (Petrini 1991). So defined, endophytes are a polyphyletic group of microorganisms inhabiting the roots, leaves, vascular tissues, and reproductive structures of plants. Endophytes may exhibit more than one type of life history at distinct life stages (e.g., latent pathogens) and may exist simultaneously or over time upon and within plant tissues (Petrini 1986).

Because endophytes occur within healthy tissue, their presence is unapparent: there is no external manifestation of symptoms on plant structures. Hyphae can only rarely be visualized without chemical fixation, and when observed *in planta* they cannot be identified. Therefore, endophytes are typically studied by culturing: plant tissues are surface-sterilized using chemical agents (e.g., ethanol, sodium hypochlorite, hydrogen peroxide; Petrini 1986, Schulz *et al.*

1993), and small pieces are plated under sterile conditions on a nutrient medium (e.g., malt extract agar, sometimes supplemented to restrict bacterial growth and slow the "weedy" fungi). Emergent hyphae are transferred to axenic cultures, which are examined for reproductive structures, grouped to morphospecies, and/or used for sequencing.

An unknown proportion of endophytes cannot be cultured, and/or are only recovered using environmental polymerase chain reaction (Arnold *et al.* 2006). These unculturable species may include obligate symbionts, poor competitors, or fungi that cannot exploit particular media. Gallery *et al.* (2007) sequenced fungi directly from apparently uninfected, surface-sterilized seeds of *Cecropia insignis*, but these methods have not yet been applied to other tropical trees. Data are needed to determine whether seedborne fungi persist in seedlings – or whether seeds, much like xylem, bark, roots, and leaves, represent a discrete substrate for endophytes, and contain a distinctive fungal community.

Box 15.2 Taxonomy of tropical endophytes.

Tropical endophytes are primarily filamentous Ascomycota (euascomycetes, or Pezizomycotina), including many species of Sordariomycetes and Dothideomycetes (Figure 15.1; Arnold and Lutzoni 2007, Van Bael *et al.* 2005). Common genera include *Xylaria*, *Colletotrichum/Glomerella*, *Botryosphaeria*, *Diaporthe/Phomopsis*, *Pestalotia*, *Rhizoctonia*, *Fusarium/Nectria*, *Trichoderma*, *Phoma*, *Phyllosticta*, *Alternaria*, *Nodulisporium/Hypoxylon*, and *Daldinia*, among others (Rodrigues 1994, Lodge *et al.* 1996, Fröhlich and Hyde 1999, Arnold 2002). Whether these genera are truly the most common in terms of biomass or infection density within leaves – or are simply most adept at growing in the nutrient-rich and benign conditions afforded by culturing – remains to be seen, and should be addressed using quantitative molecular methods. The most common genera associated with tropical angiosperms also are frequently recovered in the temperate

zone (e.g., *Xylaria*, *Colletotrichum*, and *Botryosphaeria*). However, molecular data suggest only limited overlap of species between the temperate zone and the tropics (Arnold and Lutzoni 2007).

Many of the genera shared between the temperate zone and the tropics represent fast-growing taxa that are rapidly and easily isolated using standard media. When in-depth surveys are conducted, especially with active selection against "weedy" taxa, less common genera are recovered. These often represent the Eurotiomycetes (Eurotiomycetidae and Chaetothyriomycetidae; Figure 15.1), Leotiomycetes, and various Basidiomycota. These less-common endophytes may represent the most interesting species in terms of ecological interactions, biochemical diversity, and use in bioprospecting. In particular, the Eurotiomycetes are noted for their chemical diversity and antibiotic activity, and contain a large number of species pathogenic to animals (Alexopoulos *et al.* 1996, Lutzoni *et al.* 2001).

and 15.2; Figure 15.1). Endophytes are known from mosses and other non-vascular plants, ferns and their allies, conifers, and angiosperms (Stone *et al.* 2000), and have been recovered in ecosystems ranging from hot deserts (Suryanarayanan *et al.* 2005, Hoffman and Arnold 2008) to tundra (Fisher *et al.* 1995, Higgins *et al.* 2007), mangroves (Kumaresan and Suryanarayanan 2002, Gilbert *et al.* 2002a), and temperate croplands (Arnold and Lewis 2005).

Most endophyte research has focused on a single family (Clavicipitaceae), some of whose members occur within above-ground tissues of some temperate grasses (Saikkonen *et al.* 1998, Clay and Schardl 2002). These vertically transmitted, systemic endophytes infect at least 300 species of grasses, and are recognized for their ability to produce secondary compounds, including alkaloids, which benefit hosts by deterring or sickening herbivores (e.g., Clay *et al.* 1985, Siegel *et al.* 1990, Wilkinson *et al.* 2000, but see Faeth and Sullivan 2003). In contrast, fungal endophytes of tropical trees are transmitted primarily by contagious spread (horizontal transmission), rather than by maternal inheritance (Arnold and Herre 2003,

Arnold *et al.* 2003, Herre *et al.* 2005b). These endophytes accumulate after leaf flush, growing intercellularly and subsisting on carbon in the apoplast (see Clay 2001). Tropical endophytes are highly localized within leaves, with individual infections typically occupying only ca. 2 mm^2 in area (Lodge *et al.* 1996).

In tropical forests, foliar endophytes reproduce by hyphal fragmentation and/or by the production of sexual or asexual spores on dead or senescent tissue (see Herre *et al.* 2005b). Spores and hyphal fragments may be released passively, or are liberated by physical disturbance from wind, rain, or tree- or branch-fall events. Insect herbivores also may transmit fungal propagules (see below). Many fungi, including the endophyte-rich genus *Phyllosticta*, produce slimy spores that rely at least in part on rain for dispersal (Kirk *et al.* 2001). Although heavy wind and rain are especially effective in moving spores, even light precipitation can disperse conidia of *Colletotrichum*, a common genus of pathogenic and endophytic fungi (Guyot *et al.* 2005). Similarly, light wind and the currents produced by diurnal cycles of heating and cooling are significant for dispersal of dry propagules,

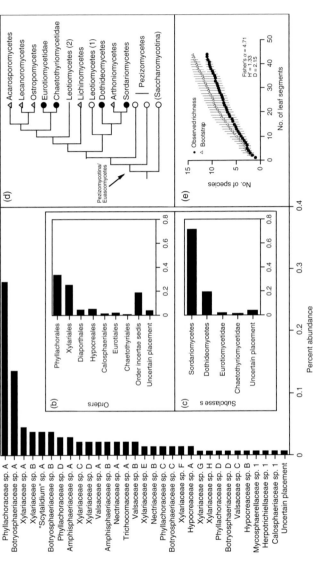

Figure 15.1 Taxonomic diversity and relative abundance of fungal endophytes inhabiting healthy leaves of tropical trees at Barro Colorado Island, Panama (BCI). (a) Relative abundance and taxonomic placement of the 31 most common species isolated from three mature leaves of *Laetia thamnia* (Flacourtiaceae), *Trichilia tuberculata* (Meliaceae), and *Gustavia superba* (Lecythidaceae). Taxonomic placements are based on BLAST matches in the NCBI GenBank database for sequence data (ca. 600 base pairs) from the nuclear ribosomal ITS regions, including the 5.8S gene (N = 127 isolates), coupled with phylogenetic analyses (Arnold and Lutzoni 2007). Most species are rare, and very few species are common. Panels (b) and (c) summarize these data at the ordinal (b) and subclass (c) levels, demonstrating the dominance of the Sordariomycetes (especially Phyllachorales and Xylariales). Panel (d) shows the current phylogenetic hypothesis for relationships of the Euascomycota (after Lutzoni *et al.* 2004), showing the phylogenetic breadth of endophytic isolates from only nine leaves at BCI. Black circles indicate lineages in which endophytes were found in the present study; white circles indicate lineages in which endophytes are known from other studies, and gray triangles indicate lineages in which the majority of (Arthoniomycetes) or all known species (e.g., Lichinomycetes) are lichenized, rather than free-living. With the exception of the clade of decomposer fungi known as Leotiomycetes (2), all major lineages that are not lichenized contain endophytic fungi. Panel (e) indicates the accumulation of endophyte species, defined as indicated in (a), as a function of the number of leaf segments sampled from leaves of *Laetia thamnia*, *Trichilia tuberculata*, and *Gustavia superba* (N = 3 leaves/species, 15 tissue segments/leaf, 5 leaves/species, 15 tissue segments/leaf). Even when singletons are excluded from the analysis, the accumulation of observed richness (black circles) and estimated richness based on bootstrap analyses (triangles) continues to rise, and diversity values (Fisher's α, Shannon index [H'], and Simpson's index [D]) remain high.

such as those of xylariaceous species and various Eurotiomycetidae (see Holb *et al.* 2004).

Inoculum volume plays an important role in determining the infection success of plant-associated fungi (Agrios 1997), but until recently, the rate of fungal propagule deposition in tropical forests was not known. Gilbert (2002), Arnold (2002), and Arnold and Herre (2003) found that typical leaves in the forest understory at Barro Colorado Island (BCI), Panama, receive ≥ 10–15 viable fungal propagules per cm^2 per hour during the mid- and late wet seasons. When adjusted for mean leaf area and extrapolated to 24 hours, these data suggest that the average leaf receives more than 15,000 viable fungal propagules per day (average based on mean leaf area for 28 tree species in the understory at BCI; Arnold 2002). These values consider only fungi capable of growing on one nutrient medium (malt extract agar) and likely underestimate total deposition.

Fungi in the forest air column represent pathogenic, saprophytic, and endophytic species, as well as numerous species of unknown ecological roles. The proportion that can form endophytic symbioses has not been quantified, but the abundance of viable propagules in air is positively associated with the frequency of endophyte infections (Arnold and Herre 2003).

Propagules of fungi at BCI are highly abundant in the air column immediately following rainfall events, and then decrease as a curvilinear function with increasing time since significant rainfall (Figure 15.2). Both ultraviolet (UV) radiation and desiccation play an important role in the mortality of fungal propagules: Arnold and Herre (2003) found that the deposition and persistence of living propagules on leaves was greater under the forest canopy than in the laboratory clearing at BCI. After 1 week of exposure, sterile plants placed in the lab clearing accumulated significantly fewer endophyte infections than plants placed in the forest understory (Arnold and Herre 2003).

ENDOPHYTE COLONIZATION AND ABUNDANCE IN TROPICAL LEAVES

The high abundance of inoculum in the air column, and the apparently universal receptivity of tropical plants to colonization by endophytic fungi (Arnold 2002, Van Bael *et al.* 2005), leads to high infection rates in mature foliage of tropical trees. Arnold (2002) recovered endophytic fungi from 100% of mature leaves sampled

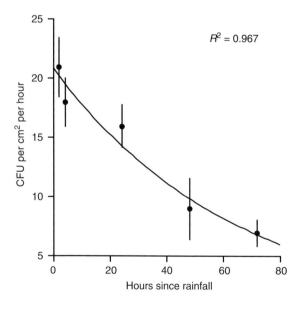

Figure 15.2 Deposition of living fungal colony-forming units (CFU; hyphal fragments and both sexual and asexual spores) per cm^2 of surface area per hour, as a function of hours since significant rainfall in the understory of secondary forest at BCI. Each point represents the mean ($\pm$ standard error) of six sampling stations (data from Arnold 2002) and underestimates total deposition: only fungi capable of growing on a single medium (2% malt extract agar) at a given temperature ($23°C$) are represented.

from 28 species of woody plants representing 24 families and 14 orders of angiosperms at BCI ($N = 9$ leaves from three individuals per species in the understory of late secondary forest). Studies in Puerto Rico and Guyana have had similar results (Gamboa and Bayman 2001, Cannon and Simmons 2002). The proportion of endophyte-infected leaves appears to increase from the arctic to the tropics (Arnold and Lutzoni 2007), although most plant communities have not yet been sampled. In the temperate zone, the frequency of endophyte infections is influenced by precipitation, humidity, elevation, irradiance, and air pollution, but the roles of these factors have not been fully assessed in the tropics. In particular, tropical savannas and dry forests – as well as the forest canopy in moist or wet forests – represent unique environmental conditions imposed by high irradiance, high temperature, and geographic congruence with endophyte-rich forests. Plants in these communities may offer a wealth of novel endophyte species.

Neither seedlings nor leaves of tropical trees typically contain culturable endophytes at emergence, but colonization proceeds rapidly given the presence of airborne inoculum and high relative humidity or wetting of leaf surfaces by dew, rain, or fog (Arnold and Herre 2003). At BCI, infection rates (defined here as the proportion of leaves containing endophytic fungi) increase to nearly 100% of leaves as foliage matures. Field experiments have shown that endophytes are present in more than 80% of *Theobroma cacao* leaves within 2 weeks of leaf emergence during the early wet season at BCI (Arnold and Herre 2003). Leaf toughness does not influence endophyte colonization: young and mature leaves can be colonized with equal frequency (Arnold and Herre 2003).

Endophyte infections within leaves are typically quantified by determining the proportion of small leaf fragments (typically ca. 2 mm^2) that yield endophytes in culture (Box 15.1). Proportions of leaf area colonized by endophytes differ among tropical sites and host species, although all tree species examined to date at BCI have consistently high densities of endophyte infection (>95% of tissue segments; Arnold 2002). Rodrigues (1994) found that 25% of leaf pieces

were colonized by endophytes in fronds of *Euterpe oleracea* (Arecaceae) in seasonally inundated sites in Amazonia. Gilbert *et al.* (2002a) recovered endophytes from 20%, 92%, and 80% of 3 mm^2 leaf fragments from three mangrove species in Panama (*Avicennia, Rhizophora,* and *Laguncularia,* respectively), with infection frequencies paralleling the salinity of water on leaf surfaces. Lodge *et al.* (1996) found a high infection rate in Puerto Rican *Manilkara bidentata* (Sapotaceae; 90–95% of 2 mm^2 leaf fragments), as did Gamboa and Bayman (2001) for leaves of *Guarea guidonia* (Meliaceae) in Puerto Rico (>95%). Apparently healthy leaves contain numerous, independent infections, rather than systemic or otherwise extensive growth of hyphae (Lodge *et al.* 1996). The biomass resulting from any given infection is very low, such that each leaf represents a densely packed mosaic of diverse endophyte species (Lodge *et al.* 1996). Mature leaves have a higher infection density than do younger leaves, reflecting the accumulation of numerous, independent infections as leaves age, and the differential proliferation of favored species as leaves approach senescence (Arnold *et al.* 2003). Synthesizing the results from several field studies in central Panama, Herre *et al.* (2005b) suggested that most leaves are saturated by endophytic fungi (i.e., contain endophytes in 100% of 2 mm^2 tissue segments) within 3–4 weeks after emergence.

Unfortunately, methodological artifacts often prevent comparisons among studies, limiting our ability to assess the role of abiotic factors or microhabitat characteristics in influencing endophyte abundance. For example, Gamboa *et al.* (2002) demonstrated a strong, inverse relationship between the size of leaf fragments used in culture and the number of fungi isolated from leaves. Only studies with similar leaf fragment sizes, media, culturing conditions, and surface-sterilization protocols can be compared, and standardization of these methods is needed.

TROPICAL ENDOPHYTE DIVERSITY

Hawksworth (1991) estimated global diversity of fungi at 1.5 million species, drawing from a ratio of six species of fungi per vascular plant species in Great Britain. However, by sampling only endophytic and saprophytic fungi, Fröhlich and Hyde (1999) suggested that a ratio of 1:33, rather than 1:6, is more appropriate for tropical plants. This conclusion is supported by the richness of endophyte communities in individual tropical leaves. Gamboa *et al.* (2002) recovered 11.7 ± 3.4 morphospecies of endophytes per leaf in a survey of five tropical plant species. Lodge *et al.* (1996) recovered 17 species from a leaf blade of *Manilkara bidentata* in Puerto Rico, and Gilbert *et al.* (2002a) recovered up to 13.1 ± 3.4 morphospecies per leaf in the mangrove *Rhizophora* in Panama. At BCI, Arnold *et al.* (2003) isolated up to 13 morphospecies from 32 mm^2 of leaf tissue from *Theobroma cacao* (i.e., 16 leaf fragments, each 2 mm^2), consistent with findings at the same site for *Heisteria concinna* and *Ouratea lucens* (11–14 morphospecies per 48 mm^2 of leaf tissue; Arnold *et al.* 2000).

To date, no study has exhaustively sampled the foliar endophyte community associated with a single tropical plant or plant species. However, it appears that individual plants can harbor a tremendous richness of species. Arnold

et al. (2003) found 47.5 ± 4.9 morphospecies of endophytes associated with only nine leaves from individual *Theobroma cacao* trees in Panama (i.e., 288 mm^2 of leaf tissue). Interestingly, total richness per tree did not differ significantly among individuals in agroecosystems, primary forest, or secondary forest, although species turnover was high among the endophyte species recovered in each site. Although leaves on the same tree will have a subset of endophytes in common, each leaf will harbor distinctive endophyte communities, leading to high rates of species turnover among leaves. In a study of *Laetia thamnia* in Panama, leaf fragments of a consistent size were used to sample increasing leaf area on a single individual. The number of morphospecies accumulated as a function of cumulative leaf area with a coefficient of 0.504, which is consistent with samples among biological provinces at a landscape scale (Figure 15.3; Rosenzweig 1995).

These and similar studies (e.g., Gamboa and Bayman 2001, Cannon and Simmons 2002, Suryanarayanan *et al.* 2002, Arnold *et al.* 2003) are providing baseline data for understanding the diversity of tropical endophyte communities. However, three major challenges still need to be

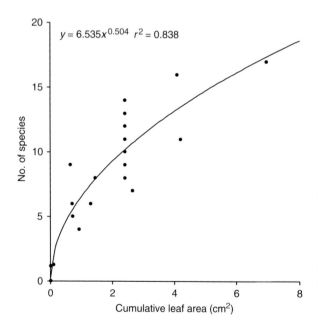

$$y = 6.535x^{0.504} \quad r^2 = 0.838$$

Figure 15.3 Species–area relationship of endophytic fungi in spatially nested samples of leaves of a tropical treelet (*Laetia thamnia*) at BCI. Richness was assessed as a function of leaf area (determined as the number of 2 mm^2 leaf segments sampled per leaf) for leaves on one tropical tree. As sampling area increased, the number of species recovered increased as a curvilinear function, with a coefficient (0.504) consistent with the species-area relationships for inter-provincial samples at macroscopic scales (as per Rosenzweig 1995).

Box 15.3   Designating functional taxonomic units.

Species boundaries in fungi are traditionally defined by reproductive morphology. Such approaches are important: they require implicit knowledge of the organisms, provide a taxonomic context, and can be compared across studies. However, many endophytes do not reproduce in culture (Guo *et al.* 2003), and existing keys (primarily based on temperate fungi) are of limited use in identifying those that do sporulate. Moreover, morphological characteristics are at times misleading, and cannot be used to identify novel anamorph/teleomorph connections (Hawksworth 1994, Rossman 1996). Accordingly, endophytic fungi are often grouped as morphospecies based on mycelial characteristics (Arnold *et al.* 2000, Guo *et al.* 2000, Gilbert *et al.* 2002a). While useful for organizing large numbers of isolates, the mechanics of such groupings are rarely detailed. How many characters are used? What is their relative weight? Are cultural characteristics labile within a putative species, and how much lability is allowed?

Arnold (2002) showed that morphospecies designations were not sensitive to a variety of factors (e.g., incubation temperature, illumination during incubation, nutrient medium, or identity of the host plant), but that grouping cultures on different media or using cultures of different ages will lead to different morphospecies boundaries and estimates of diversity. Comparisons with species defined by molecular data indicate that morphospecies can split or lump putative species: for example, Guo *et al.* (2003) recovered at least five genera of Ascomycota among 18 isolates of a single morphospecies from China. On the other hand, eight morphospecies

represented the same species of *Lophodermium* in one recent study (Arnold *et al.* 2007). Morphospecies clearly "work" better for some clades than others.

Basic Local Alignment and Search Tool (BLAST) searches of the National Center for Biotechnology Information (NCBI) GenBank database are frequently used to identify endophytes, primarily using data from the nuclear ribosomal internal transcribed spacers and 5.8S gene (ITS) (e.g., Guo *et al.* 2000, 2003). ITS data are valued for species-level systematics, but they may obscure cryptic species, and variation within, versus among, species can be unpredictable (Jacobs and Rehner 1998, Lieckfeldt and Seifert 2000, Kim and Breuil 2002). In turn, BLAST matches are based on non-phylogenetic criteria, are subject to error due to misidentified sequences, can be difficult to interpret when all top matches are unidentified isolates, and are limited to those fungi present in GenBank (21,075 ITS sequences in 2004; Lutzoni *et al.* 2004). Moreover, top BLAST matches for a given endophyte will change over time, reflecting growth of the database (Arnold and Lutzoni 2007). These issues argue for caution in assigning identities to endophytes based on BLAST searches, and underscore the need for (1) phylogenetic analyses using multiple informative loci, and (2) where possible, induction of sporulation. Leaf pieces may provide a sporulation substrate, and some culturing conditions can induce sporulation (e.g., molasses yeast medium and UV radiation). Finally, it is critical to deposit voucher specimens at recognized culture collections, including herbaria or universities in the country of origin or international depositories that will accept unidentified/sterile mycelia (e.g., Robert L. Gilbertson Mycological Herbarium, University of Arizona).

addressed before a clear understanding of tropical endophyte diversity can emerge. First, consistent and biologically meaningful species concepts are needed that incorporate sterile (non-sporulating) fungi, which typically dominate endophyte cultures (Box 15.3). Second, extensive sampling within sites is needed to reliably estimate host affinity. Third, comparative studies are needed to assess spatial structure and beta-diversity.

Although several studies have addressed one or more of these goals, none has fully satisfied these criteria. Moreover, uncertainty exists regarding the potential diversity of unculturable species, the degree of overlap between endophytes and other fungal guilds (i.e., pathogens, saprophytes; Box 15.4), and the appropriate spatial scale at which to estimate species turnover (among leaves, trees, or forests?). The number of endophyte

---

Box 15.4   Are endophytes a distinct group?

Endophytic fungi of tropical trees are distinct from arbuscular mycorrhizal (AM) fungi on the basis of habitat (aerial tissues versus rhizosphere), taxonomy (primarily Ascomycota versus Glomeromycota), and diversity (greater in endophytes than in AM fungi). However, the occurrence of entomopathogens as endophytes raises the question: How distinct are endophytes relative to other guilds of fungi? This has yet to be resolved, reflecting the lack of knowledge regarding pathogenic and saprophytic fungi in tropical forests (see Lodge 1997). Understanding where endophytes fall on the pathogen-to-mutualist continuum, and whether they exploit non-living tissues of plants, is key to interpreting endophyte ecology.

Several researchers have suggested that endophytes are saprotrophs waiting to happen: endophytes undergo a latent phase as a prelude to rapid growth following leaf death. In this way, endophytes might resemble tropical pioneer species – but with persistence in leaves instead of a soil seed bank. Four observations support the latent-saprophyte hypothesis: (1) the recovery of putatively saprotrophic taxa such as *Xylaria*, *Colletotrichum*, *Alternaria*, and *Aureobasidium* as endophytes (Bills and Polishook 1994, Fröhlich and Hyde 1999); (2) the ability of many endophytes to grow on plant-derived media; (3) the close phylogenetic relationship of many endophytes to saprophytic species; and (4) the development

of reproductive structures on surface-sterilized leaves. Interestingly, fungi that are closely related to endophytes of tropical plants, including members of *Cercospora* and *Fusarium*, can produce gibberellins and abscisic acid (Phelan and Stinner 1992). Could sapro-endophytes hasten leaf death or leaf drop, and/or accelerate the transition of leaves from carbon sources to sinks as a means to further their reproductive success? The role of endophytes in leaf hormone status and leaf lifetimes should be explored.

Similarly, it has been suggested that a large proportion of endophytes are latent pathogens. Their close relationship to known pathogens (Carroll 1986) is echoed by the observation that many pathogenic species have long latent periods: endophytes may simply represent the long end of the latent-period spectrum. Under this scenario, endophytes await environmental cues that allow them to manifest virulence, and then reproduce via the formation of necrotic lesions or other symptoms. Such cues could be abiotic (e.g., water stress due to drought), intrinsic to the host (e.g., tissue age), or could reflect an interaction with a property of the endophyte (e.g., accumulation of sufficient biomass to induce symptom development). One intriguing possibility is that plants serve as alternate hosts of one another's pathogens. The potential for a tree to harbor, at little cost (and perhaps at benefit) to itself, microbial agents detrimental to its neighbors raises a series of interesting questions regarding the cryptic roles of endophytes in tropical forest dynamics.

---

species in tropical forests remains an open question, and one that is more than academic as we attempt to understand the ecological importance and potential applications of these cryptic symbionts.

BEYOND ALPHA-DIVERSITY: HOST AFFINITY AND SPATIAL STRUCTURE

The host affinity and spatial heterogeneity of tropical endophytes are much debated. One challenge lies in the prevalence of singleton species: even

large-scale surveys, such as those at BCI, typically recover 50–65% of species only once (see Arnold *et al.* 2000, Arnold and Lutzoni 2007). Similar values have been observed for endophytes in other forest types in Panama (e.g., mangroves: 62.4% of morphospecies were singletons; Gilbert *et al.* 2002a) and for macroscopic fungi: at BCI, Gilbert *et al.* (2002b) recovered 58.1% of polypore species (shelf- and bracket-fungi) only once. Because singletons must be excluded from analyses of spatial structure or host affinity, analyses frequently consider less than half of the observed species. Consequently, conclusions regarding spatial and host specificity are based on only the most common – and perhaps most generalist – species (Arnold and Lutzoni 2007). It is not surprising that evidence for host affinity and spatial structure among endophytes has been conflicting, and at times contradictory. Arnold *et al.* (2000, 2003) found strong evidence for both host affinity and spatial structure of endophyte communities within and among forests in Panama, but Cannon and Simmons (2002) found little structure to endophyte communities in a non-quantitative study in Guyana. Suryanarayanan *et al.* (2002) found a high degree of overlap among communities of endophytes in many different hosts and forest types in India, but Arnold and Lutzoni (2007) found little overlap in endophyte genotypes among different forests along a latitudinal gradient. In the latter study, some endophyte genotypes were isolated from numerous hosts at BCI, regardless of the phylogenetic placement, leaf defenses, or phenology of those hosts; however, others were found in only a single host species. Tropical endophyte communities appear to contain a mix of generalist and site- and host-specific species. The challenge remains to infer these ecological parameters for very rare taxa, and to determine whether apparently rare species are truly rare or simply compete poorly given a set of culturing conditions (see Box 15.1).

Distribution data alone do not provide insight into the mechanisms that underlie apparent host affinity. Because leaves of tropical trees tend to be well defended against pathogens (Coley and Barone 1996), it is plausible that chemical defenses of leaves may influence host affinity of endophytes. To test this hypothesis, Arnold and Herre (2003) incorporated leaf homogenates as the nutrient source into water agar and assessed growth rates of endophytes *in vitro*. Eighty-six percent of endophytes from *Theobroma cacao*, when tested on media containing extracts from each of three host species, grew faster on extracts from *T. cacao* than on extracts from two co-occurring tree species (Arnold *et al.* 2003). To ensure that this result did not reflect a greater nutritive value in *T. cacao* extracts, Arnold *et al.* (2003) grew endophytes from three host species on leaf-extract media from all hosts, and found that endophytes grew more rapidly on extracts

of the host in which they were most frequently recovered in the field. With new methods for raising sterile seedlings and inoculating them with endophytes now available, the stage is set for much-needed assessments of host affinity *in planta*.

Unfortunately, assessing spatial structure of endophyte communities – critical for determining beta-diversity (Chave Chapter 2, this volume) – does not lend itself to such straightforward laboratory experiments, although reciprocal transplant experiments could be useful in this regard. Fungal spores have long been thought to spread long distances as aerial plankton; classic examples include wheat leaf rust (*Puccinia triticina*), which overwinters in Mexico and blows north over the Great Plains of the USA in mid-spring (Agrios 1997). Whether endophytes move about at a similar scale is not known. Genotype data are especially important for comparing fungal assemblages in different sites – as is recognized for many pathogenic fungi (Agrios 1997).

FUNGAL ENDOPHYTES AND TROPICAL FOREST COMMUNITY ECOLOGY

Given that plants in the dark forest understory are carbon limited, why do they host such a large number of obligate heterotrophs in their leaves? Given that endophytes are often closely related to pathogens, and that plants in tropical forests are well defended against pathogenic fungi (relative to temperate species; Coley and Barone 1996), why do plants host such a diversity of fungal species in their tissues?

These questions have yet to be answered, as studies assessing the ecological roles of tropical endophytes are yet in their infancy. In general, however, there are three main hypotheses regarding the roles of endophytic fungi: that they are (1) neutral inhabitants, (2) parasites, or (3) mutualists of their hosts. Given the tremendous phylogenetic diversity of tropical endophytes (Figure 15.1), endophytes as a whole likely include species with the capacity to play each of these roles, or to change roles over time or under certain conditions (see Box 15.4). Moreover, it is plausible that the most important ecological roles of endophytes are manifested with regard not to the plants they inhabit, but instead to insects or pathogens that attack those plants. Here, I present evidence for and against these general models of endophyte–host interactions as a foundation for emergent questions regarding the ecological roles of tropical endophytes.

ENDOPHYTES AS NEUTRAL INHABITANTS OF THEIR HOSTS

Over the past two decades, some authors have suggested that endophytes simply inhabit their hosts without interacting directly (e.g., Carroll 1988). Under this scenario, endophytism is an incidental part of the life cycle of fungi whose primary ecological role lies elsewhere. However, endophytes selectively colonize particular hosts (Arnold and Herre 2003), implying an interaction between endophytes and host defenses and/or other traits (see also Arnold *et al.* 2003). Further, endophytes actively penetrate leaf cuticles during colonization, and only rarely enter leaves in a passive manner (i.e., through open stomata; Mejia *et al.* 2003, Herre *et al.* 2005b). Endophytes remain metabolically active during the intercellular colonization phase, and grow slowly but actively within host foliage following infection (Deckert *et al.* 2001, Arnold *et al.* 2003). In each of these stages, fungi exude the organic molecules needed for cuticular penetration and absorptive nutrition (Van Schöll *et al.* 2006). Given the close phylogenetic relationship between endophytes and pathogens, it is likely that tropical plants are sensitive to such exudates. Can endophytes avoid inducing host defenses during colonization? If so, then how do plants in the forest understory tolerate carbon use by these heterotrophic colonists?

ENDOPHYTES AS PARASITES

Based in part on the observation that endophytes subsist on carbon from the host (Clay 2001), the potential role of endophytes as plant parasites has long been recognized. The evolutionary transiency between endophytism and

pathogenicity also underscores the possibly negative roles of endophytes (Arnold 2007). Yet leaf area, plant growth rates, and total biomass do not differ given the presence or absence of endophytes in seedlings of tropical angiosperms such as *Theobroma cacao*, *Gustavia superba*, and *Faramea occidentalis* (Arnold unpublished data). Similarly, Arnold and Engelbrecht (2007) found that endophyte infection did not influence leaf fresh weight, dry weight, or water content under well-watered conditions. While apparently supporting the neutralism hypothesis, these studies raise two questions: (1) to what degree are such outcomes sensitive to the makeup of endophyte communities in particular leaves; and (2) how are endophyte–host interactions shaped by pressure from natural enemies or abiotic stressors?

Few data are available to address the first question, but several studies have started to explore the second. For example, Arnold and Engelbrecht (2007) found that endophyte-infected leaves of *Theobroma cacao* lose water two times faster than endophyte-free leaves under severe drought conditions. In addition to the immediate consequences of desiccation, plants suffering from increased water stress may be more susceptible to fungal and bacterial pathogens (e.g., *Botryosphaeria dothidea*, Ma *et al.* 2001; *Xylella fastidiosa*, McElrone *et al.* 2001, McElrone and Forseth 2004) and abiotic stresses (Thaler and Bostock 2004). Effects on host water relations are likely most important in strongly seasonal tropical forests. In turn, beneficial effects such as anti-pathogen defense (see below) may be more important during wet seasons or in everwet forests. Similarly, apparently symptomless infections also can influence photosynthetic activity. For example, Pinto *et al.* (2000) found that infections by two endophytic Ascomycota (*Colletotrichum musae* and *Fusarium moniliforme*) reduced photosynthetic capacity in maize and banana. Because photosynthetic capacity is associated with plants' tolerance of herbivory (Agrawal 2000), endophyte infections may restrict the ability of plants to cope with damage. In the carbon-limited environment of the forest understory, the combined cost of reduced photosynthesis and a decrease in damage tolerance may be especially problematic.

ENDOPHYTES AS MUTUALISTS

A third and non-exclusive possibility is that plants benefit from the presence of endophytes in their tissues. Evolutionary theory provides little support for this hypothesis, as highly diverse symbionts and contagious spread are typically associated with parasitic or pathogenic lifestyles (Bull 1994). However, there are numerous examples of diverse, horizontally transmitted organisms that interact mutualistically with hosts (e.g., pollinating insects, mycorrhizal fungi, and root-nodulating bacteria; see Herre *et al.* 1999). The potential for endophytes to improve the fitness of hosts, and to do so in a host-specific manner, raises a series of questions of interest to community ecologists. I address a few of these below.

DO ENDOPHYTES ACT AS ENVIRONMENTALLY ACQUIRED IMMUNE SYSTEMS?

Herbivores and pathogens are important agents of density dependence (Carson *et al.* Chapter 13, this volume) and have played an important role in plant evolution, as demonstrated by the diversity and variation in chemical and structural plant defenses in tropical plants (Coley and Barone 1996). Both herbivores and pathogens have the potential to interact closely with foliar endophytes through the host plants that they share. Could endophytes provide a cryptic defense against antagonists? This hypothesis has been presented in various forms by numerous authors (e.g., Carroll 1991), yielding four possibilities regarding modes by which endophytes may contribute to host protection.

Do endophytes provide novel chemical defenses for hosts?

Endophytes may impart direct chemical defense to plants by producing secondary compounds that deter insects and inhibit pathogenic organisms (see Saikkonen *et al.* 1998). The ability of endophytes to secrete substances *in vitro* that limit the growth of other microbial species, including pathogens, has contributed to current enthusiasm regarding bioprospecting and biological control with endophytic fungi (e.g., Strobel and Daisy 2003, Gunatilaka 2006). In the context of herbivory, this mode of defense is exemplified by the alkaloids produced by clavicipitaceous endophytes of temperate grasses (Clay and Schardl 2002), and has been demonstrated in a few horizontally transmitted endophytes of woody plants (e.g., endophytes that produce compounds toxic to spruce budworms; Findlay *et al.* 2003). Arnold *et al.* (2003) suggested that endophytes of tropical trees serve as acquired immune systems, acting in concert with intrinsic leaf defensive chemistry when young, and in place of those defenses in mature leaves. However, the potential for these low-biomass infections to manifest major chemical signatures in foliage has not been assessed.

One intriguing hypothesis is that large quantities of chemical output per endophyte may not be needed to defend host tissues. Carroll (1991) proposed that endophytes protect hosts via a mosaic effect, whereby endophytes create a heterogeneous chemical landscape within and among leaves. As a result, parts of a genetically uniform plant would differ unpredictably in terms of palatability or quality for herbivores, and in terms of infectivity for pathogens. This hypothesis is compelling but has not yet been explored.

Do endophytes activate host defenses?

Systemic acquired resistance has long been recognized in plants (Agrios 1997) but there is currently no evidence for systemic protection of tropical plants as a function of endophyte infection. Arnold *et al.* (2003) raised endophyte-free seedlings, inoculated a subset of leaves on each seedling with endophytes, and then inoculated endophyte-infected and endophyte-free leaves with a virulent foliar pathogen (*Phytophthora*). In that study, the presence of endophytes in some leaves did not protect other leaves on the same plants from severe pathogen damage. That study did not examine pathogen damage in seedlings that had no endophytes, raising the possibility that some systemic defense occurred but was not detected. However, the systemic defense hypothesis is generally not supported by field observations: new endophyte infections accumulate in tissues following initial colonization, with strong evidence suggesting that early colonists do not deter later infections (until leaves are saturated with endophytes; see Arnold and Herre 2003). Instead, these data raise the possibility that endophytes evade or otherwise do not activate plant defenses.

Do endophytes interact directly with pathogens?

Endophytes have the capacity to interact directly with pathogens within the leaves they share. When Arnold *et al.* (2003) found that the anti-pathogen effects of endophytes were apparently restricted to the leaves that bore those endophytes, they concluded that direct or indirect interactions between endophytes and the *Phytophthora* pathogen were responsible for limiting the pathogen's spread. Under this scenario, the metaphor of "leaf as landscape" is apt: either through direct or indirect competition, or perhaps mycoparasitism, a robust endophyte community may limit the ability of invading pathogens to grow rapidly or extensively within leaves. Whether endophyte communities are more resistant to invasion when more diverse or more fully packed with individuals remains to be assessed, and lends itself to straightforward experiments.

Do endophytes serve as entomopathogens?

Several major pathogens of insects, including *Beauveria bassiana* (Lewis and Bing 1991), *Aspergillus* sp. (Cao *et al.* 2002), and *Paecilomyces* sp. (Arnold 2002) have been isolated as endophytes from temperate and tropical plants. In agroecosystems, endophyte infections by *Beauveria bassiana* have been successful in limiting damage to maize by the European corn borer, a major pest (Lewis and Bing 1991). Entomopathogenic infections of insects generally occur via cuticular penetration, rather than by consumption of infected plant tissues (Rawlins

1984). Thus, plants harboring entomopathogenic endophytes benefit from the production of many fungal propagules on senescent tissues. The frequency of entomopathogens among tropical endophytes has not been assessed, but is worth exploring with bioassays (to identify novel entomopathogens) and in terms of screening unnamed cultures and foliage samples with specific primers to recover known species of entomopathogenic fungi. More generally, the potential for plants to harbor entomopathogenics as symbionts is worth exploring in tropical forests (see Elliott *et al.* 2000).

ENDOPHYTES: MUTUALISTS OF INSECT HERBIVORES?

Inasmuch as endophytes may act to protect plants against insects, it is also possible that they serve as attractants of folivores, and/or may improve forage quality. Herbivorous insects are directly implicated in pathogen movement and/or infection success. At Los Tuxtlas, Mexico, García-Guzman and Dirzo (2001) showed that folivory and visible symptoms of pathogen damage were positively associated: 43% of surveyed leaves were damaged by both herbivores and pathogens, whereas 16% were damaged only by herbivores, and less than 2% were damaged only by pathogens. The authors concluded that cuticular wounding by insects is important for infection by pathogenic fungi. The same appears to be true for endophytes: Arnold (unpublished data) found that parts of *Gustavia superba* leaves that were damaged by hesperiid larvae had significantly higher rates of endophyte infection, higher endophyte species richness, and a different community of endophytes relative to undamaged areas of the same leaves (Figure 15.4). Similarly, Faeth and Wilson (1996) surface-sterilized and then artificially herbivorized living leaves of *Quercus emoryi* in Arizona, pairing each herbivore-damaged leaf with an undamaged leaf of similar age and position. At the end of the growing season, endophytes were more common in damaged leaves, and damaged areas of leaves, than in undamaged tissue.

Leaf damage by chewing or scraping insects, including Lepidoptera, Orthoptera, Coleoptera,

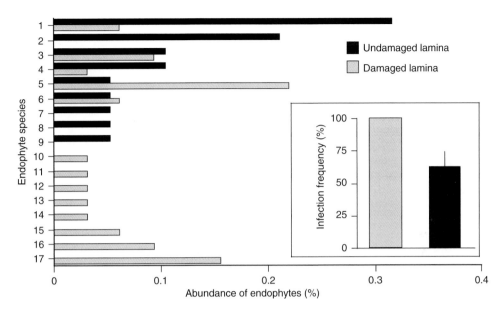

Figure 15.4 Effects of damage by hesperiid larvae on the endophyte community associated with *Gustavia superba* at BCI. Endophyte species composition differs markedly between the undamaged lamina and along edges cut by caterpillars. Inset: Caterpillar damage increases infection of leaf segments in young leaves of *Gustavia superba*.

and Hymenoptera, disrupts the leaf cuticle, thereby opening leaf interiors to ambient fungi. Insects with piercing or sucking mouthparts (e.g., Hemiptera) are more frequently associated with the transmission of viruses than with fungi, but exceptions do exist. For example, two species of aphids (*Aphis fabae* and *Uroleucon cirsii*) are capable of transmitting the fungal rust pathogen *Puccinia punctiformis* (Kluth *et al.* 2002). Moreover, sugar-rich exudates produced by piercing or sucking insects may increase the prevalence of, or survival of, fungi on leaf surfaces, increasing infection success. For example, honeydew provides a critical nutrient source for germination and survival of *Septoria nodorum*, *Uromyces vitis-fabae*, and *Botrytis fabae* on their respective host plants, which facilitates infection (reviewed in Hatcher 1995).

The observation that folivory can increase endophyte colonization, coupled with the observation that propagules of endophytic fungi remain viable following passage through the orthopteran gut (Monk and Samuels 1990), suggests that some endophytes may benefit from folivory. Such a benefit would have two components: folivores

would open new substrates by compromising leaf cuticles, and would provide a means by which endophytes could reproduce (from frass) more rapidly than if trapped within long-lived leaves. This incidental mycophagy should be evaluated: do chemical changes in leaf tissues, or the presence of fungal tissues, increase attractiveness to herbivores? Does the presence of fungal amino acids or other products improve forage quality? Could compounds generated by endophytes play a role in protecting herbivores from parasitoids? In general, the role of insect-mediated transmission of endophytes remains little explored in the tropics, but may be critical for understanding the complex interplay of endophytic and pathogenic fungi, herbivorous insects, and tropical plants.

SANTA ROSALIA'S FUNGAL BLESSINGS

A tremendous number of fungal species are capable of colonizing living plant tissues in tropical forests. Myriad species coexist within a landscape defined by only a few square centimeters of a leaf lamina, and leaves may in turn serve as provinces within the complex geography of individual branches, trees, and forests. The abundance and diversity of little-known endophytes have led some authors to characterize tropical plants as chimaera (Herre *et al.* 2005b), wherein plant tissue is interlaced with attendant fungal hyphae – or as "inside-out lichens" (Atsatt 1988), whereby plants are functionally inseparable from the genetic, structural, and evolutionary contributions of their fungal symbionts. The ecological roles of these hidden symbionts are just beginning to emerge, promising many decades of research at the interface of endophyte biology and tropical ecology.

It has been estimated that the vast majority of microfungi in tropical forests represent undescribed species (95%; see Arnold 2002), and some authors have suggested that as many as 1 million species of endophytes may exist. If working estimates are correct, then diversity of endophytes is likely many times higher than that of tropical plants. Why are there so many species of tropical endophytes? The answer may be rooted in rates of speciation and extinction, and in the case of symbionts, a subsequently complex interplay of specificity and generalism that cannot yet be reliably estimated. As indicated by Leigh (1999), the more apt question may be: What factors facilitate coexistence of such diverse species? The multiplicity of species sharing similar substrates in tropical forests may push for a continuous process of character displacement in tropical endophyte communities (see Kitajima and Poorter Chapter 10, this volume). Moreover, the nearly infinite combinations of genotypes, chemical exudates, and interactions are likely to create endophyte communities with distinctive emergent properties that are even more diverse than the fungi themselves.

ACKNOWLEDGMENTS

I thank Lissy Coley, Tom Kursar, Greg Gilbert, and especially Lucinda McDade and François Lutzoni for guidance and helpful discussion, as well as two anonymous reviewers and Stefan Schnitzer for improving the manuscript. I gratefully acknowledge funding support from the National Science Foundation (DEB-9902346 to L. McDade; DEB-0200413; DEB-0343953 to J.W. Dalling and AEA; and DEB-0640956), the College of Agriculture and Life Sciences at the University of Arizona, and the Smithsonian Tropical Research Institute (short-term fellowship under the guidance of E. Allen Herre and Egbert Leigh).

REFERENCES

Agrawal, A. (2000) Overcompensation of plants in response to herbivory and the by-product benefits of mutualism. *Trends in Plant Sciences* 5, 309–313.

Agrios, G.N. (1997) *Plant Pathology.* Academic Press, San Diego, USA.

Alexopoulos, C.J., Mims, C.W., and Blackwell, M. (1996) *Introductory Mycology.* John Wiley and Sons, New York.

Arnold, A.E. (2002) *Neotropical fungal endophytes: diversity and ecology.* PhD dissertation, University of Arizona.

Arnold, A.E. (2007) Understanding the diversity of foliar fungal endophytes: progress, challenges, and frontiers. *Fungal Biology Reviews* 21, 51–66.

Arnold, A.E. and Engelbrecht, B.M.J. (2007) Fungal endophytes nearly double minimum leaf conductance in seedlings of a tropical tree. *Journal of Tropical Ecology* 23, 369–372.

Arnold, A.E., Henk, D.A., Eells, R.L. *et al.* (2007) Diversity and phylogenetic affinity of foliar endophytes associated with loblolly pine inferred by culturing and direct PCR. *Mycologia* 99, 185–206.

Arnold, A.E. and Herre, E.A. (2003) Canopy cover and leaf age affect colonization by tropical fungal endophytes: ecological pattern and process in *Theobroma cacao* (Malvaceae). *Mycologia* 95, 388–398.

Arnold, A.E. and Lewis, L.C. (2005) Ecology and evolution of fungal endophytes, and their role against insects. In F. Vega and M. Blackwell (eds), *Ecological and Evolutionary Advances in Insect–Fungus Associations.* Oxford University Press, Oxford, pp. 74–96.

Arnold, A.E. and Lutzoni, F. (2007) Diversity and host range of foliar fungal endophytes: are tropical leaves biodiversity hotspots? *Ecology* 88, 541–549.

Arnold, A.E., Maynard, Z., Gilbert, G.S. *et al.* (2000) Are tropical fungal endophytes hyperdiverse? *Ecology Letters* 3, 267–274.

Arnold, A.E., Mejía, L.C., Kyllo, D. *et al.* (2003) Fungal endophytes limit pathogen damage in a tropical tree. *Proceedings of the National Academy of Sciences of the United States of America* 100, 15649–15654.

Atsatt, P.N. (1988) Are vascular plants inside-out lichens? *Ecology* 69, 17–23.

Augspurger, C.A. (1983) Seed dispersal of the tropical tree, *Platypodium elegans*, and the escape of its seedlings from fungal pathogens. *Journal of Ecology* 71, 759–771.

Bills, G.F. and Polishook, J.D. (1994) Abundance and diversity of microfungi in leaf litter of a lowland rain forest in Costa Rica. *Mycologia* 86, 187–198.

Bull, J.J. (1994) Perspective: virulence. *Evolution* 48, 1423–1437.

Cannon, P.F. and Simmons, C.M. (2002) Diversity and host preference of leaf endophytic fungi in the Iwokrama Forest Reserve, Guyana. *Mycologia* 94, 210–220.

Cao, L.X., You, J.L., and Zhou, S.N. (2002) Endophytic fungi from *Musa acuminata* leaves and roots in South China. *World Journal of Microbiology and Biotechnology* 18, 169–171.

Carroll, G.C. (1986) The biology of endophytism in plants with particular reference to woody perennials. In N.J. Fokkema and J. van den Huevel (eds), *Microbiology of the Phyllosphere*. Cambridge University Press, Cambridge, pp. 205–222.

Carroll, G.C. (1988) Fungal endophytes in stems and leaves: from latent pathogen to mutualistic symbiont. *Ecology* 69, 2–9.

Carroll, G.C. (1991) Beyond pest deterrence. Alternative strategies and hidden costs of endophytic mutualisms in vascular plants. In J.H. Andrews and S.S. Hirano (eds), *Microbial Ecology of Leaves*. Springer-Verlag, New York, pp. 358–375.

Clay, K. (2001) Symbiosis and the regulation of communities. *American Zoologist* 41, 810–824.

Clay, K., Hardy, T.N., and Hammond, A.M. Jr. (1985) Fungal endophytes of grasses and their effects on an insect herbivore. *Oecologia* 66, 1–6.

Clay, K. and Schardl, C. (2002) Evolutionary origins and ecological consequences of endophyte symbiosis with grasses. *American Naturalist* 160, S99–S127.

Coley, P.D. and Barone, J.A. (1996) Herbivory and plant defenses in tropical forests. *Annual Review of Ecology and Systematics* 27, 305–335.

Deckert, R.J., Melville, L.H., and Peterson, R.L. (2001) Structural features of a *Lophodermium* endophyte during the cryptic life-cycle phase in the foliage of *Pinus strobus*. *Mycological Research* 105, 991–997.

Dobson, A. and Crawley, M. (1994) Pathogens and the structure of plant communities. *Trends in Ecology and Evolution* 9, 393–398.

Elliott, S.L., Sabelis, M.W., van der Geest, L.P.S. *et al.* (2000) Can plants use entomopathogens as bodyguards? *Ecology Letters* 3, 228–235.

Faeth, S.H. and Sullivan, T.J. (2003) Mutualistic asexual endophytes in a native grass are usually parasitic. *American Naturalist* 161, 310–325.

Faeth, S.H. and Wilson, D. (1996) Induced responses in trees: mediators of interactions between macro- and micro-herbivores. In A.C. Grange (ed.), *Multitrophic Interactions in Terrestrial Systems*. Blackwell Scientific, New York, pp. 201–215.

Findlay, J.A., Li, G.Q., Miller, J.D. *et al.* (2003) Insect toxins from spruce endophytes. *Canadian Journal of Chemistry* 81, 284–292.

Fisher, P.J., Graf, F., Petrini, L.E. *et al.* (1995) Fungal endophytes of *Dryas octopetala* from a high arctic semidesert and from the Swiss Alps. *Mycologia* 87, 319–323.

Fröhlich, J. and Hyde, K.D. (1999) Biodiversity of palm fungi in the tropics: are global fungal diversity estimates realistic? *Biodiversity and Conservation* 8, 977–1004.

Gallery, R.E., Dalling, J.W., and Arnold, A.E. (2007) Diversity, host affinity, and distribution of seed-infecting fungi: a case study with neotropical *Cecropia*. *Ecology* 88, 582–588.

Gamboa, M.A. and Bayman, P. (2001) Communities of endophytic fungi in leaves of a tropical timber tree (*Guarea guidonia*: Meliaceae). *Biotropica* 33, 352–360.

Gamboa, M.A., Laureano, P., and Bayman, P. (2002) Measuring diversity of endophytic fungi in leaf fragments: Does size matter? *Mycopathologia* 156, 41–45.

Garcia-Guzman, G. and Dirzo, R. (2001) Patterns of leaf-pathogen infection in the understory of a Mexican rain forest: incidence, spatiotemporal variation, and mechanisms of infection. *American Journal of Botany* 88, 634–645.

Gilbert, G.S. (2002) Evolutionary ecology of plant diseases in natural ecosystems. *Annual Review of Phytopathology* 40, 13–43.

Gilbert, G.S., Mejia-Chang, M., and Rojas, E. (2002a) Fungal diversity and plant disease in mangrove forests: salt excretion as a possible defense mechanism. *Oecologia* 132, 278–285.

Gilbert, G.S., Ferrer, A., and Carranza, J. (2002b) Polypore fungal diversity and host density in a moist tropical forest. *Biodiversity and Conservation* 11, 947–957.

Gunatilaka, A.A.L. (2006) Natural products from plant-associated microorganisms: distribution, structural diversity, bioactivity, and implications of their occurrence. *Journal of Natural Products* 69, 509–526.

Guo, L.D., Huang, G.R., Wang, Y. *et al.* (2003) Molecular identification of white morphotype strains of endophytic fungi from *Pinus tabulaeformis*. *Mycological Research* 107, 680–688.

Guo, L.D., Hyde, K.D., and Liew, E.C.Y. (2000) Identification of endophytic fungi from *Livistona chinensis* based on morphology and rDNA sequences. *New Phytologist* 147, 617–630.

Guyot, J., Omanda, E.M., and Pinard, F. (2005) Some epidemiological investigations on *Colletotrichum* leaf disease on rubber tree. *Crop Protection* 24, 65–77.

Hatcher, P.E. (1995) Three-way interactions between plant-pathogenic fungi, herbivorous insects, and their host plants. *Biological Reviews of the Cambridge Philosophical Society* 70, 639–694.

Hawksworth, D.L. (1991) The fungal dimension of biodiversity: magnitude, significance, and conservation. *Mycological Research* 95, 641–655.

Hawksworth, D.L. (1994) *Ascomycete Systematics: Problems and Perspectives in the Nineties.* Plenum Press, New York.

Herre, E.A., Knowlton, N., Mueller, U.G. *et al.* (1999) The evolution of mutualisms: exploring the paths between conflict and cooperation. *Trends in Ecology and Evolution* 14, 49–53.

Herre, E.A., Kyllo, D.A., Mangan, S.A. *et al.* (2005a) In D.F.R.P. Burslem, M.A. Pinard, and S.E. Hartley (eds), *An Overview of Arbuscular Mycorrhizal Fungi: Composition, Distribution, and Host Effects from a Moist Tropical Forest.* Cambridge University Press, Cambridge, pp. 204–225.

Herre, E.A., Van Bael, S.A., Maynard, Z. *et al.* (2005b) Tropical plants as chimera: some implications of foliar endophytic fungi for the study of host plant defense, physiology, and genetics. In D.F.R.P. Burslem, M.A. Pinard, and S.E. Hartley (eds), *Biotic Interactions in the Tropics.* Cambridge University Press, Cambridge, pp. 226–237.

Higgins, L., Arnold, A.E., Miadlikowska, J. *et al.* (2007) Phylogenetic relationships, host affinity, and geographic structure of boreal and arctic endophytes from three major plant lineages. *Molecular Phylogenetics and Evolution* 42, 543–555.

Hoffman, M. and Arnold, A.E. (2008) Geographic locality and host identity shape fungal endophyte communities in cupressaceous trees. *Mycological Research*, in press.

Holb, I.J., Heijne, B., Withagen, J.C.M. *et al.* (2004) Dispersal of *Ventura inaequalis* ascospores and disease gradients from a defined inoculum source. *Journal of Phytopathology* 152, 639–646.

Hyde, K.D. (1997) *Biodiversity of Tropical Microfungi.* Hong Kong University Press, Hong Kong.

Husband, R., Herre, E.A., and Young, J.P.W. (2002) Temporal variation in the arbuscular mycorrhizal communities colonising seedlings in a tropical forest. *FEMS Microbiology Ecology* 42, 131–136.

Jacobs, K.A. and Rehner, S.A. (1998) Comparison of cultural and morphological characters and ITS sequences in anamorphs of *Botryosphaeria* and related taxa. *Mycologia* 90, 601–610.

Kiers, E.T., Lovelock, C.E., Krueger, E.L. *et al.* (2000) Differential effects of tropical arbuscular mycorrhizal fungal inocula on root colonization and tree seedling growth: implications for tropical forest diversity. *Ecology Letters* 3, 106–113.

Kim, S.H. and Breuil, C. (2002) Common nuclear ribosomal internal transcribed spacer sequences occur in the sibling species *Ophiostoma piceae* and *O. quercus*. *Mycological Research* 105, 331–337.

Kirk, P.M., Cannon, P.F., David, J.C. *et al.* (2001) *Dictionary of the Fungi.* 9th Edition. CABI Publishing, Wallingford.

Kluth, S., Kruess, A., and Tscharntke, T. (2002) Insects as vectors of plant pathogens: mutualistic and antagonistic interactions. *Oecologia* 133, 193–199.

Kumaresan, V. and Suryanarayanan, T.S. (2002) Occurrence and distribution of endophytic fungi in a mangrove community. *Mycological Research* 105, 1388–1391.

Leigh, E.G. Jr. (1999) *Tropical Forest Ecology.* Oxford University Press, Oxford.

Lewis, L.C. and Bing, L.A. (1991) *Bacillus thuringiensis* Berliner and *Beauveria bassiana* (Balsamo) Vuillimen for European Corn Borer control: program for immediate and season-long suppression. *Canadian Entomologist* 123, 387–393.

Lieckfeldt, E. and Seifert, K.A. (2000) An evaluation of the use of ITS sequences in the taxonomy of the Hypocreales. *Studies in Mycology* 45, 35–44.

Lodge, D.J. (1997) Factors related to diversity of decomposer fungi in tropical forests. *Biodiversity and Conservation* 6, 681–688.

Lodge, D.J., Fisher, P.J., and Sutton, B.C. (1996) Endophytic fungi of *Manilkara bidentata* leaves in Puerto Rico. *Mycologia* 88, 733–738.

Lutzoni, F., Kauff, F., Cox, C. *et al.* (2004) Assembling the fungal tree of life: progress, classification, and the evolution of subcellular traits. *American Journal of Botany* 91, 1446–1480.

Lutzoni, F., Pagel, M., and Reeb, V. (2001) Major fungal lineages are derived from lichen symbiotic ancestors. *Nature* 411, 937–940.

Ma, Z., Morgan, D.P., and Michailides, T.J. (2001) Effects of water stress on Botryosphaeria blight of pistachio caused by *Botryosphaeria dothidea*. *Plant Disease* 85, 745–749.

Mangan, S.A., Eom, A.H., Adler, G.H. *et al.* (2004) Diversity of arbuscular mycorrhizal fungi across a fragmented forest in Panama: insular spore communities differ from mainland communities. *Oecologia* 141, 687–700.

McElrone, A.J. and Forseth, I.N. (2004) Photosynthetic responses of a temperate liana to *Xylella fastidiosa* infection and water stress. *Journal of Phytopathology* 152, 9–20.

McElrone, A.J., Sherald, J.L., and Forseth, I.N. (2001) Effects of water stress on symptomatology and growth of *Parthenocissus cinquefolia* infected by *Xylella fastidiosa*. *Plant Disease* 85, 1160–1164.

Mejia, L.C., Rojas, E., Maynard, Z. *et al.* (2003) Inoculation of beneficial endophytic fungi into *Theobroma cacao* tissues. In *Proceedings of the 14th International Cocoa Research Conference*, Accra, Ghana.

Monk, K.A. and Samuels, G.J. (1990) Mycophagy in grasshoppers (Orthoptera, Acrididae) in Indo-Malayan rain forest. *Biotropica* 22, 16–21.

Petrini, O. (1986) Taxonomy of endophytic fungi of aerial plant tissues. In N.J. Fokkema and J. van den Huevel (eds), *Microbiology of the Phyllosphere*. Cambridge University Press, Cambridge, pp. 175–187.

Petrini, O. (1991) Fungal endophytes of tree leaves. In J.H. Andrews and S.S. Hirano (eds), *Microbial Ecology of Leaves*. Springer-Verlag, New York, pp. 179–197.

Phelan, P.L. and Stinner, B.R. (1992) Microbial mediation of plant-herbivore interactions. In G.A. Rosenthal and M.R. Berenbaum (eds), *Herbivores: Their Interactions with Secondary Plant Metabolites* Academic Press, New York, pp. 279–315.

Pinto, L.S.R.C., Azevedo, J.L., Pereira, J.O. *et al.* (2000) Symptomless infection of banana and maize by endophytic fungi impairs photosynthetic efficiency. *New Phytologist* 147, 609–615.

Rawlins, J.E. (1984) Mycophagy in Lepidoptera. In Q. Wheeler and M. Blackwell (eds), *Fungus–Insect Relationships: Perspectives in Ecology and Evolution*. Columbia University Press, New York, pp. 382–423.

Rodrigues, K.F. (1994) The foliar fungal endophytes of the Amazonian palm *Euterpe oleracea*. *Mycologia* 86, 376–385.

Rossman, A.Y. (1996) Morphological and molecular perspectives on systematics of the Hypocreales. *Mycologia* 88, 1–19.

Rosenzweig, M. (1995) *Species Diversity in Space and Time*. Cambridge University Press, Cambridge.

Saikkonen, K., Faeth, S.H., Helander, M. *et al.* (1998) Fungal endophytes: a continuum of interactions with host plants. *Annual Review of Ecology and Systematics* 29, 319–343.

Schulz, B., Wanke, U., Draeger, S. *et al.* (1993) Endophytes from herbaceous plants and shrubs – effectiveness of surface sterilization methods. *Mycological Research* 97, 1447–1450.

Siegel, M.R., Latch, G.C.M., Bush, L.P. *et al.* (1990) Fungal endophyte-infected grasses: alkaloid accumulation and aphid response. *Journal of Chemical Ecology* 16, 3301–3315.

Stone, J.E., Bacon, C.W., and White, J.F. Jr. (2000) An overview of endophytic microbes: endophytism defined. In C.W. Bacon and J.F. White (eds), *Microbial Endophytes*. Marcel Dekker, New York, pp. 3–29.

Strobel, G.A. and Daisy, B. (2003) Bioprospecting for microbial endophytes and their natural products. *Microbiology and Molecular Biology Reviews* 67, 491–502.

Suryanarayanan, T.S., Murali, T.S., and Venkatesan, G. (2002) Occurrence and distribution of fungal endophytes in tropical forests across a rainfall gradient. *Canadian Journal of Botany* 80, 818–826.

Suryanarayanan, T.S., Wittlinger, S.K., and Faeth, S.H. (2005) Endophytic fungi associated with cacti in Arizona. *Mycological Research* 109, 635–639.

Thaler, J.S. and Bostock, R.M. (2004) Interactions between abscisic-acid-mediated responses and plant resistance to pathogens and insects. *Ecology* 85, 48–58.

Van Bael, S.A., Maynard, Z., Rojas, E. *et al.* (2005) Emerging perspectives on the ecological roles of endophytic fungi in tropical plants. In J.F. White, Jr, J. Dighton, and P. Oudemans (eds), *The Fungal Community: Its Organization and Role in the Ecosystem*, 3rd Edition. Marcel-Dekker, New York, pp. 181–191.

Van Schöll, L., Hoffland, E., and van Breemen, N. (2006) Organic anion exudation by ectomycorrhizal fungi and *Pinus sylvestris* in response to nutrient deficiencies. *New Phytologist* 170, 153–163.

Wilkinson, H.H., Siegel, M.R., Blankenship, J.D. *et al.* (2000) Contribution of fungal loline alkaloids to protection from aphids in a grass–endophyte mutualism. *Molecular Plant–Microbe Interactions* 13, 1027–1033.

Wills, C., Condit, R., Foster, R.B. *et al.* (1997) Strong density- and diversity-related effects help to maintain tree species diversity in a tropical forest. *Proceedings of the National Academy of Sciences of the United States of America* 94, 1252–1257.

ANIMAL COMMUNITY ECOLOGY AND TROPHIC INTERACTIONS

Chapter 16

TROPICAL TRITROPHIC INTERACTIONS: NASTY HOSTS AND UBIQUITOUS CASCADES

Lee A. Dyer

OVERVIEW

In the tropics, the high diversity of species at all trophic levels combined with increased chemical defense and predation intensity create ideal opportunities for interesting research in community ecology. Two particularly useful themes in the realm of tropical tritrophic interactions are trophic cascades and coevolution, and prominent hypotheses generated by these ideas should continue to provide guidance to empirical studies in tropical communities. Trophic cascades and coevolutionary interactions are expected to be different in tropical communities simply because of the increased diversity for most taxa at all trophic levels. However, many of the assumptions about how tropical communities are different from their temperate counterparts are not well tested and could be incorrect. Thus, a major goal of understanding tropical tritrophic interactions is to thoroughly document latitudinal patterns in community attributes such as consumer specialization, plant chemical defense, and intensity of predation.

There are no adequate syntheses of trophic cascades and coevolutionary hypotheses for the tropics due to a lack of focused research programs. To explicitly test these hypotheses, tropical ecologists should focus on model systems and must utilize phylogenetic data combined with creative experimental, correlational, observational, and modeling approaches. Myrmecophytes are good candidates as model systems for such a synthetic approach, given the diversity and importance of ant plants in most tropical communities. Tritrophic interactions in tropical communities are usually part of a more complex web with highly variable interaction strengths, yet with the right approaches and study systems, we can determine which interactions are the strongest for particular taxa and ecosystems.

INTRODUCTION

The interactions between myrmecophytes and their associated arthropods are perhaps the most distinctively tropical of all documented tritrophic interactions. These diverse tropical plants, which have evolved in over 100 genera (Heil and McKey 2003), are likely the result of millions of years of strong tritrophic interactions (e.g., Itino *et al.* 2001, Quek *et al.* 2004, Tepe 2004, McKey *et al.* 2005) and are just one of the many genres of intricate tritrophic stories that have yet to be fully investigated. Tropical ant plants have provided a convincing affirmative answer to the question of whether or not natural enemy impact on herbivores exerts strong enough selection pressure to modify plant traits, which is a central question for tritrophic studies. In fact, a thorough research program that utilizes a tropical myrmecophyte as a model system should produce advances for major issues in tritrophic interactions, including trophic cascades (Schmitz *et al.* 2000), evolution of specialization (Yu and Davidson 1997), multi-trophic mutualism (Gastreich and Gentry 2004), interactions between living and detrital food webs (Dyer and Letourneau 2003), induced defenses (Fiala *et al.* 1989), genetic variation (Dalecky *et al.* 2002), plant defense theory (Heil *et al.* 2002), and chemical ecology (Rehr *et al.* 1973).

In this chapter I present challenges for understanding tritrophic interactions in the tropics. Because the field is quickly growing into an unwieldy topic worthy of its own volume (e.g., Tscharntke and Hawkins 2002, Burslem *et al.* 2005), I focus on two particularly important issues: trophic cascades and the evolution of feeding specialization. The trophic cascades hypothesis is a focus because the regulation of prey populations by natural enemies and the indirect effects on other trophic levels are focal research topics for community ecologists, population ecologists, conservation biologists, and applied scientists in agriculture and forestry. Specialization as a consequence of coevolution between hosts and parasites (which include herbivorous insects) is a key concept in tropical community ecology, and coevolutionary interactions could potentially generate a large percentage of the great diversity of plants and animals in tropical communities (Ehrlich and Raven 1964, Raven 1977, Farrell *et al.* 1992, Scott *et al.* 1992). But 40 years of theoretical development and hundreds of empirical studies still have not produced a comprehensive theoretical framework, and as a result there are no cohesive research approaches, especially for tropical taxa. In particular, coevolutionary theory has rarely considered the roles of other selective forces that could modify or enhance coevolution, such as predators and parasitoids of the herbivores (Singer and Stireman 2005).

Are tritrophic interactions in a tropical forest or agricultural system empirically distinguishable from temperate tritrophic interactions? Trophic cascades and coevolutionary interactions are expected to be different in tropical communities simply because of the increased diversity for most taxa at all trophic levels. Increased diversity at a given trophic level can weaken the effect of consumption on lower trophic levels, due to increases in interference competition (including intra-guild predation), diet shifts, omnivory, and other buffering mechanisms that are enhanced by greater complexity (Polis and Strong 1996).

Increased diversity can also weaken the effect of resource availability on upper trophic levels due to increases in exploitation competition, decreased host availability for specialists, and changes in chemical defenses (Hunter and Price 1992, Dyer and Coley 2001). If true, these ecological changes could also make coevolution a less likely outcome, since top-down and bottom-up selective forces could be weakened via the same mechanisms that weaken cascades. There are plenty of additional attributes specific to tropical communities that lead to different predictions about selective pressures between trophic levels and associated indirect effects (e.g., increased primary productivity, see Oksanen *et al.* 1981). But how many of these additional attributes are rigorously documented and how many are simply part of tropical lore? Before examining coevolution and trophic cascades in the tropics, it is worth reviewing some of the assumptions about how tropical communities are different from their temperate counterparts.

TOUGHER PREDATORS, NASTIER PLANTS, MORE SPECIALIZED CONSUMERS?

At the heart of all multitrophic issues in tropical community ecology are many assumptions that remain largely untested. Aside from obvious correlates of the increases in diversity, such as more reticulate food webs, the most prominent assumptions for tropical communities are: (1) tropical consumers are more specialized (Dobzhansky 1950, Pianka 1966, MacArthur and Wilson 1967); (2) predation is more intense in the tropics (Paine 1966, Janzen 1970); (3) chemical defenses are more abundant and toxic in the tropics (reviewed by Dyer and Coley 2001); and (4) multitrophic mutualisms are more important for tropical communities (Price 1991). It may seem that the only tenable generalization about latitudinal gradients in community ecology is the gradient in species richness, but a close examination of complex trophic interactions should reveal other strong gradients. The first job for tropical ecologists is to determine the taxa, ecosystems, and conditions for which the tropical paradigms of specialization, strong predation, toxic food, and indirect mutualisms are true. This is a prerequisite to addressing any questions in tritrophic interactions, including hypotheses on the evolution of specialization and trophic cascades. The best way to accomplish this is to ensure that improved natural history is a priority in all research on tropical tritrophic interactions.

Within the tropics, there are also complex patterns of tritrophic interactions. For example, altitudinal gradients create ecosystems in close proximity and extreme differences in overall diversity, productivity, and ant abundance – all of which decline with altitude (Janzen 1967). However, altitudinal gradients in ecological interactions have not been formally examined in the tropics (Novotny and Basset 2005). Perhaps the most striking pattern of interactions within the tropics is seen along the climate gradient from dry deciduous to wet evergreen forests in the tropics. As total annual rainfall increases and climatic variability decreases, tropical forests have higher plant diversity (Hall and Swaine 1976, Huston 1980, Gentry 1982, 1988), greater primary productivity and stem turnover (Philips *et al.* 1994), and lower seasonal production of new foliage and reproductive parts (Opler *et al.* 1980, van Schaik *et al.* 1993). In addition, plants living in wetter tropical forests appear to be better defended against herbivores, because their leaves are typically tougher, with higher concentrations of secondary compounds and lower nutritional value (Coley and Aide 1991). These changes in plant characteristics along tropical rainfall gradients should have important effects on tritrophic interactions. For example, it is possible that both the top-down impact of natural enemies and the bottom-up effect of plant defenses increase with greater rainfall and climate variability, leading to lower annual herbivore densities in wetter tropical forests (Coley and Barone 1996, Stireman *et al.* 2005). To document such a relationship between climate and herbivory based on the differences between tropical dry and wet forests, three general hypotheses should be tested: (1) in dry forests herbivore populations are limited by the bottom-up effect of plant availability (since leaves are largely deciduous and absent during the dry season) and direct abiotic effects of the severe

dry season (Janzen 1988, 1993); (2) parasitoids, predators, and plant secondary compounds have a relatively low impact on herbivore populations in climatically variable dry forests; and (3) the effect of the dry season is small in wetter, less seasonal, tropical forests but herbivore populations are strongly influenced by the bottom-up effect of greater plant defenses and the top-down impact of higher enemy densities.

Tropical community ecologists have failed to provide sufficient support for the generalizations about differences between tropical and temperate communities and have not tested any hypotheses about tritrophic trends across tropical forests. There are a number of reasons that appropriate investigations have not been completed, perhaps the most significant being lack of resources to support the necessary research. Assuming that funding is available for such work in the future, careful tests of hypotheses that examine the evolution of specialization and trophic cascades will generate data that help establish the strength of these putative patterns and the relative importance of underlying mechanisms.

EVOLUTION OF DIETARY SPECIALIZATION

Tritrophic view of feeding specialization

Most current studies on tritrophic interactions are directly or indirectly influenced by the coevolution paradigm, in which the evolution of dietary specialization is a result of increasingly specialized adaptations for secondary metabolites in one plant taxon (Dethier 1954, Fraenkel 1959, Ehrlich and Raven 1964). This hypothesis was preceded by an explicitly tritrophic idea that specialized diets represent enemy free space for herbivores, because monophagous insects are better able to utilize chemical, morphological, and phenological attributes of their host plants to defend against predators and parasitoids (Brower 1958). Multiple authors have proposed the hypothesis that plant availability/apparency (*sensu* Feeny 1976, Rhoades and Cates 1976, including plant chemistry) and pressure from enemies shape herbivore diet breadth together (Hassell and

Southwood 1978, Dyer 1995, Camara 1997a, Singer *et al.* 2004a). In tropical wet forests, if herbivores really are trapped between more toxic plants and higher rates of predation and parasitism, it is likely that herbivore specialization has evolved in response to one or both of these strong forces and is maintained by one or both. The trick is trying to determine the relative roles of these selective forces. For a given herbivore clade, did specialization evolve as herbivores developed mechanisms to enhance plant availability (e.g., overcoming chemical defenses via specific enzymes) and then enemies maintained that specialization, or vice versa? Or were plant availability and natural enemies irrelevant? These questions can only be addressed by combining strong phylogenetic approaches with experimental approaches (e.g., Farrell and Mitter 1990, Futuyma and Mccafferty 1990, also see Blackburn 2004) that examine the effects of herbivore diet on levels of parasitism and predation (reviewed by Hunter 2003). The high concordance between the cladogram of the chrysomelid genus *Phyllobrotica* and that of its host plants (Farrell and Mitter 1990) implies diversification in parallel as envisioned by the coevolutionary scenario, but is it possible that enemies were an additional selective force maintaining specialization in these beetles?

Studies of feeding specialization are actually focused on the "realized niche" of an animal's diet – the suite of resources that it is known to consume under natural conditions. Feeding efficiency is an additional component of specialized consumption (at any trophic level) and consumers that can efficiently consume only a narrow range of resources are referred to as "functional" specialists (Ferry-Graham *et al.* 2002, Irschick *et al.* 2005). Many ecologists have assumed that such functional specialization should be positively correlated with narrow diet breadths observed in nature. This assumption, however, is not appropriate (Fox and Morrow 1981, Camara 1997b) because enhanced feeding performance can evolve independently of dietary specialization. Furthermore, Fox and Morrow (1981) found that specialist insects effectively metabolize plant chemicals from plants that they rarely use in nature. The dichotomy between ecological specialization and functional specialization is more obvious when

herbivores specialize on plants with defenses that decrease feeding efficiency (bad for the herbivore) while simultaneously deterring enemies (good for the herbivore). Despite the fact that the herbivore is a specialist, it does not perform better (physiologically) on its diet of choice, but it may enjoy lower mortality. In such a case, classic laboratory and field rearing experiments designed to detect trade-offs between feeding performance and diet (e.g., Camara 1997b) do not successfully detect negative genetic correlations because the herbivores are not functional specialists – in other words there are no genotypes that perform better on one diet versus another, but they are still limited to one diet due to pressure from enemies.

Tests of the tritrophic view in temperate and tropical systems

Explicit tests of the coevolutionary (bottom-up) scenario for dietary specialization in herbivores have been conducted primarily with temperate taxa (e.g., at this writing, only 23 of 750 studies that cite Ehrlich and Raven's 1964 paper are focused on tropical taxa). A prominent exception is the well-documented synchronous evolution of *Blepharida* beetles and their host plants, *Bursera* spp. (e.g., Becerra and Venable 1999). The leaf beetles in this relationship have developed a wide array of behavioral and physiological mechanisms for circumventing each new defense of the host leaves, including squirting resins and complex mixtures of terpenes. There is no reason to assume that this and other well-documented examples of strong coevolution between host plant and herbivores are the rule in tropical communities, especially in light of the fact that several studies have also found low congruence between plant and herbivore phylogenies (e.g., Anderson 1993, Funk *et al.* 1995, Weintraub *et al.* 1995, Brandle *et al.* 2005). A rigorous coevolutionary theory for tropical systems awaits more tests of parallel phylogenies following the examples of existing work (Farrell and Mitter 1990, Mitter *et al.* 1991, Farrell *et al.* 1992, Futuyma *et al.* 1995, Becerra and Venable 1999) that encompass only a few clades.

The specific top-down view outlined above and also described by Singer and Stireman (2005) has been tested in the tropics with predators (Dyer and Floyd 1993, Dyer 1995, 1997) and parasitoids (Gentry and Dyer 2002). The results for predators mirrored temperate studies (Bernays and Graham 1988, Bernays and Cornelius 1989), with specialists being better protected against predators than generalists. However, patterns of parasitism are very different. Gentry and Dyer (2002) found that tropical specialists were not better protected than generalists, and in fact some parasitoid taxa (e.g., Braconidae) prefer specialists and chemically defended caterpillars (Dyer 2001), perhaps because these hosts represent enemy free space, since chemically defended specialists are avoided by many distinct guilds of predators (Dyer 1997). This is in striking contrast to studies in temperate systems that demonstrate anti-parasitoid defensive value of sequestered secondary compounds (Barbosa et al. 1986, 1991, Turlings and Benrey 1998, Sime 2002) and it provides evidence against the "nasty host hypothesis" (Gauld et al. 1992), which argues that parasitic hymenopterans are less diverse in the tropics because their hosts have high levels of chemical defense. For all of these studies, a major problem with comparing defenses of specialist versus generalist herbivores against their enemies is that the original selective advantages of specializing could be lost, especially for anti-parasitoid mechanisms, since parasitoids could evolve mechanisms that allow them to overcome chemical defenses sequestered by herbivores (Duffey et al. 1986, Barbosa 1988, Hunter 2003). Whenever possible, a phylogenetic approach should be utilized to examine the evolution of diet breadth in association with adaptations that allow use of phytochemicals as anti-predator and anti-parasitoid defenses (Termonia et al. 2001, Kuhn et al. 2004).

Temporal scales: from over 100 million years ago to current communities

The tropical patterns of parasitism on lepidopterans of varying diet breadths are potentially compatible with the view that enemies contributed to patterns of specialization, since pressure from parasitoids is relatively new compared with the long histories of specialized plant–herbivore relationships. For example, the Tachinidae is an estimated 20–40 million years old (Evenhuis 1994), and this family is usually the dominant source of lepidopteran mortality (Dyer and Gentry 2002, Gentry and Dyer 2002, Janzen and Hallwachs 2002, Stireman et al. 2005). In contrast, some genera of plants and herbivores have associations that go back almost 100 million years (Labandeira et al. 1994, Becerra 2003). Does this mean that selective pressures from tachinids that have arisen over the last 20 million years are driving diet breadths of herbivorous insects towards polyphagy because tachinids attack specialized herbivores? The question of evolution of generalized diet has been examined only sparingly in a phylogenetic context (for aphids, Moran 1988; for parasitoids, Stireman 2002). Singer and colleagues (2004a,b) have taken an interesting approach to understanding diet breadth of generalist arctiids. For two generalist arctiids (*Estigmene acrea* and *Grammia geneura*), a mixed diet provides benefits of increased growth due to including a high quality plant in the diet and increased defense due to including a toxic plant in the diet. For both of these arctiids, the value of enemy free space supersedes the value of enhanced larval performance due to better food quality.

Regardless of how the specialization evolved at any trophic level and whether or not it is adaptive, narrow consumer diet breadth should modify its ecological role in a community (e.g., herbivores of different diet breadth respond differently to resources, Long et al. 2003). Specialist herbivores are far more likely to present a consistent regulatory force on plants than are individual species of generalists (Strong et al. 1984, Carson and Root 2000, Dyer et al. 2004), and specialist parasitoids are traditionally thought to be more effective regulators of herbivores than generalist predators (Myers et al. 1989, Hawkins et al. 1997, Denoth et al. 2002). Putting diet breadth into a coherent ecological context should be an important goal of tropical community ecologists, given that many hypotheses about the origin and maintenance of tropical diversity make assumptions about the prevalence and consequences of consumer specialization (reviewed by Wright 2002).

TROPHIC CASCADES

Definitions

The term "trophic cascades" has been defined in many ways, which has created problems (Hunter 2001), but the most restrictive definition is: a measurable increase in primary productivity due to negative effects of predators on herbivore biomass (Paine 1980, Power 1990, Carpenter and Kitchell 1993). Defined as such, the trophic cascade hypothesis is also known as the "green world hypothesis" (Polis 1999) and was first proposed by Hairston *et al.* (1960; HSS, for Hairston, Smith, and Slobodkin). Here I use the HSS definition of a trophic cascade, but there are many other types of trophic cascades hypotheses that are potentially important forces in terrestrial systems and they fall under a more general definition provided in theoretical and empirical studies (Hunter and Price 1992, Carpenter and Kitchell 1993, Polis 1999, Halaj and Wise 2001, Dyer and Letourneau 2003, Letourneau *et al.* 2004, Schmitz 2004, Schmitz *et al.* 2004) – indirect effects of one trophic level on a non-adjacent level. This includes indirect effects among individual species or entire trophic levels, with the effects acting on densities, traits, or community parameters, such as species richness (Figure 16.1). Two additional cascades hypotheses that I consider here are the "trait-mediated cascade" and the "diversity cascade," both of which could be important in tropical communities. A trait-mediated trophic cascade is a change in plant biomass caused by modifications in herbivore foraging behavior in the presence of predators (Schmitz *et al.* 2004). A diversity cascade is an indirect effect of diversity at one trophic level on a non-adjacent trophic level (Dyer and Letourneau 2003). No trophic cascade hypothesis has been fully tested in a tropical system (Dyer and Coley 2001).

Trophic cascades hypotheses have been extended to the ecosystem exploitation hypothesis (EEH), which incorporates variation in primary productivity and generalizes predictions for even and odd numbers of trophic levels that might result along a productivity gradient (Fretwell 1977, 1987, Oksanen *et al.* 1981, Oksanen 1991, Hairston and Hairston 1997). The HSS and EEH

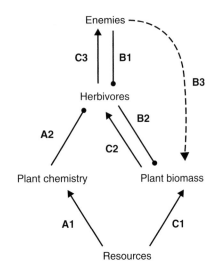

Figure 16.1   A simplified path diagram of selected direct and indirect effects that are examined in tropical food webs. Solid lines are direct effects, dashed lines are indirect effects, arrowheads are positive effects, and circle heads are negative effects. The letters next to the lines could be path coefficients or any other statistic of effect size, allowing comparisons between magnitudes of direct and indirect effects. The top-down cascade model predicts that B will be an important pathway. Plant defense (Moen *et al.* 1993) and resource availability (reviewed by Stamp 2003) hypotheses predict that A will be an important pathway. Resource limitation or bottom-up cascade models (Lindeman 1942, Slobodkin 1960, Hunter and Price 1992) predict that C will be an important pathway. This chapter focuses on pathways A and B.

models of multitrophic interactions have endured numerous attacks. Some ecologists have dismissed trophic cascades as one of many indirect effects that are unlikely to be of great importance in terrestrial systems (Polis and Strong 1996, Menge 2000, Halaj and Wise 2001). The criticism most relevant to tropical systems is that diverse terrestrial systems are unlikely to contain linear trophic levels, thus direct effects of one trophic level on another are never likely to be strong enough to cascade in any direction (Polis and Strong 1996). Omnivory, intra-guild predation, interference competition, spatial heterogeneity, prey refugia, and other factors that putatively buffer ecological systems from strong top-down effects of predators (Strong 1992, Polis and Strong 1996, Polis *et al.* 2000) are found in most tropical communities. Both Strong (1992) and Polis (1999) have argued that trophic cascades should only be expected in systems characterized by low within-trophic level diversity, simple food webs, discrete habitats, and little spatial heterogeneity. These authors assert that complex communities contain "species cascades," where the indirect positive effect of predators is demonstrated only for one species of plant, not for an entire community. In this view, predation can be important in diverse communities for particular imbedded food chains, but trophic cascades are not predicted to be important for an entire complex community.

Trait-mediated cascades

It is likely that the mechanism of trophic cascades is often trait mediated rather than density mediated (Schmitz *et al.* 2004), thus a distinction has been made between trait-mediated and density-mediated indirect interactions (TMII and DMII, respectively; Werner and Peacor 2003). In DMII, the cascade is mediated by a change in abundance of the intervening species or trophic level, while in TMII, the indirect effect is mediated by a change in behavior or defensive attributes of the intervening species (Gastreich 1999, Schmitz *et al.* 2004). DMII and TMII are not mutually exclusive; in fact it is likely that in trophic cascades, trait-mediated interactions are the most important mechanistic explanation for strong indirect effects on density (Schmitz *et al.* 2004). The best tropical example of a trait-mediated trophic cascade is reported by Gastreich (1999), who studied spiders, ants, and caterpillars associated with the ant plant *Piper obliquum*. Theridiid spiders altered the foraging of mutualist *Pheidole bicornis* ants, causing increased levels of caterpillar herbivory, while ant densities were unchanged (Gastreich 1999). Gastreich and Gentry (2004) argue that spiders are generally useful predators for examining DMII versus TMII, because they are ubiquitous enemies and have been shown to alter the density and behavior of their prey in many contexts (e.g., MacKay 1982,

Gastreich 1999, Dukas and Morse 2003, reviewed by Wise 1994).

Diversity cascades

Diversity cascades are a complex set of interactions that are particularly relevant to tropical systems. The response variables in diversity cascades can be diversity indices, species richness, abundance, or some other metric related to diversity. The most straightforward diversity cascade involves the indirect effect of plant diversity on overall consumer diversity via increased herbivore richness and abundance (Figure 16.2, path A). This bottom-up cascade hypothesis is a subset of the major hypotheses explaining the latitudinal gradient in species diversity, and it is well tested, with results indicating that plant diversity usually explains a measurable portion of consumer diversity for many different ecosystems,

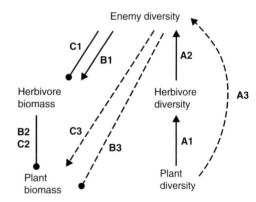

Figure 16.2   Selected diversity cascades. Solid lines are direct effects, dashed lines are indirect effects, arrowheads are positive effects, and circle heads are negative effects. The letters next to the lines could be path coefficients or any other statistic of effect size, allowing comparisons between magnitudes of direct and indirect effects. Path A is the bottom-up diversity cascade. Paths B and C represent two possible top-down diversity cascades; path B is also an important component of the argument that more diverse food webs are less likely to exhibit strong positive effects of predators on plant biomass. Several other possible diversity cascades (e.g., pathways from enemy diversity to herbivore diversity) are not depicted here.

study taxa, and scales of study (reviewed by Rohde 1992, Waide *et al.* 1999, Mittelbach *et al.* 2001). Top-down diversity cascades are less intuitive and depend on the particular assemblage of species. For example, an increase in predator diversity can cause an overall increase in herbivore abundance due to greater intra-guild predation and omnivory among predators (Hochberg 1996, Denoth *et al.* 2002); in turn, this can cause decreases in plant abundance (Figure 16.2, path B). In this scenario, as more species of predators are added, overall predation rates on herbivores decline because predators are consuming each other, herbivory increases, and primary productivity declines. However, increased enemy diversity may be just as likely to cause decreases in herbivory if the enemies are more specialized – such as parasitic Hymenoptera. In this case, complementarity between predators or "sampling error" (i.e., the "right predator" is more likely to be sampled from a more diverse community of enemies) causes an increase in overall enemy-induced mortality of herbivores (Stireman *et al.* 2004). Two examples of more complex top-down diversity cascades are presented in Dyer and Stireman (2003) and Dyer and Letourneau (2003). In the latter example, addition of a predatory beetle that specialized on one ant species caused increases in diversity of predacious ants living in a tropical understory shrub. This increase in ant diversity caused lower diversity of herbivores but overall higher levels of herbivory and lower plant biomass (Dyer and Letourneau 2003).

Diversity cascades have been examined indirectly by ecologists testing the assertion that more diverse systems are less likely to exhibit classic top-down cascades. For example, Finke and Denno (2004) demonstrate in a temperate (*Spartina* dominated) marsh that increasing diversity of predators (spiders and mirid bugs) leads to an increase in herbivore (planthoppers) density and a decrease in plant biomass (Figure 16.2, path B), mostly via intra-guild predation – in the "higher predator diversity" treatments, the spiders consumed the mirid bugs. Their experiments were conducted to demonstrate that trophic cascades are weaker in more diverse communities, but the experiments were not a valid test of cascades along a diversity gradient, since only enemy diversity

was manipulated; herbivore and plant diversity were low for all treatments. Natural ecosystems do not follow such a gradient, and increases in overall arthropod diversity, including herbivores, result in very different community dynamics than only small increases in predator richness (Dyer and Stireman 2003). Nevertheless, their results provide empirical evidence for one type of diversity cascade leading to a decline in plant biomass (Figure 16.2, path B). It is interesting to note that they were manipulating generalist predators – had the enemies in their studies been specialists, they may have found a diversity cascade that leads to an increase in plant biomass (Figure 16.2, path C).

Problems and adjustments to trophic cascades theory

Trophic cascades theory is still spinning its wheels. There have been three major reviews (Polis 1994, Pace *et al.* 1999, Persson 1999) and four meta-analyses (Schmitz *et al.* 2000, Dyer and Coley 2001, Halaj and Wise 2001, Shurin *et al.* 2002), while the number of direct empirical tests of cascades in terrestrial systems is still relatively low (fewer than 50 by December 2005), with very few studies in tropical systems. The criticisms and interpretive modifications of putative trophic cascades (or species cascades) in terrestrial systems warrant closer inspection of the methods utilized to study tritrophic interactions in diverse communities. One reason why progress has been stymied is because trophic cascades have never been tested properly in terrestrial systems. Below I outline the major faults in empirical tests of trophic cascades in terrestrial systems.

Entire trophic levels have not been deleted in diverse systems

In a complete, diverse community, it is not possible to experimentally remove all predators and parasitoids, nor is it possible to find a diverse terrestrial community from which all enemies have been removed. Thus, there are no direct experimental or correlational tests of trophic cascades in terrestrial communities, which is why the existing evidence is weak and easy to criticize. The exceptions, where a large percentage of individuals in an upper trophic level is removed (e.g., Carson and Root 2000), usually result in large indirect effects. One such exception is the case noted by Terborgh *et al.* (2001), where islands formed by hydroelectric impoundments in Venezuela were devoid of vertebrate predators. These islands experienced dramatic levels of herbivory, 10 to 100 times greater than comparable areas on the mainland, with corresponding reductions in plant seedlings and saplings. However, even in this example of "ecological meltdown" important enemies of invertebrates were not excluded; perhaps such an additional exclusion would lead to "ecological catastrophe," or the first convincing demonstration that trophic cascades are very strong stabilizing forces in tropical forests.

Insufficient natural history

All syntheses of trophic cascades and trophic interactions point to the lack of detailed knowledge of food webs as major limitations in testing hypotheses. Thus, some authors have made a strong argument that future studies need to document more details about species associations, strengths of connections between species, and other basic natural history of food webs (Cohen *et al.* 1993, Wootton 1997, Schmitz *et al.* 2000). The situation is exacerbated in tropical systems, since natural history is typically scant. A perfect example of this lack of information is the fact that actual trophic levels of predators or parasitoids are unknown. In fact, some authors acknowledge that they have lumped fourth trophic levels with third trophic levels for analyses (Halaj and Wise 2001), which results in a smaller effect size for cascades.

Temporal and spatial scales are very small

In the meta-analysis by Schmitz *et al.* (2000), almost all of the 60 studies examined were done for only one season using individual plants or very small plots (0.1–0.5 m^2). Unsurprisingly, there were no effects of study duration on the magnitude of the trophic cascade. Valid tests of

indirect predator regulation of plant populations would require decades or even centuries of study (Holt 2000, Hunter 2001), but even tests that simply demonstrate density effects consistent with regulation or control may require a large number of years (Letourneau and Dyer 1998, Carson and Root 2000). The existing experimental spatial scales are also biased towards showing no traditional trophic cascade, since none of the very small-scale manipulations that are usually conducted could conceivably cause a change in ecosystem productivity. Furthermore, this bias towards only studying smaller spatial scales is unlikely to lead to a unified understanding of tritrophic community patterns (Levin 1992).

Meta-analyses are incomplete

A simple tabulation of all the literature utilized in recent meta-analyses that test similar hypotheses allows one to calculate percent overlap of studies used in pairs of meta-analyses. The mean literature overlap between current paired meta-analysis publications (Schmitz *et al.* 2000, Dyer and Coley 2001, Halaj and Wise 2001, Shurin *et al.* 2002) is 5.7 ± 1.8%, which means that each analysis left out most of the studies that other authors deemed important. In addition, most meta-analyses do not restrict the number of studies used from single papers to avoid effect size biases, which results in amplified effect sizes for studies that report more results. This practice meets the criteria outlined by Hurlbert (1984) for pseudoreplication, since multiple results from a single study are used as independent observations in calculating the effect size statistic, increasing the relative contribution and associated biases of the selected studies. Before meta-analysts produce the complete, properly replicated quantitative summary, many more thorough empirical studies are necessary at appropriate spatial and temporal scales, especially in tropical systems where they are lacking (Dyer and Coley 2001).

The current paradigm is premature

Many authors have concluded that trophic cascades are not important in more diverse terrestrial systems (reviewed by Schmitz *et al.* 2000, Dyer and Coley 2001, Shurin *et al.* 2002, Letourneau *et al.* 2004, Stireman *et al.* 2004), and Halaj and Wise (2001) concluded this particular indirect effect is simply a trickle in most terrestrial systems. It is now assumed by many ecologists that only the simpler communities are likely to demonstrate cascades – aquatic versus terrestrial systems, grasslands versus forests, agricultural versus natural systems, and temperate versus tropical systems. While there is some limited support for this paradigm, trophic cascades, diversity cascades, trait-mediated indirect effects, and species cascades cannot be rejected as major forces in determining diversity, primary productivity, and number of trophic levels in tropical communities. In fact, the trophic cascade is one of the most useful theoretical frameworks for testing hypotheses about regulation of herbivore populations. Through tests of these and related hypotheses, ecologists will uncover the degree to which trophic cascades are weaker in more diverse terrestrial ecosystems and are likely to discover important community processes.

FUTURE DIRECTIONS

Research on tritrophic interactions in the tropics is still in its infancy. One problem that could prevent significant progress is a trend towards jumping from one hot topic to another. In fact, it has become popular to declare hypotheses "dead" without an appropriate arsenal of tests (e.g., coevolution: Rausher 1988; carbon–nutrient balance hypothesis: Hamilton *et al.* 2001; terrestrial trophic cascades: Polis and Strong 1996). This gives a false sense of progress. For tritrophic interactions in the tropics, the focus should be on utilizing a combination of the best available methods to create a broader synthesis and an improved understanding of important mechanisms behind trophic cascades and coevolutionary interactions. For example, Irschick *et al.* (2005) review studies of the evolution of specialization and provide a solid framework for future investigation, using a combination of modern approaches. On the other hand, ecologists should avoid the temptation to conduct short-term experiments at spatial scales that fail to rigorously

test the relevant hypotheses, are often contradictory, and yield few theoretical advances.

Here I propose hypotheses relevant to coevolution and trophic cascades. I also provide recommendations for approaches to testing these and related hypotheses. Since interaction strengths and corresponding statistics can vary a good deal (reviewed by Wootton and Emmerson 2005), it is relevant to differentiate between strong and weak effects within a community. For example, Halaj and Wise (2001) argue that trophic cascades are actually "trickles," which are weak effects as quantified by meta-analyses. Wootton and Emmerson (2005) provide important guidance on how to detect "strong" interactions in a community utilizing experimental, correlational, and modeling approaches. Here, I use "strong effects" to indicate where persistent additions (Yodzis 1988) or deletions (Paine 1980) of a population cause statistically significant and biologically important changes in major community parameters: productivity, diversity, number of functional trophic levels, and presence or absence of keystone species. In quantitative summaries of empirical data, strong effects would include all those that are mathematically equivalent to "large" meta-analysis effect sizes (*sensu* Gurevitch and Hedges 2001).

Future research: Hypotheses

1 Top-down and bottom-up forces have had strong effects on the evolution of diet breadth.
2 Strong consumer–resource relationships can lead to tight coevolution.
3 Diversity cascades are a strong component of tropical systems.
4 As consumer specialization increases, the strength of cascades and other indirect effects increase.
5 Top-down forces are more effective at controlling specialist herbivores while bottom-up forces are more important for generalists.

There are many appropriate alternatives to these general hypotheses. For example, in many systems narrow diet breadth may be a result of genetic drift or other non-adaptive forces, but a research program designed to test Hypothesis 1 above (e.g., as outlined by Irschick *et al.* 2005) should obviously consider this alternative. Each investigative approach outlined below allows for thorough tests of these and other alternative hypotheses.

Natural history

While ecologists have made great advances in recent decades by focusing on experimental approaches and utilizing cutting edge molecular techniques, it is still natural history and correlational data that form the basis of our most important theories and new hypotheses. Perhaps the greatest contemporary tropical tritrophic dataset is the Janzen–Hallwachs plant–caterpillar–parasitoid dataset, which provides basic natural history data for lepidopteran host plant affiliations and the parasitoid fauna that they support (Janzen and Hallwachs 2002). These data provide the raw materials for a thorough approach to interesting questions in community ecology and evolution and have already been used to guide experimental (Sittenfeld *et al.* 2002) and molecular (Hebert *et al.* 2004) approaches to testing complex hypotheses. Similar databases are being developed throughout the tropics (e.g., Dyer and Gentry 2002, Novotny *et al.* 2002). These databases have been used to test and generate numerous hypotheses, including subsets of the general hypotheses outlined above (see Lill *et al.* 2002, Barbosa and Caldas 2004, Janzen *et al.* 2005, Novotny and Basset 2005, Singer and Stireman 2005, Stireman *et al.* 2005).

Phylogenetic approaches

Advances in molecular systematics and comparative methods have provided a relatively new set of tools for ecologists to examine classic ecological questions. Farrell *et al.* (1992) present a useful outline for using phylogenies to test for tight coevolution between plants and herbivores. There are other research foci within tropical tritrophic interactions that would benefit from a phylogenetic approach. For example, Heil *et al.* (2004) used a phylogenetic approach to demonstrate that herbivore-induced extrafloral nectar in *Acacia*

myrmecophytes is a plesiomorphic state while constitutive flow of nectar is derived. Thus, plant rewards became more readily available for specialized ants that kill herbivores, indicating that a tritrophic interaction has driven recent coevolutionary relationships between plants and insects; this result is directly relevant to Hypotheses 2 and 3 above.

The phylogenetic approach is clearly necessary for testing hypotheses about specialization at any trophic level (e.g., Hypothesis 1 above) by utilizing phylogenetically controlled comparisons between specialist and generalist consumers and examining phylogenetic trends within taxa towards narrower or broader diet breadths (outlined by Irschick *et al.* 2005). This approach should be combined with a concerted effort to document ecological and functional specializations. First, tropical ecologists must establish the actual diet breadth of different species within a clade, despite the large amount of descriptive work involved. Second, a better integration of field and laboratory observations within the same taxon will allow for rigorous tests of how resource use is related to performance, thus differentiating between ecological and functional specialization (Irschick *et al.* 2005). For example, experiments could reveal the relative performance of specialized consumers when placed on more generalized diets or when exposed to alternative food items.

Large-scale, long-term experiments

The temporal and spatial scales of many experiments in the tropics are generally very small (Schmitz *et al.* 2000, Halaj and Wise 2001). Larger and longer experiments may cast light on the generality of the copious studies done in small plots for 1 year or less. Eventually, meta-analyses will provide direct quantitative comparison. Since some large-scale experiments are not possible, mensurative experiments, such as the formation of islands or fragments free of vertebrate predators (Crooks and Soule 1999, Terborgh *et al.* 2001), provide a viable alternative. Experiments should also be integrated in cohesive research programs that utilize or acknowledge models, correlational data, observational data, and phylogeny.

Modeling approach

Many theoretical components of trophic cascades in diverse communities have not been examined. Lotka–Volterra models have been used to demonstrate that adding a third trophic level to a community with four species (i.e., going from three plants and one herbivore to two plants, one herbivore and one predator) decreases herbivore population growth (Pimm and Lawton 1977). Does increasing the number of players at each trophic level (i.e., constructing a complex terrestrial community) alter this important predator control? If so, by what mechanism does it alter the response of herbivore populations to predators, and is there a threshold of this effect of diversity on predator–prey dynamics? Although these are not explicit tests of diversity cascades, such theoretical investigations could generate hypotheses and guide experimental, correlational, and observational studies. For example, several models suggest that the overall impact of parasitism on herbivore population size in biocontrol should decline with the number of parasitoid species (Kakehashi et al. 1984, Hassell and May 1986, Godfray and Waage 1991, Briggs 1993), whereas Hochberg (1996) showed that if individual parasitoids attack different hosts (as is typically the case with predation versus parasitism), multiple parasitoids should increase overall natural enemy impact. Hence, the specific assumptions employed in complex food-web models can radically alter predictions and warrant more careful consideration than in models of simpler systems.

Mesocosm/component-community approach

A mesocosm is a contained, usually experimental, assemblage of species with known physical and biotic dimensions that is a subset of a larger ecosystem (Odum 1984). Relative to microcosm studies, mesocosm studies typically utilize semi-controlled aquatic environments, and more natural assemblages that are designed to mimic natural communities (Boyle and Fairchild 1997). A similar concept is the component community (Root 1973), which is an assemblage of species associated with a particular resource; this is an example of a natural terrestrial mesocosm. If the component community is contained and easy to manipulate, it is a useful terrestrial mesocosm for testing hypotheses in community ecology. In natural terrestrial mesocosms, such as the endophytic insect fauna associated with a particular plant species, multitrophic manipulations are relatively easy. An entire trophic level can be deleted to test for trophic cascades – this is analogous to how mesocosms have been utilized in numerous studies of aquatic trophic cascades (Carpenter and Kitchell 1993). Some component communities have more than 50 species and thousands of individuals of interacting animals distributed among fewer than 100 discrete replicates (Dyer and Letourneau 2003). Fragments and islands are also mesocosms where it is possible to delete trophic levels or find systems with very few predators (Schoener and Spiller 1995, 1996, Terborgh et al. 2001). Small agricultural fields may also be treated as mesocosms if they provide enough complexity, such as an alfalfa field (Dyer and Stireman 2003). Are the cascades in these communities strong or are they trickles because of the buffering of other interactions? The limited number of studies to date suggest they are pervasive forces (Terborgh et al. 2001, Dyer and Letourneau 2003).

Species cascade approach

One could easily take experimental data that focus on a trophic chain and extrapolate out to the full web. The trophic cascades uncovered by Schoener and Spiller (Spiller and Schoener 1994, Schoener and Spiller 1999) focus on lizards and spiders as predators and on the relatively narrow trophic structure associated with them. If the same studies were conducted on the same islands with other taxa of predators and the results were consistent, it would provide strong evidence for effects of the entire predator trophic level on primary productivity of the island. This simple approach could be incorporated into any existing research program. A fixed number of tree species from a tropical forest could be selected randomly from the list of all available trees. For each tree species, all predators could be excluded using established methods (e.g., Floyd 1996) and the leaf biomass monitored for an appropriate period of time (Holt 2000). If a species cascade is demonstrated for the majority of species, it suggests that top-down forces of natural enemies are likely to result in community-wide effects. In addition, this approach provides a theoretical link to the issues about specialization discussed earlier in this chapter. Tropical communities could be modeled as parallel chains of species cascades with coevolved specialized consumers and their host plants. These chains could be connected by generalist consumers and detrital webs. Such a modeling approach would provide a framework for connecting these two bodies of tritrophic research.

CONCLUSION

Specialization and trophic cascades hypotheses should continue to provide guidance to empirical studies in tropical communities. To test these hypotheses, tropical ecologists must utilize solid phylogenetic data combined with creative experimental, correlational, observational, and modeling approaches. A concerted effort by tropical research programs that utilize these approaches to study focal communities or other model systems (e.g., Schoener and Spiller 1999, Schoener et al. 2001, Janzen and Hallwachs 2002, Dyer and Palmer 2004) will allow for fruitful synthesis and development of a useful theoretical framework for tropical specialization and trophic cascades. Such a synthetic approach would be an improvement over the existing cacophony of experiments, observations, and phylogenetic work across the geographic and taxonomic landscape of the tropics. Tritrophic interactions in tropical communities are usually part of a convoluted web with highly variable interaction strengths, yet with the right approaches and study systems we can determine which interactions are the strongest for particular taxa and ecosystems.

ACKNOWLEDGMENTS

I thank W. Carson for inviting me to put these ideas into writing. Research reported in this chapter was made possible by funding from the National Science Foundation (DEB-0344250 and DEB-0346729), Department of Energy (Southcentral Regional Center of NIGEC), Tulane University (Lathrop Fund), and Earthwatch Institute. Ideas, text, and data that helped me write this chapter were provided by D. Letourneau, R. Matlock, G. Gentry, J. Stireman, D. Janzen, W. Hallwachs, B. Braker, R. Marquis, P. Barbosa, and P. Coley. Parts of the manuscript were improved by comments from G. Gentry, A. Smilanich, T. Massad, R. Matlock, T. Floyd, O. Schmitz, W. Carson, and J. Cronin.

REFERENCES

Anderson, R.S. (1993) Weevils and plants – phylogenetic versus ecological mediation of evolution of host plant associations in Curculioninae (Coleoptera, Curculionidae). *Memoirs of the Entomological Society of Canada* 165, 197–232.

Barbosa, P. (1988) Natural enemies and herbivore–plant interactions: influence of plant allelochemicals and host specificity. In P. Barbosa and D.K. Letourneau (eds), *Novel Aspects of Insect–Plant Interactions*. John Wiley and Sons, New York, pp. 201–229.

Barbosa, P. and Caldas, A. (2004) Patterns of parasitoid-host associations in differentially parasitized macrolepidopteran assemblages on black willow *Salix nigra* (Marsh) and box elder *Acer negundo* L. *Basic and Applied Ecology* 5, 75–85.

Barbosa, P., Gross, P., and Kemper, J. (1991) Influence of plant allelochemicals on the tobacco hornworm and its parasitoid, *Cotesia congregata*. *Ecology* 72, 1567–1575.

Barbosa, P., Saunders, J.A., Kemper, J., Trumbule, R., Olechno, J., and Martinat, P. (1986) Plant allelochemicals and insect parasitoids: effects of nicotine on *Cotesia congregata* (Hymenoptera: Braconidae) and *Hyposter annulipes* (Cresson) (Hymenoptera: Ichneumonidae). *Journal of Chemical Ecology* 12, 1319–1328.

Becerra, J.X. (2003) Synchronous coadaptation in an ancient case of herbivory. *Proceedings of the National Academy of Sciences of the United States of America* 100, 12804–12807.

Becerra, J.X. and Venable, D.L. (1999) Macroevolution of insect–plant associations: The relevance of host biogeography to host affiliation. *Proceedings of the National Academy of Sciences of the United States of America* 96, 12626–12631.

Bernays, E.A. and Cornelius, M.L. (1989) Generalist caterpillar prey are more palatable than specialists for the generalist predator *Iridomyrmex humilis. Oecologia* 79, 427–430.

Bernays, E. and Graham, M. (1988) On the evolution of host specificity in phytophagous arthropods. *Ecology* 69, 886–892.

Blackburn, T.M. (2004) Method in macroecology. *Basic and Applied Ecology* 5, 401–412.

Boyle, T.P. and Fairchild, J.F. (1997) The role of mesocosm studies in ecological risk analysis. *Ecological Applications* 7, 1099–1102.

Brandle, M., Knoll, S., Eber, S., Stadler, J., and Brandl, R. (2005) Flies on thistles: support for synchronous speciation? *Biological Journal of the Linnean Society* 84, 775–783.

Briggs, C.J. (1993) Competition among parasitoid species on a stage-structured host and its effect on host suppression. *American Naturalist* 141, 372–397.

Brower, L.P. (1958) Bird predation and foodplant specificity in closely related procryptic insects. *American Naturalist* 92, 183–187.

Burslem, D.F.R.P., Pinard, M.A., and Hartley, S.E. (2005) *Biotic Interactions in the Tropics. Their Role in the Maintenance of Species Diversity.* Cambridge University Press, Cambridge.

Camara, M.D. (1997a) A recent host range expansion in *Junonia coenia* Hubner (Nymphalidae): oviposition preference, survival, growth, and chemical defense. *Evolution* 51, 873–884.

Camara, M.D. (1997b) Physiological mechanisms underlying the costs of chemical defence in *Junonia coenia* Hubner (Nymphalidae): a gravimetric and quantitative genetic analysis. *Evolutionary Ecology* 11, 451–469.

Carpenter, S.R. and Kitchell, J.F. (1993) *The Trophic Cascade in Lakes.* Cambridge University Press, New York.

Carson, W.P. and Root, R.B. (2000) Herbivory and plant species coexistence: community regulation by an outbreaking phytophagous insect. *Ecological Monographs* 70, 73–99.

Cohen, J.E., Pimm, S.L., Yodzis, P., and Saldana, J. (1993) Body sizes of animal predators and animal prey in food webs. *Journal of Animal Ecology* 62, 67–78.

Coley, P.D. and Aide, T.M. (1991) Comparison of herbivory and plant defenses in temperate and tropical broad-leaved forests. In P.W. Price, T.M. Lewinsohn, G.W. Fernandes, and W.W. Benson (eds), *Plant–Animal Interactions: Evolutionary Ecology in Tropical and Temperate Regions.* John Wiley & Sons, New York, pp. 25–49.

Coley, P.D. and Barone, J.A. (1996) Herbivory and plant defenses in tropical forests. *Annual Review of Ecology and Systematics* 27, 305–335.

Crooks, K.R. and Soule, M.E. (1999) Mesopredator release and avifaunal extinctions in a fragmented system. *Nature* 400, 563–566.

Dalecky, A., Debout, G., Mondor, G., Rasplus, J.Y., and Estoup, A. (2002) PCR primers for polymorphic microsatellite loci in the facultatively polygynous plant–ant *Petalomyrmex phylax* (Formicidae). *Molecular Ecology Notes* 2, 404–407.

Denoth, M., Frid, L., and Myers, J.H. (2002) Multiple agents in biological control: improving the odds? *Biological Control* 24, 20–30.

Dethier, V.G. (1954) Evolution of feeding preferences in phytophagous insects. *Evolution* 8, 33–54.

Dobzhansky, T. (1950) Evolution in the tropics. *American Scientist* 38, 209–221.

Duffey, S.S., Bloem, K.A., and Campbell, B.C. (1986) Consequences of sequestration of plant natural products in plant–insect–parasitoid interactions. In D.J. Boethel and R.D. Eidenbary (eds), *Interactions of Plant Resistance and Parasitoids and Predators.* Ellis Horwood, Chichester, pp. 31–60.

Dukas, R. and Morse, D.H. (2003) Crab spiders affect flower visitation by bees. *Oikos* 101, 157–163.

Dyer, L.A. (1995) Tasty generalists and nasty specialists? A comparative study of antipredator mechanisms in tropical lepidopteran larvae. *Ecology* 76, 1483–1496.

Dyer, L.A. (1997) Effectiveness of caterpillar defenses against three species of invertebrate predators. *Journal of Research on the Lepidoptera* 35, 1–16.

Dyer, L.A. (2001) In defense of caterpillars. *Natural History* 110, 42–47.

Dyer, L.A. and Coley, P.D. (2001) Latitudinal gradients in tritrophic interactions. In T. Tscharntke and B.A. Hawkins (eds), *Multitrophic Level Interactions.* Cambridge University Press, Cambridge, pp. 67–88.

Dyer, L.A. and Floyd, T. (1993) Determinants of predation on phytophagous insects: the importance of diet breadth. *Oecologia* 96, 575–582.

Dyer, L.A. and Gentry, G.L. (2002) Caterpillars and parasitoids of a Costa Rican tropical wet forest. http://www.caterpillars.org.

Dyer, L.A. and Letourneau, D. (2003) Top-down and bottom-up diversity cascades in detrital vs. living food webs. *Ecology Letters* 6, 60–68.

Dyer, L.A. and Palmer, A.D.N. (2004) *Piper: A Model Genus for Studies of Phytochemistry, Ecology, and Evolution.* Kluwer Academic/Plenum, New York.

Dyer, L.A. and Stireman, J.O. (2003) Community-wide trophic cascades and other indirect interactions in an agricultural community. *Basic and Applied Ecology* 4, 423–432.

Dyer, L.A., Letourneau, D.K., Dodson, C.D., Tobler, M.A., Stireman, J.O., and Hsu, A. (2004) Ecological causes and consequences of variation in defensive chemistry of a neotropical shrub. *Ecology* 85, 2795–2803.

Ehrlich, P.R. and Raven, P.H. (1964) Butterflies and plants: a study in coevolution. *Evolution* 18, 568–608.

Evenhuis, N.L. (1994) *Catalogue of Fossil Flies of the World (Insecta Diptera)*. Backhuys Publishers, Leiden.

Farrell, B. and Mitter, C. (1990) Phylogenesis of insect/plant interactions: Have *Phyllobrotica* leaf beetles (Chrysomelidae) and the Lamiales diversified in parallel? *Evolution* 44, 1389–1403.

Farrell, B.D., Mitter, C., and Futuyma, D.J. (1992) Diversification at the insect–plant interface. Insights from phylogenetics. *BioScience* 42, 34–42.

Feeny, P. (1976) Plant apparency and chemical defense. In J.W. Wallace and R.L. Mansell (eds), *Biochemical Interactions between Plants and Insects. Recent Advances in Phytochemistry*. Plenum, New York, pp. 1–40.

Ferry-Graham, L.A., Bolnick, D.I., and Wainwright, P.C. (2002) Using functional morphology to examine the ecology and evolution of specialization. *Integrative and Comparative Biology* 42, 265–277.

Fiala, B., Maschwitz, U., Pong, T.Y., and Helbig, A.J. (1989) Studies of a South East Asian ant–plant association – protection of *Macaranga* trees by *Crematogaster borneensis*. *Oecologia* 79, 463–470.

Finke, D.L. and Denno, R.F. (2004) Predator diversity dampens trophic cascades. *Nature* 429, 407–410.

Floyd, T. (1996) Top-down impacts on creosotebush herbivores in a spatially and temporally complex environment. *Ecology* 77, 1544–1555.

Fox, L.R. and Morrow, P.A. (1981) Specialization: species property or local phenomenon? *Science* 211, 887–893.

Fraenkel, G. (1959) The raison d'etre of secondary plant substances. *Science* 121, 1466–1470.

Fretwell, S.D. (1977) The regulation of plant communities by food chains exploiting them. *Perspectives in Biological Medicine* 20, 169–185.

Fretwell, S.D. (1987) Food chain dynamics: the central theory of ecology? *Oikos* 50, 291–301.

Funk, D.J., Futuyma, D.J., Ortí, G., and Meyer, A. (1995) A history of host associations and evolutionary diversification for *Ophraella* (Coleoptera: Chrysomelidae): new evidence from mitochondrial DNA. *Evolution* 49, 1008–1017.

Futuyma, D.J. and Mccafferty, S.S. (1990) Phylogeny and the evolution of host plant associations in the leaf beetle genus *Ophraella* (Coleoptera, Chrysomelidae). *Evolution* 44, 1885–1913.

Futuyma, D.J., Keese, M.C., and Funk, D.J. (1995) Genetic constraints on macroevolution – the evolution of host affiliation in the leaf beetle genus *Ophraella*. *Evolution* 49, 797–809.

Gastreich, K.R. (1999) Trait-mediated indirect effects of a theridiid spider on an ant–plant mutualism. *Ecology* 80, 1066–1070.

Gastreich, K.R. and Gentry, G.L. (2004) Faunal studies in model *Piper* spp. systems, with a focus on spider-induced indirect interactions and novel insect–*Piper* mutualisms. In L.A. Dyer and A.D.N. Palmer (eds), *Piper: A Model Genus for Studies of Phytochemistry, Ecology, and Evolution*. Kluwer Academic/Plenum Publishers, New York, pp. 97–116.

Gauld, I.D., Gaston, K.J., and Janzen, D.H. (1992) Plant allelochemicals, tritrophic interactions and the anomalous diversity of tropical parasitoids: the "nasty" host hypothesis. *Oikos* 65, 353–357.

Gentry, A.H. (1982) Patterns of neotropical plant-species diversity. *Evolutionary Biology* 15, 1–85.

Gentry, A.H. (1988) Changes in plant community diversity and floristic composition on environmental and geographical gradients. *Annals of the Missouri Botanical Garden* 75, 1–34.

Gentry, G. and Dyer, L.A. (2002) On the conditional nature of neotropical caterpillar defenses against their natural enemies. *Ecology* 83, 3108–3119.

Godfray, H.C.J. and Waage, J.K. (1991) Predictive modeling in biological-control – the mango mealy bug (*Rastrococcus invadens*) and its parasitoids. *Journal of Applied Ecology* 28, 434–453.

Gurevitch, J. and Hedges, L.V. (2001) Meta-analysis: combining the results of independent experiments. In S.M. Scheiner and J. Gurevitch (eds), *Design and Analysis of Ecological Experiments*. Chapman and Hall, New York, pp. 346–369.

Hairston, J. and Hairston, S. (1997) Cause–effect relationships in energy flow, trophic structure, and interspecific interactions. *American Naturalist* 142, 379–411.

Hairston, N.G., Smith, F.E., and Slobodkin, L.B. (1960) Community structure, population control, and competition. *American Naturalist* 94, 421–424.

Halaj, J. and Wise, D.H. (2001) Terrestrial trophic cascades: how much do they trickle? *American Naturalist* 157, 262–281.

Hall, J.B. and Swaine, M.D. (1976) Classification and ecology of closed-canopy forest in Ghana. *Journal of Ecology* 64, 913–951.

Hamilton, J.G., Zangerl, A.R., DeLucia, E.H., and Berenbaum, M.R. (2001) The carbon-nutrient balance hypothesis: its rise and fall. *Ecology Letters* 4, 86–95.

Hassell, M.P. and May, R.M. (1986) Generalist and specialist natural enemies in insect predator prey interactions. *Journal of Animal Ecology* 55, 923–940.

Hassell, M.P. and Southwood, T.R.E. (1978) Foraging strategies of insects. *Annual Review of Ecology and Systematics* 9, 75–98.

Hawkins, B.A., Cornell, H.V., and Hochberg, M.E. (1997) Predators, parasitoids, and pathogens as mortality agents in phytophagous insect populations. *Ecology* 78, 2145–2152.

Hebert, P.D.N., Penton, E.H., Burns, J.M., Janzen, D.H., and Hallwachs, W. (2004) Ten species in one: DNA barcoding reveals cryptic species in the neotropical skipper butterfly *Astraptes fulgerator*. *Proceedings of the National Academy of Sciences of the United States of America* 101, 14812–14817.

Heil, M. and McKey, D. (2003) Protective ant–plant interactions as model systems in ecological and evolutionary research. *Annual Review of Ecology, Evolution and Systematics* 34, 425–453.

Heil, M., Delsinne, T., Hilpert, A. *et al.* (2002) Reduced chemical defence in ant–plants? A critical reevaluation of a widely accepted hypothesis. *Oikos* 99, 457–468.

Heil, M., Greiner, S., Meimberg, H. *et al.* (2004) Evolutionary change from induced to constitutive expression of an indirect plant resistance. *Nature* 430, 205–208.

Hochberg, M.E. (1996) Consequences for host population levels of increasing natural enemy species richness in classical biological control. *American Naturalist* 147, 307–318.

Holt, R.D. (2000) Trophic cascades in terrestrial ecosystems. Reflections on Polis *et al*. *Trends in Ecology and Evolution* 15, 444–445.

Hunter, M.D. (2001) Multiple approaches to estimating the relative importance of top-down and bottom-up forces on insect populations: experiments, life tables, and time-series analysis. *Basic and Applied Ecology* 2, 295–309.

Hunter, M.D. (2003) Effects of plant quality on the population ecology of parasitoids. *Agricultural and Forest Entomology* 5, 1–8.

Hunter, M.D. and Price, P.W. (1992) Playing chutes and ladders: heterogeneity and the relative roles of bottom-up and top-down forces in natural communities. *Ecology* 73, 724–732.

Hurlbert, S.H. (1984) Pseudoreplication and the design of ecological field experiments. *Ecological Monographs* 54, 187–211.

Huston, M. (1980) Soil nutrients and tree species richness in Costa Rican forests. *Journal of Biogeography* 7, 147–157.

Irschick, D., Dyer, L.A. and Sherry, T.W. (2005) Phylogenetic methodologies for studying specialization. *Oikos* 110, 404–408.

Itino, T., Davies, S.J., Tada, H. *et al.* (2001) Cospeciation of ants and plants. *Ecological Research* 16, 787–793.

Janzen, D.H. (1967) Why migration passes are higher in the tropics. *American Naturalist* 101, 233–249.

Janzen, D.H. (1970) Herbivores and the number of tree species in tropical forests. *American Naturalist* 104, 521–528.

Janzen, D.H. (1988) Ecological characterization of a Costa Rican dry forest caterpillar fauna. *Biotropica* 20, 120–135.

Janzen, D.H. (1993) Caterpillar seasonality in a Costa Rican dry forest. In N.E. Stamp and T.M. Casey (eds), *Caterpillars: Ecological and Evolutionary Constraints on Foraging*. Chapman and Hall, New York, pp. 448–477.

Janzen, D.H., Hajibabaei, M., Burns, J.M., Hallwachs, W., Remigio, E., and Hebert, P.D.N. (2005) Wedding biodiversity inventory of a large and complex Lepidoptera fauna with DNA barcoding. *Philosophical Transactions of the Royal Society B-Biological Sciences* 360, 1835–1845.

Janzen, D.H. and Hallwachs, W. (2002) Philosophy, navigation and use of a dynamic database ("ACG Caterpillars SRNP") for an inventory of the macrocaterpillar fauna, and its food plants and parasitoids, of the Area de Conservacion Guanacaste (ACG), northwestern Costa Rica. http://janzen.sas.upenn.edu.

Kakehashi, N., Suzuki, Y., and Iwasa, Y. (1984) Niche overlap of parasitoids in host parasitoid systems – its consequence to single versus multiple introduction controversy in biological control. *Journal of Applied Ecology* 21, 115–131.

Kuhn, J., Pettersson, E.M., Feld, B.K. *et al.* (2004) Selective transport systems mediate sequestration of plant glucosides in leaf beetles: A molecular basis for adaptation and evolution. *Proceedings of the National Academy of Sciences of the United States of America* 101, 13808–13813.

Labandeira, C.C., Dilcher, D.L., Davis, D.R., and Wagner, D.L. (1994) 97-Million years of Angiosperm-insect association – paleobiological insights into the meaning of coevolution. *Proceedings of the National Academy of Sciences of the United States of America* 91, 12278–12282.

Letourneau, D.K. and Dyer, L.A. (1998) Experimental test in lowland tropical forest shows top-down effects through four trophic levels. *Ecology* 79, 1678–1687.

Letourneau, D.K., Dyer, L.A., and Vega, G.C. (2004) Indirect effects of a top predator on a rain forest understory plant community. *Ecology* 85, 2144–2152.

Levin, S.A. (1992) The problem of pattern and scale in ecology. *Ecology* 73, 1943–1967.

Lill, J.T., Marquis, R.J., and Ricklefs, R.E. (2002) Host plants influence parasitism of forest caterpillars. *Nature* 417, 170–173.

Lindeman, R.L. (1942) The trophic-dynamic aspect of ecology. *Ecology* 23, 399–418.

Long, Z.T., Mohler, C.L., and Carson, W.P. (2003) Extending the resource concentration hypothesis to plant communities: effects of litter and herbivores. *Ecology* 84, 652–665.

MacArthur, R.H. and Wilson, E.O. (1967) *The Theory of Island Biogeography*. Princeton University Press, Princeton.

Mackay, W.P. (1982) The effect of predation of western widow spiders (Araneae, Theridiidae) on harvester ants (Hymenoptera, Formicidae). *Oecologia* 53, 406–411.

McKey, D., Gaume, L., Brouat, C. *et al.* (2005) The trophic structure of tropical ant–plant–herbivore interactions: community consequences and coevolutionary dynamics. In D.F.R.P. Burslem, M.A. Pinard, and S.E. Hartley (eds), *Biotic Interactions in the Tropics. Their Role in the Maintenance of Species Diversity*. Cambridge University Press, Cambridge, pp. 386–413.

Menge, B.A. (2000) Top-down and bottom-up community regulation in marine rocky intertidal habitats. *Journal of Experimental Marine Biology and Ecology* 250, 257–289.

Mittelbach, G.G., Steiner, C.F., Scheiner, S.M. *et al.* (2001) What is the observed relationship between species richness and productivity? *Ecology* 82, 2381–2396.

Mitter, C., Farrell, B.M., and Futuyma, D.J. (1991) Phylogenetic studies of insect–plant interactions: insights into the genesis of diversity. *Trends in Ecology and Evolution* 6, 290–293.

Moen, J.H., Oksanen, L., Ericson, L., and Ekerholm, P. (1993) Grazing by food-limited microtine rodents on a productive experimental plant community: does the "green desert" exist? *Oikos* 68, 401–413.

Moran, N.A. (1988) The evolution of host-plant alternation in aphids: evidence for specialization as a dead end. *American Naturalist* 132, 681–706.

Myers, J.H., Higgins, C., and Kovacs, E. (1989) How many insect species are necessary for the biological control of insects? *Environmental Entomology* 18, 541–547.

Novotny, V. and Basset, Y. (2005) Review – host specificity of insect herbivores in tropical forests. *Proceedings of the Royal Society of London Series B-Biological Sciences* 272, 1083–1090.

Novotny, V., Miller, S.E., Basset, Y. *et al.* (2002) Predictably simple: assemblages of caterpillars (Lepidoptera) feeding on rainforest trees in Papua New Guinea. *Proceedings of the Royal Society of London Series B-Biological Sciences* 269, 2337–2344.

Odum, E.P. (1984) The mesocosm. *BioScience* 34, 558–562.

Oksanen, L. (1991) Trophic levels and trophic dynamics: a consensus emerging? *Trends in Ecology and Evolution* 6, 58–60.

Oksanen, L., Fretwell, S., Arruda, J., and Niemela, P. (1981) Exploitation ecosystems in gradients of primary productivity. *American Naturalist* 118, 240–261.

Opler, P.A., Frankie, G.W., and Baker, H.G. (1980) Comparative phenological studies of treelet and shrub species in tropical wet and dry forests in the lowlands of Costa Rica. *Journal of Ecology* 68, 167–188.

Pace, M.L., Cole, J.J., Carpenter, S.R., and Kitchell, J.F. (1999) Trophic cascades revealed in diverse ecosystems. *Trends in Ecology and Evolution* 14, 483–488.

Paine, R.T. (1966) Food web complexity and species diversity. *American Naturalist* 100, 65–75.

Paine, R.T. (1980) Food webs: linkage, interaction strength and community infrastructure. *Journal of Animal Ecology* 49, 667–685.

Persson, L. (1999) Trophic cascades: abiding heterogeneity and the trophic level concept at the end of the road. *Oikos* 85, 385–397.

Philips, O.L., Hall, P., Gentry, A.H., Sawyer, S.A., and Vasquez, R. (1994) Dynamics and species richness of tropical rain forests. *Proceedings of the National Academy of Sciences of the United States of America* 91, 2805–2809.

Pianka, E.R. (1966) Latitudinal gradients in species diversity: a review of concepts. *American Naturalist* 100, 33–46.

Pimm, S.L. and Lawton, J.H. (1977) Number of trophic levels in ecological communities. *Nature* 268, 329–331.

Polis, G.A. (1994) Food webs, trophic cascades and community structure. *Australian Journal of Ecology* 19, 121–136.

Polis, G.A. (1999) Why are parts of the world green? Multiple factors control productivity and the distribution of biomass. *Oikos* 86, 3–15.

Polis, G.A., Sears, A.L.W., Huxel, G.R., Strong, D.R., and Maron, J. (2000) When is a trophic cascade a trophic cascade? *Trends in Ecology and Evolution* 15, 473–475.

Polis, G.A. and Strong, D.R. (1996) Food web complexity and community dynamics. *American Naturalist* 147, 813–846.

Power, M.E. (1990) Effect of fish in river food webs. *Science* 250, 411–415.

Price, P.W. (1991) Patterns in communities along latitudinal gradients. In P.W. Price, T.M. Lewinsohn, G.W. Fernandes, and W.W. Benson (eds), *Plant–Animal Interactions: Evolutionary Ecology in Tropical and Temperate Regions*. John Wiley and Sons, New York, pp. 51–69.

Quek, S.P., Davies, S.J., Itino, T., and Pierce, N.E. (2004) Codiversification in an ant–plant mutualism: stem texture and the evolution of host use in *Crematogaster* (Formicidae: Myrmicinae) inhabitants of *Macaranga* (Euphorbiaceae). *Evolution* 58, 554–570.

Rausher, M.D. (1988) Is coevolution dead? *Ecology* 69, 898–901.

Raven, P.H. (1977) A suggestion concerning the Cretaceous rise to dominance of the angiosperms. *Evolution* 31, 451–452.

Rehr, S.S., Feeny, P.P., and Janzen, D.H. (1973) Chemical defence in Central American non-ant acacias. *Journal of Animal Ecology* 42, 405–416.

Rhoades, D.F. and Cates, R.G. (1976) Toward a general theory of plant antiherbivore chemistry. *Recent Advances in Phytochemistry* 10, 168–213.

Rohde, K. (1992) Latitudinal gradients in species diversity: the search for the primary cause. *Oikos* 65, 514–527.

Root, R.B. (1973) Organization of a plant–arthropod association in simple and diverse habitats: the fauna of collards (*Brassica oleracea*). *Ecological Monographs* 95–124.

Schmitz, O.J. (2004) Perturbation and abrupt shift in trophic control of biodiversity and productivity. *Ecology Letters* 7, 403–409.

Schmitz, O.J., Hamback, P.A., and Beckerman, A.P. (2000) Trophic cascades in terrestrial systems: a review of the effects of carnivore removals on plants. *American Naturalist* 155, 141–153.

Schmitz, O.J., Krivan, V., and Ovadia, O. (2004) Trophic cascades: the primacy of trait-mediated indirect interactions. *Ecology Letters* 7, 153–163.

Schoener, T.W. and Spiller, D.A. (1995) Effect of predators and area on invasion: an experiment with island spiders. *Science* 267, 1811–1813.

Schoener, T.W. and Spiller, D.A. (1996) Devastation of prey diversity by experimentally introduced predators in the field. *Nature* 381, 691–694.

Schoener, T.W. and Spiller, D.A. (1999) Indirect effects in an experimentally staged invasion by a major predator. *American Naturalist* 153, 347–358.

Schoener, T.W., Spiller, D.A., and Losos, J.B. (2001) Predators increase the risk of catastrophic extinction of prey populations. *Nature* 412, 183–186.

Scott, A., Stephenson, J., and Chaloner, W.G. (1992) Interactions and coevolution of plants and arthropods during the Paleozoic and Mesozoic. *Philosophical Transactions of the Royal Society of London Series B-Biological Sciences* 335, 129–165.

Shurin, J.B., Borer, E.T., Seabloom, E.W. *et al.* (2002) A cross-ecosystem comparison of the strength of trophic cascades. *Ecology Letters* 5, 785–791.

Sime, K. (2002) Chemical defence of *Battus philenor* larvae against attack by the parasitoid *Trogus pennator*. *Ecological Entomology* 27, 337–345.

Singer, M.S., Carriere, Y., Theuring, C., and Hartmann, T. (2004a) Disentangling food quality from resistance against parasitoids: diet choice by a generalist caterpillar. *American Naturalist* 164, 423–429.

Singer, M.S., Rodrigues, D., Stireman, J.O., and Carriere, Y. (2004b) Roles of food quality and enemy-free space in host use by a generalist insect herbivore. *Ecology* 85, 2747–2753.

Singer, M.S. and Stireman, J.O. (2005) The tritrophic niche concept and adaptive radiation of phytophagous insects. *Ecology Letters* 8, 1247–1255.

Sittenfeld, A., Uribe-Lorio, L., Mora, M., Nielsen, V., Arrieta, G. and Janzen, G.H. (2002) Does a polyphagous caterpillar have the same gut microbiota when feeding on different species of food plants? *Revista de Biologia Tropical* 50, 547–560.

Spiller, D.A. and Schoener, T.W. (1994) Effects of top and intermediate predators in a terrestrial food web. *Ecology* 75, 182–196.

Stamp, N. (2003) Out of the quagmire of plant defense hypotheses. *Quarterly Review of Biology* 78, 23–55.

Stanton, M.L., Palmer, T.M., and Young, T.P. (2002) Competition–colonization trade-offs in a guild of African Acacia-ants. *Ecological Monographs* 72, 347–363.

Stireman, J.O. (2002) Phylogenetic relationships of tachinid flies in subfamily Exoristinae (Tachinidae: Diptera) based on 28S rDNA and elongation factor-1 alpha. *Systematic Entomology* 27, 409–435.

Stireman, J.O., Dyer, L.A., Janzen, D.H. *et al.* (2005) Climatic unpredictability and parasitism of caterpillars: implications global warming. *Proceedings of the National Academy of Sciences of the United States of America* 102, 17384–17387.

Stireman, J.O., Dyer, L.A., and Matlock, R.B. (2004) Top-down forces in managed versus unmanaged habitats. In P. Barbosa and I. Castellanos (eds), *Ecology of Predator–Prey Interactions*. Oxford University Press, Oxford, pp. 303–323.

Strong, D.R. (1992) Are trophic cascades all wet? Differentiation and donor control in speciose ecosystems. *Ecology* 73, 747–754.

Strong, D.R., Lawton, J.H., and Southwood, T.R.E. (1984) *Insects on Plants: Community Patterns and Mechanisms*. Harvard University Press, Cambridge, MA.

Tepe, E., Vincent, M., and Watson, L. (2004) Phylogenetic patterns, evolutionary trends, and the origin of ant–plant associations in *Piper* section *Macrostachys*: Burger's hypothesis revisited. In L.A. Dyer and A.D.N. Palmer (eds), *Piper: A Model Genus for Studies of Phytochemistry, Ecology, and Evolution*. Kluwer Academic Publishing, New York, pp. 156–178.

Terborgh, J., Lopez, L., Nunez, P. *et al.* (2001) Ecological meltdown in predator-free forest fragments. *Science* 294, 1923–1926.

Termonia, A., Hsiao, T.H., Pasteels, J.M., and Milinkovitch, M.C. (2001) Feeding specialization and host-derived chemical defense in Chrysomeline leaf beetles did not lead to an evolutionary dead end. *Proceedings of the National Academy of Sciences of the United States of America* 98, 3909–3914.

Tscharntke, T. and Hawkins, B.A. (2002) *Multitrophic Level Interactions*. Cambridge University Press, Cambridge.

Turlings, T.C.J. and Benrey, B. (1998) Effects of plant metabolites on the behavior and development of parasitic wasps. *Ecoscience* 5, 321–333.

van Schaik, C.P., Terborgh, J.W., and Wright, S.J. (1993) Phenology of tropical forests: adaptive significance and consequences for primary consumers. *Annual Review of Ecology and Systematics* 24, 353–377.

Waide, R.B., Willig, M.R., Steiner, C.F. *et al.* (1999) The relationship between productivity and species richness. *Annual Review of Ecology and Systematics* 30, 257–300.

Weintraub, J.D., Lawton, J.H., and Scoble, M.J. (1995) Lithinine moths on ferns: a phylogenetic study of insect–plant interactions. *Biological Journal of the Linnean Society* 55, 239–250.

Werner, E.E. and Peacor, S.D. (2003) A review of trait-mediated indirect interactions in ecological communities. *Ecology* 84, 1083–1100.

Wise, D.H. (1994) *Spiders in Ecological Webs*. Cambridge University Press, New York.

Wootton, J.T. (1997) Estimates and tests of per capita interaction strength: diet, abundance, and impact of intertidally foraging birds. *Ecological Monographs* 67, 45–64.

Wootton, J.T. and Emmerson, M. (2005) Measurement of interaction strength in nature. *Annual Review of Ecology and Systematics* 36, 419–444.

Wright, S.J. (2002) Plant diversity in tropical forests: a review of mechanisms of species coexistence. *Oecologia* 130, 1–14.

Yodzis, P. (1988) The indeterminacy of ecological interactions as perceived through perturbation experiments. *Ecology* 69, 508–515.

Yu, D.W. and Davidson, D.W. (1997) Experimental studies of species-specificity in *Cecropia*–ant relationships. *Ecological Monographs* 67, 273–294.

Chapter 17

VARIATION IN TREE SEEDLING AND ARBUSCULAR MYCORRHIZAL FUNGAL SPORE RESPONSES TO THE EXCLUSION OF TERRESTRIAL VERTEBRATES: IMPLICATIONS FOR HOW VERTEBRATES STRUCTURE TROPICAL COMMUNITIES

Tad C. Theimer and Catherine A. Gehring

OVERVIEW

Vertebrates can impact species-rich, tropical plant communities by acting both as dispersers of seeds and spores and as agents of seed and seedling mortality. When we excluded terrestrial vertebrates from 14 small (6 m × 7.5 m) plots of Australian tropical rainforest, we found significantly higher seedling recruitment and survival, resulting in higher seedling species richness and diversity, on exclosure plots. These results contrasted with those for arbuscular mycorrhizal fungal (AMF) spores, in which vertebrate exclusion led to decreased abundance, richness, and diversity. In this chapter, we develop a conceptual model that explains this difference between AMF spore and tree seedling responses to vertebrate exclusion. We hypothesize that when vertebrates act primarily as agents of seed or spore dispersal, as they do for AMF spores in our system, they increase local species richness by increasing the rate of local colonization. When vertebrates act primarily as agents of random mortality, as they do for seeds and seedlings in our system, they increase the rate of local extinction and depress local species richness. In the latter case, vertebrates increase recruitment limitation and could potentially maintain diversity on larger spatial scales, although this was not the case at the scales we measured. Overall, our study suggests that (1) the relative proportion of tree or fungal species in a community for which vertebrate dispersal significantly increases the seed/spore shadow determines the magnitude of the difference in species pools available in the presence or absence of vertebrates, and thereby the rate of species accumulation and potential for terrestrial vertebrates to alter species diversity; (2) if seedling or spore mortality due to vertebrates is high enough, and not overall more strongly density dependent than in their absence, the net effect of terrestrial vertebrates will be to reduce local species richness due to increased extinction rates and thereby increase local dispersal limitation by reducing the probability that species arriving at a site will successfully establish there; and (3) indirect effects of terrestrial vertebrates, like that of altering AMF spore species richness, could alter seedling community dynamics, but these effects may take considerable time to be expressed.

INTRODUCTION

How vertebrates affect the diversity of tropical rainforest tree communities is of critical and growing importance, given the sensitivity of many species of vertebrates to human exploitation and forest fragmentation (e.g., Tabarelli *et al.* 1999, Wright *et al.* 2000, Peres 2001, Roldan and Simonetti 2001). Many studies have documented the important role individual vertebrate species can play in dispersal and recruitment of selected tree species by acting as seed dispersers and seed and seedling predators (e.g., Forget and Sabatier 1997, Wenny 2000, Wyatt and Silman 2004). The loss of these animals due to anthropogenic effects like poaching can alter the recruitment of certain tree species (e.g., Asquith *et al.* 1999, Wright *et al.* 2000, Roldan and Simonetti 2001). However, the overall effect of a larger suite of terrestrial vertebrates on rainforest plant communities has rarely been addressed (Dirzo and Miranda 1990, 1991). Most studies have focused on the impacts of vertebrates as seed and seedling predators or seed dispersers, but vertebrates may affect seedling regeneration and competitive interactions in other ways as well. For example, vertebrates could indirectly alter plant community dynamics by dispersing mycorrhizal fungal spores (e.g., McGee and Baczocha 1994, Janos *et al.* 1995, Reddell *et al.* 1997, Mangan and Adler 2000) or by changing plant–fungi interactions through herbivory (e.g., Frank *et al.* 2003).

In these ways, terrestrial vertebrates could also impact the abundance and diversity of mycorrhizal fungi. Most terrestrial plant species form associations with mycorrhizal fungi that enhance the uptake of mineral nutrients, frequently resulting in improved plant growth and survival (Smith and Read 1997). Variation in the dynamics of this symbiosis can affect community and ecosystem properties such as plant competition (Allen and Allen 1984, Hetrick *et al.* 1989, Bever *et al.* 1997), plant diversity (Grime *et al.* 1987, van der Heijden *et al.* 1998a,b, Hartnett and Wilson 1999), ecosystem productivity (Klironomos *et al.* 2000), and succession (Janos 1980). Terrestrial vertebrates may therefore affect both plant and mycorrhizal fungal communities which could in turn feed back to affect one another's performance.

Few studies have simultaneously examined the impacts of terrestrial vertebrates on the diversity of both plants and mycorrhizal fungi. We did so as part of a larger study examining the effects of experimentally excluding terrestrial vertebrates from small areas of Australian rainforest and found that vertebrate exclusion had opposite effects for seedlings and arbuscular mycorrhizal fungal (AMF) spores over the short term (4.5 years). Exclusion significantly reduced the abundance and species richness of AMF spores (Gehring *et al.* 2002), while it increased tree seedling abundance and species richness (Table 17.1). In this chapter, we explore why these communities responded differently and the

Table 17.1 Seedling community attributes (mean ± SE) for 14 exclosure–open pairs before and 4.5 years after vertebrate exclusion, and fungal spore community attributes for 13 of those pairs after 4.5 years of terrestrial vertebrate exclusion.

	Tree seedling community before exclosure		Tree seedling community 4.5 years post-exclosure		Fungal spore community 4.5 years post-exclosure	
	Open	Exclosure	Open	Exclosure	Open	Exclosure
Abundance	104 ± 14	96 ± 14	120 ± 18*	197 ± 37*	54 ± 10*[a]	28 ± 6*
Species richness	22.7 ± 1.5	21.6 ± 1.5	22.8 ± 1.5*	28.9 ± 1.4*	7.4 ± 0.5*	5.5 ± 0.6*
Evenness	0.71 ± 0.06	0.70 ± 0.05	0.78 ± 0.03	0.74 ± 0.03	0.58 ± 0.06	0.58 ± 0.06
Shannon's diversity	2.13 ± 0.18	2.04 ± 0.18	2.20 ± 0.17	2.24 ± 0.17	1.13 ± 0.1*	0.46 ± 0.06*

Note: Asterisks indicate that the difference between exclosure and open pairs was significant ($P < 0.05$) using paired *t*-tests.

[a] Fungal spore abundance is number of spores per 1 g soil sample.

implications that difference may have for how we view the effects of vertebrates on community attributes. We begin with a brief overview of how Australian tropical rainforests may differ from those in other parts of the world, followed by a summary of our results on the effect of vertebrate exclusion on seedling and AMF spore abundance and diversity. We then propose a simple conceptual model that argues AMF spores and seedlings differed in their response to vertebrates because vertebrates acted primarily to promote colonization of experimental plots by mycorrhizal fungal spores while they acted primarily as agents of local extinction for seedlings without promoting seedling colonization. We end by reiterating that the effects of a terrestrial vertebrate community on a seedling community depend on the relative importance in that plant community of (1) vertebrate seed dispersal, (2) vertebrate mycorrhizal fungal spore dispersal, and (3) the strength of vertebrate-induced density-dependent seed and seedling mortality. We suggest that the interplay of these factors will determine the net effect on community structure, and that this effect may differ from that observed or predicted based on studies of a subset of species, if that subset is not representative of the community as a whole.

COMPARATIVE ATTRIBUTES OF AUSTRALIAN RAINFORESTS

The native vertebrate fauna of Australian rainforests is depauperate compared with that of other tropical rainforests (Eisenberg 1981). Especially lacking are large (>10 kg) herbivorous mammals like deer, peccaries, and tapirs, though many areas are increasingly impacted by feral pigs (*Sus scrofa*). The largest herbivorous terrestrial mammal is the red-legged pademelon, a small (3.5–6.8 kg) kangaroo. The larger tree kangaroos and diverse possums are arboreal folivores that rarely feed on the ground. The largest frugivore (1.5–2 m tall) is the southern cassowary, a flightless bird, with no parallel in other tropical faunas except New Guinea. Several species of rodents act as seed predators and potential seed dispersers (Harrington *et al.* 1997, Theimer 2001), while a small, frugivorous kangaroo roughly approximates neotropical agoutis

and acouchies in size and behavior (Dennis 2003). As in many other forests, these terrestrial vertebrates operate at differing spatial scales, with cassowaries potentially moving seeds and spores over hundreds of meters (Westcott *et al.* 2006), while rodents and smaller marsupials move most propagules less than 50 m (Harrington *et al.* 1997, Theimer 2001, Dennis 2003). Given the relatively depauperate fauna of Australian rainforests, the effects of terrestrial vertebrates on seedling dynamics may be relatively weak compared with other, more species-rich forests. The mycorrhizal fungal communities of Australian rainforests also have similarities and differences when compared with other rainforests. Arbuscular mycorrhizae dominate (e.g., Hopkins *et al.* 1996, Gehring 2003, Gehring and Connell 2006), as is typical of many rainforests (e.g., St John 1980, Bereau and Garbaye 1994). However, seedlings of more than one third of the dominant species on our study plot are rarely or never colonized by AMF (Gehring and Connell 2006). This high proportion of non-mycorrhizal species contrasts with data from many rainforests (St John 1980, Bereau and Garbaye 1994), but is similar to that in tropical forests in Brazil and China (Zangaro *et al.* 2000, Zhao *et al.* 2001). AMF spore abundance and diversity are comparable to other rainforests, with an average of 54 spores per gram of soil and dominance of the community by members of the genera *Acaulospora* and *Glomus* (Gehring *et al.* 2002, Lovelock *et al.* 2003, Mangan *et al.* 2004). Several species of terrestrial vertebrates common in Australian rainforests carry AMF spores in their feces, including the native rodents, the musky rat kangaroo, two species of bandicoots, and several species of ground-frequenting birds (Reddell *et al.* 1997).

RESULTS OF TERRESTRIAL VERTEBRATE EXCLUSION FROM AN AUSTRALIAN RAINFOREST

In autumn 1996, we erected fourteen 6.5 m × 7.0 m vertebrate exclosures in the area surrounding Connell *et al.*'s (1984) long-term study plot. For the following 4.5 years, we monitored seedling survival and recruitment on exclosures and nearby open plots at 6-month intervals. In 2001, we assessed AMF spore abundance and diversity from soil cores collected in 13 of the 14 exclosure–open pairs (Gehring et al. 2002). Faunal censuses indicated that our study site included three species of marsupials, the long-nosed bandicoot (*Perameles nasuta*), musky rat kangaroo (*Hypsiprymnodon moschatus*), and red-legged pademelon (*Thylogale stigmatica*), and three species of eutherian rodents, the bush rat (*Rattus fuscipes*), fawn-footed melomys (*Melomys cervinipes*), and giant white-tailed rat (*Uromys caudimaculatus*). Ground-frequenting birds included the southern cassowary (*Casuarius casuarius*), brush turkey (*Alectura lathami*), and chowchilla (*Orthonyx spaldingii*). Of the birds, the cassowary was completely excluded by the fencing, while brush turkeys and chowchillas only occasionally flew over the fences. Monitoring of tracks and trapping indicated that the ground-dwelling marsupials were completely excluded from the exclosures. Bush rats were captured only once in the exclosures in 300+ trap nights versus 58 captures in open plots. Access by white-tailed rats was reduced to one third that of the open plots and *Melomys* access was reduced by more than 50%.

After 4.5 years of terrestrial vertebrate exclusion, seedling abundance averaged 40% more on exclosure plots than on open plots (Table 17.1). These differences were due to both reduced seedling recruitment and increased seedling mortality on unfenced plots. Seedling species richness averaged 27% more on exclosure plots than on open plots. However, we found no significant differences in evenness or Shannon's diversity of the seedling community (Table 17.1).

Over roughly the same time period, vertebrates promoted AMF spore abundance, species richness, species diversity, and inoculum potential in the soil (Table 17.1). Mean AMF spore abundance was 51% higher and spore species richness was 28% higher in soils from open plots than from terrestrial vertebrate exclosure plots (Gehring et al. 2002). The species composition of the spore communities varied between exclosure and open plots, with *Glomus rubiforme*, a sporocarp-forming species that might be selectively consumed by terrestrial vertebrates, serving

as the strongest indicator of treatment differences (Gehring et al. 2002). Another sporocarp-forming species, *Glomus fasciculatum*, also was markedly more abundant on open plots than on exclosure plots (Gehring et al. 2002). These differences in spore communities were not associated with indirect effects of terrestrial vertebrates on seedlings or the soil environment (Gehring et al. 2002). Differences in the AMF spore communities of open and exclosure plots also were not associated with consistent differences in tree seedling communities (Figure 17.1). We performed the same ordination analysis on the seedling communities that we had done for the spore communities (blocked multi-response permutation procedure [MRPP], Gehring et al. 2002) and found no significant difference between exclosure and open plots in seedling community composition ($A = -0.004$, $P = 0.497$).

The differences in the spore communities that resulted from vertebrate exclusion were associated with differences in the AMF inoculum potential of the soil. Both a standard bioassay plant (corn) and a rainforest seedling (*Flindersia brayleana*) had higher levels (42–50%) of AMF colonization when grown in soil cores from open plots than when grown in soil cores from exclosure plots, suggesting that the effects of vertebrates on spore communities had consequences for formation of associations with AMF (Gehring et al. 2002). Subsequent comparisons with seedlings of another rainforest species, *Cryptocarya angulata*, indicated that levels of colonization by AMF also differed between seedlings naturally establishing in exclosure versus open plots in the field, demonstrating that the spore differences are important at early stages of seedling development even when other sources of inoculum, such as intact hyphal networks, are present (Figure 17.2).

POTENTIAL REASONS FOR THE DIFFERENT RESPONSE OF SEEDLINGS AND AMF SPORES

Our study plots can be viewed as relatively small islands of space where rates of local colonization and extinction differed due to the impacts

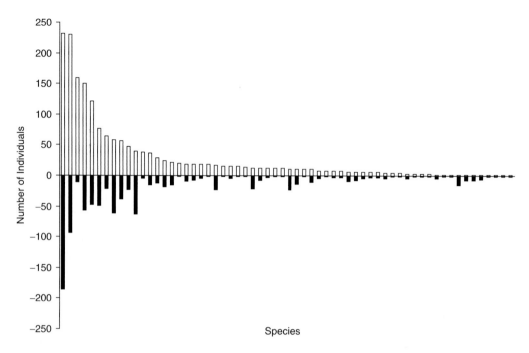

Figure 17.1   Total number of individuals of 62 seedling species that recruited on both exclosure plots (above the abscissa) and open plots during the 4.5 years after exclosures were erected.

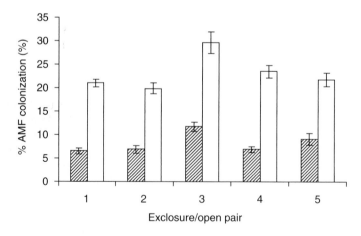

Figure 17.2   Mean levels of colonization by arbuscular mycorrhizal fungi (AMF) of *Cryptocarya angulata* seedlings germinating naturally in five paired exclosure (hatched bars) and open (open bars) plots. The bars represent the mean (±1 SE) of 3–4 seedlings from each plot. Seedlings were from the same germination cohort and were approximately 5 months old at the time of sampling. Seedlings from open plots had higher average levels of AMF colonization than seedlings from exclosure plots (means ±1 SE for open and exclosure seedlings are 23.18 ± 1.58 and 8.24 ± 0.89, respectively, paired t-test, $t = -14.51, P < 0.001$).

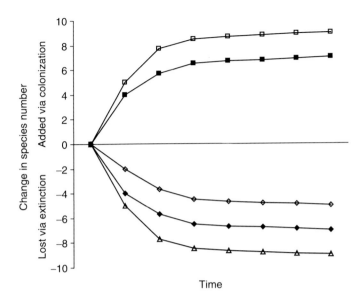

Figure 17.3 Hypothetical rate of species accumulation or loss on plots open to terrestrial vertebrates (open symbols) or protected from them (solid symbols). In this example, the species pool available to plots open to vertebrates would be higher than that available to plots where vertebrates are excluded because vertebrates carry seeds/spores to plots where they would otherwise not arrive. As a result, the rate of species addition would be greater on open plots (open squares) than on exclosure plots (solid squares). If terrestrial vertebrates increased overall mortality rates and that mortality was density independent, species loss on open plots should be more rapid (open triangles) than on exclosure plots (solid diamonds). However, if mortality due to terrestrial vertebrates was more strongly density dependent than on exclosure plots, rate of species loss would be lower on open plots than on protected plots (open diamonds) because more rare species would survive in the presence of vertebrates.

of terrestrial vertebrates. In the following paragraphs, we develop a simple conceptual model based on this analogy to explain how terrestrial vertebrates could potentially alter species richness, and discuss how that model could explain why tree seedling and AMF spore communities differed in their response to vertebrate exclosure.

Species colonizing our plots could be drawn from one of two species pools, a local pool of those species with propagules that would arrive at the site without dispersal by terrestrial vertebrates, and an enhanced pool that would include the addition of species with propagules that were dispersed by terrestrial vertebrates but that would not arrive by other means. Our exclosure plots would be limited to the local species pool while the open plots would be limited by the enhanced species pool, and in both cases the probability

that a new species would be added as species accumulated would decline to zero as the total species pool available was approached. The difference between exclosure and open plots in the number and rate of plant or AMF species accumulated would depend on the relative abundance of plant or fungal species with propagules dispersed by terrestrial vertebrates beyond their seed or spore shadow (Figure 17.3).

The probability of species extinction on any plot would depend on both the overall level of mortality and how that mortality was distributed among species. In the case of seedlings, if the overall level of seedling mortality was increased in the presence of terrestrial vertebrates, and that mortality fell on seedlings of different species without regard to their relative abundance (density-independent mortality), then the overall extinction rate on open plots would be greater than that on exclosures, and the difference in rates would be proportional to the difference in mortality rates between open and protected plots. If seedling mortality due to vertebrates was more strongly density dependent than that experienced by seedlings in their absence, then terrestrial vertebrates could potentially decrease the rate of species extinction on open plots relative to protected plots by allowing rare species to survive by differentially removing more abundant, superior competitors (Figure 17.3). The net change in tree seedling or AMF spore richness over time would depend on the difference in colonization and extinction rates.

We argue that this simple model offers an explanation for the differing responses of AMF spores and tree seedlings to terrestrial vertebrate exclosure and has important implications for how terrestrial vertebrates could alter community dynamics. In the case of tree seedlings on our site, we found little evidence that terrestrial vertebrates increased the local species pool available to colonize plots. The total number of species recorded on exclosure plots was actually higher than that on open plots. Although most seedling species were found in similar relative abundances on the two plots (Figure 17.1), there was a small suite of species that were found on exclosure plots and never recorded on open plots and vice versa. However, the relative rarity, as well as the fruit and seed characteristics, of these non-overlapping species suggested their absence was most likely due to sampling error or sensitivity to seed or seedling mortality in the presence of vertebrates. Only eight species (total of 19 individuals out of roughly 4000 seedlings recorded) occurred only on open plots; one species produced papery, winged seeds that we considered primarily wind dispersed (*Cardwellia sublimis*), six had seeds we considered primarily dispersed by arboreal birds, and only one (*Castanospora alphandi*) produced fruit and seeds that would potentially be moved primarily by terrestrial vertebrates. In contrast, 12 species (47 individuals) occurred only on exclosure plots, and again included species with a variety of putative seed dispersal modes, including wind, arboreal bird, and terrestrial vertebrates. We interpret these results as an indication that there

was no significant increase in the species pool available to colonize open plots due to increased seed dispersal by terrestrial vertebrates.

We hypothesize that this lack of effect was due to characteristics of the vertebrate community and the tree community at our site. First, two important terrestrial vertebrate seed dispersers, southern cassowaries and musky rat kangaroos, were relatively uncommon on our plots. We recorded only five cassowary droppings on the 5 ha we frequently traveled over 5 years of study, and none of these fell on a study plot (interestingly, the only one that yielded seedlings produced a clump of the most common canopy tree on the site, *Chrysophyllum* sp. nov.). White-tailed rats were common on our plot, but seedlings of one species for which white-tailed rats have been argued to be important seed dispersers, *Beilschmiedia bancroftii* (Harrington *et al.* 1997, Theimer 2001), were never recorded on an open plot but did occur in three exclosure plots, most likely because seed predation and cache recovery by white-tailed rats resulted in extremely low densities of surviving seedlings outside of exclosures. Perhaps more importantly, the majority of tree species on our plot did not depend on these animals as dispersal agents, and were instead effectively dispersed by wind (e.g., many Proteaceae) or canopy birds and bats (e.g., many Lauraceae).

In contrast, both the number of terrestrial vertebrate species that contributed to seed and seedling mortality, and therefore local extinction rates, and the number of plant species with seeds and seedlings susceptible to terrestrial vertebrate mortality were relatively large. The two most common rodents, the bush rat and *Melomys*, have not been reported to disperse seeds, but both have been documented to eat a variety of rainforest tree seeds offered to them in captive feeding trials (Harrington *et al.* 1997). In addition, several relatively common terrestrial vertebrates on our study plot (e.g., birds like chowchillas and brush turkeys, and the marsupial bandicoots) acted as important agents of seedling mortality, primarily by uprooting and burying seedlings while foraging in leaf litter, for some seedling species accounting for 70–90% of seedling cohort mortality (Theimer and Gehring 1999).

Table 17.2 Logistic regression analysis of seedling survival versus conspecific density and total seedling densitySat the scale of 1 m × 4 m subplots within exclosure and open experimental plots.

Model of seedling survival versus	Treatment	Model fit				Parameter estimates		
		d.f.	No. of seedlings	χ^2	Probability	Intercept	Slope	1 SE of slope
Conspecific density	Exclosure	1	3895	5.77	0.016	0.007	−0.002	0.001
	Open	1	2861	0.73	0.373	−0.313	−0.001	0.001
Total density	Exclosure	1	3895	0.07	0.782	−0.045	0.0001	0.0001
	Open	1	2861	3.25	0.073	−0.514	0.002	0.001

Note: Analysis was based on seedlings present in 1996 plus those recruiting between 1996 and 2000, with ultimate fate (survived versus died) based on the 2001 seedling census.

Although strong density-dependent mortality could slow the loss of uncommon species, we found little evidence for these effects. For example, logistic regression analysis of seedling survival versus conspecific density and total density at the scale of 1 m × 4 m plots showed relatively small but significant effects of conspecific density on survival in exclosure plots but not on open plots (Table 17.2). Instead, the higher seedling mortality on open plots apparently led to a more rapid rate of species extinction.

Taken together, the effects of terrestrial vertebrates on the dynamics of seedling colonization and extinction can be summarized as no significant increase in the rate of species accumulation due to an increased species pool, combined with increased seedling mortality that was not strongly density dependent. The result was a lower net rate of species accumulation on plots open to terrestrial vertebrates (Figure 17.4).

In contrast, the significantly higher abundance and richness of AMF spores on open plots suggests that vertebrates acted to enhance the species pool available on these plots and thereby increase species accumulation. AMF spores have been shown to pass through the guts of many of the species on our study plot in viable form (all of the rodents, cassowaries, bandicoots, and musky rat kangaroo; Reddell *et al.* 1997). Consistent with this hypothesis, we found that AMF species that produced spores in macroscopic sporocarps (and were therefore more likely to be attractive to mycophagous vertebrates) were more likely to occur on plots open to terrestrial vertebrates (Gehring *et al.* 2002). For example, spores of *Glomus rubiforme* were found in all open plots but in only half of the vertebrate exclosures, suggesting that these spores were more widely distributed in the presence of vertebrates. Lacking in the case of spores, however, is any indication that terrestrial vertebrates act as predators that would reduce spore abundance and thereby reduce colonization of plant roots. While vertebrates might indirectly negatively impact AMF through either herbivory, changes in soil nutrient status, or altered host plant abundance and/or species composition, we found no evidence of this (Gehring *et al.* 2002). As a result, in contrast to the effects of terrestrial vertebrates on seedling extinction, we hypothesize that terrestrial vertebrates had little impact on local extinction of species of AMF. As a result, we hypothesize that terrestrial vertebrates enhanced the species pool available to open plots through spore dispersal and thereby increased the rate of species accumulation while having little impact on spore mortality and species extinction (Figure 17.5).

IMPLICATIONS FOR THE ROLE OF TERRESTRIAL VERTEBRATES IN TROPICAL SYSTEMS

The results outlined above suggest that community response to terrestrial vertebrates will be complex, depending to a large extent on the

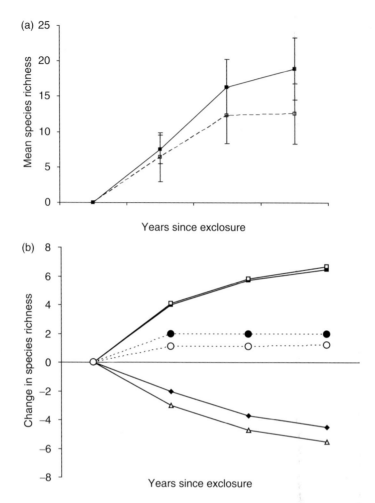

Figure 17.4 (a) The mean (± standard deviation, SD) number of seedling species recorded on open and exclosure plots over the 3 years after exclosures were erected. Paired *t*-tests showed significantly higher (*P* < 0.005) species richness on exclosure plots in all but the first year after exclosure. (b) These observed rates of species accumulation are hypothesized to result from increased mortality and subsequent higher species extinction rates on open plots (open triangles, solid line) versus exclosure plots (closed diamonds, solid line), with little to no increase in the species pool available to colonize open plots (open squares, solid line) compared with exclosure plots (closed squares, solid line) and therefore an overall lower net rate of species accumulation for open plots (open circles, dotted line) versus exclosure plots (closed circles, dotted line).

relative abundance of community members with differing attributes and the time over which those communities are monitored. We describe below what we believe are four major implications of this view for studies of vertebrate impacts on tropical rainforest community dynamics.

First, the relative proportion of species in a community for which vertebrate dispersal significantly increases the seed/spore shadow determines the magnitude of the difference in species pools available in the presence or absence of vertebrates, and thereby the rate of species accumulation.

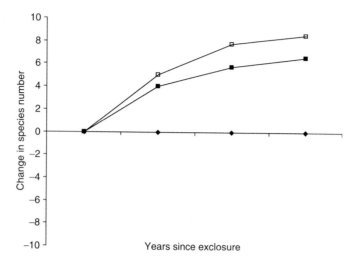

Figure 17.5 Hypothesized relationship for accumulation of mycorrhizal spore species through time on open plots (open squares) and exclosure plots (closed squares). Loss of spore species due to extinction caused by terrestrial vertebrates is assumed to be zero for both open and exclosure plots (closed diamonds). Vertebrates increase the species pool of spores arriving on open plots, resulting in the net rate of species accumulation being higher on those plots.

The tree community at our site had relatively few species with seeds dispersed primarily by terrestrial vertebrates, and therefore exclusion of those vertebrates had little overall effect on species accumulation. Tree communities with a larger proportion of species with seeds dispersed by terrestrial vertebrates will have a larger difference in the species pool available to colonize a site, and therefore loss of terrestrial vertebrates would be predicted to have a greater effect. For example, seeds of many tree species on our plots are dispersed by arboreal birds and bats, and we predict that excluding input from these animals would more strongly decrease the species pool available to exclosure plots and thereby cause a greater difference in species accumulation between treatments.

Second, even in those communities with a large number of species dependent on vertebrate dispersal, if seedling mortality due to vertebrates is high enough, and not overall more strongly density dependent than in their absence, the net effect of vertebrates may still be to reduce local species richness due to increased extinction rates. Increased local species extinction may be more likely when terrestrial vertebrates act as random mortality

agents, destroying seeds and seedlings by uprooting or trampling. On our plots, this mortality often impacted wind- and arboreal bird-dispersed species that produced small seedlings with shallow root systems or those with large cotyledons that made the seedling more susceptible to toppling or breakage. Combined with vertebrate predation on seeds and seedlings, overall mortality due to vertebrates was thereby spread across almost all seedling species on our site, resulting in higher rates of species extinction. In communities where more tree species have seeds or seedlings that are directly preyed upon by vertebrates, the potential for vertebrates to act in density-dependent ways would be greater and the overall impact of vertebrates on extinction rates may not be as great as it was on our site. Even so, this suggests that several seedling species in a community can show strong density-dependent mortality in the presence of vertebrates, but if many more species in the community experience higher, non-density-dependent mortality, the net effect of vertebrates could still be to reduce species richness.

Third, our results suggest that terrestrial vertebrates increase local dispersal limitation by reducing the probability that species arriving at a site will successfully establish there. Factors that increase dispersal limitation could potentially maintain diversity by allowing inferior competitors to survive in some areas where superior competitors never successfully establish (Tilman 1994, Hurt and Pacala 1995, Hubbell 1997, Tilman 1999). However, if seedling densities are consistently below levels necessary for strong competitive effects, dispersal limitation similar to that we documented could simply reduce local species richness. Our results support Wright's (2002) hypothesis that in many cases vertebrates maintain seedling densities at levels that preclude strong competitive effects. For example, even in the exclosure plot with highest seedling density, we detected no reduction in recruitment or survival by the fourth year after exclosure, as would be expected if strong competitive interactions were at play. Our results also underscore that even in systems in which seed arrival may not limit dispersal (e.g., Webb and Peart 2001), terrestrial vertebrates could act to increase dispersal limitation through their effects on seeds and seedlings.

Fourth, indirect effects of terrestrial vertebrates, like that of increasing AMF spore species richness, could potentially alter seedling community dynamics, but these effects may take considerable time to be expressed. The importance of AMF spore dispersal to rainforest seedling establishment and survival remains poorly understood, but there is growing evidence from other types of plant communities that some species are of greater benefit to a given plant species than others (e.g., van der Heijden *et al.* 1998a, Bever 2002, Klironomos 2003) and that AMF diversity may contribute to plant diversity (van der Heijden *et al.* 1998b). Limited distributions of some fungal species in the absence of vertebrates thus may alter plant communities. Similarly, Mangan *et al.* (2004) hypothesized that differences in the AMF communities of island and mainland sites in tropical Central America could contribute to the differences in plant communities observed at those sites. Lower AMF inoculum potential in the absence of vertebrate dispersers, as we observed, also may favor seedling species that do not form associations with AMF or that do not depend on them for growth or nutrient acquisition, leading to changes in plant communities and potentially further

reducing the populations of mycorrhizal fungi. For example, we observed that one third of the common seedling species on our rainforest study site were never or rarely observed to form associations with AMF or other types of mycorrhizal fungi (Gehring and Connell 2006). These non-mycorrhizal seedling species would be expected to become more common in the absence of terrestrial vertebrates, potentially leading to further reductions in fungal diversity and inoculum potential. Although these seedling species showed similar abundance in exclosure and open plots when AMF spore data were collected in 2001, a preliminary analysis of these seedling species in 2003 showed that the percentage of seedlings of non-mycorrhizal plant species was marginally higher in exclosure plots than in open plots (mean for open plots $= 34.8 \pm 4.5$ SE [standard error] and for exclosure plots $= 41.8 \pm 5.8$ SE, $t = 1.418$, $P = 0.08$). The diversity of microsites available for plant establishment has been recognized as an important factor maintaining tree diversity (e.g., Grubb 1977, Denslow 1980). We suggest that the effects of terrestrial vertebrates on the presence, diversity, and species composition of mycorrhizal fungal inoculum could be a relatively unexplored, but potentially important, aspect of this microsite diversity.

SUGGESTIONS FOR FUTURE RESEARCH

We suggest three avenues of research that would be fruitful in testing the ideas presented here and in expanding our knowledge of the potential interaction of vertebrates, seedlings, and AMF spores. First, comparison of seedling and spore communities that differ both in the proportion of species dependent on vertebrates for dispersing propagules and in the susceptibility of propagules to vertebrate mortality would elucidate how different communities may respond to vertebrate loss or addition. In the case of tree seedlings, such studies should distinguish between vertebrates that act primarily as seed dispersal agents that potentially reduce dispersal limitation, versus those that act as agents of seed and seedling mortality and thereby increase it. Determining how seedlings respond to spores dispersed by terrestrial vertebrates under varying conditions is critical to understanding the importance of spore dispersal by these animals.

Second, for AMF, it is important to gain a better understanding of how important spores are as propagules. Recent molecular techniques that allow the determination of the presence, absence, and abundance of AMF in root tissue allow estimates of the diversity of mycorrhizal species colonizing roots to be assessed (e.g., Husband *et al.* 2002, Rosendahl and Stukenbrock 2004). This technology allows direct comparison of the fungal community that has established on plant roots with the fungal community present in the soil or in the feces of terrestrial vertebrates.

Third, given that in many systems seedlings remain in the understory for years or decades, the full suite of vertebrate effects may not be expressed for very long periods of time, so long-term studies, or studies comparing sites with intact faunas with sites which have been defaunated for long periods, are required to understand the ultimate fate of defaunated forests. For example, if vertebrates maintain seedling densities below levels where competitive interactions are important, then the effect of defaunation may not be expressed until seedling densities have recovered to the point where these interactions come into play. Even though seedling densities across all our exclosure plots increased in the absence of vertebrates, most had still not approached the initial density of our most densely populated plot after 4 years of exclosure. Likewise, vertebrate-induced changes in the AMF community available to colonize plants may take relatively long periods to alter seedling abundance and composition. Unfortunately, the current rate of loss of many vertebrate species from tropical rainforests worldwide means that many forests will lose important vertebrate components before we fully understand their potential role in influencing diversity.

ACKNOWLEDGMENTS

We thank S. Schnitzer and W. Carson for the opportunity to contribute a chapter to this volume, two anonymous reviewers for their helpful comments, J. Connell and P. Green for stimulating discussions and invaluable assistance in the field, the CSIRO Tropical Forest Research Centre for generous access to facilities, and the National Science Foundation (grants DEB95-03217 and DEB98-06310) for financial support.

REFERENCES

Allen, E.B. and Allen, M.F. (1984) Competition between plants of different successional stages: mycorrhizae as regulators. *Canadian Journal of Botany* 62, 2625–2629.

Asquith, N.M., Terborgh, J., Arnold, A.E., and Riveros, C.M. (1999) The fruits the agouti ate: *Hymenaea courbaril* seed fate when its disperser is absent. *Journal of Tropical Ecology* 15, 229–235.

Bereau, M. and Garbaye, J. (1994) First observations on the root morphology and symbioses of 21 major tree species in the primary tropical rain forest of French Guyana. *Annales des Sciences Forestieres* 51, 407–416.

Bever, J.D. (2002) Negative feedback within a mutualism: host specific growth of mycorrhizal fungi reduces plant benefit. *Proceedings of the Royal Society of London Series B Biological Sciences* 69, 2595–2601.

Bever, J.D., Westover, K.M., and Antonovics, J. (1997) Incorporating the soil community into plant population dynamics: the utility of the feedback approach. *Journal of Ecology* 85, 561–573.

Connell, J.H. Tracey, J.G., and Webb, L.J. (1984) Compensatory recruitment, growth, and mortality as factors maintaining rain forest tree diversity. *Ecological Monographs* 54, 141–164.

Dennis, A.J. (2003) Scatterhoarding by musky rat-kangaroos, *Hypsiprymnodon moschatus*, a tropical rain-forest marsupial from Australia: implications for seed dispersal. *Journal of Tropical Ecology* 19, 619–627.

Denslow, J. (1980) Gap partitioning among tropical rain forest trees. *Biotropica* 12, 47–55.

Dirzo, R. and Miranda, A. (1990) Contemporary neotropical defaunation and forest structure, function and diversity – a sequel to John Terborgh. *Conservation Biology* 4, 444–447.

Dirzo, R. and Miranda, A. (1991) Altered patterns of herbivory and diversity in the forest understory: a case study of the consequences of contemporary defaunation. In P.W. Price, T.M. Lewisohn, G.W. Fernandes *et al.* (eds), *Plant–Animal Interactions: Evolutionary Ecology in Tropical and Temperate Regions*. John Wiley and Sons, New York, pp. 273–287.

Eisenberg, J.F. (1981) *The Mammalian Radiations*. University of Chicago Press, Chicago.

Forget, P.-M. and Sabatier, D. (1997) Dynamics of the seedling shadow of a frugivore-dispersed tree species in French Guiana. *Journal of Tropical Ecology* 13, 767–773.

Frank, D.A., Gehring, C.A., Machut, L., and Philips, M. (2003) Soil community composition and the regulation of grazed temperate grassland. *Oecologia* 442, 603–609.

Gehring, C.A. (2003) Growth responses to arbuscular mycorrhizas by rain forest seedlings vary with light intensity and tree species. *Plant Ecology* 167, 127–139.

Gehring, C.A. and Connell, J.H. (2006) The occurrence and potential importance of arbuscular mycorrhizal fungi in tree seedlings of two Australian rain forests. *Mycorrhiza* 16, 89–98.

Gehring, C.A., Wolf, J.E., and Theimer, T.C. (2002) Terrestrial vertebrates promote arbuscular mycorrhizal fungal diversity and inoculum potential in a rain forest soil. *Ecology Letters* 5, 540–548.

Grime, J.P., Mackey, J.M.L., Hillier, S.H., and Read, D.J. (1987) Floristic diversity in a model system using experimental microcosms. *Nature* 328, 420–422.

Grubb, P.J. (1977) The maintenance of species richness in plant communities: the importance of the regeneration niche. *Biological Reviews* 52, 107–145.

Harrington, G.N., Irvine, A.K, Crome, F.H.J., and Moore, L.A. (1997) Regeneration of large-seeded trees in Australian rain forest fragments: a study of higher-order interactions. In W.F. Laurance and R.O. Bierregaard (eds), *Tropical Forest Remnants*. University of Chicago Press, Chicago, pp. 292–303.

Hartnett, D.C. and Wilson, G.W.T. (1999) Mycorrhizae influence plant community structure and diversity in tallgrass prairie. *Ecology* 80, 1187—1195.

Hetrick, B.A.D., Wilson, G.W.T., and Hartnett, D.C. (1989) Relationship between mycorrhizal dependence and competitive ability of two tall grass prairie grasses. *Canadian Journal of Botany* 67, 2608–2615.

Hopkins, M.S., Reddell, P., Hewett, R.K., and Graham, A.W. (1996) Comparison of root and mycorrhizal characteristics in primary and secondary rain forest on a metamorphic soil in North Queensland, Australia. *Journal of Tropical Ecology* 12, 871–885.

Hubbell, S.P. (1997) A unified theory of biogeography and relative species abundance and its application to tropical rain forest and coral reefs. *Coral Reefs* 16(Suppl.), 9–21.

Husband, R., Herre, E.A., Turner, S.L., Gallery, R., and Young, J.P.W. (2002) Molecular diversity of arbuscular mycorrhizal fungi and patterns of host association over time and space in a tropical forest. *Molecular Ecology* 11, 2669–2678.

Janos, D.P. (1980) Vesicular-arbuscular mycorrhizae affect lowland tropical rainforest plant growth. *Ecology* 61, 151–162.

Janos, D.P., Sahley, C.T., and Emmons, L.H. (1995) Rodent dispersal of vesicular-arbuscular mycorrhizal fungi in Amazonian Peru. *Ecology* 76, 1852–1858.

Klironomos, J.N. (2003) Variation in plant response to native and exotic arbuscular mycorrhizal fungi. *Ecology* 84, 2292–2301.

Klironomos, J.N., McCune, J., Hart, M., and Neville, J. (2000) The influence of arbuscular mycorrhizae on the relationship between plant diversity and productivity. *Ecology Letters* 3, 137–141.

Lovelock, C.E., Andersen, K., and Morton, J.M. (2003) Host tree and environmental control on arbuscular mycorrhizal spore communities in tropical forests. *Oecologia* 135, 268–279.

Mangan, S.A. and Adler, G.H. (2000) Consumption of arbuscular mycorrhizal fungi by terrestrial and arboreal small mammals in a Panamanian cloud forest. *Journal of Mammalogy* 81, 563–570.

Mangan, S.A., Ahn-Heum, E., Adler, G.H, Yavitt, J.B., and Herre, E.A. (2004) Diversity of arbuscular mycorrhizal fungi across a fragmented forest in Panama: insular spore communities differ from mainland communities. *Oecologia* 141, 687–700.

McGee, P.A. and Baczocha, N. (1994) Sporocarpic endogonales and glomales in the scats of *Rattus* and *Perameles*. *Mycological Research* 98, 246–249.

Peres, C.A. (2001) Synergistic effects of subsistence hunting and habitat fragmentation on Amazonian vertebrates. *Conservation Biology* 15, 1490–1505.

Reddell, P., Spain, A.V., and Hopkins, M. (1997) Dispersal of spores of mycorrhizal fungi in scats of native mammals in tropical forests of northeastern Australia. *Biotropica* 29, 184–192.

Roldan, A.I. and Simonetti, J.A. (2001) Plant mammal interactions in tropical Bolivian forests with different hunting pressures. *Conservation Biology* 15, 617–623.

Rosendahl, S. and Stukenbrock, E.H. (2004) Community structure of arbuscular mycorrhizal fungi in undisturbed vegetation revealed by analyses of LSU rDNA sequences. *Molecular Ecology* 13, 3179–3186.

Smith, S.E. and Read, D.J. (1997) *Mycorrhizal Symbiosis*, 2nd Edition. Academic Press, London.

St John T.V. (1980) A survey of mycorrhizal infection in an Amazonian rain forest. *Acta Amazonica* 10, 527–533.

Tabarelli, M., Mantovani, W., and Peres, C.A. (1999) Effects of habitat fragmentation on plant guild structure in the montane Atlantic forests of southeastern Brazil. *Biological Conservation* 981, 119–127.

Theimer, T.C. (2001) Seed scatterhoarding by white-tailed rats: consequences for seedling recruitment by an Australian rain forest tree. *Journal of Tropical Ecology* 17, 177–189.

Theimer, T.C. and Gehring, C.A. (1999) Effects of a litter-disturbing bird on tree seedling germination ad survival in an Australian tropical rain forest. *Journal of Tropical Ecology* 15, 737–749.

Tilman, D. (1994) Competition and biodiversity in spatially structured habitats. *Ecology* 75, 2–16.

Tilman, D. (1999) Diversity by default. *Science* 283, 495–496.

van der Heijden, M.G.A., Boller, T., Wiemken, A., and Sanders, I.R. (1998a) Different arbuscular mycorrhizal fungal species are potential determinants of plant community structure. *Ecology* 79, 2082–2091.

van der Heijden, M.G.A, Klironomos, J.N., Ursic, M. et al. (1998b) Mycorrhizal fungal diversity determines plant biodiversity, ecosystem variability and productivity. *Nature* 396, 69–72.

Webb, C.O. and Peart, D.R. (2001) High seed dispersal rates in faunally intact tropical rain forest: theoretical and conservation implications. *Ecology Letters* 4, 491–499.

Wenny, D.G. (2000) Seed dispersal, seed predation, and seedling recruitment of a neotropical montane tree. *Ecological Monographs* 70, 331–351.

Westcott, D.A., Bentrupperbauemer, J., Bradford, M., and McKeown, A. (2006) Incorporating patterns of disperser behaviour into models of seed dispersal and its effects on estimated dispersal curves. *Oecologia* 146, 57–67.

Wright, S.J. (2002) Plant diversity in tropical forests: a review of mechanisms of species coexistence. *Oecologia* 130, 1–14.

Wright, S.J., Zeballos, H., Dominguez, I., Gallardo, M.M., Moreno, M.C., and Ibanez, R. (2000) Poachers alter mammal abundance, seed dispersal and seed predation in a neotropical rainforest. *Conservation Biology* 14, 227–239.

Wyatt, J.L. and Silman, M.R. (2004) Distance-dependence in two Amazonian palms: effects of spatial and temporal variation in seed predator communities. *Oecologia* 140, 26–35.

Zangaro, W., Bononi, V.L.R., and Trufen, S.B. (2000) Mycorrhizal dependency, inoculum potential and habitat preference of native woody species in South Brazil. *Journal of Tropical Ecology* 16, 603–622.

Zhao, Z.-W., Xia, Y.-M., Qin, X.-Z. et al. (2001) Arbuscular mycorrhizal status of plants and the spore density of arbuscular mycorrhizal fungi in the tropical rain forest of Xishuangbanna, southwest China. *Mycorrhiza* 11, 159–162.

Chapter 18

ECOSYSTEM DECAY IN CLOSED FOREST FRAGMENTS

John Terborgh and Kenneth Feeley

OVERVIEW

We summarize a long-term study of forest fragments isolated since 1986 as islands in Lago Guri, a vast hydroelectric impoundment in Venezuela. We studied replicate islands of two classes: small (≥ 0.5 ha, <2 ha) and medium (≥ 4 ha, <15 ha). Islands of both classes lacked predators of vertebrates and were deficient in pollinators and seed dispersers, but supported hyperdense populations of predators of invertebrates (birds, lizards, anurans, spiders), rodents, and especially herbivores (howler monkeys, common iguanas, and leaf-cutter ants). Medium islands differed from small islands in supporting a predator of leaf-cutter ants (armadillo), a scatterhoarder (agouti), and often a mesopredator (capuchin monkey), while supporting herbivores at densities intermediate between those of small islands and the large landmasses that served as controls. Large landmasses supported intact faunas and served as controls.

Our results strongly supported the hypothesis of Hairston, Smith, and Slobodkin that consumers will increase in the absence of predators to levels that result in damage to vegetation. Tree and sapling mortality was higher, and sapling recruitment dramatically lower, on small islands supporting hyperabundant herbivores. However, trophic cascades on predator-free islets were more complex than predicted by contemporary food-web theory because consumers belonged to various functional groups (e.g., pollinators, seed predators, folivores). Moreover, hyperabundant herbivores generated indirect bottom-up effects mediated via nutrient cycling. Indirect effects included enhanced tree growth, density overcompensation in bird communities, and decreased rates of avian extinction on islands supporting hyperabundant howler monkeys.

On small islands, sapling mortality exceeded recruitment for all species. This suggests that herbivore defenses of common plant species are most effective in the presence of top-down regulation. Hyperabundant herbivores appeared to be under bottom-up regulation. For example, tortoises on islands exhibited reduced growth rates and howler monkeys were below normal weight and reproduced slowly with respect to controls. We did not find any direct impact of edge exposure on tree demography.

Overall, we conclude that terrestrial trophic cascades are far more complex than implied by simple, three-level trophic models. The multiplicity of pathways suggests a complex interaction web. This web can be distorted in myriad ways with consequences that most would regard as undesirable.

INTRODUCTION

An initial effect of habitat fragmentation is to distort or disrupt many landscape-scale processes, including both biological processes, such as predation, pollination, and seed dispersal, and physical processes, such as fire and the moderating effects of a continuous habitat on microclimate (Kapos 1989, Aizen and Feinsinger 1994, Asquith *et al.* 1999, Cochrane and Laurance 2002, Laurance *et al.* 2002, Chapman and Chapman 2003). The multiplicity of processes that are altered by fragmentation has impeded efforts to understand the consequences of fragmentation from a mechanistic standpoint.

Further levels of complexity and ambiguity are inherent in the common use of "open" fragments (here defined as habitat patches in a land-use matrix) in the vast majority of fragmentation studies. The character of the matrix surrounding open patches is rarely homogeneous in terms of vegetation and distances among patches, and in many cases will be perceived in different ways by the various species inhabiting the landscape. For some species, the matrix may be freely porous, whereas for others, it may serve as an impermeable barrier (Malcolm 1994, Levey et al. 2005). Since the influence of matrix character has seldom been considered in studies of fragmentation (but see Gascon et al. 1999), the use of open fragments entails serious ambiguities, as it is rarely known which species are using different patches or moving amongst them.

The clearest interpretation of results can be obtained from "closed" fragments that are severely isolated from the surrounding matrix. Colonization by plants and animals of islands as remote as the Hawaiian archipelago demonstrates that no fragment on earth is absolutely closed, so the permeability of a matrix is a relative concept (Rose and Polis 2000). Nevertheless, water constitutes a less heterogeneous and permeable barrier for non-volant terrestrial species than a mainland habitat matrix (Gascon et al. 2000).

Since closed fragments can be thoroughly inventoried (Terborgh et al. 1997a,b), one can accurately determine which species are present and which are not. Knowing which species are absent can be as important as knowing which species are present. For example, many predators are wide-ranging and may routinely traverse unsuitable habitat to hunt in high quality patches. Because predation is difficult to detect, it is likely to go unnoticed in large complex habitats. With closed fragments, one has better knowledge of which predators are present, and consequently a greater ability to determine their influence on community dynamics.

Another advantage of closed fragments is that density-dependent emigration rarely influences the dynamics of resident animal populations. When possible, many species will emigrate in response to high densities. Animal populations of open fragments thus rarely exhibit increased

density, here termed "hyperabundance," a common feature of animal populations on islands or closed fragments (Crowell 1962, Grant 1965, Krebs et al. 1969, MacArthur et al. 1972, Morse 1977, Case et al. 1979, Emlen 1979, Wright 1980, George 1987, Blondel et al. 1988, Polis and Hurd 1995, Adler 1996).

Our purpose in this chapter is to synthesize the results of a 10-year investigation of closed forest fragments isolated as islands in Lago Guri. Lago Guri is a vast hydroelectric impoundment created in 1986 with the completion of the Raul Leoní dam on the lower Caroní River in the state of Bolívar, Venezuela. Lago Guri contains hundreds of land-bridge islands, all formerly interconnected as parts of a continuous forested landscape. The islands range in size from less than 0.1 ha to more than 1000 ha (Morales and Gorzula 1986), constituting a giant experiment on the role of spatial scale in ecology (Diamond 2001).

A crucial feature of the islands we studied is that they were essentially devoid of predators of vertebrates during the period of our research (1993–2003; Terborgh et al. 1997a, 2001). In 1960, Hairston, Smith, and Slobodkin (HSS) proposed a simple trophic cascade model which could be used to predict the consequences of predator removal. According to the HSS model, predators regulate consumers to low population densities, thereby allowing plants to escape damaging levels of herbivory. Therefore, predator loss or removal should allow herbivores to increase to levels at which damage to vegetation would be manifest. The HSS model was subsequently generalized by Oksanen et al. (1981) who proposed that a perturbation at one trophic level would propagate downward through a food web with alternating positive and negative effects at lower levels (Paine 1980, Carpenter and Kitchell 1993, Terborgh 2005). Based on these theoretical models, our expectation for predator-free Guri islands was to find higher-than-normal densities of consumers and evidence of damage to vegetation.

Both these expectations were abundantly affirmed. However, at least initially, we did not appreciate the complexity of the trophic cascade on predator-free Guri islands. The combination of restricted area, isolation, and the absence of predators of vertebrates (and some invertebrates) resulted in the distortion of many biological interactions. Since vertebrate and invertebrate consumers interact with plants as pollinators, seed dispersers, herbivores, etc., distortions in the structure of the animal community can propagate to the producer level via multiple pathways. As we investigated some of these pathways, it became clear that the strength of the different pathways was highly dependent on spatial scale in a discontinuous fashion as key species (pollinators, dispersers, predators, etc.) entered or dropped out of the system. We shall now present a synopsis of the evidence that led us to these conclusions.

METHODS

Physical setting

Lago Guri is located in east-central Venezuela on the lower Caroní River near its confluence with the Orinoco River. The dam was completed in 1986 and raised the water to 270 m above sea level from a base of 120 m, inundating an area of 4300 km^2 (Morales and Gorzula 1986). Due to the hilly terrain, inundation resulted in the transformation of hundreds of forested hilltops into isolated land-bridge islands (<0.1 ha to >1000 ha; Alvarez *et al.* 1986).

Lago Guri and a broad watershed protection zone surrounding it are administered by EDELCA (Electrificación del Caroní), the Venezuelan energy company that operates the hydroelectric station. Access to the lake is strictly controlled and no hunting is allowed, though some illegal poaching still occurs, primarily for capybara.

Mean annual rainfall at the dam is 1100 mm, but is somewhat higher 60–80 km to the south where we conducted our research. The habitat of all study islands was tropical dry forest supporting 50–70 species of trees per hectare (Peetz 2001, Feeley *et al.* 2005). Much of the canopy is facultatively deciduous, such that many crowns keep their leaves in wet years but may lose them for several months during dry years (Huber 1986).

Our research built upon two salient features of the Guri experiment: the availability of replicate islands in each of three discrete size

classes – small (S), medium (M), and large (L) – and the high degree of consistency in animal community composition within each size class (Table 18.1). Thus, islands were not assigned to classes on the basis of area *per se*, but by the consistency with which they retained less complete to more complete animal communities (Terborgh *et al.* 1997a). Nearly all of the study islands had gentle topography formed of well-drained, clay-rich, oxisols (three islands were rocky and relatively steep).

Animal communities

Our research began in 1993, some 7 years after the lake level reached its final stage. We surmise that non-volant species inhabiting the Guri landscape prior to inundation were forced to swim or migrate upslope as the water level rose, thereby becoming concentrated on hilltops that eventually became islands. After 7 years, most individual animals that survived the inundation would have died, except for long-lived species such as monkeys and tortoises. The populations of short-lived species either became self-sustaining or died out soon after isolation. This inference is supported by the fact that three-quarters or more of all vertebrate species present on the mainland were already absent from S and M islands when we undertook the first surveys in 1993 (Terborgh *et al.* 1997a,b). Although populations of some species subsequently disappeared from some islands (Feeley 2005a), most non-volant species recorded on the S and M islands in 1993 were still present in 2003 when the project ended. Thus, community composition remained reasonably consistent throughout the period of the research.

Small islands ranged in size from 0.25 to 2.5 ha and supported predators of invertebrates (birds, lizards, amphibians, spiders, etc.), pollinators (bees, wasps, lepidoptera, hummingbirds), seed predators (rodents), and generalist herbivores (common iguanas, *Iguana iguana*; red howler monkeys, *Alouatta seniculus* – several islands; leaf-cutter ants, *Atta* spp., *Acromyrmex* sp.), but few, if any, seed dispersers and no predators of vertebrates.

Table 18.1 Relative abundance of some animals on Lago Guri islands, by functional group.

Functional group	Landmass category			
	Small	**Medium**	**Large**	**Mainland**
Predators of vertebrates				
Jaguar, puma, harpy			tr.	+
Ocelot		tr.	+	+
Raptors			+	+
Boa, rattlesnake			+	+
Predators of invertebrates				
Birds	++	−	+	+
Amphibians				
Dendrobates leucomelos	+++	++	+	+
Bufo spp.	++	+	+	+
Lizards				
Ameiva ameiva	+++	++	+	+
Spiders	+++	+	+	+
Pollinators				
Social bees	−	+	+	+
Butterflies	−	+	+	+
Hummingbirds	++	+	+	+
Seed dispersers				
Capuchin monkey		++	++	+
Guans		#	+	+
Toucans		#	+	+
Cotingidae		−	+	+
Pipridae		+	++	++
Thraupidae	#	+	+	+
Turdidae		#	+	+
Seed predators				
Collared peccary		tr.	++	+
Bearded saki monkey		#	+	+
Agouti		++	++	+
Paca			+	+
Spiny rat			+	+
Other small rodents	+++	+++	+	+
Generalist herbivores				
Tortoise		++	++	+
Deer		#	+	+
Tapir				+
Howler monkey	+++	++	+	+
Iguana	+++	++	+	+
Porcupine	#	?	+	+
Leaf-cutter ants	+++	++	+	+
Social insect predators				
Armadillo, nine-banded		++	+	+
Armadillo, large		#	+	+
Tamandua		#	+	+
Giant anteater			tr.	+
Army ant (*E. hammatum*)			+	+

Note: +, abundance equivalent to mainland; ++, supranormal abundance; +++, hyperabundant; −, abundance less than on mainland; #, present on one or more islands of class; tr., transient; ?, presence uncertain.

Medium islands ranged in area from 4 to 25 ha. In addition to the species present on small islands, M islands supported a scatterhoarding rodent (agouti, *Dasyprocta* spp.), one or sometimes two predators of social insects (nine-banded armadillo, *Dasypus novemcinctus* – all medium islands; tamandua, *Tamandua tetradactylus* – one island), and an additional generalist herbivore, the tortoise (*Geochelone carbonaria*). A mesopredator (capuchin monkey, *Cebus olivaceus*) was present on three M islands.

Large landmasses served as controls. So that we could benefit from multiple large landmass sites while limiting pseudoreplication, we used two large islands (88 and 190 ha) and two sites 2 km apart on the nearby mainland. The large landmasses supported complete or nearly complete faunas, including all primates known to occur in the region, mesopredators (opossum, *Didelphis marsupialis*; tayra, *Eyra barbara*; coati, *Nasua nasua*), predators of vertebrates (felids, raptors, snakes), and large terrestrial mammals such as deer (*Mazama americana*), collared peccary (*Tayassu tajacu*), and tapir (*Tapirus terrestris*). Jaguar, puma, and harpy eagle were all confirmed for the mainland and the 190 ha island, but were not observed on the 88 ha island or any smaller island.

Vegetation dynamics

We investigated the prediction that the vegetation of predator-free S and M islands would suffer herbivore-induced damage by censusing seedlings and monitoring woody stems of several size classes on islands and the mainland. Trees ≥ 10 cm diameter at breast height (dbh) were monitored for survival, recruitment, mortality, and growth in plots on 24 islands and at two mainland sites. Within 14 of these plots (located at 5 S, 4 M, 5 L sites) we subsampled small (≥ 1 m tall, <1 cm dbh) and large (≥ 1 cm dbh and <10 cm dbh) saplings in 15 m × 15 m (225 m^2) subplots, two in each tree plot. In addition, we censused seedlings (<1 m tall) in a total of 136, 2 m × 2 m plots located on 17 S islands. Tree plots were recensused after either 1 or 5 years, and sapling plots after 5 years (seedling plots were not recensused). Overall, we tagged,

mapped, measured, and identified 5609 adult trees, 7027 saplings, and 4086 seedlings representing more than 320 woody species (exclusive of lianas).

RESULTS

Hyperdensity of persistent populations and functional imbalances

Faunal surveys revealed that most animal populations persisting on S and M islands displayed increased density (hyperabundance) compared with mainland/large landmass populations. Hyperabundance was documented for some species of birds (Terborgh *et al.* 1997b, Feeley and Terborgh 2006), various mammals, reptiles, amphibians, and spiders (Terborgh *et al.* 1997a), leaf-cutter ants (Rao 2000), rodents (Lambert *et al.* 2003), tortoises (Aponte *et al.* 2003), and howler monkeys (Terborgh *et al.* 2001; Table 18.1). In addition, some islands presumably supported high numbers of arthropods (other than leaf-cutter ants) judging from the hyperabundance of their predators (lizards, anurans, birds, and spiders; Terborgh *et al.* 1997a, Feeley and Terborgh 2006). Common iguanas were present on S and M islands at roughly 10 times mainland density; howler monkeys and rodents at 20–30 times, and leaf-cutter ants at up to 100 times. Importantly, species loss/persistence was strongly non-random with regard to guild membership such that some functional groups were over-represented (predators of invertebrates, rodents), whereas others were under-represented (pollinators, seed predators) or altogether absent (predators of vertebrates; Terborgh *et al.* 1997a, Feeley *et al.* 2007).

Causes of hyperabundance

It is tempting to attribute the hyperabundance of birds, lizards, amphibians, rodents, and herbivores to the absence of predators. Although this causal link can only be inferred, the circumstances are highly suggestive. We found primary consumers and lesser vertebrates to be hyperabundant only on S and M islands that lacked predators of vertebrates. In contrast, we did not observe increases in the abundance of these same species on somewhat larger islands that held populations of mesopredators (ocelot, raptors, snakes, etc.).

A more direct test of top-down regulation was conducted by taking advantage of the contrasting densities of leaf-cutter ants on S islands (4.5 mature colonies per hectare) versus M islands (0.2 mature colonies per hectare; Terborgh et al. 2001). Leaf-cutter density was negatively associated with the presence of armadillos, a major predator: none of the S islands supported armadillos whereas all M islands did. Cages placed over the entrances to young *Atta* colonies on M islands greatly prolonged their survival, but had no effect on S islands (Rao 2000). Where armadillos were present, incipient *Atta* colonies were systematically dug out and destroyed. Telltale claw marks left no doubt that the predators were armadillos. These results strongly suggest that the loss of predators caused the extreme hyperabundance of *Atta* colonies on small islands (Rao 2000).

An alternative hypothesis sometimes evoked to explain the hyperabundance of animals on islands is "ecological release" or "density compensation." These terms refer to increases in the mean abundance of members of a guild on species-poor islands in comparison with the nearby mainland (Crowell 1962, MacArthur et al. 1972, Yeaton and Cody 1974). Density compensation could potentially result from a number of mechanisms, but is most frequently attributed to the utilization of novel resources or habitats in the absence of competitors (Case et al. 1979, Wright 1980, Vassallo and Rice 1982, Cody 1983, Anjos and Bocon 1999, Rodda and Dean-Bradley 2002).

Density compensation is an unlikely explanation for the hyperabundance of many species at Lago Guri since their numbers far exceed those that could be expected in compensation of missing potential competitors. For example, the biomass of howler monkeys on some islands was equivalent to almost 4000 $kg\,km^{-2}$, more than the total biomass of any known primate community in the world (Terborgh 1983, Peres 1997,

Peres and Dolman 2000). Such extreme "density *over*compensation" cannot be explained by the same mechanisms as density compensation. Even with niche expansions and niche shifts, compensatory increases are not expected to surpass mainland density levels (Diamond 1970, MacArthur et al. 1972). Incomplete compensation is predicted because resource use efficiency is expected to be reduced in novel or "inappropriate" habitats/food resources (Diamond 1970, Conner et al. 2000, Rodda and Dean-Bradley 2002).

Effects at the producer level

The presence of hyperabundant consumer populations on S and M islands implied that the vegetation of such islands would be under intense top-down pressure (Hairston et al. 1960). In 1997, we found that the density of small saplings on S islands was only one third of those on L landmasses, in accord with the expectation of strong negative top-down effects. In a recensus conducted 5 years later, the densities of small saplings had decreased further to less than a quarter of the density on L landmasses. On M islands, sapling densities in 1997 were indistinguishable from those found on L landmasses but by 2002 had declined modestly (Terborgh et al. 2006).

The vegetation of S islands already had a conspicuously battered appearance by 1997 when vegetation monitoring began. By the end of the 5-year monitoring period in 2002, standing dead trees were sprinkled through the canopies of these islands, dead branches and vines littered the forest floor, and the understory was empty. The recruitment of small saplings (stems ≥1 m tall and <1 cm dbh) on S islands was only 20% of that observed on L landmasses, whereas stems of all size classes experienced greater mortality than on large landmasses. Sapling recruitment was similarly depressed to 20% of control values on M islands, but in keeping with reduced herbivore densities on this class of islands, the decline in sapling stem numbers was retarded by about a decade in comparison with S islands (Terborgh et al. 2006).

Multiple pathways to vegetation decline

Although damaging levels of herbivory, especially by leaf-cutter ants, appear to offer the most likely explanation for the low recruitment and high mortality of saplings on S islands, other processes could have been involved. For example, decreased abundance or diversity of pollinators could lower seed set (Aizen and Feinsinger 1994); lack of seed dispersers could prevent adequate seed distribution (Leigh *et al.* 1993, Asquith *et al.* 1997, 1999, Asquith and Mejia-Chang 2005); hyperabundant seed predators such as small rodents could reduce seed survivorship (Asquith *et al.* 1997, Lopez and Terborgh 2007, Asquith and Mejia-Chang 2005); and/or high herbivory on adult trees could result in reduced allocation of resources to reproduction (Ruess and McNaughton 1984, Belsky 1986). Distortions in any or all of these processes could contribute to suppressed recruitment of tree saplings.

Pollination

Despite a steady cross-water flow of immigrant butterflies (a representative pollinator group) to many S and M islands, the overall density of butterflies on these islands was low relative to L landmasses (Shahabuddin and Terborgh 1999). Immigrant or experimentally introduced butterflies left most S islands within 24 hours. These islands often lacked either adult food resources and/or larval food plants (Shahabuddin and Terborgh 1999). Similarly, dung beetles emigrated quickly from small islands (Larsen *et al.* 2005). Based on these results, we surmise that many flying insects do not turn back when they fly out over open water, so that individuals tend to emigrate from smaller islands faster than they immigrate. Fragmentary evidence (unpublished) suggests that the reproduction of some tree species on S and M islands may have been hindered by a deficiency of pollinators.

Seed predation

We tested the hypothesis that S and M islands would have higher rates of seed predation due to high rodent abundances by setting out arrays of 16 species of seeds on S, M, and L landmasses and monitoring removal rates. Contrary to expectation, removal rates did not differ between large landmasses and islands. This seemingly paradoxical result held for seeds set out on the forest floor as well as for lightly buried seeds (Lopez and Terborgh 2007). Apparently the species of rodents able to persist on S and M islands were inefficient seed predators. Abnormally high rates of seed predation thus did not seem to be contributing to low sapling recruitment rates.

Herbivory

Finally, we tested the role of herbivory in repressing recruitment rates by setting out seedlings of six species of common forest trees under three types of cages: impermeable (closed to arthropods and all larger animals), skirted (open only to arthropods but not larger animals), and gated (open to both arthropods and rodents). After 4 months, 38% of the seedlings in open cages on S islands had been lost to herbivory, whereas losses at medium and large landmass sites were lower at 18% and 14%, respectively ($P = 0.001$). Survivorship in skirted and gated cages did not differ significantly and was lower than in impermeable cages ($P < 0.005$; Lopez and Terborgh 2007). These results suggest that arthropod herbivory and not seedling predation accounted for differences in survivorship.

Indirect effects and nutrient cycling

Tree growth rates were over six times faster on S islands supporting hyperdense howler monkeys than on islands lacking howlers ($P < 0.01$). This seemingly counterintuitive association could result from an accelerated return of plant available nutrients via monkey urine and feces (recall that howler monkey biomass was ca. 4000 $kg\,km^{-2}$ on some islands) as opposed to slower pathways such as via leaf litter, throughfall, etc. (Feeley 2005b, Feeley and Terborgh 2005). However, the enhancement of tree growth is likely to be transitory. Intense herbivore pressure has already resulted in significant shifts in tree community composition on S islands, such that there is now a significant positive relationship between the density of herbivores and the relative abundance of "unpreferred" tree species ($r^2 = 0.24$, $P < 0.05$; Rao et al. 2001, Feeley and Terborgh 2005, Orihuela-Lopez et al. 2005). These unpreferred tree species tend to have nutrient-poor/highly defended leaves and as a result their increased predominance has led to a decrease in leaf litter quality ($r^2 = 0.23$, $P < 0.05$). Consequently, soil nutrient availability is now decreasing on S islands supporting dense herbivore populations (for example, there is a significant positive relationship between C:N [high C:N indicates low soil fertility] and the density of howler monkeys, $r^2 = 0.28$–0.50, $P < 0.05$; Feeley and Terborgh 2005). This decrease in soil nutrient availability will likely lead to slower growth rates, further stressing these declining forests (Pastor and Naiman 1992, Pastor et al. 1993).

Unlike howler monkeys, leaf-cutter ants sequester nutrients underground in their fungal gardens and refuse chambers. The deepest chambers in mature leaf-cutter colonies can be 5 m below the surface at a level reached by few roots. During the rainy season, when the water table rises, the nutrients concentrated in underground chambers can be leached away, leading one investigator studying nutrient fluxes in a Venezuelan rainforest to conclude that "leaf-cutting ants bleed mineral elements out of a rainforest" (Haines 1983). This may explain why tree growth was slower on S islands supporting hyperabundant leaf-cutter ants but lacking howler monkeys than on large landmasses supporting much lower densities of leaf-cutter ants.

The presence of hyperabundant herbivores can carry important implications for other major faunal groups, such as birds. For example, the density and diversity of birds on small islands is positively correlated with the density of howlers ($r^2 = 0.59$, $P < 0.001$; Feeley and Terborgh 2006). In contrast, islands lacking howlers have experienced accelerated rates of avian extinctions during 10 years of observation ($r^2 = 0.78$, $P < 0.005$), perhaps in part as an indirect consequence of nutrient

sequestration by hyperabundant leaf-cutter ants (Feeley 2005a).

Evidence for such bottom-up effects on bird communities was limited to S islands. On some M and L islands, hyperabundant mesopredators (capuchin monkeys) imposed a negative top-down effect on bird communities via nest predation (Terborgh et al. 1997a). Olive capuchins on the Venezuelan mainland utilize a 200 ha home range (Robinson 1986), but groups of this species persisted on some Lago Guri islands of less than 15 ha. The greatly reduced home ranges of these trapped capuchins are likely to be searched at a much higher frequency than the vastly larger home ranges of unrestricted groups. Accordingly, breeding bird densities on M islands harboring capuchins were markedly less than those on similar-sized islands lacking capuchins and in some cases were depressed to such a degree that total numbers of breeding birds were less than on small predator-free islands despite the order of magnitude difference in area (Terborgh et al. 1997b). In another situation, capuchin densities approximately doubled on the 190 ha L island between 1993 and 2003, during which time the island lost 57% of its avian species (Feeley 2005a). The disappearance of birds on M and L islands due to mesopredator release may eventually have important implications for the vegetation, potentially causing altered rates of pollination, seed predation, seed dispersal, or even insect herbivory (Marquis and Whelan 1994, Van Bael et al. 2003).

Bottom-up forces prevail in the absence of top-down regulation

Current ecological theory predicts that consumers will increase in the absence of top-down regulation, but does not predict the amount of increase (Oksanen and Oksanen 2000). One of the most unexpected findings of our research was that some species increased by many-fold (up to 100 times in the case of leaf-cutter ants), implying that the carrying capacity of the Guri forest for primary consumers is potentially much greater than suggested by animal densities on the nearby mainland.

Consumer populations regulated from the bottom up (i.e., by food resources) should show evidence of nutritional stress, a prediction we were able to test with two species: tortoises and howler monkeys. Tortoises belonging to a hyperdense island population grew at roughly half the rate of uncrowded mainland tortoises (Aponte et al. 2003); likewise, hyperdense howler monkeys weighed a third less as adults and reproduced at lower rates than their large landmass counterparts (K. Glander unpublished data, Terborgh et al. 2001). Howlers confined to a small island lacked many food resources available to howlers living on a large island and fed heavily on tree species that were ignored or consumed only sparingly by the latter (Orihuela-Lopez et al. 2005).

Possible effects of edge and increased exposure

"Edge effects," particularly exposure to desiccating convectional drafts, are commonly invoked to explain vegetation changes in other tropical forest fragment systems (Laurance et al. 2002, Laurance 2004). Forest margins around Lago Guri are exposed to persistent trade winds (Feeley 2004), suggesting the possibility that die-back and/or recruitment failure along exposed edges was contributing to the demographic decline of tree stands. We conducted a series of tests to investigate this possibility. In comparisons of sapling plots situated on windward versus leeward slopes of islands, we found no discernible effect of exposure on the number, mortality, or recruitment of either small or large saplings. We also failed to find significant effects on the density, species richness, or composition of seedling cohorts growing on the windward versus leeward margins of small islands. Similarly, we found no effects of proximity to exposed edges on either the growth or mortality of trees ≥10 cm dbh (Terborgh et al. 2006).

DISCUSSION

We tested the top-down hypothesis of Hairston et al. (1960) by taking advantage of a mega-experiment initiated in 1986 with the creation of Lago Guri in the lower Caroní Valley of Bolívar State, Venezuela. Lago Guri islands smaller than ca. 15 ha provided predator-free habitats supporting hyperabundant densities of species belonging to a non-random selection of functional groups. Previous predator-exclusion experiments have produced less dramatic results because the spatial scale of the experiment was small (usually only a few square meters; reviewed in Schmitz et al. 2000), the exclusion of predators was only partial (Sinclair et al. 2000), or the duration of exclusion was brief (usually <1 year; Schmitz et al. 2000). The Lago Guri experiment largely overcame these difficulties, providing replicate isolates of natural habitat over spatial scales ranging from less than 1 ha to more than 100 ha and isolated for more than a decade.

The presence of water barriers effectively trapped populations of non-volant animals in emergent patches of a previously continuous tropical dry forest. A majority of vertebrate species apparently disappeared from islands of less than 15 ha within the first few years of isolation, but some species survived. Among these, predators of vertebrates were conspicuously absent. Released from predation and unable to disperse or emigrate, trapped populations of several vertebrates and some invertebrates increased in density by as much as one to two orders of magnitude.

Our research compared small (≥0.5 ha, <2 ha), medium (≥4 ha, <15 ha), and large (≥80 ha) landmasses, the faunas of which conformed to nearly perfect nested subsets, presumably a consequence of the differing area requirements of persistent species (Terborgh 1992, Terborgh et al. 1997a, Feeley 2003). The loss of some ecological functions (predation, seed dispersal), reduction of others (pollination), and exaggeration of still others (folivory) resulted in pronounced functional imbalances in the residual animal communities of small and medium islands, unleashing a complex trophic cascade with both top-down and bottom-up components.

Hyperabundant consumers marginally increased tree and sapling mortality on S islands compared with M or L landmasses. However, recruitment of small saplings on S and M islands was only about 20% of that on the L landmasses and insufficient to offset mortality (Terborgh *et al.* 2006). Exclosure experiments revealed that arthropod herbivory was a major factor in the failure of seedlings to survive (Lopez and Terborgh 2007). Strikingly, no tree species exhibited positive population growth on the S or M islands, demonstrating that the trophic cascade affected the entire tree community. These results suggest that the effectiveness of plant anti-herbivore defenses is limited in the absence of top-down control.

In all likelihood, multiple effects of isolation and hyperabundant consumers contributed to the high mortality and low recruitment of tree saplings on S and M islands. However, we strongly suspect that elevated herbivory was the main driver, particularly herbivory by leaf-cutter ants since they forage from the ground to the canopy, whereas iguanas and howler monkeys forage only in the canopy. A leading role of leaf-cutter ants is supported by three observations. First, M islands supported lower densities of leaf-cutter ants than S islands. Sapling stem numbers were higher on M islands and mortality rates lower than on S islands. Consistent with this, we found a strong negative relationship between the density of leaf-cutter ants and sapling recruitment rates ($r^2 = 0.58, P < 0.05$; unpublished data). Second, on the S islands, mortality exceeded recruitment for all tree species. It is unlikely that deficiencies of pollinators or dispersers would negatively impact all members of a tree community because a multiplicity of agents is involved and fragmentation does not impede them all (e.g., wind; also, hummingbirds were frequent on S islands; Feeley 2003). Third, S islands tended to be dominated by species carrying foliage that was determined through independent tests to be unpreferred by either howler monkeys or leaf-cutter ants (Rao *et al.* 2001, Feeley and Terborgh 2005, Orihuela-Lopez *et al.* 2005). The most obvious explanation for this bias is the selective mortality of preferred species under the pressure of intense herbivory.

Early in the decade over which the research unfolded, we wondered whether large fragments would experience the same changes as small fragments, only more slowly (Levin 1992). In general, we now feel that the answer is "no" because of the stepwise addition/loss of ecologically important species with spatial scale. Leaf-cutter ants offer

a good example. On predator-free S islands they occurred at a mean density of 4.5 mature colonies per hectare, whereas on M islands, in the presence of armadillos, colony density was only 0.2 per hectare. On the mainland, where colonies are exposed to army ants (*Eciton hammatum*) as well as armadillos and giant anteaters (*Myrmecophaga tridactyla*), colony density dropped to approximately 0.05 per hectare (Vasconcelos and Cherrett 1997, Terborgh *et al.* 2001). Similarly, rodent hyperabundance is likely to disappear at a fragment size large enough to sustain ocelots, raptors, and/or snakes. At large enough scales, ungulates will replace leaf-cutter ants as herbivores and agoutis and peccaries will complement small rodents as seed predators (Wright *et al.* 2000). Thus, important consequences of fragmentation result from the presence/absence of individual species, a fact that raises questions about the use of area as a continuous variable in fragmentation studies (e.g., Wardle *et al.* 1997). The broad consequences of fragmentation – community change accompanied by biodiversity loss – will occur over a wide range of spatial scales, but the details (e.g., the mortality and recruitment schedules of individual species) will certainly differ.

An important question for conservation is whether there is an upper limit to the scale effects of fragmentation. Research on ungulate predator–prey systems in North America strongly implies that the presence of top carnivores (wolves or the equivalent) is necessary to avert top-down trophic cascades (Mclaren and Peterson 1994, Ripple *et al.* 2001). The areas needed to sustain populations of top carnivores are in the thousands of square kilometers (Woodroffe and Ginsberg 1998).

A related question of conservation relevance is whether the processes observed in closed fragments extend to open fragments in a mainland land-use matrix. The question has no simple answer because mainland systems are typically more complex and less controlled than those on forested islets. In open mainland systems, non-resident animals are free to move through the matrix and resident animals have the option of density-dependent emigration. Moreover, human activities such as hunting, roadkill, agricultural chemicals, domestic animals, fire control (or the lack of it), and the qualities of the matrix can all influence the presence and/or abundance of non-volant vertebrates in fragments (Malcolm 1994, Gascon *et al.* 1999, Laurance *et al.* 2002, Forman *et al.* 2003). Thus, comparisons of results from open systems with Lago Guri are difficult. But some general principles do emerge. The loss of top predators predictably releases herbivores and mesopredators (Soulé and Terborgh 1999). These groups then impose top-down pressure on plants, smaller vertebrates, and arthropods. Accelerated nutrient cycling may enhance productivity, but excessive browsing by herbivores leads to structural as well as compositional changes in the vegetation, while hyperdense mesopredators exert negative pressures on birds, lizards, amphibians, snakes, and doubtless some arthropods (Crooks and Soulé 1999, McShea and Rappole 2000). Eventually, herbivore pressure will result in a more herbivore-resistant vegetation that in turn will suppress herbivore numbers from the bottom up (Pastor and Naiman 1992, Pastor *et al.* 1993).

The results from Lago Guri open a window on the operation of trophic cascades in terrestrial systems by demonstrating that a perturbation at one level (predator removal) can propagate to the producer level via multiple pathways. These pathways are both direct, such as the effects of altered densities of pollinators, seed dispersers, and herbivores on plant recruitment, and indirect, such as the effects of hyperabundant herbivores on nutrient cycling, tree growth, and bird diversity (Davidson *et al.* 1985). Contemporary theory (e.g., Oksanen and Oksanen 2000) is primarily concerned with direct effects exerted through herbivory. Our results from Lago Guri reveal a much deeper reality that finds parallels in the intricate interaction webs of the marine realm (Yodzis 2001). In fact, the very complexity of the Guri trophic cascade, with its multiplicity of pathways, implies that plant composition is established and maintained by a balance of processes representing the numerous interaction links, both direct and indirect, between plants and animals. From a scientific standpoint, these new insights are both exciting and challenging, but the conservation implications are unsettling. The existence of multiple pathways, operating with various interaction strengths and strongly sensitive to

spatial scale as well as the character of the matrix and the encompassing landscape, implies that almost any change in the structure/composition of an animal community may lead to instability in the composition of the plant community. Changes in plant community composition, in turn, will inevitably feed back to the animal community with consequences we are as yet unable to predict.

ACKNOWLEDGMENTS

We thank EDELCA for allowing us to conduct the research in the Guri impoundment and especially Luis Balbas for many kinds of advice and support. The work described here is the result of a long-term collaborative effort involving the work of many students and researchers, all of whom we gratefully acknowledge. K. Glander is thanked for data on the body mass of howler monkeys. We are grateful to Walter Carson and two anonymous reviewers for many helpful suggestions on the manuscript. Financial support was provided by the John T. and Katherine D. MacArthur Foundation and the National Science Foundation (DEB97-09281, DEB01-08107).

REFERENCES

Adler, G.H. (1996) The island syndrome in isolated populations of a tropical forest rodent. *Oecologia* 108, 694–700.

Aizen, M.A. and Feinsinger, P. (1994) Forest fragmentation, pollination, and plant reproduction in a Chaco dry forest, Argentina. *Ecology* 75, 330–351.

Alvarez, E., Balbas, L., Massa, I., and Pacheco, J. (1986) Aspectos ecológicas del embalse Guri. *Interciencia* 11, 325–333.

Anjos, L.D. and Bocon, R. (1999) Bird communities in natural forest patches in southern Brazil. *Wilson Bulletin* 111, 397–414.

Aponte, C., Barreto, G.R., and Terborgh, J. (2003) Consequences of habitat fragmentation on age structure and life history in a tortoise population. *Biotropica* 35, 550–555.

Asquith, N.M. and Mejia-Chang, M. (2005) Mammals, edge effects, and the loss of tropical forest diversity. *Ecology* 86, 379–390.

Asquith, N.M., Terborgh, J., Arnold, A.E., and Riveros, C.M. (1999) The fruits the agouti ate: *Hymenaea courbaril* seed fate when its disperser is absent. *Journal of Tropical Ecology* 15, 229–235.

Asquith, N.M., Wright, S.J., and Clauss, M.J. (1997) Does mammal community composition control recruitment in neotropical forests? Evidence from Panama. *Ecology* 78, 941–946.

Belsky, A.J. (1986) Does herbivory benefit plants? A review of the evidence. *American Naturalist* 127, 870–892.

Blondel, J., Chessel, D., and Frochot, B. (1988) Bird species impoverishment, niche expansion, and density inflation in Mediterranean island habitats. *Ecology* 69, 1899–1917.

Carpenter, S.R. and Kitchell, J.F. (1993) *The Trophic Cascade in Lakes*. Cambridge University Press, Cambridge, UK.

Case, T.J., Gilpin, M.E., and Diamond, J.M. (1979) Overexploitation, interference competition, and excess density compensation in insular faunas. *American Naturalist* 113, 843–854.

Chapman, C.A. and Chapman, L.J. (2003) Fragmentation and the alteration of seed dispersal processes: an initial evaluation of dung beetles, seed fate, and seedling diversity. *Biotropica* 35, 382–393.

Cochrane, M.A. and Laurance, W.F. (2002) Fire as a large-scale edge effect in Amazonian forests. *Journal of Tropical Ecology* 18, 311–325.

Cody, M.L. (1983) Bird diversity and density in South African forests. *Oecologia* (Berlin) 59, 201–215.

Connor, E.F., Courtney, A.C., and Yoder, J.M. (2000) Individuals–area relationships: the relationship between animal population density and area. *Ecology* 81, 734–748.

Crooks, K.R. and Soulé, M.S. (1999) Mesopredator release and avifaunal extinctions in a fragmented system. *Nature* 400, 563–566.

Crowell, K.J. (1962) Reduced interspecific competition among the birds of Bermuda. *Ecology* 43, 75–88.

Davidson, D.W., Samson, D.A., and Inouye, R.S. (1985) Granivory in the Chihuahuan Desert: interactions within and between trophic levels. *Ecology* 67, 486–502.

Diamond, J. (2001) Dammed experiments! *Science* 294, 1847–1848.

Diamond, J.M. (1970) Ecological consequences of island colonization by Southwest Pacific birds, II. The effect of species diversity on total population density. *Proceedings of the National Academy of Sciences of the United States of America* 67, 1715–1721.

Emlen, J.T. (1979) Land bird densities on Baja California islands. *The Auk* 96, 152–167.

Feeley, K. (2003) Analysis of avian communities in Lake Guri, Venezuela, using multiple assembly rule models. *Oecologia* 137, 104–113.

Feeley, K.J. (2004) The effects of forest fragmentation and increased edge exposure on leaf litter accumulation. *Journal of Tropical Ecology* 20, 709–712.

Feeley, K.J. (2005a) *The effects of tropical dry forest fragmentation on floral and faunal communities as mediated through trophic interactions.* Doctoral dissertation, Duke University, Durham, NC.

Feeley, K.J. (2005b) The role of clumped defecation in the spatial distribution of soil nutrients and the availability of nutrients for plant uptake. *Journal of Tropical Ecology* 21, 99–102.

Feeley, K.J., Gillespie, T.W., and Terborgh, J.W. (2005) The utility of spectral indices from Landsat ETM+ for measuring the structure and composition of tropical dry forests. *Biotropica* 37, 508–519.

Feeley, K.J. and Terborgh, J.W. (2005) The effects of herbivore density on soil nutrients and tree growth in tropical forest fragments. *Ecology* 86, 116–124.

Feeley, K.J. and Terborgh, J.W. (2006). Habitat fragmentation and effects of herbivore (red howler monkey) abundances on bird species richness. *Ecology* 87, 144–150.

Feeley, K.J., Gillespie, T.W., Lebbin, D.J., and Walter, H.S. (2007) Species characteristics associated with extinction vulnerability and nestedness rankings of birds in tropical forest fragments. *Animal Conservation* 10, 493–501.

Forman, R.T.T., Sperling, D., Bissonette, J.A. *et al.* (2003) *Road Ecology: Science and Solutions*. Island Press, Washington, DC.

Gascon, C., Lovejoy, T.E., Bierregaard, R.O. Jr. *et al.* (1999) Matrix habitat and species richness in tropical forest remnants. *Biological Conservation*, 91, 1–7.

Gascon, C., Malcolm, J.R., Patton, J.L. *et al.* (2000) Riverine barriers and the geographic distribution of Amazonian species. *Proceedings of the National Academy of Sciences of the United States of America* 97, 13672–13677.

George, T.L. (1987) Greater land bird densities on island versus mainland: relation to nest predation level. *Ecology* 68, 1393–1400.

Grant, P.R. (1965) The density of land birds on the Tres Marias islands in Mexico. *Canadian Journal of Zoology* 44, 391–399.

Haines, B. (1983) Leaf-cutting ants bleed mineral elements out of rainforest in southern Venezuela. *Tropical Ecology* 24, 85–93.

Hairston, N.G., Smith, F.E., and Slobodkin, L.B. (1960) Community structure, population control, and competition. *American Naturalist* 94, 421–424.

Huber, O. (1986) La vegetación de la cuenca del Rio Caroní. *Interciencia* 11, 301–310.

Kapos, V. (1989) Effects of isolation on the water status of forest patches in the Brazilian Amazon. *Journal of Tropical Ecology* 5, 173–185.

Krebs, C., Keller, B., and Tamarin, R. (1969) *Microtus* population biology. *Ecology* 50, 587–607.

Lambert, T.D., Adler, G.H., Riveros, C.M., Lopez, L., Ascanio, R., and Terborgh, J. (2003) Rodents on tropical land-bridge islands. *Journal of Zoology (London)* 260, 179–187.

Larsen, T.H., Williams, N.M., and Kremen, C. (2005) Extinction order and altered community structure rapidly disrupt ecosystem functioning. *Ecology Letters* 8, 538–547.

Laurance, W.F. (2004) Forest–climate interactions in fragmented tropical landscapes. *Philosophical Transactions of the Royal Society of London B* 359, 345–352.

Laurance, W.F., Lovejoy, T.E., Vasconcelos, H.L. *et al.* (2002) Ecosystem decay of Amazonian forest fragments: a 22-year investigation. *Conservation Biology* 16, 605–618.

Leigh, E.G. Jr., Wright, S.J., Herre, E.A., and Putz, F.E. (1993) The decline of tree diversity on newly isolated tropical islands: a test of a null hypothesis and the implications. *Evolutionary Ecology* 7, 76–102.

Levey, D.J., Boiker, B.M., Tewksbury, J.J., Sargent, S., and Haddad, N.M. (2005) Effects of landscape corridors on seed dispersal by birds. *Science* 309, 146–148.

Levin, S.A. (1992) The problem of space and scale in ecology. *Ecology* 73, 1943–1967.

Lopez, L. and Terborgh, J. (2007) Seed predation and seedling herbivory as factors in tree recruitment failure on predator-free forested islands. *Journal of Tropical Ecology* 21, 129–137.

MacArthur, R.H., Diamond, J.M., and Karr, J.R. (1972) Density compensation in island faunas. *Ecology* 53, 330–342.

Malcolm, J.R. (1994) Edge effects in central Amazonian forest fragments. *Ecology* 75, 2438–2445.

Marquis, R.J. and Whelan, C.J. (1994) Insectivorous birds increase growth of white oak through consumption of leaf-chewing insects. *Ecology* 75, 2007–2014.

Mclaren, B.E. and Peterson, R.O. (1994) Wolves, moose and tree rings on Isle Royale. *Science* 266, 1555–1558.

McShea, W.J. and Rappole, J.H. (2000) Managing the abundance and diversity of breeding bird populations through manipulations of deer populations. *Conservation Biology* 14, 1161–1170.

Morales, L.C. and Gorzula, S. (1986) The interrelations of the Caroní River Basin ecosystems and hydroelectric power projects. *Interciencia* 11, 272–277.

Morse, D.H. (1977) Occupation of small islands by passerine birds. *The Condor* 79, 399–412.

Oksanen, L., Fretwell, S.D., Arruda, J., and Niemelä, P. (1981) Exploitation ecosystems in gradients of primary productivity. *American Naturalist* 118, 240–261.

Oksanen, L. and Oksanen, T. (2000) The logic and realism of the hypothesis of exploitation. *American Naturalist* 155, 703–723.

Orihuela-Lopez, G., Terborgh, J., and Ceballos, N. (2005) Food selection by a hyperdense population of red howler monkeys (*Alouatta seniculus*). *Journal of Tropical Ecology* 21, 445–450.

Paine, R.T. (1980) Food webs: linkage, interaction strength, and community infrastructure. *Journal of Animal Ecology* 49, 667–685.

Pastor, J., Dewey, B., Naiman, R.J., McInnes, P.F., and Cohen, Y. (1993) Moose browsing and soil fertility in the boreal forests of Isle Royale National Park. *Ecology* 74, 467–480.

Pastor, J. and Naiman, R.J. (1992) Selective foraging and ecosystem processes in boreal forests. *American Naturalist* 139, 690–705.

Peetz, A. (2001) Ecology and social organization of the bearded saki *Chiropotes satanas* (Primates: Pitheciinae) in Venezuela. *Ecotropical Monographs* 1, 1–170.

Peres, C.A. (1997) Primate community structure at twenty western Amazonian flooded and unflooded forests. *Journal of Tropical Ecology* 13, 381–405.

Peres, C.A. and Dolman, P.M. (2000) Density compensation in neotropical primate communities: evidence from 56 hunted and nonhunted Amazonian forests of varying productivity. *Oecologia* 122, 175–189.

Polis, G.A. and Hurd, S.D. (1995) Extraordinarily high spider densities on islands – flow of energy from marine to terrestrial food webs and the absence of predation. *Proceedings of the National Academy of Sciences of the United States of America* 92, 4382–4386.

Rao, M. (2000) Variation in leaf-cutter ant (*Atta* sp.) densities in forest isolates: the potential role of predation. *Journal of Tropical Ecology* 16, 209–225.

Rao, M., Terborgh, J., and Nuñez, P. (2001) Increased herbivory in forest isolates: implications for plant community structure and composition. *Conservation Biology* 15, 624–633.

Ripple, W.J., Larsen, E.J., Renkin, R.A., and Smith, D.W. (2001) Trophic cascades among wolves, elk, and aspen on Yellowstone National Park's northern range. *Biological Conservation* 102, 227–234.

Robinson, J.G. (1986) Seasonal variation in use of time and space by the wedge-capped capuchin monkey, *Cebus olivaceus*: implications for foraging theory. *Smithsonian Contributions to Zoology* 431, 1–60.

Rodda, G.H. and Dean-Bradley, K. (2002) Excess density compensation of island herpetofaunal assemblages. *Journal of Biogeography* 29, 623–632.

Rose, M.D. and Polis, G.A. (2000) On the insularity of islands. *Ecography* 23, 693–701.

Ruess, R.W. and McNaughton, S.J. (1984) Urea as a promotive coupler of plant–herbivore interactions. *Oecologia* 63, 331–337.

Schmitz, O.J., Hambäck, P.A., and Beckerman, A.P. (2000) Trophic cascades in terrestrial systems: a review of the effects of carnivore removal on plants. *American Naturalist* 155, 141–153.

Shahabuddin, G. and Terborgh, J.W. (1999) Frugivorous butterflies in Venezuelan forest fragments: abundance, diversity and the effects of isolation. *Journal of Tropical Ecology* 15, 703–722.

Sinclair A.R.E., Krebs, C.J., Fryxell, J.M. *et al.* (2000) Testing hypotheses of trophic level interactions: a boreal forest ecosystem. *Oikos* 89, 313–328.

Soulé, M.E. and Terborgh, J. (eds) (1999) *Continental Conservation: Scientific Foundations of Regional Reserve Networks*. Island Press, Washington, DC.

Terborgh, J. (1983) *Five New World Primates: A Study in Comparative Ecology*. Princeton University Press, Princeton.

Terborgh, J. (1992) Maintenance of diversity in tropical forests. *Biotropica* 24, 283–292.

Terborgh, J. (2005) The green world hypothesis revisited. In J. Ray, J. Ginsberg, and K. Redford (eds), *Large Carnivores and Biodiversity: Does Saving one Conserve the Other?* Island Press, Washington, DC, pp. 82–99.

Terborgh, J., Lopez, L., Tello, J., Yu, D., and Bruni, A.R. (1997a) Transitory states in relaxing ecosystems of land bridge islands. In W.R. Laurance and R.O. Bierregaard, Jr. (eds), *Tropical Forest Remnants*. University of Chicago Press, Chicago, pp. 256–274.

Terborgh, J., Lopez, L., and Tello, J. (1997b) Bird communities in transition: the Lago Guri islands. *Ecology* 78, 1494–1501.

Terborgh, J., Lopez, L., Nuñez V. *et al.* (2001) Ecological meltdown in predator-free forest fragments. *Science* 294, 1923–1926.

Terborgh, J., Feeley, K., Nuñez V.P., Balukjian, B., and Silman, M. (2006) Vegetation dynamics of predator-free land-bridge islands. *Journal of Ecology* 94, 253–263.

Van Bael, S.H., Brawn, J.D., and Robinson, S.K. (2003) Birds defend trees from herbivores in a Neotropical forest canopy. *Proceedings of the National Academy of Sciences of the United States of America* 100, 8304–8307.

Vasconcelos, H.L. and Cherrett, J.M. (1997) Leafcutting ants and early forest regeneration in central Amazonia: effects of herbivory on tree seedling establishment. *Journal of Tropical Ecology* 13, 357–370.

Vassallo M.I. and Rice, J.C. (1982) Ecological release and ecological flexibility in habitat use and foraging of an insular avifauna. *Wilson Bulletin* 94, 139–155.

Wardle, D.A., Zackrisson, O., Hörnberg, G., and Gallet, C. (1997) Influence of island area on ecosystem properties. *Science* 277, 1296–1299.

Woodroffe, R. and Ginsberg, J.R. (1998) Edge effects and the extinction of populations inside protected areas. *Science* 280, 2126–2128.

Wright, S.J. (1980) Density compensation in island avifaunas. *Oecologia* 45, 385–389.

Wright, S.J., Zeballos, H., Dominguez, I., Gallardo, M.M., Moreno, M. and Ibañez, R. (2000) Poachers alter mammal abundance, seed dispersal, and seed predation in a Neotropical forest. *Conservation Biology* 14, 227–239.

Yeaton, R.I. and Cody, M.L. (1974) Competitive release in island song sparrow populations. *Theoretical Population Biology* 5, 42–58.

Yodzis, P. (2001) Must top predators be culled for the sake of fisheries? *Trends in Ecology and Evolution* 16: 78–84.

Chapter 19

RESOURCE LIMITATION OF INSULAR ANIMALS: CAUSES AND CONSEQUENCES

Gregory H. Adler

OVERVIEW

Populations of herbivorous animals may be limited by many factors, some of which may be grouped as either bottom-up or top-down processes. Bottom-up processes include both quantity and quality of food, while top-down processes include predation. Ecologists have long been divided over the importance of these two groups of processes in limiting herbivore populations. Herbivores in tropical forests appear to be at least seasonally limited by bottom-up processes, as evidenced by seasonal fluctuations in abundance, reproductive activity, and general health, but there is also strong evidence that populations of such animals are limited by predators. Support for the importance of top-down processes has been derived from studies of herbivores isolated on islands with few predators, where population densities reach much higher levels than on adjacent mainland areas. In the absence of strong top-down limitation, herbivore populations may be limited by food availability, even in seasons of resource abundance, and a trophic cascade may result. Such cascades may ramify throughout a tropical forest and have implications for plant recruitment and ultimately forest composition and structure. In this chapter, I outline a series of studies on spiny rats (*Proechimys semispinosus*) isolated on small islands in the Panama Canal that either directly or indirectly lend both descriptive and experimental evidence for the importance of bottom-up and top-down processes in limiting population growth and densities. Insular populations of spiny rats were censused over a 9-year period to evaluate fluctuations in demography (e.g., population density, reproductive output, survival, body weight, and age structure). Selected populations also were subjected to various manipulations to test the importance of food availability in limiting population growth and density. In the insular setting, spiny rats conform to the general trend of higher densities and reduced reproductive output. I suggest that spiny rats on the islands largely have been relieved of top-down limitation and instead are more strongly limited by food availability than their counterparts living within intact forest.

INTRODUCTION

Biologists have expended enormous effort to identify factors that influence population growth and ultimately limit population density, either by density-independent or density-dependent means. Population limitation includes any factor, either density independent or density dependent, that curbs population growth, while population regulation includes only those factors that have a density-dependent effect. In this chapter,

I speak primarily of population limitation because, in most cases, the relationship of limiting factors to density has not been established. In the case of herbivores (including frugivores and granivores), no consensus on the factors that are most important in limiting density has yet emerged. Indeed, two major opposing viewpoints have gained traction as the most likely limiting factors. One viewpoint argues that bottom-up processes ultimately limit herbivore populations (White 1993) and that top-down "community regulation" is relatively rare in complex terrestrial ecosystems (Polis and Strong 1996). According to this viewpoint, such populations are limited by the quantity and quality of food from primary producers. Thus, not only is the quantity of available plant material important but also the extent to which plants are protected mechanically and chemically. The second viewpoint argues that herbivore populations are limited primarily by top-down processes (Hairston *et al.* 1960, Terborgh 1988). Thus, predators ultimately limit herbivore populations (either in density-independent or density-dependent ways), with bottom-up processes playing only a secondary role because population densities are depressed by predators and virtually never at the level where food would become limiting.

INSULAR ENVIRONMENTS

Herbivores isolated on islands frequently show consistent differences when compared with conspecifics on the adjacent mainland. In the case of mammals, population densities are higher and often more stable on islands (e.g., Adler and Levins 1994). Individuals often have better short-term survival probabilities and longer lifespans, reduced aggressive behavior, and reduced reproductive output (Adler and Levins 1994). Body size tends to increase in the case of small mammals and to decrease in the case of large mammals (the island rule; Van Valen 1973, Lomolino 1985). The entire suite of population- and individual-level changes following isolation has been called the island syndrome (Adler and Levins 1994). Such differences often occur very rapidly (i.e., within a single generation), greatly preceding speciation, and are likely due initially to reaction norms (the entire complement of phenotypes expressed by a single genotype across a complete range of environmental conditions) rather than to immediate genetic changes (Adler and Levins 1994). However, initial genetic changes also could occur concomitantly through the founder effect and genetic drift. Thus, in the novel environmental setting of higher densities, body size increases and reproductive output decreases, even in the absence of immediate genetic changes. Over longer time frames, however, genetic changes, driven by directional selection, may occur that provide better adaptation to the novel insular setting (Adler and Levins 1994).

The cause of the island syndrome is unknown, but two major hypotheses have been offered (reviewed by Adler and Levins 1994). First, predation is greatly reduced on most islands, and the increased densities are the result of release from top-down limitation. Because top-down limitation is absent or at least reduced, herbivore populations are limited by food and intrinsic regulatory (density-dependent) mechanisms. Second, interspecific competitors are often absent from islands, and the higher island densities result from competitive release and greater resource availability for the species that remain. Thus, the bottom-up versus top-down controversy can be examined in an insular setting in which there are fewer predators and competitors. In either case, individual-level changes (e.g., increased survival probabilities, increased body size, and reduced reproductive output) might be expected to occur in response to increased densities and greater crowding, and bottom-up limitation, perhaps in concert with intrinsic regulatory mechanisms (e.g., density-dependent recruitment and reproductive output), should assume primary importance for herbivores.

TROPICAL ENVIRONMENTS

Throughout most of the tropics, seasonal fluctuations in rainfall and concomitant fluctuations in other climatic variables such as irradiance, relative humidity, and soil water potentials impose a seasonal rhythm on the activities of organisms living within the tropics. Thus, tropical organisms in both mainland and insular settings generally show seasonal changes in reproductive output, behavior, and vitality in response to seasonal fluctuations in resource abundance (Fleming 1971, 1992, Bonaccorso 1979, Glanz *et al.* 1982, Milton 1982, Russell 1982, Gliwicz 1984, Hentry 1994, Adler and Beatty 1997). Seasonal fluctuations also occur with great regularity (just as temperate regions exhibit seasonal regularity), albeit with some variability among years, and provide environmental cues that allow organisms to prepare physiologically and behaviorally in advance of the impending seasonal changes or to respond quickly once those changes have occurred.

In tropical forests that experience a seasonal climate, most plants reproduce seasonally (e.g., Foster 1982a, Van Schaik et al. 1993) and commonly use irradiance or some measure of moisture availability, such as drought, as a cue to initiate reproduction (e.g., Van Schaik et al. 1993, Wright et al. 1999). An 18-year record from old-growth forest in central Panama reveals that fruit production varies greatly throughout the year and is greatest in April at the end of the dry season and least in November, December, and January at the end of the rainy season and beginning of the dry season (Wright and Calderon 2006).

The seasonal availability of fruits and seeds that results from plant reproduction greatly affects frugivorous and granivorous animals. Thus, populations of such animals appear to be limited by seasonal shortage of fruits and seeds (e.g., Van Schaik et al. 1993), and there are abundant descriptive data on mammals to support this proposition (e.g., Russell 1982, Smythe et al. 1982, Terborgh 1986). During the season of resource scarcity, fruit production is insufficient to support the biomass of frugivorous mammals (Smythe et al. 1982, Terborgh 1986), and frugivores often manifest signs of food deprivation, including weight loss, increased time spent foraging over wider areas, increased foraging risk, and reproductive quiescence (Foster 1982b, Russell 1982, Smythe et al. 1982). When fruits and seeds are most abundant during the year, much of the fruit that is produced is not consumed by mammals (Hladik and Hladik 1969, personal observations). In contrast, fruits and seeds that fall to the forest floor during the season of resource scarcity are consumed nearly as rapidly as they fall. Diets of frugivorous and granivorous mammals often overlap extensively when resources are abundant, but overlap is reduced when resources are scarce (e.g., Smythe 1978, Smythe et al. 1982, Terborgh 1983, 1986).

The demography of frugivorous and granivorous mammals also reflects the impact of seasonal changes in resource abundance. Reproduction is often seasonal, and mammals time their maximal reproductive effort to coincide with the season of greatest resource abundance (e.g., Fleming 1971, Bonaccorso 1979, Glanz et al. 1982, Milton 1982, Russell 1982, Gliwicz 1984). When fruit is scarce, mortality rates are higher (Smythe 1978, Milton 1982). Because birth and mortality rates are not balanced throughout a year, population densities fluctuate widely and demonstrate an annual pattern of peaks and nadirs. Thus, density reaches a peak following the season of resource abundance, declines as mortality exceeds fecundity during the season of resource scarcity, reaches a nadir at the end of that season, and begins to increase again when fruit is abundant and fecundity again exceeds mortality.

Irregular fluctuations such as storms and El Niño Southern Oscillation (ENSO) events frequently are superimposed on seasonal fluctuations and may impose additional constraints on the activities of organisms. Such fluctuations cannot be anticipated in advance of the impending changes by organisms using environmental cues because of their irregular periodicity and consequent unpredictability. In seasonally dry tropical forests, such as in central Panama, an abnormally wet dry season apparently causes famine conditions for frugivores and granivores, apparently because many plants require a protracted dry period or increased irradiance (such as normally occurs in the dry season) to initiate flowering (e.g., Foster 1982b, Wright 1999). Thus, few resources are available for frugivores and granivores later in the year because of poor plant reproduction. In contrast, ENSO events, which contribute most to inter-annual variation in tropical climates (Wright and Calderon 2006), cause abnormally dry conditions in much of the tropics, which apparently stimulate massive plant reproduction, thereby providing hyperabundant resources (e.g., Wright et al. 1999). Although it is not clear why plant reproduction increases during ENSO events, the most likely explanation is the increase in irradiance (Wright and Calderon 2006). Frugivores and granivores respond to the increased abundance of resources by reaching extremely high but unsustainable densities and subsequently decline, sometimes catastrophically, when fruit production assumes more normal levels (Davis 2001, Adler unpublished data).

Clearly, frugivorous and granivorous mammals regularly experience bottom-up limitation by seasonal scarcity of resources, even in intact forests. However, there is compelling evidence that frugivorous and granivorous mammals that live in mainland tropical forests also are subjected to strong top-down limitation (e.g., Terborgh *et al.* 2001). In the absence of top predators, a trophic cascade apparently results, leading to greatly increased population densities of animal consumers (Terborgh *et al.* 2001), including rodents (Lambert *et al.* 2003). Therefore, mammals isolated on islands with depauperate predator communities and where densities are much higher than on mainland areas are expected to be affected to an even greater degree by seasonal shortage of resources and stronger bottom-up limitation. To what extent are populations of herbivorous mammals, when largely released from top-down limitation, limited by bottom-up processes?

SPINY RATS AS A CASE STUDY

Introduction and methods

Since 1989, my students and I have been studying the Central American spiny rat (*Proechimys semispinosus*) on islands in the Panama Canal as a model system to examine population processes, including resource limitation. There are over 200 islands available for study, and the spiny rat is distributed widely on the islands (Adler and Seamon 1991). Even tiny islands less than 0.1 ha in area frequently contain rats if the islands are close to larger landmasses, even though such islands are too small to support persistent populations (Adler and Seamon 1991). Most islands of more than 1 ha support persistent populations, and the spiny rat is the only terrestrial mammal that maintains such populations on those small islands (Adler 1996). My students and I have studied spiny rats on 76 islands ranging in size from less than 0.1 to 1500 ha, and we selected 12 islands ranging in size from 1.7 to 3.9 ha for long-term study to examine population-level patterns and processes. We have conducted both

short-term studies (i.e., within a single season or year) on geographical factors that influence spiny rat distributions on islands, seed predation and dispersal by spiny rats, mycorrhizal fungal spore dispersal by spiny rats, and spacing patterns of spiny rats, and long-term studies (i.e., spanning several years) on demography and factors that limit spiny rat populations, particularly resource abundance. These long-term studies will be the focus of this review.

Long-term study islands were selected because they (1) were known or predicted to maintain persistent populations, (2) were of a size that permitted regular and thorough censuses, (3) were of roughly similar size and isolation, and (4) differed in tree species composition and forest structure and therefore were likely to harbor populations that differed in density and demography (Adler 2000). Because the long-term study islands maintain persistent and generally large populations of spiny rats (often more than 100 or 200 individuals), they (1) are amenable to both descriptive studies and manipulative experiments to address the question of population limitation by resource abundance, (2) are likely to yield large sample sizes of individuals, and (3) are therefore appropriate for robust statistical analysis. Because the islands are isolated, they represent essentially closed systems from the perspective of spiny rat demography and therefore ideal experimental systems. Although there is circumstantial evidence that spiny rats occasionally swim among islands (Adler and Seamon 1991), and we indeed recently documented several inter-island movements (Lopez and Adler unpublished data), we never recorded such movements on the long-term study islands during the study period. We therefore assume that colonization events are rare along an ecological time scale and have a negligible impact on the demography of spiny rat populations.

We conducted monthly censuses of spiny rats on each of the islands for 9 years (eight islands, January 1991 through March 2000) or 7 years (four islands, February 1993 through March 2000). For this purpose, we established permanent sampling grids that covered the whole of each island. Sampling points on the grids were 20 m apart, and each such point was occupied by a live-trap. Live-trapping was conducted for four nights each month, and all captured spiny rats were uniquely marked for individual identification. Standard data were recorded for each rat upon first capture each month (Adler 1994). We assumed that each rat captured on an island was born on that island, and we estimated the month of birth of all captured rats based on growth curves of individuals captured shortly after birth (Adler 1994).

We also conducted monthly censuses of fruiting trees and lianas on each island to search for patterns in the relationships between spiny rat demography and resource abundance. To facilitate these censuses, we marked, measured, and identified all trees ≥ 10 cm in diameter at breast height (Adler 2000), and censuses were conducted each month by walking the entirety of each island and recording all trees and lianas that were producing ripe fruit. We included in the censuses only those trees and lianas whose fruits and seeds are consumed by spiny rats, based upon feeding trials of captive individuals (Adler 1995). For animals that occupy small home ranges (generally <0.2 ha in the case of spiny rats; Endries and Adler 2005), censuses that include all fruiting individuals within an animal's home range are more accurate indicators of fruit availability than sampling along transects or using fruit traps (Chapman *et al.* 1994).

For the remainder of this chapter, I will summarize those studies that are relevant to resource limitation of spiny rats in an insular setting. First, I will address descriptive studies that provide the framework for implicating resources in limiting spiny rat populations on the islands. These studies rely solely on the monthly censuses of spiny rats and fruit production. I will conclude by describing two studies that experimentally test the role of resources in limiting populations of the spiny rat. These studies rely on the monthly censuses of spiny rats and fruit production and food-provisioning experiments.

Natural history of spiny rats

Proechimys semispinosus is a large echimyid rodent that is distributed widely throughout lowland tropical forests from southern Honduras to northwestern South America (Woods 1993). This rodent is sufficiently generalized that it is able to live in many types of forests, including dry and wet or primary and secondary forests (Gonzalez and Alberico 1993, Tomblin and Adler 1998, Adler 2000). Demography varies not only temporally but also spatially, and this demographic flexibility presumably promotes persistence in heterogeneous environments (Adler 1996). Spiny rats reach their greatest abundance in seasonally dry secondary forests and are associated statistically with treefall gaps (Lambert and Adler 2000). Mean adult male body weight is approximately 350 g, with exceptional males reaching over 700 g (Adler 1996, 2000, Adler *et al.* 1998).

This rodent is primarily frugivorous and granivorous and consumes a wide variety of fruits and seeds (Adler 1995). Mycorrhizal fungi also constitute an important component of its diet (Mangan and Adler 1999, 2002). Spiny rats actively consume subterranean sporocarps of such fungi even when fruits and seeds are abundant. Provisioning spiny rats with food during the season of fruit scarcity does not reduce the importance of fungi in their diet, and spiny rats apparently consume the fungi according to their availability in the soil (Mangan and Adler 2002). The spiny rat is strictly terrestrial (Seamon and Adler 1999, Lambert and Adler 2000) and therefore is able to consume only subterranean (hypogeous) sporocarps of mycorrhizal fungi and those fruits and seeds that fall to the ground.

In both mainland and insular forests, *Proechimys semispinosus* demonstrates seasonal fluctuations in density, reproductive output, and recruitment (Fleming 1971, Gliwicz 1984, Adler 1994, Adler and Beatty 1997), but isolated populations frequently are asynchronous in their fluctuations, despite the similar seasonal climatic changes (Adler 1994). This asynchrony presumably reflects differences in floristic composition among isolated forest patches, which fosters differences in overall fruiting phenology and consequent differences in food availability (Adler 1994). Such asynchrony contrasts sharply with synchrony often observed with temperate rodent populations over large spatial scales (e.g., Van Horne 1981, Adler 1987). In temperate forests, virtually all plant reproduction occurs during warm months, and availability of resources is entrained by environmental conditions that prevail across wide areas, despite differing floristic composition.

Mammalian predators almost never visit the study islands, and avian predators such as owls are transient and usually absent for long periods of time (Adler 1996). The only predators of spiny rats that are regularly present on the islands are snakes, but their low metabolic demands no doubt render them much less important than mammalian and avian predators in top-down limitation, as in temperate regions (e.g., Lin and Batzli 1995). Insular populations consequently demonstrate a pronounced island syndrome, with densities sometimes reaching at least an order of magnitude greater than on adjacent mainland areas (Adler 1996). Body size is greater on islands, and the largest individual ever recorded (720 g) was captured on one of the long-term study islands. Reproductive output also is often greatly reduced on islands, where the number of births per adult female in a year is less than 3 (Adler 1996), compared with a maximum reproductive capacity of 11 births per adult female per year in mainland populations (Fleming 1971). The breeding season may be as short as 2 months in a year (Adler and Beatty 1997) but is often continuous in mainland populations (Fleming 1971). Most insular populations show a period of reproductive quiescence each year, but that period often varies among islands in timing and duration (Adler and Beatty 1997).

Anecdotal evidence for resource limitation

Anecdotal observations of spiny rats on the islands strongly suggest that they are seasonally food-stressed. As fruiting activity subsides towards the end of the rainy season and beginning of the dry season, spiny rats lose weight rapidly (routinely as much as 25% in 1 month), reproductive activity ceases, and the vitality of individuals declines noticeably. Mite infestations and fur loss are substantial, and many individuals are reduced to skeletal, depilated ghosts of their former robust states (personal observations). Once fruit begins falling again in large quantities in the mid- to late

dry season, the physical decline is reversed, and individuals rapidly regain their lost weight and fur and begin breeding.

Statistical patterns further indicate that spiny rats on the islands are seasonally food-stressed. Time series analysis reveals a negative relationship between spiny rat trappability (the proportion of individuals known to be present that are captured during a monthly census) and the density of fruiting trees and lianas (Adler and Lambert 1997). Thus, spiny rats more readily enter baited traps when fruit availability is lower. Mean annual population density on small islands is positively related to the mean annual density of fruiting trees and lianas (Figure 19.1; Adler and Beatty 1997). There is no relationship between the density of spiny rats and any single species of tree that produces large crops of edible fruit (Adler 2000), indicating that spiny rats are not dependent on a single species of fruiting tree. Similarly, there is no relationship between spiny rat density and groups of species of such trees (e.g., palms), with one notable exception.

The density of spiny rats is positively related to the density of all species of fig trees (*Ficus* spp.). This relationship may be biologically relevant because of the fruiting phenology of fig trees. These trees reproduce asynchronously, and individuals can produce figs at any time of the year, even during the season of resource scarcity when few or no other species of trees are fruiting. Spiny rats also avidly consume figs (Adler 1995). Interestingly, not only spiny rat density but also demography in general closely mirrors fig tree density and species composition on the islands (Adler 2000). Thus, islands with higher densities of fig trees, particularly those species that frequently produce large figs, had higher densities of spiny rats, better survival rates, and longer breeding seasons. Fig trees long have been known to provide important resources for a wide variety of animal consumers (Janzen 1979). More recently, figs have been considered to be keystone plant species for animal consumers and of conservation concern because of the large area occupied by single breeding units (Nason *et al.* 1998).

Resource abundance appears to be a major force in influencing both reproduction and density and in driving population trajectories through

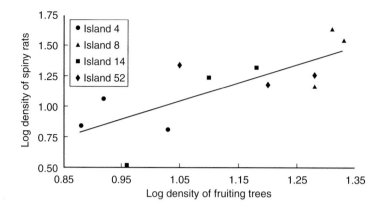

Figure 19.1 Relationship between mean annual spiny rat density and mean annual density of fruiting trees and lianas (from Adler and Beatty 1997).

reproduction–density space, that is, the statistical space resulting when a measure of reproductive output is plotted versus population density (Adler and Beatty 1997). Thus, plots of reproductive output, expressed at the individual level as births per adult female or at the population level as length of the breeding season, versus density demonstrate clockwise trajectories through reproduction–density space. The clockwise trajectories conform to theoretical predictions and strongly implicate intrinsic regulation of population density by density-dependent changes in reproductive output (Schaffer and Tamarin 1973). Thus, reproductive effort was adjusted with changes in density, apparently as individuals attempted to maximize their fitness in a seasonal environment with attendant changes in resource abundance.

Reproductive output therefore is influenced both directly and indirectly by climatic seasonality and resource abundance. The role of resource abundance may be summarized in a simple flow diagram (Figure 19.2). According to this scenario, climatic seasonality influences the timing of plant reproduction and therefore community-wide fruiting phenology and seasonal fruit and seed abundance. In turn, resource abundance determines to a great extent annual and longer-term changes in the demography of *Proechimys semispinosus*. Thus, resource abundance influences density, which drives reproductive effort

(albeit with a time lag because reproductive responses to changing density and resource abundance are not instantaneous), and reproductive effort feeds back directly onto density by either increased effort and more recruitment when densities are low or reduced effort and less recruitment when densities are high. Resource abundance also may directly influence reproductive effort irrespective of density when per capita resource abundance is insufficient to support the physiological demands of reproduction.

Experimental tests of resource limitation

To test the role of food resources in limiting populations of *Proechimys semispinosus*, I conducted two food-provisioning experiments. The first experiment tested the hypothesis that insular populations that are largely released from top-down limitation are limited by food resources even during the season of resource abundance (Adler 1998). The second experiment tested the hypothesis that such populations are limited during the season of resource scarcity (Adler unpublished data). By using small islands as experimental replicates, we could control for the "commuter effect," whereby individuals living near the experimental area temporarily increase density in that area (Adler 1998). This effect is common in

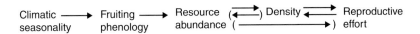

Figure 19.2 Conceptual diagram showing the putative influence of climatic seasonality on spiny rat population density (from Adler and Beatty 1997).

experimental provisioning studies and renders the utility of such studies dubious because, although densities increase in response to provisioning, the increase is often due largely to nearby individuals temporarily visiting the provisioned area rather than to any organic response such as increased survival or reproductive success. In the case of relatively closed systems such as islands, any substantive increase in density would have to be promulgated by increased survival or reproductive output because the rate of immigration over water would be too low to explain such an increase. We also could census each spiny rat population regularly to obtain reliable estimates of density, reproductive output, and survival. If the first hypothesis were supported by the experiment, it would seem that the second hypothesis was unnecessary to test. However, we were interested not only in the effect on density but also in the processes (e.g., increased survival, increased reproductive output, or both) by which density might change.

To test the first hypothesis, we selected four islands as controls, whereby we simply censused spiny rats and fruiting trees and lianas every month, and four islands as experimentals. Criteria for island designations as controls and experimentals are given in Adler (1998). The experimental populations were provisioned with fresh native fruit from four species of trees known to be eaten by spiny rats (Adler 1995; the palms *Astrocaryum standleyanum, Attalea butyracea*, and *Bactris major*, and the fig *Ficus insipida*) from May through October 1992, and spiny rat populations and fruiting trees and lianas were censused every month. Each experimental population was provisioned with 315.2 kg ha^{-1} over the 6-month experimental period. We provisioned the populations every week, except during the week in which rat censuses were conducted. For this purpose, we placed fruit into semipermeable exclosures to exclude larger frugivorous rodents such as agoutis

and pacas, if such rodents were present (both species were occasionally seen on two experimental islands for several months but did not establish populations). Exclosures were situated permanently and spaced evenly across each island at a density of 10 per hectare. Each month, spiny rats had access to 52.5 kg ha^{-1} of extra food, or roughly 1 to 3 kg per spiny rat (depending on the island and time of year), which is approximately 3–9 times the mean body weight of a spiny rat. Thus, all spiny rats had access to reliable and predictable additional resources throughout the provisioning period.

Densities of spiny rats and of known births were analyzed by repeated measures analysis of covariance, with the density of fruiting trees and lianas as the covariate. Thus, we controlled for the availability of natural fruit, which certainly had an impact on population densities even in the absence of the provisioned fruit (Adler and Beatty 1997). Both density (Figure 19.3) and density of known births (Figure 19.4) increased on the experimental islands relative to the controls while controlling for the availability of natural fruits. The experimental increase in density could have been due to improved survival or increased reproductive output, and the increase in the density of births could have been due to either increased per capita reproductive output or simply an increase in the number of adults and therefore in the number of young that they produced. We therefore analyzed monthly survival rates and the number of known births per adult female using linear analysis of count data (Lindsey 1995). Survival rates did not differ between control and experimental populations, but per capita reproductive output was greater within the experimental populations. Thus, the increase in density was due to increased per capita reproductive output by females rather than to improved survival, and spiny rat populations were food-limited even during the season of greatest resource abundance.

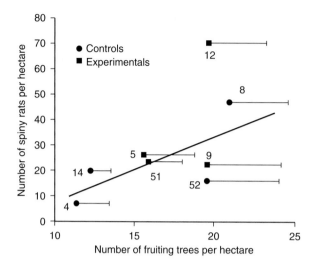

Figure 19.3 Effects of provisioned food on mean population density of spiny rats compared with control populations in relation to natural fruit availability (from Adler 1998). Numbers represent island designations.

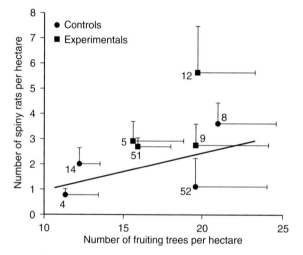

Figure 19.4 Effects of provisioned food on mean density of known births of spiny rats compared with control populations in relation to natural fruit availability (from Adler 1998). Numbers represent island designations.

To test the second hypothesis regarding responses during periods of resource scarcity, I used similar methods but increased the number of experimental populations to eight and provisioned the populations from November 1997 through January 1998 and from November 1998 through January 1999 when fruit production typically was lowest. I provisioned each population with cracked corn (which is eaten by spiny rats) because sufficient fresh native fruit was unavailable. I provisioned each experimental population every week (except during the week in which rats were censused) at the rate of 50 kg ha^{-1}, again using permanently placed exclosures at a density of 10 per hectare.

Of the eight experimental populations, six reached higher densities than would have been expected based on the quantity of natural fruit available when compared with the controls. The two experimental populations that did not reach higher densities were at extremely low densities at the beginning of the study but increased steadily throughout the 2-year study period, but it is not possible to determine if the increase was due to a natural recovery from low population densities or to the provisioning. Apparently, their densities were so low at the beginning of the study that even extra food did not allow them to reach densities comparable to those expected based on natural fruit availability. Therefore, those two populations either were not limited by food or I could not detect such limitation with my approach because their population densities were so low, but the other six populations were strongly limited by the availability of natural fruit. Reproductive activity was very low in all populations during the provisioning period, but survival was enhanced. Experimental populations consequently began each season of abundant resources at higher densities relative to controls and to natural fruit availability, and each of those six populations reached relatively higher peak densities. Thus, food provisioning during the season of resource scarcity increased density via increased survival rather than via increased reproductive output during the provisioning period.

CONSEQUENCES AND GENERAL IMPLICATIONS

Descriptive and experimental evidence demonstrates that frugivore and granivore populations can be food-limited in seasonal tropical forests and that bottom-up limitation is important. I suggest that the role of bottom-up limitation is particularly strong in insular settings where predators are scarce or absent and frugivore and granivore populations largely are released from top-down pressures. I further suggest that attempts to categorize herbivore populations as being limited solely by either bottom-up or top-down processes are likely to fail because both are no doubt important, and their relative importance is likely to fluctuate seasonally and in response to irregular climatic fluctuations and therefore is context-specific. Furthermore, populations are likely at least partly regulated by density-dependent intrinsic processes such as recruitment and reproductive output, particularly when released from top-down limitation. Such regulation is common within rodent populations (e.g., Wolff 1997) and

would not be unexpected within dense spiny rat populations.

Bottom-up and top-down processes have implications far beyond merely limiting population growth and densities of herbivores. Tropical forest fragmentation is creating landscapes of insular patches of forest that lie within a matrix of less hospitable habitat such as agricultural areas, pasturage, urban areas, and water from the construction of hydroelectric projects. Predator populations are reduced in or eliminated from many patches, which increases the importance of bottom-up limitation, thereby unleashing a trophic cascade, and the resulting cascades have important implications for plant recruitment and diversity (Terborgh *et al.* 2001). Thus, in the absence of predators, herbivores as diverse as leafcutter ants, monkeys, and rodents reach abnormally high densities and greatly influence plant diversity, species composition, and forest structure (Terborgh *et al.* 2001).

ACKNOWLEDGMENTS

I thank Craig Bernstein, Nicole Casteel, Jorge Castillo, Mark Endries, Jennifer Freund, Gregory Hoch, Thomas Lambert, Scott Mangan, Felix Matias, Andrew Roper, Gregory Ruthig, Jill Stahlman, David Tomblin, and Paul Trebe for field assistance, the Smithsonian Tropical Research Institute for logistical support, and two anonymous reviewers for comments on a previous draft of the manuscript. Studies described in this chapter were supported by the US National Science Foundation, the National Geographic Society, the Smithsonian Institution, the Smithsonian Tropical Research Institute, and the University of Wisconsin – Oshkosh.

REFERENCES

Adler, G.H. (1987) Influence of habitat structure on demography of two rodent species in eastern Massachusetts. *Canadian Journal of Zoology* 65, 903–912.
Adler, G.H. (1994) Tropical forest fragmentation and isolation promote asynchrony among populations of a frugivorous rodent. *Journal of Animal Ecology* 63, 903–911.

Adler, G.H. (1995) Fruit and seed exploitation by Central American spiny rats, *Proechimys semispinosus*. *Studies on Neotropical Fauna and Environment* 30, 237–244.

Adler, G.H. (1996) The island syndrome in isolated populations of a tropical forest rodent. *Oecologia* 108, 694–700.

Adler, G.H. (1998) Impacts of resource abundance on populations of a tropical forest rodent. *Ecology* 79, 242–254.

Adler, G.H. (2000) Tropical tree diversity, forest structure and the demography of a frugivorous rodent, the spiny rat (*Proechimys semispinosus*). *Journal of Zoology* 250, 57–74.

Adler, G.H. and Beatty, R.P. (1997) Changing reproductive rates in a Neotropical forest rodent, *Proechimys semispinosus*. *Journal of Animal Ecology* 66, 472–480.

Adler, G.H. and Lambert, T.D. (1997) Ecological correlates of trap response of a Neotropical forest rodent, *Proechimys semispinosus*. *Journal of Tropical Ecology* 13, 59–68.

Adler, G.H. and Levins, R. (1994) The island syndrome in rodent populations. *Quarterly Review of Biology* 69, 473–490.

Adler, G.H. and Seamon, J.O. (1991) Distribution and abundance of a tropical rodent, the spiny rat, on islands in Panama. *Journal of Tropical Ecology* 7, 349–360.

Adler, G.H., Tomblin, D.C., and Lambert, T.D. (1998) Ecology of two species of echimyid rodents (*Hoplomys gymnurus* and *Proechimys semispinosus*) in central Panama. *Journal of Tropical Ecology* 14, 711–717.

Bonaccorso, F.J. (1979) Foraging and reproductive ecology in a Panamanian bat community. *Bulletin of the Florida State Museum, Biological Sciences* 24, 359–408.

Chapman, C.A., Wrangham, R., and Chapman, L.J. (1994) Indices of habitat-wide fruit abundance in tropical forests. *Biotropica*, 26, 160–171.

Davis, S.L. (2001) *Environmental variation and the demography of a tropical forest rodent.* Unpublished Masters thesis, University of Wisconsin – Oshkosh.

Endries, M.J. and Adler, G.H. (2005) Spacing patterns of a tropical forest rodent, the spiny rat (*Proechimys semispinosus*), in Panama. *Journal of Zoology* 265, 147–155.

Fleming, T.H. (1971) Population ecology of three species of Neotropical rodents. *Miscellaneous Publications of the Museum of Zoology, University of Michigan* 143, 1–77.

Fleming, T.H. (1992) How do fruit- and nectar-feeding birds and mammals track their food resources? In *Effects of Resource Distribution on Animal–Plant Interactions.* Academic Press, London, pp. 355–391.

Foster, R.B. (1982a) The seasonal rhythm of fruitfall on Barro Colorado Island. In E.G. Leigh, Jr., A.S. Rand, and D.M. Windsor (eds), *The Ecology of a Tropical Forest: Seasonal Rhythms and Long-Term Changes.* Smithsonian Institution Press, Washington, DC, pp. 151–172.

Foster, R.B. (1982b) Famine on Barro Colorado Island. In E.G. Leigh, Jr., A.S. Rand, and D.M. Windsor (eds), *The Ecology of a Tropical Forest: Seasonal Rhythms and Long-Term Changes.* Smithsonian Institution Press, Washington, DC, pp. 201–212.

Glanz, W.E., Thorington, R.W. Jr., Giacalone-Madden, J., and Heaney, L.R. (1982) Seasonal food use and demographic trends in *Sciurus granatensis.* In E.G. Leigh, Jr., A.S. Rand, and D.M. Windsor (eds), *The Ecology of a Tropical Forest: Seasonal Rhythms and Long-Term Changes.* Smithsonian Institution Press, Washington, DC, pp. 239–252.

Gliwicz, J. (1984) Population dynamics of the spiny rat *Proechimys semispinosus* on Orchid Island (Panama). *Biotropica* 16, 73–78.

Gonzalez, M.A. and Alberico, M. (1993) Seleccion de habitat en unacomunidad de mamiferos pequenos en la costa Pacifica de Colombia. *Caldasia* 17, 313–324.

Hairston, N.G. Sr, Smith, F.E., and Slobodkin, L.B. (1960) Community structure, population control, and competition. *American Naturalist* 94, 421–425.

Henry, O. (1994) Saisons de reproduction chez trios Rongeurs et un Artiodactyle en Guyane francaise, en function des facteurs du milieu et de l'alimentation. *Mammalia* 58, 183–200.

Hladik, A. and Hladik, C.M. (1969) Rapports trophiques entre vegetation et primates dans la foret de Barro Colorado (Panama). *La Terre et la Vie* 23, 25–117.

Janzen, D.H. (1979) How to be a fig. *Annual Review of Ecology and Systematics* 10, 13–51.

Lambert, T.D. and Adler, G.H. (2000) Microhabitat use by a tropical forest rodent, *Proechimys semispinosus*, in central Panama. *Journal of Mammalogy* 81, 70–76.

Lambert, T.D., Adler, G.H., Mailen Riveros, C., Lopez, L., Ascanio, R., and Terborgh, J. (2003) Rodents on tropical land-bridge islands. *Journal of Zoology* 260, 179–187.

Lin, Y.K. and Batzli, G.O. (1995) Predation on voles: an experimental approach. *Journal of Mammalogy* 76, 1003–1012.

Lindsey, J.K. (1995) *Modelling Frequency and Count Data.* Clarendon Press, Oxford.

Lomolino, M.V. (1985) Body size of mammals on islands: the island rule reexamined. *American Naturalist* 125, 310–316.

Mangan, S.A. and Adler, G.H. (1999) Consumption of arbuscular mycorrhizal fungi by spiny rats

(*Proechimys semispinosus*) in eight isolated populations. *Journal of Tropical Ecology* 15, 779–790.

Mangan, S.A. and Adler, G.H. (2002) Seasonal dispersal of arbuscular mycorrhizal fungi by spiny rats in a neotropical forest. *Oecologia* 131, 587–597.

Milton, K. (1982) Dietary quality and demographic regulation in a howler monkey population. In E.G. Leigh, Jr., A.S. Rand, and D.M. Windsor (eds), *The Ecology of a Tropical Forest: Seasonal Rhythms and Long-Term Changes*. Smithsonian Institution Press, Washington, DC, pp. 273–289.

Nason, J.D., Herre, E.A., and Hamrick, J.L. (1998) The breeding structure of a tropical keystone plant resource. *Nature* 391, 685–687.

Polis, G.A. and Strong, D.R. (1996) Food web complexity and community dynamics. *American Naturalist* 147, 813–846.

Russell, J.K. (1982) Timing of reproduction by coatis (*Nasua narica*) in relation to fluctuations in food. In E.G. Leigh, Jr., A.S. Rand, and D.M. Windsor (eds), *The Ecology of a Tropical Forest: Seasonal Rhythms and Long-Term Changes*. Smithsonian Institution Press, Washington, DC, pp. 413–431.

Schaffer, W.M. and Tamarin, R.H. (1973) Changing reproductive rates and population cycles in lemmings and voles. *Evolution* 27, 111–124.

Seamon, J.O. and Adler, G.H. (1999) Short-term use of space by a Neotropical forest rodent, *Proechimys semispinosus*. *Journal of Mammalogy* 80, 899–904.

Smythe, N. (1978) The natural history of the Central American agouti (*Dasyprocta punctata*). *Smithsonian Contributions to Zoology* 257, 1–52.

Smythe, N., Glanz, W.E., and Leigh, E.G. Jr. (1982) Population regulation in some terrestrial frugivores. In E.G. Leigh, Jr., A.S. Rand, and D.M. Windsor (eds), *The Ecology of a Tropical Forest: Seasonal Rhythms and Long-Term Changes*. Smithsonian Institution Press, Washington, DC, pp. 227–238.

Terborgh, J. (1983) *Five New World Primates*. Princeton University Press, Princeton.

Terborgh, J. (1986) Community aspects of frugivory in tropical forests. In A. Estrada and T.H. Fleming (eds), *Frugivores and Seed Dispersal*. W. Junk, Dordrecht, pp. 371–384.

Terborgh, J. (1988) The big things that run the world: a sequel to E.O. Wilson. *Conservation Biology* 2, 402–403.

Terborgh, J., Lopez, L., Nunez, P. *et al.* (2001) Ecological meltdown in predator-free forest fragments. *Science* 294, 1923–1926.

Tomblin, D.C. and Adler, G.H. (1998) Differences in habitat use between two morphologically similar tropical forest rodents. *Journal of Mammalogy* 79, 953–961.

Van Horne, B. (1981) Demography of *Peromyscus maniculatus* populations in seral stages of coastal coniferous forest in southeast Alaska. *Canadian Journal of Zoology* 59, 1045–1061.

Van Schaik, C.P., Terborgh, J.W., and Wright, S.J. (1993) The phenology of tropical forests: adaptive significance and the consequences for primary consumers. *Annual Review of Ecology and Systematics* 24, 353–377.

Van Valen, L. (1973) Pattern and the balance of nature. *Evolutionary Theory* 1, 31–49.

White, T.C.R. (1993) *The Inadequate Environment: Nitrogen and the Abundance of Animals*. Springer-Verlag, Barcelona.

Wolff, J.O. (1997) Population regulation in mammals: an evolutionary perspective. *Journal of Animal Ecology* 66, 1–13.

Woods, C.A. (1993) Suborder Hystricognathi. In D.E. Wilson and D.M. Reeder (eds), *Mammal Species of the World: A Taxonomic and Geographic Reference*. Smithsonian Institution Press, Washington, DC, pp. 771–806.

Wright, S.J. and Calderon, O. (2006) Seasonal, El Nino and longer term changes in flower and seed production in a moist tropical forest. *Ecology Letters* 9, 35–44.

Wright, S.J., Carrasco, C., Calderon, O., and Paton, S. (1999) The El Nino Southern Oscillation, variable fruit production, and famine in a tropical forest. *Ecology* 80, 1632–1647.

Chapter 20

TROPICAL ARBOREAL ANTS: LINKING NUTRITION TO ROLES IN RAINFOREST ECOSYSTEMS

Diane W. Davidson and Steven C. Cook

OVERVIEW

Arboreal ants are extraordinarily abundant and functionally important in tropical rainforests worldwide. With carbohydrate-rich, nitrogen (N)-poor diets, many such ant taxa invest "excess carbon" in chemical weaponry, wide-ranging and high tempo foraging, spatial territoriality, support of populous colonies, and nutrition of N-contributing microsymbionts. Despite such investments, and their potential contributions to nitrogen gain and/or conservation, feeding assays document greater average N-limitation in arboreal exudate-feeders than in predatory/scavenging species.

Better anti-herbivore protection may accrue to extrafloral nectar (EFN) plants attracting the most N-starved, ecologically dominant, and territorial arboreal ant species. Such taxa should take sugars only at high concentration but amino acids even at low concentrations. Not surprisingly, EFNs average higher total sugar concentrations and lower amino acid concentrations, compared with minimum sugar and amino acid concentrations acceptable to foraging workers of arboreal taxa as a whole.

Rankings of exudate-feeders by increasing dietary N contributions from predation and scavenging (increasing δ^{15}N) generally correspond to expectations of myrmecologists and may be reasonable predictors of the relative capacities of different ant taxa to deter plant herbivores. Exceptionally low values for certain "herbivorous" taxa may be explained in part by differential recycling of light N by microsymbionts.

In the most prominent exudate-feeding taxa (Formicinae and Dolichoderinae), foraging functional groups may be defined by a combination of digestive anatomy and body size. Among formicines, derived proventriculi and large, size-adjusted crop capacities and liquid uptake rates correlate with solitary leaf-foraging, visual navigation, diurnal activity, diverse diets, and high species richness. In contrast, large-bodied dolichoderines (*Dolichoderus* species) have plesiomorphic proventriculi, poorer foraging performances, uniform diets of trophobiont honeydew, and comparatively low species richness. Wide-ranging leaf-foragers should be good exploitative competitors, whereas trophobiont-tenders may be better interference competitors. Foraging in aggregate, the latter taxa may also encounter fewer herbivorous prey, and often damage hosts via resource parasitism and disease transmission. Small-bodied taxa (including some myrmicines) are generally highly N-limited and accepting of lower quality resources. Finer distinctions are inevitable within all functional groups. Although foraging functional groups correlate strongly with taxonomy, it remains uncertain how they "map" to guilds defined by structure and intensity of interspecific interactions.

INTRODUCTION

Measured in terms of both numbers and biomass in canopy fogging samples, ants (Formicidae: Hymenoptera) are the dominant arthropods of tropical rainforest canopies. Insecticidal fogging may oversample cursorial insects and undersample some sap-feeders (Dejean *et al.* 2000), but the abundance of ants is so striking (regularly half of all sampled arthropods; Tobin 1995, Davidson and Patrell-Kim 1996) that their pre-eminence should survive correction for these biases. Relative to ant communities in temperate forests, tropical communities are remarkable more for high ant abundance than for high species richness (Stork 1988); exceptional abundance is typical of one or a few species and is principally a phenomenon of the arboreal zone (Stork 1988, Tobin 1994, Yanoviak and Kaspari 2000). Many of the abundant arboreal taxa are spatially territorial and/or behaviorally dominant in interspecific encounter competition (Yanoviak and Kaspari 2000). Together, superabundance and behavioral dominance define various ant species as ecological dominants (Davidson 1998).

Based on high abundance, behavioral dominance, and perhaps the lack of evolutionary opposition to risky behaviors in reproductively sterile workers, ants are rivaled only by bees as functionally important canopy arthropods. They can deter insect and vertebrate herbivores (e.g., Janzen 1972, Bentley 1976, Dejean and Corbara 2003), magnify damage from sap-feeding herbivores (Queiroz and Oliveira 2001, Dejean and Corbara 2003), interfere with plant reproductive services (Davidson and Epstein 1989, Willmer and Stone 1997, Raine *et al.* 2002), and provide enemy free space for nesting birds (e.g., Koepke 1972, Young *et al.* 1990, and *Cyphorhinus aradus* on myrmecophytic *Triplaris*). By one estimate, fully a third of tropical woody dicots and herbaceous vines produce extrafloral nectar (EFN) and/or pearl bodies to attract ants for anti-herbivore protection (Schupp and Feener 1991, see also Blüthgen and Reifenrath 2003). Still other plants, principally herbs, epiphytes, and primary hemiepiphytes, use ants to disperse their seeds (e.g., Davidson and Epstein 1989, Kaufmann *et al.* 1991, Horvitz and Schemske 1994, Kaufman 2003); some such plant taxa may be keystone resource species for rainforest frugivores (Terborgh 1986). Ants also interact strongly with other arthropods, especially sap-feeding Hemiptera (Auchenorhyncha and Sternorhyncha [e.g., Buckley 1987]) and other ant species (Jeanne and Davidson 1984).

Wilson (1959) first noted an important asymmetry between arboreal and terrestrial ants. Although few terrestrial taxa forage in the arboreal zone, many arboreal species forage terrestrially. Coupled with larger mean colony sizes in many arboreal species (Tobin 1994, Yanoviak and Kaspari 2000), this observation suggests that arboreal ants may often be competitively superior

to their terrestrial counterparts. Numerous recent studies identify specific arboreal taxa as ecological dominants, defending mutually exclusive territories ranging from treetops to the ground (e.g., Majer 1993, Dejean and Corbara 2003, Blüthgen *et al.* 2004b). Nevertheless, territorial defense is selective, occurring against certain other ant species and not others, though the basis for selectivity remains poorly understood.

Why are arboreal ants so abundant? Why may they outcompete terrestrial species, and what accounts for the competitive superiority of ecological dominants? These questions may have common answers, related in part to how resource imbalances have molded ant biology (Tobin 1991, Davidson 1997, 1998, 2005, Yanoviak and Kaspari 2000). Extraordinary dependence on carbohydrate (CHO)-rich, nitrogen (N)-poor exudates of associated plants and sap-feeding insects appears to have strongly influenced the ecology and evolution of arboreal ant taxa, together with their functional roles in tropical ecosystems. In turn, effects of dietary resource ratios on the elemental and ecological stoichiometry of ants have likely driven evolution of these associates. Here, we briefly highlight some theoretical aspects of the nascent fields of elemental and ecological stoichiometry, and then illustrate how such principles may play out in the biology and ecology of tropical arboreal ants and other closely interacting species. We then consider how digestive anatomy and function help to define the nature of ant interactions with one another, plants, and trophobionts. Our focus throughout is on ant nutrition, but we do not dismiss the importance of other aspects of ant ecology (e.g., nest site limitation and top-down forces), already better understood and summarized elsewhere (Herbers 1985, 1989, Fonseca 1999, Orr *et al.* 2003, Foitzik *et al.* 2004, Lebrun 2005).

ELEMENTAL AND ECOLOGICAL STOICHIOMETRY OF TROPICAL ARBOREAL ANTS

The parallel theories of elemental/ecological stoichiometry and the geometric framework explicitly recognize mass balance of energy and nutrient flows as essential for understanding both the functional biology of individual organisms and higher-level processes in populations, communities, and ecosystems (Sterner and Elser 2002, Raubenheimer and Simpson 2004). Food storage aside, consumers act to maintain elemental homeostasis in body composition, and they respond ecologically and evolutionarily to dietary resource ratios. All organisms regularly harvest energy and matter from their environments, convert a fraction of these resources to biomass, and return energy and materials back to their surroundings. Natural selection should have molded each of these processes by producing regulatory mechanisms that both enhance acquisition or retention of limiting macronutrients and either dispose of nutrient excesses or channel them to useful functions (Davidson 1997, Raubenheimer and Simpson 1998, 2004). Selection for such regulation should be strongest in organisms with nutritional imbalances, and given the remarkably broad range of dietary resource ratios utilized across the ants, these insects represent a compelling study system for research into the effects of elemental and ecological stoichiometry on insect ecology and evolution (Davidson 1997, Cook and Davidson 2006). Importantly also, workers can be assayed directly for evidence of resource imbalances (Kay 2002, Davidson 2005).

Earliest ants were stinging huntresses, provisioning prey to brood in terrestrial nests (Hölldobler and Wilson 1990). For taxa adapted to forage on highly connected planar terrestrial surfaces, colonization of the three-dimensional, poorly connected, and vertically structured arboreal zone must have been accompanied by marked increases in foraging costs. Ancestors of contemporary arboreal taxa appear to have solved this problem by exploiting energy-rich plant and insect secretions (EFNs, plant wound sap, and trophobiont honeydews) to fund energetically expensive foraging. (Present-day examples of terrestrial taxa transitioning to arboreal life include some *Myrmicaria*, *Pheidole*, and *Anoplolepis*, as well as *Paraponera clavata*.) Recent isotopic studies suggest that many exudate-feeders also obtain substantial N from exudates (Blüthgen *et al.* 2003, Davidson *et al.* 2003, but see below). If this is so, ants must be ingesting and processing very large volumes

of CHO-rich exudates for the sparse N they contain. Digestive systems specialized to process liquid foods in volume (Eisner 1957, Davidson *et al.* 2004), perhaps together with nutrient contributions from microsymbionts (see below), may have enabled these taxa to persist and even thrive on highly imbalanced, N-poor diets.

High CHO:N dietary ratios of exudate-feeders should be correlated with availability of "excess CHOs," that is, CHOs exceeding those funding growth and reproduction through primary metabolism (Davidson 1997). Natural selection is postulated to have turned these energy sources to good use in supporting large populations of CHO-dependent workers (Tobin 1994) and subsidizing pursuit of N-rich prey by one or more of three mechanisms (Davidson 1997): (1) increased activity "tempos" (velocities), correlated with higher "dynamic densities" of foraging workers; (2) defense of absolute spatial territories and associated prey; and (3) N-free or N-poor defensive/offensive chemical weaponry. A review of existing information provided at least circumstantial evidence for each of these mechanisms, all of which contribute to ecological dominance (Davidson 1997, see also Yanoviak and Kaspari 2000).

If arboreal exudate-feeders have evolved to invest excess CHOs in mechanisms ensuring ecological dominance (and thus access to limiting N), to what extent do these taxa remain N-deprived? Answering this question could also help to resolve the degree to which various ant taxa convey anti-herbivore protection to plants. Recently, Davidson (2005) evaluated N-deprivation in a behavioral assay of 54 arboreal or terrestrial Amazonian ant taxa with diets ranging from carnivorous to highly herbivorous, as evaluated by worker $\delta^{15}N$ ratios (Davidson *et al.* 2003). Relative N-deprivation was quantified as an exchange ratio (ER), defined as SUCmin/AAmin, or the minimum sucrose concentration divided by the minimum amino acid concentration, accepted as food by $\geq 50\%$ of tested workers (see Kay 2002, pioneering this approach). ER values averaged almost five-fold higher (corresponding to greater N-deprivation) for "N-omnivores" and "N-herbivores" (N-OH taxa) than for "N-carnivores," that is, taxa which gain no detectable N from plant or trophobiont exudates (Davidson *et al.* 2003). At a range of taxonomic levels (species/species groups, genera, and tribes), N-deprivation declined marginally with increasing trophic level (i.e., higher $\delta^{15}N$ ratio) and significantly with increasing body size. Overall, use of N-free or less "N-dense" chemical weaponry by N-OH taxa appeared to reduce N-deprivation. (However, among relatives with similarly N-free weaponry, the pattern with trophic level was reversed, that is, the most N-limited taxa were often more carnivorous.) Groups with proteinaceous venoms tended to be more carnivorous and less N-deprived than did those with alkaloidal venoms or N-free chemical weaponry, and most small-bodied ants fell into the latter category.

In general then, exudate-foraging ants tend to have highly imbalanced and N-poor diets, and despite deploying excess CHOs to obtain protein, many such taxa remain N-deprived. This is particularly true of several small-bodied taxa with relatively carnivorous diets for their subfamilies. Perhaps not coincidentally, such taxa include several genera rich in specialized plant-ants (e.g., Davidson and McKey 1993): *Azteca* and *Crematogaster* (Davidson 2005), and also *Myrmelachista* (with just a single plant-ant surveyed to date; Davidson unpublished data). Together, these observations suggest that myrmecophytes could have evolved filters to selectively attract the most carnivorous of N-limited ant taxa that had previously responded to N-poor diets by reducing N investment in chemical weaponry. The next section describes how myrmecophilous plants, interacting more casually and opportunistically with ants, might vary CHO:N ratios of ant rewards to similar effect.

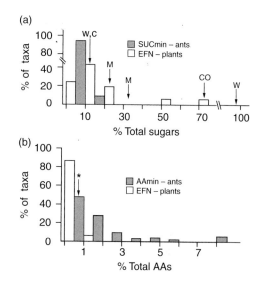

Figure 20.1 Comparisons of SUCmin and AAmin for 54 Amazonian and 33 Bornean ant taxa (Davidson 2005 and unpublished) versus (a) total sugars and (b) total amino acids (AAs) in EFNs of 16 Australian tropical rainforest plant taxa (Blüthgen and Fiedler 2004a; all EFNs used mainly by ants). Arrows designate (a) individual values (not included in columns) for wound secretions (W) or trophobiont honeydews from the same study: C, Cicadelidae; CO, Coccidae; M, Membracidae; and (b) all values for these same wound and trophobiont secretions (asterisk). All concentrations are wt/vol. Median SUCmin values were identical (4.00) in Brunei and Peru, and lumped values were significantly lower than % total sugars in EFN (12.90, $X_1^2 = 26.67$, $P < 0.0001$ in a Wilcoxon test). Median AAmin values did not differ between Amazonian and Bornean ants (1.00 and 0.21, respectively, $P \gg 0.05$), and lumped values significantly exceeded the median for EFN (0.10, $X_1^2 = 18.28$, $P < 0.0001$ in a Wilcoxon test).

THE ROLE OF STOICHIOMETRY IN OPPORTUNISTIC ANT–PLANT INTERACTIONS

In contrast to dimensionless ERs, threshold sugar and amino acid values for feeding cannot be compared independently of one another among taxa because they are influenced by such factors

as proximity to nest sites (Davidson 1978) and natural enemies (Nonacs and Dill 1991, Blüthgen and Fiedler 2004b). Nonetheless, we can ask how the distribution of minimum acceptable concentrations across all assayed ant taxa compares with total sugar and amino acid concentrations in EFN and honeydew for tropical plants and trophobionts. Figure 20.1 presents such comparisons for total sugars (a) and amino acids (b). Reflecting mainly sucrose, glucose, and fructose content (Engel *et al.* 2001, Blüthgen *et al.* 2003), the median value for total sugars in EFNs exceeds that of SUCmin across Amazonian and Bornean ants (Davidson 2005 and unpublished data) by a factor of about five (19.85% versus 4.00%). In contrast, the median amino acid concentration in EFN is just a tenth that of AAmin (0.1% versus 1.0%). Opposing patterns for sugars and amino acids suggest that sugar concentrations in EFN were not simply biased to high values by evaporation during nectar accumulation after ant exclusion. Overall, it appears that both EFN plants and trophobionts offer ants sugars at concentrations much greater than the minimum requirements of many arboreal ant taxa, but amino acids at much lower concentrations. Predictably, the ratio of % total sugars to % total amino acids in EFN far exceeds that of SUCmin to AAmin (Figure 20.2). The few available values for sugar and amino acid concentrations in honeydews (arrows in figures) hint at a similar conclusion for those secretions.

"Oversupply" of sugars relative to amino acids in EFN may have both multiple causes and important consequences. It could be favored, in part, because myrmecophilous plants themselves are more N-limited than C-limited (Schupp and Feener 1991). Additionally, it may alter the outcome of ant–ant competition in ways favorable to deterrence of the plant's natural enemies. The most N-starved, behaviorally dominant, territorial, widely foraging, and populous ant taxa (e.g., Rocha and Bergallo 1992, see also below) should provide better anti-herbivore protection, and these same taxa should accept sugars only at high concentration. Recent work by Blüthgen and Fiedler (2004a,b and online material) in the Australian tropics has established that the aggressive, ecologically dominant *Oecophylla smaragdina* (authors' unpublished ER = 8) regularly monopolizes the highest quality exudates: EFN and hemipteran honeydews with high concentrations of both total sugars and total amino acids, and/or high amino acid diversity. Distributions of just four other ant species were correlated or marginally correlated with solute concentrations in exudates; these taxa most often used nectars of low or

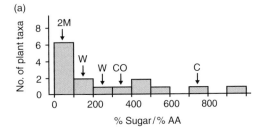

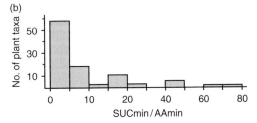

Figure 20.2 Distributions of SUCmin/AAmin ratios for Amazonian and Bornean ants (bottom panel; this study), and ratios of % total sugars/% total amino acids for 16 Australian EFN plants (top panel; Blüthgen and Fiedler 2004; lettered arrows as in Figure 20.1). Note the different scales of the abscissas in the two panels. Unequal variances again required non-parametric comparison: median SUCmin/AAmin ratios were identical (4.00) for Amazonia and Borneo, but combined ratios were significantly lower than the sugar/amino acid ratio in EFNs of Australian rainforest plants (median = 110, $X_1^2 = 34.92$, $P < 0.0001$ in a Wilcoxon test; additional details in Figure 20.1 legend).

low relative total sugar concentration and were unaffected by amino acid concentrations. Eleven remaining taxa were non-selective. Although high quality resources may have been included in the fundamental niches of the latter 15 taxa, they were under-represented in their realized niches.

Because plants have limited control over the recipients of ant rewards, all optimality models of EFN composition must take into account multiple and potentially competing ant species. Nevertheless, rather than focusing on all potential attending ant species individually, future studies might profitably address reward composition in relation to ant functional groups, as determined from feeding ecology (see below) and/or N-deprivation (Blüthgen and Fiedler 2004a,b, Blüthgen *et al.*

2004b, Davidson *et al.* 2004, Davidson 2005). Ideally also, future tests of the "manipulated competition hypothesis" should compare EFN and trophobiont data with dietary assays of ants from the same communities (*contra* Figure 20.1, see legend). Further, some ant taxa respond to mixtures of amino acids and sugars in individualistic and non-additive ways (e.g., Lanza *et al.* 1993, Blüthgen and Fiedler 2004b), and rigorous tests of the hypothesis must eventually take such complications into account. Finally, in at least some EFN plants, changes in nectar quality can be induced by herbivory (Stephenson 1982, Smith *et al.* 1990), and the same may be true of honeydew (Fischer and Shingleton 2001, Fischer *et al.* 2002). In the single case where sugars and amino acids were monitored simultaneously, amino acid content but not sugar content increased following simulated herbivory (Smith *et al.* 1990). However, more studies are needed to determine whether constitutive EFNs are proportionally richer in sugars and poorer in N than are nectars produced after herbivory.

ANT ASSOCIATIONS WITH ENDOSYMBIONTS

In general, predatory ants might be anticipated to be more effective than are herbivorous or omnivorous ants in deterring herbivorous insects. Although stable isotope technology has proven useful in ranking arboreal ant taxa from lesser to greater dependence on predation and scavenging (Blüthgen *et al.* 2003, Davidson *et al.* 2003), $\delta^{15}N$ values are likely imperfect predictors of trophic levels. Thus, although the very low $\delta^{15}N$ values of some arboreal exudate-feeders, especially formicines and cephalotines, suggest that these taxa feed as herbivores or highly herbivorous omnivores, an alternative or additional explanation is possible. Isotopic fractionation during putative N-recycling by symbiotic bacteria may produce differential retention of light N, preferentially released during biochemical reactions and then transaminated to convert non-essential to essential amino acids. Because most ants are omnivores, the potential exists for gradually magnifying colony N reserves by recycling N acquired

through consumption of fungi (authors' unpublished data), and/or hunted or scavenged prey, including hemolymph (Zientz *et al.* 2005). Given the opportunity, selection to counteract stoichiometric imbalances should have favored such relationships in N-deprived taxa, as in the N-limited sap-feeding trophobionts often tended by ants (e.g., Douglas *et al.* 2001). Recently published evidence shows N-recycling and upgrading by microsymbionts of ants (Feldhaar *et al.* 2007). Moreover, coupled with autocoprophagy and anal trophalaxis, the considerable urate stores in the *Dolichoderus* fat body suggest possible urate recycling in this genus (Cook and Davidson 2006). If any or all of these groups do recycle or fix N, and those processes measurably lower $\delta^{15}N$, then exudate-feeding, omnivorous ants in these taxa would be less "herbivorous" than isotopes indicate, and therefore potentially more beneficial and less harmful to the plants on which they forage.

An important goal of continued research in this area is to understand what fractions of ants' N budgets come from plant secretions (including those filtered through trophobionts), as opposed to carnivory followed by N-recycling. This goal will likely prove difficult to achieve in anything other than "closed systems" of ants and myrmecophytes. Moreover, identifying ant taxa with the greatest potential for providing anti-herbivore protection to plants would require distinguishing N acquired via predation versus scavenging. General effects of ants on the forest may be easiest to define by direct tests, that is, ant removal experiments, and would be most useful if focused on particular taxonomic and functional groups (see below).

FORAGING FUNCTIONAL GROUPS

Many studies of ant–ant, ant–trophobiont, and ant–plant interactions are formulated in terms of "generic" ants, differing perhaps in body size and/or colony size, but with few or no other defining features. Nevertheless, achieving a predictive understanding of interspecific interactions in ant communities, as well as of the roles these insects play in tropical ecosystems, will likely depend more on advancements in our knowledge of their functional biology than on attempts to model behaviors and interactions of ants as "black boxes." Within communities of arboreal exudate-foragers, for example, we have just begun to identify functional groups, correlated in some cases with deep phylogenetic disparities in ant digestive anatomy (Davidson *et al.* 2004). Here we review structural and functional variation in the ant proventriculus (gizzard) and then relate this variation to the activities and roles of arboreal ants in rainforest ecosystems.

Form and function of the ant proventriculus

The proventriculus controls food flow between the worker crop (the colony's "social stomach") and the midgut, where digestion occurs, and it is remarkably diverse in form and function (Hölldobler and Wilson 1990, p. 290 and figure 1 in Davidson *et al.* 2004). As early as the 1950s, the biomechanics of the ant proventriculus were inferred from histological sections, in elegant studies by some of the foremost insect biologists of the past century (Eisner and Wilson 1952, Eisner 1957, Eisner and Brown 1958). This early work tied anatomical structure of derived proventriculi to both reliance on liquid foods and capacity for trophalaxis (regurgitative food sharing), hypothesized as key to colony social integration. Proventricular anatomy came to play a determining role in the designation of tribes within subfamilies Formicinae and Dolichoderinae (Baroni-Urbani *et al.* 1992, Shattuck 1992). However, in recent work based on anatomical characters, Bolton (2003) recognized no tribes within Dolichoderinae, and by his proposed revision of formicines, the derived or "sepalous" proventriculus is either homoplasious in various tribes or has been lost repeatedly following a single origin. Thus, although linkages likely exist between derived proventriculi and liquid diets, the case for such associations must be made anew based on current phylogenetic hypotheses.

Formicines and dolichoderines are most prominent among the exudate-foragers from which canopy dominants are drawn. Other arboreal exudate-foragers include Ectatomminae (*Ectatomma* and *Gnamptogenys*), the monotypic

Paraponerinae (*Paraponera*), some Ponerinae (*Pachycondyla* and *Diacamma*), and Myrmicinae (especially *Crematogaster*, *Myrmicaria*, *Cephalotes*, and *Cataulacus*). Liquids are transported internally (in the crop) in all but the Ectatomminae, Paraponerinae, and Ponerinae, which carry food droplets in the mandibles. Early investigators (especially Eisner 1957) noted that several independent lineages specializing in liquid foods, and carrying them internally, exhibited "passive damming" of crop fluids against posterior flow, by virtue of sclerotization of the anterior proventriculus and/or evolution of an "occlusory tract." Simultaneous with each of these innovations, loss of key muscle groups provided energetic savings and prevented dilution from compromising enzyme function. The most highly derived proventriculi include the sepalous formicine organ and the unique filtering shield associated with pollen-feeding in *Cephalotes* (Cephalotini, Myrmicinae; see Baroni-Urbani and Andrade 1997, Roche and Wheeler 1997, Andrade and Baroni-Urbani 1999). Additionally, several small-bodied genera in one or more clades of the Dolichoderinae have highly modified proventriculi that, in their most derived state, appear functionally convergent with the sepalous formicine proventriculus (Eisner and Wilson 1952). In both cases, passive damming is so complete as to require canals to pass liquid foods posteriorly from the crop to the proventricular bulb, which, in turn, delivers liquids to the midgut.

Although early investigators mentioned only intranidal functions of derived proventriculi in liquid-feeding ants, novel approaches and tools have recently raised new questions about how proventricular anatomy has influenced foraging behaviors. Recent field measures of liquid-feeding performances in tropical and temperate ant taxa reveal associations between passive damming and two body size-adjusted measures of liquid-feeding performance: relatively large maximum load sizes in all such taxa, and rapid drinking rates in formicines and dolichoderines with the most derived proventriculi (Davidson *et al.* 2004). *In vivo* biomechanical studies of feeding ants, using synchrotron radiation, are currently defining determinants of feeding rates more precisely (Cook 2008).

Defining foraging functional groups

Correlations between proventricular anatomy and diet suggest potential utility of these structures in defining taxonomically based niche differences among the ants. Foraging performances of relatively large-bodied formicines and dolichoderines correlate with both proventricular anatomy and (more loosely) membership in one of two broad foraging functional groups defined by predominant worker activities (Davidson *et al.* 2004). "Trophobiont-tenders" include all studied dolichoderines and a small subset of formicines with atypically slow uptake rates. In contrast, most formicines are "leaf-foragers," defined operationally by regular searching of leaf laminae by solitary workers. Although ants in the two categories certainly overlap in their activities, foraging modes differ on average (Davidson *et al.* 2004, supplementary online dataset). With the most rapid, body size-adjusted uptake rates and the largest load sizes, sepalous formicines are better equipped to forage solitarily than are comparably sized species of *Dolichoderus* with plesiomorphic proventriculi and much slower drinking rates. In contrast, *Dolichoderus* species specialize in aggregate trophobiont-tending (Figure 20.3), with nestmates present to assist one another in handling honeydew production that, where measured, is gradual (Tjallingii 1995, Yao and Akimoto 2002). Workers may not then have been selected for rapid liquid uptake, and a requirement for microbial assistance in food processing

in the hindgut may also slow digestion (Cook and Davidson 2006). In both *Dolichoderus* and highly trophobiont-tending formicines (e.g., *Oecophylla* spp.), workers do not stray far from nestmates (Dill *et al.* 2002), and unlike most formicines, are not systematic leaf searchers (Davidson *et al.* 2004). Guarding trophobionts both day and night, workers regularly commute over chemical trails (Hölldobler and Wilson 1990). Compared with the monotonous lifestyles of *Dolichoderus* species, foraging modes of sepalous formicines are many and diverse, perhaps due to their capacity for independent foraging (Davidson *et al.* 2004).

Additional foraging functional groups almost certainly remain to be defined within both leaf-foraging and trophobiont-tending exudate-foragers. Trophobiont-tenders might specialize on certain categories of sap-feeders based on ease of controlling these associates, and/or on resource quality (including CHO:N ratios) and the plant parts from which they feed. For example, Coccidae are mobile only as tiny crawlers and cannot later be relocated to new, relatively N-rich growth. Nevertheless, they might be more easily controlled by small-bodied ants than are Membracidae, mobile at all developmental stages and sizes. Leaf-foragers exhibit diverse habits, searching leaf laminae for EFNs, cast-off honeydews and wound secretions, pollen and fungal spores (Wheeler and Bailey 1920, Andrade and Baroni-Urbani 1999), prey, and even bird droppings (potentially recyclable urate N). Several of

Figure 20.3 Crematogaster ants tending hemipteran egg mass from which nymphs will emerge and be tended individually.

these food types are dispersed and unpredictable in space and time, favoring their location and collection by widely and independently searching workers; for example, EFN production often varies with leaf age (McKey 1984). Interspecific variation in search intensity suggests additional functional group structure among leaf-foragers (Cook and Davidson in preparation). Predatory taxa like New World *Camponotus sericeiventris* and *Paraponera clavata* (isotopic evidence in Davidson *et al.* 2003 and Tillberg and Breed 2004) visit leaves in a coarse-grained way, either out to the tip and back, or a trip around the leaf edge. Fine-grained searchers, which directly contact a large fraction of each visited leaf, include other New World *Camponotus* (e.g., members of subgenera *Myrmobrachys* and *Myrmaphaenus*), most *Cephalotes* and *Pseudomyrmex* species, and many Old World *Polyrhachis*, *Echinopla*, *Camponotus* (*Colobopsis*), and *Tetraponera*.

Kaspari and Weiser (2000) have noted a bimodal distribution of body sizes in a Panamanian community of arboreal and terrestrial ants, and this finding is consistent with our experience in western Amazonia (authors' unpublished data). Smaller-bodied taxa comprising the lower mode are abundant and important elements of the tropical arboreal fauna but do not fall neatly into either of the above two major functional groups. Most small-bodied exudate-foragers, including many *Crematogaster*, *Wasmannia*, *Azteca*, and *Technomyrmex* species, tend to forage with nestmates, in accord with their size-limited forage capacities. However, liquid-feeding performances vary, and consistent with their plesiomorphic proventriculi (DeMoss 1973), tested *Crematogaster* species exhibit relatively poor feeding performances (both load sizes and uptake rates; Davidson *et al.* 2004). Nevertheless, they can dominate resources to the exclusion of much larger ants by virtue of powerful contact toxins (Daloze *et al.* 1986). The same may be true of *Wasmannia*, defended by a potent sting (Howard *et al.* 1982), though its proventriculus has apparently not been studied. In contrast, two dolichoderines with moderately derived proventriculi (Eisner 1957) exhibit relatively rapid liquid uptake (*Technomyrmex*) or large load capacities (*Azteca*) for their body sizes (Davidson *et al.* 2004),

as well as volatile alarm/defense secretions that can quickly attract nestmates (Do Nascimento *et al.* 1998 and Brophy *et al.* 1993, respectively). Future research may resolve the question of whether these genera differ on average in their capacities for resource defense versus exploitation, as well as whether small-bodied taxa are constrained to forage nearer to nest sites or pavilions.

Foraging functional groups and ant community structure

Based on emerging knowledge of foraging functional groups in tropical arboreal exudate-foragers, can we predict how membership in particular groups should affect interspecific interactions and community structure? First, some have argued that dominant status and true spatial territoriality (see Hölldobler and Wilson 1990) are mainly a feature of trophobiont-tenders (Blüthgen *et al.* 2004b). However, our own studies reveal that many leaf-foragers defend spatial territories centered around live nest trees (Jones *et al.* 2004, for Asian *Camponotus* (*Colobopsis*) in the species-rich *cylindricus* group, and Davidson *et al.* 2007, for neotropical *Camponotus sericeiventris*). Therefore, territoriality appears to correlate more strongly with expectation of long-term gain from spatially defined resources than with foraging functional group *per se*.

In contrast, and correlated with differences in proventricular anatomy, large-bodied formicines and dolichoderines may have diverged early on in ways that differentiate both their contemporary roles as exploitative versus interference competitors and their potential roles as predators and scavengers in tropical forests. With greater numbers of independently searching workers and disproportionately large crop capacities and rapid uptake rates, leaf-foragers may be superior exploitative competitors for both non-honeydew exudates and prey. In contrast, densely populous foraging groups of trophobiont-tenders (Figure 20.3) are likely better adapted than are leaf-foragers for interference competition over locally concentrated resources (see especially Blüthgen *et al.* 2004b). Thus, various *Dolichoderus* species dominate EFNs on plants where they tend trophobionts (authors' observations), and a phalanx of *Do. quadridenticulatus* workers can wrest control of EFN plants from aggressive *Ca. sericeiventris* workers (Davidson personal observation). Nevertheless, lacking wide-ranging foraging, trophobiont-tenders may encounter fewer potential prey (apart from tended trophobionts) than do leaf-foraging formicines. Consistent with this conjecture, larval formicines have the anatomical structures to process solid food, whereas dolichoderine larvae do not (Wheeler and Wheeler 1976), though they may consume insect haemolymph, particularly that of trophobionts (Dill *et al.* 2002).

Small-bodied taxa may be more likely than their larger counterparts to accept low quality resources, because absolute locomotory costs are greater for large-bodied workers (although offset somewhat by longer stride lengths; Fewell *et al.* 1996). Moreover, if they forage regularly over shorter distances, this could also allow them to be less selective (Davidson 1978). Consistent with these hypotheses, SUCmin declines marginally, and AAmin significantly, with log of ant body size ($P = 0.06$ and $P = 0.03$, respectively). Additionally, small-bodied taxa tend to have populous colonies that potentially both increase the colony's capacity for intensive and extensive search (e.g., Swain 1980 for *Crematogaster*) and confer numerical advantage in battles of attrition (McGlynn 1999, Palmer 2004). The aforementioned attributes should contribute to both exploitative and interference competitive ability, making these taxa difficult to dislodge from controlled resources. However, except where nesting polydomously, and due to their limited foraging ranges, smaller taxa may prevail more often at localized resources than over widely distributed ones, on a scale, for example, of whole trees.

Guild structure, or the structure of interactions within communities, is a key determinant of local species diversity (May 1972), and once foraging functional groups are more finely elucidated for arboreal taxa, it will be important to determine whether interspecific interactions are stronger within or between those groups. It is obvious, however, that we currently know little about the interaction structure within

communities of opportunistic arboreal rainforest exudate-feeders, and about how that structure maps to phylogenetic structure. Such questions must be resolved if we are ever to fully understand one of the earliest and most commonly noted patterns in such communities, that is, associations between particular territorial dominants and the specific non-dominant ants capable of coexisting within their territories (e.g., Leston 1978, Majer 1993, Dejean and Corbara 2003, Blüthgen *et al.* 2004b, but see Davidson *et al.* 2007).

Foraging functional groups and plant defense

We return briefly to the matter of which free-living ant taxa might be associated with reduced damage to plants on which they forage. Table 20.1 summarizes attributes of the two broad functional groups as they bear on that subject. By 7 of 10 criteria, leaf-foragers appear more likely than trophobiont-tenders to exert a net positive effect on plants where they forage. Of these seven, criteria not previously discussed include alteration of plant metabolism and development, frequent transmission of plant pathogens through stylets of tended trophobionts (e.g., Buckley 1987), and plant resource losses to ants and trophobionts. Although two criteria apparently show greater benefits from trophobiont-tenders, they could actually be associated with higher resource losses if the principal effect of trophobiont-tenders per unit time and worker number are negative, as the first seven criteria suggest. By the final criterion, effects of leaf-foragers and trophobiont-tenders may be equivalent on average, with spatially territorial, carnivorous, and small-bodied taxa within each group exhibiting the greatest N-deprivation (Davidson 2005). We do not mean to suggest that trophobiont-tenders are always less desirable associates of plants than are leaf-foragers, but rather argue from first principles why that should be so on average for free-living ant taxa. Against natural enemies deterred only by large numbers of foraging workers, trophobiont-tenders could be the "preferred" associates, and plant protection has been demonstrated for at least some trophobiont-tenders,

Table 20.1 Comparison (on average) of leaf-foragers versus trophobiont-tenders with respect to traits hypothesized to confer efficacy in defense against plant herbivores and pathogens.

Trait	LF relative to TT
Potential for pathogen transmission (through trophobionts or wounding)	LF < TT[+]
Parasitism of plant resources	LF < TT[+]
Alteration of plant metabolism and development	LF < TT[+]
Percentage of workers searching independently	LF > TT[+]
Amount of plant surface (leaves, stems) covered per foraging worker	LF > TT[+]
Prevention of pathogen development in wounds	LF > TT[+]
Potential for larvae to consume solid foods (prey)[a]	LF > TT[+]
Length of daily activity period	LF < TT[−]
Total numbers of workers on plant surfaces	LF < TT[−]
Average N-deprivation of ant taxa[b]	LF = TT[=]

Notes: Leaf-foragers (LF): Formicinae, Ponerinae, Ectatomminae, Paraponerini, Pseudomyrmecinae, Cephalotini and Cataulacini, following Bolton (2003). Trophobiont-tenders (TT): mainly dolichoderines, a few formicines, and perhaps *Cephalotes attratus*; see Blüthgen *et al.* (2000) and Davidson *et al.* (2004). Superscripts "+", "−" and "=" highlight, respectively, cases where LFs should provide greater protection (or do less harm) than do TTs, where the reverse likely holds, and where effects of LFs and TTs are apparently equivalent.
[a] From Wheeler and Wheeler (1976).
[b] See Davidson (2005).

including Old World *Dolichoderus thoracicus* (Khoo and Ho 1992) and *Oecophylla smaragdina* (e.g., Offenberg *et al.* 2004), and New World *Azteca chartifex* (De Medeiros *et al.* 1999). Nevertheless, just a few existing studies document plant resource losses to tended trophobionts and ants (but see Kay *et al.* 2004), and convincingly demonstrate even intermittent and context-dependent net positive effects of ants (Messina 1981, Horvitz and Schemske 1984, Gaume *et al.* 1998, Oliveira and Del-Claro 2005).

If leaf-foragers are generally more desirable associates than are trophobiont-tenders, plants should have been selected to favor the former species over the latter. In this light, Davidson *et al.* (2004) have resurrected Becerra and Venable's (1989) hypothesis that EFNs evolved in part to entice trophobiont-tenders to desert their associates and feed directly from plants, effectively short-circuiting sap-feeders from the interaction. In its simplest form, this hypothesis is refuted by evidence that ant colonies respond numerically to tend both EFNs and trophobionts on the same plants (Buckley 1983), and by refutation of the theory's correlate, that nutritive values of EFNs should exceed those of honeydews (Fiala 1990,

Blüthgen and Fiedler 2004a). However, a more complicated revision of the theory might suggest that EFN plants have evolved disproportionately CHO-rich nectars (Figure 20.2) to support the wide-ranging and energy-demanding activities of leaf-foragers. Members of that guild could both convey greater plant protection (Table 20.1) and, by virtue of better exploitative competitive abilities, keep the resource too low to attract takeovers by large-bodied trophobiont-tenders, especially *Dolichoderus* species. Because total sugar concentration is directly correlated with total amino acid content in EFN (Blüthgen and Fiedler 2004a), and since SUCmin and AAmin are also positively correlated ($P = 0.0008$ in Spearman rank test), this suggestion need not contradict Blüthgen *et al.*'s (2004a) assertion that identities of ant associates respond to the amino acid component of EFN. EFN plants may attract aggressive, territorial species with cheap and abundant sugars but increase amino acid production only when damaged by herbivores (Smith *et al.* 1990).

Finally, because some small-bodied ants tend to exhibit individually low resource requirements, high N-limitation, and carnivory (see above; Davidson *et al.* 2003, Davidson 2005), as well as colonies finely divided into many searchers, they may be particularly attractive partners for plants and trophobionts. Small-bodied ant taxa (e.g., *Azteca*, *Crematogaster*, *Allomerus*, and *Myrmelachista* spp.) are the most common inhabitants of myrmecophytes (true "ant-plants," e.g., Davidson and McKey 1993), which should have been selected consistently to obtain good protection for minimal reward.

NOTE

Results of *in vivo* synchrontron x-ray imaging of feeding *Camponotus* workers have now revealed no direct effect of proventricular activity on liquid-feeding performances. Other anatomical features (e.g., glossal morphology, buccal volume, and mass of cibarial pump musculature) likely directly determine liquid uptake rates of ants generally.

ACKNOWLEDGMENTS

For facilitating our work in the Manu National Park of western Amazonia, we gratefully acknowledge Peru's office of Áreas Naturales Protegidas (ANP-INRENA), Manu N.P. officials, and the Museo de Entomología, Universidad Nacional Agraria La Molina. We also thank administrations and personnel of the Universiti Brunei Darussalam, Kuala Belalong Field Studies Center, and the Brunei Museums, for facilitating our work in Borneo. Our contribution benefited greatly from a critical review and from NSF support (award IBN-9707932).

REFERENCES

Andrade M.L. de and Baroni-Urbani, C. (1999) *Cephalotes*: diversity and adaptation in the ant genus *Cephalotes*, past and present. *Stuttgarter Beiträge zur Naturkunde, Ser. B (Geologie und Paläontologie)* 271.

Baroni-Urbani, C. and Andrade, M.L. (1997) Pollen eating, storing, and spitting in ants. *Naturwissenschaften* 84, 256–258.

Baroni-Urbani, C., Bolton, B., and Ward, P.S. (1992) The internal phylogeny of ants (Hymenoptera: Formicidae). *Systematic Entomology* 17, 301–329.

Becerra, J.X. and Venable, D.L. (1989) Extrafloral nectaries: a defense against ant–Homoptera mutualisms? *Oikos* 55, 276–280.

Bentley, B.L. (1976) Plants bearing extrafloral nectaries and the associated ant community: interhabitat differences in the reduction of herbivore damage. *Ecology* 57, 815–820.

Blüthgen, N. and Fiedler, K. (2004a) Competition for composition: lessons from nectar-feeding ant communities. *Ecology* 85, 1479–1485.

Blüthgen, N. and Fiedler, K. (2004b) Preferences for sugars and amino acids and their conditionality in a diverse, nectar-feeding ant community. *Journal of Animal Ecology* 73, 155–156.

Blüthgen, N., Gebauer, G., and Fiedler, K. (2003) Disentangling a rainforest food web using stable isotopes: dietary diversity in a species-rich ant community. *Oecologia* 137, 426–435.

Blüthgen, N., Gottsberger, G., and Fiedler, K. (2004a) Sugar and amino acid composition of ant-attended nectar and honeydew sources from an Australian rainforest. *Australian Ecology* 29, 418–429.

Blüthgen, N. and Reifenrath, K. (2003) Extrafloral nectaries in Australian rainforest: structure and distribution. *Australian Journal of Botany* 51, 515–527.

Blüthgen, N., Stork, N.G., and Fiedler, K. (2004) Bottom-up control and co-occurrence in complex communities: honeydew and nectar determine a rainforest ant mosaic. *Oikos* 106, 344–358.

Blüthgen, N., Verhaagh, M., Goitia, W., Jaffe, K., Morawetz, W., and Barthlott, W. (2000) How plants shape the ant community in the Amazonian rainforest canopy: the key role of extrafloral nectaries and homopteran honeydew. *Oecologia* 125, 229–240.

Bolton, B. (2003) Synopsis and classification of Formicidae. *Memoirs of the American Entomological Institute* 71.

Brophy, J.J., Clezy, P.S., Leung, C.W.F., and Robertson, P.L. (1993) Secondary amines isolated from venom gland of dolichoderine ant, *Technomyrmex albipes*. *Journal of Chemical Ecology* 19, 2183–2192.

Buckley, R. (1983) Interaction between ants and membracid bugs decreases growth and seed set of host plant bearing extrafloral nectaries. *Oecologia* 58, 132–136.

Buckley, R.C. (1987) Interactions involving plants, homoptera, and ants. *Annual Review of Ecology and Systematics* 12, 111–135.

Cook, S.C. (2008) *Functional biology of exudate-feeding ants*. Ph.D. thesis, University of Utah, USA (in press).

Cook, S.C. and Davidson, D.W. (2006) Nutritional and functional biology of exudates-feeding ants (Hymenoptera: Formicidae). *Entomologia Experimentalis et Applicata* 118, 1–10.

Daloze, D., Braekman, J.C., Vanhecke, P., Boevé, J.L., and Pasteels, J.M. (1986) Long chain electrophilic contact poisons from the Dufour's gland of the ant *Crematogaster scutellaris* (Hymenoptera, Myrmicinae). *Canadian Journal of Chemistry* 65, 432–436.

Davidson, D.W. (1978) Experimental tests of optimal diet predictions in a social insect. *Sociobiology* 4, 35–41.

Davidson, D.W. (1997) The role of resource imbalances in the evolutionary ecology of tropical arboreal ants. *Biological Journal of the Linnean Society* 61, 153–181.

Davidson, D.W. (1998) Resource discovery versus resource domination in ants: a functional mechanism for breaking the trade-off. *Ecological Entomology* 23, 484–490.

Davidson, D.W. (2005) Ecological stoichiometry of ants in a New World rain forest. *Oecologia* 142, 221–231.

Davidson, D.W., Cook, S.C., Snelling, R.R., and Chua, T.H. (2003) Explaining the abundance of ants in lowland tropical rainforest canopies. *Science* 300, 969–972.

Davidson, D.W., Cook, S.C., and Snelling, R.R. (2004) Liquid-feeding performances of ants (Formicidae): ecological and evolutionary implications. *Oecologia* 139, 255–266.

Davidson, D.W. and Epstein, W.W. (1989) Epiphytic associations with ants. In U. Lüttge (ed.), *Vascular Plants as Epiphytes*. Springer-Verlag, New York, pp. 200–233.

Davidson, D.W., Lessard, J.P., Bernau, C.R., and Cook, S.C. (2007) The tropical ant mosaic in a primary Bornean rain forest. *Biotropica* 39, 468–475.

Davidson, D.W. and McKey, D. (1993) The evolutionary ecology of symbiotic ant–plant interactions. *Journal of Hymenoptera Research* 2, 13–83.

Davidson, D.W. and Patrell-Kim, L. (1996) Tropical arboreal ants: why so abundant? In A.C. Gibson (ed.), *Neotropical Biodiversity and Conservation*. UCLA Botanical Garden, Publ. No. 1, Los Angeles.

Dejean A. and Corbara, B. (2003) Review on mosaics of dominant ants in rainforests and plantations. In Y. Basset, V. Novotny, S.E. Miller, and R.L. Kitching (eds), *Arthropods of Tropical Forests: Spatio-temporal Dynamics and Resource Use in the Canopy*. Cambridge University Press, Cambridge, pp. 341–347.

Dejean, A., McKey, D., Gibernau, M., and Belin, M. (2000) The arboreal ant mosaic in a Cameroonian rainforest. *Sociobiology* 35, 403–423.

De Medeiros, M.A., Delabie, J.H.C., and Fowler, H.G. (1999) Predatory potential of the ant *Azteca chartifex spiriti* (Hymenoptera: Formicidae) in cocoa plantations of Bahia Brazil. *Cientif. Jaboticabal* 27, 41–46.

DeMoss, G.L. (1973) *Phylogenetic relationships among selected genera of North American Myrmicinae as evidenced by comparative morphological studies of proventriculi and Malpighian tubules*. PhD dissertation, University of Tennessee.

Dill, M., Williams, D.J., and Maschwitz, U. (2002) Herdsmen ants and their mealybug partners. *Abh. Senckenberg. Naturforsch. Ges.* 557, i–iii, 373.

Do Nascimento, R.R., Billen, J., Sant'ana, A.E.G., Morgan, E.D., and Harada, A.Y. (1998) Pygidia gland of *Azteca* nr. *bicolor* and *Azteca chartifex*: morphology and chemical identification of volatile components. *Journal of Chemical Ecology* 24, 1629–1637.

Douglas, A.E., Minto, L.B., and Wilkinson, T.L. (2001) Quantifying nutrient production by the microbial symbionts in an aphid. *Journal of Experimental Biology* 204, 349–358.

Eisner, T. (1957) A comparative morphological study of the proventriculus of ants (Hymenoptera: Formicidae). *Bulletin of the Museum of Comparative Zoology* 116, 441–490.

Eisner, T. and Brown, W.L. Jr. (1958) The evolution and social significance of the ant proventriculus. *Proceedings of the 10th International Congress of Entomology* 2, 503–508.

Eisner, T. and Wilson, E.O. (1952) The morphology of the proventriculus of a formicine ant. *Psyche* 59, 47–60.

Engel, V., Fischer, M.K., Wäckers, F.L., and Voelkl, W. (2001) Interactions between extrafloral nectaries, aphids and ants: are there competition effects between plant and homopteran sugar sources? *Oecologia* 129, 577–584.

Feldhaar, H., Straka, J., Krischke, M. *et al.* (2007) Nutritional upgrading for omnivorous carpenter ants by the endosymbiont *Blochmannia*. *BMC Biology* 5, 48.

Fewell, J.H., Harrison, J.F., and Lighton, J.R.B. (1996) Foraging energetics of the ant, *Paraponera clavata*. *Oecologia* 105, 419–427.

Fiala, B. (1990) Extrafloral nectaries vs ant–Homoptera mutualisms: a comment on Becerra and Venable. *Oikos* 59, 281–282.

Fischer, M.K. and Shingleton, A.W. (2001) Host plant and ants influence the honeydew sugar composition of aphids. *Functional Ecology* 15, 544–550.

Fischer, M.K., Völkl, W., Schopf, R., and Hoffmann, K. (2002) Age-specific patterns in honeydew production and honeydew composition in the aphid *Metopeurum fuscoviride*: implications for ant-attendance. *Journal of Insect Physiology* 48, 319–326.

Foitzik, S., Backus, V.L., Trindl, A., and Herbers, J.M. (2004) Ecology of *Leptothorax* ants: impact of food, nest sites, and social parasites. *Behavioral Ecology and Sociobiology* 55, 484–493.

Fonseca, C.R. (1999) Amazonian ant–plant interactions and the nesting space limitation hypothesis. *Journal of Tropical Ecology* 15, 807–825.

Gaume, L. McKey, D. and Terrin, S. (1998) Ant–plant–homopteran mutualism: how the third partner affects the interaction between a plant-specialist ant and its myrmecophyte host. *Proceedings of the Royal Society of London Series B* 265, 569–575.

Herbers, J.M. (1985) Nest site limitation and facultative polygyny in the ant *Leptothorax longispinosus. Behavioral Ecology and Sociobiology* 19, 115–122.

Herbers, J.M. (1989) Community structure in north temperate ants: temporal and spatial variation. *Oecologia* 81, 201–211.

Hölldobler, B. and Wilson, E.O. (1990) *The Ants.* Harvard University Press, Cambridge, MA.

Horvitz, C.C. and Schemske, D.W. (1984) Effects of ants and ant-tended herbivores on seed production of a neotropical herb. *Ecology* 65, 1369–1378.

Horvitz, C.C. and Schemske, D.W. (1994) Effects of dispersers, gaps and predators on dormancy and seedling emergence in a tropical herb. *Ecology* 75, 1949–1958.

Howard, D.F., Blum, M.S., Jones, T.H., and Tomalski, M.D. (1982) Behavioral responses to an alkylpyrazine from the mandibular gland of the ant *Wasmannia auropunctata. Insectes Sociaux* 29, 369–374.

Janzen, D.H. (1972) Protection of *Barteria* (Passifloraceae) by *Pachysima* ants (Pseudomyrmecinae) in a Nigeria rain forest. *Ecology* 53, 885–892.

Jeanne, R.L. and Davidson, D.W. (1984) Population regulation in social insects. In C.B. Huffaker and R.L. Rabb (eds), *Ecological Entomology.* Wiley-Interscience, New York, pp. 559–587.

Jones, T.H., Clark, D.A., Edwards, A.A., Davidson, D.W., Spande, T.F., and Snelling, R.S. (2004) The chemistry of exploding ants, *Camponotus* spp. (*cylindricus* complex). *Journal of Chemical Ecology* 30, 1479–1492.

Kaspari, M. and Weiser, M.D. (2000) Ant activity along moisture gradients in a neotropical forest. *Biotropica* 32, 703–711.

Kauffmann, E. (2003) *South-east Asian ant gardens: diversity, ecology, ecological significance, and evolution of mutualistic ant–epiphyte interactions.* PhD dissertation, J.W. Goethe-University, Frankfurt.

Kauffmann, S, McKey, D.B., Hossaert-McKey, M., and Horvitz, C.C. (1991) Adaptations for a two-phase seed dispersal system involving vertebrates and ants in a hemiepiphytic fig (*Ficus microcarpa*: Moraceae). *American Journal of Botany* 78, 971–977.

Kay, A. (2002) Applying optimal foraging theory to assess nutrient availability ratios for ants. *Ecology* 83, 1935–1944.

Kay, A.D., Scott, S.E., Schade, J.D., and Hobbie, S.E. 2004. Stoichiometric relations in an ant–treehopper mutualism. *Ecology Letters* 7, 1024–1028.

Khoo, K.C. and Ho, C.T. (1992) The influence of *Dolichoderus thoracicus* (Hymenoptera: Formicidae) on losses due to *Helopeltis theivora* (Heteroptera: Miridae), black pod disease, and mammalian pests in cocoa in Malaysia. *Bulletin of Entomologic Research* 82, 485–491.

Koepke, M. (1972) Über die Resistenzformen de Vogelnester in einem begrenzten Gebeit des tropischen Regenwaldes in Peru. *Journal of Ornithology* 113,138–160.

Lanza, J., Vargo, E.L., Sandeep, P., and Yu, Z.C. (1993) Preferences of the fire ants *Solenopsis invicta* and *S. geminata* (Hymenoptera: Formicidae) for amino acid and sugar components of extrafloral nectars. *Environmental Entomology* 22, 411–417.

Lebrun, E. (2005) Who is the top dog in ant communities? Resources, parasitoids, and multiple competitive hierarchies. *Oecologia* 142, 642–653.

Leston, D. (1978) A neotropical ant mosaic. *Annals of the Entomological Society of America* 71, 649–653.

Majer, J.D. (1993) Comparison of the arboreal ant mosaic in Ghana, Brazil, Papua New Guinea and Australia – its structure and influence on arthropod diversity. In J. LaSalle and I.D. Gauld (eds), *Hymenoptera and Biodiversity.* CAB International, Wallingford, pp. 115–141.

May, R.M. (1972) Will a large complex system be stable? *Nature* 238, 413–414.

McGlynn, T. P. (1999) Non-native ants are smaller than related native ants. *American Naturalist* 154, 690–699.

McKey, D. (1984) Interaction of the ant-plant *Leonardoxa africana* (Caesalpiniaceae) with its obligate inhabitants in a rainforest in Cameroon. *Biotropica* 16, 81–99.

Messina, F.J. (1981) Plant protection as a consequence of an ant–membracid mutualism: interactions on goldenrod (*Solidago* sp.) *Ecology* 62, 1433–1440.

Nonacs, P. and Dill, L.M. (1991) Mortality risk versus food quality trade-offs in ants' patch use over time. *Ecological Entomology* 16, 73–80.

Offenberg, J., Havanon, S., Aksornkoae, S, Macintosh, D.J., and Nielsen, M.G. (2004) Observations on the ecology of weaver ants (*Oecophylla smaragdina* Fabricius) in a Thai mangrove ecosystem and their effect on herbivory of *Rhizophora mucronata* Lam. *Biotropica* 36, 345–351.

Oliveira, P.S. and Del-Claro, K. (2005) Multitrophic interactions in a neotropical savanna: ant–hemipteran systems, associated insect herbivores, and host plants. In D.F.R.P. Burslem, M.A. Pinard, and S.E. Hartley (eds) *Biotic Interactions in the Tropics.* Cambridge University Press, Cambridge, pp. 415–456.

Orr, M.R., Dahlsten, E.L., and Benson, W.W. (2003) Ecological interactions among ants in the genus *Linepithema*, their phorid parasitoids, and ant competitors. *Ecological Entomology* 28, 203–210.

Palmer, T.M. (2004) Wars of attrition: colony size determine competitive outcomes in a guild of African acacia-ants. *Animal Behaviour* 68, 993–1004.

Queiroz, J.M. and Oliveira, P.S. (2001) Tending ants protect honeydew-producing white flies (Homoptera: Aleyrodidae). *Environmental Entomology* 30, 295–297.

Raine, N.E., Willmer, P., and Stone, G.N. (2002) Spatial structure and floral avoidance behavior prevent ant-pollinator conflict in a Mexican ant-acacia. *Ecology* 83, 3086–3096.

Raubenheimer, D. and Simpson, S.J. (1998) Nutrient transfer functions: the site of integration between feeding behaviour and nutritional physiology. *Chemoecology* 8, 61–68.

Raubenheimer, D. and Simpson, S.J. (2004) Organismal stoichiometry: quantifying non-independence among food components. *Ecology* 85, 1203–1216.

Roche, C.F.D. and Bergallo, H.G. (1992) Bigger ant colonies reduce herbivory and herbivore residence time on leaves of an ant-plant: *Azteca muelleri* vs. *Coelomera ruficornis* on *Cecropia pachystrachya*. *Oecologia* 91: 249–252.

Roche, R.K and Wheeler, D.E. (1997) Morphological specializations of the digestive tract of *Zacryptocerus rohweri* (Hymenoptera: Formicidae). *Journal of Morphology* 234, 253–262.

Schupp, E.W. and Feener, D.H. (1991) Phylogeny, life-form and habitat dependence of ant-defended plants in a Panamanian forest. In C.R. Huxley and D.F. Cutler (eds), *Ant–Plant Interactions*. Oxford University Press, Oxford, pp. 175–197.

Shattuck, S.O. (1992) Higher classification of the ant subfamilies Aneuretinae, Dolichoderinae and Formicinae (Hymenoptera: Formicidae). *Systematic Entomology* 17, 199–206.

Smith, L.L., Lanza, J., and Smith, G.C. (1990) Amino acid concentrations in extrafloral nectar of *Impatiens sultani* increase after simulated herbivory. *Ecology* 71, 107–115.

Stephenson, A.G. (1982) The role of the extrafloral nectaries of *Catalpa speciosa* in limiting herbivory and increasing fruit production. *Ecology* 63, 663–669.

Sterner, R.W. and Elser, J.J. (2002) *Ecological Stoichiometry: The Biology of Elements from Molecules to the Biosphere*. Princeton University Press, Princeton, NJ.

Stork, N.E. (1988) Insect diversity: facts, fiction and speculation. *Biological Journal of the Linnean Society* 35, 321–337.

Swain, R.B. (1980) Trophic competition among parabiotic ants. *Insectes Sociaux* 27, 377–390.

Terborgh, J.T. (1986) Keystone plant resources in the tropical forest. In M.E. Soulé (ed.), *Conservation Biology*. Sinauer, Sunderland, MA, pp. 330–344.

Tillberg, C.V. and Breed, M.D. (2004) Placing an omnivore in a complex food web. *Biotropica* 36, 266–272.

Tjallingii, W.F. (1995) Regulation of phloem sap feeding by aphids. In R.F. Chapman and G. de Boer (eds), *Regulatory Mechanisms in Insect Feeding*. Chapman and Hall, New York, pp. 120–209.

Tobin, J.E. (1991) A neotropical rainforest ant community: some ecological considerations. In C.R. Huxley and D.F. Cutler (eds), *Ant–Plant Interactions*. Oxford Press, Oxford, pp. 536–538.

Tobin, J.E. (1994) Ants as primary consumers: diet and abundance in the Formicidae. In J.H. Hunt and C.A. Nalepa (eds), *Nourishment and Evolution in Insect Societies*. Westview, Boulder, CO, pp. 279–307.

Tobin, J.E. (1995) Ecology and diversity of tropical forest canopy ants. In M.D. Lowman and N.M. Nadkarni (eds) *Forest Canopies*. Academic Press, New York, pp. 129–147.

Wheeler, G.C. and Wheeler, J. (1976) Ant larvae: review and synthesis. *Memoirs of the Entomological Society of Washington* 7. The Entomological Society of Washington, DC.

Wheeler, W.M. and Bailey, I.W. (1920) The feeding habits of pseudomyrmecine and other ants. *Transactions of the American Philosophical Society* 22, 253–279.

Willmer, P.G. and Stone, G.N. (1997) How aggressive ant-guards assist seed-set in Acacia flowers. *Nature* 388, 165–167.

Wilson, E.O. (1959) Some ecological characteristics of ants in New Guinea rainforests. *Ecology* 40, 437–446.

Yanoviak, S.P. and Kaspari, M. (2000) Community structure and the habitat templet: ants in the tropical forest canopy and litter. *Oikos* 89, 259–266.

Yao, I. and Akimoto, S.I. (2002) Flexibility in the composition and concentration of amino acids in honeydew of the drepanosiphid aphid *Tuberculatus quercicola*. *Ecological Entomology* 27, 745–752.

Young, B.E., Kaspari, M., and Martin, T.E. (1990) Species-specific nest site selection by birds in ant-acacia trees. *Biotropica* 22, 310–315.

Zientz, E., Feldhaar, H., Stoll, S., and Gross, R. (2005) Insights into the microbial world associated with ants. *Archives of Microbiology* 184, 199–206.

Chapter 21

SOIL FERTILITY AND ARBOREAL MAMMAL BIOMASS IN TROPICAL FORESTS

Carlos A. Peres

OVERVIEW

Tropical forests have been characterized across a wide range of primary plant productivity, which is partly dependent on soil properties. Yet the relationships between soil fertility and plant productivity and herbivore biomass remain poorly understood in tropical forests. Here I review the evidence regarding the relationship between soil fertility and mammal assemblage biomass from the perspective of primates spanning a wide dietary spectrum. I also present new data based on a comprehensive compilation of available community-wide estimates of neotropical primate biomass density in Amazonian and Guianan forests. A composite index of soil fertility, based on both chemical and physical properties of soils, explained 37% of the variation in total diurnal primate biomass in a set of 60 undisturbed Amazonian forest sites that had not been affected by hunting pressure and anthropogenic habitat disturbance. I discuss the mechanisms by which tropical forest soil nutrient availability may constrain bottom-up trophic cascades from green plant producers to primary and secondary consumers. These include edaphic effects on the quality and amount of resources produced by individual food plants, as well as wholesale changes in floristic composition. Measures of soil fertility and other environmental gradients affecting forest productivity can serve as an efficient framework for predicting the diversity and population sizes of vertebrate species that can be protected by potential reserve polygons.

INTRODUCTION

Early perceptions that evergreen tropical forests must be sustained by fertile soils (Wallace 1853) – which is reinforced by the paradoxical high phytomass supported by highly efficient nutrient capture and cycling in nutrient-poor soils – have been unequivocally demystified by a vast body of evidence on the limited agropastoral potential of the humid tropics (Goodland and Irwin 1975, Irion 1978, Sioli 1980). Yet the physical and chemical properties of unfertilized tropical forest soils are remarkably variable (Projeto RADAMBRASIL 1972–1978, Sanchez 1981, Uehara and Gilman 1981, Cochrane and Sanchez 1982, Vitousek and Sanford 1986,

Jordan 1989, Furley 1990, Richter and Babbar 1991). The local to regional scale heterogeneity in soil age, texture, drainage, depth, parent materials, pH, and macro- and micronutrient content therefore presents a highly diverse set of consequences to forest primary productivity.

Most soils in the humid tropics are highly weathered and relatively nutrient poor (Irion 1978, Uehara and Gilman 1981). Tropical forest plants are often able to prevent nutrient loss through leaching and herbivory using a number of above- and below-ground strategies. However, retention of scarce nutrients may be less imperative in fertile soils that can replace nutrients lost through leaching and herbivory. Given the fundamental laws of thermodynamics,

bottom-up constraints on green plant producers often reverberate via successive nodes in a
food chain onto the size and dynamics of herbivore and carnivore populations, and through
the structure of whole forest ecosystems. Yet the
role of soil fertility as a factor regulating vertebrate populations remains poorly investigated. In
particular, the diverse relationships between soil
nutrient limitation and forest composition, forest
phytochemistry, and ultimately the amount and
quality of digestible resources available to invertebrate and vertebrate consumers have been poorly
explored in tropical forests (but see Janzen 1974,
McKey *et al.* 1978, Coley *et al.* 1985, Chapin
et al. 1986, Vitousek and Sanford 1986, Oates
et al. 1990, Coley and Aide 1991). If rainfall
and light are not limiting, food quality for herbivores will depend on the rate of nutrient uptake
by food plants, which is ultimately a function of
soil fertility and the underlying geological parent
material.

Soil texture and nutrient status have major
effects on the distribution and abundance of plant
communities in terms of both understory shrubs
and herbs (Tuomisto and Poulsen 1996, Tuomisto
et al. 2003, Costa *et al.* 2005) and trees (Huston
1980, Gentry 1988, Clark *et al.* 1999, Givnish
1999, Potts *et al.* 2002, Phillips *et al.* 2003).
The composition and density of food plants, the
productivity and growth rate of preferred food
items, and foliage levels of defensive secondary
metabolites may therefore be more favorable to
herbivores in higher fertility soils (Coley *et al.*
1985, Chapin *et al.* 1986, Vitousek and Sanford
1986, Waterman and Mole 1989). In more fertile
soils, this may lead to higher folivore and frugivore densities as documented in tropical forests
of South America (Emmons 1984, Peres 2000a),
central Africa (Barnes and Lahm 1998), and
northern Australia (Kanowski *et al.* 2001), as well
as in tropical savannas (Bell 1982, du Preez *et al.*
1983, Runyoro *et al.* 1995, Augustine *et al.* 2003)
and temperate regions (Jones and Hanson 1985,
Pastor *et al.* 1993, Recher *et al.* 1996). In Amazonian forests, studies on the relationship between
soil chemistry and wildlife abundance strongly
suggest that population densities of large vertebrates can be depressed under conditions of low
fertility (Emmons 1984, Peres 1997a,b, 1999a,b,

2000a,b, Peres and Dolman 2000, Haugaasen
and Peres 2005a,b, Palacios and Peres 2005).

In general, species richness should increase
with the size of the resource base because higher
population densities result from greater energy
availability, thereby enabling more species to
attain viable population sizes within a given area
(Wright 1983, Rosenzweig and Abramsky 1993).
This species–energy relationship predicts that the
species richness of an area will be positively
correlated with the aggregate population density of all taxa (but see Srivastava *et al.* 1998,
Mittelbach *et al.* 2001), although this correlation
can also emerge from other mechanisms (Evans
et al. 2005). If soil nutrient availability limits
primary consumer population sizes, and therefore species richness, one would expect that, at
regional scales, not only will there be a positive relationship between soil fertility and species
richness, but these species should have greater
biomass densities in high productivity areas.

In this chapter, I review the relationship
between soil fertility and mammal biomass in
tropical forests. I explore this relationship from
the perspective of arboreal mammals based on
community-wide estimates of platyrrhine primate biomass in Amazonian and Guianan forests.
Finally, I discuss the mechanisms by which tropical forest soil nutrient availability may constrain
bottom-up trophic cascades from green plant
producers to primary and secondary consumers.

SOIL INFERTILITY IN TROPICAL FORESTS

The structure of any community may be largely
determined by its primary productivity (Fretwell
1977, Hunter and Price 1992, Power 1992,
Rosenzweig 1995), although this claim remains
contentious against much empirical evidence
(e.g., Crête 1999, Howe and Brown 1999, Seagle
and Liang 2002). Primary productivity in moist
tropical forests is often constrained by a limited
supply of nutrients and trace elements in low
pH soils, which tend to be poorer than those in
the temperate zone due to a long and repeated
history of intensive leaching and weathering
(Hacker 1982, Jordan 1989). A low soil pH may

result in increased toxicity caused by H^+, Al, and Mn and a reduced uptake of most nutrients (Marshner 1991).

If tropical lowland forests are nutrient limited, one might predict that they would respond to nutrient enrichment treatments which could result in higher vegetative or reproductive productivity, litterfall, and soil organic matter accumulation. Although responses to fertilization experiments have been variable, most tropical forest plots fertilized on a meaningful scale show increased above-ground net primary production, radiation conversion efficiency, leaf area index, and nutrient content of leaf litter (Harrington et al. 2001). Other studies in primary forests have shown that tree girths, litterfall mass, and litter nutrient content increase in fertilized plots (e.g., Tanner et al. 1998, Mirmanto et al. 1999). In regenerating secondary forests, tree biomass can increase significantly following N-only and N + P treatments (Davidson et al. 2004), indicating that secondary productivity and recovery of above-ground biomass is often constrained by soil fertility and texture across regions and soil types within a region (Chazdon 2003). Crucially, quantitative allocation to reproductive plant parts – that are important to consumers of flowers, nectar, and whole unripe/mature fruits or seeds – also appears to increase. In a Sumatran forest, production of leaf litter and fruit increased along a natural gradient of increasing soil fertility (van Schaik and Mirmanto 1985). Nutrient enrichment enhances allocation to reproduction in other tropical ecosystems like dwarf mangrove stands in Panama (Lovelock et al. 2004). But as far as I am aware, the only experimental fertilization study in a tropical forest where litterfall was fractioned into both leaf litter and reproductive components (at Barito Ulu, central Kalimantan, Borneo) shows a significant increase in reproductive parts (flowers and fruits) in most tree species within fertilized plots (Mirmanto et al. 1999, J. Proctor, personal communication). Further studies are however necessary to confirm whether nutrient enrichment augments plant reproductive productivity at the community level in a range of soil types.

Only 7% of the soils under forest or agriculture in the Amazon basin show no sign of fertility limitation (Cochrane and Sanchez 1982), and most of Brazilian Amazonia consists of nutrient-poor, acidic soils that are often associated with aluminum toxicity (Nicholaides et al. 1984). Agricultural production is severely constrained by nitrogen, phosphorus, potassium, and calcium deficiency in 62–90% of the Amazon. In much of the basin, soil nutrients exported through leaching and runoff are replaced not so much from weathering of the parent material but from long-range export of approximately 40 million tons $year^{-1}$ of atmospheric dust particles and dissolved material carried by wind and rainfall from the Sahara alone (Swap et al. 1992, Koren et al. 2006). Exogenous nutrient inputs from rainfall at a remote site in the state of Amazonas are in the order of 0.34, 0.9, 0.3, and 1.32 $kg\,ha^{-1}\,year^{-1}$ for P, K, Mg, and Ca, respectively (Williams and Fisher 1997). In fact, water flowing out of crystal-clear forest streams in many upland parts of the basin is more distilled and may contain only half of the elemental concentrations (e.g., P, Ca, and Mg) of rainwater, obviously attesting to the net efficiency with which nutrients are retained by the vegetation (Irion 1978, Furch 1984, Bruijnzeel 1991).

However, all regional-scale soil maps available for the Amazon show a diversity of pedological processes and a highly variable macromosaic of soil types of varying fertility (Sombroek 1966, EMBRAPA 2002). While vast upland tracts of lowland Amazonia consist of highly weathered soils of ancient pre-Cambrian origin, young soils along white-water rivers are mainly Quaternary (Pleistocene and Holocene) deposits that are renewed annually by a prolonged flood pulse (Junk 1997). Most of central Amazonia both north and south of the Amazon River consists of the so-called Barreira formation characterized by extremely low clay fractions of key inorganic nutrients which were radically impoverished during the formation of the kaolinitic topsoil. Total amounts of sodium, calcium, magnesium, and potassium are often in the range of 100–300 ppm and the cation exchange capacity barely exceeds 5 mval 100 g^{-1} (Irion 1978). By contrast, the Cretaceous to Tertiary fine-grained sediments that formed much of the soils of southwestern Amazonia are often relatively fertile.

BASIN-WIDE PATTERNS OF PRIMATE BIOMASS

Primates are ideal candidates for a regional-scale test of bottom-up effects of soil fertility because (1) they represent one of the most important biomass components of arboreal vertebrate assemblages; (2) they consume a significant but unknown proportion of the primary vegetative and reproductive productivity of neotropical forests (Eisenberg 1980, Terborgh 1983, Peres 1999a, 2000a, Haugaasen and Peres 2005b); (3) they are strictly arboreal and therefore have priority of access to food items produced in the forest understory and canopy before they fall to the ground and become available to terrestrial vertebrates; (4) they often form highly conspicuous and observable groups, and are amenable to highly standardized population surveys that can be replicated in any tropical forest (Peres 1999c); and (5) they consequently have attracted a disproportionately large amount of interest from field ecologists and behavioral biologists,

resulting in a strong cadre of primatologists in most habitat-countries.

I compiled data on the population density and biomass for all diurnal primate species from 96 spatially independent undisturbed forest sites in lowland Amazonia and the Guianan shield (Figure 21.1). This excludes only night monkeys (*Aotus* spp.), which are rarely censused by primatologists. Over half of these sites ($N = 52$) resulted from our own long-term series (1987–2004) of standardized line-transect censuses of mid-sized to large-bodied vertebrate assemblages throughout lowland Amazonia (Peres 1997a, 2000a,b, Peres and Dolman 2000, Haugaasen and Peres 2005b, Palacios and Peres 2005, Peres and Nascimento 2006, Peres and Palacios 2007). Data compilation for all other sites was updated from Peres (1999b) and included an exhaustive survey of published and unpublished reports of primate population densities obtained through line-transect census techniques. However, I excluded from the final dataset any study that either failed to report densities for one or more diurnal

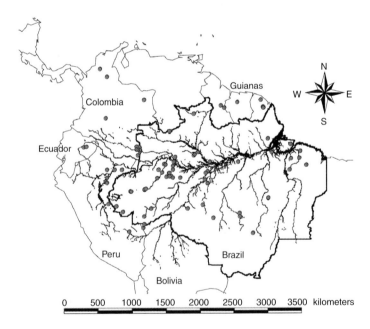

Figure 21.1 Location of 96 forest sites within eight South American countries where community-wide primate surveys considered in this analysis were conducted. The boundary polygon of Brazilian Amazonia is indicated by a thick line.

primate species occurring at a given site, or was considered to be based on an insufficient census effort (<100 km of census walks). All survey sites were part of continuous tracts of primary forest that may have been selectively hunted to a varying extent but otherwise had not been subjected to structural habitat disturbance due to selective logging, slash-and-burn agriculture, surface wildfires, and forest fragmentation. In most of the analyses, I excluded sites that had been hunted to a moderate or persistent extent (see Peres and Palacios 2007) because subsistence game hunting profoundly affects the size structure and aggregate biomass of Amazonian primate assemblages (Peres 1990, 1999b, Peres and Dolman 2000). Non-hunted and lightly hunted sites, on the other hand, showed no significant differences in total biomass and size distribution of the primate assemblage, and were therefore pooled together. Conversions of population density to biomass estimates relied on mean body mass values for adult males and females available from the literature, which were then corrected using a factor of 0.8 to account for juveniles in the population (see Peres 2000a). Descriptive details on all but the most recently surveyed study areas (2001–2004), forest site classification according to levels of hunting pressure, and procedures used during line-transect censuses and data analysis can be obtained elsewhere (Peres 1999b, 2000a, Peres and Palacios 2007 and references therein) or from the author.

SOIL FERTILITY

I used a classification of the agricultural potential of the Amazon basin based on key physical and chemical indicators of soil fertility. These data are based on a 1:3,000,000-scale digital soil map of Brazilian Amazonia that was produced in the 1970s by the Soils Division of the Brazilian Institute for Agricultural Research (EMBRAPA 2002). This is regarded as the best available soil map for the Amazon (Laurance *et al.* 2002), containing 17 major soil types that are further subdivided into over 100 subtypes, using the Brazilian soil taxonomy (cf. Beinroth 1975). The different soil subtypes were classified using published sources (especially Sombroek 1984, 2000).

The index of soil fertility ranged from 1 (poorest soils) to 5 (best soils) with nine class-intervals of 0.5. Although soil chemistry was considered, a slightly greater weight was given to physical properties of the soil (e.g., soil depth, texture, stoniness, waterlogging) that cannot be easily enhanced by agricultural inputs. Soil fertility classes 4.0–5.0 have the highest agricultural potential. These include nutrient-rich alluvial soils in várzea forests (seasonally inundated by white-water rivers of Andean origin), terra roxa soils (nutrient-rich, well-structured upland soils formed on base-rich rock), eutrophic Cambisols (young, relatively unweathered soils with high activity clay and high nutrient status), and Vertisols (clay soils with high activity clay minerals and high nutrient content). These soil types collectively encompass only 1.8% of the Brazilian Amazon (EMBRAPA 2002). Soil classes 2.5–3.5, comprising 53.4% of the Brazilian Amazon, have some agricultural potential but also important limitations, such as high acidity, low nutrient availability, shallowness, waterlogging, and concretionary status. Soil classes 1.5–2.0 are suitable mainly for cattle pasture or undemanding tree crops, and encompass 34.8% of the Brazilian Amazon. They include the intensively weathered Xanthic Ferralsols of central Amazonia, very stony and shallow soils, nutrient-poor waterlogged soils, and Plinthosols (soils that become hardened laterite when exposed to wetting and drying cycles). Soil class 1 encompasses 7.8% of Brazilian Amazonia, has no potential for agriculture, and largely consists of very sandy soils, including podzols and quartz sands, some of which are waterlogged. Because 31 of the 96 sites were located outside Brazilian Amazonia, I assigned fertility classes based on available soil-type information and applied the same criteria used in the EMBRAPA soil map.

EDAPHIC DETERMINANTS OF PRIMATE BIOMASS

The assemblage biomass of all sympatric primate species, estimated for 60 non-hunted to lightly hunted forest sites and 36 moderately to heavily hunted sites, was highly variable. In non-hunted to lightly hunted sites, it ranged from as low as 20.1 $kg\,km^{-2}$ to as high as 953.1 $kg\,km^{-2}$ (mean $\pm$ SD $=$ 248.6 $\pm$ 156.7). The lowest biomass estimates were very similar across major levels of hunting pressure (20.1 versus 26.3 $kg\,km^{-2}$) but the highest estimate in moderately to heavily hunted sites was 626.8 $kg\,km^{-2}$ (mean $\pm$ SD $=$ 148.1 $\pm$ 127.7), and there was a significant difference in primate biomass between hunted and non-hunted sites ($t = 3.25$, d.f. $= 94$, $P_{adj} = 0.002$). By contrast, there was no significant difference in the total primate density between non-hunted to lightly hunted sites (mean $\pm$ SD $=$ 106.4 $\pm$ 66.9; range $=$ 9.9 $-$ 355.2 ind. km^{-2}) and moderately to heavily hunted sites (mean $\pm$ SD $=$ 106.6 $\pm$ 75.1; range $= 13.1 - 357.8$ ind. km^{-2}; $t = -0.014$, d.f. $= 94$, $P_{adj} = 0.989$), partly because of density compensation in hunted sites by small-bodied species (Peres and Dolman 2000).

Considering only non-hunted to lightly hunted sites, primate biomass tended to increase away from the equator towards the Guianan and Guaporé shield (north and south of the Amazon, respectively), but especially towards seasonally dry parts of southwestern Amazonia (Figure 21.2). Soil fertility alone explained one third of the variation in the log-transformed estimates of total primate biomass ($R^2 = 0.368$, $F_{1,58} = 31.4, P < 0.001, N = 60$; Figure 21.3a). This is roughly equivalent to a mean primate biomass increment of 47 $kg\,km^{-2}$ across consecutive classes of soil fertility, or a nearly six-fold increase in biomass from the least to the most fertile soils. Soil fertility also had an appreciable effect on the overall primate density in non-hunted sites ($R^2 = 0.342$, $F_{1,58} = 30.19$, $P < 0.001$, $N = 60$), but not on the mean individual body mass of all co-occurring species ($R^2 = 0.002$, $P = 1.0$), which ranged from 1436 to 4874 g (mean $\pm$ SD $= 2469 \pm 781$g, $N = 60$). There was no significant interaction between levels of hunting pressure and soil fertility, and combining these two variables further improved a minimum regression model explaining nearly half the total variation in primate biomass across all sites ($R^2 = 0.452$, $F_{2,93} = 38.4, P < 0.001, N = 96$).

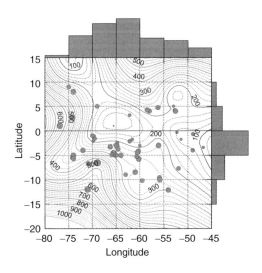

Figure 21.2 Geographic patterns of primate biomass in Amazonian and Guianan forests. Sizes of shaded circles are scaled according to the log_{10} total diurnal primate biomass estimates whereas contour lines indicate interpolations of untransformed biomass values. Border histograms indicate the total number of non-hunted to lightly hunted sites within 5-degree bands for which data are available. Solid line represents the equator.

Once the effects of hunting pressure and soil fertility were taken into account (in an analysis of covariance), the total primate biomass was still affected by the local primate species richness, which ranged from 2 to 13 species (mean $\pm$ SD $= 7.87 \pm 2.67$ species, $N = 96$). I therefore examined the effects of different environmental variables on the mean primate biomass per species co-occurring at any given site. Soil fertility again had a significantly positive effect on the primate biomass per species richness, explaining 43% of the variation in this ratio considering only the 60 non-hunted and lightly hunted sites (Figure 21.3b).

None of the other environmental variables associated with each forest site, including total rainfall and strength of the dry season, had a significant effect on primate biomass. Rainfall gradients are often closely correlated with levels of soil fertility because cumulative leaching and runoff of soil and plant nutrients are more likely if ancient

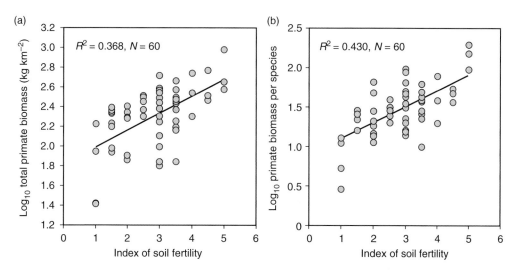

Figure 21.3 Relationships between a composite index of soil fertility and (a) the aggregate biomass density (kg km^{-2}) and (b) the mean biomass density per co-occurring species of all diurnal primate species in non-hunted to lightly hunted forest sites of Amazonia and the Guianan shield. R^2 values of each relationship are indicated in each plot.

soils have been subjected to a long and repeated history of heavy rainfall. For the 96 forest sites considered here, there was a significantly negative correlation between total annual rainfall and soil fertility ($r = -0.434$, $P_{adj} < 0.001$), so this relationship appears to hold at a pan-Amazonian scale despite the wide mesoscale variation in soil fertility within regions sharing the same rainfall regime. However, rainfall alone was a very weak correlate of total primate biomass ($r = -0.079$, $P_{adj} = 1.0$) or the primate biomass per species ratio ($r = -0.118$, $P_{adj} = 0.906$).

All sites surveyed had a full complement of terrestrial and aerial predator species, which are known to affect herbivore abundance (Hairston *et al.* 1960, Dyer and Letourneau 1999, Halaj and Wise 2001). However, primate biomass was considerably higher in nutrient-rich forests despite the concomitant higher abundance of predators that habitually or occasionally take primates, such as harpy eagles, ornate hawk eagles, and a range of scansorial mammalian carnivores (Peres unpublished data). Arboreality does not necessarily confer immunity to predation, and none of the primate species surveyed grow to a complete

size refuge. Indeed, predation pressure shapes a range of behavioral adaptations and the ecology of group living in Amazonian primates (Terborgh and Janson 1986, Peres 1993, Isbell 1994). However, natural predators do not appear to play a major role in limiting primate population density along the productivity gradient experienced by undisturbed Amazonian forests. Moreover, human predation (hunting) of medium- to large-bodied primate species tends to be heavier in nutrient-rich forests, but is unlikely to significantly mitigate the impact of natural predation through predator control. These lines of evidence suggest that regulation of primate abundance, at least in vast tracts of continuous Amazonian forest, is primarily a bottom-up rather than a top-down process.

SOIL NUTRIENT LIMITATION AND HABITAT PRODUCTIVITY

The importance of soil nutrient limitation on the cost-effectiveness of a plant's anti-herbivore arsenal has become fairly well established in plant physiological ecology (e.g., Coley *et al.* 1985, Chapin *et al.* 1986, Fine *et al.* 2004). But essentially two arguments can be distinguished in Janzen's (1974) proposal that soil nutrient status should affect the secondary metabolism of plants adapted to impoverished soils. First, plant tissues lost to herbivores cannot be inexpensively replaced in such habitats, making a heavy investment in defensive chemistry cost effective. Second, the spatial distribution of plant defenses is partly governed by phylogenetic inertia in that plant families colonizing and forming low diversity stands in nutrient-poor soils are predisposed to produce high levels of secondary metabolites. Adequate tests of these hypotheses in terms of the phytochemistry of entire plant communities in heterogeneous soil mosaics would require a comprehensive analysis of biochemical profiles and to a large extent this has not been done. However, there appears to be a fairly tight relationship between soil nutrient availability, plant chemistry and digestibility, and the abundance of mammalian herbivores in undisturbed tropical forests (McKey *et al.* 1978, Waterman and McKey 1988, Waterman *et al.* 1988, Chapman and Chapman 1999, Chapman *et al.* 2002, but see Oates *et al.* 1990) and tropical savannas (Bell 1982). Nutrient-deficient environments often lead to an emphasis on secondary metabolites derived from "carbon-overflow" pathways, whereas nutrient-rich environments are often characterized by a greater production of nitrogen-based metabolites or enhanced growth rates (Waterman and Mole 1989).

Levels of plant reproductive investment (e.g., production of flowers, fruits, and seeds) relative to somatic investment (e.g., energy storage, survival, and morphological or chemical defense) is also likely to be determined by the uptake of macronutrients and trace elements. Higher per capita investments into large crop sizes of reproductive parts (flowers, fruits, or seeds), of higher densities of large-crowned trees that can afford to produce large fruit crops would favor nectivores, frugivores, and seed predators. In Madre de Dios, Peru, for example, the annual yield of fresh edible fruits in nutrient-rich alluvial soils ($592 \ kg \, ha^{-1} \, year^{-1}$), that are replenished by a flood pulse from the Tambopata River once

every decade, is nearly twice that of terra firme forest on clay soils, and six-fold greater than that of terra firme forest on nutrient-poor sandy soils (Phillips 1993). Likewise, production rates of young leaves by both saplings and mature trees are also likely to be affected by soil fertility (Mirmanto *et al.* 1999, Harrington *et al.* 2001). Despite the apparent hyperabundance of green foliage in evergreen tropical forests, soil fertility can affect the resource base available to strict and facultative folivores, which are often highly selective and limited by the amount and quality of palatable leaves (Ganzhorn 1980, Peres 1997b, Gupta and Chivers 1999). Moreover, the trade-off between leaf growth rate and anti-herbivore defenses (Coley 1988, Kitajima 1994) enforces edaphic specialization among tropical trees (Fine *et al.* 2004), further increasing community-wide differences in foliage quality between nutrient-rich and nutrient-poor soils.

MAMMAL BIOMASS AND SOIL FERTILITY IN TROPICAL FORESTS

The positive effects of soil fertility on the total output and quality of plant food items may seem obvious, especially if light and moisture are not limiting. However, few tropical forest studies have demonstrated this relationship despite over 30 years of mammal surveys following Janzen's (1974) seminal discussion on this topic. The multitrophic consequences of soil nutrient limitation to vertebrate communities are even less well understood, and no tropical forest study has been able to demonstrate a direct link between soil fertility and mammal biomass at regional scales. I have shown that key indicators of soil fertility, including soil chemistry and texture, can affect the biomass of primate assemblages in Amazonian forests, and that this relationship is significant even when differences in species richness are taken into account.

The relatively strong bottom-up effect of soil fertility on primate biomass is quite remarkable given the variation in other environmental variables that can also affect primate abundance, including forest type, forest structure, floristic composition, total fruit production, and the density and patch size of keystone plant resources (Peres 1997a, 1999b, 2000c, Chapman and Chapman 1999, Stevenson 2001). Some of these variables may be partly nested within the effects of large-scale edaphic gradients considered here but a more robust multivariate model could explain a larger proportion of the variation in primate abundance over such a vast region.

Primary forest productivity in terms of total litterfall is a strong predictor of primate species richness in neotropical forests (Kay *et al.* 1997), both increasing with rainfall up to approximately 2500 mm year^{-1}. This relationship can now be extended to total primate biomass in that soil fertility is likely to be correlated with the total above-ground turnover of the forest biomass. In general, both primate species richness (Peres and Janson 1999) and forest biomass turnover (Malhi *et al.* 2004) increase from eastern to western Amazonian forests, and this geographic pattern also holds for total primate biomass and primate biomass per species. Above-ground coarse wood productivity in Amazonian forests increases with soil fertility, particularly towards the eastern flanks of the Andes (Malhi *et al.* 2004), which mirrors the geographic pattern of primate biomass. I predict that the relationship between forest productivity (e.g., as indexed by litterfall) and the biomass of a group of primary consumers such as primates would be even tighter if flower and fruit production were considered separately, but few studies distinguish the vegetative and reproductive fractions of litterfall. I also predict that mammal biomass in tropical forests is a strong positive correlate of the above- to below-ground ratio in forest phytomass, and that this relationship will hold at most meso- to large scales, depending on the extent to which wide-ranging mammals integrate local edaphic constraints at the landscape scale.

Fittkau (1973, 1974) was one of the first to show a severe deficiency of some nutrients essential to plant growth in central Amazonia, particularly calcium, phosphorus, nitrogen, potassium, and a number of trace elements. He attributed the paucity of snails and mussels to the notorious calcium deficiency of Amazonian forest streams. Both terrestrial and aquatic food webs are affected by regional differences in geochemistry and soil fertility. The net productivity of Amazonian nutrient-rich white-water lakes can be 15- to 19-fold greater than that of nutrient-poor black-water lakes, where fish can show signs of severe nutrient deficiency in their vertebrae (Geisler and Schneider 1976, Smith 1979). For example, the fish production of sediment-rich rivers of Andean origin such as the Madeira or the Purús (52 kg ha^{-1} year^{-1}) is much greater than that of rivers draining primarily nutrient-poor podzols and spodosols such as the Negro–Casiquiare–Guainia (6.6–13.2 kg ha^{-1} year^{-1}; Goulding 1979, Clark and Uhl 1987, Goulding *et al.* 1988).

Previous studies had already shown large differences in total primate biomass between eutrophic soils in seasonally inundated Amazonian várzea forests and mesotrophic or oligotrophic soils in upland terra firme forests (Peres 1997a,b, 1999b, Peres and Dolman 2000). These patterns are consistent with those found for Amazonian assemblages of small canopy mammals (Malcolm *et al.* 2005), large terrestrial and arboreal mammals (Emmons 1984, Haugaasen and Peres 2005), and mid-sized to large-bodied vertebrates (Peres 2000a,b). Primate communities in seasonally flooded and terra firme forests consistently show a reverse abundance–diversity relationship with high biomass, species-poor assemblages occurring in the most nutrient-rich soils (Peres 1997a). Variation in primate biomass throughout the western Amazon can also be explained by regional differences in soil types and geochemistry. Primate densities in southeastern Colombia, eastern and southern Peru, western Brazilian Amazonia, and northern Bolivia are consistently higher in white-water than black-water drainages (Freese *et al.* 1982, Peres 1997a, Palacios and Peres 2005), despite sediment overflow from white-water rivers to adjacent black-water drainages in exceptionally high inundation years. For example, primate biomass along the black-water Rio Nanay, upriver of Iquitos, Peru, tends to be particularly low (Freese *et al.* 1982), reflecting the nutrient-poor status of the soils in this region (Kauffman *et al.* 1998). On the basis of 300 km of census effort conducted at seven Amazonian forest sites of varying productivity, Emmons (1984) suggested that mammal abundance in Amazonian forests is largely a function of soil fertility. She noted that densities of non-volant mammals gradually decreased from forests on fertile alluvial or volcanic soils in western Amazonia, through those on upland latosols, to very nutrient-poor white sands of the Guianan shield. Abrupt declines in population densities of howler monkeys (Peres 1997b) and other arboreal folivores (Peres 1999a) can be observed throughout Amazonian forests and across a gradient of soil fertility from annually flooded várzea forests, to supra-annually flooded floodplain forests, to Paleovárzea forests, to mesotrophic terra firme forests, and finally oligotrophic terra firme forests. The present analysis confirms the positive effect of soil fertility on mammal biomass on a much larger scale. Salovaara (2005) also showed that primate and ungulate biomass in a non-flooded forest landscape of eastern Peruvian Amazonia was considerably higher in more fertile soils. In this study, major soil formations were classified using estimates of soil cation content (Ca^+ K^+ Mg^+ Na) based on the composition of pteridophyte species (ferns and allies) with known optimal cation requirements (Salovaara et al. 2004).

The relationship between large vertebrate population abundance and soil fertility can be generalized to other continental mammal faunas. Once the effects of altitude were taken into account, the combined abundance of folivorous marsupials in Australian rainforests was significantly higher in sites on nutrient-rich basalts than in those on nutrient-poor acid igneous or metamorphic rocks (Kanowski et al. 2001). Barry (1984) showed that infertile podzol soils in rainforest sites of southeast Queensland supported significantly fewer small mammals than fertile krasnozem soils. There is also conclusive evidence that the richest and most abundant Australian vertebrate (or mammal) faunas occur in sites with the greatest degree of soil fertility (Barry 1984, Recher et al. 1996, Woinarski et al. 1999, Claridge and Barry 2000), although these studies may be confounded by the effects of rainfall.

Although the high species richness of the large mammal fauna of East Africa may be largely due to the sheer size of its savanna biome (Cristoffer and Peres 2003), the exceptionally high native ungulate biomass (e.g., Runyoro et al. 1995,

Caro 1999) can be largely attributed to the rich volcanic soil originating from the Great Rift. Low concentrations of essential mineral elements may limit the distribution of some species. The spectacularly large mixed-species herds of East African ungulates have been spatially correlated with high concentrations of Na, Mg, and P in grasses (McNaughton 1988). This is consistent with the striking differences in large mammal biomass between savannas on fertile and infertile soils (East 1984, Fritz and Duncan 1994), a pattern that can be extended to North American savannas and forests (Jones and Hanson 1985). The distribution of elephant and rhinos in Borneo may be limited by mineral-rich soils in salt licks (Davies and Payne 1982). Conversely, the remarkably low mammal biomass sustained by even relatively undisturbed South American savannas of the Brazilian cerrado (Marinho-Filho et al. 2003, personal observations) can be partly attributed to its highly weathered latosols that are particularly poor in key plant nutrients, especially P and Ca. Although large mammal assemblages of the cerrado were far more species rich in the Plio-Pleistocene (Simpson 1980), there is no evidence to suggest that the megafaunal abundance of this biome was ever analogous to extant nutrient-rich African or Asian savannas.

Low biomass of folivorous lemurs in Malagasy evergreen forests has been attributed to the relatively high fiber content of mature leaves (Ganzhorn 1992), which in plants on nutrient-poor soils tends to be associated with slower growth rates and higher leaf replacement costs (Janzen 1974, McKey et al. 1978, Coley et al. 1985). In central African forests, both the biomass of wild herbivores and densities of humans exploiting them increase steeply in sites characterized by medium to high soil nutrient availability primarily due to greater deposits of sediments from volcanic, marine, and sedimentary rocks (Barnes and Lahm 1998). Compared with most upland Amazonian forests, central African forests of the Congo basin are more nutrient rich and in general can usually sustain a much higher biomass of diurnal primates and game vertebrates (Fa and Peres 2001), and these differences do not take into account the more prominent nocturnal African primate fauna which is rarely censused. In fact, from an intercontinental perspective, mammal biomass in a typical undisturbed terra firme forest of central Amazonia is more analogous to that of nutrient-poor forests of central Borneo where primate and ungulate densities can be extremely low (Bodmer *et al.* 1991, McConkey 1999).

CONCLUDING REMARKS

The effect of soil fertility on the geographic variation in terrestrial vertebrate biomass at different spatial scales is a reminder of the powerful influence of bottom-up forces regulating the structure of tropical forest communities. Baseline densities of wildlife populations can be properly investigated only in continents and regions that remain relatively unadulterated by large-scale anthropogenic disturbance, including structural habitat changes and direct or atmospheric inputs of industrial fertilizers. There is no reason why this relationship should not apply to temperate forests and other biomes, but sadly the opportunities to understand the distribution and movements of large vertebrates in pre-agricultural Europe and North America are no longer with us. Several questions remain wide open for future investigation, including the consequences of soil fertility on the availability of vertebrate-mediated seed dispersal services to plant taxa bearing fleshy fruits, which may have a positive feedback effect on the density of fruiting plants in nutrient-rich soils. Large-scale edaphic constraints on tropical forest habitat productivity should also be explicitly considered in increasingly overhunted landscapes because productivity–abundance relationships are likely to affect the size, recovery rate, and source–sink dynamics of game vertebrate populations (Joshi and Gadgil 1991). For example, sustainable harvest rates of different vertebrate prey species in Amazonian forests are profoundly affected by soil fertility largely because this boosts standing population densities of game-birds and large mammals (Peres 2000b). Finally, edaphic constraints on habitat productivity should be considered in regional-scale conservation planning, particularly in terms of the size of herbivore populations that can be sustained in forest polygons to be set aside as nature reserves. Yet few

community ecologists have linked soil processes to vertebrate populations and assemblages at large spatial scales in the tropics. It is to be hoped that our understanding of soil–productivity–abundance–diversity relationships will improve while they can still be unraveled in the world's remaining tracts of relatively undisturbed tropical forests.

ACKNOWLEDGMENTS

I thank Götz Schroth for making the Amazon-wide soil data available to me. Primate surveys conducted over the years 1987–2004 throughout the Brazilian Amazon were funded by Conservation International, the Bay Foundation, Wildlife Conservation Society, the Wellcome Trust (UK), National Geographic Society, and World Wildlife Fund-US. Unpublished primate biomass estimates for several Colombian sites were made available by Erwin Palacios. I am deeply indebted to all field assistants who have helped during the survey work, and all local communities for their unreserved collaboration and hospitality during this study. The manuscript benefited from comments by Walter Carson and two anonymous reviewers.

REFERENCES

Augustine, D.J., McNaughton, S.J., and Frank, D.A. (2003) Feedbacks between soil nutrients and large herbivores in a managed savanna ecosystem. *Ecological Applications* 13, 1325–1337.

Barnes, R.F.W. and Lahm, S.A. (1998) An ecological perspective on human densities in the Central African forests. *Journal of Applied Ecology* 34, 245–260.

Barry, S.J. (1984) Small mammals in a south-east Queensland rainforest: the effects of soil fertility and past logging disturbance. *Australian Wildlife Research* 11, 31–39.

Beinroth, F.H. (1975) Relationships between US Soil Taxonomy, the Brazilian system and FAO/UNESCO units. In E. Bornemisza and A. Alvarado (eds), *Soil Management in Tropical America*. North Carolina State University, Raleigh, NC, pp. 97–108.

Bell, R.H.V. (1982) The effect of soil nutrient availability on community structure in African ecosystems. In G.J. Huntley and B.H. Walker (eds), *Ecology of*

Tropical Savannas. Springer-Verlag, New York, Berlin, pp. 183–216.

Bodmer, R., Sidik, I., and Iskander, S. (1991) *Mammalian biomass shows that uneven densities of orang-utans are caused by variation in fruit availability.* Project Barito Ulu Report, Cambridge, pp. 1–18.

Bruijnzeel, L.A. (1991) Nutrient input–output budgets of tropical forest ecosystems: a review. *Journal of Tropical Ecology* 7, 1–24.

Caro, T.M. (1999) Abundance and distribution of mammals in Katavi National Park, Tanzania. *African Journal of Ecology* 37, 305–313.

Chapin, F.S., Vitousek, P.M., and van Cleve, K. (1986) The nature of nutrient limitation in plant communities. *American Naturalist* 127, 48–58.

Chapman, C.A. and Chapman, L.J. (1999) Implications of small scale variation in ecological conditions for the diet and density of red colobus monkey. *Primates* 40, 215–231.

Chapman, C.A., Chapman, L.J., Bjorndal, K.A., and Onderdonk, D.A. (2002) Application of protein to fiber ratios to predict colobine abundance on different spatial scales. *International Journal of Primatology* 23, 283–310.

Chazdon, R.L. (2003) Tropical forest recovery: legacies of human impact and natural disturbances. *Perspectives in Plant Ecology Evolution and Systematics* 6, 51–71.

Claridge, A.W. and Barry, S.C. (2000) Factors influencing the distribution of medium-sized ground-dwelling mammals in southeastern mainland Australia. *Australian Ecology* 25, 676–688.

Clark, D.B., Palmer, M.W., and Clark, D.A. (1999) Edaphic factors and the landscape-scale distributions of tropical rain forest trees. *Ecology* 80, 2662–2675.

Clark, K. and Uhl, C. (1987) Farming, fishing, and fire in the history of the upper Rio Negro region of Venezuela. *Human Ecology* 15, 1–26.

Cochrane, T.T. and Sanchez, P.A. (1982) Land resources, soils and their management in the Amazon region: a state of knowledge report. In S.B. Hecht (ed.), *Amazonia: Agriculture and Land Use Research.* Centro Internacional de Agricultura Tropical, Cali, Colombia, pp. 137–209.

Coley, P.D. (1988) Effects of plant growth rate and leaf lifetime on the amount and type of anti-herbivore defense. *Oecologia* 74, 531–536.

Coley, P.D. and Aide, T.M. (1991) Comparison of herbivory and plant defenses in temperate and tropical broad-leaved forests. In P.W. Price, T.M. Lewinsohn, G.W. Fernandes, and W.W. Benson (eds), *Plant–Animal Interactions: Evolutionary Ecology in Tropical and Temperate Regions.* Wiley and Sons, New York, pp. 25–49.

Coley, P.D., Bryant, J.P., and Chapin III, F.S. (1985) Resource availability and plant anti-herbivore defense. *Science* 230, 895–899.

Costa, F.R., Magnusson, W.E., and Luizão, R.C. (2005) Mesoscale distribution patterns of Amazonian understorey herbs in relation to topography, soil and watersheds. *Journal of Ecology* 92, 863–878.

Crête, M. (1999) The distribution of deer biomass in North America supports the hypothesis of exploitation ecosystems. *Ecology Letters* 2, 223–227.

Cristoffer, C. and Peres, C.A. (2003) Elephants vs. butterflies: the ecological role of large herbivores in the evolutionary history of two tropical worlds. *Journal of Biogeography* 30, 1357–1380.

Davidson, E.A., de Carvalho, C.J.R., Vieira, I.C.G. *et al.* (2004) Nitrogen and phosphorus limitation of biomass growth in a tropical secondary forest. *Ecological Applications* 14, S150–S163.

Davies, A.G. and Payne, J. (1982) *A Faunal Survey of Sabah.* WWF Malaysia, Kuala Lumpur.

du Preez, D.R., Gunton, C., and Bate, G.C. (1983) The distribution of macronutrients in a broad leaf savanna. *South African Journal of Botany* 2, 236–242.

Dyer, L.A. and Letourneau, D.K. (1999). Relative strengths of top-down and bottom-up forces in a tropical forest community. *Oecologia* 119, 265–274.

East, R. (1984) Rainfall, soil nutrient status and biomass of large African savanna mammals. *African Journal of Ecology* 22, 245–270.

Eisenberg, J.F. (1980) The density and biomass of tropical mammals. In M.E. Soulé and B.A. Wilcox (eds), *Conservation Biology: An Evolutionary–Ecological Perspective.* Sinauer, Sunderland, MA, pp. 35–55.

EMBRAPA (2002) *Mapa de Solos do Brasil, escala 1:3,000,000.* Serviço Nacional de Levantamento e Conservação de Solos, Rio de Janeiro.

Emmons, L.H. (1984) Geographic variation in densities and diversities of non-flying mammals in Amazonia. *Biotropica* 16, 210–222.

Evans, K.L., Warren, P.H., and Gaston, K.J. (2005) Species–energy relationships at the macroecological scale: a review of the mechanisms. *Biological Reviews* 80, 1–25.

Fa, J.E. and Peres, C.A. (2001) Game vertebrate extraction in African and Neotropical forests: an Intercontinental Comparison. In J.D. Reynolds, G.M. Mace, K.H. Redford, and J.G. Robinson (eds), *Conservation of Exploited Species.* Cambridge University Press, Cambridge, pp. 203–241.

Fine, P.V.A., Mesones, I., and Coley, P.D. (2004) Herbivores promote habitat specialization by trees in Amazonian forests. *Science* 305, 663–665.

Fittkau, E.J. (1973) Artenmannigfaltigkeit amazonischer Lebensräume aus ökologischer Sicht. *Amazoniana* 4, 321–340.

Fittkau, E.J. (1974) Zur Okologischen Gliederung Amazoniens I. *Amazoniana* 5, 77–134.

Freese, C.H., Heltne, P.G., Castro, N., and Whitesides, G. (1982) Patterns and determinants of monkey densities in Peru and Bolivia with notes on distributions. *International Journal of Primatology* 3, 53–90.

Fretwell, S.D. (1977) The regulation of plant communities by food chains exploiting them. *Perspectives in Biology and Medicine* 20, 169–185.

Fritz, H. and Duncan, P. (1994) On the carrying capacity for large ungulates of African savanna ecosystems. *Proceedings of the Royal Society of London B* 256, 77–82.

Furch, K. (1984) Water chemistry of the Amazon basin: the distribution of chemical elements among freshwaters. In H. Sioli (ed.), *The Amazon: Limnology and Landscape Ecology of a Mighty Tropical River and Its Basin*. W. Junk Publishers, The Netherlands, pp. 167–199.

Furley, P.A. (1990) The nature and sustainability of Brazilian Amazon soils. In D. Goodman and A. Hall (eds), *The Future of Amazonia*. Macmillan, London, pp. 309–359.

Ganzhorn, J.U. (1992) Leaf chemistry and the biomass of folivorous primates in tropical forests. *Oecologia* 91, 540–547.

Geisler, R. and Schneider, J. (1976) The element matrix of Amazon waters and its relationship with the mineral content of fishes. *Amazoniana* 6, 47–65.

Gentry, A.H. (1988) Changes in plant community diversity and floristic composition on environmental and geographical gradients. *Annals of the Missouri Botanical Garden* 75, 1–34.

Givnish, T.J. (1999) On the causes of gradients in tropical tree diversity. *Journal of Ecology* 87, 193–210.

Goodland, R.J.A. and Irwin, H.S. (1975) *Amazon Jungle: Green Hell to Red Desert*. Elsevier, New York.

Goulding, M. (1979) *Ecologia de pesca do Rio Madeira*. Manaus: CNPq-Conselho Nacional do Desenvolvimento Cientifico e Tecnologico. Instituto Nacional de Pesquisas da Amazonia-INPA.

Goulding, M., Carvalho, M.L., and Ferreira, E.G. (1988) *Rio Negro, Rich Life in Poor Water*. SPB Academic Publication, The Hague.

Gupta, A.K. and D.J. Chivers (1999). Biomass and use of resources in south and south-east Asian primate communities. In J.G. Fleagle, C. Janson, and C.K. Read (eds), *Primate Communities*. Cambridge University Press, Cambridge, pp. 38–54.

Hacker, J.B. (ed.) (1982) *Nutritional Limits to Animal Production from Pastures*. Commonwealth Agricultural Bureau, Slough.

Hairston, N.G., Smith, F.E., and Slobodkin, L.B. (1960). Community structure, population control, and competition. *American Naturalist* 44, 421–425.

Halaj, J. and Wise, D.H. (2001). Terrestrial trophic cascades: how much do they trickle? *American Naturalist* 157, 262–281.

Harrington, R.A., Fownes, J.H., and Vitousek, P.M. (2001) Production and resource Use efficiencies in N- and P-limited tropical forests: a comparison of responses to long-term fertilization. *Ecosystems* 4, 646–657.

Haugaasen, T. and Peres, C.A. (2005a) Mammal assemblage structure in Amazonian flooded and unflooded forests. *Journal of Tropical Ecology* 21, 133–145.

Haugaasen, T. and Peres, C.A. (2005b) Primate assemblage structure in Amazonian flooded and unflooded forests. *American Journal of Primatology* 67, 243–258.

Howe, H.F. and Brown, J.S. (1999) Effects of birds and rodents on synthetic tallgrass communities. *Ecology* 80, 1776–1781.

Hunter, M.D. and Price, P. (1992) Playing chutes and ladders: bottom-up and top-down forces in natural communities. *Ecology* 73, 733–746.

Huston, M. (1980) Soil nutrients and tree species richness in Costa Rican forests. *Journal of Biogeography* 7, 147–157.

Irion, G. (1978) Soil infertility in the Amazonian rain forest. *Naturwissenschaften* 65, 515–519.

Isbell, L.A. (1994) Predation on primates: ecological patterns and evolutionary consequences. *Evolutionary Anthropology* 3, 61–71.

Janzen, D.H. (1974) Tropical blackwater rivers, animals, and mast fruiting by the Dipterocarpaceae. *Biotropica* 6, 69–103.

Jones, R.L. and Hanson, H.C. (1985) *Mineral Licks, Geophagy, and Biogeochemistry of North American Ungulates*. Iowa State University Press, Ames, IA.

Jordan, C.F. (1989) Are process rates higher in tropical forest ecosystems? In J. Proctor (ed.), *Mineral Nutrients in Tropical Forest and Savanna Ecosystems*. Blackwell Scientific Publications, Oxford, pp. 217–240.

Joshi, N.V. and Gadgil, M. (1991) On the role of refugia in promoting prudent use of biological resources. *Theoretical Population Biology* 40, 211–229.

Junk, W.J. (1997) *The Central Amazon Floodplain: Ecology of a Pulsing System*. Springer, New York.

Kanowski, J., Hopkins, M.S., Marsh, H., and Winter, J.W. (2001) Ecological correlates of folivore abundance in north Queensland rainforests. *Wildlife Research* 28, 1–8.

Kauffman, S.G., Paredes Arce, G., and Marquina, R. (1998) Suelos de la zona de Iquitos. In R. Kalliola and S. Flores Paitán (eds), *Geoecología y desarrollo Amazónico: Estudio integrado en zona de Iquitos, Perú.* Annales Universitatis Turkuensis, Ser A II, vol. 114, University of Turku, Finland, pp. 139–229.

Kay, R.F., Madden, R.H., van Schaik, C., and Higdon, D. (1997) Primate species richness is determined by plant productivity: implication for conservation. *Proceedings of the National Academy of Sciences of the United States of America* 94, 13023–13027.

Kitajima, K. (1994) Relative importance of photosynthetic traits and allocation patterns as correlates of seedling shade tolerance of 13 tropical trees. *Oecologia* 98, 419–428.

Koren, I., Kaufman, Y.J., Washington, R. *et al.* (2006) The Bodélé depression: a single spot in the Sahara that provides most of the mineral dust to the Amazon forest. *Environmental Research Letters* 1, 1–5.

Laurance, W.F., Albernaz, A.K.M., Gotz, S. *et al.* (2002) Predictors of deforestation in the Brazilian Amazon. *Journal of Biogeography* 29, 737–748.

Lovelock, C.E., Feller, I.C., Mckee, K.L., Engelbrecht, B.M.J., and Ball M.C. (2004) The effect of nutrient enrichment on growth, photosynthesis and hydraulic conductance of dwarf mangroves in Panamá. *Functional Ecology* 18, 25–33.

McConkey, K.R. (1999) *Gibbons as Seed Dispersers in the Rain-forests of Central Borneo.* University of Cambridge, Cambridge.

McKey, D.D., Waterman, P.G., Mbi, C.N., Gartlan, J.S., and Struhsaker, T.T. (1978) Phenolic content of two African rain forests: ecological implications. *Science* 202, 61–64.

McNaughton, S.J. (1988) Mineral nutrition and spatial concentrations of African ungulates. *Nature* 334, 343–345.

Malcolm, J.R., Patton, J.L., and da Silva, M.N.F. (2005) Small mammal communities in upland and floodplain forests along an Amazonian white-water river. In E.A. Lacey and P. Myers (eds), *Mammalian Diversification: From Chromosomes to Phylogeography*, University of California Publications 133, pp. 335–391.

Malhi, Y., Baker, T.R., Phillips, O.L. *et al.* (2004) The above-ground coarse wood productivity of 104 Neotropical forest plots. *Global Change Biology* 10, 563–591.

Marinho-Filho, J., Rodrigues, F.H.G., and Juarez, K.M. (2003) The cerrado mammals: diversity, ecology, and natural history. In P.S. Oliveira and R.J. Marquis (eds), *The Cerrados of Brazil: Ecology and Natural History of a Neotropical Savanna.* Columbia University Press, New York, pp. 266–284.

Marshner, H. (1991) Mechanisms and adaptation of plants to acid soils. *Plant and Soil* 134, 1–20.

Mirmanto, E., Proctor, J., Green, J., Nagy, L., and Suriantata (1999) Effects of nitrogen and phosphorus fertilization in a lowland evergreen rainforest. *Philosophical Transactions of the Royal Society of London Series B-Biological Sciences* 354, 1825–1829.

Mittelbach, G.G., Steiner, C.F., Scheiner, S.M. *et al.* (2001) What is the observed relationship between species richness and productivity? *Ecology* 82, 2381–2396.

Nicholaides III, J.J., Bandy, D.E., Sanchez, P.A., Villachienea, J.H., Coutou, A.J., and Valverde, C. (1984) From migratory to continuous agriculture in the Amazon basin. In *Improved Production Systems as an Alternative to Shifting Cultivation. Soils Bulletin* 53. FAO, Rome, pp. 141–168.

Oates, J.F., Whitesides, G.H., Davies, A.G. *et al.* (1990) Determinants of variation in tropical forest primate biomass: new evidence from West Africa. *Ecology* 71, 328–343.

Palacios, E. and Peres, C.A. (2005) Primate population densities in three nutrient-poor Amazonian terra firme forests of south-eastern Colombia. *Folia Primatologica* 76, 135–145.

Pastor, J., Dewey, B., Naiman, R.J., MacInnes, P.F., and Cohen, Y. (1993) Moose browsing and soil fertility in the boreal forests of Isle Royale National Park. *Ecology* 74, 467–480.

Peres, C.A. (1990) Effects of hunting on western Amazonian primate communities. *Biological Conservation* 54, 47–59.

Peres, C.A. (1993) Antipredation benefits in a mixed-species group of Amazonian tamarins. *Folia Primatologica* 61, 61–76.

Peres, C.A. (1997a) Primate community structure at twenty western Amazonian flooded and unflooded forests. *Journal of Tropical Ecology* 13, 381–405.

Peres, C.A. (1997b) Effects of habitat quality and hunting pressure on arboreal folivore densities in neotropical forests: a case study of howler monkeys (*Alouatta* spp.). *Folia Primatologica* 68, 199–222.

Peres, C.A. (1999a) The structure of nonvolant mammal communities in different Amazonian forest types. In J.F. Eisenberg and K.H. Redford (eds), *Mammals of the Neotropics, the Central Neotropics. Ecuador, Peru, Bolivia, Brazil* 3. University of Chicago Press, Chicago, pp. 564–580.

Peres, C.A. (1999b) Effects of subsistence hunting and forest type on the structure of Amazonian primate communities. In J.G. Fleagle, C. Janson, and K.E. Reed (eds), *Primate Communities.* Cambridge University Press, Cambridge, pp. 268–283.

Peres, C.A. (1999c) General guidelines for standardizing line-transect surveys of tropical forest primates. *Neotropical Primates* 7, 11–16.

Peres, C.A. (2000a) Effects of subsistence hunting on vertebrate community structure in Amazonian forests. *Conservation Biology* 14, 240–253.

Peres, C.A. (2000b) Evaluating the impact and sustainability of subsistence hunting at multiple Amazonian forest sites. In J.G. Robinson and L.E. Bennett (eds), *Hunting for Sustainability in Tropical Forests*. Columbia University Press, New York, pp. 31–56.

Peres, C.A. (2000c) Identifying keystone plant resources in tropical forests: the case of gums from *Parkia* pods. *Journal of Tropical Ecology* 16, 287–317.

Peres, C.A. and Dolman, P. (2000). Density compensation in neotropical primate communities: evidence from 56 hunted and nonhunted Amazonian forests of varying productivity. *Oecologia* 122, 175–189.

Peres, C.A. and Janson, C. (1999) Species coexistence, distribution, and environmental determinants of neotropical primate richness: a community-level zoogeographic analysis. In J.G. Fleagle, C. Janson, and K.E. Reed (eds), *Primate Communities*. Cambridge University Press, Cambridge, pp. 55–74.

Peres, C.A. and Nascimento, H.S. (2006) Impact of game hunting by the Kayapó of southeastern Amazonia: implications for wildlife conservation in tropical forest indigenous reserves. *Biodiversity and Conservation* 15, 2627–2653.

Peres, C.A. and Palacios, E. (2007) Basin-wide effects of game harvest on vertebrate population densities in Amazonian forests: implications for animal-mediated seed dispersal. *Biotropica* 39(3), 304–315.

Phillips, O. (1993) The potential for harvesting fruits in tropical rainforests: new data from Amazonian Peru. *Biodiversity and Conservation* 2, 18–38.

Phillips, O.L., Nuñez, P.V., Monteagudo, A.L. *et al.* (2003) Habitat association among Amazonian tree species: a landscape-scale approach. *Journal of Ecology* 91, 757–775.

Potts M.D., Ashton, P.S., Kaufman, L.S., and Plotkin, J.B. (2002) Habitat patterns in tropical rain forests: a comparison of 105 plots in Northwest Borneo. *Ecology* 83, 2782–2797.

Power, M.E. (1992) Top-down and bottom-up forces in food webs – do plants have primacy. *Ecology* 73, 733–746.

Recher, H.F., Majer, J.D., and Ganesh, S. (1996) Eucalypts, arthropods and birds: on the relation between foliar nutrients and species richness. *Forest Ecology and Management* 85, 177–195.

Richter, D.D. and Babbar, L.I. (1991) Soil diversity in the tropics. *Advances in Ecological Research* 21, 315–389.

Rosenzweig, M.L. (1995) *Species Diversity in Space and Time*. Cambridge University Press, Cambridge.

Rosenzweig, M.L. and Abramsky, Z. (1993) How are diversity and productivity related? In R.E. Ricklefs and D. Schluter (ed.), *Species Diversity in Ecological Communities*. University of Chicago Press, Chicago, pp. 52–65.

Runyoro, V.A., Hofer, H., Chausi, E.B., and Moehlman, P.D. (1995). Long-term trends in the herbivore populations of the Ngorongoro Crater, Tanzania. In A.R.E. Sinclair and P. Arcese (eds), *Serengeti II*. University of Chicago Press, Chicago, IL, pp.146–168.

Salovaara, K.J. (2005) *Habitat heterogeneity and the distribution of herbivorous mammals in Peruvian Amazonia*. PhD thesis, Turku University, Turku, Finland.

Salovaara, K.J., Cardenas, G.G., and Tuomisto, H. (2004) Forest classification in an Amazonian rainforest landscape using pteridophytes as indicator species. *Ecography* 27, 689–700.

Sanchez, P.A. (1981) Soils of the humid tropics. In *Blowing in the Wind: Deforestation and Long-term Implications*, Department of Anthropology, College of William and Mary, Williamsburg, pp. 347–410.

van Schaik, C.P. and Mirmanto, E. (1985) Spatial variation in the structure and litterfall of a Sumatran rain forest. *Biotropica* 17, 196–205.

Seagle, S.W. and Liang, S.Y. (2002) Browsing and microhabitat effects on riparian forest woody seedling demography. *Ecology* 83, 212–227.

Simpson, G.G. (1980) *Splendid Isolation: The Curious History of South American Mammals*. Yale University Press, New Haven, CT.

Smith, N. (1979) *A pesca no Rio Amazonas*. INPA, Manaus.

Sioli, H. (1980) Foreseeable consequences of actual development schemes and alternative ideas. In F. Barbira-Scazzocchio (ed.), *Land, People and Planning in Contemporary Amazonia*. Centre of Latin American Studies Occasional Paper No. 3. Cambridge University, Cambridge, pp. 257–268.

Sombroek, W.G. (1966) *Amazon Soils, a Reconnaissance of the Soils of the Brazilian Amazon*. Pudoc, Wageningen, The Netherlands.

Sombroek, W.G. (1984) Soils of the Amazon region. In H. Sioli (ed.), *The Amazon – Limnology and Landscape Ecology of a Mighty Tropical River and Its Basin*. Dr W. Junk, Dordrecht, The Netherlands, pp. 521–535.

Sombroek, W.G. (2000) Amazon landforms and soils in relation to biological diversity. *Acta Amazonica* 30, 81–100.

Srivastava, D.S. and Lawton, J.H. (1998) Why more productive sites have more species: an experimental test of theory using tree-hole communities. *American Naturalist* 152, 510–529.

Stevenson, P.R. (2001) The relationship between fruit production and primate abundance in Neotropical communities. *Biological Journal of the Linnean Society* 72, 161–178.

Swap, R., Garstang, M., Greco, S., and Talbot, R. (1992) Saharan dust in the Amazon basin. *Tellus* 44B, 133–149.

Tanner, E.V.J., Vitousek, P.M., and Cuevas, E. (1998) Experimental investigation of nutrient limitation of forest growth on wet tropical mountains. *Ecology* 79, 10–22.

Terborgh, J. (1983) *Five New World Primates – a Study in Comparative Ecology*. Princeton University Press, Princeton.

Terborgh, J.W. and Janson, C.H. (1986) The sociocology of primate groups. *Annual Review of Ecology Systematics* 17, 111–135.

Tuomisto, H. and Poulsen, A. (1996) Influence of edaphic specialization on pteridophyte distribution in neotropical rain forest. *Journal of Biogeography* 23, 283–293.

Tuomisto, H., Poulsen, A.D., Ruokolainen, K. *et al.* (2003) Linking floristic patterns with soil heterogeneity and satellite imagery in Ecuadorian Amazonia. *Ecological Applications* 13, 352–371.

Uehara, G. and Gilman, G. (1981) *The Mineralogy, Chemistry and Physics of Tropical Soils with Variable Charge Clays*. Westview Press, Boulder, CO.

Vitousek, P.M. and Sanford, R.L. Jr. (1986) Nutrient cycling in moist tropical forest. *Annual Review of Ecology and Systematics* 17, 137–167.

Wallace, A.R. (1853) *A Narrative of Travels on the Amazon and Rio Negro, with an Account of the Native Tribes and Observations on the Climate, Geology, and Natural History of the Amazon Valley*. Reeve and Co., London.

Waterman, P.G. and McKey, D.B. (1988) Secondary compounds in rain-forest plants: patterns of distribution and ecological implications. In H. Leith and O.R. Werger (eds), *Ecosystems of the World, Tropical Rain Forests*, 14b. Elsevier, The Hague, pp. 513–536.

Waterman, P.G. and Mole, S. (1989) Soil nutrients and plant secondary compounds. In J. Proctor (ed.), *Mineral Nutrients in Tropical Forest and Savanna Ecosystems*. Blackwell Scientific Publications, Oxford, pp. 241–254.

Waterman P.G., Ross, J.A.M., Bennett, E.L., and Davies, A.G. (1988) A comparison of the floristics and leaf chemistry of the tree flora in two Malaysian rain forests and the influence of leaf chemistry on populations of colobine monkeys in the Old World. *Biological Journal of the Linnean Society* 34, 1–32.

Williams, M.R. and Fisher, T.R. (1997) Chemical composition and deposition of rain in the central Amazon, Brazil. *Atmospheric Environment* 31, 207–217.

Woinarski J.C.Z., Fisher, A., and Milne, D. (1999) Distribution patterns of vertebrates in relation to an extensive rainfall gradient and variation in soil texture in the tropical savannas of the Northern Territory, Australia. *Journal of Tropical Ecology* 15, 381–398.

Wright, D.H. (1983) Species–energy theory: an extension of species–area theory. *Oikos* 41, 496–506.

SECONDARY FOREST SUCCESSION, DYNAMICS, AND INVASION

Chapter 22

PROCESSES CONSTRAINING WOODY SPECIES SUCCESSION ON ABANDONED PASTURES IN THE TROPICS: ON THE RELEVANCE OF TEMPERATE MODELS OF SUCCESSION

Chris J. Peterson and Walter P. Carson

OVERVIEW

We review the major constraints on woody species succession in abandoned pastures in the tropics and ask whether conceptual models developed primarily in temperate regions are useful in tropical habitats. We found that the majority of tropical post-agricultural succession research (>60 studies) was not typically focused on testing broad general hypotheses or conceptual models, particularly those developed to explain patterns of early succession in temperate regions. Instead, the studies focused more on evaluating general constraints on woody species recruitment. Among the studies that do consider the major successional models, the three mechanisms of interaction described by Connell and Slatyer (1977) are the most frequently considered. Empirical studies suggest that woody species succession in abandoned agricultural fields is constrained primarily by the availability of woody propagules, though studies rarely simultaneously evaluate the relative importance of other potential processes. Additional constraints on succession are competition with residual or resident herbaceous vegetation (e.g., graminoids) and seed and seedling predation. Most models of succession in temperate regions were unsatisfactory because they failed to place enough emphasis on propagule limitation and facilitation. Nonetheless, one conceptual model, the nucleation model, appears to provide a general conceptual framework that is robust enough to encapsulate post-agricultural succession in many tropical habitats. Under the nucleation model, a few successful early woody colonists, residual trees, or key microsites present in the pasture early in succession facilitate the establishment and survival of woody species in the immediate proximity of these early woody residents or on these microsites. Succession proceeds as these patches of woody vegetation that form around recruitment foci spread and coalesce. The repeated observation of enhanced seed input and woody species recruitment near surviving remnant trees and on unique microsites is consistent with this model. We urge tropical succession researchers to evaluate general models of succession and further quantify the degree to which propagule limitation, facilitation, seed predation, and life-history trade-offs interact to determine the rate of woody species succession into abandoned agricultural habitats. The refinement and further development of the nucleation model in combination with quantitative models of dispersal holds much promise.

INTRODUCTION

Is post-agricultural succession in the tropics fundamentally different than in the temperate zone? A number of broad models of succession have been developed for and tested in temperate systems since Clements (1916) presented the first general theory nearly a century ago. Can we apply these models to tropical systems? Because tropical systems typically have greater species diversity, putatively more benign environmental conditions, and more complex and intimate plant–animal interactions, processes underlying succession in the tropics could be qualitatively different than processes within temperate communities. Here, we briefly describe nine prominent conceptual and life-history based models that were developed for temperate systems but appear to be applicable to tropical succession. We ask whether these models have provided a conceptual basis for or motivated studies of post-agricultural succession in the tropics. We then identify the empirical basis for the major constraints on woody species recruitment in abandoned tropical pastures. We suggest that the facilitation mechanism encapsulated within the nucleation model (Yarranton and Morrison 1974) provides a potential robust foundation for new conceptual models of succession in post-agricultural communities in the tropics. Finally, we conclude with a description of five areas where additional research will help provide a foundation for both the testing and building of general models of post-agricultural succession in the tropics.

SCOPE OF THE CHAPTER

The scope of this chapter includes studies of early succession in tropical habitats that were previously in agriculture. We focus on the first 40–50 years of succession. We do not consider succession following logging without agriculture (e.g., Swaine and Hall 1983), succession after removal of exotic tree plantations (Duncan and Chapman 2003), or succession in different-sized gaps in intact forests (e.g., Schnitzer and Carson 2001, Pearson *et al.* 2003). Although we have examined 61 publications, this chapter is not

a comprehensive review of patterns of structural and compositional change (for this see Finegan 1996, Guariguata and Ostertag 2001, Meli 2003).

OVERVIEW OF PROMINENT MODELS OR CONCEPTUAL FRAMEWORKS OF SUCCESSION

Here we briefly introduce nine prominent general models of secondary succession, all of which may apply to post-agricultural succession in the tropics. Typically, these models are not mutually exclusive, often have different foci, vary in their mechanistic detail, or attempt to describe different components of succession.

Relay floristics

Under relay floristics, succession proceeds as early colonists and their respective communities modify the environment in such a way that they facilitate and thereby hasten their own replacement by later colonists and communities that are not present at the beginning of succession. The result is that whole communities arise and decline in near unison, rather than individualistic species dynamics; thus the synchronicity of turnover of species is very high. This captures the essence of Clements' (1916) views of succession. The relay floristics model is not spatial and does not consider the influence of animals (e.g., dispersers, herbivores, seed predators; Table 22.1). Nor is variation in propagule availability considered as a major influence on successional trajectories. Among the life-history traits that are considered important, tolerance of environmental extremes is probably foremost (Table 22.1). Early colonizers are thought to moderate these harsh conditions and thereby facilitate the arrival and survival of a suite of later successional species.

Initial floristic composition

In 1954, Egler proposed the initial floristic composition (IFC) model of succession and contrasted it with relay floristics. Under the original

Table 22.1 Summary of similarities and differences among major models of secondary succession.

Model traits	Relay floristics	Initial floristic composition	Nucleation	Gradient in time	Interaction categories	Vital attributes	Resource ratio	Shade tolerance	Hierarchical framework
Spatial	No	No	Explicit	Implicit	No	Implicit	No	No	Implicit
Propagule availability	Not considered	Explicit	Indirect and patchy	Explicit but species specific	Explicit but species specific	Explicit but species specific	Not considered	Not considered	Explicit
Interactions important	Yes	No	Yes	Somewhat	Yes	Somewhat	Yes	Yes	Varies
Dominant interactions	Facilitation	None	Facilitation	Facilitation, inhibition	Inhibition	None	Inhibition	Inhibition	All
Important life-history traits	Shade tolerance	Longevity, growth rate	Environmental stress tolerance	Longevity, growth rate	Longevity, stress tolerance, low resource tolerance	Dispersal ability, longevity, timing of reproduction	Low resource tolerance	Shade tolerance, size, growth rate, longevity	Many
Animal effects	None	None	Dispersers	Herbivores, seed predators	None	Dispersers	None	None	Dispersers, herbivores, seed predators
Inhibition by herbaceous species	No	No	No	Yes	Usually	No	Sometimes	No	Varies
Regional repeatability	High	Low	Moderate	Moderate	Moderate	Varies	High	Moderate	Varies

Notes: Interactions are defined as either inhibition, tolerance, or facilitation. "Explicit" signifies that the trait was directly considered in the model. "Implicit" signifies that in the presentation of the model the trait was important, even if not stated directly by the authors. "Varies" signifies that whether a trait is important for that model is contingent on a number of factors, thus no general statement can be made.

interpretation of initial floristics, all or the vast majority of species that will eventually become prominent later in succession are present at the outset. Species that dominate early do so because of rapid growth rates or because a large supply of propagules were present in the first year of succession, or both (Table 22.1). Late successional species that were present at the outset become dominant thereafter because they develop more slowly and ultimately outlive the early occupants. Thus the primary determinant of species turnover is differences in life-history characteristics, especially growth rate and lifespan. The broad trends in growth rate and lifespan observed in early tropical secondary succession (from early successional grasses, herbs, and shrubs, to later successional short-lived pioneer trees, and finally to longer-lived pioneer trees) appear to be consistent with this model (Guariguata and Ostertag 2001), but not if late successional and primary tree species are considered. Still, Finegan (1996) maintained that all long-lived pioneer trees, as well as many mature forest trees, typically establish in the first few years of succession, thus offering support for the initial floristics model in post-agricultural habitats. Wilson *et al.* (1992) pointed out that researchers are not consistent in their use of the IFC model, and have actually used two variants. The first, described above, is closer to Egler's original concept and the second is called pre-emptive initial floristics. In this latter variant, the species that colonize a new site prevent the establishment of later successional species, until after the initial colonists die. However, this expanded definition of IFC is not consistent with Egler's (1954) original model. We argue that the defining characteristic of IFC is the presence of many of the primary forest or late successional species during the first few years of succession.

Nucleation

Yarranton and Morrison (1974) developed a facilitation-based nucleation model for stressful habitats that have low resource availability; these habitats are typically inimical to colonization. This model is a spatial extension of the Clementsian relay floristics model, whereby early colonizers, typically woody species, facilitate and promote the colonization of additional woody species. It was originally applied to succession on temperate dunes. Under the nucleation model, a few successful early colonists ameliorate harsh local conditions (e.g., enhancing local nutrient and water status via shading and litter), thereby facilitating the establishment and survival of later arrivals in the immediate proximity of the early residents (Table 22.1). The early residents also serve as foci for enhanced localized seed dispersal and accumulation (e.g., via serving as perches for birds). Spatially, the model predicts that succession will proceed outward from some number of recruitment foci of early colonists, which grow and coalesce as additional late arrivals establish in the vicinity of the early colonists. Thus the nucleation model is spatially explicit. In its most general form it makes no particular predictions about future species composition though it could easily be refined to do so based upon differential species responses to stressful habitats and colonization ability and dispersal mode. The nucleation model differs substantially from all of the other models in explicitly predicting high levels of neighborhood-scale heterogeneity in vegetation structure and composition during secondary succession. Under this model, as under relay floristics, the critical differences among species are in their tolerance of stressful conditions that occur in recently abandoned habitats. Here, pioneer species are defined by their ability to both arrive in these habitats and more importantly cope with stressful conditions (e.g., low resource availability; Table 22.1). Animal effects were given only passing mention as dispersers that bring late successional propagules into the "zone of influence" of early successional nuclei.

Gradient in time

Pickett (1976) and Drury and Nisbet (1973) developed explanations for species turnover during succession that focused on the contrasting life-history traits (particularly differential growth rates, dispersal ability, lifespan, etc.) of species that dominated early versus late during succession. Both perspectives viewed secondary succession as a gradient in time whereby temporally changing biotic and abiotic conditions favored species with one suite of life-history traits early in succession versus different life-history traits later on. For both of these conceptual models of succession, species turnover occurred due to known correlations among plant size, longevity, and slow growth (Table 22.1). Thus, succession was a population process that could be understood only by knowing the interspecific differences in species life-history traits. In addition, Pickett (1976) argued that early occupants would inhibit the establishment and growth of later occupants. These two gradient-in-time models explicitly identified dispersal ability as a critical life-history attribute of pioneers. For example, Pickett (1976) noted that long-range dispersal would be required to reach the interior of large disturbed patches. Unfortunately these approaches failed to incorporate specific mechanisms of resource competition that permitted residents to resist displacement or what factors caused residents to decline in abundance other than shorter lifespans.

Interaction categories

In a compelling review of previous work and theory, Connell and Slatyer (1977) described how succession might occur via three categories of interaction: facilitation, tolerance, and inhibition (Table 22.1). They pointed out that the interaction among earlier and later species could be positive (facilitation), neutral (tolerance), or negative (inhibition). They highlighted how some existing models focused on resident species *inhibiting* later successional species (e.g., pre-emptive initial floristics) while some focused on resident species *facilitating* later successional species (e.g., the nucleation model). Pickett *et al.* (1987) pointed out several problems with these models, noting that they were actually broad categories of mechanisms (e.g., inhibition could occur via resource competition or allelopathy) that underlie species change and that each model could operate simultaneously during any succession even at the same site. Also, the interaction between any given pair of species could be both facilitative and inhibitory. Nonetheless, if interactions could be summed among numerous pairs of species, these three categories would allow successional interactions to be classified jointly as either

positive via facilitation or negative via inhibition or neutral. The interaction categories of Connell and Slatyer (1977) did not explicitly consider space or impacts of herbivores, although herbivores were briefly considered. Connell and Slatyer did point out that wide dispersal and numerous propagules would cause pioneer species to be the earliest dominants but subsequently focused on successional dynamics thereafter. They concluded that inhibition by earlier dominants predominated during secondary succession. Thus they suggested that longevity, stress tolerance, and low resource tolerance should be crucial life-history traits among later successional species.

Vital attributes

Noble and Slatyer (1980) developed the vital attributes model whereby particular disturbances produce conditions that can be best exploited by species with appropriate life-history traits. Species were classified according to three key traits: mode of dispersal, ability to regenerate after disturbance, and timing of reproduction and senescence. Like the work of Pickett (1976) and Drury and Nisbet (1973) this model took a population-based approach focusing on the life-history traits that led to early arrival at a site and those that led to later arrival and allowed for long-term persistence. Dispersal ability was granted a more prominent role in influencing species availability than in the gradient-in-time models. Once again, however, the model was not explicitly spatial and the actual underlying mechanisms that caused species displacement and turnover were not explicitly considered (Table 22.1). This model did not consider enemies, and competitive interaction among plant species was de-emphasized relative to the other models.

Resource ratios

Tilman (1985, 1988) developed the resource ratio model of succession whereby species turnover during succession is driven by interspecific competition that depends upon the shift from high light/low nutrient conditions early in succession to low light/high nutrient conditions later in succession. Tilman (1985) concluded that patterns of succession were highly repeatable and very similar within broad regions (see also MacMahon 1981). Tilman's model assumed that light and soil resources (particularly nitrogen) were inversely correlated during succession. Thus dominant species were superior competitors at equilibrium for the ratio of light and nutrients that occurs at a given period during succession. As the ratio of these two resources changes, different species would gain a competitive advantage and become dominant. In this model, space is implicit, while adequate propagule availability is assumed among all species (Table 22.1). The most important life-history traits are the requirements of a species for light and soil resources and how this defines their competitive ability. Tilman (1985) did not consider effects of dispersers or herbivores.

Shade tolerance

Huston and Smith (1987) developed an individualistic model mainly for secondary forest succession that relied on inversely correlated life-history traits. In particular, they focused on longevity, sapling establishment rate, and maximum size, age, and growth rates (Table 22.1). They argued that population-level models are too simplistic to explain successional dynamics even when population-level models produce predictions that are confirmed in nature. They emphasized variation in conditions, particularly light levels, at the scale of individuals, and showed that when competition for light is strong, the trade-off between shade tolerance and other life-history traits may be the primary mechanism that underlies species replacement (Table 22.1). Huston and Smith demonstrated through the use of computer simulations that hypothetical species that differed in growth rate or rate of sapling establishment, and shade tolerance or maximum size, would reproduce actual successional dynamics. As with several of the previous models, this model assumed adequate propagule availability and did not consider any animal impacts. Although they mentioned variation in dispersal ability among species, their model was not spatial. Similar to the resource ratio model, this model focused on

competitive interactions (primarily for light) as major drivers of successional change.

Hierarchical causes

Pickett et al. (1987) developed a hierarchical successional framework that explicitly considered nearly all conceivable causes of turnover during succession. Unlike the other successional models, this approach focused far greater attention on site characteristics, rather than only species characteristics. Pickett et al. established that there are three fundamental causes of compositional change and its variation during succession: (1) site availability, (2) differential species availability (e.g., identity and number of propagules of potential colonists), and (3) differential species performance (e.g., differential competitive ability of species that actually colonized the site). This approach emphasized neglected aspects of succession such as how initial site size and landscape configuration could alter successional processes. The advantage of this framework was that one could choose among an exhaustive list of processes that could influence succession (e.g., colonization limitation) at any given site and pick ones that applied to a give locale. The drawback of this approach was that all mechanisms were implicitly given equal importance within each position of the hierarchy and succession was viewed as relatively site specific, leaving little room for generality. Regardless, this exhaustive hierarchical framework undoubtedly applies to tropical systems. We suggest that this paper should be read by anyone studying succession because it is comprehensive and considers nearly all possible causes for species turnover.

DO STUDIES OF POST-AGRICULTURAL SUCCESSION INCORPORATE OR TEST TEMPERATE MODELS?

Overall we found that few studies of tropical post-agricultural succession were designed to test the above models or were motivated by these models. Indeed, among 61 publications reviewed none identified that their objectives included examining the models of Pickett (1976), Noble and Slatyer (1980), Tilman (1985, 1988), Huston and Smith (1987), or Pickett *et al.* (1987). Three studies evaluated Egler's IFC model (Harcombe 1977, Myster 2003a, Pena-Claros 2003) and three studied the role of remnant trees in pastures and used the nucleation model as their conceptual foundation (Holl *et al.* 2000, Slocum 2001, Carriere *et al.* 2002). In an approach similar to that of Egler (1954), Finegan (1996) and Gómez-Pompa and Vásquez-Yanes (1981) presented models where most of the species from all successional stages establish very early in succession (for a test of these models see Pena-Claros 2003). Nearly a dozen papers mentioned the interaction categories of Connell and Slatyer (1977) though these studies were not designed a priori to distinguish among these categories. Furthermore, a search of citations in tropical post-agricultural succession papers on Web of Science showed only one citation of Tilman's work (Ganade and Brown 2002), one of Huston and Smith (Nepstad *et al.* 1996), and none of Pickett, or Noble and Slatyer. Overall, few studies were undertaken to test general conceptual models and most studies were not theory driven. Most researchers report that their primary motivation was either to identify the processes limiting woody species establishment early in succession (e.g., seed dispersal, germination, predation) or to add to the existing case studies of succession in the tropics and to aid restoration efforts. As a result, several broad generalizations have emerged regarding the constraints on woody species establishment in tropical old fields and pastures (e.g., Holl *et al.* 2000). Nonetheless, it was clear that no widely adopted theory is currently central to studies of tropical post-agricultural succession regardless of whether these theories were developed in temperate or tropical regions.

THE TWO MAJOR CONSTRAINTS ON TROPICAL POST-AGRICULTURAL SUCCESSION

Two major constraints appear to strongly influence the pattern of early succession in tropical habitats: recruitment limitation and inhibition by resident vegetation, particularly perennial graminoids that are present at the time of abandonment.

Recruitment limitation

Recruitment limitation caused by a sparse seed bank

Very few seeds of forest species are present in the soil seed bank of old fields and abandoned pastures (Uhl *et al.* 1982, Zahawi and Augspurger 1999, Wijdeven and Kuzee 2000, Zimmerman *et al.* 2000, Cubina and Aide 2001, Slocum 2001, Myster 2004). For example, the seed bank density of woody species was nearly five times greater in intact forest than in abandoned pastures in eastern Amazonia (679 m^{-2} versus 144 m^{-2}), and viable seeds of primary forest tree species were essentially absent (Nepstad *et al.* 1996). Similarly, in southern Mexico, Guevara *et al.* (2004) found that primary forest tree species made up only 0.3% of the soil seed bank around remnant figs in pastures. Thus it seems unlikely that succession typically starts with woody species establishing, in even small numbers, from a seed bank.

Recruitment limitation caused by low seed rain

Seed rain into pastures is typically low, concentrated near forest edges, and enhanced when fields have structural complexity. Although there are few studies on the behavior of animal dispersal vectors (predominantly birds and bats), avian movements appear to be more common in pastures that have scattered shrubs and trees versus those devoid of structural complexity. Avian movements typically occur within 80 m of the forest edge (Sisk 1991, Da Silva *et al.* 1996, but see Puyravaud 2003). Seed rain into abandoned agricultural lands (independent of remnant trees) declines rapidly across short distances from the forest edge (e.g., Martinez-Garza and Gonzalez-Montagut 2002). Dosch *et al.* (2007) found that 45 m from a forest edge seed rain in five pastures averaged less than 1% of that at the forest–pasture interface. No species with seeds larger than 1 mm arrived more than 10 m from the forest during

13 months of monitoring. Holl (1999) found that seeds of animal-dispersed woody species arrived in open pasture at a density of only $3 \, \text{m}^{-2} \, \text{year}^{-1}$ and only one genus (*Solanum*) had seed input into seed traps more than 5 m from forest. Cubina and Aide (2001) found that only three species and 0.3% of seeds dispersed more than 4 m from the forest edge. Those seeds that do arrive in pastures are generally small-seeded pioneer species that are dispersed by wind (e.g., *Trema micrantha*) or bats (e.g., *Cecropia* spp.) (Martinez-Garza and Gonzalez-Montagut 1999, 2002). Ingle (2003) studied the seed rain into successional land adjacent to lower montane rainforest in the Philippines. Wind-dispersed seeds were 15 times more common than vertebrate-dispersed seeds and more than 95% of the total seed rain was within 40 m of the forest edge. Seed input appears to be high only in very small pastures surrounded by forest (e.g., Zahawi and Augspurger 1999, Wijdeven and Kuzee 2000, Myster 2004) or where seeds originate from numerous remnant trees growing within the pasture (Guevara and Laborde 1993, Slocum and Horvitz 2000). Overall, woody colonist density and diversity decays rapidly with distance from forest edge and this likely reflects very low seed input (Peterson and Haines 2000, Chinea 2002, Myster 2003a), although exceptions exist (Aide *et al.* 1996, Duncan and Duncan 2000).

Recruitment limitation caused by seed predation

Rates of seed predation as measured via seed removal studies are typically high in early successional communities. For example, in Costa Rican pastures Holl and Lulow (1997) found that 66% of all seeds were removed within the first 30 days for 10 different woody species. In the eastern Amazon, Nepstad *et al.* (1996) documented more than 80% seed removal within 20 days for 6 of 11 woody species. Duncan and Duncan (2000) estimated that nearly 50% of seeds of six species were removed within 11 days at a site near Kibale National Park in Uganda. Removal rates in these early successional communities are typically greater than those found in nearby intact forest or in treefall gaps (Nepstad *et al.* 1996). It is especially important to note that removal rates typically decrease as seed size increases (Figure 22.1; Osunkoya 1994, Nepstad *et al.* 1996, Sarmiento 1997, Duncan and Duncan 2000, Jones *et al.* 2003, Myster 2003b). When Zimmerman *et al.* (2000) placed the very small seeds of *Cecropia schreberiana* in Puerto Rican pasture, 100% of seeds were removed within 8 hours. This may suggest that large seeds have a size refuge from the smaller seed predators that are typically present in pastures (Figure 22.1), and that in the rare instances when large seeds arrive in pastures, they may escape predation. Overall, most studies

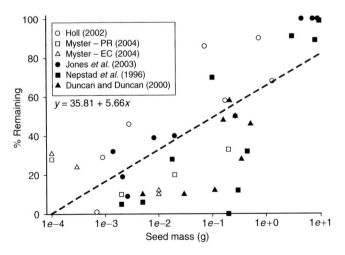

Figure 22.1 Percent of original number of seeds remaining versus seed mass (g), for 21 species, pooled across five studies ($r^2 = 0.41$, $P < 0.001$). From Nepstad *et al.* (1996), Duncan and Duncan (2000), Holl (2002), Jones *et al.* (2003), and Myster (2004; PR, Puerto Rico; EC, Ecuador). Four of the five within-study regressions of seeds remaining versus seed mass were also significant.

(Osunkoya 1994, Sarmiento 1997, Duncan and Duncan 2000) confirm a moderate to high level of seed removal though these rates can vary widely among sites, species, seed sizes, and distance from forest edge (Nepstad *et al.* 1996, Holl and Lulow 1997, Pena-Claros and De Boo 2002, Jones *et al.* 2003, Myster 2003b). Regardless, it appears likely that seed predation may strongly limit rates of woody species colonization and potentially filter out species whose seeds are highly preferred. We caution that these results are based primarily on seed removal trials where seed disappearance may not always equal seed death; future studies should document the fates of removed seeds (e.g., Forget 1997).

Recruitment limitation caused by seedling predation

Small mammals are fairly abundant in tropical pastures (Nepstad *et al.* 1996, Jones *et al.* 2003) and the few studies available suggest they can cause substantial seedling mortality. For example, Holl and Quiros-Nietzen (1999) found that rabbits clipped more than 50% of transplanted seedlings of four native tree species in a lower montane pasture in southern Costa Rica. Predation did not differ between open pasture and under-tree locations (Holl and Quiros-Nietzen 1999) and some species were clipped more than others (high in *Sideroxylon portoricense* and *Vochysia allenii*, low in *Ocotea galucosericea* and *Ocotea whitei*). In Honduras, Zahawi and Augspurger (2006) found levels of seedling predation similar to those of Holl and Quiros-Nietzen (1999), although only at one of three sites. In Puerto Rico, Zimmerman *et al.* (2000) observed substantial herbivory on three small-seeded species and suggested that small-seeded species were more vulnerable than large-seeded species. Duncan and Duncan (2000) found that rodent predation caused at least 29% mortality among woody seedlings introduced as seeds into East African pastures.

Leaf-cutting ants can be major seedling predators and herbivores. For example, *Atta sexdens* removed more than one third of the foliage within 16 days of planting of seedlings of four tree species in eastern Amazonia (Nepstad *et al.* 1996).

In a 2-year-old Brazilian pasture Vasconcelos and Cherrett (1997) found that small seedlings were particularly vulnerable to leaf-cutters and at least 65% of the individuals of six species were attacked. Because some species are far more vulnerable than others, it is likely that leaf-cutter attacks, like seed predation, can delay succession as well as alter species composition.

Recruitment limitation can be ameliorated by resident trees and shrubs

Seed input can increase dramatically beneath pasture trees and shrubs, regardless of whether these pasture trees were remnants or new colonists (e.g., Sarmiento 1997, Slocum and Horvitz 2000, Carriere *et al.* 2002). For example, both Nepstad *et al.* (1996) and Holl *et al.* (2000) found seed deposition for animal-dispersed seeds was more than two orders of magnitude higher beneath resident trees compared with open pasture (see also Slocum and Horvitz 2000). Not surprisingly, wind-dispersed seeds show similar rates of input both beneath trees and in open pasture (Holl *et al.* 2000).

Seedling abundance is much greater beneath isolated trees. In the southern Costa Rican pre-montane zone, tree seedlings of animal-dispersed species were more concentrated beneath shrubs and trees than in open pasture (Holl 1999); similar trends were documented in lowland pastures of northeastern Costa Rica (Cusack and Montagnini 2004). In lower montane Ecuadorian pastures, woody recruitment was essentially nil in open pastures, whereas recruitment steadily increased in pastures with abundant guava trees (Zahawi and Augspurger 1999). Zanne and Chapman (2001) found that grasslands converted to tree plantations had greater understory woody stem density and species richness than unplanted grasslands. The identity of resident trees appears to influence colonization. For example, in pastures in Costa Rica, Slocum (2001) found more recruits were present beneath *Cordia* and *Cecropia* trees than beneath *Ficus* trees. Also, the woody species composition beneath both *Cordia* and *Cecropia* trees was distinct from that under *Ficus* or *Pentaclethra* trees.

Holl (2002) demonstrated that while woody species (shrubs) overall facilitated tree establishment into pastures in Costa Rica, this was the product of positive, negative, and neutral effects at different stages of tree regeneration. Specifically, shrubs enhanced seed input and seedling survival of animal-dispersed species, but seed predation was greater beneath shrubs. Germination did not vary between habitats. This work confirms the conceptual problems pointed out by Pickett *et al.* (1987) with trying to apply the models of Connell and Slatyer (1977) in a mutually exclusive way; nonetheless net effects (i.e., facilitation) were still apparent. Regardless, there appears to be severe dispersal and colonization limitation into these early successional habitats and resident woody species generally ameliorate this constraint and likely alter rates of woody species succession as well as composition.

Inhibition caused by resident vegetation

Following abandonment, all types of agricultural land uses leave some residual herbaceous vegetation. Typically, this residual vegetation reduces woody species germination and establishment (Nepstad *et al.* 1996, Slocum 2000), though less so for large-seeded species (Figure 22.2). For example, Holl *et al.* (2000) found that in montane pastures in Costa Rica, species richness and cover of broad-leaved species was up to five-fold greater after just 6 months in sites where graminoids had been removed versus where they were left intact (see also Peterson and Haines 2000). In African post-agricultural grasslands and plantations, Zanne and Chapman (2001) reported significant negative correlations between the number and richness of new woody stems and the abundance of grasses and forbs (see also Ferguson *et al.* 2003). The impact of pasture graminoids, however, on woody species establishment is not uniformly negative and can be species specific. For instance, Zimmerman *et al.* (2000) introduced seeds of 11 woody species into a Puerto Rican pasture; germination across all species went from 32% in natural vegetation to only 17% in removal plots. Four species had significantly lower germination where the graminoids had been removed. From the same experiment, survivorship of two species was significantly higher

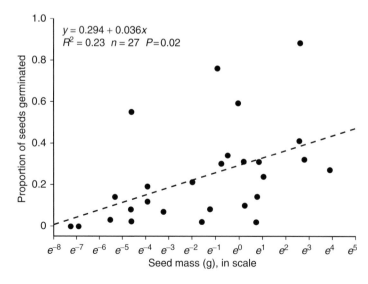

Figure 22.2 Relationship between seed mass (natural log scale) and germination for 27 species sown into intact grassland or pasture vegetation. Pooled from Zimmerman *et al.* (2000) and Hooper *et al.* (2002).

in removal plots, and the survival of other species did not differ among treatments. Notably, three of the four species with higher germination had survival that did not differ among treatments, so for these species, removal of pasture vegetation had a net negative effect on establishment.

In post-agricultural *Saccharum* grasslands in Panama, Hooper *et al.* (2002) compared the effect of two cutting treatments (cut once or twice), shading plus cutting (75% and 95% light reduction), and fire on the establishment of 20 woody species. Cutting significantly increased light at ground level, while cutting plus shading significantly increased soil moisture while reducing light. In general, species with larger seeds tended to have higher germination in the presence of dense *Saccharum*, although these findings were not analyzed statistically. These results and other studies (e.g., Holl *et al.* 2000, Myster 2004) suggest that resident grasses suppress smaller-seeded woody colonists more than large-seeded species, perhaps because grasses ameliorate detrimental light and heat levels for large-seeded species (Figure 22.2). For the 16 species that germinated, removing graminoids did not affect germination; however, shading led to substantially improved germination for 10 species. Thus shade may ameliorate harsh conditions (e.g., heat or desiccation) that inhibit germination. For 19 of 20 species, survival was highest in one of the shaded treatments versus the control, fire, or cutting treatments (Hooper *et al.* 2002). Overall, they concluded that cutting plus shading eliminated below-ground interference between *Saccharum* and tree seedlings. They suggested that the major constraint on survival and growth of seedlings is below-ground interference, that it is worse for smaller-seeded tree species, and that high light conditions could be inimical to germination.

Findings are similarly complex for survival and growth of transplants. We summarize the results of four studies (Gerhardt 1993, Hardwick *et al.* 1997, Zimmerman *et al.* 2000, Hooper *et al.* 2002) that examined the survival of tree seedlings transplanted into experimental plots where resident vegetation was left intact and where it had been removed (Figure 22.3). Survival was greater within intact vegetation for 16 species and greater in the cleared treatments for 10 species.

We arranged species in rank order of seed size from smallest to largest (Figure 22.3). The results suggest greater negative effects of grassy vegetation for species with smaller seeds but there was no statistically significant relationship between seed size and response to vegetation removal. Thus, it is premature to draw any firm conclusions from research to date.

Still, when taking all studies into account, the general trend is that high abundance of non-woody vegetation in old fields or abandoned pastures retards succession. For example, several types of non-woody vegetation can hinder woody plant colonization in post-agricultural sites: *Bacharis trinervis* in Ecuador (Zahawi and Augspurger 1999); the fern *Dicranopteris linearis* in Sri Lanka (Cohen *et al.* 1995); the fern *Nephrolepis multiflora* in the Caribbean (Rivera *et al.* 2000); a weedy *Bidens* species in Guatemala (Ferguson *et al.* 2003); *Melampodium divaricatum*, *Bidens pilosa*, and *Paspalum conjugatum* in Mexico (Purata 1986); *Phytolacca rivinoides* in Costa Rica (Harcombe 1977); and bamboos in southeastern Peru (Griscom and Ashton 2003).

Favorable microsites may ameliorate inhibition by pioneer vegetation

Pasture that is dominated by graminoids is often inimical to woody species recruitment. Consequently, microsites within pasture where grasses are less abundant may be foci for tree recruitment. For example, Peterson and Haines (2000) found much higher densities of woody species on rotting logs versus unbroken graminoid cover in a pasture in Costa Rica (see also Lack 1991). Similarly, Slocum (2000) found that woody recruits were five and eight times more abundant on rotting logs and within fern patches, respectively, versus open pasture dominated by grasses. In southern Venezuela, in the first year following slash-and-burn agriculture, Uhl *et al.* (1982) found significantly more tree seedlings on logs and slash versus areas away from these microsites. Apparently, these sites reduced competition with grasses and moderated harsh environmental conditions, but such sites may also be a refuge from seed and seedling predators (e.g., Long *et al.* 1998).

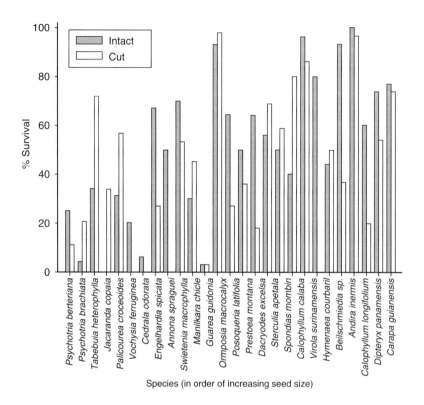

Figure 22.3 Survival of tree seedlings transplanted into or germinating in intact grassland or pasture vegetation, and into plots where vegetation had been cut, for 27 species. Pooled from Gerhardt (1993), Hardwick et al. (1997), Zimmerman et al. (2000), and Hooper et al. (2002). Species are arranged in order of increasing seed size, from left to right.

IS THE NUCLEATION MODEL A VIABLE CONCEPTUAL FRAMEWORK FOR TROPICAL SUCCESSION?

Relative to succession following agriculture in temperate regions, facilitation appears to play a more prominent role in the tropics. The colonization of woody species beneath both remnant trees and early colonizing shrubs and within plantations (Vieira et al. 1994, Slocum and Horvitz 2000, Slocum 2001, Zanne and Chapman 2001, Carriere et al. 2002, Holl 2002) are clear examples of facilitation operating to produce expanding patches as predicted by the nucleation model. Thus, much of the variation in the woody species composition, richness, and rate of succession may be explained by the initial abundance and diversity of trees that extend above the herbaceous layer. In addition, favorable microsites such as rotting logs may also provide foci for woody species establishment and thus the formation of distinct patches where woody vegetation becomes established. Early successional communities with few or no remnant trees or a paucity of key microsites may have very slow rates of succession. This strongly supports predictions of the nucleation model where patches of woody vegetation become established within a matrix of graminoids and eventually coalesce. Thus, while it is incomplete as a full conceptual framework, the nucleation model does describe the general dynamics of early post-agricultural succession, though more research is clearly needed. Combining the nucleation model with realistic models of seed dispersal and propagule supply (e.g., Nuttle and Haefner 2005) could provide a robust conceptual foundation for understanding early succession in the tropics.

The frequently dense graminoid vegetation in post-agricultural sites is clearly able to influence establishment of woody species. Indeed, many reports of arrested succession have been attributed to inhibition caused by competition with resident vegetation (Cohen *et al.* 1995, Zahawi and Augspurger 1999, Rivera *et al.* 2000). Still, as reported above, the presence of intact vegetation in some cases facilitates woody species establishment (Zimmerman *et al.* 2000, Hooper *et al.* 2002). It is likely that the effect of resident vegetation is neither uniformly negative nor positive, but that the effect varies predictably with the abundance of resident vegetation. Thus experiments are needed that extend removal studies to a focus on how variation in the abundance and identity of resident herbaceous vegetation influences woody species recruitment among contrasting life-history types. The degree to which the presence of resident vegetation facilitates or inhibits subsequent colonization of woody species needs to be integrated into the nucleation model.

LIMITATIONS OF EXISTING MODELS

Two of the major functions of models and hypotheses about succession are to organize and make sense of divergent findings, and to predict dynamics in new settings. We suggest that existing succession models are to varying degrees inadequate though they remain for the most part untested and they deserve greater consideration even if they are rejected. Existing models of secondary succession have overwhelmingly concentrated on life-history traits (e.g., Egler 1954, Drury and Nisbet 1973) and competitive interactions (Tilman 1985, Huston and Smith 1987). These models need to be expanded or refined in order to apply to tropical habitats. Two simple examples are illustrative. First, the least surprising spatial pattern documented in empirical studies is a much greater density and rate of colonization near

forest edges (e.g., Myster 2003a). Despite the ubiquity of this pattern, it cannot be explained via relay floristics, initial floristic composition, resource ratios, shade tolerance, or interaction categories. Second, several well-known examples illustrate the potential for multiple constraints on tropical succession (e.g., Aide and Cavelier 1994, Nepstad *et al.* 1996, Chapman and Chapman 1999, Holl *et al.* 2000), resulting in very slow or non-existent structural and compositional change. The explanations from existing models are unsatisfactory. The gradient-in-time models would imply that no species is adapted to do well in these sites, a conclusion challenged by the use of transplants. The resource ratio and shade tolerance models would imply that sites had reached a final stage with the best competitor occupying the site, a conclusion seed predation and dispersal studies show to be incorrect. The inhibition mechanism under Connell and Slatyer's interaction categories would imply that the current vegetation is simply preventing later-stage species from establishing, when the cause is at least partly a lack of propagules of later successional species.

AN EMPIRICAL RESEARCH PROGRAM

We recommend research that will contribute to laying a foundation for improved models and synthesis. We argue that it is important to undertake research that more fully quantifies the following five processes.

1 *Propagule input as a function of distance from source areas.* Succession requires that propagules arrive into abandoned pastures and a lack of propagules no doubt delays tropical succession. Models that attempt to predict colonization based upon interspecific differences in fecundity, seed size, modes of dispersal, and source populations will no doubt pay large dividends (e.g., Nuttle and Haefner 2005).

2 *The degree to which resident herbaceous vegetation inhibits or facilitates woody species recruitment.* The essential element is to evaluate the response of potential woody colonists to natural variation in the abundance and identity of the herbaceous

vegetation present in the pasture. Hypotheses should be developed regarding which life-history traits or functional groups are likely to be facilitated or inhibited by varying amounts of residual vegetation and then experiments conducted that vary the relevant factors (e.g., propagule number, resident biomass, functional group, litter mass). Overall, we suggest that a moderate amount of extant vegetation or litter may serve to facilitate germination and seedling survivorship by ameliorating local microclimatic conditions. As this vegetation or litter becomes very dense, it will likely become inimical to woody species recruitment (e.g., Carson and Peterson 1990). This may be why microsites that are devoid of grasses but have favorable microclimates (rotting logs) are areas of relatively dense woody seedling recruitment.

3 *The degree that resident trees facilitate woody species recruitment.* If the nucleation model applies broadly, then resident trees and shrubs should typically promote the establishment of woody vegetation in their proximity and also result in the spread of woody vegetation around these recruitment foci. Studies need to be conducted across a range of both abiotic (e.g., soil moisture, fertility) and biotic (e.g., graminoid abundance and composition) conditions to test the robustness of this model. These studies need to be combined with studies that manipulate the number of trees, structural complexity, or key microsites early in succession to fully evaluate the degree to which early succession in tropical systems is constrained by variation in these putative recruitment foci.

4 *The spatial and temporal variability in the intensity of predation on seeds and seedlings.* Greater seed predation appears to occur on smaller-seeded species; otherwise few generalizations have emerged due to a paucity of studies. This is problematic in terms of testing models of succession. For example, if seed predators congregate beneath trees or shrubs (due to enhanced cover or elevated food resources), seed predators could make areas beneath trees and shrubs inimical to woody species recruitment, in contrast to the predictions of the nucleation model.

5 *The correlations among life-history traits such as seed size, growth rate, fecundity, and shade and drought tolerance.* These correlations have been widely examined for temperate woody species but less so for tropical species (but see Garwood 1983). Almost all of the major models assume various correlations and trade-offs among plant traits, and the validity of predictions of these models rests on the validity of these assumptions. For example, the resource ratio model assumes a negative correlation between competitive ability for light and competitive ability for nutrients with relevant differences in allocation to shoots versus roots.

CONCLUSIONS

A substantial number of models or conceptual frameworks have been developed to explain or predict patterns or causes of species turnover in early successional habitats in temperate regions. We have presented a brief review of these models and found that most were limited in their usefulness when applied to post-agricultural succession in the tropics. However, these models have rarely been tested in tropical habitats and at least one, the nucleation model (Yarranton and Morrison 1974), shows real promise and should be tested, expanded to incorporate quantitative models of dispersal, and revised where appropriate. Another conceptual model, a hierarchical framework that provides an exhaustive list of causes for species turnover during succession (Pickett *et al.* 1987), was also considered highly relevant for studies of succession in any habitat. We suggest that future studies of succession should be designed to test robust general models that can then be refined and applied across larger geographic regions.

ACKNOWLEDGMENTS

We thank Brian McCarthy, Stefan Schnitzer, Henry Stevens, and especially John Paul for comments on earlier drafts of this manuscript. This work was supported by NSF grant DEB 94-24606 to C.J.P. The NSF grant to Peterson and Haines was named "Research toward sustainable land use and biodiversity in a mosaic of agriculture and tropical forest".

REFERENCES

Aide, T.M. and Cavelier, J. (1994) Barriers to lowland tropical forest restoration in the Sierra Nevada de Santa Marta, Colombia. *Restoration Ecology* 2, 219–229.

Aide, T.M., Zimmerman, J.K., Rosario, M., and Marcano, H. (1996) Forest recovery in abandoned cattle pastures along an elevational gradient in northeastern Puerto Rico. *Biotropica* 28, 537–548.

Carriere, S.M., Andre, M., Letourmy, P., Olivier, I., and McKey, D.B. (2002) Seed rain beneath remnant trees in a slash-and-burn agricultural system in southern Cameroon. *Journal of Tropical Ecology* 18, 353–374.

Carson, W.P. and Peterson, C.J. (1990) The role of litter in an old-field community: impact of litter quantity in different seasons on plant species richness and abundance. *Oecologia* 85, 8–13.

Chapman, C.A. and Chapman, L.J. (1999) Forest restoration in abandoned agricultural land: a case study from East Africa. *Conservation Biology* 13, 1301–1311.

Chinea, J.D. (2002) Tropical forest succession on abandoned farms in the Humacao Municipality of eastern Puerto Rico. *Forest Ecology and Management* 167, 195–207.

Clements, F.E. (1916) *Plant Succession: An Analysis of the Development of Vegetation*. Carnegie Institution of Washington Publication 242. Carnegie Institute, Washington, DC.

Cohen, A.L., Singhakumara, B.M.P., and Ashton, P.M.S. (1995) Releasing rain forest succession: a case study in the *Dicranopteris linearis* fernlands of Sri Lanka. *Restoration Ecology* 3, 261–270.

Connell, J.H. and Slatyer, R.O. (1977) Mechanisms of succession in natural communities and their role in community stability and organization. *American Naturalist* 111, 1119–1144.

Cubina, A. and Aide, T.M. (2001) The effect of distance from forest edge on seed rain and soil seed bank in a tropical pasture. *Biotropica* 33, 260–267.

Cusack, D. and Montagnini, F. (2004) The role of native species plantations in recovery of understory woody diversity in degraded pasturelands of Costa Rica. *Forest Ecology and Management* 188, 1–15.

Da Silva, J.M.C., Uhl, C., and Murray, G. (1996) Plant succession, landscape management, and the ecology of frugivorous birds in abandoned Amazonian pastures. *Conservation Biology* 10, 491–503.

Dosch, J.J., Peterson, C.J., and Haines, B.L. (2007) Seed rain during initial colonization of replicate abandoned pastures in the premontane wet forest zone of southern Costa Rica. *Journal of Tropical Ecology* 23, 1–9.

Drury, W.H. and Nisbet, I.C.T. (1973) Succession. *Journal of the Arnold Arboretum* 54, 331–368.

Duncan, R.S. and Chapman, C.A. (2003) Consequences of plantation harvest during tropical forest restoration in Uganda. *Forest Ecology & Management* 173, 235–250.

Duncan, R.S. and Duncan, V.E. (2000) Forest succession and distance from forest edge in an Afro-Tropical grassland. *Biotropica* 32, 33–41.

Egler, F.E. (1954) Vegetation science concepts. 1. Initial floristic composition, a factor in old-field vegetation development. *Vegetatio* 4, 412–417.

Ferguson, B.G., Vandermeer, J., Morales, H., and Griffith, D.M. (2003) Post-agricultural succession in El Peten, Guatemala. *Conservation Biology* 17, 818–828.

Finegan, B. (1996) Pattern and process in neotropical secondary rain forests: the first 100 years of succession. *Trends in Ecology and Evolution* 11, 119–124.

Forget, P.-M. (1997) Effect of microhabitat on seed fate and seedling performance in two rodent-dispersed tree species in rainforest in French Guiana. *Journal of Ecology* 85, 693–703.

Ganade, G. and Brown, V.K. (2002) Succession in old pastures of central Amazonia: role of soil fertility and plant litter. *Ecology* 83, 743–754.

Garwood, N.C. (1983) Seed germination in a seasonal tropical forest in Panama: a community study. *Ecological Monographs* 53, 159–181.

Gerhardt, K. (1993) Tree seedling development in tropical dry abandoned pasture and secondary forest in Costa Rica. *Journal of Vegetation Science* 4, 95–102.

Gómez-Pompa, A. and Vásquez-Yanes, C. (1981) Successional studies of a rain forest in Mexico. In D.C. West, H.H. Shugart, and D.B. Botkin (eds), *Forest Succession*. Springer-Verlag, New York, pp. 246–266.

Griscom, B.W. and Ashton, P.M.S. (2003) Bamboo control of forest succession: *Guadua sarcocarpa* in southeastern Peru. *Forest Ecology and Management* 175, 445–454.

Guariguata, M.R. and Ostertag, R. (2001) Neotropical secondary forest succession: changes in structural and functional characteristics. *Forest Ecology and Management* 148, 185–206.

Guevara, S. and Laborde, J. (1993) Monitoring seed dispersal at isolated standing trees in tropical pastures: consequences for local species availability. *Vegetatio* 107/108, 319–338.

Guevara, S., Laborde, J., and Sanchez-Rios, G. (2004) Rain forest regeneration beneath the canopy of fig trees isolated in pastures of Las Tuxtlas, Mexico. *Biotropica* 36, 99–108.

Harcombe, P.A. (1977) The influence of fertilization on some aspects of succession in a humid tropical forest. *Ecology* 58, 1375–1383.

Hardwick, K., Healey, J., Elliott, S., Garwood, N., and Anusarnsunthorn, V. (1997) Understanding and assisting natural regeneration processes in degraded seasonal evergreen forests in northern Thailand. *Forest Ecology and Management* 99, 203–214.

Holl, K.D. (1999) Factors limiting tropical rain forest regeneration in abandoned pasture: seed rain, seed germination, microclimate, and soil. *Biotropica* 31, 229–242.

Holl, K.D. (2002) Effect of shrubs on tree seedling establishment in an abandoned tropical pasture. *Journal of Ecology* 90, 179–187.

Holl, K.D., Loik, M.E., Lin, E.H.V., and Samuels, I.A. (2000) Tropical montane forest restoration in Costa Rica: overcoming barriers to dispersal and establishment. *Restoration Ecology* 8, 339–349.

Holl, K.D. and Lulow, M.E. (1997) Effects of species, habitat, and distance from edge on post-dispersal seed predation in a tropical rain forest. *Biotropica* 29, 459–468.

Holl, K.D. and Quiros-Nietzen, E. (1999) The effect of rabbit herbivory on reforestation of an abandoned pasture in southern Costa Rica. *Biological Conservation* 87, 391–395.

Hooper, E., Condit, R., and Legendre, P. (2002) Responses of 20 native tree species to reforestation strategies for abandoned farmland in Panama. *Ecological Applications* 12, 1626–1641.

Huston, M. and Smith, T. (1987) Plant succession: life history and competition. *American Naturalist* 130, 168–198.

Ingle, N.R. (2003) Seed dispersal by wind, birds, and bats between Philippine montane rainforest and successional vegetation. *Oecologia* 134, 251–261.

Jones, F.A., Peterson, C.J., and Haines, B.L. (2003) Seed predation in neotropical pre-montane pastures: site, distance, and species effects. *Biotropica* 35, 219–225.

Lack, A.J. (1991) Dead logs as a substrate for rainforest trees in Dominica. *Journal of Tropical Ecology* 7, 401–405.

Long, Z.T., Carson, W.P., and Peterson, C.J. (1998) Treefall mounds as refugia from herbivory. *Journal of the Torrey Botanical Society* 125, 165–168.

MacMahon, J.A. (1981) Successional processes: comparisons among biomes with special reference to probable roles of and influences on animals. In D.C. West, H.H. Shugart, and D.B. Botkin (eds), *Forest Succession*. Springer-Verlag, New York, pp. 277–304.

Martinez-Garza, C. and R. Gonzalez-Montagut (1999) Seed rain from forest fragments into tropical pastures in Los Tuxtlas, Mexico. *Plant Ecology* 145, 255–265.

Martinez-Garza, C. and Gonzalez-Montegut, R. (2002) Seed rain of fleshy-fruited species in tropical pastures in Los Tuxtlas, Mexico. *Journal of Tropical Ecology* 18, 457–462.

Meli, P. (2003) Tropical forest restoration: twenty years of academic research. *Intersciencia* 28, 581–589.

Myster, R.W. (2003a) Vegetation dynamics of a permanent pasture plot in Puerto Rico. *Biotropica* 35, 422–428.

Myster, R.W. (2003b) Effects of species, density, patch-type, and season on post-dispersal seed predation in a Puerto Rican pasture. *Biotropica* 35, 542–546.

Myster, R.W. (2004) Regeneration filters in post-agricultural fields of Puerto Rico and Ecuador. *Plant Ecology* 172, 199–209.

Nepstad, D.C., Uhl, C., Pereira, C.A., and Cardoso da Silva, J.M. (1996) A comparative study of tree establishment in abandoned pasture and mature forest of eastern Amazonia. *Oikos* 76, 25–39.

Noble, I.R. and Slatyer, R.O. (1980) The use of vital attributes to predict successional changes in plant communities subject to recurrent disturbances. *Vegetatio* 43, 5–21.

Nuttle, T. and Haefner, J.W. (2005) Seed dispersal in heterogeneous environments: bridging the gap between mechanistic dispersal and forest dynamics models. *American Naturalist* 165, 336–349.

Osunkoya, O.O. (1994) Postdispersal survivorship of north Queensland rainforest seeds and fruits: effects of forest, habitat and species. *Australian Journal of Ecology* 19, 52–64.

Pearson, T.R.H., Burslem, D.F.R.P., Goeriz, R.E., and Dalling, J.W. (2003) Regeneration niche partitioning in neotropical pioneers: effects of gap size, seasonal drought and herbivory on growth and survival. *Oecologia* 137, 456–465.

Pena-Claros, M. (2003) Changes in forest structure and species composition during secondary forest succession in the Bolivian Amazon. *Biotropica* 35, 450–461.

Pena-Claros, M. and De Boo, H. (2002) The effect of forest successional stage on seed removal of tropical rain forest tree species. *Journal of Tropical Ecology* 18, 261–274.

Peterson, C.J. and Haines, B.L. (2000) Early successional patterns and potential facilitation of woody plant colonization by rotting logs in premontane Costa Rican pastures. *Restoration Ecology* 8, 361–369.

Pickett, S.T.A. (1976) Succession: an evolutionary interpretation. *American Naturalist* 110, 107–119.

Pickett, S.T.A., Collins, S.L., and Armesto, J.J. (1987) Models, mechanisms, and pathways of succession. *Botanical Review* 53, 335–371.

Purata, S.E. (1986) Floristic and structural changes during old-field succession in the Mexican tropics in relation to site history and species availability. *Journal of Tropical Ecology* 2, 257–276.

Puyravaud, J.-P. (2003) Rain forest expansion mediated by successional processes in vegetation thickets in the Western Ghats of India. *Journal of Biogeography* 30, 1067–1080.

Rivera, L.W., Zimmerman, J.K., and Aide, T.M. (2000) Forest recovery in abandoned agricultural lands in a karst region of the Dominican Republic. *Plant Ecology* 148, 115–125.

Sarmiento, F.O. (1997) Arrested succession in pastures hinders regeneration of Tropandean forests and shreds mountain landscapes. *Environmental Conservation* 24, 14–23.

Schnitzer, S.A. and Carson, W.P. (2001) Gap dynamics and the maintenance of diversity in a tropical forest. *Ecology* 82, 913–919.

Sisk, T.D. (1991) *Distribution of birds and butterflies in heterogeneous landscapes*. PhD dissertation, Stanford University, Stanford, CA.

Slocum, M.G. (2000) Logs and fern patches as recruitment sites in a tropical pasture. *Restoration Ecology* 8, 408–413.

Slocum, M.G. (2001) How tree species differ as recruitment foci in a tropical pasture. *Ecology* 82, 2547–2559.

Slocum, M.G. and Horvitz, C.C. (2000) Seed arrival under different genera of trees in a neotropical pasture. *Plant Ecology* 149, 51–62.

Swaine, M.D. and Hall, J.B. (1983) Early succession on cleared forest land in Ghana. *Journal of Ecology* 71, 601–627.

Tilman, D. (1985) The resource-ratio hypothesis of plant succession. *American Naturalist* 125, 827–852.

Tilman, D. (1988) *Plant Strategies and the Dynamics and Structure of Plant Communities*. Princeton University Press, Princeton.

Uhl, C., Clark, H., Clark, K., and Maquirino, P. (1982) Successional patterns associated with slash-and-burn agriculture in the Upper Rio Negro region of the Amazon basin. *Biotropica* 14, 249–254.

Vasconcelos, H.L. and Cherrett, J.M. (1997) Leaf-cutting ants and early forest regeneration in central Amazonia: effects of herbivory on tree seedling establishment. *Journal of Tropical Ecology* 13, 357–370.

Vieira, I.C.G., Uhl, C., and Nepstad, D. (1994) The role of the shrub *Cordia multispicata* Cham. as a "succession facilitator" in an abandoned pasture, Paragominas, Amazonia. *Vegetatio* 115, 91–99.

Wijdeven, S.M.J. and Kuzee, M.E. (2000) Seed availability as a limiting factor in forest recovery processes in Costa Rica. *Restoration Ecology* 8, 414–424.

Wilson, J.B., Gitay, H., Roxburgh, S.H., King, W.M., and Tangney, R.S. (1992) Egler's concept of "Initial Floristic Composition" in succession – ecologists citing it don't agree what it means. *Oikos* 64, 591–593.

Yarranton, G.A. and Morrison, R.G. (1974) Spatial dynamics of a primary succession: nucleation. *Journal of Ecology* 62, 417–428.

Zahawi, R.A. and Augspurger, C.K. (1999) Early plant succession in abandoned pastures in Ecuador. *Biotropica* 31, 540–552.

Zahawi, R.A. and Augspurger, C.K. (2006) Tropical forest restoration: tree islands as recruitment foci in degraded lands of Honduras. *Ecological Applications* 16, 464–478.

Zanne, A.E. and Chapman, C.A. (2001) Expediting reforestation in tropical grasslands: distance and isolation from seed sources in plantations. *Ecological Applications* 11, 1610–1621.

Zimmerman, J.K., Pascarella, J.B., and Aide, T.M. (2000) Barriers to forest regeneration in an abandoned pasture in Puerto Rico. *Restoration Ecology* 8, 350–360.

Chapter 23

CHANCE AND DETERMINISM IN TROPICAL FOREST SUCCESSION

Robin L. Chazdon

OVERVIEW

Based on chronosequence studies and permanent plot studies, I describe successional changes in vegetation structure, population dynamics, species richness, and species composition in tropical forests. Tropical secondary forests initially increase rapidly in structural complexity and species richness, but the return to pre-disturbance species composition may take centuries or longer – or may never occur. Vegetation dynamics during secondary tropical forest succession reflect a complex interplay between deterministic and stochastic processes. As more studies of succession are carried out, the importance of stochastic factors is becoming more evident. Features of the local landscape, such as proximity to forest fragments or large areas of diverse, mature forest, strongly impact the nature and timing of species colonization. Disturbance history and previous land use strongly determine the extent to which resprouts or remnant mature-forest vegetation dominate during secondary forest regeneration and, together with soil fertility, strongly influence the composition of dominant pioneer species. Community assembly processes during succession appear to be strongly affected by dispersal limitation at all stages. Initial community composition, often dominated by long-lived pioneer species, changes extremely slowly over time. Tree seedlings that colonize only after the stem exclusion stage of succession may take decades or longer to recruit as trees, thus contributing to a slow rate of change in tree species composition. Long-term studies within individual sites do not support the notion that secondary succession in tropical forests leads to convergence in species composition, as suggested by some chronosequence studies. Predictable, directional changes do occur in vegetation during tropical forest succession, but convergent trends are more apparent for structural features, life-forms, and functional groups than for species composition. Although relay floristics may well describe changes in species dominance early in succession, there is little evidence to support this model during later phases of succession. Clearly, there is much more work to be done, with a particular need to avoid biasing initial site selection and to use experimental approaches in combination with long-term studies. Through these research approaches, we will be better able to identify the effects of deterministic versus stochastic processes in tropical forest succession.

INTRODUCTION: SUCCESSIONAL THEMES AND VARIATIONS

Secondary succession is the long-term directional change in community composition following a disturbance event, often at a large (>1 ha) spatial scale. Hurricanes, floods, landslides, windstorms, cyclones, and fires are examples of major natural disturbances that can initiate the successional process (Waide and Lugo 1992, Whitmore and Burslem 1998, Chazdon 2003). Human impacts are responsible for most of the world's secondary forests, however (Brown and Lugo 1990, Guariguata and Ostertag 2001). The relationship between land use and forest succession is complex; the type and intensity of land use, soil fertility, and the surrounding landscape matrix all strongly influence the nature and rate of successional processes (Purata 1986, Hughes *et al.* 1999, Johnson *et al.* 2000, Moran *et al.* 2000, Pascarella *et al.* 2000, Silver *et al.* 2000, Ceccon *et al.* 2003, Ferguson *et al.* 2003, Myster 2004).

A major challenge in studies of tropical forest successional dynamics is to reveal the relative importance of deterministic versus stochastic processes affecting species composition, spatial distributions, and their rates of change. Niche-based processes, such as the competition–colonization trade-off and successional niche theory, generate predictable transitions between early and late successional species with distinct sets of life-history traits (Rees *et al.* 2001). But the rate of these transitions and the particular species involved can vary widely across forests within the same region and climate. Although these theoretical predictions reflect an underlying theme of successional change observed in many temperate and tropical forests, the overall importance of stochastic factors during vegetation succession remains poorly understood. Deterministic successional processes are defined as orderly and predictable changes in species abundance determined by climate, soils, and species life history (Clements 1904, 1916), whereas stochastic processes are influenced by random events that are not predictable in nature.

The most direct way to study succession is to follow changes in structure and composition over time. Yet in tropical forests, few studies have examined changes in vegetation structure and composition over time for more than a few years (Chazdon *et al.* 2007). Consequently, our knowledge of successional processes derives almost exclusively from chronosequence studies (Pickett 1989, Guariguata and Ostertag 2001). Space-for-time substitutions make (often) unrealistic assumptions, such as similar environmental conditions, site history, and seed availability across sites as well as over time. Moreover, sites are often carefully selected to minimize variation in abiotic conditions, and site selection may favor stands that conform to preconceived models of successional development of vegetation. Successional areas available for study may also represent a biased sample of the landscape due to underlying differences in soil fertility, slope, elevation, or drainage – these environmental factors often influence patterns of land use and abandonment. Ideally, chronosequence studies should be based on a series of replicated plots of different ages selected using objective criteria (land-use records,

soil type). Ruiz *et al.* (2005), for example, randomly selected 59 forests in six age classes (based on aerial photographs and satellite imagery) in a 56-year tropical dry forest chronosequence on Providencia Island, Colombia. Long-term studies within individual sites, however, are more effective in providing a mechanistic understanding of succession, population dynamics, and effects of local site factors on recruitment, growth, and mortality of different growth forms and size classes (Foster and Tilman 2000, Sheil 2001). These aspects are poorly understood for most tropical secondary forests, but are essential for a complete understanding of successional dynamics and their local, regional, or geographic variations. Knowledge of successional processes is also critically needed to develop ecologically sound tropical forest management and restoration programs.

Successional studies in tropical forests have generally emphasized the tree component, ignoring the community dynamics of tree seedlings and saplings and non-tree life-forms. Moreover, few studies have examined non-arboreal life-forms, such as herbs, shrubs, and lianas (Dewalt *et al.* 2000, Martin *et al.* 2004, Capers *et al.* 2005). Thus, we have a limited understanding of how the plant community as a whole is changing during succession within forests of known history. Successional forests are embedded within a dynamic regional landscape that determines the pool of species available for colonization, the genetic diversity of seed sources, the availability of pollinators, herbivores, seeds, dispersal agents, and pathogens, and the likelihood of repeated human perturbations. Finally, within successional as well as mature tropical forests, climate fluctuations and human-induced environmental changes simultaneously exert pressure on a wide range of ecological processes (Ramakrishnan 1988, Clark 2004, Laurance *et al.* 2004, Malhi and Phillips 2004). All of these factors lead to a highly complex set of interactions that ultimately drive community dynamics, and thus seriously challenge our ability to distinguish the relative importance of niche-based versus neutral processes (Vandermeer 1996).

In this chapter, I examine patterns and processes of vegetation dynamics during secondary tropical forest succession. First, I present a brief discussion of successional theory, as it applies to forest succession. I then describe the basic framework of tropical forest succession, based on chronosequence studies in wet, dry, and montane tropical forests. I discuss the few long-term studies that describe successional change in vegetation structure, population dynamics, composition, and species richness within individual forests. Finally, I consider the question of whether secondary tropical forests ever reach a stable climax community. Throughout, I re-examine the role of deterministic versus stochastic processes during different phases of tropical forest succession and across different types of landscapes.

THEORETICAL BACKGROUND

Successional theory has a long history, originating at the beginning of the twentieth century with studies by Cowles (1899), Clements (1904, 1916), Gleason (1926), and Tansley (1935). Clements viewed succession as a highly orderly, deterministic process in which the community acts as an integrated unit, analogous to the development of an individual organism. The endpoint of succession is a stable climax community (homeostasis), which exists in equilibrium with the contemporary climatic conditions. This deterministic view was later emphasized by Odum (1969) in his pioneering studies of ecosystem development.

Critics challenged this view of communities as highly integrated units and stressed the importance of chance events and the role of individualistic behavior of species during succession. Gleason (1926) viewed succession as a largely stochastic process with communities reflecting individualistic behavior of component species, whereas Tansley (1935) argued that regional climate alone does not determine the characteristics of climax vegetation. Watt (1919, 1947) examined successional processes in small-scale disturbances within forests, emphasizing the unstable spatial mosaic created by patch dynamics. Egler (1954) maintained that the initial floristic composition of an area was a strong determinant of later vegetation composition, emphasizing the role of site pre-emption and the long-term legacy of chance colonization events.

During the 1970s, ecologists replaced equilibrium paradigms with alternative non-equilibrium theories and began to emphasize the mechanistic basis of ecological processes. Vegetation dynamics during succession were viewed as emerging from properties of component species (*sensu* Gleason 1926), rather than from a holistic, organismal concept of community development (Pickett *et al.* 1987). The mechanistic approach of Drury and Nisbet (1973), Pickett (1976), Connell and Slatyer (1977), Bazzaz (1979), and Noble and Slatyer (1980) emphasized changes in resource availability during succession in relation to the life-history characteristics of dominant species. These works led to the predominant contemporary view that vegetation change emerges from the interactions of component populations as they ebb and flow in response to changing environmental conditions (Rees *et al.* 2001). The intermediate disturbance hypothesis (Connell 1978, 1979) also grew out of this non-equilibrium thinking, and predicted that species diversity would reach a peak during intermediate phases of succession and would decrease to low levels (approaching monodominance) in a late successional community in the absence of disturbance. The notion that competitive exclusion is prevented by disturbance events, thus permitting more species to coexist, has now become well accepted (Huston 1979, Wilkinson 1999, Sheil and Burslem 2003).

Early studies of tropical vegetation showed strong evidence of non-equilibrium viewpoints. In his studies of forests of Ivory Coast, Aubréville (1938) questioned the concept of stable "climax" vegetation, replacing it with a concept of gap-phase dynamics that Richards (1952) termed the "mosaic theory" and Watt (1947) termed the "cyclical theory of regeneration" (Burslem and Swaine 2002). Studies by Eggeling (1947) and Jones (1956) on old secondary forests of Uganda and Nigeria, respectively, were used to provide detailed empirical support for Connell's non-equilibrium theory. Webb *et al.* (1972), in one of the first experimental studies of tropical forest regeneration, emphasized the importance of chance dispersal events and highly patchy spatial distributions in early phases of secondary succession.

Three conceptual frameworks have been applied to studies of vegetation dynamics during tropical forest succession. The first framework examines the role of deterministic (predictable) versus stochastic (unpredictable) factors in vegetation dynamics. If deterministic forces predominate (as viewed by Clements), successional communities that share the same climate should exhibit predictable convergence in community composition over time, regardless of differences in initial composition (Christensen and Peet 1984). This view also holds that mature forests within a region should maintain stable and similar species composition (Terborgh *et al.* 1996). Environmental variation across sites, among other factors, can create divergence in species composition during succession, rather than convergence (Leps and Rejmánek 1991).

A second framework is based on the timing of colonization of species during succession and contrasts the models of initial floristic composition versus relay floristics (Egler 1954). Relay floristics involves colonization by later successional species well after the initial disturbance, whereas initial floristic composition applies when species from all stages colonize early following disturbance but reach peak abundances at different times according to their growth rates and longevities (Gómez-Pompa and Vázquez-Yanes 1974, Bazzaz and Pickett 1980, Finegan 1996).

A third framework focuses on the relative importance of species life-history traits and species interactions in determining the balance among mechanisms of tolerance, inhibition, and facilitation during succession (Connell and Slatyer 1977, Rees *et al.* 2001). Later successional species may establish due to facilitation or release from inhibition by earlier successional species, or due to intrinsic life-history characteristics such as arrival time, growth rate, and longevity with no direct interaction with early species.

These conceptual frameworks also apply to community assembly processes in mature forests (Young *et al.* 2001), in the study of gap-phase dynamics (Whitmore 1978), in assessing the role of random drift versus environment in determining spatial variation in species composition (Hubbell and Foster 1986), and in developing neutral models of community composition based on source pools and dispersal limitation (Hubbell *et al.* 1999). Moreover, the relative importance of successional processes may change over time (Connell and Slatyer 1977, Walker and Chapin 1987).

AN OVERVIEW OF TROPICAL SECONDARY FOREST SUCCESSION

Phases of succession

In its general outline, tropical forest succession is similar to temperate forest succession (Oliver and Larson 1990), but the recovery of forest structure can be particularly rapid in tropical wet climates (Ewel 1980). The sequence and duration of successional phases may vary substantially among tropical forests, depending upon the nature of the initializing disturbance and the potential for tree colonization and forest structural development. Vegetation succession following hurricanes follows a different trajectory than post-agricultural succession in the same region (Boucher *et al.* 2001, Chazdon 2003). Similarly, post-extraction secondary forests follow different successional trajectories than swidden fallows (Riswan *et al.* 1985, Chokkalingam and de Jong 2001, Chazdon 2003).

The first phase of secondary succession is often dominated by herbaceous species (grasses or ferns in abandoned pastures), vines, shrubs, and woody lianas (Budowski 1965, Kellman 1970, Gómez-Pompa and Vázquez-Yanes 1981, Ewel 1983, Toky and Ramakrishnan 1983, Finegan 1996). This building phase is termed the "stand initiation stage" by Oliver and Larson (1990). Dramatic changes in vegetation structure and composition occur during the first decade of succession in tropical regions, as woody species rapidly colonize abandoned fields (see reviews by Brown and Lugo 1990 and Guariguata and Ostertag 2001). Rapid growth of early colonizing trees ("pioneers") can bring about canopy closure in only 5–10 years after abandonment. Early woody regeneration consists of new seedling recruits from seed rain and the seed bank (Benitez-Malvido *et al.* 2001) as well as resprouts; the latter often dominate the early woody community (Uhl *et al.* 1981,

Kammesheidt 1998). Resprouting is the most common form of early plant establishment in swidden fallows (Uhl *et al.* 1981, Kammesheidt 1998, Perera 2001, Schmidt-Vogt 2001), and may lead to the development of uneven cover and clumped tree distributions during the first phase of regrowth (Schmidt-Vogt 2001).

Following hurricanes, logging, and superficial fires, resprouting residual trees dominate early regenerating woody vegetation, often bypassing the stand initiation phase. Studies of forest regeneration following Hurricane Joan in southeastern Nicaragua provide a detailed description of this "direct regeneration" process, where the post-hurricane forest composition was similar to that of the mature, pre-disturbance forest due to extensive resprouting of damaged stems (Yih *et al.* 1991, Vandermeer *et al.* 1995, 1996, Boucher *et al.* 2001).

Following abandonment of intensive agriculture, such as cattle pastures, the first seedling shrub and tree recruits emerge from the seed bank or newly dispersed seed and tend to be wind-, bird-, or bat-dispersed species with small seeds that require direct light or high temperatures to germinate (Uhl and Jordan 1984, Vázquez-Yanes and Orozco-Segovia 1984). Rotting logs (Peterson and Haines 2000) and remnant trees (Elmqvist *et al.* 2001, Slocum 2001, Guevara *et al.* 2004) facilitate colonization of bird- and bat-dispersed tree species in abandoned pastures, whereas the aggressive growth and clonal spread of shrubs, vines, and lianas can inhibit seedling recruitment of light-demanding tree species (Schnitzer *et al.* 2000, Schnitzer and Bongers 2002). In Sri Lanka, dense growth of bamboo can suppress tree regeneration during early succession following swidden agriculture (Perera 2001).

The stand initiation phase of succession is the most vulnerable to invasion by exotic species (Fine 2002). In many tropical regions, particularly on islands, exotic pioneer species form dense, monospecific stands in early phases of succession, such as *Lantana camara* in Australia, *Piper aduncum* in eastern Malesia, and *Leucaena leucocephala* in Vanuatu (Whitmore 1991). Invasive plant species can have long-lasting effects on tropical forest succession. Invasive grasses such as *Saccharum spontaneum* in Panama and

Imperata cylindrica in Indonesia can inhibit regeneration of woody species (D'Antonio and Vitousek 1992, Otsamo *et al.* 1995, Hooper *et al.* 2004). Young secondary forests in the Caribbean islands of Puerto Rico and the Dominican Republic are often dominated by exotic species (Rivera and Aide 1998, Aide *et al.* 2000, Lugo 2004, Lugo and Helmer 2004, Martin *et al.* 2004). In moist forests of Madagascar that were logged (50 years earlier) or cleared for subsistence agriculture (150 years earlier), populations of invasive species persisted throughout the successional trajectory, with a lasting effect on woody species richness and composition (Brown and Gurevitch 2004). Inhibitory effects of invasive species are not limited to tropical islands, however. In subtropical northwestern Argentina, native tree recruitment in 5–50-year-old secondary forests was negatively related to the dominance (% basal area) of the invasive tree *Ligustrum lucidum* (Oleaceae; Lichstein *et al.* 2004).

Canopy closure signals the beginning of the second phase, termed the "stem exclusion phase" by Oliver and Larson (1990). As early colonizing trees increase rapidly in basal area and height, understory light availability decreases dramatically. These changes are associated with decreasing woody seedling density and high seedling mortality of shade-intolerant species of shrubs, lianas, and canopy trees (Capers *et al.* 2005). Low light availability in the understory favors establishment of shade-tolerant tree and palm species that are dispersed into the site from surrounding vegetation by birds and mammals (particularly bats). By 10–20 years after abandonment, the stage is set for a shift in the abundance and composition of tree species that gradually plays out over decades, if not centuries. This constitutes the third and longest phase of forest succession.

This third phase of forest succession corresponds to the "understory reinitiation stage" of Oliver and Larson (1990) and is characterized by a gradual turnover of species composition in canopy and subcanopy layers. The advance regeneration in the understory often contains species characteristic of mature old-growth forests (Guariguata *et al.* 1997, Chazdon *et al.* 1998, Denslow and Guzman 2000). Eventually, the death of canopy trees creates gaps, increasing resource availability for new recruits. Over long periods of time, perhaps several hundred years, the canopy will consist of mixed cohorts of tree species that were not present early in succession, thus initiating the "old-growth stage" of forest dynamics (Oliver and Larson 1990). Old-growth forests are characterized by a complex vertical and horizontal structure, presence of large, living, old trees, large woody debris, and highly diverse canopy and understory vegetation (Budowski 1970).

Ecological processes affecting vegetation dynamics and species composition vary among successional phases. During the stand initiation phase of succession, stochastic processes of dispersal and colonization are likely to influence community composition most strongly, whereas later in succession, deterministic processes, such as species

fidelity to environment, may become more powerful factors (Walker and Chapin 1987). Thus, processes of dispersal, seed germination, resprouting, and rapid growth of shade-intolerant species determine early species composition (Table 23.1). Some studies show that rates of seed predation are highest during this stage of tropical forest succession (Hammond 1995, Peña-Claros and de Boo 2002, Andresen *et al.* 2005), but these patterns may be species- and site-specific (Holl and Lulow 1997). After canopy closure, forest dynamics in the stem exclusion phase (phase 2) reflect high mortality of shade-intolerant shrubs and lianas, suppression and mortality of shade-intolerant tree species within the subcanopy, and high recruitment of shade-tolerant species that are primarily dispersed by birds and bats (Table 23.1). These processes have been described in detail by Chazdon *et al.* (2005) and Capers *et al.* (2005)

Table 23.1 Vegetation dynamics processes across successional phases in tropical forests.

Phase 1: Stand initiation phase (0–10 years)
Germination of seed-bank and newly dispersed seeds
Resprouting of remnant trees
Colonization by shade-intolerant and shade-tolerant pioneer trees
Rapid height and diameter growth of woody species
High mortality of herbaceous old-field colonizing species
High rates of seed predation
Seedling establishment of bird- and bat-dispersed, shade-tolerant tree species

Phase 2: Stem exclusion phase (10–25 years)
Canopy closure
High mortality of lianas and shrubs
Recruitment of shade-tolerant seedlings, saplings, and trees
Growth suppression of shade-intolerant trees in understory and subcanopy
High mortality of short-lived, shade-intolerant pioneer trees
Development of canopy and understory tree strata
Seedling establishment of bird- and bat-dispersed, shade-tolerant tree species
Recruitment of early colonizing, shade-tolerant tree and palm species into the subcanopy

Phase 3: Understory reinitiation stage (25–200 years)
Mortality of long-lived, shade-intolerant pioneer trees
Formation of canopy gaps
Canopy recruitment and reproductive maturity of shade-tolerant canopy and subcanopy tree and
   palm species
Increased heterogeneity in understory light availability
Development of spatial aggregations of tree seedlings

Notes: Names of phases are derived from Oliver and Larson (1990). Dispersal remains a key process throughout, but shifts from predominantly long-distance dispersal initially to predominantly local dispersal towards the end of phase 3. Ages reflect approximate rates of succession as observed in the Caribbean lowlands of Costa Rica.

for secondary forests in northeastern Costa Rica. Over time, these processes lead to the long understory reinitiation phase, characterized by mortality of long-lived pioneer tree species, formation of canopy gaps, and reproductive maturity of shade-tolerant tree species and their continued recruitment into the canopy. The relatively homogeneous, low light conditions in the understory of phase 2 forests act as a strong filter for recruitment of the shade-tolerant tree species that will later recruit in the canopy. Understory light conditions become more heterogeneous during later stages of succession and create more diverse opportunities for seedling and sapling recruitment than in phase 2 forests (Nicotra et al. 1999). Thus, the understory reinitiation phase (phase 3) is associated with increasing species richness and evenness in all vegetation size classes. Successional phases do not correspond strictly with age classes, however, as actual rates of succession are known to vary widely with climate, soils, previous land use, and landscape configuration (Arroyo-Mora et al. 2005).

Successional patterns of tree colonization

Studies of vegetation dynamics in mature tropical forests emphasize two divergent life-history modes of trees: pioneer and shade-tolerant species (Swaine and Whitmore 1988). Yet studies of successional forests clearly suggest a far greater complexity in regeneration modes and life histories. For example, Budowski (1965, 1970), Knight (1975), and Finegan (1996) noted the distinction between short- and long-lived pioneer tree species in lowland forests of Mesoamerica. Secondary forest in phase 2 (stem exclusion phase) in northeastern Costa Rica is actually composed of three groups of pioneer tree species: (1) short-lived shade-intolerant species, (2) long-lived shade-intolerant species, and (3) long-lived shade-tolerant species. All of these species colonize early, but the "short-lived" species (which tend to be smaller in stature as well) generally do not persist in the canopy beyond the first 10–15 years (Budowski 1965, 1970). The inheritors of the canopy are two groups of "long-lived"

trees that grow to large stature and persist for many decades or longer. One group of these secondary forest trees lacks seedling or sapling recruits in older secondary forests (Figure 23.1a), whereas a second group shows abundant recruitment of seedlings and saplings (Figure 23.1b). This second group of "shade-tolerant pioneers" has been recognized in only one previous study (Dalling et al. 2000), but plays an important role in wet forest succession, at least in northeastern Costa Rica. These species are common or dominant species in mature forests of the region, such as *Pentaclethra macroloba*, *Hernandia didymantha*, and *Inga thibaudiana* (Figure 23.1). Canopy individuals of these species appear to reach reproductive maturity within 15–20 years during secondary forest succession (personal observation).

Although many shade-tolerant tree species (and canopy palm species) colonize during the stand initiation phase (e.g., Kenoyer 1929, Knight 1975, Peña-Claros 2003), other species do not appear in the seedling community until decades have passed, and these tend to occur in low abundance and frequency. Finegan (1984) maintained that forest species generally do not colonize during the stand initiation phase of succession and that some facilitation is required for their establishment. He proposed a composite mechanism of succession, whereby short- and long-lived pioneers establish early and forest species colonize later, during the stem exclusion and understory reinitiation phases (phases 2 and 3). Later establishment could reflect limited seed dispersal, differences in abundance of mature trees in surrounding communities, or specific regeneration requirements that are met only during later stages of succession. We have little detailed information on patterns of tree colonization within individual sites in the second and third phases of succession, as most studies have emphasized vegetation dynamics during the stand initiation phase (Finegan 1996, Myster 2004).

Gómez-Pompa and Vásquez-Yanes (1981) first proposed that tropical forest succession follows a relay floristics model (*sensu* Egler 1954), where species achieve their greatest abundance at different times, such that dominant species shift temporally across a successional sere. A study

(a)

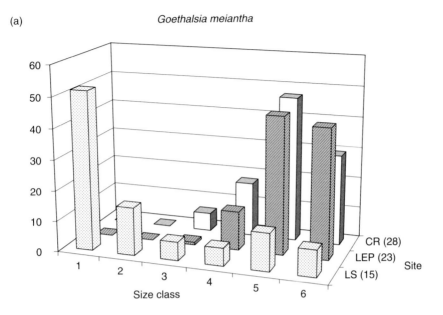

(b)

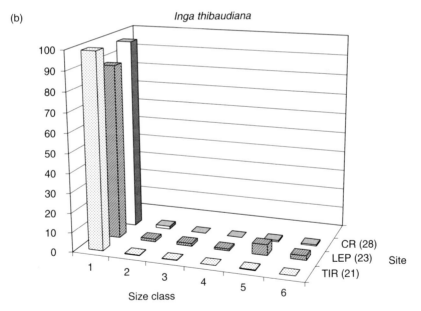

Figure 23.1 Size distributions of (a) a shade-intolerant pioneer (*Goethalsia meiantha*) and (b) a shade-tolerant pioneer (*Inga thibaudiana*) in three secondary forests of different age since abandonment in northeastern Costa Rica. Forest sites are abbreviated as follows: CR = Cuatro Rios, LEP = Lindero el Peje; LS = Lindero Sur; TIR = Tirimbina. Site ages are in parentheses. Size classes are defined as follows: 20 cm ht > class 1 < 1 cm dbh ≤ class 2 < 2.5 cm dbh ≤ class 3 < 5 cm dbh ≤ class 4 < 10 cm dbh ≤ class 5 < 25 cm dbh ≤ class 6.

of forest succession in the Bolivian Amazon, based on two chronosequences following shifting cultivation, generally supported this model. Peña-Claros (2003) distinguished four different groups of tree species: (1) species that reach maximum abundance during phase 1 of succession; (2) species that dominate in phase 2; (3) species that reach their peak abundance in phase 3 or old-growth forests; and (4) mid-successional species that showed no trend in abundance with stand age. Species in the third group varied in their period of first colonization; some species were present in 2–3-year-old stands, whereas others first appeared in stands 20 years old (Peña-Claros 2003). Few data are available to test this model during the late phases of succession.

Forest structure

The most striking changes that occur during tropical forest succession are structural changes, such as the increase in canopy height, density of trees ≥ 10 cm diameter at breast height (dbh), basal area, and above-ground biomass. In wet lowland areas of northeastern Costa Rica, these structural changes cause a reduction of understory light availability to below 1% transmittance of diffuse photosynthetically active radiation within 15–20 years after abandonment (Nicotra et al. 1999). Leaf area index increases rapidly and often reaches a peak before other components of forest structure (Brown and Lugo 1990). Mean photosynthetic light availability near the forest floor was not significantly different between young secondary forest (15–20 years old) and mature forest stands in wet tropical regions of Costa Rica (Nicotra et al. 1999). Light availability in young secondary forests, however, is more spatially homogeneous than in mature forests due to even-aged canopy cover and absence of treefall gaps (Nicotra et al. 1999). Structural changes during tropical forest succession are well documented in reviews by Brown and Lugo (1990), Guariguata and Ostertag (2001), Chazdon (2003), and Chazdon et al. (2007). Tropical secondary forests often show rapid structural convergence with mature forests (Saldarriaga et al. 1988, Guariguata et al.

1997, Ferreira and Prance 1999, Aide et al. 2000, Denslow and Guzman 2000, Kennard 2002, Peña-Claros 2003, Read and Lawrence 2003). Rates of structural convergence depend strongly on soil fertility (Moran et al. 2000), soil texture (Johnson et al. 2000, Zarin et al. 2001), and the duration and intensity of land use prior to abandonment (Uhl et al. 1988, Nepstad et al. 1996, Hughes et al. 1999, Steininger 2000, Gehring et al. 2005).

Species richness and diversity

Many chronosequence studies have also documented rapid recovery of species richness and species diversity during tropical succession, but these trends are strongly influenced by soil fertility and land-use history (Brown and Lugo 1990, Guariguata and Ostertag 2001, Chazdon 2003, Chazdon et al. 2007). Inconsistent methods, use of different stem size classes, and presentation of biased diversity measures confound accurate comparisons of species abundance and richness across study plots. Moreover, many chronosequence studies lack replication of age classes and use small plots 0.1 ha or less in size. Finally, the ability to identify and locate forest areas that have remained undisturbed for over a century has proven challenging in many tropical areas (Clark 1996, Willis et al. 2004). These problems help to explain the inconsistent patterns in species diversity found across chronosequence studies.

Species richness and stem density are positively correlated in virtually all vegetation samples (Denslow 1995, Condit et al. 1996, Chazdon et al. 1998, Sheil 2001, Howard and Lee 2003), confounding comparisons of species number among sites that differ in overall stem density or area sampled (Gotelli and Colwell 2001). Thus, the best way to compare species richness among sites is to use rarefaction techniques to compare the accumulation of species within a site as a function of the cumulative number of individuals sampled (Chazdon et al. 1998). It is not appropriate to use sample data to compare species per stem, because species accumulation is a nonlinear function of the number of individuals in a sample (Chazdon *et al.* 1999, Gotelli and Colwell 2001). Indices of species diversity, such as Shannon–Weiner or Simpson indices, that emphasize evenness or dominance, respectively, are less biased by density than simple species counts per unit area (species density). Species richness estimation techniques can also be useful in correcting for sample-size bias (Colwell and Coddington 1994, Chazdon *et al.* 1998), although no method (including Fisher's α; Condit *et al.* 1996) can overcome limitations of sparse data due to small sample areas or small numbers of stems. Here, I restrict my comparisons to studies based on diversity indices or that have incorporated rarefaction techniques or species richness estimators to compare species richness across stands within a chronosequence.

A variety of temporal patterns have been observed in successional studies of tropical forests. Eggeling (1947) conducted the first study of species composition across a tropical forest chronosequence, based on a series of 10 plots in Budongo Forest, Uganda. He concluded that there was an initial rise in species numbers (species density) during succession, reaching a peak at intermediate phases of succession, followed by a decline during late succession. His analysis, however, did not take into account differences in tree density among the plots. Sheil (2001) applied the rarefaction method of Hurlbert (1971) to these data, and confirmed that the plots of intermediate age indeed had the highest species richness of trees ≥ 10 cm dbh, whereas late successional plots had the lowest species richness. In a comparison of early, intermediate, and late successional tropical dry forests in Costa Rica, Kalacska *et al.* (2004) also found higher species richness of trees ≥ 5 cm dbh in sites of intermediate age. This trend was further supported by the Shannon diversity index and an incidence-based, non-parametric species richness estimator (Kalacska *et al.* 2004). In northwest Guyana, 60-year-old secondary forest had higher species richness (Fisher's α) of trees ≥ 10 cm dbh than mature forests (van Andel 2001).

Other studies have documented continuously increasing species diversity with stand age, but these studies often lack comparative data for older secondary forests or "primary" forests. In swidden

fallow succession in northeastern India, Toky and Ramakrishnan (1983) found a linear increase in species diversity (Shannon index) with fallow age during the first 15–20 years. Chinea (2002) found that the Shannon diversity index for trees ≥ 2.5 cm dbh increased with age since abandonment in sites from 1 to 45 years old in eastern Puerto Rico. In a 56-year chronosequence in tropical dry forest on Providencia Island, Colombia, Ruiz *et al.* (2005) found that species richness, based on rarefaction of stems ≥ 2.5 cm dbh, increased steadily with increasing age of abandonment; abundance-based, non-parametric species richness estimators confirmed this trend. Peña-Claros (2003) found a similar pattern for two 40-year chronosequences in Bolivian Amazon forest; Shannon diversity index increased with stand age for understory, subcanopy, and canopy vegetation layers. Along a chronosequence in Argentinian subtropical montane forests, Grau *et al.* (1997) also found that Shannon diversity of trees ≥ 10 cm dbh increased in young stands and by 45–50 years reached values similar to mature forests in the region. In the upper Rio Negro of Colombia and Venezuela, Saldarriaga *et al.* (1988) found similar values of Shannon and Simpson's indices for stems ≥ 1 cm dbh between 40-year-old stands and mature forests.

Several studies in the Old World tropics suggest that species richness recovers very slowly, even in older secondary forests. Shannon diversity for trees ≥ 10 cm dbh in a 55-year-old secondary rainforest in central Kalimantan, Indonesia was significantly lower compared with adjacent mature forest (Brearley *et al.* 2004). In Singapore, Turner *et al.* (1997) also found significantly lower Shannon diversity for trees ≥ 30 cm dbh in approximately 100-year-old secondary forest compared with primary forest. Even after 150 year of recovery following clearing for subsistence agriculture, moist forests of Ranomafana National Park, Madagascar showed significantly lower species richness (estimated number of species/250 stems) than uncleared forests (Brown and Gurevitch 2004).

In general, canopy trees (≥ 30 cm dbh) show slower recovery of species richness during succession compared with seedlings and saplings due to the longer time required for shade-tolerant species to reach these size classes. Using sample-based rarefaction curves, Guariguata *et al.* (1997) found that species richness of trees ≥10 cm dbh was consistently lower in young secondary stands (15–20 years) compared with mature forest stands in wet lowland forest of Costa Rica, but these differences were less pronounced or absent for woody seedlings and saplings. Similarly, Denslow and Guzman (2000) found that estimates and indices of seedling species richness did not vary with stand age across a 70-year tropical moist forest chronosequence in Panama.

Species composition

Species composition appears to vary independently of species richness across a chronosequence (Finegan 1996, Guariguata and Ostertag 2001, Chazdon 2003). Even where species richness and forest structure of secondary forests are not significantly different from those of mature forests, species composition remains quite distinct in secondary forests for periods up to centuries (Finegan 1996). Early differences in colonizing vegetation and land use can impact the successional trajectory of a particular site (Janzen 1988, Mesquita *et al.* 2001). In 6–10-year-old forest of the Brazilian Amazon, *Cecropia*-dominated logging clear-cuts were considerably more diverse than *Vismia*-dominated stands on abandoned pastures (Mesquita *et al.* 2001), reflecting facilitation of recruitment by residual vegetation following logging (Chazdon 2003). Variation in species composition due to site history and environmental heterogeneity creates a major challenge in comparing floristic composition of secondary forests with mature forests within a single landscape (Whitmore 1973, 1974, Ashton 1976, Duivenvoorden 1996, Swaine 1996, Clark *et al.* 1998). First, this variability makes it difficult to select representative mature forest areas for robust comparisons of species composition between secondary and mature forests. Second, land-use history may interact with environmental conditions, such as elevation, soil fertility, slope, and drainage. Third, in many instances, remaining mature forest areas have been exposed to human and natural disturbances of variable spatial and temporal impact

(Whitmore and Burslem 1998) or may continue to be influenced by disturbances that occurred centuries ago or longer (Denevan 1992, Brown and Gurevitch 2004, Wardle *et al.* 2004, Willis *et al.* 2004). Consequently, the use of nearby mature forests as a benchmark can be problematic.

Tropical dry forests tend to exhibit fewer successional stages and faster recovery of species composition compared with wet forests (Ewel 1980, Murphy and Lugo 1986, Perera 2001, Kennard 2002). In tropical dry forests, late successional species are tolerant of hot and dry conditions, resprouting is common (Denslow 1996), and wind dispersal is more common than in wet forests. Furthermore, due to the higher frequency of large-scale fire, even the oldest, least disturbed dry forests in the landscape may be undergoing late stages of secondary succession (Kennard 2002). Most of the present closed-canopy mature forests in dry regions of Sri Lanka, for example, are secondary forests on abandoned formerly irrigated cultivated land (Perera 2001). Fire tends to damage small stems more than large stems, and frequent fires may therefore retard succession (Goldammer and Seibert 1990, Cochrane and Schulze 1999).

Although we do not yet know what processes influence the rate of change of species richness during tropical wet forest succession, three factors are probably involved. First, long-lived pioneer species persist well into the understory reinitiation stage, pre-empting space and slowing the rate of species turnover. Second, low light availability in young and intermediate aged second-growth forests and the rarity or absence of canopy gaps may restrict establishment and recruitment of gap-requiring tree species (Nicotra *et al.* 1999, Dupuy and Chazdon 2006). Third, low seed availability may limit colonization of tree species. Dispersal limitation is high in recently abandoned clearings and in secondary as well as mature tropical forests (Dalling *et al.* 1988, Holl 1999, Wijdeven and Kuzee 2000, Muller-Landau *et al.* 2002, Hooper *et al.* 2004, Svenning and Wright 2005). The extent of dispersal limitation may be greatest for species with animal-dispersed seeds. Following logging in lowland rainforests of eastern Borneo, seed addition increased seedling recruitment for five animal-dispersed species, but not for two wind-dispersed species (Howlett and Davidson 2003). Even when secondary forests are close to mature forests, seed dispersal can be a major limitation (Gorchov *et al.* 1983, Corlett 1992, Turner *et al.* 1997, Wunderle 1997, Duncan and Chapman 1999, Holl 1999, Ingle 2003). Martinez-Garza and Gonzalez-Montagut (1999) found that dispersal limitation of forest interior species resulted in pioneer dominance for 30–70 years in abandoned pastures of lowland tropical regions of Mexico.

Under ideal conditions, the early arrival and establishment of some shade-tolerant canopy tree species (including palms) can increase the rate of succession, as these species often grow rapidly in height and reach reproductive maturity within 20–30 years, when they begin to produce their own seedling cohorts (Sezen *et al.* 2005). Many of these species are capable of recruitment into canopy tree size classes (≥ 25 cm dbh) in the absence of gaps (Chazdon unpublished data). If seedlings of shade-tolerant and slow-growing species colonize later, during the stem exclusion or understory reinitiation phase, their recruitment to the canopy may require several decades or longer (Finegan and Chazdon unpublished data).

Few studies have statistically compared species composition across a tropical forest chronosequence. Terborgh *et al.* (1996) compared species composition in early, middle, and late successional floodplain vegetation with mature floodplain forests of the Manu River in Peru. In this study, cluster analysis showed that the five mature floodplain forests were most similar to each other in species composition and that they differed considerably from successional forests. A detrended correspondence analysis suggested a clear directionality to species compositional changes during floodplain succession in this region. A similar approach was used by Sheil (1999) to compare canopy tree species composition for the 10 sites in Eggeling's (1947) study of forests in Budongo, Uganda. This analysis indicated a consistent compositional progression across the plot series, with the ranking of plots conforming precisely to Eggeling's original successional sequence. Within this set of plots, there was strong evidence

for compositional convergence towards a species-poor forest dominated by *Cynometra alexandri* (ironwood; Sheil 1999). An alternative interpretation, suggested by Sheil (1999), is that Eggeling originally selected the plots to fit his preconceived model of an ordered developmental successional series in Budongo Forest.

Life-forms, functional groups, and life-history traits

During succession, life-form composition shifts dramatically, particularly during the stand initiation phase. During the first 5 years of post slash-and-burn succession, Ewel and Bigelow (1996) documented decreases in herbaceous vines, increases in shrubs and trees, and a dramatic increase in epiphytes between 30 and 36 months. Grass and forb dominance peaked after 3 years in an abandoned pasture in Puerto Rico (Myster 2003). Vines, ferns, and persistent grasses can impede establishment and growth of woody shrubs and trees in abandoned pastures (Holl *et al.* 2000, Hooper *et al.* 2004). More often, however, early dominance of large-leaved herbaceous species facilitates establishment of shade-tolerant woody species (Denslow 1978, Ewel 1983). Across a sequence of stands from 20 to over 100 years old in Barro Colorado Island and surrounding areas, liana abundance decreased as a function of stand age (Dewalt *et al.* 2000). Liana size increased during succession, however, resulting in a lack of correspondence between stand age and liana basal area. Liana diversity (as measured by Fisher's α) was higher in young stands than in older stands, up to 70 years in age (Dewalt *et al.* 2000).

Considering only woody life-forms in wet tropical forests of northeastern Costa Rica, Guariguata *et al.* (1997) found that shrub abundance was significantly higher whereas understory palm abundance was significantly lower in young secondary stands (15–20 years old) compared with old-growth stands. Mature canopy palms (stems ≥ 10 cm dbh) were also significantly more abundant in old-growth stands. Woody seedlings in second-growth permanent plots in this region showed decreasing abundance of shrubs and lianas and increasing abundance of canopy and understory palms over a 5-year period, mirroring chronosequence trends (Capers *et al.* 2005).

Several studies have documented successional changes in leaf phenology and wood characteristics. Tropical dry forests are a mix of deciduous and evergreen species, but early successional communities tend to be dominated by deciduous species, with increasing abundance of evergreen species later in succession. A trend towards increasing leaf lifespan with succession is well established for tropical wet and seasonal forests (Reich *et al.* 1992). Another well-established trend is that of increasing wood density from early to late succession (Whitmore 1998, Suzuki 1999, Muller-Landau 2004).

Successional trends have also been observed in seed dispersal modes and other reproductive traits. During the first few months of succession following clear-cutting in northeastern Costa Rica, nearly all newly establishing plants were of wind-dispersed species (Opler *et al.* 1977). This fraction decreased over time, while the percentage of fleshy-fruited species increased. Within 3 years, animal-dispersed species composed 80% of the species, similar to values in mature forest. Self-compatibility is more prevalent among species in early successional stages, whereas out-crossing is more common in later stages as dioecy and self-incompatibility increase (Opler *et al.* 1980). Chazdon *et al.* (2003) compared the distribution of reproductive traits in woody vegetation in relation to successional stage in forests of northeastern Costa Rica. In second-growth trees, relative abundance of species with explosive seed dispersal, hermaphroditic flowers, and insect pollination was higher, whereas relative abundance of species with animal dispersal and mammal pollination was lower compared with old-growth forests (Chazdon *et al.* 2003). In the same study area, Kang and Bawa (2003) examined variation in flowering time, duration, and frequency in relation to successional status. Supra-annual flowering was proportionately less common in early successional species than in species of later successional stages, but flowering time did not vary consistently with successional status (Kang and Bawa 2003).

SUCCESSIONAL DYNAMICS WITHIN INDIVIDUAL FORESTS

Few studies have examined successional dynamics within individual tropical forests over time. Here, I highlight these studies and examine whether the trends observed within individual forests are similar to those trends described from chronosequence studies. This topic is discussed in more detail by Chazdon *et al.* (2007), based on case studies from northeastern Costa Rica and Chiapas, México. Sheil (1999, 2001) and Sheil *et al.* (2000), examined long-term changes in species richness and composition in five plots (1.5–1.9 ha) originally studied by Eggeling (1947) in Budongo, a semi-deciduous forest in Uganda. In plot 15, a former grassland at the forest margin, the number of tree species more than 10 cm dbh increased from 25 to 74 over 48 years and rarefaction revealed an increase in species per 200 individuals from 22 to 45 (Sheil 2001). But few shade-tolerant stems or species were present over these years (Sheil *et al.* 2000), suggesting a strong influence of savanna species. In plot 7, which was a late successional stand in the 1940s, species richness increased and the number of smaller stems increased. Over 54 years, there was a relative increase in shade-tolerant stems, but a decrease in the proportion of shade-tolerant species (Sheil *et al.* 2000). Larger stems in this plot showed lower average mortality rates (1% per year) than those reported for other tropical forests. Considering all of the plots in Eggeling's study that were also monitored over 54 years (several had silvicultural interventions), Sheil (2001) found that each plot showed increases in species richness, exceeding the richness found within Eggeling's original chronosequence. The peak in species richness observed for intermediate successional sites in Eggeling's original series was not observed in the time series data, however. Using a size-structured approach, Sheil (1999) compared temporal trends in species composition within plots. In the time series analysis, only one plot supported Eggeling's model, but overall the temporal changes within plots did not support the model of convergent vegetation composition during succession or a mid-successional peak in species richness.

Vandermeer *et al.* (2000) monitored annual changes in species richness of forests severely damaged by Hurricane Joan in eastern Nicaragua. Over a 10-year period, species richness of stems ≥3.2 cm dbh increased two- to three-fold. After only 10 years of recovery, the hurricane-damaged forests had higher species richness than undisturbed forests within the region (Vandermeer *et al.* 2000). These same six hurricane-damaged forests were subjected to an analysis of species compositional trajectories, including 12 years of data (Vandermeer *et al.* 2004). Analyses of multidimensional distance were used to assess whether these sites were becoming more similar over time, as predicted by deterministic (Clementsian) models of succession. Results indicated that three of these six pairwise comparisons showed increasingly divergent vegetation, two showed increasingly similar vegetation, and one showed no significant trend. Thus, Vandermeer *et al.* (2004) concluded that successional pathways were not convergent among these different plots, perhaps due to differences in initial conditions or to later successional dynamics.

Lang and Knight (1983) followed changes in tree growth and dynamics over a 10-year period in a 60-year-old secondary forest on Barro Colorado Island, Panama. All species ≥2.5 cm dbh were followed in a single 1.5 ha plot. During this period, mortality exceeded recruitment and net tree density declined by 11%. Trees above 10 cm dbh increased in density and stand basal area and biomass increased accordingly. Species varied widely in mortality rates and in diameter growth rates. The overall turnover rate of stems in the plot was 7.3%, with pioneer species showing overall declines in abundance and species typical of older forest recruiting into the canopy (Lang and Knight 1983).

Few studies have monitored vegetation dynamics of second-growth forests on an annual basis (Breugel *et al.* 2006, Chazdon *et al.* 2007). Chazdon *et al.* (2005) monitored mortality and recruitment annually for 6 years for trees ≥5 cm dbh in four 1 ha plots in wet second-growth, lowland rainforests in northeastern Costa Rica. In 12–15-year-old stands, abundance decreased 10–20% in the small size class (5–10 cm dbh), but increased 49–100% in the large size class (≥25 cm dbh) over 6 years. Common species changed dramatically in abundance over 6 years within plots, reflecting high mortality of early colonizing tree species and high rates of recruitment of shade-tolerant tree and canopy palm species. Mortality rates of small trees (5–9.9 cm dbh) were higher in younger than in older stands, but large trees (≥25 cm dbh) showed low rates of mortality, averaging 0.89% per year across stands and years. Most tree mortality occurred in overtopped individuals and therefore did not lead to the formation of canopy gaps. Tree mortality in these young secondary forests (particularly for trees ≥25 cm dbh) was highly sensitive to dry season rainfall, even during non-El Niño Southern Oscillation years (Chazdon *et al.* 2005). Woody seedling density in these four secondary forest plots declined over 5 years, whereas Shannon diversity and the proportion of rare species increased (Capers *et al.* 2005). Among plots, seedling species composition showed no tendency towards convergence over this period.

These studies support the hypothesis that successional dynamics are being driven by high mortality of light-demanding species (mainly in small size classes) and simultaneous recruitment of shade-tolerant trees into the canopy (Rees *et al.* 2001), with low mortality of long-lived pioneer species in the canopy. Vandermeer *et al.* (2004) documented high rates of mortality of suppressed trees beneath the canopy in 10–14-year-old forests recovering from hurricane damage. In secondary forests developing after pasture abandonment in Costa Rica, mortality rates of large trees appear to be lower compared with mature tropical forests, whereas recruitment of trees into canopy size classes is high (Chazdon *et al.* 2005). Thus, trees recruit to canopy positions in the absence of canopy gaps in these second-growth forests. Whereas canopy gaps are thought to drive much of the dynamics of canopy tree recruitment in mature tropical forests, the *absence* of canopy gaps seems to drive species turnover in secondary forests during the transition from phase 2 to phase 3.

The few studies conducted to date suggest that rates of recruitment, mortality, growth, and species turnover are particularly high within smaller size classes (<10 cm dbh) during the stand initiation phase and decrease as stands enter the understory reinitiation phase of succession (Breugel *et al.* 2006). The decreased rates of change in species and stem turnover over time reflect an increased relative abundance of slow-growing, shade-tolerant species overall, but particularly in smaller size classes (Chazdon *et al.* 2007).

RECRUITMENT LIMITATION DURING SUCCESSION

Many studies have examined the relative importance of biotic and abiotic factors that affect seedling establishment and recruitment during tropical forest succession. These factors, such as light availability, seed predation, and non-local seed dispersal, vary in importance across successional stages (Figure 23.2). During early stages of succession in abandoned fields and pastures, for example, seedling recruits originate from the seed bank and from non-local seed rain, and these factors assume high importance in controlling seedling establishment (Young *et al.* 1987, Dupuy and Chazdon 1998, Benitez-Malvido *et al.* 2001). Seed predation rates are initially high in abandoned fields (Uhl 1987), and several

studies suggest that rates of mammalian seed predation (post-dispersal) decrease during succession (Hammond 1995, Notman and Gorchov 2001).

Light availability is uniformly high in abandoned fields, but becomes increasingly limiting for seedling recruitment as forest cover increases during succession. Gaps are small and relatively uncommon in young secondary forests (Yavitt *et al.* 1995, Nicotra *et al.* 1999, Denslow and Guzman 2000), but increase in size and frequency in later stages of succession. In a 1.5 ha plot, Lang and Knight (1983) documented 13 new canopy gaps created by treefalls during a 10-year observation period in a 60-year-old secondary forest on Barro Colorado Island, whereas no gaps had been observed in this site previously. Gap creation in young secondary forests (phase 2) resulted in increased abundance and species richness of woody seedlings (Dupuy and Chazdon 2006). It is therefore likely that canopy gaps are associated with increased abundance and species richness of regenerating seedlings during the understory reinitiation phase of secondary forest succession as well as during the old-growth phase (Nicotra *et al.* 1999; Figure 23.2). Canopy gaps promote increases in tree species richness through increasing overall levels of recruitment as well as permitting establishment and

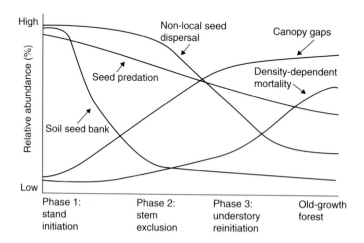

Figure 23.2 Shifting relative importance of biotic and abiotic factors that affect seedling recruitment and mortality across tropical forest secondary succession following abandonment of cultivated fields or pastures.

recruitment of light-demanding species (Denslow 1987, Hubbell *et al.* 1999, Brokaw and Busing 2000).

During early stages of succession, shade-tolerant tree species are not yet reproductively mature and therefore seeds must be dispersed from nearby or distant mature forests or forest fragments, if remnant trees are not present (Guevara *et al.* 1986). As shade-tolerant species recruit to canopy positions and become reproductively mature, local seed shadows increase the potential for density-dependent effects on seedling recruitment and growth (Janzen 1970, Connell 1971; Figure 23.2). Mean seed dispersal distances also are expected to decrease. Ultimately, these successional trends in seedling recruitment and spatial distribution of reproductive trees influence the abundance, species composition, and genetic composition of saplings and trees (Sezen *et al.* 2005).

During the end of the stand initiation phase of succession, when the forest canopy begins to close, fast-growing, shade-intolerant colonizing tree species are present as canopy trees and are also found as smaller individuals in the understory, as seedlings and saplings. As time progresses and the understory becomes more shaded, however, these shade-intolerant tree species are eliminated from the seedling and sapling pool and shade-tolerant species not present in the canopy colonize these small size classes (Guariguata *et al.* 1997). Chao *et al.* (2005) predicted that, as secondary forests mature, compositional similarity between tree species and seedlings or saplings would initially be high (phase 1), but would quickly decline to a minimum during intermediate stages of succession (phase 2) followed by an increase later in succession as shade-tolerant trees reach reproductive maturity and produce seedlings that can establish, grow, and survive (phase 3). Using an abundance-based estimator of the Jaccard index, Chao *et al.* (2005) found that compositional similarity between seedling and tree assemblages and between sapling and tree assemblages was, indeed, initially high in the youngest (12-year-old) stand, as predicted. As the forest matures, tree seedling and sapling pools gradually become enriched by shade-tolerant species not represented as canopy trees, resulting

in a decrease in compositional similarity that reached a minimum in the 23-year-old stand. This minimum similarity represents a point in forest succession of maximum recruitment limitation for both seedlings and saplings. In the 28-year-old second-growth plot, the abundance-based Jaccard index increased, reflecting recruitment of shade-tolerant species in all three size classes. The similarity index continued to increase and stabilized at 0.4–0.5 in two old-growth stands.

IS THERE AN ENDPOINT TO SUCCESSION?

The distinction between old secondary forests and mature forests is often blurry. Budowski (1970) pointed out several features that distinguish "climax" from old secondary forests in the Neotropics, including abundant regeneration of dominant shade-tolerant canopy tree species, slow-growing species, trees with dense wood and large gravity- or animal-dispersed seeds, lower abundance of shrubs, highly diverse and abundant epiphytes, and abundant large woody lianas.

Does succession ever reach a stable climax? This is a difficult question to address because the process of succession can occur over broad spatial scales. The successional framework described here applies to large-scale disturbances that lead to relatively homogeneous regenerating stands. As stands develop and spatial heterogeneity increases due to canopy gaps or other disturbances, small-scale patch dynamics and dispersal limitation begin to exert a strong influence on community composition and organization. Thus, different late successional forest stands are likely to show divergence in species composition due to exogenous disturbances or endogenous heterogeneity, even if they shared a similar early successional trajectory. For tropical forests, there is much reason to question the notion of a stable climax (Clark 1996).

Just as responses to disturbance can move forests off a late successional trajectory, historical legacies of human disturbance can influence long-term patterns of species composition in forests that are not visibly disturbed at present. In Central Africa, the dominant tree species in many old-growth forests recruit poorly, even in canopy gaps (Aubréville 1938, Jones 1956). More than 20% of the tree species in old-growth forests of southern Cameroon showed a preference for recruitment in shifting cultivation fields. The presence of charcoal in almost a third of the areas sampled supports the view that these forests are currently undergoing late stages of succession (van Gemerden *et al.* 2003). Evidence from other studies confirms that large-scale disturbances in rainforest areas throughout the world have been caused by widespread historical human impact (Denevan 1992, White and Oates 1999, Bayliss-Smith *et al.* 2003, Willis *et al.* 2004).

If there is no stable endpoint to the successional process, we are forced to view all forests as points along a successional continuum. We must also recognize that we may never be able to reconstruct the initial (pre-disturbance) species composition of a successional forest. The challenge is then to identify how biotic and abiotic factors at a range of spatial scales influence the successional trajectory of particular forests. This task may ultimately require experimental approaches at the scale of entire communities and landscapes, but such large-scale experiments will be challenging to execute and manage over long time periods. A mixed approach involves conducting experimental studies combined with monitoring of long-term changes in vegetation dynamics in sets of replicated stands that initially span a range of successional ages but share similar abiotic conditions.

SUCCESSION IN RELATION TO LANDSCAPE PATTERN

Forest succession occurs within the context of the surrounding landscape. As tropical landscapes become more deforested and fragmented over time, these landscape patterns will influence both the pattern and the processes of secondary forest succession. In shifting cultivation fields of Belize, composition of woody and herbaceous species was significantly influenced by distance to older forest, but species richness and evenness were not significantly affected (Kupfer *et al.*

2004). Abandoned fields close to intact forest had greater densities of successional woody taxa that are common in seasonally dry, subtropical forests.

Landscape-level studies clearly show that secondary forests more frequently develop in areas close to or bordering existing forest patches and that species diversity and composition recover more quickly in areas close to large forest patches (Tomlinson *et al.* 1996). Although comparative studies are greatly needed, these trends may be more representative of neotropical regions than in East Asian forests, where mature forest species often fail to recruit, even in adjacent second-growth forests (Turner *et al.* 1997). In this case, recruitment failure may be due, at least in part, to the extinction or rarity of large frugivores, which are important dispersal agents for large-seeded mature forest species (Turner *et al.* 1997). In montane Costa Rica, secondary forests were more likely to occur near old-growth forests, at increased elevation, on steeper slopes, further from roads, in areas of lower population density, and within forest reserves (Helmer 2000). Distance to older forest was a key predictor of species richness and diversity in a landscape-scale study of secondary forests in Puerto Rico (Chinea 2002). Chinea and Helmer (2003) examined the effect of landscape pattern on species composition in secondary forests in Puerto Rico, based on a series of 167 forest inventory plots (each approximately 120 m^2) that varied in climate, land-use history, and landscape structure. Canonical correspondence analysis based on nine variables explained only 16% of the total variance in species abundances. Land use covaried with elevation and substrate, so variation in species composition of secondary forests was generated by interactions between biophysical and socioeconomic forces (Chinea and Helmer 2003). Species composition of abandoned coffee plantations (at higher elevations) remained distinct from that of abandoned pastures (at lower elevations). Distance to large forest patches (at least 3800 ha) was also a significant factor explaining variation in total and native species richness, although the effect was small in this large-scale study (Chinea and Helmer 2003).

Tropical forests are among the most complex and diverse ecosystems in the world. It should be no surprise, therefore, that the process of transformation from a massively disturbed forest or an abandoned agricultural field or pasture to a community resembling the original structure, species richness, life-form composition, and species composition is prolonged, often idiosyncratic, and strongly contingent upon history and chance events.

Clearly, there is much more work to be done, with a particular need to avoid biasing initial site selection and to use experimental approaches in combination with long-term studies. Through these research approaches, we will be better able to identify the effects of deterministic versus stochastic processes in tropical forest succession.

ACKNOWLEDGMENTS

I gratefully thank Robert Capers, Susan Letcher, Walter Carson, and three anonymous reviewers for their comments on an earlier version of this manuscript. My research on vegetation dynamics in tropical secondary forests has been supported by the Andrew Mellon Foundation, the National Science Foundation, the University of Connecticut Research Foundation, and the Organization for Tropical Studies. Many students, field assistants, and research collaborators have contributed to this effort and I thank them all.

REFERENCES

Aide, T.M., Zimmerman, J.K., Pascarella, J.B., Rivera, L., and Marcano-Vega, H. (2000) Forest regeneration in a chronosequence of tropical abandoned pastures: Implications for restoration ecology. *Restoration Ecology* 8, 328–338.

Andresen, E., Pedroza-Espino, Allen, E.B., and Perez-Salicrup, D.R. (2005) Effects of selective vegetation thinning on seed removal in secondary forest succession. *Biotropica* 37, 145–148.

Arroyo-Mora, J.P., Sanchez-Azofeifa, A., Kalacska, M.E.R., and Rivard, B. (2005) Secondary forest detection in a Neotropical dry forest landscape using Landsat 7 ETM+ and IKONOS imagery. *Biotropica* 37, 498–507.

Ashton, P.S. (1976) Mixed dipterocarp forest and its variation with habitat in the Malayan lowlands: a re-evaluation at Pasoh. *Malayan Forester* 39, 56–72.

Aubréville, A. (1938) La forêt coloniale: les forêts de l'Afrique occidentale française. *Annales Académie Sciences Coloniale* 9, 1–245.

Bayliss-Smith, T.P., Hviding, E., and Whitmore, T.C. (2003) Rain forest composition and histories of human disturbance in Solomon Islands. *Ambio* 32, 346–352.

Bazzaz, F.A. (1979) The physiological ecology of plant succession. *Annual Review of Ecology and Systematics* 10, 351–371.

Bazzaz, F.A. and Pickett, S.T.A. (1980) Physiological ecology of tropical succession: a comparative review. *Annual Review of Ecology and Systematics* 11, 287–310.

Benitez-Malvido, J., Martinez-Ramos, M., and Ceccon, E. (2001) Seed rain versus seed bank, and the effect of vegetation cover on the recruitment of tree seedlings in tropical successional vegetation. In G. Gottsberger, S. Liede, and J. Cramer (eds), *Life Forms and Dynamics in Tropical Forests*. Stuttgart, pp. 1–18.

Breugel, M. van, Bongers, F., and Martínez-Ramos, M. (2006) Community-level species dynamics during early secondary forest succession. *Journal of Tropical Ecology* 22, 1–12; 663–674. doi:10.1017/S0266467406003452

Boucher, D.H., Vandermeer, J.H., de la Cerda, I.G., Mallona, M.A., Perfecto, I., and Zamora, N. (2001) Post-agriculture versus post-hurricane succession in southeastern Nicaraguan rain forest. *Plant Ecology* 156, 131–137.

Brearley, F.Q., Prajadinata, S., Kidd, P.S., Proctor, J., and Suriantata (2004) Structure and floristics of an old secondary rain forest in Central Kalimantan, Indonesia, and a comparison with adjacent primary forest. *Forest Ecology and Management* 195, 385–397.

Brokaw, N. and Busing, R.T. (2000) Niche versus chance and tree diversity in forest gaps. *Trends in Ecology and Evolution* 15, 183–187.

Brown, K.A. and Gurevitch, J. (2004) Long-term impacts of logging on forest diversity in Madagascar. *Proceedings of the National Academy of Sciences of the United States of America* 101, 6045–6049.

Brown, S. and Lugo, A.E. (1990) Tropical secondary forests. *Journal of Tropical Ecology* 6, 1–32.

Budowski, G. (1965) Distribution of tropical American rain forest species in the light of successional processes. *Turrialba* 15, 40–42.

Budowski, G. (1970) The distinction between old secondary and climax species in tropical Central American lowland forests. *Tropical Ecology* 11, 44–48.

Burslem, D.F.R.P. and Swaine, M.D. (2002) Forest dynamics and regeneration. In R.L. Chazdon and T.C. Whitmore (eds), *Foundations of Tropical Forest Biology: Classic Papers with Commentaries*. University of Chicago Press, Chicago, pp. 577–583.

Capers, R.S., Chazdon, R.L., Redondo Brenes, A., and Vilchez Alvarado, B. (2005) Successional dynamics of woody seedling communities in tropical secondary forests. *Journal of Ecology* 93, 1071–1084.

Ceccon, E., Huante, P., and Campo, J. (2003) Effects of nitrogen and phosphorus fertilization on the survival and recruitment of seedlings of dominant tree species in two abandoned tropical dry forests in Yucatan, Mexico. *Forest Ecology and Management* 182, 387–402.

Chao, A., Chazdon, R.L., Colwell, R.K., and Shen, T.-J. (2005) A new statistical approach for assessing similarity of species composition with incidence and abundance data. *Ecology Letters* 8, 148–159.

Chazdon, R.L. (2003) Tropical forest recovery: legacies of human impact and natural disturbances. *Perspectives in Plant Ecology, Evolution and Systematics* 6, 51–71.

Chazdon, R.L., Careaga, S., Webb, C., and Vargas, O. (2003) Community and phylogenetic structure of reproductive traits of woody species in wet tropical forests. *Ecological Monographs* 73, 331–348.

Chazdon, R.L., Colwell, R.K., and Denslow, J.S. (1999) Tropical tree richness and resource-based niches. *Science* 285, 1459a.

Chazdon, R.L., Colwell, R.K., Denslow, J.S., and Guariguata, M.R. (1998) Statistical methods for estimating species richness of woody regeneration in primary and secondary rain forests of NE Costa Rica. In F. Dallmeier and J. Comiskey (eds), *Forest Biodiversity Research, Monitoring and Modeling: Conceptual Background and Old World Case Studies*. Parthenon Publishing, Paris, pp. 285–309.

Chazdon, R.L., Letcher, S.G. Breugel, M. *et al.* (2007) Rates of change in tree communities of secondary tropical forests following major disturbances. *Proceedings of the Royal Society of London, Series B* 362, 273–289.

Chazdon, R.L., Redondo Brenes, A., and Vilchez Alvarado, B. (2005) Effects of climate and stand age on annual tree dynamics in tropical second-growth rain forests. *Ecology* 86, 1808–1815.

Chinea, J.D. (2002) Tropical forest succession on abandoned farms in the Humacao Municipality of eastern Puerto Rico. *Forest Ecology and Management* 167, 195–207.

Chinea, J.D. and Helmer, E.H. (2003) Diversity and composition of tropical secondary forests recovering from large-scale clearing: results from the 1990 inventory in Puerto Rico. *Forest Ecology and Management* 180, 227–240.

Chokkalingam, U. and De Jong, W. (2001) Secondary forest: a working definition and typology. *International Forestry Review* 3, 19–26.

Christensen, N.L. and Peet, R.K. (1984) Convergence during secondary forest succession. *Journal of Ecology* 72, 25–36.

Clark, D.A. (2004) Sources or sinks? The responses of tropical forests to current and future climate and atmospheric composition. *Philosophical Transactions of the Royal Society of London Series B-Biological Sciences* 359, 477–491.

Clark, D.B. (1996) Abolishing virginity. *Journal of Tropical Ecology* 12, 735–739.

Clark, D.B., Clark, D.A., and Read, J.M. (1998) Edaphic variation and the mesoscale distribution of tree species in a neotropical rain forest. *Journal of Ecology* 86, 101–112.

Clements, F.E. (1904) *The Development and Structure of Vegetation*. Botanical Survey of Nebraska 7. Studies in the Vegetation of the State. Lincoln, NE.

Clements, F.E. (1916) *Plant Succession: An Analysis of the Development of Vegetation*. Carnegie Institute Publication 242. Washington, DC.

Clements, F.E. (1928) *Plant Succession and Indicators*. Wilson, New York.

Cochrane, M.A. and Schulze, M.D. (1999) Fire as a recurrent event in tropical forests of the eastern Amazon: effects on forest structure, biomass, and species composition. *Biotropica* 31, 2–16.

Colwell, R.K. and Coddington, J.A. (1994) Estimating terrestrial biodiversity through extrapolation. *Philosophical Transactions of the Royal Society (Series B)* 345, 101–118.

Condit, R., Hubbell, S.P., LaFrankie, J.V., *et al.* (1996) Species–area and species–individual relationships for tropical trees: a comparison of three 50-ha plots. *Journal of Ecology* 84, 549–562.

Connell, J.H. (1971) On the role of natural enemies in preventing competitive exclusion in some marine animals and in rain forest trees. In P.J. Den Boer and G.R. Gradwell (eds), *Dynamics of Populations*. PUDOC, Wageningen, pp. 298–312.

Connell, J.H. (1978) Diversity in tropical rain forests and coral reefs. *Science* 199, 1302–1310.

Connell, J.H. (1979) Tropical rain forests and coral reefs as open non-equilibrium systems. In R.M. Anderson, B.D. Turner, and L.R. Taylor (eds), *Population Dynamics*. British Ecological Society, Blackwell Publishing, Oxford, pp. 141–163.

Connell, J.H. and Slatyer, R.O. (1977) Mechanisms of succession in natural communities and their role in community stability and organization. *American Naturalist* 111, 1119–1144.

Corlett, R.T. (1992) The ecological transformation of Singapore, 1819–1990. *Journal of Biogeography* 19, 411–420.

Cowles, H.C. (1899) The ecological relations of vegetation on the sand dunes of Lake Michigan. *Botanical Gazette* 27, 95–117, 167–202, 281–308, 361–391.

Dalling, J.W., Hubbell, S.P., and Silvera, K. (1998) Seed dispersal, seedling establishment and gap partitioning among tropical pioneer trees. *Journal of Ecology* 86, 674–689.

Dalling, J.S., Winter, K., Hubbell, S.P., Hamrick, J.L., Nason, J.D., and Murawski, D.A. (2000) The unusual life history of *Alseis blackiana*: a shade-persistent pioneer tree? *Ecology* 82, 933–945.

D'Antonio, C.M. and Vitousek, P.M. (1992) Biological invasions by exotic grasses, the grass/fire cycle, and global change. *Annual Review of Ecology and Systematics* 23, 63–87.

Denevan, W.M. (1992) The pristine myth: the landscape of the Americas in 1492. *Annals of the Association of American Geographers* 82, 369–385.

Denslow, J.S. (1978) *Secondary succession in a Colombian rainforest: Strategies of species response across a disturbance gradient.* Doctoral dissertation, University of Wisconsin, Madison.

Denslow, J.S. (1987) Tropical rainforest gaps and tree species diversity. *Annual Review of Ecological Systems* 18, 431–452.

Denslow, J.S. (1995) Disturbance and diversity in tropical rain forests: The density effect. *Ecological Applications* 5, 962–968.

Denslow, J.S. (1996) Functional group diversity and responses to disturbance. In G.H. Orians, R. Dirzo, and J.H. Cushman (eds), *Biodiversity and Ecosystem Processes in Tropical Forests.* Ecological Studies 122. Springer, Berlin, pp. 127–151.

Denslow, J.S. and Guzman, G.S. (2000) Variation in stand structure, light and seedling abundance across a tropical moist forest chronosequence, Panama. *Journal of Vegetation Science* 11, 201–212.

Dewalt, S.J., Schnitzer, S.A., and Denslow, J.S. (2000) Density and diversity of lianas along a chronosequence in a central Panamanian lowland forest. *Journal of Tropical Ecology* 16, 1–19.

Drury, W.H. and Nisbet, I.C.T. (1973) Succession. *Journal of the Arnold Arboretum* 54, 331–368.

Duivenvoorden, J.F. (1996) Patterns of tree species richness in rain forests of the middle Caqueta area, Colombia, NW Amazonia. *Biotropica* 28, 142–158.

Duncan, R.S. and Chapman, C.A. (1999) Seed dispersal and potential forest succession in abandoned

agriculture in tropical Africa. *Ecological Applications* 9, 998–1008.

Dupuy, J.M. and Chazdon, R.L. (1998) Long-term effects of forest regrowth and selective logging on the seed bank of tropical forests in NE Costa Rica. *Biotropica* 30, 223–237.

Dupuy, J.M. and Chazdon, R.L. (2006) Effects of vegetation cover on seedling and sapling dynamics in secondary tropical wet forests in Costa Rica. *Journal of Tropical Ecology* 22, 1–22.

Eggeling, W.J. (1947) Observations on the ecology of the Budongo Rain Forest, Uganda. *Journal of Ecology* 34, 20–87.

Egler, F.E. (1954) Vegetation science concepts: I. Initial floristic composition – a factor in old-field vegetation development. *Vegetatio* 4, 412–417.

Elmqvist, T., Wall, M., Berggren, A.-L., Blix, L., Fritioff, A., and Rinman, U. (2001) Tropical forest reorganization after cyclone and fire disturbance in Samoa: remnant trees as biological legacies. *Conservation Ecology* 5, 10. [online] URL: http://www.consecol.org/vol5/iss2/art10/

Ewel, J. (1980) Tropical succession: manifold routes to maturity. *Biotropica* 12 (Suppl.), 2–7.

Ewel, J.J. (1983) Succession. In F.B. Golley (ed.), *Tropical Rain Forest Ecosystems.* Elsevier Scientific Publishing Co, Amsterdam, The Netherlands, pp. 217–223.

Ewel, J.J. and Bigelow, S.W. (1996) Plant life-forms and tropical ecosystem functioning. In G.H. Orians, R. Dirzo, and J.H. Cushman (eds), *Biodiversity and Ecosystem Processes in Tropical Forests*, Springer, New York, pp. 101–126.

Ferguson, B.G., Vandermeer, J., Morales, H., and Griffith, D.M. (2003) Post-agricultural succession in El Peten, Guatemala. *Conservation Biology* 17, 818–828.

Ferreira, L.V. and Prance, G.T. (1999) Ecosystem recovery in terra firme forests after cutting and burning: a comparison on species richness, floristic composition and forest structure in the Jaú National Park, Amazonia. *Botanical Journal of the Linnean Society* 130, 97–110.

Fine, P.V.A. (2002) The invisibility of tropical forests by exotic plants. *Journal of Tropical Ecology* 18, 687–705.

Finegan, B. (1984) Forest succession. *Nature* 312, 109–114.

Finegan, B. (1996) Pattern and process in neotropical secondary forests: the first 100 years of succession. *Trends in Ecology and Evolution* 11, 119–124.

Foster, B.L. and Tilman, D. (2000) Dynamic and static views of succession: testing the descriptive power of the chronosequence approach. *Plant Ecology* 146, 1–10.

Gehring, C., Denich, M., and Vlek, P.L.G. (2005) Resilience of secondary forest regrowth after slash-and-burn agriculture in central Amazonia. *Journal of Tropical Ecology* 21, 519–527.

Gleason, H.A. (1926) The individualistic concept of the plant association. *Bulletin of the Torrey Botanical Club* 53, 7–26.

Glenn-Lewin, D.C., Peet, R.K., and Veblen, T.T. (eds), (1992) *Plant Succession: Theory and Prediction.* Chapman and Hall, New York.

Goldammer, J.G. and Seibert, B. (1990) The impact of droughts and forest fires on tropical lowland rain forest of East Kalimantan. In J.G. Goldammer (ed.), *Fire in the Tropical Biota.* Ecological Studies 84. Springer, Berlin, pp. 11–31.

Gómez-Pompa, A. and Vázquez-Yanes, C. (1974) Studies on the secondary succession of tropical lowlands: the life cycle of secondary species. *Proceedings of the First International Congress of Ecology*, The Hague, pp. 336–342.

Gómez-Pompa, A. and Vázquez-Yanes, C. (1981) Successional studies of a rain forest in Mexico. In D.C. West, H.H. Shugart, and D.B. Botkin (eds), *Forest Succession: Concepts and Application.* Springer-Verlag, New York, pp. 246–266.

Gotelli, N.J and Colwell, R.K. (2001) Quantifying biodiversity: procedures and pitfalls in the measurement and comparison of species richness. *Ecology Letters* 4, 379–391.

Gorchov, D.L., Cornejo, F., Ascorra, C., and Jaramillo, M. (1993) The role of seed dispersal in the natural regeneration of rain forest after strip-cutting in the Peruvian Amazon. *Vegetatio* 107/108, 339–349.

Grau, H.R., Arturi, M.F., Brown, A.D., and Acenolaza, P.G. (1997) Floristic and structural patterns along a chronosequence of secondary forest succession in Argentinean subtropical montane forests. *Forest Ecology and Management* 95, 161–171.

Guariguata, M. and Ostertag, R. (2001) Neotropical secondary forest succession: changes in structural and functional characteristics. *Forest Ecology and Management* 148, 185–206.

Guariguata, M.R., Chazdon, R.L., Denslow, J.S., Dupuy, J.M., and Anderson, L. (1997) Structure and floristics of secondary and old-growth forest stands in lowland Costa Rica. *Plant Ecology* 132, 107–120.

Guevara, S., Laborde, J., and Sanchez-Rios, G. (2004) Rain forest regeneration beneath the canopy of fig trees isolated in pastures of Los Tuxtlas, Mexico. *Biotropica* 36, 99–108.

Guevara, S., Purata, S.E., and van der Maarel, E. (1986) The role of remnant forest trees in tropical secondary succession. *Vegetatio* 66, 77–84.

Hammond, D.S. (1995) Post-dispersal seed and seedling mortality of tropical dry forest trees after shifting agriculture, Chiapas, Mexico. *Journal of Tropical Ecology* 11, 295–313.

Helmer, E.H. (2000) The landscape ecology of tropical secondary forest in montane Costa Rica. *Ecosystems* 3, 98–114.

Holl, K.D. (1999) Factors limiting tropical rain forest regeneration in abandoned pasture: seed rain, seed germination, microclimate, and soil. *Biotropica* 31, 229–242.

Holl, K.D., Loik, M.E., Lin, E.H.V., and Samuels, I.A. (2000) Tropical montane forest restoration in Costa Rica: overcoming barriers to dispersal and establishment. *Restoration Ecology* 8, 339–349.

Holl, K.D. and Lulow, M.W. (1997) Effects of species, habitat, and distance from the edge on post-dispersal seed predation in a tropical rainforest. *Biotropica* 29, 459–468.

Hooper, E.R., Legendre, P., and Condit, R. (2004) Factors affecting community composition of forest regeneration in deforested, abandoned land in Panama. *Ecology* 85, 3313–3326.

Howard, L.F. and Lee, T.D. (2003) Temporal patterns of vascular plant diversity in southeastern New Hampshire forests. *Forest Ecology and Management* 185, 5–20.

Howlett, B.E. and Davidson, D.W. (2003) Effects of seed availability, site conditions, and herbivory on pioneer recruitment after logging in Sabah, Malaysia. *Forest Ecology and Management* 184, 369–383.

Hubbell, S.P. and Foster, R.B. (1986) Biology, chance, and history and the structure of tropical rain forest tree communities. In J. Diamond and T.J. Case (eds), *Community Ecology.* Harper and Row, New York, pp. 77–96.

Hubbell, S.P., Foster, R.B., O'Brian, S.T. *et al.* (1999) Light gap disturbances, recruitment limitation, and tree diversity in a neotropical forest. *Science* 283, 554–557.

Hughes, R.F., Kauffman, J.B., and Jaramillo, V.J. (1999) Biomass, carbon, and nutrient dynamics of secondary forests in a humid tropical region of Mexico. *Ecology* 80, 1892–1907.

Hurlbert, S.H. (1971) The non-concept of species diversity: A critique and alternative parameters. *Ecology* 52, 577–586.

Huston, M. (1979) A general hypothesis of species diversity. *American Naturalist* 113, 81–101.

Ingle, N. (2003) Seed dispersal by wind, birds, and bats between Philippine montane rainforest and successional vegetation. *Oecologia* 134, 251–261.

Janzen, D.H. (1970) Herbivores and the number of tree species in tropical forests. *American Naturalist* 104, 501–528.

Janzen, D.H. (1988) Management of habitat fragments in a tropical dry forest – growth. *Annals of the Missouri Botanical Garden* 75, 105–116.

Johnson, C.M., Zarin, D.J., and Johnson, A.H. (2000) Post-disturbance aboveground biomass accumulation in global secondary forests. *Ecology* 81, 1395–1401.

Jones, E.W. (1956) Ecological studies on the rain forest of Southern Nigeria: IV (Continued), the plateau forest of the Okomu Forest Reserve. *Journal of Ecology* 44, 83–117.

Kalacska, M., Sanchez-Azofeifa, G.A., Rivard, B. *et al.* (2004) Leaf area index measurements in a tropical moist forest: a case study from Costa Rica. *Remote Sensing of Environment* 91, 134–152.

Kammesheidt, L. (1998) The role of tree sprouts in the restorations of stand structure and species diversity in tropical moist forest after slash-and-burn agriculture in Eastern Paraguay. *Plant Ecology* 139, 155–165.

Kang, H. and Bawa, K.S. (2003) Effects of successional status, habit, sexual systems, and pollinators on flowering patterns in tropical rain forest trees. *American Journal of Botany* 90, 865–876.

Kellman, M.C. (1970) *Secondary plant succession in tropical montane Mindanao.* Australian National University, Department of Biogeography and Geomorphology, Publ. BG/2, Canberra.

Kennard, D.K. (2002) Secondary forest succession in a tropical dry forest: patterns of development across a 50-year chronosequence in lowland Bolivia. *Journal of Tropical Ecology* 18, 53–66.

Kenoyer, L.A. (1929) General and successional ecology of the lower tropical rain forest at Barro Colorado Island, Panama. *Ecology* 10, 201–222.

Knight, D.H. (1975) A phytosociological analysis of species-rich tropical forest on Barro Colorado Island, Panama. *Ecological Monographs* 45, 259–284.

Kupfer, J.A., Webbeking, A.L., and Franklin, S.B. (2004) Forest fragmentation affects early successional patterns on shifting cultivation fields near Indian Church, Belize. *Agriculture Ecosystems and Environment* 103, 509–518.

Lang, G.E. and Knight, D.H. (1983) Tree growth, mortality, recruitment, and canopy gap formation during a 10-year period in a tropical moist forest. *Ecology* 64, 1075–1080.

Laurance, W.F., Oliveira, A.A., Laurance, S.G. *et al.* (2004) Pervasive alteration of tree communities in undisturbed Amazonian forests. *Nature* 428, 171–175.

Leps, J. and Rejmánek, M. (1991) Convergence or divergence: what should we expect from vegetation succession? *Oikos* 62, 261–264.

Lichstein, J.W., Grau, H.R., and Aragón, R. (2004) Recruitment limitation in secondary forests dominated by an exotic tree. *Journal of Vegetation Science* 15, 721–728.

Lugo, A.E. (2004) The outcome of alien tree invasions in Puerto Rico. *Frontiers in Ecology and Environment* 2, 265–273.

Lugo, A.E. and Helmer, E. (2004) Emerging forests on abandoned land: Puerto Rico's new forests. *Forest Ecology and Management* 190, 145–161.

Malhi, Y. and Phillips, O.L. (2004) Tropical forests and global atmospheric change: a synthesis. *Philosophical Transactions of the Royal Society of London Series B-Biological Sciences* 359, 549–555.

Martin, P.H., Sherman, R.E., and Fahey, T.J. (2004) Forty years of tropical forest recovery from agriculture: structure and floristics of secondary and old-growth riparian forests in the Dominican Republic. *Biotropica* 36, 297–317.

Martinez-Garza, C. and Gonzalez-Montagut, R. (1999) Seed rain from forest fragments into tropical pastures in Los Tuxtlas, Mexico. *Plant Ecology* 145, 255–265.

Mesquita, R.C.G., Ickes, K., Ganade, G., and Williamson, G.B. (2001) Alternative successional pathways in the Amazon Basin. *Journal of Ecology* 89, 528–537.

Moran, E.F., Brondizio, E., Tucker, J.M., da Silva-Fosberg, M.C., McCracken, S., and Falesi, I. (2000) Effects of soil fertility and land-use on forest succession in Amazônia. *Forest Ecology and Management* 139, 93–108.

Muller-Landau, H.C. (2004) Interspecific and inter-site variation in wood specific gravity of tropical trees. *Biotropica* 36, 20–32.

Muller-Landau, H.C., Wright, S.J., Calderon, O., Hubbell, S.P., and Foster, R.B. (2002) Assessing recruitment limitation: concepts, methods and case-studies from a tropical forest. In D.J. Levey, W.R. Silva, and M. Galetti (eds), *Seed Dispersal and Frugivory: Ecology, Evolution, and Conservation.* CABI Publishing, Wallingford, pp. 35–53.

Murphy, P.G. and Lugo, A.E. (1986) Ecology of tropical dry forest. *Annual Review of Ecology and Systematics* 17, 89–96.

Myster, R.W. (2003) Vegetation dynamics of a permanent pasture plot in Puerto Rico. *Biotropica* 35, 422–428.

Myster, R.W. (2004) Post-agricultural invasion, establishment, and growth of neotropical trees. *Botanical Review* 70, 381–402.

Nepstad, D., Uhl, C., Pereira, C.A., and da Silva, J.M.C. (1996) A comparative study of tree establishment in abandoned pasture and mature forest of eastern Amazonia. *Oikos* 76, 25–39.

Nicotra, A.B., Chazdon, R.L., and Iriarte, S. (1999) Spatial heterogeneity of light and woody seedling regeneration in tropical wet forests. *Ecology* 80, 1908–1926.

Noble, I.R. and Slatyer, R.O. (1980) The use of vital attributes to predict successional changes in plant communities subject to recurrent disturbances. *Vegetatio* 43, 5–21.

Notman, E. and Gorchov, D.L. (2001) Variation in post-dispersal seed predation in mature Peruvian lowland tropical forest and fallow agricultural sites. *Biotropica* 33, 621–636.

Odum, E.P. (1969) The strategy of ecosystem development. *Science* 164, 262–270.

Oliver, C.D. and Larson, B.C. (1990) *Forest Stand Dynamics*. McGraw-Hill, New York.

Opler, P.A., Baker, H.G., and Frankie, G.W. (1977) Recovery of tropical lowland forest ecosystems. In J. Cairns, K.L. Dickson, and E.E. Herricks (eds), *Recovery and Restoration of Damaged Ecosystems*. University Press of Virginia, Charlottesville, VA, pp. 399–421.

Opler, P.A., Baker, H.G., and Frankie, G.W. (1980) Plant reproductive characteristics during secondary succession in Neotropical lowland forest ecosystems. *Biotropica (Suppl.)* 12, 40–46.

Otsamo, A., Ådjers, G., Hadi, T.S., Kuusipalo, J., Tuomela, K., and Vuokko, R. (1995) Effect of site preparation and initial fertilization on the establishment and growth of four plantation tree species used in reforestation of *Imperata cylindrica* (L.) Beauv. dominated grasslands. *Forest Ecology & Mangement* 73, 271–277.

Pascarella, J.B., Aide, T.M., Serrano, M.I., and Zimmerman, J.K. (2000) Land-use history and forest regeneration in the Cayey Mountains, Puerto Rico. *Ecosystems* 3, 217–228.

Peña-Claros, M. (2003) Changes in forest structure and species composition during secondary forest succession in the Bolivian Amazon. *Biotropica* 35, 450–461.

Peña-Claros, M., and de Boo, H. (2002) The effect of forest successional stage on seed removal of tropical rain forest tree species. *Journal of Tropical Ecology* 18, 261–274.

Perera, G.A.D. (2001) The secondary forest situation in Sri Lanka: a review. *Journal of Tropical Forest Science* 13, 768–785.

Peterson, C.J., and Haines, B.L. (2000) Early successional patterns and potential facilitation of woody plant colonization by rotting logs in premontane Costa Rican pastures. *Restoration Ecology* 8, 361–369.

Pickett, S.T.A. (1976) Succession: an evolutionary interpretation. *American Naturalist* 110, 107–119.

Pickett, S.T.A. (1989) Space-for-time substitution as an alternative to long-term studies. In G.E. Likens (ed.), *Long-term Studies in Ecology*. Springer-Verlag, New York, pp. 110–135.

Pickett, S.T.A., Collins, S.L., and Armesto, J.J. (1987) Models, mechanisms and pathways of succession. *Botanical Review* 53, 335–371.

Purata, S.E. (1986) Floristic and structural changes during old-field succession in the Mexican tropics in relation to site history and species availability. *Journal of Tropical Ecology* 2, 257–276.

Ramakrishnan, P.S. (1988) Sustainable development, climate change, and tropical rain forest landscape. *Climatic Change* 39, 583–600.

Read, L. and Lawrence, D. (2003) Recovery of biomass following shifting cultivation in dry tropical forests of the Yucatan. *Ecological Applications* 13, 85–97.

Reich, P.B., Walters, M.B., and Ellsworth, D.S. (1992) Leaf life-span in relation to leaf, plant, and stand characteristics among diverse ecosystems. *Ecological Monographs* 62, 365–392.

Rees, M., Condit, R., Crawley, M., Pacala, S., and Tilman, D. (2001) Long-term studies of vegetation dynamics. *Science* 293, 650–655.

Richards, P.W. (1952) *The Tropical Rain Forest*. Cambridge University Press, Cambridge.

Riswan, S., Kentworthy, J.B., and Kartawinata, K. (1985) The estimation of temporal processes in tropical rain forest: a study of primary mixed dipterocarp forest in Indonesia. *Journal of Tropical Ecology* 1, 171–182.

Rivera, L.W. and Aide, T.M. (1998) Forest recovery in the karst region of Puerto Rico. *Forest Ecology and Management* 108, 63–75.

Ruiz, J., Fandiño, M.C., and Chazdon, R.L. (2005) Vegetation structure, composition, and species richness across a 56-year chronosequence of dry tropical forest on Providencia Island, Colombia. *Biotropica* 37, 520–530.

Saldarriaga, J.G., West, D.C., Tharp, M.L., and Uhl, C. (1988) Long-term chronosequence of forest succession in the upper Rio Negro of Colombia and Venezuela. *Journal of Ecology* 76, 938–958.

Schmidt-Vogt, D. (2001) Secondary forests in swidden agriculture in the highlands of Thailand. *Journal of Tropical Forest Science* 13, 748–767.

Schnitzer, S.A. and Bongers, F. (2002) The ecology of lianas and their role in forests. *Trends in Ecology and Evolution* 17, 223–230.

Schnitzer, S.A., Dalling, J.W., and Carson, W. (2000) The impact of lianas on tree regeneration in tropical forest canopy gaps: evidence for an alternative pathway of gap-phase regeneration. *Journal of Ecology* 88, 655–666.

Sezen, U.U., Chazdon, R.L., and Holsinger, K.E. (2005) Genetic consequences of tropical second-growth forest regeneration. *Science* 307, 891.

Sheil, D. (1999) Developing tests of successional hypotheses with size-structured populations, and an assessment using long-term data from a Ugandan rain forest. *Plant Ecology* 140, 117–127.

Sheil, D. (2001) Long-term observations of rain forest succession, tree diversity and responses to disturbance. *Plant Ecology* 155, 183–199.

Sheil, D. and Burslem, D. (2003) Disturbing hypotheses in tropical forests. *Trends in Ecology and Evolution* 18, 18–26.

Sheil, D., Jennings, S., and Savill, P. (2000) Long-term permanent plot observations of vegetation dynamics in Budongo, a Ugandan rain forest. *Journal of Tropical Ecology* 16, 765–800.

Silver, W.L., Ostertag, R., and Lugo, A.E. (2000) The potential for carbon sequestration through reforestation of abandoned tropical agricultural and pasture lands. *Restoration Ecology* 8, 394–407.

Slocum, M.G. (2001) How tree species differ as recruitment foci in a tropical pasture. *Ecology* 82, 2547–2559.

Steininger, M.K. (2000) Secondary forest structure and biomass following short and extended land-use in central and southern Amazonia. *Journal of Tropical Ecology* 16, 689–708.

Suzuki, E. (1999) Diversity in specific gravity and water content of wood among Bornean tropical rainforest trees. *Ecological Research* 14, 211–224.

Svenning, J.C. and Wright, S.J. (2005) Seed limitation in a Panamanian forest. *Journal of Ecology* 93, 853–862.

Swaine, M.D. (1996) Rainfall and soil fertility as factors limiting forest species distributions in Ghana. *Journal of Ecology* 84, 419–428.

Swaine, M.D. and Whitmore, T.C. (1988) On the definition of ecological species groups in tropical rain forests. *Vegetatio* 75, 81–86.

Tansley, A.G. (1935) The use and abuse of vegetational concepts and terms. *Ecology* 16, 284–307.

Terborgh, J., Foster, R.B., and Nuñez V., P. (1996) Tropical tree communities: a test of the nonequilibrium hypothesis. *Ecology* 77, 561–567.

Toky, O.P. and Ramakrishnan, P.S. (1983) Secondary succession following slash and burn agriculture in north-eastern India. I. Biomass, litterfall, and productivity. *Journal of Ecology* 71, 735–745.

Tomlinson, J.R., Serrano, M.I., Lopez, T.M., Aide, T.M., and Zimmerman, J.K. (1996) Land-use dynamics in a post-agricultural Puerto Rican landscape (1936–1988). *Biotropica* 28, 525–536.

Turner, I.M., Wong, Y.K., Chew, P.T., and bin Ibrahim, A. (1997) Tree species richness in primary and old secondary tropical forest in Singapore. *Biodiversity and Conservation* 6, 537–543.

Uhl, C. (1987) Factors controlling succession following slash-and-burn agriculture in Amazonia. *Journal of Ecology* 75, 377–407.

Uhl, C., Buschbacker, R., and Serrão, E.A.S. (1988) Abandoned pastures in eastern Amazonia. I. Patterns of plant succession. *Journal of Ecology* 75, 663–681.

Uhl, C., Clark, K., Clark, H., and Murphy, P. (1981) Early plant succession after cutting and burning in the upper Rio Negro of the Amazon Basin. *Journal of Ecology* 69, 631–649.

Uhl, C. and Jordan, C.F. (1984) Successional and nutrient dynamics following forest cutting and burning in Amazonia. *Ecology* 65, 1476–1490.

van Andel, T. (2001) Floristic composition and diversity of mixed primary and secondary forests in northwest Guyana. *Biodiversity and Conservation* 10, 1645–1682.

Vandermeer, J. (1996) Disturbance and neutral competition theory in rain forest dynamics. *Ecological Modelling* 85, 99–111.

Vandermeer, J., Boucher, D., Granzow-de la Cerda, I., and Perfecto, I. (2001) Growth and development of the thinning canopy in a post-hurricane tropical rain forest in Nicaragua. *Forest Ecology and Management* 148, 221–242.

Vandermeer, J., Boucher, D., Perfecto, I., and Granzow-de la Cerda, I. (1996) A theory of disturbance and species diversity: evidence from Nicaragua after Hurricane Joan. *Biotropica* 28, 600–613.

Vandermeer, J. and de la Cerda, I.G. (2004) Height dynamics of the thinning canopy of a tropical rain forest: 14 years of succession in a post-hurricane forest in Nicaragua. *Forest Ecology and Management* 199, 125–135.

Vandermeer, J., de la Cerda, I.G., Boucher, D., Perfecto, I., and Ruiz, J. (2000) Hurricane disturbance and tropical tree species diversity. *Science* 290, 788–791.

Vandermeer, J., de la Cerda, I.G., Perfecto, I., Boucher, D., Ruiz, J., and Kaufmann, A. (2004) Multiple basins of attraction in a tropical forest: evidence for nonequilibrium community structure. *Ecology* 85, 575–579.

Vandermeer, J., Mallona, M.A., Boucher, D., Yih, K., and Perfecto, I. (1995) Three years of ingrowth following catastrophic hurricane damage on the Caribbean coast of Nicaragua: evidence in support of the direct regeneration hypothesis. *Journal of Tropical Ecology* 11, 465–471.

van Gemerden, B.S., Olff, H., Parren, M.P.E., and Bongers, F. (2003) The pristine rain forest? Remnants of historical human impacts on current tree species composition and diversity. *Journal of Biogeography* 30, 1381–1390.

Vázquez-Yanes, C. (1980) Light quality and seed germination in *Cecropia obtusifolia* and *Piper auritum* from a tropical rain forest in Mexico. *Phyton* 38, 33–35.

Vázquez-Yanes, C. and Orozco-Segovia, A. (1984) Ecophysiology of seed germination in the tropical humid forests of the world: a review. In E. Medina, H.A. Mooney, and C. Vázquez-Yanes (eds), *Physiological Ecology of Plants of the Wet Tropics*. Dr W Junk, The Hague, pp. 37–50.

Waide, R.B. and Lugo, A.E. (1992) A research perspective on disturbance and recovery of a tropical forest. In J.G. Goldammer (ed.), *Tropical Forests in Transition*. Birkhauser, Basel, Switzerland, pp. 173–189.

Walker, L.R. and Chapin, F.S.I. (1987) Interactions among processes controlling successional change. *Oikos* 50, 131–135.

Wardle, D.A., Walker, L.R., and Bardgett, R.D. (2004) Ecosystem properties and forest decline in contrasting long-term chronosequences. *Science* 305, 509–513.

Watt, A.S. (1919) On the causes of failure of natural regeneration in British oakwoods. *Journal of Ecology* 7, 173–203.

Watt, A.S. (1947) Pattern and process in the plant community. *Journal of Ecology* 35, 1–22.

Webb, L.J., Tracey, J.G., and Williams, W.T. (1972) Regeneration and pattern in the subtropical rain forest. *Journal of Ecology* 60, 675–695.

White, L.J.T. and Oates, J.F. (1999) New data on the history of the plateau forest of Okomu, southern Nigeria: an insight into how human disturbance has shaped the African rain forest. *Global Ecology and Biogeography* 8, 355–361.

Whitmore, T.C. (1973) Frequency and habitat of tree species in the rain forests of Ulu Kelantan. *Garden's Bulletin Singapore* 26, 195–210.

Whitmore, T.C. (1974) *Change with Time and the Role of Cyclones in Tropical Rain Forest on Kolombangara,*

Solomon Islands. Commonwealth Forestry Institute Paper 46. Commonwealth Forestry Institute, Oxford.

Whitmore, T.C. (1978) Gaps in the forest canopy. In P.B. Tomlinson and M.M. Zimmerman (eds), *Tropical Trees as Living Systems*. Cambridge University Press, New York, pp. 639–655.

Whitmore, T.C. (1998) *Tropical Rain Forests of the Far East*. Clarendon Press, Oxford.

Whitmore, T.C. (1991) Tropical rain forest dynamics and its implications for management. In A. Gómez-Pompa, T.C. Whitmore, and M. Hadley (eds), *Rain Forest Regeneration and Management*. UNESCO and Parthenon Publishing Group, Paris, pp. 67–89.

Whitmore, T.C. and Burslem, D.F.R.P. (1998) Major disturbances in tropical rainforests. In D.M. Newbery, H.H.T. Prins, and N.D. Brown (eds), *Dynamics of Tropical Communities*. Blackwell Science Ltd, Oxford, pp. 549–565.

Wijdeven, S.M.J. and Kuzee, M.E. (2000) Seed availability as a limiting factor in forest recovery processes in Costa Rica. *Restoration Ecology* 8, 414–424.

Willis, K.J., Gillson, L., and Brncic, T.M. (2004) How "Virgin" is virgin rainforest? *Science* 304, 402–403.

Wilkinson, D.M. (1999) The disturbing history of intermediate disturbance. *Oikos* 84, 145–147.

Wunderle, J.M. (1997) The role of animal seed dispersal in accelerating native forest regeneration on degraded tropical lands. *Forest Ecology and Management* 99, 223–235.

Yavitt, J.B., Battles, J.J., Lang, G.E., and Knight, D.H. (1995) The canopy gap regime in a secondary Neotropical forest in Panama. *Journal of Tropical Ecology* 11, 391–402.

Yih, K., Boucher, D.H., Vandermeer, J.H., and Zamora, N. (1991) Recovery of the rain forest of southeastern Nicaragua after destruction by Hurricane Joan. *Biotropica* 23, 106–113.

Young, K.R., Ewel, J.J., and Brown, B.J. (1987) Seed dynamics during forest succession in Costa Rica. *Vegetatio* 71, 157–173.

Young, T.P., Chase, J.M., and Huddleston, R.T. (2001) Community succession and assembly: comparing, contrasting and combining paradigms in the context of ecological restoration. *Ecological Restoration* 19, 5–18.

Zarin, D.J., Ducey, M.J., Tucker, J.M., and Salas, W.A. (2001) Potential biomass accumulation in Amazonian regrowth forests. *Ecosystems* 4, 658–668.

Chapter 24

EXOTIC PLANT INVASIONS IN TROPICAL FORESTS: PATTERNS AND HYPOTHESES

Julie S. Denslow and Saara J. DeWalt

OVERVIEW

In the tropics, exotic plants have been widely introduced for industrial timber, for land reclamation and forage crops, and as ornamentals. In spite of the apparent opportunity for naturalization and spread, invasive exotic plants are scarce in many continental tropical forests. We examine several conditions under which exotic species do pose substantial threats to tropical ecosystems or to their management. These include island ecosystems, open-canopied forests, fragmented or disturbed ecosystems, and forests managed for timber or crops. We explore four hypotheses to account for the scarcity of exotic species in many tropical forests: (1) tropical forests are resistant to invasions by exotic species because they are rich in species and functional groups; (2) native rainforest species competitively exclude exotic species; (3) high pest loads and high pest diversity in the tropics deter establishment and spread of exotic species; and (4) low propagule availability contributes to the rarity of exotic species in many tropical forests. While current research suggests that high species diversity *per se* is not likely to be an impediment to exotic species, functional group diversity, high competitive exclusion rates, and high pest loads all may confer a certain biotic resistance to the establishment and spread of exotic species in tropical forests. Similarly, high functional diversity and high productivity may increase the resilience of tropical forests to the kinds of ecosystem changes effected by invasive species in other ecosystems. However, we are unable to fully evaluate these hypotheses and their interactions in the absence of a better assessment of the actual exposure of tropical forests to exotic propagules and results from seed addition experiments to test the relative importance of biotic resistance and dispersal limitation in limiting the spread of exotic species into tropical forests.

INTRODUCTION

Tropical forests face myriad threats from human activities, including land conversion and habitat fragmentation, altered fire cycles, and defaunation (Sala *et al.* 2000). With some exceptions, however, few continental tropical forests appear to be affected strongly by invasive exotic plants (e.g., Ramakrishnan 1991, Whitmore 1991, Rejmánek 1996, Fine 2002). Rejmánek (1996) found only 42 exotic plant species known to invade tropical rainforests; of those, about half were known to invade forests only on islands and eight

were reported only from treefall gaps. Similarly, exotic species constitute small percentages of the floras of two tropical field stations, La Selva Biological Station in Costa Rica (7.6%) and Barro Colorado Island in Panama (21%), where exotic species are confined to pastures, clearings, or other highly disturbed sites (Foster and Hubbell 1990, Hammel 1990). A global survey of threats to biodiversity suggests that biotic exchange is secondary to other factors such as land-use change for tropical forests as it is for most forest ecosystems (Sala *et al.* 2000). The apparently low impact of exotic species on tropical forest ecosystems could reflect biotic resistance (Mack 1996) to exotic invaders and/or historically low exposure to propagules from exotic species (Fine 2002). Both biotic resistance (in the form of impact from competitors, predators, and pathogens) and propagule availability (via reproductive output, vegetative spread, and dispersal) are important components of plant community composition (e.g., Turnbull et al. 2000) and have figured in rates of biotic change throughout evolutionary time. Of interest here is their role in the spread and impacts of exotic invasive species in tropical forest ecosystems. In this chapter we examine patterns of exotic plant invasions in tropical and subtropical forests and explore four hypotheses proposed to account for these patterns.

Our perception of the vulnerability of an ecosystem to invasive species has at least two components (D'Antonio and Dudley 1995): (1) the ease with which exotic species are able to establish and spread, and (2) the tendency for exotic species to alter ecosystem and community processes. Here we will use the term "ecosystem resistance" to describe the degree to which competition, predation, and disease limit the ability of exotic species to establish reproducing populations. A community with low ecosystem resistance will be more highly invasible than a community with high resistance. "Ecosystem resilience" will be used to describe the tendency for ecosystem processes to remain unchanged following exotic invasion. Thus ecosystem processes such as disturbance frequency or resource supply rates will remain relatively unchanged following establishment of an exotic species in a resilient community. Propagule pressure – a key component of the invasion cycle (D'Antonio and Dudley 1995) – is a function of sizes of source populations, seed production, and propagule dispersal, all of which reflect the ecology and introduction history of the invader rather than attributes of the ecosystem.

Figure 24.1 illustrates some of the processes that affect the establishment and impacts of a potential exotic invader. The impact of an exotic plant species on an ecosystem will be a function both of its abundance (population size and density) and of its capacity, relative to established species, to alter ecosystem structure and

processes. Propagule pressure, resource availability, and pressure from natural enemies all influence the probability that an exotic species will establish a reproducing population and the rate of growth of that population. Habitat fragmentation increases exposure of forests to propagule pressure from exotic species in nearby disturbed or managed ecosystems. Available resources, such as light and space, also are increased by disturbance and habitat fragmentation and decreased when pre-empted by native species via competition. Rates of competitive exclusion are thought to be highest where primary productivity and growth rates are high (e.g., Rosenzweig and Abramsky 1993). Similarly, high species and functional group richness is thought to reduce resources available to newly establishing exotic species. The complexity of the invasion process and scarcity of appropriate data preclude evaluation of the relative importance of the many factors affecting the impact of exotic species on tropical forests. Many of the processes illustrated in Figure 24.1 are interdependent and most studies focus on situations in which invasive species present substantial threats to the biotic integrity of ecosystems.

Moreover, there is a strong historic component to current distributions of invasive exotic plants. For example, Wu et al. (2004) suggest that the low number of naturalized exotic plant species per log (area) in Taiwan versus Japan reflects Taiwan's shorter history of introductions. The apparent vulnerability of Hawai'i's forests to invasive species reflects, in part, a history of large-scale introductions. Between 1910 and 1960, some 1026 taxa, all exotic except for 78 native species, were out-planted into forest reserves statewide (Woodcock 2007). This enterprise, carried out to restore Hawai'i's watersheds, also provided opportunity for the establishment and spread of invasive species into native forests at an unprecedented scale, and certainly affects our perceptions today of the vulnerability of Hawai'i's forests to exotic species.

Our objective here is to consider the circumstances under which invasive exotic plants have had strong ecological impacts on tropical ecosystems and to use these examples to provide insight into the attributes of some tropical rainforests that

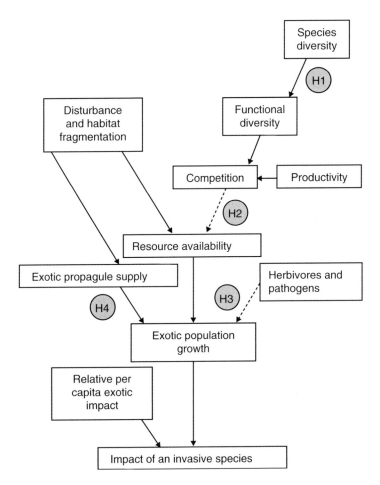

Figure 24.1 Conceptual diagram of factors affecting the impact of exotic plant invasions on a tropical forest ecosystem. Positive effects are shown with solid lines and negative effects with dashed lines. The diagram depicts a variety of interacting processes, some of which reflect attributes of the invaded community and contribute to its resistance to the establishment of new species. Others are attributes of the invading species that affect its ability to reach suitable establishment sites and to alter local ecosystem processes. H1 through H4 refer to hypotheses addressing these processes discussed in the text.

might account for the scarcity of exotic species in them. Our examples are drawn from the available literature, which necessarily addresses ecosystems that may be vulnerable to the establishment of exotics by virtue of location (islands, peninsulas) or exposure to frequent or historic disturbances. Cited examples are listed in Table 24.1 for ease of reference. We have found few examples of investigations in large tracts of intact continental forests,

so we will draw insights from examples on their fringes. For the same reason, we have defined tropical forests broadly to include forests in the tropics and subtropics under a wide range of climatic and edaphic conditions. We discuss several hypotheses that might account for scarcity of exotic plants in many tropical forest ecosystems, review the available information, and offer suggestions for future research.

Table 24.1 Exotic species cited in the text as invasive in tropical forests.

Exotic species	Growth form	Invasive range	Invaded ecosystems	Attributes of the invaded ecosystem	References
Annona glabra	Tree	Pacific Islands, Australia (Queensland), Vietnam	Fresh- or brackish-water wetlands; Melaleuca swamps (Queensland)	Open canopy, flooded soils	Humphries et al. (1991)
Ardisia elliptica	Shrub	Pacific Islands, Australia, USA (Florida)	Moist lowland forest and riparian areas	Forest understory	Horvitz et al. (1998)
Broussonetia papyrifera	Tree	Western Pacific, SE USA, Uganda	Tropical rainforest	Treefall clearings	Sheil et al. (2000)
Cassia spectabilis	Tree	Uganda	Tropical rainforest	Treefall clearings	Sheil et al. (2000)
Chromolaena odorata	Shrub	Pacific Islands, SE Asia, Australia	Tropical moist forest	Disturbances caused by ungulates, swidden agriculture	Chandrasekaran and Swamy (2000), Chandrashekara and Ramakrishnan (1994), Ghazoul (2004)
Cinchona pubescens	Tree	Pacific Islands, esp. Galápagos	Tropical moist upland forest	Island ecosystems, forest understory	MacDonald et al. (1988)
Cinnamomum zeylanicum	Tree	Pacific Islands, Seychelles	Tropical moist secondary forest	Island ecosystems, slopes, ravines	Fleischmann (1997)
Clidemia hirta	Shrub	Pacific Islands, Australia, Malaysia, Réunion	Tropical moist forest	Island ecosystems, forest understory	DeWalt et al. (2004), Wester and Wood (1977)
Falcataria moluccana	Tree	Pacific Islands, Singapore, Indonesia, Réunion, Seychelles	Tropical moist forest	Island ecosystems, open canopies, nitrogen-poor substrates	Hughes and Denslow (2005)
Hedychium gardnerianum	Herb	USA (Hawai'i), New Zealand, Réunion	Tropical moist submontane forest	Island ecosystems, forest understory	Smith (1985)
Lantana camara	Shrub	Pacific Islands, Australia, India, SE Asia, Indian Ocean	Mesic and dry forests	Disturbances due to ungulates, swidden agriculture	Chandrasekaran and Swamy (2000), Chandrashekara and Ramakrishnan (1994), Ramakrishnan and Vitousek (1989)
Melaleuca quinquenervia	Tree	Pacific Islands, USA (Florida)	Subtropical forest/marsh ecotone	Flooded soils, open canopy	Ewel (1986)

Species	Growth form	Invasive range	Forest type	Ecology	References
Melastoma candidum	Shrub/tree	USA (Hawai'i)	Tropical moist forest	Island ecosystems, disturbance, open canopy	Smith (1985)
Miconia calvescens	Tree	USA (Hawai'i), French Polynesia, Australia (Queensland)	Tropical wet and moist forests	Forest understory, disturbed areas	Meyer (1996), Conant et al. (1997)
Mikania micrantha	Vine	Pacific Islands, SE Asia, India, Australia	Tropical moist forest	Disturbed areas, open canopy	Chandrasekaran and Swamy (2000), Chandrashekara and Ramakrishnan (1994)
Morella faya	Tree	USA (Hawai'i)	Tropical moist to wet forest	Island ecosystems, nitrogen-poor substrates, forest understory	Vitousek et al. (1987)
Panicum maximum	Grass	Pantropical	Tropical dry to moist forest	Disturbed areas including forest edges and clearings	Uhl and Kauffman (1990), D'Antonio and Vitousek (1992)
Passiflora tarminiana	Vine	USA (Hawai'i)	Tropical moist forest	Island ecosystems, disturbed areas	Smith (1985)
Piper aduncum	Shrub/tree	Papua New Guinea, Fiji, Solomon Islands	Tropical moist forest	Forest edges, clearings, and secondary forest understory	Rogers and Hartemink (2000)
Pittosporum undulatum	Tree	Australian offshore islands, USA (Hawai'i), Jamaica	Tropical moist montane forest	Island ecosystems, disturbed areas, forest understory	Bellingham et al. (2005)
Psidium cattleianum	Tree	Pacific Islands, Australia (Queensland, offshore islands), Indian Ocean	Tropical moist and wet forests	Island ecosystems, disturbances from ungulates and canopy opening, forest understory	Smith (1985), Huenneke and Vitousek (1990)
Rubus alceifolius	Shrub	Australia (Queensland, Christmas Island), Réunion	Tropical moist forest	Disturbances and forest clearings	Baret et al. (2003)
Schinus terebinthifolius	Tree	USA (Florida, Hawai'i), Australia, Pacific Islands, Indian Ocean	Tropical seasonal and moist forests	Disturbed areas, clearings, open canopy	Smith (1985)
Urochloa mutica	Grass	Pacific Islands, Australia	Tropical swamps and riparian ecosystems	Flooded soils, disturbances, open canopy	Uhl and Kauffman (1990)

Notes: Information on ecology and invasive range from Space (2005). Literature references are examples only and not intended to be exhaustive; Space (2005) provides additional references.

INVASIBLE TROPICAL ECOSYSTEMS

While many tropical forests appear to be substantially weed-free, invasives can have strong impacts on mainland forest ecosystems where canopy structure is naturally open, where rainforests are fragmented or disturbed, or where forests are exploited for crops or timber, and on island ecosystems, where both disturbed and intact forest ecosystems are vulnerable.

Islands

Tropical islands are often seen as invasive-species hotspots because of both the abundance of exotic species and their impacts in those ecosystems (D'Antonio and Dudley 1995, Sax *et al.* 2002, Denslow 2003, Wu *et al.* 2004). The effects of these species are not confined to highly disturbed areas. For example, the flora of Hawai'i contains similar numbers of native (989) and naturalized exotic (1044) species (Wagner *et al.* 1999), among which are many that invade and alter native forests (Smith 1985). These include subcanopy trees, such as *Psidium cattleianum* Sabine (Myrtaceae), *Morella faya* (Ait.) Wilbur (Myricaceae), and *Schinus terebinthifolius* Raddi (Anacardiaceae); large herbs like *Hedychium gardnerianum* Ker Gawl (Zingiberaceae); shrubs such as *Clidemia hirta* (L.) D. Don and *Melastoma candidum* D. Don (Melastomataceae); and vines such as *Passiflora tarminiana* Coppens & Barney (Passifloraceae). Examples of forest invaders abound from other islands as well. *Pittosporum undulatum* Vent. (Pittosporaceae) invades montane rainforests of Jamaica (Bellingham *et al.* 2005); *Cinchona pubescens* Vahl (Rubiaceae) is a major forest conservation concern in the Galápagos highlands (MacDonald *et al.* 1988); *Rubus aceifolius* Poiret (Rosaceae) invades the forests of Christmas Island and Réunion (Baret *et al.* 2003); *Cinnamomum zeylanicum* Blume (Lauraceae) invades forest in the Seychelles (Fleischmann 1997); and the neotropical understory shrub *Piper aduncum* L. (Piperaceae) is spreading in lowland forest in Papua New Guinea (Rogers and Hartemink 2000). Sax *et al.* (2002)

note that, on average, islands have about twice as many exotic plant species as comparable mainland habitats.

However, not all tropical islands are characterized by high densities of exotic species. In their summary of 20 island floras, Wu *et al.* (2004) show that tropical islands do not have notably more naturalized exotic species per unit log (area) than islands elsewhere nor do oceanic islands have a higher species density than continental islands (see also Rejmánek 1996, Sax *et al.* 2002). These patterns suggest that factors other than isolation or latitude likely affect invasibility of island ecosystems.

Open-canopied forests

Invasive exotic species strongly affect some mainland tropical forests as well, especially those with naturally open canopies, even when relatively undisturbed. *Melaleuca quinquenervia* (Cav.) S.T. Blake (Myrtaceae) invasion is altering the structure of the Florida Everglades where it invades scrub cypress habitats in the ecotone between upland pine forests and cypress swamps (Ewel 1986). *Annona glabra* L. (Annonaceae), a native of Florida and Central America, creates dense thickets where it invades Queensland (Australia) *Melaleuca* swamp forests (Humphries *et al.* 1991). *Falcataria moluccana* (Miq.) Barneby and J.W. Grimes (Leguminosae), a large nitrogen-fixing tree, invades undisturbed but open-canopied *Metrosideros* forests on recent lava flows in Hawai'i (Hughes and Denslow 2005). These examples also emphasize that high-stress habitats, such as frequently flooded or shallow soils, are also vulnerable to invasions if exposed to exotic species with appropriate physiological tolerances.

Disturbed forests

Chronic disturbances open forest canopies and provide opportunities for the spread of aggressive exotics. Such disturbances long have been recognized to predispose plant communities to exotic species establishment, in part because of the increase in resource availability they cause (Rejmánek 1989, Kitayama and Mueller-Dombois

1995, Mueller-Dombois 1995, Davis *et al.* 2000, Mack *et al.* 2002). On the continental island of Singapore, numbers of exotic species are positively correlated with canopy openness, but intact rainforest appeared to be resistant (Teo *et al.* 2003). In Florida (Horvitz *et al.* 1995) and Jamaica (Bellingham *et al.* 2005), canopy opening following hurricanes facilitated the growth of exotic species already present in the seed and seedling pool. The forests of the Western Ghats in India support large populations of native ungulates (Bagchi *et al.* 2004) and indigenous human populations practicing swidden agriculture. The understories of these forests are dominated by dense stands of *Lantana camara* L. (Verbenaceae), *Mikania micrantha* H.B.K. (Asteraceae), and *Chromolaena odorata* (L.) R.M. King and H. Robinson (Asteraceae), all of neotropical origins (Chandrashekara and Ramakrishnan 1994, Chandrasekaran and Swamy 2000).

Browsing and rooting by exotic ungulates frequently is associated with invasions of exotic plant species. In Hawai'i, pigs contribute to tree and shrub death, churn the soil, and disperse seeds of exotic species, facilitating the spread of *Psidium cattleianum* into moist forests (Aplet *et al.* 1991). Fensham *et al.* (1994) described high densities of *Lantana camara* in dry rainforest in northern Australia following impacts of pig digging and ground fire. In Pasoh Forest Reserve in Peninsular Malaysia, the native pig, *Sus scrofa*, strongly modifies the forest understory (Ickes *et al.* 2001) which may facilitate the spread of the neotropical understory shrub *Clidemia hirta*. Thus the disturbances associated with ungulate foraging are associated with the spread of exotic plant species both where ungulates are recent introductions and where ungulates are a historic component of the forest ecosystem.

Fragmented forests

Where rainforests are highly fragmented, they are subject to edge encroachment from grass fires, penetration of wind and light into the forest interior, high rates of canopy damage, and seed rain from adjacent clearings, all of which facilitate the establishment of disturbance-adapted species

(Laurance 1997, DiStefano *et al.* 1998). Naturally fragmented riparian forests in Seychelles are heavily invaded (Fleischmann 1997). In North Queensland, Australia, forest fragments are degraded further by a suite of exotic vines which smother canopy trees and understory (Humphries *et al.* 1991), a common pattern in fragmented tropical forests (Laurance 1997). Native vines and lianas have similar impacts on fragmented forests in Brazil (Tabánez *et al.* 1997). In contrast, extensive intact Queensland rainforest appears resistant to invasions by exotic species, even when disturbed by occasional severe windstorms (Humphries and Stanton 1992). One consequence of forest fragmentation has been the alteration of successional trajectories by exotic species and the establishment on abandoned agricultural land of new forest types sometimes dominated initially by exotic species (Lugo and Helmer 2004).

Managed ecosystems

When coupled with exotic seed sources, disturbance and canopy opening due to logging and swidden agriculture also facilitate the spread of invasive species. *Chromolaena odorata* invades tropical dry forest in Thailand after extraction of *Shorea siamensis* Miq. (Dipterocarpaceae) for timber (Ghazoul 2004). In South and Central America, invasive African grasses, such as *Panicum maximum* Jacq. (Poaceae) and *Urochloa mutica* (Forssk.) T.Q. Nguyen (Poaceae), impede forest regeneration following logging or swidden agriculture and carry fire into the edges of intact forest (Uhl and Kauffman 1990, D'Antonio and Vitousek 1992). The high productivity, aggressive spread, and nutritious foliage of these exotic grasses have been important drivers of the conversion of Central and South American tropical forest to cattle pasture (Parsons 1972).

The considerable cost of controlling weeds in managed ecosystems in the tropics also is indicative of their potential to affect the course of secondary succession, forest restoration projects, and long-term forest management. For example, competition from exotic grasses and related increased fire frequencies are major impediments

to the restoration of tropical dry and mesic forest ecosystems (D'Antonio and Vitousek 1992,
Cabin *et al.* 2000). Along with nutrient depletion,
weed encroachment is a principal cause of field
abandonment in swidden agriculture (Nye and
Greenland 1960). Early fallow vegetation in tropical rainforest environments often is dominated
by native pioneer species, but Ramakrishnan and
Vitousek (1989) note that reducing the time
between clearings in northeast India increased
the dominance of invasive exotics such as *Lantana*
and *Chromolaena* and other pantropical weeds. In
tropical tree plantations, competition from both
exotic weeds and native pioneer species is an
important determinant of the success or failure
of tree establishment (Wadsworth 1997). Tropical forest ecosystems may be especially vulnerable
to the spread of exotic plants from swidden or
logging operations. These activities not only provide disturbances that facilitate the establishment
of exotic species, but are also the vehicle for the
movement of novel species and varieties into lands
adjacent to forest reserves (Denslow 2002).

This brief review suggests several generalizations: (1) exotic species are not perceived as
a threat to most continental tropical forests;
(2) nonetheless, invasive alien species do affect
some tropical forest communities severely, notably
those on islands, those with an open canopy structure, and those frequently disturbed or highly
fragmented; and (3) invasive alien species present
substantial problems in managed ecosystems in
tropical environments where they alter successional trajectories, impede restoration, and may
become propagule sources driving invasion into
intact ecosystems.

IMPACTS OF EXOTIC PLANTS ON TROPICAL FORESTS

These examples stand in strong contrast to intact
close-canopied forests where exotic plants are rare,
even in treefall gaps. We discuss several hypotheses to account for the apparently high resistance of
intact continental tropical forests to the establishment of invasive exotic plants. These hypotheses
address different processes in exotic invasions as
indicated on Figure 24.1.

Hypothesis 1. Tropical forests are resistant to invasions by exotic species because they are rich in species and functional groups

The idea that species-rich communities are less
invasible than species-poor communities dates
from the writings of Elton (1958), who suggested
that more resources were likely to be pre-empted
and more niches filled in species-rich than in
species-poor communities. It has been offered as
one of the central organizing tenets of invasion
ecology (see reviews by Levine and D'Antonio
1999, Mack *et al.* 2002) and is an often-used
example of the effects of diversity on ecosystem
processes (Hooper *et al.* 2005). The relationship
between native and exotic species diversity is
negative when plot sizes are small (e.g., Fridley
et al. 2004) and experimental manipulations of
community structure have shown that species-
rich communities resist establishment of new
species more effectively than do less rich communities (Levine and D'Antonio 1999, Levine
2000, Tilman *et al.* 2001, Kennedy *et al.* 2002).
In contrast, studies of grasslands (Stohlgren *et al.*
1999), riparian ecosystems (Levine 2000), islands
(Lonsdale 1999, Sax *et al.* 2002), and continental ecosystems (Stark *et al.* 2006) show that
at regional scales both native and exotic species
richness are similarly correlated with environmental gradients – that is, native and exotic
species richness are positively correlated and both
increase along gradients of increasing resource
supply. In the absence of direct evidence, however, these patterns are not sufficient to document
competitive exclusion or resistance of diverse
communities to exotic invaders. Two recent studies have shown that these patterns of negative
and positive correlations do not differ from that
predicted by a neutral model of no species interactions and that the relationship between exotic
and native species richness depends on the area
and/or number of individuals sampled (Fridley
et al. 2004, Herben *et al.* 2004). At small plot
sizes, native and exotic richness are negatively
correlated because the number of individuals
and species sampled is necessarily limited. At
large plot sizes, the number of individuals and
species sampled in a plot is more variable and

native and exotic species richness are positively correlated.

Thus there appears to be little support for the hypothesis that species richness makes communities more resistant to the establishment of exotic species. Diverse communities can be invaded where propagule pressure is high (Levine and D'Antonio 1999). At Semliki Forest Reserve, Uganda, high-diversity forests were no more resistant to exotic invasion than were the low-diversity plantations (Rejmánek 1996). The effectiveness with which the native community pre-empts available resources may be more important than diversity *per se* in impeding the establishment of exotic species (Davis *et al.* 1998, 2000, Shea and Chesson 2002, Denslow 2003 and see Hypothesis 2). Although key functional groups are more likely to be present in species-rich than in species-poor assemblages (Hooper *et al.* 2005), diverse ecosystems may be highly invasible when key functional groups are missing, as has been proposed for tropical islands (Kitayama 1996, Kitayama and Itow 1999, Lonsdale 1999, Fine 2002, Denslow 2003).

High diversity of functional groups may buffer continental tropical forests against the kinds of ecosystem and structural alteration caused by exotic species in other ecosystems. When invasive exotic species represent novel functional groups they are likely to alter community structure, disturbance regimes, or soil processes with ecosystem-wide consequences (Vitousek 1986). Tropical forests may be resilient to such ecosystem-altering consequences of exotic invasions when exotics do not add new functional groups to the plant community or have high per capita impacts relative to native species. Ecosystem processes such as nutrient and moisture supply rates are not easily altered in rainforest environments where moisture supply, nitrogen turnover rates, and net primary productivity are naturally high (Vitousek and Sanford 1987). The principal impact of exotic species on rainforest communities thus is likely to be through competition for space and resources rather than through alteration of ecosystem processes or disturbance regimes. Exotic vines and lianas may be an exception to these generalizations. While this is a well-represented functional group in mainland tropical forests, a heavy infestation of lianas – exotic or native – can kill or damage canopy trees, prevent sapling growth, and contribute to the gradual degradation of rainforest structure (Humphries *et al.* 1991, Tabánez *et al.* 1997, Horvitz *et al.* 1998).

Hypothesis 2. Native rainforest species competitively exclude exotic species

The ability of the native plant community to exclude potential invaders also will be a function, in part, of growth and dispersal rates of native species and of their ability to rapidly pre-empt resources. Two functional groups in particular may play important roles in reducing seedling establishment, thus contributing to invasion resistance in tropical forests (e.g., Rejmánek 1996, 1999, Fine 2002): (1) fast-growing pioneers of several growth forms that effectively occupy space and pre-empt resources in high-light environments, and (2) broad-leaved understory species that are able to persist in low-light environments.

High-light-demanding pioneers, including fast-growing trees, shrubs, large herbs, and lianas, are important components of forest regeneration processes because they quickly establish in large treefall openings or other disturbed areas, such as landslips or riparian corridors. These sites also provide establishment opportunities for exotic species in forest ecosystems (e.g., Rejmánek 1996, Knapp and Canham 2000, Webb *et al.* 2000, McDowell and Turner 2002). In Hawai'i several exotic *Rubus* species compete with the native *Rubus hawaiiensis* A. Gray for gap establishment sites (Gerrish *et al.* 1992) and, in the Budongo Forest, Uganda, spread of the exotics *Cassia spectabilis* DC. (Fabaceae) and *Broussonetia papyrifera* (L.) L'Hér. ex Vent. (Moraceae) is facilitated by gap openings (Sheil *et al.* 2000).

High rates of competitive exclusion in productive wet tropical forests have been suggested as a mechanism for the hump-shaped pattern of diversity across productivity gradients (e.g., Rosenzweig and Abramsky 1993) and high competitive exclusion rates likely reduce establishment success of exotic species as well as native species. For example, new treefall openings are colonized by a combination of large-leaved herbs, vines, palms, and fast-growing, light-demanding trees, which rapidly reduce light levels near the ground (Walker *et al.* 1996, Denslow *et al.* 1998, Schnitzer *et al.* 2000). Genera such as *Cecropia*, *Trema*, *Balsa*, and *Macaranga* exhibit some of the highest growth rates observed among woody plants (Wadsworth 1997), with life-history characteristics similar to many invasive exotics. Where they are abundant, vines and lianas suppress seedling establishment and sapling growth in gaps (Putz 1991, Schnitzer *et al.* 2000). Where such pioneer species are rare, forests may be particularly vulnerable to the establishment of exotics. Horvitz *et al.* (1995) comment on the scarcity of pioneer species in the hardwood hammock flora of south Florida and speculate that exotic species (especially vines) in these hammocks usurped that role following the passage of Hurricane Andrew in 1992. On islands such as the Galápagos and Hawai'i, species with pioneer growth characteristics are sparse, possibly contributing to the invasibility of these ecosystems (Kitayama and Itow 1999, Denslow 2003). These species are able to convert high resource availability into rapid growth and high rates of production. Where invasive species increase productivity, such as through nitrogen addition to nitrogen-poor soils, competitive exclusion rates are expected to increase as well. For example, where the nitrogen-fixing tree *Falcataria moluccana* invades Hawai'ian *Metrosideros* forest on nitrogen-poor soils, productivity increases but the native *Metrosideros* declines (Hughes and Denslow 2005).

At the other end of the shade-tolerance spectrum, understory palms, shrubs, ferns, and herbs produce dense shade at ground levels (Montgomery 2004). Where these species are common, seedling establishment is suppressed and seedling densities are low (Denslow *et al.* 1991, Farris-Lopez *et al.* 2004, Harms *et al.* 2004, Wang and Augspurger 2004). The inhibitory effect is strong on native species and likely would affect exotic species as well. The combined effect of fast-growing pioneers and shade-tolerant herbs and shrubs is to reduce site occupancy by seedlings and increase the importance of recruitment limitation in rainforest dynamics. While density

may be more critical than the number of species, redundancy within functional groups is likely to increase their distribution and impact.

Hypothesis 3. High pest loads and high pest diversity in the tropics deter establishment and spread of exotic species

In tropical forests, high diversity and abundance of natural enemies (herbivores and pathogens) occasionally may lead to high impacts of native pests on exotic species (e.g., Nair 2001, Novotny *et al.* 2003). Certainly, rates of leaf damage by herbivores and pathogens tend to be high, although variable among species, in tropical forests (Coley and Aide 1991, Coley and Barone 1996). Thus, native generalist herbivores, pathogens, and viruses may provide a biotic barrier to invasion (Mack 1996, Parker *et al.* 2006) if they are able to exploit exotic plants. High diversities of both pests and host plants, which reach their peak in the wet tropics for many taxa, may increase the probability that an exotic plant is suppressed by native herbivores, as shown by Prieur-Richard *et al.* (2002) in a Mediterranean plant community. In addition, generalist pest species may play important regulatory roles in population dynamics of tropical plants. For example, Augspurger (1984) describes the importance of damping-off fungi as a source of seedling mortality in the tropical forest understory. Oomycetes, a common group of damping-off fungi, can persist in the soil in the absence of hosts and exhibit low host specificity (Augspurger 1984, Hood *et al.* 2004). Among insect herbivores, most species are not monophagous but feed on multiple species within a genus or family (Coley and Barone 1996). Some common foliage-feeding herbivores, such as leaf-cutter (Attine) ants (Fowler *et al.* 1989, Farji-Brener 2001, Wirth *et al.* 2003) and orthoptera (Novotny *et al.* 2004), have broad diets. Leaf-cutter ants in particular are serious predators of a number of exotic crops, including citrus, eucalyptus, coffee, and cacao; Cherett (1989) suggests that the susceptibility of so many crop species to this pest is due in part to their lack of defenses. A meta-analysis by Parker *et al.* (2006) showed that exotic invaders often are repelled by native herbivores. We suggest that such generalist natural enemies may contribute to the apparent invasion resistance of tropical forests.

This biotic barriers hypothesis runs counter to one of the principal explanations of invasiveness – the enemy release hypothesis (ERH). This hypothesis proposes that the accidental or even intentional introduction of plants away from their native range is accomplished most often without concomitant introduction of the specialist herbivores, pathogens, and viruses that attack and limit their populations in their native range. If native species are limited by their own suite of natural enemies where exotics are introduced, then exotics may proliferate because of their relatively lower pest loads (Elton 1958, Maron and Vilà 2001, Keane and Crawley 2002). Indeed, there is evidence that, where introduced, some exotic tropical species have lower pest loads in their invasive than in their native range and lower pest loads than local native species in their invasive range. For example, invasive species on Mahé, the main island of the tropical Seychelles, suffered less leaf area loss to herbivores than native woody species (Dietz et al. 2004). Native pioneer species were especially susceptible to herbivores (C. Kueffer pers. comm.). In addition, a biogeographical comparison of the impact of natural enemies on the neotropical shrub *Clidemia hirta* in its native and introduced ranges found that plants were heavily attacked by insect herbivores and fungal pathogens in the native range, particularly in forest understory, but that they were relatively pest free in the introduced range (DeWalt et al. 2004). The consequences of pest-load reduction to C. hirta appear to include not only proliferation in the introduced range, but also invasion of forest understory, where it does not occur in its native range (DeWalt et al. 2004). The relative success of plantations of exotic species such as rubber (*Hevea brasiliensis* (Willd.) Muell.-Arg. [Euphorbiaceae]) and *Eucalyptus* (Myrtaceae) in the tropics is also attributable in part to their escape from heavy pest loads, particularly from specialists in their native ranges (Rosenthal et al. 1979, Gadgil and Bain 1999). Thus, some tropical plant populations may be regulated by natural enemies to the extent that release from these enemies leads to their proliferation in introduced ranges (DeWalt 2005).

Does a biotic barrier in the form of high pest loads contribute to the relative dearth of invasive exotic species in tropical forests? Does escape from natural predators give exotic species an advantage where they are introduced? Both of these hypotheses are compelling and supported by examples. Further evaluation awaits more information on the population-level effects of natural enemies on exotic species.

Hypothesis 4. Low propagule availability contributes to the rarity of exotic species in many tropical forests

The distribution and abundance of many forest plant species, in temperate as well as in tropical forests, are limited by failure to recruit seedlings to sites otherwise suitable to their establishment and growth (Clark et al. 1998, 1999a, Turnbull et al. 2000, Beckage and Clark 2003, Svenning and Wright 2005). Factors contributing to recruitment limitation include those affecting the size of the seed crop (fecundity and the density and distribution of adult trees), close and distant dispersal (Clark et al. 1999b), and post-dispersal factors such as pests and pathogens which affect germination and seedling establishment (Clark et al. 1998, Nathan and Muller-Landau 2000, Turnbull et al. 2000, Zimmerman et al. 2000). Dispersal and recruitment limitation increasingly are seen as major determinants of the relative abundances of species in forest ecosystems (Clark et al. 1999b, Harms et al. 2000, Nathan and Muller-Landau 2000, Hubbell 2001, Denslow et al. 2006).

Propagule supply also plays a major role in exotic species invasions (Von Holle and Simberloff 2005). For example, some of the strongest predictors of invasiveness are those that affect propagule distribution and abundance, including the duration, frequency, and area of exotic species introductions (Richardson 1999, Lockwood et al. 2005). Forests most likely to be free of exotic species are those with low exposure to propagules from urban or agricultural areas (Aragon and Morales 2004, Sullivan et al. 2005). Disturbed forests may be free of exotic species if propagule input is low. In Hawai'i Volcanoes National Park, montane rainforests heavily damaged by pig browsing are little affected by exotic species where human traffic is low (T. Tunison personal communication). Forest fragmentation, road construction, and other sources of human disturbance are likely to expose adjacent forests to seed rain from exotic species. Fine (2002) has suggested that the scarcity of exotic species in tropical forests may reflect a more recent history of disturbance and fragmentation in tropical than in temperate forest. Although forest clearing and fragmentation is a more recent phenomenon in tropical than in temperate forests, human populations have lived in and exploited tropical forests for centuries. Further, natural disturbance regimes in wet tropical forests are high (Denslow 1987). It is unlikely that disturbance alone accounts for the distribution of exotic species.

Low exotic species abundances may reflect a historic lack of seed availability from species with appropriate physiological characteristics, such as shade tolerance, for establishment in tropical rainforest. Exotic plants, especially woody species, have been widely introduced in the tropics, often at grand scales. Extensive forestry, agricultural, and land-reclamation projects juxtapose large populations of exotic species with native forest. Many of these planted exotic species have life-history attributes similar to those of native pioneers and gap colonizers (Wadsworth 1997). For example, fast-growing species in the genera *Pinus*, *Tectona*, *Eucalyptus*, and *Gmelina* commonly have been planted for industrial timber (Wadsworth 1997), while many species and genotypes of *Leucaena*, *Albizia*, *Acacia*, and *Calliandra* have been introduced as utility species to rehabilitate degraded soils and provide fodder and firewood (Hughes and Styles 1989, Hughes 1994). Tropical forests may not be immune to the spread of such high-light-demanding exotic species, which may appear in natural forest clearings where they can impede regeneration of native species (Rejmánek 1996).

A more likely source of shade-tolerant species is ornamental plants which have been widely introduced into tropical habitats and are the source of many invasive species (Meyer and Lavergne 2004, Wu *et al.* 2004). Daehler (personal communication) estimates that 39% of the approximately 100 naturalized exotic species that pose the greatest threat to Hawai'i's native ecosystems were introduced as ornamentals. Noteworthy examples of shade-tolerant ornamentals that have become invasive in tropical forests include *Miconia calvescens* DC. (Melastomataceae), a neotropical tree invasive in native forests of French Polynesia (Meyer 1996) and Hawai'i (Conant *et al.* 1997); *Ardisia elliptica* Thunb. (Myrsinaceae), which has become invasive in hammocks of south Florida (Horvitz *et al.* 1998); and numerous vines and lianas with shade-tolerant juvenile stages, such as *Merremia tuberosa* (L.) Rendle (Convolvulaceae) and *Jasminum dichotomum* Vahl (Oleaceae) (Horvitz *et al.* 1998), also invasive in south Florida.

In the absence of experimental additions of seeds (Turnbull *et al.* 2000), it is difficult to evaluate the role of propagule availability in limiting exotic species in tropical forests. Tropical forests, like their temperate counterparts, are not likely to be strongly affected by exotic plant invasions if the forests are buffered from exposure to propagule sources (e.g., Pyšek *et al.* 2002).

CONCLUSIONS

The scarcity of exotic plants recorded from intact continental tropical forests suggests that tropical forests may be resistant to invasions of exotic plants. High species and functional group diversity, high competitive exclusion rates, and high pest loads all may confer a certain biotic resistance to the establishment and spread of exotic species in tropical forests. Similarly, high diversity and high productivity may increase the resilience of tropical forests to the kinds of ecosystem changes effected by invasive species in other ecosystems. However, we are unable to evaluate these hypotheses fully because we cannot evaluate exposure levels of tropical forests to propagules of exotic species. There are exceptions to the general pattern of sparse exotic species in tropical forests. Prevention and control of invasive species on islands, in fragmented or disturbed ecosystems, and in managed ecosystems are all major conservation and management concerns in tropical forest environments. Lessons from islands and exploited ecosystems suggest that control of invasive exotic species will assume increasing importance in the conservation of forest preserves as habitats become fragmented and exposed to increasing varieties of exotic species.

Our review highlights several areas in which further research would be productive. For example, what is the role of natural enemies (herbivorous pests and pathogens) in regulating native and exotic plant populations? We know little about the extent to which top-down factors control tropical plant populations, much less about their role in plant invasions. Basic research on the role of pests and pathogens in regulating species abundances may provide insight into the mechanisms of invasion as well as elucidating factors structuring plant communities. These issues also are related directly to the development of safe and effective biological control agents for wildland weeds (Denslow and Johnson 2006).

What are the roles of seemingly minor species such as understory shrubs and pioneer trees in regulating resource availability in tropical forest ecosystems? Our review suggests that competition from native pioneers may be critical in invasion resistance, yet these species are relatively sparse in intact rainforests. Recent research documents the importance of understory vegetation in reducing light levels near the ground (Montgomery 2004). Thus shrubs, ferns, and understory palms could play important roles in limiting exotic species establishment.

Experimental additions of seeds, propagules, or seedlings in tropical forest environments would provide important insights into processes limiting the establishment and spread of species (e.g., Turnbull *et al.* 2000) and how those processes vary within forests and across landscapes, regions, and species.

While forests in general, and tropical rainforests in particular, often appear more resistant to the establishment of exotic species than many other ecosystem types (Rejmánek 1989, Fine 2002), global change is likely to increase their vulnerability. Changing climate, altered disturbance regimes, and increased forest fragmentation and exploitation (Sala *et al.* 2000) may open forest canopies, and increased global movement of species, biotypes, pests, and diseases will provide exposure to new species capable of taking advantage of local environmental opportunities.

ACKNOWLEDGMENTS

We are pleased to acknowledge our debt to the faculty and students of the Department of Botany of the University of Hawai'i at Manoa during J.S.D.'s tenure as G.P. Wilder Visiting Professor and to the Department of Ecology and Evolutionary Biology at Rice University during S.J.D.'s appointment as a Huxley Research Fellow. We are also grateful for comments from C. Kueffer, two ad hoc reviewers and from W. Carson which were helpful in improving the manuscript.

REFERENCES

Aplet, G.H., Anderson, S.J., and Stone, C.P. (1991) Association between feral pig disturbance and the composition of some alien plant assemblages in Hawai'i Volcanoes National Park. *Vegetatio* 95, 55–62.

Aragon, R. and Morales, J.M. (2004) Species composition and invasion in NW Argentinian secondary forests: effects of land-use history, environment and landscape. *Journal of Vegetation Science* 14, 195–204.

Augspurger, C. (1984) Seedling survival of tropical tree species: interactions of dispersal distances, light gaps, and pathogens. *Ecology* 65, 1705–1712.

Bagchi, S., Goyal, S.P., and Sankar, K. (2004) Herbivore density and biomass in a semi-arid tropical dry deciduous forest of western India. *Journal of Tropical Ecology* 20, 475–478.

Baret, S., Nicolini, E., Le Bourgeois, T., and Strasberg, D. (2003) Developmental patterns of the invasive bramble (*Rubus alceifolius* Poiret, Rosaceae) in Réunion island: an architectural and morphometric analysis. *Annals of Botany* 91, 39–48.

Beckage, B. and Clark, J.S. (2003) Seedling survival and growth of three forest tree species: the role of spatial heterogeneity. *Ecology* 84, 1849–1861.

Bellingham, P.J., Tanner, E.V.J., and Healey, J.R. (2005) Hurricane disturbance accelerates invasion by the alien tree *Pittosporum undulatum* in Jamaican montane rain forests. *Journal of Vegetation Science* 16, 675–684.

Cabin, R.J., Weller, S.G., Lorence, D.H. *et al.* (2000) Effects of long-term ungulate exclusion and recent alien species control on the preservation and restoration of a Hawaiian Tropical Dry Forest. *Conservation Biology* 14, 439–453.

Chandrasekaran, S. and Swamy, P.S. (2000) Changes in herbaceous vegetation following disturbance due to biotic interference in natural and man-made ecosystems in Western Ghats. *Tropical Ecology* 36, 213–220.

Chandrashekara, U.M. and Ramakrishnanan, P.S. (1994) Successional patterns and gap phase dynamics of a humid tropical forest of the Western Ghats of Kerala, India: ground vegetation, biomass, productivity and nutrient cycling. *Forest Ecology and Management* 70, 23–40.

Cherett, J.M. (1989) Leaf-cutting ants. In H. Lieth and M.J.A. Weiger (eds), *Ecosystems of the World.* Elsevier, New York, pp. 473–489.

Clark, J.S., Beckage, B., Camill, P. *et al.* (1999a) Interpreting recruitment limitation in forests. *American Journal of Botany* 86, 1–16.

Clark, J.S., Macklin, E., and Wood, L. (1998) Stages and spatial scales of recruitment limitation in southern Appalachian forests. *Ecological Monographs* 68, 213–235.

Clark, J.S., Silman, M., Kern, R., Macklin, E., and HilleRisLambers, J. (1999b) Seed dispersal near and far: patterns across temperate and tropical forests. *Ecology* 80, 1475–1494.

Coley, P.D. and Aide, T.M. (1991) Comparison of herbivory and plant defenses in temperate and tropical broad-leaved forests. In P.W. Price, T.M. Lewinsohn, G.W. Fernandes, and W.W. Benson (eds), *Plant–Animal Interactions: Evolutionary Ecology in Tropical and Temperate Regions.* John Wiley & Sons, New York, pp. 25–49.

Coley, P.D. and Barone, J.A. (1996) Herbivory and plant defenses in tropical forests. *Annual Review of Ecology and Systematics* 27, 305–335.

Conant, P., Medeiros, A.C., and Loope, L.L. (1997) A multiagency containment program for miconia (*Miconia calvescens*), an invasive tree in Hawaiian rain forests. In J.O. Luken and J.W. Thieret (eds), *Assessment and Management of Plant Invasions.* Springer-Verlag, New York, pp. 249–254.

D'Antonio, C.M. and Dudley, T.L. (1995) Biological invasions as agents of change on islands versus mainlands. In P.M. Vitousek, L.L. Loope, and H. Adsersen (eds), *Islands: Biological Diversity and Ecosystem Function.* Springer-Verlag, Stanford, pp. 103–121.

D'Antonio, C.M. and Vitousek, P.M. (1992) Biological invasions by exotic grasses, the grass/fire cycle, and global change. *Annual Review of Ecology and Systematics* 23, 63–87.

Davis, M.A., Wrage, K.J., and Reich, P.B. (1998) Competition between tree seedlings and herbaceous vegetation: support for a theory of resource supply and demand. *Journal of Ecology* 86, 652–661.

Davis, M.A., Grime, J.P., and Thompson, K. (2000) Fluctuating resources in plant communities: a general theory of invasibility. *Journal of Ecology* 88, 528–534.

Denslow, J.S. (1987) Tropical rainforest gaps and tree species diversity. *Annual Review of Ecology and Systematics* 18, 431–451.

Denslow, J.S. (2002) Invasive alien woody species in Pacific island forests. *Unasylva* 53, 62–63.

Denslow, J.S. (2003) Weeds in paradise: thoughts on the invasibility of tropical islands. *Annals of the Missouri Botanical Garden* 90, 119–127.

Denslow, J.S., Ellison, A., and Sanford, R.E. Jr. (1998) Treefall gap size effects on above- and below-ground processes in a tropical wet forest. *Journal of Ecology* 86, 597–609.

Denslow, J.S. and Johnson, M.T. (2006) Biological control of tropical weeds: research opportunities in plant–herbivore interactions. *Biotropica* 38, 139–142.

Denslow, J.S., Newell, E., and Ellison, A.M. (1991) The effect of understory palms and cyclanths on the growth and survival of *Inga* seedlings. *Biotropica* 23, 225–234.

Denslow, J.S., Uowolo, A.L., and Hughes, R.F. (2006) Limitations to seedling establishment in a mesic Hawaiian forest. *Oecologia* 148, 118–128.

DeWalt, S.J. (2005) Effects of natural enemies on tropical plant invasions. In D.F.R.P. Burslem, M. Pinard, and S.E. Hartley (eds), *Biotic Interactions in the Tropics.* Cambridge University Press, Cambridge, pp. 459–483.

DeWalt, S.J., Denslow, J.S., and Ickes, K. (2004) Natural-enemy release facilitates habitat expansion of the invasive tropical shrub *Clidemia hirta. Ecology* 85, 471–483.

Dietz, H., Wirth, L.R., and Buschmann, H. (2004) Variation in herbivore damage to invasive and native woody plant species in open forest vegetation on Mahé, Seychelles. *Biological Invasions* 6, 511–521.

Di Stefano, J.F., Fournier, L.A., Carranza, J., Marin, W., and Mora, A. (1998) Invasive potential of *Syzygium jambos* (Myrtaceae) in forest fragments: the case of Ciudad Colon, Costa Rica. *Revista de Biologia Tropical* 46, 567–573.

Elton, C.S. (1958) *The Ecology of Invasions of Animals and Plants.* Methuen, London.

Ewel, J.J. (1986) Invasibility: lessons from South Florida. In H.A. Mooney and J.A. Drake (eds), *Ecology of Biological Invasions of North America and Hawai'i.* Springer-Verlag, New York, pp. 214–230.

Ewel, J.J., O'Dowd, D.J., Bergelson, J. et al. (1999) Deliberate introductions of species: research needs. *BioScience* 49, 619–630.

Farji-Brener, A.G. (2001) Why are leaf-cutting ants more common in early secondary forests than in old-growth tropical forests? An evaluation of the palatable forage hypothesis. *Oikos* 92, 169–177.

Farris-Lopez, K., Denslow, J.S., Moser, B., and Passmore, H. (2004) Influence of a common palm, *Oenocarpus mapora*, on seedling establishment in a tropical moist forest in Panama. *Journal of Tropical Ecology* 20, 429–439.

Fensham, R.J., Fairfax, R.J., and Cannell, R.J. (1994) The invasion of *Lantana camara* L. in Forty Mile Scrub National Park, North Queensland. *Australian Journal of Ecology* 19, 297–305.

Fine, P.V.A. (2002) The invasibility of tropical forests by exotic plants. *Journal of Tropical Ecology* 18, 687–705.

Fleischmann, K. (1997) Invasion of alien woody plants on the islands of Mahé and Silhouette, Seychelles. *Journal of Vegetation Science* 8, 5–12.

Foster, R.B. and Hubbell, S.P. (1990) The floristic composition of the Barro Colorado Island forest. In A.H. Gentry (ed.), *Four Neotropical Rain Forests.* Yale University Press, New Haven, pp. 85–98.

Fowler, H.G., Pagani, M.I., Da Silva, D.A., Forti, L.C., da Silva, V.P., and Vasconcelos, H.L. (1989) A pest is a pest is a pest? The dilemma of Neotropical leaf cutting ants: keystone taxa of natural ecosystems. *Environmental Management* 13, 671–675.

Fridley, J.D., Brown, R.L., and Bruno, J.F. (2004) Null models of exotic invasion and scale-dependent patterns of native and exotic species richness. *Ecology* 85, 3215–3222.

Gadgil, P.D. and Bain, J. (1999) Vulnerability of planted forests to biotic and abiotic disturbances. *New Forests* 17, 227–238.

Gerrish, G., Stemmermann, L., and Gardner, D.E. (1992) The distribution of *Rubus* species in the State of Hawai'i. Cooperative National Park Resources Studies Unit, Honolulu, Hawai'i. No. 85.

Ghazoul, J. (2004) Alien abduction: disruption of native plant-pollinator interactions by invasive species. *Biotropica* 36, 156–164.

Green, P.T., O'Dowd, D.J., and Lakes, P.S. (2004) Resistance of island rainforest to invasion by alien plants: influence of microhabitat and herbivory on seedling performance. *Biological Invasions* 6, 1–9.

Hammel, B. (1990) The distribution of diversity among families, genera and habitat types in the La Selva Flora. In A.H. Gentry (ed.), *Four Neotropical Rain Forests.* Yale University Press, New Haven, pp. 75–84.

Harms, K.E., Powers, J.S., and Montgomery, R.A. (2004) Variation in small sapling density, understory cover, and resource availability in four Neotropical forests. *Biotropica* 36, 40–51.

Harms, K.E., Wright, S.J., Calderon, O., Hernández, A., and Herre, E.A. (2000) Pervasive density-dependent recruitment enhances seedling diversity in a tropical forest. *Nature* 404, 493–495.

Herben, T., Mandák, B., Bímová, K., and Münzbergová, Z. (2004) Invasibility and species richness of a community: a neutral model and a survey of published data. *Ecology* 85, 3223–3233.

Hood, L.A., Swaine, M.D., and Mason, P.A. (2004) The influence of spatial patterns of damping-off disease and arbuscular mycorrhizal colonization on tree seedling establishment in Ghanaian tropical forest and soil. *Journal of Ecology* 92, 816–823.

Hooper, D.U., Chapin III, F.S., Ewel, J.J. et al. (2005) Effects of biodiversity on ecosystem functioning: a consensus of current knowledge and needs for future research. *Ecological Monographs* 75, 3–35.

Horvitz, C.C., McMann, S., and Freedman, A. (1995) Exotics and hurricane damage in three hardwood hammocks in Dade County Parks, Florida. *Journal of Coastal Research* 21, 145–158.

Horvitz, C.C., Pascarella, J.B., McMann, S., Freedman, A., and Hofstetter, R.H. (1998) Functional roles of invasive non-indigenous plants in hurricane-affected subtropical hardwood forests. *Ecological Applications* 8, 947–974.

Hubbell, S.P. (2001) *The Unified Neutral Theory of Biodiversity and Biogeography.* Princeton University Press, Princeton, NJ.

Huenneke, L.F. and Vitousek, P.M. (1990) Seedling and clonal recruitment of the invasive tree *Psidium cattleianum*: implications for management of native Hawaiian forests. *Biological Conservation* 53, 199–211.

Hughes, C.E. (1994) Risks of species introductions in tropical forestry. *Commonwealth Forestry Review* 73, 243–252.

Hughes, R.F. and Denslow, J.S. (2005) Invasion by an N-fixing tree, *Falcataria moluccana*, alters function and structure of wet lowland forests of Hawai'i. *Ecological Applications* 15, 1615–1628.

Hughes, C.E. and Styles, B.T. (1989) The benefits and risks of woody legume introductions. In C.H. Stirton and J.L. Zarruchi (eds), *Advances in Legume Biology. Monographs in Systematic Botany*, Missouri Botanical Garden 29, pp. 505–531.

Humphries, S.E., Groves, R.H., and Mitchell, D.S. (1991) Part I. Plant invasions of Australian ecosystems: a status review and management directions. *Kowari 2*, 1–134.

Humphries, S.E. and Stanton, J.P. (1992) *Weed Assessment in the Wet Tropics World Heritage Area of North Queensland*. Report to The Wet Tropics Management Agency Report.

Ickes, K., DeWalt, S.J., and Appanah, S. (2001) Effects of native pigs (*Sus scrofa*) on woody understorey vegetation in a Malaysian lowland rain forest. *Journal of Tropical Ecology* 17, 191–206.

Keane, R.M. and Crawley, M.J. (2002) Exotic plant invasions and the enemy release hypothesis. *Trends in Ecology and Evolution* 17, 164–170.

Kennedy, T.A., Naeem, S., Howe, K.M., Knops, J.M.H., Tilman, D., and Reich, P.B. (2002) Biodiversity as a barrier to ecological invasion. *Nature* 417, 636–638.

Kitayama, K. (1996) Patterns of species diversity on an oceanic versus a continental island mountain: a hypothesis on species diversification. *Journal Vegetation Science* 7, 879–888.

Kitayama, K. and Itow, S. (1999) Above-ground biomass and soil nutrient pools of a *Scalesia pedunculata* montane forest on Santa Cruz, Galápagos. *Ecological Research* 14, 405–408.

Kitayama, K. and Mueller-Dombois, D. (1995) Biological invasion on an oceanic island mountain – do alien plant species have wider ecological ranges than native species? *Journal of Vegetation Science* 6, 667–674.

Knapp, L.B. and Canham, C.D. (2000) Invasion of an old-growth forest in New York by *Ailanthus altissima*: sapling growth and recruitment in canopy gaps. *Journal of the Torrey Botanical Society* 127, 307–315.

Laurance, W.F. (1997) Hyper-disturbed parks: edge effects and the ecology of isolated rain forest reserves in tropical Australia. In W.F. Laurance and R.O. Bierregaard, Jr. (eds), *Tropical Forest Remnants: Ecology, Management, and Conservation of Fragmented Communities*. University of Chicago Press, Chicago, pp. 71–83.

Levine, J.M. (2000) Species diversity and biological invasions: relating local process to community pattern. *Science* 288, 852–854.

Levine, J.M. and D'Antonio, C.M. (1999) Elton revisited: a review of evidence linking diversity and invasibility. *Oikos* 87, 15–26.

Lockwood, J.L., Cassey, P., and Blackburn, T.M. (2005) The role of propagule pressure in explaining species invasions. *Trends in Ecology and Evolution* 20, 223–228.

Lonsdale, W.M. (1999) Global patterns of plant invasions and the concept of invasibility. *Ecology* 80, 1522–1536.

Lugo, A.E. and Helmer, E. (2004) Emerging forests on abandoned land: Puerto Rico's new forests. *Forest Ecology and Management* 190, 145–161.

MacDonald, I.A.W., Ortiz, L., Lawesson, J.E., and Nowak, J.B. (1988) The invasion of highlands in Galápagos by the red quinine tree, *Cinchona succirubra*. *Environmental Conservation* 15, 215–220.

Mack, R.N. (1996) Biotic barriers to plant naturalization. In V.C. Moran and J.H. Hoffmann (eds), *IX International Symposium on Biological Control*. University of Cape Town, Stellenbosch, South Africa, pp. 39–46.

Mack, R.N., Simberloff, D., Lonsdale, W.M., Evans, H., and Clout, M. (2002) Biotic invasions: causes, epidemiology, global consequences and control. *Ecological Applications* 10, 689–710.

Maron, J.L. and Vilà, M. (2001) When do herbivores affect plant invasion? Evidence for the natural enemies and biotic resistance hypotheses. *Oikos* 95, 361–373.

McDowell, C.L. and Turner, D.P. (2002) Reproductive effort in invasive and non-invasive *Rubus*. *Oecologia* 133, 102–111.

Meyer, J.-Y. (1996) Status of *Miconia calvescens* (Melastomataceae), a dominant invasive tree in the Society Islands (French Polynesia). *Pacific Science* 50, 66–76.

Meyer, J.-Y. and Lavergne, C. (2004) *Beautes fatales*: Acanthaceae species as invasive alien plants on tropical Indo-Pacific Islands. *Diversity and Distributions* 10, 333–347.

Montgomery, R.A. (2004) Effects of understory foliage on patterns of light attenuation near the forest floor. *Biotropica* 36, 33–40.

Mueller-Dombois, D. (1995) Biological diversity and disturbance regimes in island ecosystems. In P.M. Vitousek, L.L. Loope, and H. Adsersen (eds), *Islands: Biological Diversity and Ecosystem Function*. Springer, Berlin, pp. 163–175.

Nair, K.S.S. (2001) *Pest Outbreaks in Tropical Forest Plantations: Is There a Greater Risk for Exotic Tree Species?* Center for International Forestry Research, Bogor, Indonesia.

Nathan, R. and Muller-Landau, H.C. (2000) Spatial patterns of seed dispersal, their determinants and consequences for recruitment. *Trends in Ecology and Evolution* 15, 278–285.

Novotny, V., Basset, Y., Miller, S.E. *et al.* (2004) Local species richness of leaf-chewing insects feeding on woody plants from one hectare of a lowland rainforest. *Biological Conservation* 18, 227–237.

Novotny, V., Miller, S.E., Cizek, L. *et al.* (2003) Colonising aliens: caterpillars (Lepidoptera) feeding on *Piper aduncum* and *P. umbellatum* in rainforests of Papua New Guinea. *Ecological Entomology* 28, 704–716.

Nye, P.H. and Greenland, D.J. (1960) *The Soil Under Shifting Cultivation.* Technical Communication No. 51, Commonwealth Bureau of Soils, Harpenden.

Parker, J.D., Burkepile, D.E., and Hay, M.E. (2006) Opposing effects of native and exotic herbivores on plant invasions. *Science* 311, 1459–1461.

Parsons, J.J. (1972) Spread of African pasture grasses in the American tropics. *Journal of Range Management* 25, 12–17.

Prieur-Richard, A.K., Lavorel, S., Linhart, Y.B. and Dos Santos, A. (2002) Plant diversity, herbivory and resistance of a plant community to invasions, in Mediterranean annual communities. *Oecologia* 130, 96–104.

Putz, F.E. (1991) Silvicultural effects of lianas. In F.E. Putz and H.A. Mooney (eds), *The Biology of Vines.* Cambridge University Press, Cambridge, pp. 493–501.

Pyšek, P., Jarošek, V., and Kučera, T. (2002) Patterns of invasion in temperate nature reserves. *Biological Conservation* 104, 13–24.

Ramakrishnan, P.S. (1991) *Ecology of Biological Invasion in the Tropics.* International Scientific Publications, New Delhi.

Ramakrishnan, P.S. and Vitousek, P.M. (1989) Ecosystem-level processes and consequences of biological invasions. In J.A. Drake, H.A. Mooney, F. Di Castri *et al.* (eds), *Biological Invasions: A Global Perspective.* Wiley, New York, pp. 281–328.

Rejmánek, M. (1989) Invasibility of plant communities. In J.A. Drake, H.A. Mooney, F. Di Castri *et al.* (eds) *Biological Invasions: A Global Perspective.* Wiley, New York, pp. 369–388.

Rejmánek, M. (1996) Species richness and resistance to invasions. In G. Orians, R. Dirzo, and J.H. Cushman (eds), *Biodiversity and Ecosystem Processes in Tropical Forests.* Springer Verlag, New York, pp. 153–172.

Rejmánek, M. (1999) Invasive plant species and invasible ecosystems. In O.T. Sandlund, P.J. Schei, and A. Viken (eds), *Invasive Species and Biodiversity Management.* Kluwer Academic Publishers, Boston, MA, pp. 79–102.

Richardson, D.M. (1999) Commercial forestry and agroforestry as sources of invasive alien trees and shrubs. In O.T. Sandlund, P.J. Schei, and A. Viken (eds), *Invasive Species and Biodiversity Management.* Kluwer Academic Publishers, Boston, MA, pp. 237–257.

Rogers, H.M. and Hartemink, A.E. (2000) Soil seed bank and growth rates of an invasive species, *Piper aduncum*, in the lowlands of Papua New Guinea. *Journal of Tropical Ecology* 16, 243–251.

Rosenthal, G.A. and Janzen, D.H. (1979) *Herbivores, Their Interaction with Secondary Plant Metabolites.* Academic Press, New York.

Rosenzweig, M.L. and Abramsky, Z. (1993) How are diversity and productivity related? In R.E. Ricklefs and D. Schluter (eds), *Species Diversity in Ecological Communities.* University of Chicago Press, Chicago, pp. 52–65.

Sala, O.E., Chapin III, F.S., Armesto, J.J. *et al.* (2000) Global biodiversity scenarios for the year 2100. *Science* 287, 1770–1774.

Sax, D.F., Gaines, S.D., and Brown, J.H. (2002) Species invasions exceed extinctions on islands worldwide: a comparative study of plants and birds. *American Naturalist* 160, 766–783.

Schnitzer, S.A., Dalling, J.W., and Carson, W.P. (2000) The impact of lianas on tree regeneration in tropical forest canopy gaps: evidence for an alternative pathway of gap-phase regeneration. *Journal of Ecology* 88, 655–666.

Shea, K. and Chesson, P. (2002) Community ecology theory as a framework for biological invasions. *Trends in Ecology and Evolution* 17, 170–176.

Sheil, D. (1994) Naturalized and invasive species in the evergreen forests of the East Usambara Mountains, Tanzania. *African Journal of Ecology* 32, 66–71.

Sheil, D., Jennings, S., and Savill, P. (2000) Long-term permanent plot observations of vegetation dynamics in Budongo, a Ugandan rain forest. *Journal of Tropical Ecology* 16, 765–800.

Smith, C.W. (1985) Impact of alien plants on Hawai'i's native biota. In C.P. Stone and J.M. Scott (eds), *Hawai'i's Terrestrial Ecosystems: Preservation and Management.* University of Hawai'i Cooperative National Park Resources Studies Unit, Honolulu, HI, pp. 180–250.

Space, J.C. (2005) *Pacific Island Ecosystems at Risk. v 5.0.* USDA Forest Service, Institute of Pacific Islands Forestry. (www.hear.org/pier).

Stark, S.C., Bunker, D.E., and Carson, W.P. (2006) A null model of exotic plant diversity tested with exotic and native species-area relationships. *Ecology Letters* 9, 136–141.

Stohlgren, T.J., Binkley, D., Chong, G.W. *et al.* (1999) Exotic plant species invade hot spots of native plant diversity. *Ecological Monographs* 69, 25–46.

Sullivan, J.J., Timmins, S.M., and Williams, P.A. (2005) Movement of exotic plants into coastal native forests from gardens in northern New Zealand. *New Zealand Journal of Ecology* 29, 1–10.

Svenning, J.C. and Wright, S.J. (2005) Seed limitation in a Panamanian forest. *Journal of Ecology* 93, 853–862.

Tabánez, A.A.J., Viana, V.M., and Dias, A.D.S. (1997) Conseqüências da fragmentação e do efeito de borda sobre a estrutura, diversidade e sustentabilidade de um fragmento de floresta de Planalto de Piracicaba, SP. *Revista Brasileira de Biologia* 57, 47–60.

Teo, D.H.L., Tan, H.T.W., Corlett, R.T., Wong, C.M., and Lum, S.K.Y. (2003) Continental rain forest fragments in Singapore resist invasion by exotic plants. *Journal of Biogeography* 30, 305–310.

Tilman, D., Reich, P.B., Knops, J., Wedin, D., Mielke, T., and Lehman, C. (2001) Diversity and productivity in a long-term grassland experiment. *Science* 294, 843–845.

Turnbull, L.A., Crawley, M.J., and Rees, M. (2000) Are plant populations seed-limited? A review of seed-sowing experiments. *Oikos* 88, 225–238.

Uhl, C. and Kauffman, J.B. (1990) Deforestation, fire susceptibility, and potential tree responses to fire in the eastern Amazon. *Ecology* 71, 437–449.

Vitousek, P.M. (1986) Biological invasions and ecosystem properties: can species make a difference? In H.A. Mooney and J.A. Drake (eds), *Ecology of Biological Invasions of North American and Hawai'i*. Springer Verlag, New York, pp. 163–176.

Vitousek, P.M. and Sanford, R.L. (1987) Nutrient cycling in moist tropical forest. *Annual Review of Ecology and Systematics* 17, 137–168.

Vitousek, P.M., Walker, L.R., Whiteaker, L.D., Mueller-Dombois, D., and Matson, P.A. (1987) Biological invasion by *Myrica faya* alters ecosystem development in Hawaii. *Science* 238, 802–804.

Von Holle, B. and Simberloff, D. (2005) Ecological resistance to biological invasion overwhelmed by propagule pressure. *Ecology* 86, 3212–3218.

Wadsworth, F.H. (1997) *Forest Production for Tropical America*. Agriculture Handbook 710. USDA Forest Service, Washington, DC.

Wagner, W.L., Herbst, D.R., and Sohmer, S.H. (1999) *Manual of the Flowering Plants of Hawaii*, 2nd Edition. University of Hawaii Press, Honolulu, HI.

Walker, L.R., Zimmerman, J.K., Lodge, D.J., and Guzman-Grajales, S. (1996) An altitudinal comparison of growth and species composition in hurricane-damaged forests in Puerto Rico. *Journal of Ecology* 84, 877–890.

Wang, Y.-H. and Augspurger, C.K. (2004) Dwarf palms and cyclanths strongly reduce Neotropical seedling recruitment. *Oikos* 107, 619–633.

Webb, S.L., Dwyer, M., Kaunzinger, C.K., and Wyckoff, P.H. (2000) The myth of the resilient forest: Case study of the invasive Norway maple (*Acer platanoides*). *Rhodora* 102, 332–354.

Wester, L.L. and Wood, H.B. (1977) Koster's curse (*Clidemia hirta*), a weed pest in Hawaiian forests. *Environmental Conservation* 4, 35–41.

Whitmore, T.C. (1991) Invasive woody plants in perhumid tropical climates. In P.S. Ramakrishnan (ed.), *Ecology of Biological Invasions in the Tropics*. International Scientific Publications, New Delhi, pp. 35–40.

Wirth, R., Beyschlag, W., Ryel, R., Herz, H., and Holldobler, B. (2003) *Herbivory of Leaf-Cutting Ants: A Case Study on Atta columbica in the Tropical Rain Forest of Panama*. Springer-Verlag, New York.

Woodcock, D. (2007) To restore the watersheds: early twentieth-century tree planting in Hawaii. *Annals of the Association of American Geographers* 93, 624–635.

Wu, S.-H., Hsieh, C.-F., Chaw, S.-M., and Rejmanek, M. (2004) Plant invasions in Taiwan: insights from the flora of casual and naturalized alien species. *Diversity and Distributions* 10, 349–362.

Zimmerman, J.K., Pascarella, J.B., and Aide, T.M. (2000) Barriers to forest regeneration in an abandoned pasture in Puerto Rico. *Restoration Ecology* 8, 350–360.

TROPICAL FOREST CONSERVATION

Chapter 25

LINKING INSIGHTS FROM ECOLOGICAL RESEARCH WITH BIOPROSPECTING TO PROMOTE CONSERVATION, ENHANCE RESEARCH CAPACITY, AND PROVIDE ECONOMIC USES OF BIODIVERSITY

Thomas A. Kursar, Todd L. Capson, Luis Cubilla-Rios,
Daniel A. Emmen, William Gerwick, Mahabir P. Gupta, Maria V. Heller,
Kerry McPhail, Eduardo Ortega-Barría, Dora I. Quiros, Luz I. Romero,
Pablo N. Solis, and Phyllis D. Coley

OVERVIEW

Bioprospecting has frequently been cited as a sustainable use of biodiversity that should also provide a motivation for conservation. Nevertheless, in the tropical, biodiversity-rich regions of the world the level of bioprospecting is much below the potential, with the result that bioprospecting has had limited impact on conservation. Our group has developed a bioprospecting program in Panama that has addressed these critical issues. The program was initiated using the insights from 20 years of basic ecological research to enhance the likelihood of finding active compounds. In addition, instead of sending samples abroad, most of the research in our program is carried out in Panamanian laboratories. As a result, many young Panamanian scientists are trained. Through this and other mechanisms Panama receives immediate benefits from investigation into the uses of its biodiversity. Over the long term, such research may lead to intellectual property that assists with establishing new industry in Panama that is based upon sustainable uses of biodiversity. Additionally, we have linked our bioprospecting efforts to conservation via transparent communication about the program's use of biodiversity, resulting in a self-evident need to promote conservation. The Panama program has also made direct conservation efforts in a newly established protected area, Coiba National Park, an area in which we also collect. Hence, beginning with insights from ecological research, both conservation and sustainable development benefit from the enhanced bioprospecting effort that we have established in Panama.

THREATS TO BIODIVERSITY, BIOPROSPECTING, AND THE CONVENTION ON BIOLOGICAL DIVERSITY

Humans have greatly impacted earth's ecosystems, resulting in large and rapid changes. To date, human activity, including grazing, clearing for agriculture, and urban development, has extensively modified 35–50% of earth's land area, with the largest future impacts predicted to occur in the tropics (Ramankutty and Foley 1999, Tilman *et al.* 2001, Defries *et al.* 2002). Freshwater and marine ecosystems also are threatened (Jackson *et al.* 2001, Pauly *et al.* 2002, Rabalais *et al.* 2002). Many believe we are experiencing a major extinction crisis (Pimm *et al.* 1995). Exacerbating these environmental threats, the human population may grow by nearly 50% between 2000 and 2030, and resource use per capita is rising fast (Myers and Kent 2003). Given that diversity is concentrated in the tropics, it is likely that extinctions will be most frequent in the tropics. While conservation can potentially mitigate many of these problems, the extent and speed of conservation efforts may not be sufficient.

How can we better motivate conservation, especially in the developing countries of the tropics that harbor a large fraction of the world's biodiversity? Clearly, all appropriate conservation approaches should be applied with maximal effectiveness. Here we focus on a strategy often termed "use it or lose it." Put simply, if humans can obtain more value from habitat left in a natural state than from conversion for human uses, then economic forces can drive conservation. While "sustainable use," such as natural forest management, often leads to habitat degradation (Oates 1999, Terborgh 1999), other economic uses of biodiversity may prove easier to sustain. Particularly promising are medicinal and horticultural products, ecosystem services, ecotourism, and bioprospecting, the investigation of biodiversity as a source of useful medicines or genes (ten Kate and Laird 1999). For areas with high biodiversity, such as the tropics, bioprospecting may be an economic use of biodiversity that effectively promotes habitat protection.

Nevertheless, the utility of bioprospecting for providing benefits for developing countries and for enhancing the protection of their biodiversity has been controversial. This controversy results, in part, from changing perceptions on who owns and who should benefit from biodiversity. Between the 17th and early 20th centuries sovereign countries and colonial powers prohibited the export of viable seed or live plants of nutmeg, *Cinchona*, and coffee in order to retain the benefits of biodiversity. However, such efforts were not successful over the long term; there were many examples of "biopiracy" such as the smuggling of rubber seedlings out of Brazil and the subsequent establishment of lucrative rubber plantations in Southeast Asia (Balick and Cox 1996). During the 1970s and 1980s, a number of international conventions provided for biotic and open ocean resources as a common heritage of humanity and promoted their shared use (Gepts 2004).

In 1992–3 the Convention on Biological Diversity (CBD; Gollin 1993) reversed this trend by recognizing that nations have sovereignty over, and hence the right to control access to, their species ("genetic resources") and by requiring equitable sharing of the benefits derived from biodiversity. More recent international agreements have provided additional support for the new legal regime, one that researchers and industry presently abide by (Gollin 1999). A key component of the CBD that is not adequately appreciated provides that both developing and developed countries should facilitate the study of the uses of biodiversity. For example, the CBD indicates that each country should provide for appropriate access to biodiversity (Article 15) and that the developed countries allow for the transfer of technology (Article 16). In short, the CBD is a wide-ranging and expansive document that lays out a very broad perspective on the use of biodiversity. This ranges from guaranteeing nations sovereign rights over their biotic resources to stating that nations should "endeavour to create conditions to facilitate access to genetic resources for environmentally sound uses" as well as "develop and carry out scientific research based on genetic resources."

Given these provisions, many expected that the CBD would promote biodiversity research, including bioprospecting, and would allow nations to capture the value of their biodiversity. In fact, many governments developed unrealistic expectations that have inhibited both basic and commercially oriented research on biodiversity (Grajal 1999, Gomez-Pompa 2004). This phenomenon resulted from the perception that biotic resources have a high value, even in a "raw," undeveloped state. In part these perceptions arose because in 1991, a 1.1 million dollar agreement between Merck and Costa Rica's National Biodiversity Institute (Aldhous 1991) fueled unrealistic hopes for substantial access payments to biodiversity-rich nations. Since then many of the academic studies and international conferences have dealt with legal issues such as defining prior informed consent for the use of traditional knowledge and specifying the nature of benefit sharing arrangements (CBD 2005). Inadequate attention has been paid to the equally important issues of promoting and streamlining scientific research on the uses of biodiversity. Furthermore, some countries have passed legislation that severely restricts basic research, such as the export of herbarium specimens as well as research by their own scientists on their own biodiversity (Grajal 1999, ten Kate and Laird 1999, p. 19). In addition, in biodiverse countries applied research by pharmaceutical and agricultural companies has been inhibited. Because the value of biotic resources can be realized only through research, the low investment in research has had the ironic effect of decreasing the value of biodiversity.

How can we return to the vision outlined in the CBD? How can we promote the study of the uses of biodiversity as well as link such studies with economic development and conservation? Prior to 1990 most bioprospecting arrangements involved shipping samples abroad with the expectation of receiving royalties or milestone payments. During the last 10 years, more emphasis has been placed on the provision of benefits that are derived from research (Laird and ten Kate 2002). Nevertheless, this approach to bioprospecting is in its infancy. In fact, no comprehensive programs to study the uses of biotic resources have been created in the biodiversity-rich countries of the world. This has left the bioprospecting-based argument, to conserve nature as a future source of medicines and

genes, without any modern examples. A separate issue is that there is no inherent link between bioprospecting and conservation; in fact creating mechanisms by which bioprospecting directly promotes conservation has been elusive (Laird and ten Kate 2002). In this chapter we provide an overview of our Panama-based bioprospecting program, one that is attempting to address many of these issues (Kursar *et al.* 2006, 2007). In particular we focus on how ecological studies can assist biodiversity prospecting, and how such research can be linked to economic development as well as to conservation.

IS BIOPROSPECTING BENEFICIAL FOR DEVELOPING COUNTRIES?

The model implicit in many bioprospecting arrangements has been that the source country provides biological materials and the developed country provides research. If a drug is commercialized, the source country would receive royalties. However, with a success rate of much less than 1 in 10,000 samples, royalties are a highly unlikely outcome (Principe 1991, McChesney 1996). Additionally, the time frame is long, perhaps 10–12 years from discovery to receiving benefits. This arrangement fails because neither bioprospecting nor conservation is promoted and because biodiversity-rich countries receive no immediate benefits. This model inhibits research on the beneficial uses of biodiversity and greatly weakens or invalidates the argument that biodiversity should be preserved because of future utility.

In order to create a bioprospecting program in which the source country receives immediate and tangible benefits, we have initiated a collaborative project based in Panama in which most of the drug discovery research is carried out in-country (Capson *et al.* 1996, Kursar *et al.* 1999). In this chapter, we describe our bioprospecting project which has been ongoing in Panama since 1995. Royalty agreements are in place with the Panamanian government. However, royalties are not the focus of the project. By conducting the research in Panama, the project provides guaranteed benefits even if a drug does not become commercialized.

THE PROCESS OF DRUG DISCOVERY RESEARCH

A key function of bioprospecting is to contribute some of the thousands of compounds that are discovered annually to have interesting structures or activities. A subset of these become new "lead compounds," that is, compounds that are promising enough to merit substantial investment and continued investigation. In a typical year, relatively few lead compounds are successful and become approved medicines. Hence, the drug discovery process can be thought of as a pyramid having a very broad base that is composed of thousands of compounds with new activities, with many of these derived from bioprospecting (Principe 1991, McChesney 1996). Higher on the pyramid are many fewer lead compounds and only 20–30 of these make it to market annually. Even though the discoveries of many research groups encompass interesting structures or activities that do not become medicines, such compounds represent the essential, initial steps of the drug discovery process. Recently it has been estimated that a third of new drugs, and perhaps more, including many of the most innovative medicines, are derived from research in academia, government, or small biotech companies (Angell 2004). Tens of billions of dollars are spent annually to support this research. Furthermore, nature is still a productive source of new medicines. Taken together, these observations indicate that bioprospecting research conducted in academia or in small companies will continue to provide both jobs and promising lead compounds. Given that much of biodiversity lies in the developing world, what mechanisms will promote the funding of bioprospecting research in these regions?

THE INTERNATIONAL COOPERATIVE BIODIVERSITY GROUPS

In 1992 the United States government (National Institutes of Health, the National Science Foundation, and the US Department of Agriculture) initiated an imaginative and ambitious program, The International Cooperative Biodiversity Groups (ICBG), with the goals of combining drug discovery from natural products with biodiversity conservation, scientific capacity-building, and economic development (Rosenthal *et al.* 1999). The motivation for the program was derived from the concerns that were outlined above regarding threats to biodiversity and the slow pace of research on uses of biodiversity. In addition, the program inception was prompted by the recognition that improvements in human health and agricultural productivity historically have depended on access to biodiversity (Grifo and Rosenthal 1997). For example, discovery of taxol in the 1960s led scientists to uncover a previously undescribed mechanism of anti-cancer activity in 1979 (Horwitz 1992). These discoveries had considerable consequences, including the development of taxol as an effective anti-cancer treatment, new assays to detect other microtubule stabilizing agents and, quite recently, the discovery of additional anti-cancer drugs that work through the same mechanism (Mani *et al.* 2004). An agricultural example is the case of the then-new, high yield rice varieties that were protected from grassy stunt virus during the 1970s by breeding with a wild species of rice from India (Plucknett *et al.* 1987). Because of many similar successes, the collection and protection of crop germplasm for use in crop breeding remains a very high priority worldwide (Biodiversity International 2007).

In many regions of the world the future development of biodiversity-based research is threatened not only by the biodiversity losses already mentioned, but also by restraints on access to biodiversity and by weak scientific infrastructure. To address these issues an experimental approach was applied to the creation of new modes of accessing and using biodiversity in pharmaceutical and agricultural discovery. The ICBG program is based upon "biodiscovery partnerships" in which systematists, chemists, cell biologists, conservationists, and lawyers from academia, business, and government in the USA and in developing countries have succeeded in promoting biodiversity-based research by developing novel institutional and legal arrangements. These programs have been or currently are based in many countries, including Panama, Peru, Surinam, Madagascar, Cameroon/Nigeria, Vietnam/Laos,

Chile/Argentina, Mexico, Papua New Guinea, Costa Rica, Peru, and Uzbekistan/Kyrgyzstan (Rosenthal *et al.* 1999, Fogarty International Center 2005). They include collections of plants, algae, microbes, and invertebrates, a wide variety of bioassays, and the isolation and structural elucidation of active compounds. The ICBG program emphasizes the training of young scientists, the enhancement of research capacity, and the promotion of conservation, particularly in developing countries. The emphasis of the Panama ICBG, the focus of this chapter, has been to assure that Panama receives immediate benefits from bioprospecting and to link bioprospecting with conservation and sustainable development.

THE USE OF ECOLOGICAL INSIGHT IN BIOPROSPECTING IN THE PANAMA ICBG

Finding compounds that lead to marketable drugs is a highly unlikely process. Although many programs make random collections, using biological insight could enhance discovery. The Panama ICBG has used over 20 years of basic research on plant–herbivore interactions to guide our collections. The research suggested that young leaves are very dependent on chemical defenses whereas mature leaves depend more on toughness (Kursar and Coley 2003). We tested this and related ecological hypotheses by making extracts from fresh young and mature leaves and comparing their activities in bioassays. We found that 10.0% of the extracts from young leaves were highly active in anti-cancer assays while only 4.5% of the extracts from mature leaves were active. The National Cancer Institute has tested hundreds of thousands of samples, primarily dried, mature leaves, and found activity in only 4.3% (Figure 25.1). Extracts from young leaves also were more active in bioassays for activity against Chagas' disease, malaria, and HIV (Coley *et al.* 2003). Out of 23 species from which active compounds were purified, 10 species had compounds of interest only in the young leaves. Four species had some compounds of interest in the young leaves and other compounds in the mature leaves. For another 10 species most of the compounds of interest were found in both

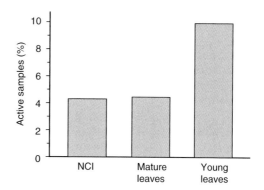

Figure 25.1 Comparison of the activity against cancer bioassays of samples from the National Cancer Institute (NCI) and the Panama ICBG. For the NCI data, about 114,000 extracts prepared during 1961–1980 from 25,000 to 35,000 species were measured for activity against lung (H-460), breast (MCF-7), and central nervous system (SF-268) cancer cell lines. NCI scored samples as "confirmed active" based upon an *in vivo* assay for anti-leukemia activity along with an *in vitro* assay for anti-mitotic activity (Suffness and Douros 1982). In our study, extracts were scored active if growth relative to the controls was inhibited by 50% or more at an extract concentration of $\leq 20\,\mu g$ of extract per ml (data from Gupta, Solis, and co-workers; bioassay methods described in Coley *et al.* 2003).

the young and mature leaves. In general, more interesting compounds were found in the young leaves, such as those isolated from *Myrospermum frutescens* and *Guatteria* spp. (Mendoza *et al.* 2003, 2004, Montenegro *et al.* 2003, Correa *et al.* 2006).

Based upon our ecological studies we predicted that, comparing the young leaves of different species, the speed of leaf expansion would show a negative correlation with activity. In short, some species invest less in secondary metabolites during leaf development and depend on rapid expansion to minimize the window when leaves are vulnerable to herbivores (Aide and Londoño 1989). In order to expand rapidly, they allocate resources from chloroplast development to growth, so the young leaves appear white or light green (Kursar and Coley 1992a–c). We classify this as an "escape" syndrome (Kursar and Coley 2003). In contrast, other species follow a "defense" syndrome in which they expand leaves more slowly, invest in effective secondary metabolites throughout development, and green normally (Coley et al. 2005). Although investment in secondary metabolites appears linked to expansion rate, expansion rate is not easy to assess rapidly in the field. Instead, we hypothesized that the extent of chloroplast development, or the "greenness" of a leaf, might be a rapid visual clue to the level of chemical defense and hence activity in our bioassays (Coley et al. 2003). Although extracts from green young leaves were more active than light green or white leaves, leaf color was not significant (non-parametric analysis of variance, $P = 0.16$; Coley et al. 2003). Thus, our data did not warrant focusing collection efforts only on young green leaves.

We also predicted that shade-tolerant species, because they are adapted to lower resource environments, should be better defended than species that require high light for establishment (Coley 1983, Coley et al. 2003). In fact, extracts from mature leaves of shade species were significantly more active than those from gap-requiring species (Coley et al. 2003). Although we would not predict a priori that growth form should influence chemical defense, it has been suggested that epiphytes (Bennett 1992) or lianas (Hegarty et al. 1991, Phillips 1991) should be active. We found no significant effect of growth form on activity among shrubs, trees, herbs, ferns, lianas, vines, and epiphytes (Coley et al. 2003). However, palms were less active, perhaps due to a greater reliance on toughness.

THE SCIENTIFIC OUTPUT OF THE PANAMA ICBG

The Panama ICBG group has collected over 1500 species of plants, made over 1000 isolates of endophytic fungi as well as collected dozens of cyanobacteria and marine invertebrates. Using ICBG funds, two laboratories were set up in Panama and several existing laboratories in Panama were enhanced. The project also acquired the first nuclear magnetic resonance (NMR) facility in Panama (Bruker Avance 300 MHz). This infrastructure has supported the isolation and structure elucidation of over 50 compounds with medicinally relevant activities, primarily from plants. These represent the majority of the published studies of the uses of Panama's biodiversity in which all or nearly all of the elucidation of the chemical structure and the development of the medicinal bioassays has been accomplished in Panama-based laboratories.

Plants have been an excellent source of anti-cancer compounds (Cragg et al. 1993) and a major biotic resource accessed by the ICBG. Since the 1970s Dr M.P. Gupta and more recently Dr P.N. Solis (Faculty of Pharmacy, University of Panama) have carried out bioassay-guided purification of active molecules from plants. As part of the ICBG they established the first cytotoxicity, or cell-based, anti-cancer assay in Panama and isolated 40 anti-cancer compounds, 13 of which are new to science (Hussein et al. 2003a,b, 2004, 2005).

Drs Ortega-Barría and Romero have developed in vitro bioassays for testing the activity of extracts and compounds against tropical diseases, including Chagas' disease, leishmaniasis, and malaria (Williams et al. 2003). These assays do not use radioactivity and are more appropriate for developing nations. The malaria assay, based upon a DNA-sensitive fluorometric probe, uses no radioactive reagents and is particularly novel (Corbett et al. 2004). The assay has been patented (Ortega-Barría et al. 2005) and scientists from Bolivia, Madagascar, and Peru have come to Panama to learn the fluorescent bioassay method.

Agricultural products, such as pesticides, also are relevant for bioprospecting. Aphids and whiteflies may constitute the worst agricultural pests in the tropics (Oerke and Dehne 1997). Drs Quiros and Emmen in the Department of Zoology at the University of Panama developed a method for the rapid, efficient assay of plant extracts for activity against aphids in a 96-well microtiter plate format as well as an improved whitefly assay.

Drs L. Cubilla and L. Romero submitted a patent application for aporphine alkaloids from young leaves of two species of Guatteria (Annonaceae) that have high in vitro toxicity against Leishmania mexicana (leishmaniasis) but 65-fold lower toxicity to the human host cell

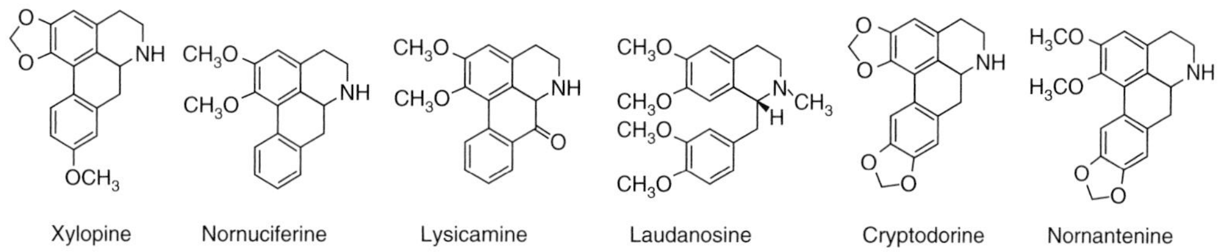

Figure 25.2 Six aporphine alkaloids isolated from crude extracts of *Guatteria amplifolia* and *G. dumetorum* and purified based upon their *in vitro* activity against *Leishmania mexicana* (Montenegro *et al.* 2003).

(Figure 25.2; Montenegro *et al.* 2003, Rios *et al.* 2004). These compounds are sufficiently promising that they should be evaluated at the next level, for safety and efficacy in a vertebrate (mouse) model.

COMBINING BIOPROSPECTING WITH TECHNOLOGY TRANSFER AND TRAINING IN THE PANAMA ICBG

The Panama ICBG is based at the Smithsonian Tropical Research Institute (STRI), the University of Panama, and the Institute of Advanced Scientific Investigations and High Technology Services or INDICASAT, all in Panama City, as well as at US universities. The sophisticated scientific, communications, and administrative infrastructure provided by STRI has been essential for meeting project goals of technology transfer and training in Panamanian laboratories. The factor that most limits research in biodiversity-rich countries is the development of laboratories that publish in international journals. The technology transfer and training with the Panama ICBG has been recognized within Panama and also internationally as a model program (Dalton 2004, *The Economist* 2005). With funding of about $500,000 per year, the Panama ICBG has contributed substantially towards enhancing infrastructure, provided research opportunities for eight Panamanian laboratory leaders, and given research experience to over 70 young scientists. Twenty are working on BS degrees in Panama, 5 working on MS degrees within Panama, and 24 having moved to other countries for MS and PhD studies. Informal comparisons suggest that in many developing

countries students have few opportunities to participate in research on the uses of their own biodiversity.

LINKING BIOPROSPECTING WITH ECONOMIC DEVELOPMENT

In the context of bioprospecting we define economic development to mean enhanced research. At a minimum developing countries should expect bioprospecting to provide new jobs. A vigorous and productive bioprospecting program also will provide developing country researchers with intellectual property, such as novel, active compounds or genes. In developed countries intellectual property can be the basis for creating new companies. In principle, the same process should operate in developing countries. Based upon the substantial amount of funding expended for pharmaceutical and biomedical research, this expectation is quite reasonable. Annual spending by the largest pharmaceutical companies on research is estimated at 27–43 billion dollars worldwide (Agnew 2000). Additional research funds are expended by government (e.g., National Institutes of Health), non-profit institutions (e.g., Howard Hughes Foundation, Medicines for Malaria Venture, Institute for OneWorld Health), and many small companies (Morel *et al.* 2005). About one third of total research in the large pharmaceutical companies (9–14 billion dollars) is similar to the initial steps of bioprospecting (ten Kate and Laird 1999). These include the discovery of active compounds through bioassay, purification, and structure elucidation, their modification to enhance activity, and their testing in vertebrate models. These all are currently employed in the

Panama ICBG and can be carried out in most developing countries. Expending a small fraction of total funding on biodiversity-based research in developing countries would have a substantial economic impact in most countries.

Pharmaceutical companies regularly collaborate with academia and small companies in developed countries. To what extent has bioprospecting in developing countries promoted growth in their economies? While some bioprospecting activity has been carried out in developing countries, extension of the model described above to biodiversity-rich countries has been slow. In other words, developing country scientists have not participated in the more advanced stages of bioprospecting research and few researchers have used biodiversity-based intellectual property in order to attract funding and establish biotechnology companies in developing countries. Serious barriers remain to be overcome in order to meet these goals.

WHAT ARE THE OBSTACLES AND THE SOLUTIONS FOR REALIZING ECONOMIC AND SCIENTIFIC DEVELOPMENT THROUGH BIOPROSPECTING?

The rapid and substantial successes of fields such as genetics and cell biology, as well as their ability to attract funding, can be assigned in large part to the premium placed on collaboration and on the sharing of materials and techniques among competing laboratories (Edwards 2004). Thus, bioprospecting will be most competitive where an open, dynamic research environment is created. Nevertheless, a barrier to collaboration is the tendency of some to view bioprospecting as a confidential activity. As far as possible, the Panama ICBG maintains open access, sharing materials and techniques.

Another barrier to collaboration is the difficulty of entering into legal agreements with academic and pharmaceutical collaborators. Lack of experience and restrictive regulations, leading to very slow and expensive legal processes, probably blocks many bioprospecting projects. At present sufficient experience exists worldwide

such that, in principle, developing countries could be provided with legal advice that is consistent with the CBD, protects the interests of all sides, and also allows negotiations to be completed rapidly. A key recommendation would be the creation of institutions which provide such assistance (e.g., Public Interest Intellectual Property Advisors, Inc. 2005).

Another limitation is that very few of the thousands of active compounds discovered in academic laboratories and published each year are investigated for safety and efficacy in vertebrate models. In effect, the research process ends before the utility of these compounds has been determined. This lack is especially critical in the case of tropical diseases; 3 billion people live in affected areas and no safe and effective treatments are available (Trouiller *et al.* 2001, Gelb and Hol 2002). By working with vertebrate models, researchers would more effectively address the need for new treatments as well as establish more substantial intellectual property.

A major barrier to linking bioprospecting to economic development is that laboratories in the developing world often are not internationally competitive. This key step, enhancing in-country scientific training, infrastructure, funding, and institutional capacity, deserves to be a focus of development efforts (Kettler and Modi 2001, Annan 2003, Holmgren and Schnitzer 2004). In order to attract established scientists or highly qualified postdoctoral associates from developed countries, it is essential to arrange not just space and set-up funds, but also a secure option for such scientists to return to an institution in the USA or Europe should they choose to do so. Once such laboratories are established in-country, they can pursue additional research funds.

In general, funding to developing country researchers should be provided on a competitive basis, with researchers and their institutions held accountable for the use of funds and productivity evaluated as part of the competitive process. Accountability should be a centerpiece since indiscriminate support by donors can impede scientific development.

Another obstacle is the failure to appreciate the breadth of bioprospecting-related research.

Narrowly defined, bioprospecting includes only collections, bioassay, and natural products chemistry, ending with the publication of novel, active compounds. In fact, chemically based interactions among organisms, particularly for tropical animals, microbes, and plants, is a vastly understudied area that could be the basis for vigorous and exciting research programs. Studies of mechanisms of action, biosynthesis, and chemical modification of active compounds followed by retesting would involve researchers from genetics to neurobiology to organic synthetic chemistry. Other important areas include agricultural applications, the safety and efficacy of medicinal plants as well as the traditional areas of biodiversity research such as ecology and systematics.

LINKING BIOPROSPECTING WITH CONSERVATION IN THE PANAMA ICBG

While excellent conservation efforts originate from large organizations that are based in developed countries, more conservation initiatives within developing countries are needed. Clearly there is a great need to develop an in-country conservation ethic. In particular, urbanites are an increasingly important fraction of the population in developing countries, with an estimated increase in Latin America and the Caribbean, for example, of 42% to 85% during the period 1950–2030 (United Nations 2003). We must address the need for conservation tools that effectively engage these citizens, especially the urban-based businesses, governments, and universities (Aide and Grau 2004). Our experience has been that bioprospecting provides a Panamanian voice in support of conservation, one that is especially unique and powerful since it originates in the urban areas.

The first link with conservation must be to assure that biodiversity-rich countries receive immediate economic benefits from bioprospecting. We argue that research must be viewed as the primary economic benefit of bioprospecting. In essence research on biodiversity provides jobs, training opportunities for young scientists, and could lead to the creation of new industries.

In addition to the indirect link between economic benefits from the Panama ICBG's studies of the uses of biodiversity, other ICBG efforts have promoted conservation. All of the investigators involved in the project have worked to create a link between bioprospecting and conservation. Members of the Panama ICBG, from the principal investigators to assistants, give up to 100 talks annually to students at schools, to the public in small towns, to the business community, to government officials, and to visitors from outside of Panama. These can be powerful since they are Panamanians giving talks to their countrymen about discoveries from their biodiversity, as well as the value of and threats to Panama's biodiversity. In addition, the newspapers and television frequently report on the Panama ICBG. These efforts have created wide public awareness of the Panama ICBG's bioprospecting efforts, giving the project a high degree of transparency. Legal agreements that provide for nearly all benefits, such as royalties, to return to Panama also meet the need for equitable sharing of benefits, as defined in the CBD, and enhance local support for the project. Efforts at transparency, the provision of immediate benefits, and equitable legal agreements promote the perception that Panama's biodiversity has direct value to Panama.

Many of these conservation efforts can be viewed as indirect, making it difficult to quantify their conservation impact. Nevertheless, if ecotourism can motivate conservation, as clearly is the case in Costa Rica, the same outcome can be foreseen for bioprospecting. Hence, we believe such efforts do have an impact and, just as conservation should be an integral aspect of ecotourism, conservation should be part of bioprospecting. Bioprospecting also has direct effects on conservation since researchers in parks and reserves have in many cases made important contributions to habitat protection. In particular, Dr Todd Capson, a chemist, pioneered a novel approach in which an ICBG scientist worked actively in conservation. He provided scientific support for the establishment of Coiba National Park, a spectacular marine and terrestrial park. Rather than representing a conservation organization or conservation biology, Dr Capson represented the interests of an applied project that is recognized within Panama for the economic and other benefits it provides. To our knowledge, Dr Capson's initiative was the first application of this highly effective conservation strategy.

THE FUTURE OF BIOPROSPECTING RESEARCH

Any discussion of bioprospecting must touch on the issue of fear of biopiracy. To what extent should the goals and organization of bioprospecting projects be determined by biopiracy issues? Due to these concerns the Panama ICBG has not worked with indigenous groups and campesinos. In our view, the absence of a single authority and the consequent uncertain legal relationships among these groups create the possibility of biopiracy claims (Dalton 2001). Otherwise our experience has been that charges of biopiracy have not taken hold. From its inception, the Panama program has focused on the concept of providing immediate benefits that would link bioprospecting, economic development, and conservation. Similarly constituted ICBG projects have been politically acceptable in other countries (Kingston *et al.* 1999, Schuster *et al.* 1999, Soejarto *et al.* 1999).

Bioprospecting continues to play an important role in the discovery of novel, active compounds. Recent experience demonstrates that natural products research continues to complement the drug discovery research of medicinal chemists and cell biologists. We define natural products to include compounds derived from plants, fungi, bacteria, corals, sponges and other invertebrate animals, and vertebrate venoms but to exclude compounds that are based upon human physiology such as interferon and testosterone. Despite the fact that all of the available analyses combine both classes of compounds, it is clear that natural products, as we have defined them, contribute considerably to new medicines. The percentage of new medicines derived from natural products, in excess of 35%, remained constant during the period 1981–2002 (Grifo and Newman 1997, Newman *et al.* 2003, Koehn and Carter 2005), suggesting that this rate of success will continue. Many of these have mechanisms of action that are novel and not previously discovered. Consequently, natural products also lead to the discovery of novel molecular targets, creating opportunities for additional innovations. Another reason that natural products research will retain its value is that new medicines will be needed due to pathogen resistance, the spread or evolution of new diseases, the failure of vaccine or public health measures to control disease, and the societal expectation that medical care should improve. Thus bioprospecting has great potential to contribute to human health. Clearly, many developing countries could carry out much exciting, high quality research on the uses of their own biodiversity that provides in-country training in laboratories and contributes to human health, economic development, and conservation. Nevertheless, new laboratories, cross-disciplinary collaborations, and a dynamic research environment must be established in more developing countries before the promise of bioprospecting can be fulfilled.

ACKNOWLEDGMENTS

Among the many who have made key contributions to the Panama ICBG we acknowledge Mirei Endara, Rodrigo Tarte, Elena Lombardo, Ira Rubinoff, Joshua Rosenthal, Flora Katz, and Yali Hallock. The project was supported by funds from the National Institutes of Health, the National Science Foundation, and the US Department of Agriculture.

REFERENCES

Agnew, B. (2000) When pharma merges, R & D is the dowry. *Science* 287, 1952–1953.

Aide, T.M. and Grau, H.R. (2004) Globalization, migration and Latin American ecosystems. *Science* 305, 1915–1916.

Aide, T.M. and Londoño, E.C. (1989) The effects of rapid leaf expansion on the growth and survivorship of a lepidopteran herbivore. *Oikos* 55, 66–70.

Aldhous, P. (1991) "Hunting license" for drugs. *Nature* 353, 290.

Angell, M. (2004) The truth about the drug companies. *The New York Review of Books*, July 15, 52–58.

Annan, K. (2003) A challenge to the world's scientists. *Science* 299, 1485.

Balick, M.J. and Cox, P.A. (1996) *Plants, People, and Culture. The Science of Ethnobotany.* Scientific American Library, New York.

Bennett, B. (1992) Use of epiphytes, lianas and parasites by Shuar people of Amazonian Ecuador. *Selbyana* 13, 99–114.

Capson, T.L., Coley, P.D., and Kursar, T.A. (1996) A new paradigm for drug discovery in tropical rainforests. *Nature Biotechnology* 14, 1200–1202.

CBD (2005) http://www.cbd.int/programmes/socio-eco/benefit/case-studies.asp (accessed 25 February 2008)

Coley, P.D. (1983) Herbivory and defense characteristics of tree species in a lowland tropical forest. *Ecological Monographs* 53, 209–233.

Coley, P.D., Heller, M.V., Aizprua, R. *et al.* (2003) Use of ecological criteria in designing plant collection strategies for drug discovery. *Frontiers in Ecology and the Environment* 1, 421–428.

Coley, P.D., Lokvam, J., Rudolph, K. *et al.* (2005) Divergent defensive strategies of young leaves in two species of Inga. *Ecology* 86, 2633–2643.

Corbett, Y., Herrera, L., Gonzalez, J. *et al.* (2004) A novel DNA-based microfluorimetric method to evaluate antimalarial drug activity. *American Journal of Tropical Medicine and Hygiene* 70, 119–124.

Correa, J.E., Ríos, C.H., Castillo, A. *et al.* (2006) Minor alkaloids from *Guatteria dumetorum* with antileishmanial activity. *Planta Medica* 72, 270–272.

Cragg, G.M., Boyd, M.R., Cardellina, I.I. *et al.* (1993) Role of plants in the National Cancer Institute drug discovery and development program. In A.D. Kinghorn and M.F. Balandrin (eds), *Human Medicinal Agents from Plants.* American Chemical Society, Washington, DC, pp. 80–95.

Dalton, R. (2001) The curtain falls. *Nature* 414, 685.

Dalton, R. (2004) Bioprospects less than golden. *Nature* 429, 598–600.

Defries, R.S., Bounoua, L., and Collatz, G.J. (2002) Human modification of the landscape and surface climate in the next fifty years. *Global Change Biology* 8, 438–458.

Edwards, J.L. (2004) Research and societal benefits of the global biodiversity information facility. *BioScience* 54, 486–487.

Fogarty International Center (2005) http://www.fic.nih.gov/programs/research_grants/icbg/index.html (accessed 25 February 2008)

Gelb, M.H. and Hol, W.G.J. (2002) Drugs to combat tropical protozoan parasites. *Science* 297, 343–344.

Gepts, P. (2004) Who owns biodiversity, and how should the owners be compensated? *Plant Physiology* 134, 1295–1307.

Gollin, M.A. (1993) The convention on biological diversity and intellectual property rights. In W.V. Reid, S.A. Laird, C.A. Meyer *et al.* (eds), *Biodiversity Prospecting: Using Genetic Resources for Sustainable Development.* World Resources Institute, Washington, DC, pp. 289–324.

Gollin, M.A. (1999) New rules for natural products research. *Nature Biotechnology* 17, 921–922.

Gomez-Pompa, A. (2004) The role of biodiversity scientists in a troubled world. *BioScience* 54, 217–225.

Grajal, A. (1999) Biodiversity and the nation state: regulating access to genetic resources limits biodiversity research in developing countries. *Conservation Biology* 13, 6–10.

Grifo, F. and Newman, D.J. (1997) The origins of prescription drugs. In F. Grifo and J. Rosenthal (eds), *Biodiversity and Human Health.* Island Press, Washington, DC, pp. 131–163.

Grifo, F. and Rosenthal, J. (eds) (1997) *Biodiversity and Human Health.* Island Press, Washington, DC.

Hegarty, M.P., Hegarty, E.E., and Gentry, A.H. (1991) Secondary compounds in vines with an emphasis on those with defensive functions. In F.E. Putz and H.A. Mooney (eds), *The Biology of Vines.* Cambridge University Press, Cambridge, pp. 287–310.

Holmgren, M. and Schnitzer, S.A. (2004) Science on the rise in developing countries. *PLoS Biology* 2, 10–13.

Horwitz, S.B. (1992) Mechanisms of action of taxol. *Trends in Pharmacological Science* 13, 134–136.

Hussein, A.A., Barberena, I., Capson, T.L. *et al.* (2004) New cytotoxic naphthopyrane derivatives from *Adenaria floribunda. Journal of Natural Products* 67, 451–453.

Hussein, A.A., Barberena, I., Correa, M., Coley, P.D., Solis, P.N., and Gupta, M.P. (2005) Cytotoxic flavonol glycosides from *Triplaris cumingiana. Journal of Natural Products* 68, 231–233.

Hussein, A.A., Bozzi, B., Correa, M. *et al.* (2003a) Bioactive constituents from three *Vismia* species. *Journal of Natural Products* 66, 858–860.

Hussein, A.A., Gomez, B., Ramos, M. *et al.* (2003b) Constituents of *Hirea reclinata* and their anti-HIV activity. *Revista Latinoamericana de Quimica* 31, 5–8.

Biodiversity International (2007) http://www.biodiversityinternational.org/index.asp (accessed 25 February 2008)

Jackson, R.B., Carpenter, S.R., Dahm, C.N. *et al.* (2001) Water in a changing world. *Ecological Applications* 11, 1027–1045.

Kettler, H.E. and Modi, R. (2001) Building local research and development capacity for the prevention and cure of neglected diseases: the case of India. *Bulletin of the World Health Organization* 79, 742–747.

Kingston, D.G.I., Abdel-Kader, M., Zhou, B.-N. et al. (1999) The Suriname International Cooperative Biodiversity Group program: lessons from the first five years. *Pharmaceutical Biology* 37 (Suppl.), 22–34.

Koehn, F.E. and Carter, G.T. (2005) The evolving role of natural products in drug discovery. *Nature Reviews* 4, 206–220.

Kursar, T.A., Caballero-George, C.C., Capson, T.L. et al. (2006) Securing economic benefits and promoting conservation through bioprospecting. *BioScience* 56, 1005–1012.

Kursar, T.A., Caballero-George, C.C., Capson, T.L. et al. (2007) Linking bioprospecting with sustainable development and conservation: the Panama case. *Biodiversity and Conservation* 16, 2789–2800.

Kursar, T.A., Capson, T.L., Coley, P.D. et al. (1999) Ecologically guided bioprospecting in Panama. *Pharmaceutical Biology* 37 (Suppl.), 114–126.

Kursar, T.A. and Coley, P.D. (1992a) Delayed development of the photosynthetic apparatus in tropical rainforest species. *Functional Ecology* 6, 411–422.

Kursar, T.A. and Coley, P.D. (1992b) The consequences of delayed greening during leaf development for light absorption and light use efficiency. *Plant, Cell, and Environment* 15, 901–909.

Kursar, T.A. and Coley, P.D. (1992c) Delayed greening in tropical leaves: an anti-herbivore defense? *Biotropica* 24, 256–262.

Kursar, T.A. and Coley, P.D. (2003) Convergence in defense syndromes of young leaves in tropical rainforests. *Biochemical Systematics and Ecology* 21, 929–949.

Laird, S.A. and ten Kate, K. (2002) Linking biodiversity prospecting and forest conservation. In S. Pagiola, J. Bishop, and N. Landell-Mills (eds), *Selling Forest Environmental Services. Market-based Mechanisms for Conservation and Development*. Earthscan Publications Limited, Sterling, Virginia, pp. 151–172.

Mani, S., Macapinlac, M. Jr., Goel, S. et al. (2004) The clinical development of new mitotic inhibitors that stabilize the microtubule. *Anti-Cancer Drugs* 15, 553–558.

McChesney, J.D. (1996) Biological diversity, chemical diversity, and the search for new pharmaceuticals. In M.J. Balick, E. Elisabetsky, and S.A. Laird (eds), *Medicinal Resources of the Tropical Forest: Biodiversity and its Importance to Human Health*. Columbia University Press, New York, pp. 11–18.

Mendoza, D.T., Ureña González, L.D., Heller, M.V., Ortega-Barría, E., Capson, T.L., and Cubilla Rios, L. (2004) New cassane and cleistanthane diterpenes with activity against *Trypanosoma cruzi* from *Myrospermum frutescens*: absolute stereochemistry of the cassane series. *Journal of Natural Products* 67, 1711–1715.

Mendoza, D.T., Ureña González, L.D., Ortega-Barría, E., Capson, T.L., and Cubilla Rios, L. (2003) Five new cassane diterpenes with activity against *Trypanosoma cruzi* from *Myrospermum frutescens*. *Journal of Natural Products* 66, 928–932.

Montenegro, H., Gutiérrez, M., Romero, L.I., Ortega-Barría, E., Capson, T.L., and Cubilla Rios, L. (2003) Aporphine alkaloids from *Guatteria* spp. with leishmanicidal activity. *Planta Medica* 69, 677–679.

Morel, C.M., Acharya, T., Broun, D. et al. (2005) Health innovation networks to help developing countries address neglected diseases. *Science* 309, 401–404.

Myers, N. and Kent, J. (2003) New consumers: The influence of affluence on the environment. *Proceedings of the National Academy of Sciences of the United States of America* 100, 4963–4968.

Newman, D.J., Cragg, G.M., and Snader, K.M. (2003) Natural products as sources of new drugs over the period 1981–2002. *Journal of Natural Products* 66, 1022–1037.

Oates, J.F. (1999) *Myth and Reality in the Rainforest: How Conservation Strategies Are Failing in West Africa*. University of California Press, Berkeley, CA.

Oerke, E.C. and Dehne, H.W. (1997) Global crop production and the efficacy of crop protection: current situation and future trends. *European Journal of Plant Pathology* 103, 203–215.

Ortega-Barría, E., Corbett, Y., Herrera, L., González, J., and Capson, T.L. (2005) Fluorescence-based bioassay for anti-parasitic drugs. International application published under the Patent Cooperation Treaty (PCT), WO 2005/035783.

Pauly, D., Christensen, V., Guénette, S. et al. (2002) Towards sustainability in world fisheries. *Nature* 418, 689–695.

Phillips, O. (1991) The ethnobotany and economic botany of tropical vines. In F.E. Putz and H.A. Mooney (eds), *The Biology of Vines*. Cambridge University Press, Cambridge, pp. 427–475.

Pimm, S.L., Russell, G.J., Gittleman, J.L., and Brooks, T.M. (1995) The future of biodiversity. *Science* 269, 347–350.

Plucknett, D.L., Smith, N.J.H., Williams, J.T., and Anishetty, N.M. (1987) A case study in rice germplasm: IR 36. *Gene Banks and the World's Food*. Princeton University Press, Princeton, NJ, pp. 171–185.

Principe, P.P. (1991) Valuing the biodiversity of medicinal plants. In O. Akerele, V. Heywood, and

H. Synge (eds), *The Conservation of Medicinal Plants*. Cambridge University Press, Cambridge, pp. 79–124.

Public Interest Intellectual Property Advisors, Inc. (PIIPA) (2005) http://www.piipa.org/ (accessed 25 February 2008)

Rabalais, N.N., Turner, R.E., and Wiseman, W.J. (2002) Gulf of Mexico hypoxia, a.k.a. "The Dead Zone." *Annual Review of Ecology and Systematics* 33, 235–263.

Ramankutty, N. and Foley, J.A. (1999) Estimating historical changes in global land cover: croplands from 1700 to 1992. *Global Biogeochemical Cycles* 13, 997–1027.

Rios, L.C., Romero, L.I., Ortega-Barría, E., and Capson T. (2004) Treatments for leishmaniasis, International application published under the Patent Cooperation Treaty (PCT). WO 2004/084801.

Rosenthal, J.P., Beck, D., Bhat, A. *et al.* (1999) Combining high risk science with ambitious social and economic goals. *Pharmaceutical Biology* 37 (Suppl.), 6–21.

Schuster, B.G., Jackson, J.E., Obijiofor, C.N. *et al.* (1999) Drug development and conservation of biodiversity in West and Central Africa: a model for collaboration with indigenous people. *Pharmaceutical Biology* 37 (Suppl.), 84–99.

Soejarto, D.D., Gyllenhaal, C., Regalado, J.C. *et al.* (1999) Studies on biodiversity of Vietnam and Laos: the UIC-based ICBG program. *Pharmaceutical Biology* 37 (Suppl.), 100–111.

Suffness, M. and Douros, J. (1982) Current status of the NCI plant and animal product program. *Journal Natural Products* 45, 1–14.

ten Kate, K. and Laird, S.A. (1999) *The Commercial Use of Biodiversity: Access to Genetic Resources and Benefit-sharing*. Earthscan Publications, London.

Terborgh, J. (1999) *Requiem for Nature*. Island Press, Covelo.

The Economist (2005) The scum of the Earth. April 7.

Tilman, D., Fargione, J., Wolff, B. *et al.* (2001) Forecasting agriculturally driven global environmental change. *Science* 292, 281–284.

Trouiller, P., Torreele, E., Olliaro, P. *et al.* (2001) Drugs for neglected diseases: a failure of the market and a public policy health failure? *Tropical Medicine and International Health* 6, 945–951.

United Nations (2003) *World Urbanization Prospects: The 2003 Revision Population Database*. United Nations Population Division. http://esa.un.org/unup/ (accessed 25 February 2008).

Williams, C., Espinosa, O.A., Montenegro, H. *et al.* (2003) Hydrosoluble formazan XTT: its application to natural products drug discovery for *Leishmania*. *Journal of Microbiological Methods* 55, 813–816.

Chapter 26

TROPICAL RAINFOREST CONSERVATION: A GLOBAL PERSPECTIVE

Richard T. Corlett and Richard B. Primack

OVERVIEW

The five major rainforest regions (Asia, Africa, Madagascar, Neotropics, and New Guinea) are distinct ecological and biogeographical entities, each with its own levels of threat from various human activities. The purpose of this chapter is to review these threats, and then to evaluate the conservation strategies being used to deal with them. Across the tropics, commercial logging is increasingly the primary driver of forest degradation and loss, with particularly heavy logging rates in Southeast Asia. Hunting now threatens large vertebrates in most accessible forest areas, with potentially major consequences for the ecosystem processes these vertebrates mediate. Uncontrolled forest fires are an expanding problem when farmers set fires following logging. Globally, rainforest destruction is still dominated by poor farmers, but large-scale commercial monocultures are an increasingly important driver. Cattle ranching is particularly important in the Neotropics. Political instability and armed conflict are a problem in several areas, but particularly in Africa. Clearance rates vary hugely within and between regions, with Southeast Asia – particularly Indonesia – the current "disaster area." Because of a rapidly rising human population and poverty, threats to rainforests will become even more severe in coming decades.

Protected areas can conserve tropical forests, but most are underfunded and therefore underprotected. Linking conservation with development in Integrated Conservation and Development Projects has a poor record of success, but some mechanism is needed to transfer the costs of establishing protected areas from local communities to the developed world. Regulating rainforest exploitation is the other key challenge. Listings in CITES (Convention on International Trade in Endangered Species) can limit international trade in overexploited species and certification schemes can support examples of best practice, but controlling the internal trade in timber, bushmeat, and other forest products is much more difficult. While efforts to restore rainforests on degraded sites will become increasingly important in future, these projects should not distract attention from protecting the rainforests that still remain.

INTRODUCTION

The world's tropical rainforests exist in five major regions that are distinct ecological and biogeographical entities, each with its own unique biota and interactions (Primack and Corlett 2005, Corlett and Primack 2006) (Figure 26.1). These differences result largely from tens of millions of years of independent evolution during the Tertiary, when wide oceanic barriers made dispersal between regions particularly difficult (Morley 2003), and have survived the more recent joining of North and South America at the Isthmus of Panama and the convergence of New Guinea with Southeast Asia (Primack and Corlett 2005). The absence of major groups from particular regions is one obvious difference: New Guinea, for example, lacks groups that are critical to other forests, such as primates, ungulates, and eutherian carnivores, and, until

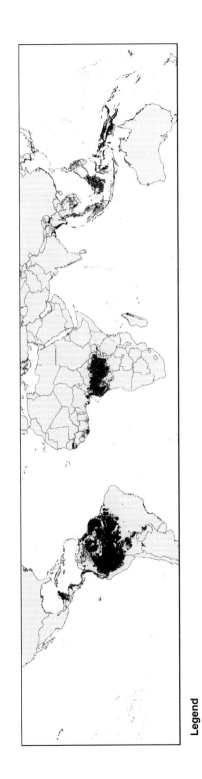

Legend

Rainforest

Figure 26.1 The current global distribution of lowland tropical rainforests. (Courtesy of UNEP World Conservation Monitoring Centre 2004.)

recently, honeybees. Conversely, some functionally important groups are found in only one region, such as leaf-cutter ants in the Neotropics and cassowaries in New Guinea. Equally significant are cases where separate evolutionary radiations have occupied superficially similar niches, such as the pteropodid fruit bats of the Old World and the very different phyllostomid bats of the New World.

Inherent biological differences between rainforest regions interact in various ways with anthropogenic threats to rainforests worldwide. In some cases, the influence of biogeography on a key driver of deforestation is direct: Southeast Asia's dipterocarp forests are logged in direct proportion to the density of these large trees. In most cases, however, the effects are less clear. In our book (Primack and Corlett 2005), we document differences in pollination, seed dispersal, folivory, and predation, but in the absence of cross-continental comparisons of the community-level consequences of these differences we can only speculate about the forests' differing responses to human impacts. One likely difference is in the impact of hunting on key ecological processes. Hunters are selective by size and ease of capture, with kills dominated by different taxa belonging to different functional groups in different rainforest regions (Robinson and Bennett 2000). For example, hunters favor pigs in Asia (Robinson and Bennett 2000), frugivorous ungulates in Africa (Fa *et al.* 2005), and large rodents in the Neotropics (Wright 2003): animals that interact with fallen fruits and seeds in very different ways, with potentially very different consequences for forest regeneration (Primack and Corlett 2005). Another potentially important difference is in the role of frugivorous vertebrates in forest succession on abandoned land. The early stages of woody succession in the Neotropics are dominated by tiny-seeded pioneers dispersed by fruit bats and emberizid birds (tanagers and their relatives), which swallow only the smallest seeds, whereas in the rainforest regions of Africa and Asia, larger-seeded pioneers are dispersed by bulbuls and other birds that swallow most seeds (Corlett 2002). Bats defecate in flight while birds defecate from perches, so we would predict differences in the spatial pattern of pioneer establishment in each region.

The message for conservation is that there are many threats and many rainforests. Although conservationists in one region can learn from experiences in another, they must acknowledge not only the obvious differences in political and social factors, but also the less obvious biological differences. These differences provide an additional motivation for saving not just "the rainforest," but the many rainforests. In this chapter we first review the major threats to rainforests, and then examine how these threats differ among rainforest regions. Finally, we evaluate various approaches to rainforest conservation.

MANY THREATS

Rainforests are threatened throughout the tropics by human activities, but the intensity of each threat varies by region.

Logging

Commercial logging is often the primary driver of forest degradation and loss (e.g., Curran *et al.* 2004). Official statistics on tropical timber production and trade, though readily available (ITTO 2004; Table 26.1), are an unreliable measure of logging impacts (Asner *et al.* 2005). Most tropical logging harvests at least some trees illegally (illegal sites, species, or tree sizes) (Barber *et al.* 2002, Curran *et al.* 2004, Ravenal *et al.* 2004, Richards 2004), and in many countries, poorly documented internal markets are much more important than exports (e.g., Laschefski and Freris 2001).

In most cases only a few species are exported, so logging intensities are low, but in Southeast Asian dipterocarp forests many species are grouped into a few market categories, resulting in more intense logging (Whitmore 1998). Domestic markets are usually far less fussy about the species, size, and quality of logs. This can greatly increase initial logging intensities in accessible forest areas and encourages re-logging for smaller, less desirable trees (Corlett personal observations).

Table 26.1 The state of the world's tropical rainforests: statistics on the forests and human populations of major rainforest countries.

Year(s)	Total forest area (1000 km²) 2005	Annual forest loss (1000 km²) 2000–2005	Annual forest loss[a] (%) 2000–2005	Industrial roundwood production (million m³) 2004	Population density (per km²) 2005	Annual population increase (%) 2000–2005	GDP[b] per capita (US$) 2005
Africa							
Cameroon	212	2.2	1.0	2	34	1.9	941
Congo	225	0.2	0.1	1	12	3.0	1740
Democratic Republic of Congo	1336	3.2	0.2	4	25	2.8	115
Gabon	218	0.1	0.0	4	5	1.7	6680
Asia							
Indonesia	885	18.7	2.0	32	117	1.3	1233
Malaysia	209	1.4	0.7	18	77	2.0	5110
Myanmar	322	4.7	1.4	4	75	1.1	107
Thailand	145	0.6	0.4	8	125	0.9	2563
America							
Brazil	4777	31.0	0.6	103	22	1.4	4297
Colombia	607	0.4	0.1	2	40	1.6	2436
Peru	687	0.9	0.1	1	22	1.5	2763
Venezuela	477	2.9	0.6	1	29	1.8	4956
Other							
Madagascar	128	0.4	0.3	0	32	2.8	276
Papua New Guinea	294	1.4	0.5	2	13	2.1	585

Notes: Note that the statistics for forests and roundwood production refer to all forests in that country, including open woodlands. Data from the online databases of the Food and Agriculture Organization of the United Nations (FAO), the United Nations Population Division, and the International Monetary Fund.

[a] As a percentage of the remaining forest.
[b] Gross domestic product per capita; a measure of personal income. These figures do not allow for differences in the cost of living.

Even in so-called "selective logging," the process of finding, cutting, preparing, and extracting rainforest logs can be devastating (Nepstad et al. 1999), yet most wildlife can survive one cycle of selective logging (Johns 1997, Fimbel et al. 2001). However, the loss of trees is a lesser impact than the construction of roads and other infrastructure. Improved access brings in hunters and encourages recurrent cycles of logging. Landless farmers often move into the area and remove the remaining trees for agriculture. Recently logged forests are also far more likely to burn than those that have not been logged or were logged long ago (Siegert et al. 2001).

Hunting

Rising human populations, the wide availability of guns, and improved transport systems connecting hunters to regional markets have transformed subsistence hunting of rainforest wildlife into a commercial enterprise (Robinson and Bennett 2000, Milner-Gulland and Bennett 2003, Walsh et al. 2003, Corlett 2007). The bushmeat trade, and more local hunting for traditional medicine (e.g., Nijman 2005) and the pet trade (e.g., Wright et al. 2001, Duarte-Quiroga and Estrada 2003, Raselimanana 2003), threatens vertebrates throughout the tropics. Hunting of forest vertebrates not only affects the survival of the harvested species, but also disrupts the web of interactions that maintains forest diversity, including seed dispersal (particularly of large seeds), seed predation, browsing of seedlings, and predation (Wright 2003).

Fire

Undisturbed rainforests do not normally burn except under extreme drought conditions (Whitmore 1998). Rainforest fires became more common over the last 25 years due to forest fragmentation, logging, and increased ignition sources (Barlow and Peres 2004). Farmers commonly use fire for clearing rainforest and preventing regrowth, but these fires are difficult to control. Uncontrolled fires burned an estimated 50–60 million ha of forest in Southeast Asia and

the Neotropics during unusually dry conditions caused by the 1997–1998 El Niño event (Nepstad et al. 2004). The open canopy and abundant fuel supply in logged forests make them particularly vulnerable to fire (Siegert et al. 2001). Forest fragments are also vulnerable because structural changes at the edges increase available fuel, while exposure to wind and sunlight reduces humidity (Laurance 2004). Single fires lead to positive feedbacks by further increasing fuel load and canopy openness (Cochrane et al. 1999). The result is a landscape that becomes vulnerable to fire after weeks, rather than months, without rain.

Deforestation

Rainforests potentially can recover from logging, hunting, and single fires, but few rainforest species can survive complete removal of the forest (Figure 26.2). Generalizing about the causes of rainforest conversion is impossible, given the variety of human systems in the tropics, but globally, most rainforest destruction still results from small-scale crop cultivation by poor farmers, typically migrants from other rural areas in the same country. Large farms and cattle ranches on the forest frontier are often formed by consolidating smaller plots opened up by the first settlers, although direct clearance by large landowners and commercial interests is the major cause of conversion in some areas. Small-scale shifting cultivation is the primary cause of deforestation in Central Africa (Zhang et al. 2005). Cattle ranching has been most important in tropical America (Fearnside 2005), while tree plantations (oil palm, rubber trees, cacao, etc.) cause most deforestation in Southeast Asia (Barber et al. 2002, Curran et al. 2004). In Brazil, mechanized soybean cultivation is an expanding threat in the Amazon region (Fearnside 2001, 2002, but see also Brown et al. 2005).

Deforestation rates vary greatly both within and between regions (Table 26.1), changing rapidly in response to factors such as El Niño, local and global economies, and political events in individual countries. Satellite imagery showed that approximately 58,000 km^2 per year (0.5%)

Figure 26.2 Agriculture is a major contributor to rainforest destruction. In this case, indigenous people in the Brazilian Amazon have cut down trees and burned them in preparation for planting their crops. Here a local chief stands in front of land that has been cleared (photograph courtesy of Milla Jung).

of the 11.5 million km^2 of humid tropical forest observed in 1990 were cleared from 1990 to 1997 (Achard *et al.* 2002, Mayaux *et al.* 2005). Another 23,000 km^2 annually (0.2%) were degraded to an extent that was visible in satellite images. Rates of both loss and degradation were twice as high in Asia (including New Guinea) as in Africa, with tropical America showing the lowest overall rates of both. These averages, however, masked huge differences within regions, with deforestation "hotspots" such as Acre in Brazilian Amazonia, parts of Madagascar, and central Sumatra experiencing deforestation of more than 4% per year. Note also that the 1990–1997 period of this study omits the 1997–1998 El Niño Southern Oscillation (ENSO) event, during which huge areas of forest were burned.

Invasive species

Invasive species are most obviously a threat on oceanic islands such as Hawai'i, where exotic species of all types are a huge problem (Loope *et al.* 2001). Continental rainforests have proven much more resistant (Teo *et al.* 2003, Denslow and DeWalt Chapter 24, this volume), but increasing numbers of cases exist where species from one continent have invaded disturbed and fragmented rainforests on other continents. Old World honeybees in the Amazonian rainforest (Roubik 2000), feral pigs in Australia and New Guinea (Heise-Pavlov and Heise-Pavlov 2003),

and neotropical pioneer plants in Old World forests (Peters 2001, Struhsaker *et al.* 2005) are just three examples.

Global climatic and atmospheric change

Abundant evidence exists that climate changes are affecting biological communities in the northern temperate zone (Parmesan 2007), but the evidence from the tropics is less clear. Empirical evidence suggests increased tree turnover in Amazonian forests in the last two decades, but there is currently no consensus on the driver(s) of these changes (Lewis *et al.* 2006). Global climate models predict changes in temperature and water balance in the tropics over the next century that will subject many species to conditions outside the range of tolerance that their current distributions indicate (e.g., Meynecke 2004, Miles *et al.* 2004). In most areas, the direct impact of these changes will probably be dwarfed by interactions with other human impacts (Corlett 2003). In particular, any increase in the frequency and/or intensity of dry periods would accelerate the synergy between logging, drought, agricultural clearance, and fires (Laurance 2004).

MANY RAINFORESTS

There are five main rainforest regions in the world, each with its own level of threat from human activities.

Asia

Southeast Asia has less than half its forest remaining and has the highest rates of forest loss and degradation in the tropics (Achard *et al.* 2002; Table 26.1). Moreover, the rate of clearance increased during the 1980s and 1990s (DeFries *et al.* 2002) and is probably still increasing. Although poverty and population growth play an important role in the destruction of the region's rainforests, the high rates seen today result from a combination of logging and conversion of forest to cash crops or industrial timber plantations (Corlett 2005, Mackinnon 2005). The vertebrate fauna of the remaining forests is threatened by unsustainable hunting pressure, driven increasingly by a massive regional trade in wild animals and their parts for luxury foods, traditional medicines, trophies, raw materials, and pets (Corlett 2007).

Indonesia has most of the region's surviving rainforests, but it has also replaced the Philippines as the region's new "rainforest disaster area" (e.g., Jepson *et al.* 2001, Curran *et al.* 2004). The rate of forest loss in Indonesia is accelerating, with recent estimates putting it at 20,000 km^2 per year, about 2% of the remaining forest (Barber *et al.* 2002). Indonesia is the world's largest supplier of plywood and other processed wood products, creating a huge demand for wood that is met largely by illegal logging (Barber *et al.* 2002). As elsewhere, logging promotes deforestation by providing access to farmers. In Indonesia, however, the same industrial conglomerates control much of the logging, wood processing, and plantation industries, so the link between logging and deforestation is often more direct, with logging just the first stage in the conversion of rainforest into a plantation monoculture (Barber *et al.* 2002, Curran *et al.* 2004). Two-thirds of the plantations on former forest land in Indonesia consist of oil palm, which covered an estimated 3 million ha in the year 2000 (Glastra *et al.* 2002). Global demand for palm oil is expected to double in the next 20 years, and half of the new plantation land required to expand production likely will be in Indonesia. Most of this will come from the conversion of lowland rainforests in Sumatra, Kalimantan, and Papua. In an additional twist, proposals for new plantations in Indonesia are often used as an excuse for logging in areas that are unsuitable for oil palm (Sandker *et al.* 2007).

Africa

The rainforest countries of Africa combine rapid population growth with extreme poverty (Table 26.1). Political instability and armed conflicts have been an additional problem in several countries, with varied, but generally negative, impacts on the rainforest (e.g., Draulans and Van Krunkelsven 2002). The rainforests of West Africa are largely gone, and the remnants are threatened by a dense, growing human population (Minnemeyer 2002). Vast areas of rainforest still remain in Central Africa, but these face a variety of threats (Zhang *et al.* 2005). Large areas of rainforest remain relatively intact in the Democratic Republic of Congo (the DRC, formerly Zaire) because the road and river networks there do not provide adequate access to loggers, commercial hunters, and landless migrants (Minnemeyer 2002). Throughout the accessible forests of the region, however, subsistence and commercial hunting of wildlife is intense; meat from wild game is an important protein source for rural populations and sometimes preferred to domestic meat even in urban areas (Wilkie *et al.* 2005). In most areas, defaunation rather than deforestation is the primary problem at present (Minnemeyer 2002). The greatest threat in the immediate future is that the expansion of logging activity will open up isolated areas to hunters and migrant farmers (Zhang *et al.* 2005). Logging concessions now cover almost half the Central African rainforest (Minnemeyer 2002). The projected human population for the DRC by the year 2050 is 200 million (United Nations Population Division 2001), so deforestation seems certain to accelerate.

Madagascar

The biodiversity of Madagascar is "extraordinarily distinctive, diverse, and endangered" (Yoder *et al.* 2005). Nearly 90 million years of isolation has resulted in endemism of more than 80% in most groups of terrestrial organisms, while extreme poverty and one of the highest population growth rates on earth has put all forests under pressure (Goodman and Benstead 2003). The original broad band of rainforest along the eastern side of the island has been largely cleared; the surviving forest is fragmented and in many places badly degraded (Dufils 2003, Du Puy and Moat 2003). Although deforestation rates have declined from their peak, the major threat continues to be clearing by poor farmers for rice and cattle. Hunting and fuelwood collection are also major problems, as is logging, although the latter is not carried out on the scale seen in other rainforest regions (Goodman and Benstead 2003). Endemic reptiles and amphibians are widely collected for the pet trade (Raselimanana 2003). Invasive plants and animals seem to be a greater problem in Madagascar than in other rainforest regions (e.g., Brown and Gurevitch 2004).

America

Half the world's remaining tropical rainforests, as well as the largest relatively intact forest blocks, are in the Amazon basin. Although percentage deforestation rates are lower than in Asia and much of Africa, the absolute quantity of forest loss is higher. The major problem is the advancement of the agricultural frontier into the rainforest, both from the margins in the southeast and west and from hotspots in the interior (Fearnside 2005). Forest is converted to crops and cattle pasture, mostly by small farmers whose access to the forest stems from the government's road expansion activities, which far outstrip those in any other major rainforest region (Laurance *et al.* 2002, 2005; but see also the correspondence on this issue in *Science* 307, pp. 1043–1047). Much of this infrastructure is justified by the production of soybeans, an export crop grown by wealthy agribusinesses that employ very few people and displace small farmers to the forest frontier (Fearnside 2002). Deforestation, logging, and forest fires are all concentrated along the new roads (Laurance *et al.* 2005). As in other

regions, logging roads extend access beyond the government road system and the total forest area affected each year by logging and accidental fires is similar to the cleared area (Asner *et al.* 2005). Environmental protection is improving in Amazonia, but the Brazilian government does not currently have the capacity to control illegal deforestation, logging, and mining across this vast area (Fearnside 2005, Laurance *et al.* 2005).

There were two other major rainforest blocks in the Neotropics: the Brazilian Atlantic Forest along the southeast coast of Brazil, and a band of forest that extended from the Pacific coast of northwest South America through Central America to southern Mexico (Primack and Corlett 2005). In both areas, the accessible forests have been completely cleared or are highly fragmented, with the area of forest remaining being inversely related to human population density (Mast *et al.* 1999, Tabarelli *et al.* 2005). As in the Amazon region, ranching and cash crops are the major causes of deforestation. On the positive side, these two areas showcase innovative conservation projects and conservation-related research.

New Guinea

New Guinea, the largest tropical island, supports the third largest contiguous block of rainforest after the Amazon and Congo basins. The lowland rainforest flora is generally similar at the family and genus levels to that of Southeast Asia, but the vertebrate fauna is very different (Primack and Corlett 2005). The western half of the island is the Indonesian province of Papua (previously Irian Jaya). Papua has the lowest population density of any Indonesian province, but as rainforest resources are depleted in western Indonesia, Papua's vast rainforests are increasingly exploited. Clearing for cash crops such as oil palm, exploitation of wildlife, and fires are all growing problems in Papua, but the biggest threat is from a huge (and technically illegal) logging industry (EIA 2005). The main target is merbau (*Intsia* spp.). A recent report documented a complex web of operations, involving mostly Malaysian logging companies on the ground, traders in Singapore, buyers in Hong Kong, and factories in China, where at least one merbau log is consumed every minute of every working day (EIA 2005). The ongoing construction of roads in previously inaccessible areas will inevitably accelerate this illegal exploitation.

Until recently, the rainforests of Papua New Guinea, which occupies the eastern half of the island, appeared to have a much brighter future. The relatively low human population density in the lowlands and a unique system of clan control of forest lands created barriers to large-scale logging or conversion to cash crops. However, over the last two decades Papua New Guinea has experienced a logging boom, causing severe, but still localized, environmental damage (Sizer and Plouvier 2000, World Bank 2002). Corruption has been a massive problem, with logging concessions awarded in return for bribes to senior officials. High birth rates pose a problem for the future, with the human population of Papua New Guinea expected to grow from its present 4.8 million to 11 million by 2050.

Australia

Although the total area of tropical rainforest in Australia is tiny by global standards, it is of great interest from both a biological and a conservation perspective. After a long period of exploitation of the forests, followed by an epic struggle between competing interest groups in the 1980s, Australia has now protected most of its remaining tropical rainforests, including all of the larger blocks, in the 8940 km^2 Wet Tropics of Queensland World Heritage Area (McDonald and Lane 2002, Stork 2005). Despite the absence of spectacular wildlife, the protected rainforests have been a hugely successful tourist attraction. The region still has many environmental problems, but most seem relatively minor in comparison with the massive threats to rainforests elsewhere. One exception is the threat from climate change, since the concentration of the Australian endemic rainforest vertebrates in upland areas makes them especially vulnerable to global warming (Hilbert et al. 2001, Meynecke 2004).

Island rainforests

Most rainforest islands are in the Pacific, but there are also a number in the Indian Ocean (the Andaman, Nicobar, Comoro, Mascarene, and Seychelles islands), a few in the Atlantic (Principe, São Tomé, and Annobon), and many in the Caribbean. The total area of these island rainforests worldwide is very small and the forests on each island are much less species rich than their continental counterparts, but the high rate of endemism means that together they support a significant proportion of all rainforest species. Their biotas are also highly endangered: around 75% of terrestrial vertebrate extinctions in the last 400 years have been on islands, although by no means all in rainforest (Primack 2006). Today, island rainforests suffer from the same problems as continental rainforests, exacerbated by tiny forest areas and population sizes, with the additional problem of an apparently much greater susceptibility to invasion by exotic plant and animal species (Loope et al. 2001, Teo et al. 2003).

SAVING THE RAINFORESTS

We have reached the point where preliminary evaluations can be made of the various strategies that have been used to conserve rainforests.

Protected areas

With a few exceptions, relatively intact tropical rainforests only survive today either in regions with very few human inhabitants or in areas set aside for their protection. Human populations in the tropics are expected to rise by 2 billion over the next 25 years (Wright 2005); this increase, coupled with expanding exploitation of rainforest resources, will ensure that isolation will not provide protection much longer. Thus, the single most important strategy for protecting intact rainforest communities is to establish – and effectively manage – protected areas (Terborgh et al. 2002, Peres 2005).

There is a great deal of variation in the success of existing parks in protecting rainforest biodiversity. Costa Rica's parks are relatively successful (Sánchez-Azofeifa *et al.* 2003), for example, while two decades of chaos in the DRC has left most parks there effectively unprotected (Inogwabini *et al.* 2005), and many of Indonesia's parks are subject to virtually uncontrolled logging, hunting, and, in some areas, forest clearance (Curran *et al.* 2004). Most rainforest protected areas lie between these extremes. They often have huge problems, including poaching and encroachment of park boundaries, but both the vegetation and fauna are usually in much better condition inside parks than outside (Bruner *et al.* 2001, DeFries *et al.* 2005, Struhsaker *et al.* 2005). Pressures on rainforest parks are mounting, however, as human populations increase. A recent satellite-based survey found that 69% of moist tropical forest parks surveyed experienced a decline in forest habitat within 50 km of their boundary over the last 20 years (DeFries *et al.* 2005).

Rainforest parks come in all different sizes and shapes (Putz and Zuidema Chapter 28, this volume). The massive, but almost completely unmanaged, national parks of the Brazilian Amazon represent one extreme (Peres and Terborgh 1995). Such huge parks are probably the only way to preserve complete rainforest ecosystems, including the full range of species, habitats, and ecological processes (Laurance 2005, Peres 2005). The window of opportunity for establishing large parks is rapidly closing, as loggers and settlers move into new areas, so the completion of a pantropical network of representative protected parks is the most urgent priority in rainforest conservation. Large parks alone will not be enough, however; smaller rainforest reserves, down to a few hectares in size, can also play a valuable role in an overall conservation strategy, protecting species and habitats not represented in the larger parks (Turner and Corlett 1996). Indeed, in much of the tropics, there are no large areas of intact rainforest left to protect, so small reserves are the only way to save what survives.

Declaring new parks is one thing; effectively protecting them is another. Most rainforest parks are chronically underfunded and therefore chronically underprotected (e.g., Inogwabini *et al.* 2005). The priorities are usually to increase the number of staff, improve their training, and increase their

mobility. Although tourism can provide a source of income in accessible parts of politically stable countries (Gossling 1999, Naidoo and Adamowicz 2005), most parks cannot be expected to generate significant income (Balmford and Whitten 2003, Inogwabini *et al.* 2005). As a result, they are a net cost to the local and national economies. Even if rainforest countries spend a similar proportion of their national budgets on protected areas as is done in Europe or North America, the bulk of the costs of most parks will have to be paid by the developed world (Balmford and Whitten 2003, Blom 2004). The key needs for funding protected areas are stability and accountability: the first is a prerequisite for long-term planning, particularly in areas with ongoing political instability, while the second is necessary to reassure donors that their money is being spent correctly (Blom 2004). Conservation trust funds are one way of achieving both of these (Balmford and Whitten 2003, Blom 2004, Kiss 2004).

In practice, the costs of establishing new protected areas in inhabited regions are currently borne by the local people, who lose access to resources within the park boundaries, and who may be displaced from their homes and farms (Ferraro 2002). Creating parks without adequate compensation and/or opportunities for participation in any benefits is both immoral and, in the long term, unworkable, since the cost of protecting a large forest area against resentful local communities is likely to be prohibitive (Balmford and Whitten 2003). In Africa, a positive attitude by the neighboring community was the best predictor of success in rainforest parks (Struhsaker *et al.* 2005).

For more than a decade, the dominant approach to biodiversity conservation in developing countries was the Integrated Conservation and Development Project (ICDP), which linked conservation of biodiversity with the economic development of neighboring communities. Huge amounts of both conservation and development funding have been sunk into such projects, despite little evidence for success in either objective, never mind in reconciling the two (Terborgh *et al.* 2002, Christensen 2004, McShane and Wells 2004). There has recently been a backlash against ICDPs, but the problems they aimed to address have not gone away. The "fences and fines" approach to conservation that has worked so well in the developed world does not transfer easily to large tropical parks surrounded by desperately poor people. One option currently being developed is to make direct payments to individual landowners and local communities, either for environmental services (water, carbon fixation) or for the protection of the rainforest and its fauna (Ferraro and Kiss 2002, Kiss 2004, Primack 2006, Putz and Zuidema Chapter 28, this volume). Direct payments have a much shorter record than ICDPs, but their relative simplicity makes their effectiveness potentially testable with an experimental approach.

Regulating exploitation

Parks will not be enough. However successful we are in expanding the present coverage of protected areas and ensuring their proper management, they will inevitably be too few, too small, and too unrepresentative to preserve all of the rainforest biodiversity. Most rainforest regions will continue to have a larger area of forest outside the parks, so regulating its exploitation can make a major contribution to the protection of rainforest diversity. Even the unregulated exploitation of timber and wildlife, as long as it maintains forest cover, protects much more biodiversity than clearance for pasture or crops.

Logging damage can be reduced by using methods known collectively as reduced impact logging or RIL. These involve guidelines designed to minimize damage to soils and the next generation of commercial trees, as well as non-target species of plants and animals (Putz et al. 2000). A number of studies have now shown that the application of RIL can potentially benefit everyone, reducing not just environmental damage, but also the financial costs of logging (Putz et al. 2000, Boltz et al. 2003, Pearce et al. 2003). However, most rainforest logging is either illegal or involves only a short-term concession, so the logger derives no financial benefit from protecting soils and future generations of trees. Some aspects of RIL, such as the training of workers, careful planning of roads, and directional felling of trees, make sense in any logging operation, but others, such as the

exclusion of steep slopes and streamside forests, merely cut profits (Putz et al. 2000). The forest owner – the state in most rainforest countries – would undoubtedly benefit from the strict application of RIL guidelines, but enforcing them requires a well-trained, adequately paid, and highly motivated team of forest officers, which few rainforest countries have.

In many places, logging activity is focused on a single species, thus providing, at least in theory, a relatively simple target for control. In 2002, the Conference of Parties to CITES (Convention on International Trade in Endangered Species of Wild Fauna and Flora) voted to list the big-leaf mahogany, *Swietenia macrophylla*, on Appendix II – the first such listing for a major timber species (Chen and Zain 2004). Other species have followed, including ramin (*Gonystylus* spp.), from the peat-swamp forests of Indonesia and Malaysia. The criteria for trade in Appendix II species focus on sustainability, with exporting countries having to provide permits verifying that the shipment was legally acquired and that export will not be detrimental to the survival of the species. Although CITES listing is certainly not the perfect answer to controlling the legal and illegal overexploitation of rainforest timber, it has the advantage of being rapid and of using laws already in force in most of the producer and consumer nations.

CITES has already had a large impact on the international trade in endangered rainforest animals, such as primates (Chapman and Peres 2001) and parrots (Wright et al. 2001), but listing will not affect the huge internal trade in endangered species in many rainforest countries (e.g., Chapmen and Peres 2001, Wright et al. 2001, Duarte-Quiroga and Estrada 2003), nor does it help with the much broader threat from the bushmeat trade. In some areas, the threat from bushmeat hunting is so urgent that only an immediate and massive investment in enforcement can save viable populations of large vertebrates (Walsh et al. 2003). Other strategies include working with logging companies to stop the transport of hunters and wild meat on logging trucks, promoting affordable alternative sources of protein, education, and banning the commercial trade while still allowing subsistence hunting (Robinson and Bennett 2000).

Certification

Another approach to encouraging the sustainable exploitation of rainforest resources is "eco-certification" or "eco-labeling," based on the fact that at least some consumers will prefer to buy products that they know have been produced in an environmentally friendly manner (Nunes and Riyanto 2005, Putz and Zuidema Chapter 28, this volume). Certification of forests, and the wood products harvested from them, has the longest track record (Rametsteiner and Simula 2003). An increasing number of both individual consumers and industrial buyers of timber and wood products in the developed world now insist on certification. However, only a tiny proportion of tropical rainforest is currently certified because the costs of meeting certification standards are rarely justified by the premium paid for certified products.

Many products directly linked to rainforest loss are invisible in the end-products purchased by the consumer, so mobilizing consumer support is a lot more difficult. Developed countries import huge amounts of palm oil and soybean, for instance, but few people knowingly buy either product at the supermarket. Palm oil is used in soap, cosmetics, candles, and a variety of processed foods, while rainforest soybeans reach consumers as chicken, pork, or beef. In such cases, private companies that import the raw products must be pressured to ensure that they have been produced in a way that does not contribute to deforestation. As with certified timber, the direct impact is likely to be small, but certification can help maintain examples of biodiversity-friendly practices.

Restoring the rainforest

In vast areas of the humid tropics, including much of tropical Asia, it is already too late to preserve the large, continuous tracts of little-disturbed rainforest needed to ensure the long-term survival of rainforest biotas. In such areas we need to learn how to restore forests on abandoned pastures and eroded hillsides. Some conservationists fear that focusing on ecological restoration takes attention from more urgent problems of saving the last viable tracts of intact rainforest. However,

there is also a "time tax" on ecological restoration, since species are inexorably lost from unrestored landscapes (Martínez-Garza and Howe 2003). The technical problems of large-scale restoration are huge and the processes slow (Florentine and Westbrooke 2004, Lamb *et al.* 2005), so we must start trials now if we are to have any hope of resolving a problem that will inevitably expand.

CONCLUSIONS

The situation is critical but not yet hopeless. Even in the worst hit regions, the majority of the rainforest biota still survives in small protected areas, in fragments of primary forest on sites that are too steep, too wet, or too infertile to be worth clearing, in logged forests, in secondary forest on abandoned land, and in woody regrowth along streams and fences (Corlett 2000). More species will survive if existing parks are fully protected, gaps in the protected area system are filled, and unprotected areas are managed sustainably. International support on a massive scale is needed to ensure that financing is available for protected area systems and to encourage practices such as RIL in exploiting unprotected areas. It makes no sense to expect some of the world's poorest countries to pay for the protection of the world's richest ecosystems, when the benefits are global.

ACKNOWLEDGMENTS

This chapter builds on our recent book, *Tropical Rain Forests: An Ecological and Biogeographical Comparison*, and all the many people who helped us with the book have indirectly contributed to this chapter. We would also like to thank Kamal Bawa, Peter Feinsinger, Elizabeth Platt, and Billy Hau for constructive comments on an earlier draft.

REFERENCES

Achard, F., Eva, H.D., Stibig, H.-J. *et al.* (2002) Determination of deforestation rates of the world's humid tropical forests. *Science* 297, 999–1002.

Asner, G.P., Knapp, D.E., Broadbent, E.N., Oliveira, P.J.C., Keller, M., and Silva, J.N. (2005) Selective logging in the Brazilian Amazon. *Science* 310, 480–482.

Balmford, A. and Whitten, T. (2003) Who should pay for tropical conservation, and how should the costs be met? *Oryx* 37, 238–250.

Barber, C.V., Matthews, E., Brown, D. *et al.* (2002) *State of the Forest: Indonesia.* World Resources Institute, Washington, DC.

Barlow, J. and Peres, C.A. (2004) Ecological responses to El Nino-induced surface fires in central Brazilian Amazonia: management implications for flammable tropical forests. *Philosophical Transactions of the Royal Society of London Series B-Biological Sciences* 359, 367–380.

Blom, A. (2004) An estimate of the costs of an effective system of protected areas in the Niger Delta – Congo Basin Forest Region. *Biodiversity and Conservation* 13, 2661–2678.

Boltz, F., Holmes, T.P., and Carter, D.R. (2003) Economic and environmental impacts of conventional and reduced-impact logging in tropical South America: a comparative review. *Forest Policy and Economics* 5, 69–81.

Brown, K.A. and Gurevitch, J. (2004) Long-term impacts of logging on forest diversity in Madagascar. *Proceedings of the National Academy of Sciences of the United States of America* 101, 6045–6049.

Brown, J.C., Koeppe, M., Coles, B., and Price, K.P. (2005) Soybean production and conversion of tropical forest in the Brazilian Amazon: the case of Vilhena, Rondônia. *Ambio* 34, 462–469.

Bruner, A.G., Gullison, R.E., Rice, R.E., and da Fonseca, G.A.B. (2001) Effectiveness of parks in protecting tropical biodiversity. *Science* 291, 125–128.

Chapman, C.A. and Peres, C.A. (2001) Primate conservation in the new millennium: the role of scientists. *Evolutionary Anthropology* 10, 16–33.

Chen, H.K. and Zain, S. (2004) CITES, trade and sustainable forest management. *Unasylva* 219, 44–45.

Christensen, J. (2004) Win–win illusions. *Conservation in Practice* 5, 12–19.

Cochrane, M.A., Alencar, A., Schulze, M.D. *et al.* (1999) Positive feedbacks in the fire dynamics of closed canopy tropical forests. *Science* 284, 1832–1835.

Corlett, R.T. (2000). Environmental heterogeneity and species survival in degraded tropical landscapes. In M.J. Hutchings, E.A. John and A. Stewart (eds), *The Ecological Consequences of Environmental Heterogeneity.* Blackwell Science, Oxford, UK, pp. 333–355.

Corlett, R.T. (2002). Frugivory and seed dispersal in degraded tropical East Asian landscapes. In D.J. Levey, W.R. Silva and M.Galetti (eds), *Seed Dispersal and Frugivory: Ecology, Evolution and Conservation.* CABI Publishing, Wallingford, pp. 451–465.

Corlett, R.T. (2003) Climate change and biodiversity in tropical East Asian forests. In R.E. Green, M. Harley, L. Miles, J. Scharlemann, A. Watkinson, and O. Watts (eds), *Global Climate Change and Biodiversity.* RSPB, Bedford, pp. 9–10.

Corlett, R.T. (2005) Vegetation. In A. Gupta (ed.), *The Physical Geography of Southeast Asia.* Oxford University Press, Oxford, pp. 105–119.

Corlett, R.T. (2007). The impact of hunting on the mammalian fauna of tropical Asian forests. *Biotropica* 39, 292–303.

Corlett, R.T. and Primack, R.B. (2006) Tropical rainforests: why cross-continental comparisons are needed. *Trends in Ecology and Evolution* 21, 104–110.

Curran, L.M., Trigg, S.N., McDonald, A.K. *et al.* (2004) Lowland forest loss in protected areas of Indonesian Borneo. *Science* 303, 1000–1003.

DeFries, R.S., Hansen, A., Newton, A.C., and Hansen, M.C. (2005) Increasing isolation of protected areas in tropical forests over the past twenty years. *Ecological Applications* 15, 19–26.

DeFries, R.S., Houghton, R.A., Hansen, M.C., Field, C.B., Skole, D., and Townshend, J. (2002) Carbon emissions from tropical deforestation and regrowth based on satellite observations for the 1980s and 1990s. *Proceedings of the National Academy of Sciences of the United States of America* 99, 14256–14261.

Draulans, D. and Van Krunkelsven, E. (2002) The impact of war on forest areas in the Democratic Republic of Congo. *Oryx* 36, 35–40.

Duarte-Quiroga, A. and Estrada, A. (2003) Primates as pets in Mexico City: an assessment of the species involved, source of origin, and general aspects of treatment. *American Journal of Primatology* 61, 53–60.

Dufils, J.-M. (2003) Remaining forest cover. In S.M. Goodman and J.P. Benstead (eds), *The Natural History of Madagascar.* University of Chicago Press, Chicago, pp. 88–96.

Du Puy, D.J. and Moat, J. (2003) Using geological substrate to identify and map primary vegetation types in Madagascar and the implications for planning biodiversity conservation. In S.M. Goodman and J.P. Benstead (eds), *The Natural History of Madagascar.* University of Chicago Press, Chicago, pp. 51–67.

EIA (2005) *The Last Frontier: Illegal Logging in Papua and China's Massive Timber Theft.* Environmental Investigation Agency and Telepak, Washington, DC.

Fa, J.E., Ryan, S.F. and Bell, D.J. (2005). Hunting vulnerability, ecological characteristics and harvest rates of bushmeat species in Afrotropical forests. *Biological Conservation* 121, 167–176.

Fearnside, P.M. (2001) Soybean cultivation as a threat to the environment in Brazil. *Environmental Conservation* 28, 23–38.

Fearnside, P.M. (2002) Avanca Brasil: environmental and social consequences of Brazil's planned infrastructure in Amazonia. *Environmental Management* 30, 735–747.

Fearnside, P.M. (2005) Deforestation in Brazilian Amazonia: history, rates, and consequences. *Conservation Biology* 19, 680–688.

Ferraro, P.J. (2002) The local costs of establishing protected areas in low-income nations: Ranomafana National Park, Madagascar. *Ecological Economics* 43, 261–275.

Ferraro, P.J. and Kiss, A. (2002) Direct payments to conserve biodiversity. *Science* 298, 1718–1719.

Fimbel, R.A., Grajal, A., and Robinson, J.G. (2001) *The Cutting Edge: Conserving Wildlife in Logged Tropical Forest.* Columbia University Press, New York.

Florentine, S.K. and Westbrooke, M.E. (2004) Evaluation of alternative approaches to rainforest restoration on abandoned pasturelands in tropical north Queensland, Australia. *Land Degradation & Development* 15, 1–13.

Glastra, R., Wakker, E., and Richert, W. (2002) *Oil Palm Plantations and Deforestation in Indonesia: What Role do Europe and Germany Play?* WWF Schweiz, Zurich, Switzerland.

Goodman, S.M. and Benstead, J.P. (2003) *The Natural History of Madagascar.* University of Chicago Press, Chicago.

Gossling, S. (1999) Ecotourism: a means to safeguard biodiversity and ecosystem function? *Ecological Economics* 29, 303–320.

Heise-Pavlov, P.M. and Heise-Pavlov, S.R. (2003) Feral pigs in tropical lowland rainforest of northeastern Australia: ecology, zoonoses and management. *Wildlife Biology* 9 (Suppl 1), 21–27.

Hilbert, D.W., Ostendorf, B., and Hopkins, M.S. (2001) Sensitivity of tropical forests to climate change in the humid tropics of north Queensland. *Austral Ecology* 26, 590–603.

Inogwabini, B.-I., Ilambu, O., and Gbanzi, M.A. (2005) Protected areas of the Democratic Republic of Congo. *Conservation Biology* 19, 15–22.

ITTO (2004) *Annual Review and Assessment of the World Timber Situation 2003.* International Tropical Timber Organization, Yokohama, Japan.

Jepson, P., Jarvie, J.K., MacKinnon, K., and Monk, K.A. (2001) The end for Indonesia's lowland forests? *Science* 292, 859–861.

Johns, A.G. (1997) *Timber Production and Biodiversity Conservation in Tropical Rain Forests.* Cambridge University Press, New York.

Kiss, A. (2004) Making biodiversity conservation a land-use priority. In T.O. McShane and M.P. Wells (eds), *Getting Biodiversity Projects to Work: Towards More Effective Conservation and Development.* Columbia University Press, New York, pp. 98–123.

Lamb, D., Erskine, P.D., and Parrotta, J.A. (2005) Restoration of degraded tropical forest landscapes. *Science* 310, 1628–1632.

Laschefski, K. and Freris, N. (2001) Saving the wood ... from the trees. *Ecologist* 31, 40–43.

Laurance, W.F. (2004) Forest–climate interactions in fragmented landscapes. *Philosophical Transactions of the Royal Society of London Series B-Biological Sciences* 359, 345–352.

Laurance, W.F. (2005) When bigger is better: the need for Amazonian mega-reserves. *Trends in Ecology and Evolution* 20, 645–648.

Laurance, W.F., Albernaz, A.K.M., Schroth, G. et al. (2002) Predictors of deforestation in the Brazilian Amazon. *Journal of Biogeography* 29, 737–748.

Laurance, W.F., Bergen, S., Cochrane, M.A. et al. (2005) The future of the Amazon. In E. Bermingham, C.W. Dick, and C. Moritz (eds), *Tropical Rainforests: Past, Present, and Future.* University of Chicago Press, Chicago, pp. 583–609.

Lewis, S.L., Phillipps, O.L., and Baker, T.T. (2006) Impacts of global atmospheric change on tropical forests. *Trends in Ecology and Evolution* 21, 173–174.

Loope, L.L., Howarth, F.G., Kraus, F., and Pratt, T.K. (2001) Newly emergent and future threats of alien species to Pacific birds and ecosystems. *Studies in Avian Biology* 22, 291–304.

Mackinnon, K. (2005) Parks, people, and policies: conflicting agendas for forests in Southeast Asia. In E. Bermingham, C.W. Dick, and C. Moritz (eds), *Tropical Rainforest: Past, Present & Future.* University of Chicago Press, Chicago, pp. 558–582.

Martinez-Garza, C. and Howe, H.F. (2003) Restoring tropical diversity: beating the time tax on species loss. *Journal of Applied Ecology* 40, 423–429.

Mast, R.B., Mahecha, J.V.R., Mittermeier, R.A., Hemphill, A.H., and Mittermeier, C.G. (1999) Choco-Darien-Western Ecuador. In R.A. Mittermeier, N. Miers, P.R. Gil, and C.G. Mittermeier (eds), *Hotspots: Earth's Biologically Richest and Most Endangered Terrestrial Ecoregions.* Mittermeier, C.G.CEMEX, Mexico City, Mexico, pp. 122–131.

Mayaux, P., Holmgren, P., Achard, F., Eva, H., Stibig, H., and Branthomme, A. (2005) Tropical forest cover change in the 1990s and options for future monitoring. *Philosophical Transactions of the Royal Society of London Series B-Biological Sciences* 360, 373–384.

McDonald, G. and Lane, M. (2002) *Securing the Wet Tropics?* Federation Press, Leichhardt, NSW.

McShane, T.O. and Wells, M.P. (2004) *Getting Biodiversity Projects to Work: Towards More Effective Conservation and Development.* Columbia University Press, New York.

Meynecke, J.O. (2004) Effects of global climate change on geographic distributions of vertebrates in North Queensland. *Ecological Modelling* 174, 347–357.

Miles, L., Grainger, A., and Phillips, O. (2004) The impact of global climate change on tropical forest biodiversity in Amazonia. *Global Ecology and Biogeography* 13, 553–565.

Milner-Gulland, E.J. and Bennett, E.L. (2003) Wild meat: the bigger picture. *Trends in Ecology & Evolution* 18, 351–357.

Minnemeyer, S. (2002) *An Analysis of Access into Central Africa's Rainforests.* World Resources Institute, Washington, DC.

Morley, R.J. (2003) Interplate dispersal paths for megathermal angiosperms. *Perspectives in Plant Ecology and Evolution* 6, 5–20.

Naidoo, R. and Adamowicz, W.L. (2005) Economic benefits of biodiversity exceed the costs of conservation at an African rainforest reserve. *Proceedings of the National Academy of Sciences of the United States of America* 102, 16712–16716.

Nepstad, D., Lefebvre, P., da Silva, U.L. *et al.* (2004) Amazon drought and its implications for forest flammability and tree growth: a basin-wide analysis. *Global Change Biology* 10, 704–717.

Nepstad, D., Verissimo, A., Alencar, A. *et al.* (1999) Large-scale impoverishment of Amazonian forests by logging and fire. *Nature* 398, 505–508.

Nijman, V. (2005) Decline of the endemic Hose's langur *Presbytis hosei* in Kayan Mentarang National Park, East Borneo. *Oryx* 39, 223–226.

Nunes, P.A.L.D. and Riyanto, Y.E. (2005) Information as a regulatory instrument to price biodiversity benefits: certification and ecolabeling policy practices. *Biodiversity and Conservation* 14, 2009–2027.

Parmesan, C. (2007) Influences of species, latitudes and methodologies on estimates of phenological response to global warming. *Global Change Biology* 13, 1860–1872.

Pearce, D., Putz, F.E., and Vanclay, J.K. (2003) Sustainable forestry in the tropics: panacea or folly. *Forest Ecology and Management* 172, 229–247.

Peres, C.A. (2005) Why we need megareserves in Amazonia. *Conservation Biology* 19, 728–733.

Peres, C.A. and Terborgh, J. (1995) Amazonian nature reserves: an analysis of the defensibility status of existing conservation units and design criteria for the future. *Conservation Biology* 9, 34–46.

Peters, H.A. (2001) *Clidemia hirta* invasion at the Pasoh Forest Reserve: an unexpected invasion in an undisturbed tropical forest. *Biotropica* 33, 60–68.

Primack, R.B. (2006) *Essentials of Conservation Biology,* 4th Edition. Sinauer Associates, Sunderland, MA.

Primack, R.B. and Corlett, R.T. (2005) *Tropical Rain Forests: An Ecological and Biogeographical Comparison.* Blackwell Science, Oxford.

Putz, F.E., Dykstra, D.P., and Heinrich, R. (2000) Why poor logging practices persist in the tropics. *Conservation Biology* 14, 951–956.

Rametsteiner, E. and Simula, M. (2003) Forest certification – an instrument to promote sustainable forest management? *Journal of Environmental Management* 67, 87–98.

Raselimanana, A.P. (2003) Trade in reptiles and amphibians. In S.M. Goodman and J.P. Benstead (eds), *The Natural History of Madagascar.* University of Chicago Press, Chicago, pp. 1564–1568.

Ravenal, R.M., Granoff, I.M.E., and Magee, C. (2004) *Illegal Logging in the Tropics: Strategies for Cutting Crime.* Haworth Press, Binghamton, NY.

Richards, M. (2004) Forest trade policies: how do they affect forest governance? *Unasylva* 219, 39–43.

Robinson, J.G. and Bennett, E.L. (2000) *Hunting for Sustainability in Tropical Forests.* Columbia University Press, New York.

Roubik, D.W. (2000). Pollination system stability in Tropical America. *Conservation Biology* 14, 1235–1236.

Sánchez-Azofeifa, G.A., Daily, G.C., Pfaff, A.S.P., and Busch, C. (2003) Integrity and isolation of Costa Rica's national parks and biological reserves: examining the dynamics of land-cover change. *Biological Conservation* 109, 123–135.

Sandker, M., Suwarno, A., and Campbell, B.M. (2007) Will forests remain in the face of oil palm expansion? Simulating change in Malinau, Indonesia. *Ecology and Society* 12, 37.

Siegert, F., Rucker, G., Hinrichs, A., and Hoffmann, A. (2001) Increased damage from fires in logged forests during droughts caused by El Niño. *Nature* 414, 437–440.

Sizer, N. and Plouvier, D. (2000) *Increased Investment and Trade by Transnational Logging Companies in Africa, the Caribbean and the Pacific.* WWF Belgium, Brussels.

Stork, N.E. (2005) The theory and practice of planning for long-term conservation of biodiversity in the Wet Tropics rainforests of Australia. In E. Bermingham, C.W. Dick, and C. Moritz (eds), *Tropical Rainforest: Past,*

Present and Future. University of Chicago Press, Chicago, pp. 507–526.

Struhsaker, T.T., Struhsaker, P.J., and Siex, K.S. (2005) Conserving Africa's rain forests: problems in protected areas and possible solutions. *Biological Conservation* 123, 45–54.

Tabarelli, M., Pinto, L.P., Silva, J.M.C., Hirota, M., and Bedê, L. (2005) Challenges and opportunities for biodiversity conservation in the Brazilian Atlantic Forest. *Conservation Biology* 19, 695–700.

Teo, D.H.L., Tan, H.T.W., Corlett, R.T., Wong, C.M., and Lum, S.K.Y. (2003) Continental rain forest fragments in Singapore resist invasion by exotic plants. *Journal of Biogeography* 30, 305–310.

Terborgh, J., Van Schaik, C.P., Davenport, L., and Rao, M. (2002) *Making Parks Work: Strategies for Preserving Tropical Nature*. Island Press, Washington, DC.

Turner, I.M. and Corlett, R.T. (1996) The conservation value of small, isolated fragments of lowland tropical rainforest. *Trends in Ecology and Evolution* 11, 330–333.

United Nations Population Division (2001) *World Population Prospects: The 2000 Revision: Highlights*. United Nations, New York.

Walsh, P.D., Abernethy, K.A., Bermejo, M. *et al.* (2003) Catastrophic ape decline in western equatorial Africa. *Nature* 422, 611–614.

Whitmore, T.C. (1998) *An Introduction to Tropical Rain Forests*, 2nd Edition. Oxford University Press, Oxford.

Wilkie, D.S., Starkey, M., Abernathy, K., Effa, E.N., Telfer, P. and Godoy, R. (2005) Role of prices and wealth in consumer demand for bushmeat in Gabon, West Africa. *Conservation Biology* 19, 268–274.

World Bank (2002) *Papua New Guinea Environmental Monitor*. World Bank, Washington, DC.

Wright, S.J. (2003) The myriad consequences of hunting for vertebrates and plants in tropical forests. *Perspectives in Plant Ecology Evolution and Systematics* 6, 73–86.

Wright, S.J. (2005) Tropical forests in a changing environment. *Trends in Ecology and Evolution* 20, 553–560.

Wright, T.F., Toft, C.A., Enkerlin-Hoeflich, E. *et al.* (2001) Nest poaching in Neotropical parrots. *Conservation Biology* 15, 710–720.

Yoder, A.D., Olson, L.E., Hanley, C. *et al.* (2005) A multidimensional approach for detecting species patterns in Malagasy vertebrates. *Proceedings of the National Academy of Sciences of the United States of America* 102, 6587–6594.

Zhang, Q., Devers, D., Desch, A., Justice, C.O., and Townshend, J. (2005) Mapping tropical deforestation in Central Africa. *Ecological Monitoring and Assessment* 101, 69–83.

Chapter 27

ENVIRONMENTAL PROMISE AND PERIL IN THE AMAZON

William F. Laurance

OVERVIEW

The Amazon basin sustains about half of the world's remaining tropical forests, and is being destroyed and degraded at alarming rates. About one fifth of the Amazon has been deforested and perhaps another third degraded by selective logging, surface fires, habitat fragmentation, and edge effects. Hunting and illegal gold mining have also altered large expanses of the region, even in many remote areas. The rapid pace of forest conversion may accelerate in the near future because of a major planned expansion of transportation infrastructure, which greatly facilitates forest colonization, predatory logging, and land speculation. If such projects continue unabated, much of the basin's forests could be fragmented on a large spatial scale, sharply increasing the vulnerability of surviving forest tracts to a range of exploitative activities.

However, the conservation prognosis is not entirely negative. In parts of the Amazon, regenerating forest on abandoned land provides habitat for certain wildlife and is far superior to cattle pastures in its hydrological functions and carbon storage. The greatest cause for optimism is the prospect of a substantial expansion of protected and semi-protected areas, particularly in Brazilian Amazonia. In addition, a growing network of indigenous lands is helping to reduce forest exploitation in some areas. Unfortunately, many reserves are poorly managed and protected, and a key challenge is to establish basic staffing and infrastructure for planned and existing parks. Improving the enforcement of environmental legislation in remote frontier areas is also a daunting challenge for Amazonian nations.

INTRODUCTION

In the biblical book of Revelation, the dawning of the Apocalypse sees four dark horsemen – famine, war, pestilence, and disease – raining down horror on humanity. Some believe that the Amazon could face its own kind of apocalypse in the coming century. Its horsemen will be different: not famine but the rapid expansion of agriculture; not war but industrial logging; not pestilence but wildfires; and not disease but widespread forest fragmentation. Others, however, believe that the analogy of an apocalypse is too pessimistic (e.g., see Putz and Zuidema Chapter 28, this volume). In this chapter I briefly describe the most important threats to the Amazon, and suggest how the basin's forests might be altered in coming decades. The Amazon, I conclude, faces a dynamic combination of environmental promise and peril.

The Amazon basin sustains well over half of the world's remaining tropical rainforest (Whitmore 1997) and includes some of the most biologically rich ecosystems ever encountered. Closed-canopy forests in the basin encompass about 5.3 million km^2, an area the size of western Europe (Sarre *et al.* 1996). By far the most extensive forest type is *terra firme* – forests that are not seasonally flooded. There also are large areas of seasonally flooded forest along rivers and in floodplains (termed *várzea* if they are flooded by relatively nutrient-rich white waters, and *igapó* if inundated by nutrient-poor black waters), and limited areas of bamboo forest and vine forest. In addition, scattered savannas and open forests occur in drier areas of the basin, where narrow strips of rainforest vegetation (termed "gallery forest") often persist along permanent rivers and streams (IBGE 1997).

Most of the Amazon is flat or undulating, at low elevation (<300 m), and overlays very poor soils. Roughly four-fifths of the Amazon's soils are classified as latosols (Brown 1987, Sarre *et al.* 1996), which are heavily weathered, acidic, high in toxic aluminum, and poor in nutrients (Richter and Babbar 1991). Somewhat more productive soils in the Amazon are concentrated along the basin's western margin, in the Andean foothills and their adjoining floodplains. These areas are much more recent geologically than the rest of the basin and thus their soils are less heavily weathered.

Rainfall varies markedly across the Amazon. In general, forests in the basin's eastern and southern portions are driest, with the strongest dry season. Although evergreen, these forests are near the physiological limits of tropical rainforest, and can persist only as a result of having deep root systems that access groundwater during the dry season (Nepstad *et al.* 1994). The wettest and least seasonal forests are in the northwestern Amazon, with the central Amazon being intermediate; forests in these areas do not require deep roots.

DIRECT THREATS TO THE AMAZON

Agriculture

Historically, Amazonian development has been limited by the basin's poor soils, remoteness from major population centers, and diseases such as malaria and yellow fever. This is rapidly changing. In the Brazilian Amazon, which comprises two-thirds of the basin, more forest was destroyed during the last 30 years than in the previous 450 years since European colonization (Lovejoy 1999). Losses of Amazonian forests in Bolivia, Ecuador, Colombia, and Peru have also risen substantially in recent decades (e.g., Sarre *et al.* 1996, Viña and Cavalier 1999, Steininger *et al.* 2001a,b).

Deforestation rates in the Amazon average roughly 3–4 million ha per year – an area larger than Belgium. The most reliable deforestation statistics are for the Brazilian Amazon (Figure 27.1). These statistics have been produced annually since 1989 (except 1993) by Brazil's national space agency based on interpretation of satellite imagery (INPE 2005). Despite various initiatives to slow forest loss, deforestation in Brazilian Amazonia has accelerated substantially since 1990 ($F_{1,14} = 11.17$, $R^2 = 44.4\%$, $P = 0.005$; linear regression with log-transformed deforestation data). Considerable year-to-year variation in deforestation rates (Figure 27.1) results from changing economic trends (such as fluctuating commodity prices and international currency-exchange rates, which affect timber, beef, and soy exports); evolving government policies (such as stabilization of Brazilian hyperinflation in 1994 that freed pent-up funds for development, ongoing infrastructure expansion, periodic crackdowns on illegal logging, and the designation of new protected areas); and climatic conditions (particularly droughts, which strongly influence forest burning) (Laurance 2005a). Rates of deforestation have been especially high in recent years; from 2002 to 2004, nearly 2.5 million ha of forest was destroyed annually – equivalent to 11 football fields a minute. This increase mostly resulted from rapid destruction of seasonal forest types in the southern and eastern parts of the basin; relative to preceding years (1990–2001), forest loss shot up by 48% in the states of Pará, Rondônia, Mato Grosso, and Acre (Laurance *et al.* 2004a).

The most important proximate drivers of deforestation in the Amazon today are directly related to agriculture. The greatest cause of forest loss is large-scale cattle ranching, typically by relatively wealthy landowners. Ranchers commonly use bulldozers to extract timber prior to felling and burning the forest (Uhl and Buschbacher 1985). Large- and medium-scale ranchers may cause as much as three-quarters of all deforestation in the Brazilian Amazon (Fearnside 1993, Nepstad *et al.* 1999a) and also account for much forest loss elsewhere in Latin America (e.g., Viña and Cavalier 1999). From 1990 to 2005, the number of cattle in Brazilian Amazonia nearly tripled, from about 22 million to 60 million head. Brazilian beef exports rose sharply during this period both because of favorable exchange rates

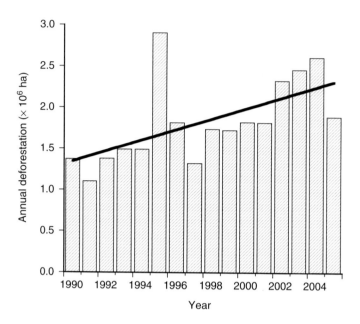

Figure 27.1 Estimated deforestation rates in Brazilian Amazonia from 1990 to 2005 (data from INPE 2005). The regression line shows the overall trend.

and because Brazil is free from hoof-and-mouth disease (Kaimowitz *et al.* 2004, Laurance 2005a).

Second in importance is slash-and-burn farming by landowners who clear small (typically 1–2 ha) areas of forest each year to plant manioc, corn, bananas, and other crops (Fearnside 1993). The forest's understory is slashed with machetes and the debris is ignited during the dry season. The ash from the burned vegetation provides a pulse of plant nutrients, which supports crops for a few years before the area is left to fallow and the farmer is forced to clear more forest. Slash-and-burn farming occurs both opportunistically (often illegally) and as a result of government-sponsored colonization programs that allocate small forest tracts (usually <100 ha) to individual families. Brazil has hundreds of Amazonian colonization projects involving at least half a million people (Homma *et al.* 1992), initiated in part to help divert population flows that would otherwise further overcrowd Brazil's major cities (Fearnside 1990, 1993).

The third cause of deforestation, industrial agriculture, is growing rapidly in importance in drier areas of the Amazon basin and in adjoining transitional forests and *cerrado* woodlands and savannas. Most of these farms are devoted to soy, which involves clearing large expanses of relatively flat land for crop production. Soy farming has been a major cause of deforestation in northern and eastern Bolivia (Steininger *et al.* 2001a,b) and is rapidly increasing in Pará, Maranhão, and especially Mato Grosso states in Brazil (Fearnside 2001). In 2004 nearly half of all deforestation in Brazilian Amazonia occurred in Mato Grosso (INPE 2005), largely as a result of the explosive growth of industrial soy farms (Laurance *et al.* 2004a).

Logging

In recent decades, industrial logging (Figure 27.2) has increased sharply in the Amazon, and now affects 1–2 million ha of forest each year in Brazilian Amazonia alone (Nepstad *et al.* 1999b, Asner *et al.* 2005). In the tropics, logging is normally selective, in that only a relatively small

Figure 27.2 Industrial logging, like this operation in northern Bolivia, creates labyrinths of roads that promote forest colonization and overhunting (photograph by W.F. Laurance).

percentage of all trees are harvested. However, the number of harvested species varies considerably among regions. In new frontiers, only 5–15 species are typically harvested (1–3 trees ha^{-1}), but in older frontiers up to 100–150 species are harvested (5–10 trees ha^{-1}) (Uhl et al. 1997). Valuable timbers such as mahogany (*Swietenia* spp.) are overexploited and play a key role in making logging operations profitable (Fearnside 1997).

The immediate impacts of logging mostly arise from the extensive networks of roads, tracks, and small clearings created during cutting operations (Figure 27.2), which cause collateral tree mortality, soil erosion and compaction, vine and grass invasions, and microclimatic changes associated with disruption of the forest canopy (Uhl and Vieira 1989, Veríssimo et al. 1992, 1995, Johns 1997). Many sensitive wildlife species decline in logged forests (Johns 1997 and references therein). In addition, logging has important indirect effects; by creating labyrinths of forest roads, logging opens up areas for colonization by migrant settlers and ranchers who often use destructive slash-and-burn farming methods (Uhl and Buschbacher 1985, Veríssimo et al. 1995). Logging often leads to an increase in hunting, which can seriously affect some wildlife species. In the Malaysian state of Sarawak, for example, a single large logging camp was estimated to consume over 30,000 kg of wildlife meat each year (Bennett and Gumal 2001).

Logging is a multi-billion dollar business in the Amazon. Brazil currently has about 400 domestic timber companies operating in the Amazon, which operate from 6000 to 7000 timber mills, whereas Bolivia has about 150 domestic companies (Laurance 1998). In addition, multinational timber corporations from Malaysia, Indonesia, China, South Korea, and other Asian nations have moved rapidly into the Brazilian Amazon by acquiring control of large forest tracts, often by purchasing interests in local timber firms. In Guyana, Suriname, and Bolivia, these corporations have obtained extensive long-term forest leases (termed "concessions"; Colchester 1994, Sizer and Rice 1995). In 1996 alone, Asian corporations invested more than 500 million dollars in the Brazilian timber industry (Muggiati and Gondim 1996). Asian multinationals now own or control at least 13 million ha of Amazonian forest (Laurance 1998).

A striking feature of the Amazonian timber industry is that illegal logging is rampant. A 1997 study by Brazil's national security agency concluded that 80% of Amazonian logging was illegal, and recent raids have netted massive stocks of stolen timber (Abramovitz 1998). Aside from widespread illegal cutting, most legal operations from the hundreds of domestic timber companies in the Amazon are poorly managed. A government inspection of 34 operations in Paragominas, Brazil, for example, concluded that "the results were a disaster," and that not one was using accepted practices to limit forest damage (Walker 1996). In the late 1990s, in a controversial attempt to gain better control over Amazonian logging operations, Brazil opened 39 of its National Forests, totaling 14 million ha, to logging, arguing that concessions would not be granted to companies with poor environmental records (Anon. 1997). Brazil plans greatly to expand its system of National Forests in the Amazon, adding 50 million ha of new logging reserves by the year 2010 (Veríssimo et al. 2002).

Forest fragmentation

The rapid pace of deforestation is causing forest fragmentation on many spatial scales. On a basin-wide scale, major new highways, roads, and transportation projects are now penetrating deep into the heart of the basin, promoting forest colonization, logging, mining, and deforestation in areas once considered too remote for development (Laurance 1998, 2005a, Carvalho et al.

2001, Laurance et al. 2001a,b, 2002a, 2004a). By 1988, the area of forest in Brazilian Amazonia that was fragmented (<100 km^2 in area) or prone to edge effects (<1 km from forest edge) was more than 150% larger than the area that had actually been deforested (Skole and Tucker 1993). Because over 18% of the region's forests have now been cleared (INPE 2005), the total area affected by fragmentation, deforestation, and edge effects could constitute one third or more of the Brazilian Amazon today (Laurance 1998).

On a landscape scale, different land uses tend to generate distinctive patterns of fragmentation. Cattle ranchers destroy large, rectangular blocks of forest, and habitat fragments in such landscapes are often moderately regular in shape (Figure 27.3, right). Forest-colonization projects, however, result in more complex patterns of fragmentation (Figure 27.3, left), creating very irregularly shaped fragments with a high proportion of forest edge (Dale and Pearson 1997, Laurance et al. 1998b). Remote-sensing studies suggest that, as a result of rapid habitat fragmentation, nearly 20,000 km of new forest edge is being created each year in the Brazilian Amazon (W. Chomentowski, D. Skole, and M. Cochrane personal communication).

Habitat fragmentation has myriad effects on Amazonian forests (reviewed in Laurance et al. 2002b), such as altering the diversity and composition of fragment biota, and changing ecological processes like pollination, nutrient cycling, and carbon storage (Lovejoy et al. 1986, Bierregaard et al. 1992, Didham et al. 1996, Laurance and Bierregaard 1997). Edge effects – ecological changes associated with the abrupt, artificial edges of forest fragments – penetrate at least 300 m into Amazonian forests (Figure 27.4; Laurance et al. 1997, 1998a, 2000, 2002b). Moreover, forest fragmentation appears to interact synergistically with ecological changes such as hunting, fires, and logging (Laurance and Cochrane 2001, Peres 2001, Cochrane and Laurance 2002, Laurance and Peres 2006), collectively posing an even greater threat to the rainforest biota.

As a result of such changes, many faunal groups, including insectivorous understory birds, most primates, and larger mammals,

Tailândia   Paragominas

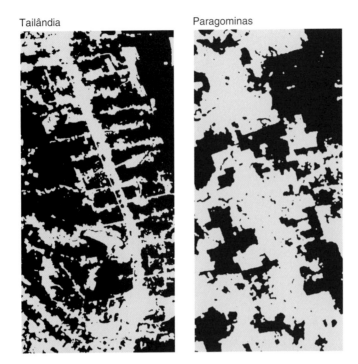

Figure 27.3   Different land uses in the Brazilian Amazon produce distinctive patterns of forest fragmentation. Government-sponsored colonization projects in Tailândia result in a "fishbone" pattern of fragmentation, which differs from the fragmentation pattern caused by cattle ranching near Paragominas. Each image shows an area of about 600 km^2.

decline in abundance or disappear in fragmented forests (Lovejoy *et al.* 1986, Schwartzkopf and Rylands 1989, Bierregaard *et al.* 1992, Stouffer and Bierregaard 1995). Numerous invertebrate species, such as certain ants, beetles, butterflies, and termites, also respond negatively to fragmentation and edge effects (Klein 1989, Didham *et al.* 1996, Brown and Hutchings 1997, Carvalho and Vasconcelos 1999). Remarkably, many arboreal mammals, understory birds, and invertebrates are unable or unwilling to cross even small (30–80 m wide) forest clearings (Laurance *et al.* 2002b, Laurance, S.G. *et al.* 2004).

Wildfires

Under natural conditions, large-scale fires are evidently very rare in Amazonian rainforests,

perhaps occurring only once or twice every thousand years during exceptionally severe El Niño droughts (Sanford *et al.* 1985, Saldariagga and West 1986, Meggers 1994, Piperno and Becker 1996). Closed-canopy tropical forests are poorly adapted to fire (Uhl and Kauffman 1990), and even light ground-fires kill many trees and virtually all vines (Kauffman 1991, Barbosa and Fearnside 1999, Cochrane and Schulze 1999, Cochrane *et al.* 1999, Nepstad *et al.* 1999a,b).

The incidence of fire has increased radically in the Amazon, for two reasons. First, the number of ignition sources has increased by orders of magnitude since European colonization. Fire is used commonly in the Amazon today, to clear forests, destroy slash piles, and help control weeds in pastures. Over a 4-month period in 1997, satellite images revealed nearly 45,000 separate fires in the Amazon (P. Brown 1998), virtually all of

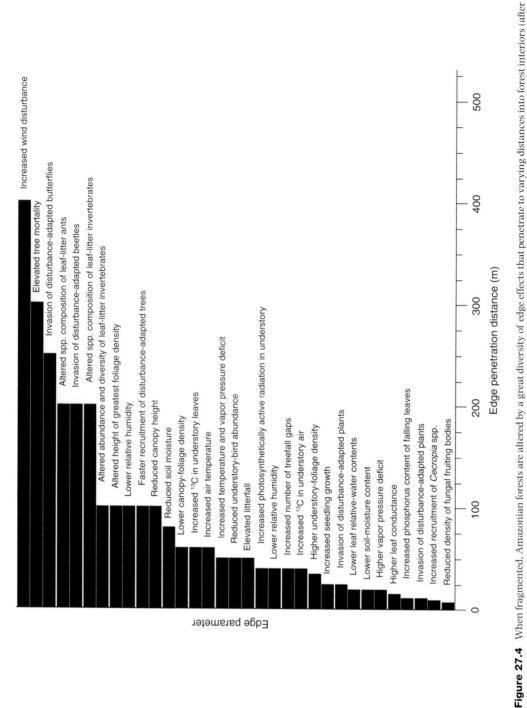

Figure 27.4 When fragmented. Amazonian forests are altered by a great diversity of edge effects that penetrate to varying distances into forest interiors (after Laurance *et al.* 2002b).

Figure 27.5 Ground-fires can penetrate several kilometers into forests, killing many trees and vines and making forests vulnerable to even more devastating wildfires in the future (photograph by M.A. Cochrane).

them human-caused. During the 1997–1998 El Niño drought, wildfires lit by farmers and ranchers swept through an estimated 3.4 million ha of fragmented and natural forest, savanna, regrowth, and farmlands in the northern Amazonian state of Roraima (Barbosa and Fearnside 1999), and there were many large fires in other locations (Cochrane and Schulze 1998). Smoke from forest burning becomes so bad during strong droughts that regional airports must be closed and hospitals report large increases in the incidence of respiratory problems (Laurance 1998).

Second, human land uses increase the vulnerability of tropical forests to fire. Logged forests

are far more susceptible to fires, especially during droughts. Logging increases forest desiccation and woody debris (Uhl and Kauffman 1990), and greatly increases access to slash-and-burn farmers and ranchers, which are the main sources of ignition (Uhl and Buschbacher 1985). The combination of logging, migrant farmers, and droughts was responsible for the massive fires that destroyed millions of hectares of Southeast Asian forests in 1982–1983 and 1997–1998 (Leighton 1986, Woods 1989, N. Brown 1998).

Fragmented forests are also exceptionally vulnerable to fire (Figure 27.5), especially in more seasonal areas of the basin. This is because

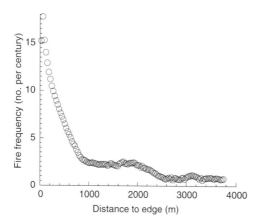

Figure 27.6 Fragmented forests are often extremely vulnerable to fire. Shown is the mean estimated fire frequency as a function of distance from forest edge, for 419 forest fragments in a 2500 km² landscape in eastern Amazonia (adapted from Cochrane and Laurance 2002).

fragment edges are prone to desiccation (Kapos 1989) and contain large amounts of flammable litter and wood debris (Nascimento and Laurance 2004), and because forest remnants are juxtaposed with fire-prone pastures, farmlands, and regrowth forests (Gascon *et al.* 2000). Ground-fires originating in nearby pastures can penetrate thousands of meters into fragmented forests (Figure 27.6; Cochrane and Laurance 2002). These low-intensity fires kill many trees and increase canopy openings and fuel loads, making the forest far more prone to catastrophic wild-fires in the future (Cochrane and Schulze 1999, Cochrane *et al.* 1999). Roughly 45 million ha of forests in Brazilian Amazonia (13% of the total area) are currently vulnerable to edge-related fires (Cochrane 2001).

Additional pressures

Today, even the remotest areas of the Amazon are being influenced by human activities. Illegal gold mining is widespread, with wildcat miners polluting streams with mercury (used to separate gold from sediments) and degrading stream basins with pressure hoses. Illegal miners have

also threatened indigenous Amerindians through intimidation and introduction of new diseases (Christie 1997). In addition, increasing numbers of major oil, natural gas, and mineral developments (iron ore, bauxite, gold, copper) are being sanctioned by Amazonian governments (Nepstad *et al.* 1997, Laurance 1998); such projects provide the economic impetus for construction of roads, highways, and transportation networks, which greatly increase forest loss and fragmentation. Finally, hunting pressure is growing throughout the Amazon because of greater access to forests and markets and the common use of shotguns (Alvard *et al.* 1997, Peres 2001). Intensive hunting can alter the structure of animal communities, extirpate species with low reproductive rates, and exacerbate the effects of habitat fragmentation on exploited species (Robinson and Redford 1991).

The magnitude of the human footprint in the Amazon is illustrated by a recent study. Barreto *et al.* (2005) used extensive spatial data on deforestation, urban centers, agrarian reform settlements, hotspots indicating forest fires, areas licensed for mining and mineral reserves, and positions of authorized logging operations to estimate the extent of human activities in the Brazilian Amazon. By 2002, they found, an estimated 47% of the region was under direct human pressure. Their study was conservative because it did not include illegal logging, which is very extensive (e.g., Asner *et al.* 2005), as well as insidious changes such as overhunting that are largely not detectable using available remote-sensing techniques.

FUTURE THREATS

Pressures on Amazonian forests will almost certainly increase in the future. Ultimately, the rapid expansion of the Amazonian population, which rose in Brazil from about 2.5 million in 1960 to over 20 million today (IBGE 2000), is increasing pressures on forests. Such striking growth has mainly resulted from long-term government policies designed to accelerate immigration and economic development in the region, including large-scale colonization schemes, a tax-free development zone, and credit incentives to attract private capital (Moran 1981, Smith 1982, Fearnside 1987, Laurance 2005a). As a result, the Amazon has the highest rate of immigration of any region in Brazil, and has often been characterized as an "escape valve" for reducing overcrowding, social tensions, and displacement of agricultural workers in other parts of the country (Anon. 2001).

Of more immediate importance is that several Amazonian countries have ambitious, near-term plans to develop major infrastructure projects encompassing large expanses of the basin. These projects are intended to accelerate economic development and exports, especially in the industrial agriculture, timber, and mining sectors of the economy. In the Brazilian Amazon, unprecedented investments, on the order of 20 billion dollars, are being fast-tracked to facilitate construction of new highways, roads, railroads, gas lines, hydroelectric reservoirs, power lines, and river-channelization projects (Laurance *et al.* 2001b, Fearnside 2002). Under current schemes, about 7500 km of new paved, all-weather highways will be created. Key environmental agencies, such as the Ministry of the Environment, are being largely excluded from the planning of these developments (Laurance and Fearnside 1999).

The new infrastructure projects have the potential to cause unprecedented forest loss and degradation (Figure 27.7). The once-remote northern Amazon, for example, has been bisected by the BR-174 highway, which spans some 800 km between Manaus and the Venezuelan border, greatly increasing physical access for logging and colonization projects. Other large highways, such as the BR-319 and BR-163, will soon bisect the central-southern Amazon along a north–south axis. Permanent waterways are being constructed that involve channelizing thousands of kilometers of the Madeira, Xingu, Tocantins, and Araquaia rivers, to allow river barges to transport soybeans from rapidly expanding agricultural areas in central Brazil (Fearnside 2001). In addition, planned road projects will traverse large expanses of the southern Amazon and ascend the Andes to reach the Pacific coast, passing through Bolivia, Peru, and northern Chile. A 3000 km natural-gas line

is also under construction between Santa Cruz, Bolivia and São Paulo, Brazil (Soltani and Osborne 1994).

If they proceed as currently planned, the new infrastructure projects will be one of the most serious threats to Amazonian forests (Laurance *et al.* 2001a, 2004a). By criss-crossing the basin and greatly increasing physical access to forests, the new projects will open up expansive frontiers for colonization and encourage further immigration into a region that is already experiencing rapid population growth. Forest loss and fragmentation are expected to increase considerably (Figure 27.7). In the future, the resulting forest remnants will be far more vulnerable than are large expanses of intact forest to predatory logging, wildfires, and other degrading activities.

A final concern is that Amazonian forests could be subjected to major environmental alterations as a result of global warming, changes in atmospheric composition, or large-scale land-cover changes that reduce evapotranspiration and alter land–atmosphere interactions (e.g., Laurance 2004, Laurance *et al.* 2004b, Malhi and Phillips 2005 and references therein). Reductions in future precipitation are especially likely to have important impacts on forests. For example, several (but not all) of the leading global circulation models suggest that global warming and increasing deforestation will collectively lead to substantial future declines in Amazonian rainfall (Costa and Foley 2000, Cox *et al.* 2000, Zhang *et al.* 2001). These declines are likely to be most damaging in the large expanses of Amazonian forest that experience strong dry seasons and are already at or near the physiological limit of tropical rainforest. In such areas, the incidence of intentional or unplanned forest fires could rise sharply.

AN EXPANDING NETWORK OF RESERVES

Despite the growing panoply of environmental threats, this is also a period of unparalleled opportunity for conservation in the Amazon. Most notably, Brazil, via various federal and state initiatives, is currently designating many

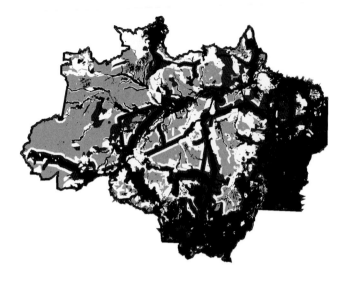

Figure 27.7 Optimistic (above) and non-optimistic (below) scenarios for the Brazilian Amazon, showing predicted forest degradation by the year 2020 (black is deforested or heavily degraded, including savannas and other non-forested areas; dark gray is moderately degraded; white is lightly degraded; and light gray is pristine) (after Laurance *et al.* 2001b).

new Amazonian protected areas and sustainable-use forests (Laurance 2005b, Peres 2005). For example, with an area of nearly 3.9 million ha, the recently designated Tumucumaque Mountains National Park in northeastern Brazil is the largest tropical forest reserve in the world (Mitchell 2002).

The new conservation units in Brazilian Amazonia vary in the kinds of resource uses that are legally permitted (Rylands and Brandon 2005). Intensive uses including industrial

logging are permitted in some reserves, such as National Forests and Environmental Protection Areas, whereas other units, such as National Parks, nominally allow only limited uses that include tourism and scientific research. Yet other conservation units, such as Extractive Reserves, permit intermediate activities such as hunting, rubber tapping, and traditional swidden farming (Laurance 2005).

Although less than 5% of the Brazilian Amazon is currently in strict-protection reserves such as

National Parks, this figure will rise in coming years. Via the Amazon Regional Protected Area (ARPA) initiative, the Brazilian federal government has committed to establish a total of 10% of forests in the region (50 million ha) in strict-protected areas (Rylands and Brandon 2005). ARPA is also promoting new "sustainable-use" reserves that allow various types of extractive activities, from rubber tapping to industrial logging, and in which biodiversity conservation is a secondary priority. Although many new reserves have been designated since ARPA's inception in 2002, most are still "paper parks" that as yet have little staffing or infrastructure.

In addition to ARPA, some forward-looking states in the Brazilian Amazon, especially Amapá and Amazonas, are currently establishing many new conservation units, mostly smaller sustainable-use reserves. The Brazilian Amazon also contains several hundred indigenous lands and territories that are controlled by Amerindian tribes. Although not formally considered conservation units, these lands encompass one fifth of the Brazilian Amazon and often have an important role in protecting forests from predatory logging and land development (Schwartzman and Zimmerman 2005). To provide territories for additional Amerindian groups, the network of indigenous lands is likely to increase in the future (Rylands and Brandon 2005).

Strategies for locating reserves in Amazonia have evolved over time. During the 1970s, the initial emphasis was on protecting putative Pleistocene forest refugia, major vegetation formations, suggested phytogeographical regions, and areas with little economic potential (Rylands and Brandon 2005). Today, however, reserve locations are being influenced by three concepts that arose during the mid- to late 1990s. One of these is ARPA, which is focusing on establishing reserves within 23 Amazonian ecoregions, identified by WWF, that encompass major river drainages and vegetation types (Ferreira 2001). Another is a series of expert workshops initiated by Brazil's Ministry for the Environment, which identified 385 priority areas for conservation in Amazonia (MMA 2002). The third is the biodiversity corridor concept, which proposes to link conservation units of various types into several large chains,

to help maintain forest connectivity (Ayres *et al.* 1997). Several of the proposed corridors span major rainfall gradients and might, if adequately secured and protected, limit the impacts of future climate change, by enabling species to shift their ranges in response to changing conditions.

CONCLUSIONS

As discussed above, the Amazon has already been substantially altered by human activities, with roughly one fifth of all its forests having been destroyed to date, and larger expanses – perhaps another third of the remaining forest – having been degraded by selective logging, surface fires, habitat fragmentation, and edge effects. Moreover, even many of the remotest areas of the Amazon have been altered to some degree by hunting and by other forms of exploitation such as illegal gold mining. The rapid pace of Amazon forest loss could easily accelerate in the future given current plans for major expansion of transportation infrastructure, with a number of new projects slated to penetrate deep into intact forest tracts. Especially alarming is the prospect that the basin's forests could be fragmented on a large spatial scale, which could dramatically increase the vulnerability of remaining forests to a range of exploitative activities.

Nonetheless, the conservation prognosis is not entirely negative. As has occurred in the past, especially in areas with infertile soils, large expanses of exploited land in the Amazon will be abandoned, usually after cattle ranching, leading to regeneration of secondary forests. These secondary forests are clearly superior to pastures in terms of their hydrological functions and carbon storage. They also provide some habitat for wildlife, but their benefits for old-growth forest species are usually limited where regrowth is young or does not adjoin primary forest (a source of seeds and animal seed dispersers) (Uhl *et al.* 1988, Lamb *et al.* 2005). In the Amazon, many areas of secondary forest are burned after one to several decades to create new pastures (Fearnside 2000).

Perhaps the greatest cause for optimism in the Amazon is the prospect of a major expansion of the current system of protected and semi-protected areas. Although many of these new conservation units will be under multiple-use management and thus can be subjected to intensive uses such as industrial logging, they clearly afford some degree of protection to forests. The growing network of indigenous lands will also help to limit the extent of forest exploitation. The great challenges for the near future are to rapidly expand the existing protected-area network, and to establish direly needed staffing and infrastructure for park management. Such initiatives will be crucial, because pressures on protected areas will increase rapidly in the future as highways and other transportation infrastructure ramify throughout the basin, bringing conservation units and the expanding Amazonian population into ever-closer contact.

ACKNOWLEDGMENTS

I thank Walter Carson, Stefan Schnitzer, and two anonymous reviewers for commenting on the manuscript. Support was provided by the NASA-LBA program, Andrew W. Mellon Foundation, and Conservation, Food and Health Foundation. This is publication number 505 in the BDFFP technical series.

REFERENCES

Abramovitz, J. (1998) *Taking a Stand: Cultivating a New Relationship with the World's Forests*. World Watch Institute, Washington, DC.

Alvard, M.S., Robinson, J.G., Redford, K.H., and Kapland, H. (1997) The sustainability of subsistence hunting in the neotropics. *Conservation Biology* 11, 977–982.

Anon. (1997) Controle sobre florestas exige a reforma do IBAMA. *A Crítica Newspaper*, Manaus, Brazil, January 19.

Anon. (2001) Amazônia cede as terras e o governo se esquece das verbas. *O Liberal*, Belém, Brazil, April 4.

Asner, G.P., Knapp, D., Broadbent, E., Oliveira, P., Keller, M., and Silva, J. (2005) Selective logging in the Brazilian Amazon. *Science* 310, 480–482.

Ayres, J.M., da Fonseca, G.A.B., Rylands, A.B., Quieroz, H.L., Pinto, I.P., and Cavalcanti, R. (1997) *Abordagens*

inovadoras para conservação da biodiversidade no Brasil: os corridores das florestas neotropicais. Ministério do Meio Ambiente, Brasília, Brazil.

Barbosa, R.I. and Fearnside, P.M. (1999) Incêndios na Amazônia brasileira: estimativa da emissão de gases do efeito estufa pela queima de diferentes ecossistemas de Roraima na passagem do evento "El Niño" (1997/98). *Acta Amazonica* 29, 513–534.

Barreto, P., Souza, C., Anderson, A., Salomão, R., Wiles, J., and Noguerón, R. (2005) *Human Pressure in the Brazilian Amazon*. Technical report, Institute of People and the Environment in the Amazon (IMAZON), Belém, Brazil.

Bennett, E.L. and Gumal, M.T. (2001) The interrelationships of commercial logging, hunting, and wildlife in Sarawak: recommendations for forest management. In R. Fimbel, A. Grajal, and J.G. Robinson (eds), *The Cutting Edge: Conserving Wildlife in Logged Tropical Forests*. Columbia University Press, New York, pp. 359–374.

Bierregaard, R.O. Jr., Lovejoy, T.E., Kapos, V., dos Santos, A., and Hutchings, R.W. (1992) The biological dynamics of tropical rainforest fragments. *BioScience* 42, 859–866.

Brown, K.S. (1987) Soils and vegetation. In T.C. Whitmore and G.T. Prance (eds), *Biogeography and Quaternary History in Tropical America*. Oxford Monographs in Biogeography 3. Oxford, pp. 19–45.

Brown, K.S. Jr. and Hutchings, R.W. (1997) Disturbance, fragmentation, and the dynamics of diversity in Amazonian forest butterflies. In W.F. Laurance and R.O. Bierregaard, Jr. (eds), *Tropical Forest Remnants: Ecology, Management, and Conservation of Fragmented Communities*. University of Chicago Press, Chicago, pp. 91–110.

Brown, N. (1998) Out of control: fires and forestry in Indonesia. *Trends in Ecology and Evolution* 13, 41.

Brown, P. (1998) Forest fires: setting the world ablaze. *The Guardian*, London, England, March 20.

Carvalho, G., Barros, A.C., Moutinho, P., and Nepstad, D.C. (2001) Sensitive development could protect the Amazon instead of destroying it. *Nature* 409, 131.

Carvalho, K.S. and Vasconcelos, H.L. (1999) Forest fragmentation in central Amazonia and its effects on litter-dwelling ants. *Biological Conservation* 91, 151–158.

Christie, M. (1997) Yanomami Indians appeal for help against invaders. *Reuters News Service*, August 31.

Cochrane, M.A. (2001) In the line of fire: understanding the impacts of tropical forest fires. *Environment* 43, 28–38.

Cochrane, M.A., Alencar, A., Schulze, M.D. *et al.* (1999) Positive feedbacks in the fire dynamics of closed canopy tropical forests. *Science* 284, 1832–1835.

Cochrane, M.A. and Laurance, W.F. (2002) Fire as a large-scale edge effect in Amazonian forests. *Journal of Tropical Ecology* 18, 311–325.

Cochrane, M.A. and Schulze, M.D. (1998) Forest fires in the Brazilian Amazon. *Conservation Biology* 12, 948–950.

Cochrane, M.A. and Schulze, M.D. (1999) Fire as a recurrent event in tropical forests of the eastern Amazon: effects on forest structure, biomass, and species composition. *Biotropica* 31, 2–16.

Colchester, M. (1994) The new sultans: Asian loggers move in on Guyana's forests. *Ecologist* 24, 45–52.

Costa, M. and Foley, J. (2000) Combined effects of deforestation and doubled atmospheric CO_2 concentrations on the climate of Amazonia. *Journal of Climate* 13, 18–34.

Cox, P.M., Betts, R.A., Jones, C.D., Spall, S.A. and Totterdell, I.J. (2000) Acceleration of global warming due to carbon cycle feedbacks in a coupled climate model. *Nature* 408, 184–187.

Dale, V.H. and Pearson, S.M. (1997) Quantifying habitat fragmentation due to land-use change in Amazonia. In W.F. Laurance and R.O. Bierregaard, Jr. (eds), *Tropical Forest Remnants: Ecology, Management, and Conservation of Fragmented Communities*. University of Chicago Press, Chicago, pp. 400–409.

Didham, R.K., Ghazoul, J., Stork, N.E., and Davis, A.J. (1996) Insects in fragmented forests: a functional approach. *Trends in Ecology and Evolution* 11, 255–260.

Fearnside, P.M. (1987) Causes of deforestation in the Brazilian Amazon. In R.F. Dickinson (ed.), *The Geophysiology of Amazonia: Vegetation and Climate Interactions*. John Wiley, San Francisco, pp. 37–61.

Fearnside, P.M. (1990) Environmental destruction in the Amazon. In D. Goodman and A. Hall (eds), *The Future of Amazonia: Destruction or Sustainable Development?* Macmillan, London, pp. 179–225.

Fearnside, P.M. (1993) Deforestation in the Brazilian Amazon: the effect of population and land tenure. *Ambio* 8, 537–545.

Fearnside, P.M. (1997) Protection of mahogany: a catalytic species in the destruction of rain forests in the American tropics. *Environmental Conservation* 24, 303–306.

Fearnside, P.M. (2000) Global warming and tropical land-use change: greenhouse gas emissions from biomass burning, decomposition and soils in forest conversion, shifting cultivation and secondary vegetation. *Climatic Change* 46, 115–158.

Fearnside, P.M. (2001) Soybean cultivation as a threat to the environment in Brazil. *Environmental Conservation* 28, 23–38.

Fearnside, P.M. (2002) Avança Brasil: environmental and social consequences of Brazil's planned infrastructure in Amazonia. *Environmental Management* 30, 735–747.

Ferreira, L.V. (2001) Identifição de áreas prioritárias para a conservação da biodiversidade por meio da representatividade das unidades de conservação e tipos de vegetação nas ecorregiões da Amazônia brasiliera. In A. Verissimo, A. Moreira, D. Sawyer, I. dos Santos, L. Pinto, and J. Capobianco (eds), *Biodiversidade na Amazônia brasiliera: avaliação e ações prioritárias para conservação, uso sustentável e repartição de benefícios*. Editora Estação Liberdade, São Paulo, Brazil, pp. 211–245.

Gascon, C., Williamson, G.B., and da Fonseca, G.A.B. (2000) Receding forest edges and vanishing reserves. *Science* 288, 1356–1358.

Homma, A.K.O., Walker, R.T., Scatena, F. *et al.* (1992) *A dinâmica dos desmatamentos e das queimadas na Amazônia: uma análise microeconômica*. Unpublished manuscript, EMBRAPA, Belém, Brazil.

IBGE (1997) *Diagnóstico ambiental da Amazônia Legal. CD-ROM with GIS version of RADAM maps*. Institute for Geography and Statistics (IBGE), Brasília, Brazil.

IBGE (2000) *Censo Demográfico, Dados Distritais: XI Recenseamento Geral do Brasil 2000*. Brazilian Institute of Geography and Statistics, Brasília.

INPE (2005) *Deforestation Estimates in the Brazilian Amazon*. National Institute for Space Research (INPE), São Jose dos Campos, Brazil.

Johns, A. (1997) *Timber Production and Biodiversity Conservation in Tropical Rain Forests*. Cambridge University Press, Cambridge.

Kaimowitz, D., Mertens, B., Wunder, S., and Pacheco, P. (2004) *Hamburger connection fuels Amazon destruction*. Technical report, Center for International Forest Research, Bogor, Indonesia.

Kapos, V. (1989) Effects of isolation on the water status of forest patches in the Brazilian Amazon. *Journal of Tropical Ecology* 5, 173–185.

Kauffman, J.B. (1991) Survival by sprouting following fire in tropical forests of the eastern Amazon. *Biotropica* 23, 219–224.

Klein, B.C. (1989) Effects of forest fragmentation on dung and carrion beetle communities in central Amazonia. *Ecology* 70, 1715–1725.

Lamb, D., Erskine, P.D., and Parotta, J.A. (2005) Restoration of degraded tropical forest landscapes. *Science* 310, 1628–1632.

Laurance, S.G.W., Stouffer, P.C., and Laurance, W.F. (2004) Effects of road clearings on movement patterns of understory rainforest birds in central Amazonia. *Conservation Biology* 18, 1099–1109.

Laurance, W.F. (1998) A crisis in the making: responses of Amazonian forests to land use and climate change. *Trends in Ecology and Evolution* 13, 411–415.

Laurance, W.F. (2004) Forest–climate interactions in fragmented tropical landscapes. *Philosophical Transactions of the Royal Society B* 359, 345–352.

Laurance, W.F. (2005a) Razing Amazonia. *New Scientist*, October 15, pp. 34–39.

Laurance, W.F. (2005b) When bigger is better: the need for Amazonian megareserves. *Trends in Ecology and Evolution* 20, 645–648.

Laurance, W.F., Albernaz, A.K.M., and Da Costa, C. (2001a) Is deforestation accelerating in the Brazilian Amazon? *Environmental Conservation* 28, 305–311.

Laurance, W.F., Albernaz, A.K.M., Fearnside, P.M., Vasconcelos, H.L., and Ferreira, L.V. (2004a) Deforestation in Amazonia. *Science* 304, 1109.

Laurance, W.F., Albernaz, A.K.M., Schroth, G., Fearnside, P.M., Ventincinque, E., and Da Costa, C. (2002a) Predictors of deforestation in the Brazilian Amazon. *Journal of Biogeography* 29, 737–748.

Laurance, W.F. and Bierregaard, R.O. Jr. (eds) (1997) *Tropical Forest Remnants: Ecology, Management and Conservation of Fragmented Communities.* University of Chicago Press, Chicago.

Laurance, W.F. and Cochrane, M.A. (2001) Synergistic effects in fragmented landscapes. *Conservation Biology* 15, 1488–1489.

Laurance, W.F., Cochrane, M.A., Bergen, S. *et al.* (2001b) The future of the Brazilian Amazon. *Science* 291, 438–439.

Laurance, W.F., Delamonica, P., Laurance, S.G., Vasconcelos, H., and Lovejoy, T.E. (2000) Rainforest fragmentation kills big trees. *Nature* 404, 836.

Laurance, W.F. and Fearnside, P.M. (1999) Amazon burning. *Trends in Ecology and Evolution* 14, 457.

Laurance, W.F., Ferreira, L.V., Rankin-de Merona, J.M., and Laurance, S.G. (1998a) Rain forest fragmentation and the dynamics of Amazonian tree communities. *Ecology* 79, 2032–2040.

Laurance, W.F., Laurance, S.G., and Delamonica, P. (1998b) Tropical forest fragmentation and greenhouse gas emissions. *Forest Ecology and Management* 110, 173–180.

Laurance, W.F., Laurance, S.G., Ferreira, L.V., Rankin-de Merona, J., Gascon, C., and Lovejoy, T.E. (1997) Biomass collapse in Amazonian forest fragments. *Science* 278, 1117–1118.

Laurance, W.F., Lovejoy, T.E., Vasconcelos, H.L. *et al.* (2002b) Ecosystem decay of Amazonian forest fragments: a 22-year investigation. *Conservation Biology* 16, 605–618.

Laurance, W.F., Oliveira, A.A., Laurance, S.G. *et al.* (2004b) Pervasive alteration of tree communities in undisturbed Amazonian forests. *Nature* 428, 171–175.

Laurance, W.F. and Peres, C.A. (eds) (2006) *Emerging Threats to Tropical Forests.* University of Chicago Press, Chicago.

Leighton, M. (1986) Catastrophic drought and fire in Borneo tropical rain forest associated with the 1982–1983 El Niño Southern Oscillation Event. In G.T. Prance (ed.), *Tropical Rain Forests and the World Atmosphere.* Cambridge University Press, Cambridge, England, pp. 75–102.

Lovejoy, T.E. (1999) Preface. *Biological Conservation* 91, 100.

Lovejoy, T.E., Bierregaard, R.O., Rylands, A. *et al.* (1986) Edge and other effects of isolation on Amazon forest fragments. In M.E. Soule (ed.), *Conservation Biology: The Science of Scarcity and Diversity.* Sinauer, Sunderland, MA, pp. 257–285.

Malhi, Y. and Phillips, O.L. (eds) (2005) *Tropical Forests and Global Atmospheric Change.* Oxford University Press, Oxford.

Meggers, B.J. (1994) Archeological evidence for the impact of mega-Niño events on Amazonian during the past two millennia. *Climatic Change* 28, 321–338.

Mitchell, A. (2002) Brazil unveils project to save rain forest. *The Globe and Mail*, UK, September 7.

MMA (2002) *Biodiversidade brasileira: avaliação e identificação de áreas e acões prioritárias para conservação, utilização sustentável e repartição de benefícios da biodiversidade brasileira.* Ministério do Meio Ambiente, Brasília, Brazil.

Moran, E.F. (1981) *Developing the Amazon.* University of Indiana Press, Bloomington.

Muggiati, A. and Gondim, A. (1996) Madeireiras. *O Estado de S. Paulo*, São Paulo, Brazil, September 16.

Nascimento, H.E.M. and Laurance, W.F. (2004) Biomass dynamics in Amazonian forest fragments. *Ecological Applications* 14, S127–S138.

Nepstad, D.C., Carvalho, C., Davidson, E. *et al.* (1994) The role of deep roots in the hydrological cycles of Amazonian forests and pastures. *Nature* 372, 666–669.

Nepstad, D.C., Klink, C., Uhl, C. *et al.* (1997) Land-use in Amazonia and the cerrado of Brazil. *Ciencia e Cultura* 49, 73–86.

Nepstad, D.C., Moreira, A.G., and Alencar, A.A. (1999a) *Flames in the rain forest: origins, impacts, and alternatives to Amazonian fires.* Pilot Program to Conserve the Brazilian Rain Forest, Brasília, Brazil.

Nepstad, D.C., Verissimo, A., Alencar, A. *et al.* (1999b) Large-scale impoverishment of Amazonian forests by logging and fire. *Nature* 398, 505–508.

Peres, C.A. (2001) Synergistic effects of subsistence hunting and habitat fragmentation on Amazonian forest vertebrates. *Conservation Biology* 15, 1490–1505.

Peres, C.A. (2005) Why we need megareserves in Amazonia. *Conservation Biology* 19, 728–733.

Piperno, D.R. and Becker, P. (1996) Vegetational history of a site in the central Amazon Basin derived from phytolith and charcoal records from natural soils. *Quaternary Research* 45, 202–209.

Richter, D.D. and Babbar, L.I. (1991) Soil diversity in the tropics. *Advances in Ecological Research* 21, 315–389.

Robinson, J. and Redford, K. (eds) (1991) *Neotropical Wildlife Use and Conservation.* University of Chicago Press, Chicago.

Rylands, A.B. and Brandon, K. (2005) Brazilian protected areas. *Conservation Biology* 19, 612–618.

Saldariagga, J. and West, D.C. (1986) Holocene fires in the northern Amazon basin. *Quaternary Research* 26, 358–366.

Sanford, R.L., Saldariagga, J., Clark, K., Uhl, C., and Herrera, R. (1985) Amazon rain-forest fires. *Science* 227, 53–55.

Sarre, A., Sobral Filho, M., and Reis, M. (1996) The amazing Amazon. *ITTO Tropical Forest Update* 6, 3–7.

Schwartzkopf, L. and Rylands, A.B. (1989) Primate species richness in relation to habitat structure in Amazonian rainforest fragments. *Biological Conservation* 48, 1–12.

Schwartzman, S. and Zimmerman, B. (2005) Conservation alliances with indigenous peoples of the Amazon. *Conservation Biology* 19, 721–727.

Sizer, N. and Rice, R. (1995) *Backs to the Wall in Suriname: Forest Policy in a Country in Crisis.* World Resources Institute, Washington, DC.

Skole, D. and Tucker, C.J. (1993) Tropical deforestation and habitat fragmentation in the Amazon: satellite data from 1978 to 1988. *Science* 260, 1905–1910.

Smith, N.J.H. (1982) *Rainforest Corridors: The Transamazon Colonization Scheme.* University of California Press, Berkeley, CA.

Soltani, A. and Osborne, T. (1994) *Arteries for Global Trade, Consequences for Amazonia.* Amazon Watch, Malibu, CA.

Steininger, M.K., Tucker, C., Ersts, P., Killeen, T., and Hecht, S. (2001a) Clearance and fragmentation of semi-deciduous tropical forest in the Tierras Bajas zone, Santa Cruz, Bolivia. *Conservation Biology* 15, 856–866.

Steininger, M.K., Tucker, C., Townshend, J. *et al.* (2001b) Tropical deforestation in the Bolivian Amazon. *Environmental Conservation* 28, 127–134.

Stouffer, P.C. and Bierregaard, R.O. (1995) Use of Amazonian forest fragments by understory insectivorous birds. *Ecology* 76, 2429–2445.

Uhl, C., Barreto, P., Verissimo, A. *et al.* (1997) Natural resource management in the Brazilian Amazon. *BioScience* 47, 160–168.

Uhl, C. and Buschbacher, R. (1985) A disturbing synergism between cattle ranch burning practices and selective tree harvesting in the eastern Amazon. *Biotropica* 17, 265–268.

Uhl, C., Buschbacher, R., and Serrão, E. (1988) Abandoned pastures in eastern Amazonia. I. Patterns of plant succession. *Journal of Ecology* 76, 633–681.

Uhl, C. and Kauffman, J.B. (1990) Deforestation, fire susceptibility, and potential tree responses to fire in the eastern Amazon. *Ecology* 71, 437–449.

Uhl, C. and Vieira, I.C.G. (1989) Ecological impacts of selective logging in the Brazilian Amazon: a case study from the Paragominas region of the state of Pará. *Biotropica* 21, 98–106.

Veríssimo, A., Barreto, P., Mattos, M., Tarifa, R., and Uhl, C. (1992) Logging impacts and prospects for sustainable forest management in an old Amazonian frontier: the case of Paragominas. *Forest Ecology and Management* 55, 169–199.

Veríssimo, A., Barreto, P., Tarifa, R., and Uhl, C. (1995) Extraction of a high-value natural resource in Amazonia: the case of mahogany. *Forest Ecology and Management* 72, 39–60.

Veríssimo, A., Cochrane, M.A., and Souza, C. Jr. (2002) National forests in the Amazon. *Science* 297, 1478.

Viña, A. and Cavalier, J. (1999) Deforestation rates (1938–1988) of tropical lowland forests on the Andean foothills of Colombia. *Biotropica* 31, 31–36.

Walker, G. (1996) Kinder cuts. *New Scientist* 151, 40–42.

Whitmore, T.C. (1997) Tropical forest disturbance, disappearance, and species loss. In W.F. Laurance and R.O. Bierregaard, Jr. (eds), *Tropical Forest Remnants: Ecology, Management, and Conservation of Fragmented Communities.* University of Chicago Press, Chicago, pp. 3–12.

Woods, P. (1989) Effects of logging, drought, and fire on tropical forests in Sabah, Malaysia. *Biotropica* 21, 290–298.

Zhang, H., Henderson-Sellers, A., and McGuffie, K. (2001) The compounding effects of tropical deforestation and greenhouse warming on climate. *Climate Change* 49, 309–338.

Chapter 28

CONTRIBUTIONS OF ECOLOGISTS TO TROPICAL FOREST CONSERVATION

Francis E. Putz and Pieter A. Zuidema

OVERVIEW

Given that tropical forest conservation is not solely an ecological problem, ecologists can only hope to provide partial solutions. Despite this fundamental limitation, ecological insights are needed and ecologists can also help to build environmental awareness. But, they should also address the causes of destruction by considering the challenges facing the people who determine forest fates. Efforts at reducing the technical impediments to conservation in the tropics are likely to be successful only if researchers recognize the complex social, economic, and political contexts in which conservation happens or fails to happen in developing countries. Ecologists working in these real landscapes need to be careful when making assumptions about the concerns and values they share with local stakeholders. Caution is also warranted when extrapolating from small-scale and narrowly focused research carried out in purportedly pristine protected areas to the complex landscapes in which reasonable conservation interventions are needed. Furthermore, ecologists should realize that conservation solutions for tropical forests vary in size, land use history, and socioeconomic context, as well as that protecting depopulated parks is just one of a large variety of conservation options. If more than a small portion of tropical forest biodiversity is to be conserved, more ecologists need to work outside protected areas and focus on maximizing biodiversity conservation in the vast remaining areas of multifunctional and semi-natural landscapes.

INTRODUCTION

The ongoing destruction of tropical forests is of great concern to environmentalists and ecologists in the West and to many people in the rest of the world as well. School children are instructed about the evils of deforestation, the media bombard us with graphic accounts of widespread biodiversity losses, and ecologists document in ever-increasing detail and decry with stridency the deleterious consequences of logging, fragmentation, farming, and fires (e.g., Laurance Chapter 27, this volume). While these clarion calls play important roles in the politics of conservation and in general awareness-building, most ecologists and conservation biologists fail to respond to them with viable alternatives to forest conversion because they disregard the social, economic, and political contexts in which tropical forests are destroyed or maintained. A consequence of this failure to recognize the complex reality of tropical forest conservation is that ecologists are increasingly alienated from potential conservation allies whose principal foci are peace, justice, poverty alleviation, indigenous rights, sustainable resource use, and policy reform (e.g., Colchester 1996, Neumann 1997, Slater 2003, Chapin 2004, Terborgh 2005).

In this chapter we briefly describe the complex and mostly human-dominated dimensions in which tropical forest conservation happens or fails to happen and explore some options for conserving forests and promoting social welfare. We then focus on how ecologists might substantially increase their impacts on the fates of tropical forests. We make these suggestions in full recognition of our own failures, as ecologists, to understand conservation challenges fully or to alter the fates of many tropical forests.

Unfortunately, there is no overarching ecological theory to guide our efforts to conserve tropical forests. Although ecological insights, theory based and otherwise, are needed to address the technical impediments to effective conservation, the cumulative experience of successful conservation practitioners suggests that conservation solutions are more than ecological and need to be negotiated locally with the assistance of interdisciplinary teams of very patient and culturally sensitive individuals (Putz 2000, Sayer and Campbell 2004, Colfer 2005). While the authoritarian, simplistic, and "expertocratic" approaches generally associated with the demarcation and defense of protected areas against incursions by local people can succeed over the short term (Rice *et al.* 1997, Terborgh 2000, 2004), and may actually serve their ecological purpose under some conditions (e.g., Peres 2005), long-term conservation solutions will usually need to reflect and respond to the complex local realities (Hutton and Leader-Williams 2003, Bray 2004, Andrade 2005, Kaimowitz and Sheil 2007). Many ecologists find collaborating or even communicating with social scientists challenging, and working collaboratively with local people during project design, implementation, and dissemination nearly impossible given their personal proclivities and time constraints (but see Sheil and Lawrence 2004). Often we fail to recognize the differences in our worldviews from those of other tropical forest stakeholders (Kaimowitz and Sheil 2007), a challenge made greater when local people adopt politically expedient language and concepts learned from earlier generations of visiting environmentalists (e.g., Brosius 1997, Dove *et al.* 2003). More fundamentally, few ecologists are willing to make the transition from being problem describers (e.g., the effects of logging on x, y, or z) to problem solvers (e.g., financially feasible methods for minimizing the effects of logging on x, y, or z).

For many people living outside of the tropics, part of the allure of tropical forests is that they are far away, exotic, and inhabited by strange and wonderful organisms but few people (Slater 2003). Many perceive tropical forests as aseasonal, primeval, and extremely sensitive to human interventions. Even ecologists often distinguish themselves as *tropical* ecologists, suggesting that they recognize distinctive attributes of tropical ecosystems. But what is it that renders tropical ecology and conservation different from ecology and conservation anywhere else? Perhaps it is the phenomenal species diversity of tropical forest ecosystems that makes them distinctive, but many other hyper-diverse ecosystems in temperate regions do not draw so much global attention (e.g., the "fynbos" woodlands of South Africa or the scrublands of Western Australia). Perhaps it is the alarming rates at which tropical ecosystems are being destroyed (e.g., Laurance Chapter 27, this volume, Corlett and Primack Chapter 26, this volume), but many ecosystems in temperate and boreal regions are suffering similar fates at similar rates (Millennium Ecosystem Assessment 2005). For example, the hyper-diverse pine savannas of the southeastern coastal plain in the USA were reduced to less than 3% of their historical range during the past century but are still being destroyed at an alarming rate (e.g., Croker 1987). Whatever the reasons for separating out tropical forests for special consideration, the social, political, and economic contexts of conservation are certainly very different from those in developed countries. It is the failure to consider these contexts during project selection, design, implementation, and publication that limits the conservation value of many well-intended conservation-motivated ecological research efforts (Robbins 2004, Sayer and Campbell 2004).

THE ECOLOGICAL FOUNDATION FOR TROPICAL FOREST CONSERVATION

Conservation is not solely an ecological or technical issue, especially where most local stakeholders are poor, plagued by violence, and unsure of their continued access to the resources they need to survive. Nevertheless, ecologists dominate the field of tropical forest conservation and most conservation recommendations are grounded primarily in ecological theory (e.g., Pickett *et al.* 1997, McCool and Stankey 2004, Groom *et al.* 2006). For example, if Brazilians and Indonesians and Cameroonians are destroying tropical forests by building roads and opening farms, then island biogeography theory calls for the establishment of large inviolate preserves in which such activities are prohibited. Similarly, widespread forest fragmentation provokes calls for connecting protected areas with corridors of natural forest to allow gene flow and thereby avoid the extinctions that will otherwise occur. And if human-induced disturbances such as grazing, logging, and burning practices exceed the natural range of disturbances to which forest organisms are presumed to be adapted, then herd reductions, logging bans, and fire-use restrictions are obviously justified. As we hope to show below, all of these well-intentioned and ecological theory-based recommendations can be inappropriate at certain

scales and in many social, economic, and political contexts (Figure 28.1).

The island-biogeographical model of MacArthur and Wilson (1967) has influenced conservation science perhaps more than any other concept in ecology. Concerns about species losses in forest fragments, for example, are explained by this theory. Unfortunately, while the theory may be used to make robust predictions for oceanic islands, it does not work as well for forest fragments embedded in landscapes with edge-buffering plantations or secondary forests across which many taxa readily move (e.g., Malcolm 1994, Gascon *et al.* 1999, Barlow *et al.* 2007). Another problem is that most of the numerous studies documenting changes in fragment microclimate, ecosystem functions, and biodiversity were conducted in small fragments of 1–100 ha, and much less often in fragments of more than 1000 ha (Zuidema *et al.* 1996, Laurance and Bierregaard 1997). Furthermore, fragmentation effects on

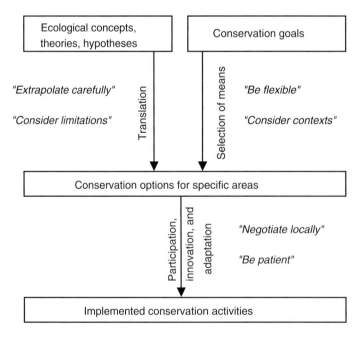

Figure 28.1 From ecological concepts and conservation goals to implemented conservation activities. Quoted remarks pertain to the steps in the process and are the main messages of this chapter.

populations of tree species reportedly vary, from rapid losses (e.g., Leigh *et al.* 1993) to long-term persistence (Thomas 2004) in even small fragments. And although edge effects undoubtedly also occur in large fragments in Brazil (Nepstad *et al.* 1999, Laurance 2000, 2004, Laurance *et al.* 2002, Silva Dias *et al.* 2002), Phillips *et al.* (2006) failed to find the expected extensive edge effects in Amazonian Peru. Furthermore, there is little evidence that forest fragments of more than 10,000 ha lose a large portion of species if hunting is controlled and wildfires are avoided. Nevertheless, results of fragmentation research are often used to justify pleas for parks much larger ($>10^6$ ha) than any studied forest fragments (e.g., Terborgh *et al.* 1997, Peres 2005, Tabarelli *et al.* 2004, Tabarelli and Gascon 2005). Although the "Single Large Or Several Small" (SLOSS) reserves debate has been described by some critics as a purely theoretical exercise (e.g., Simberloff 1997), the "larger is better" argument seems to prevail in some conservation and scientific circles (e.g., Laurance 2005, Peres 2005). Ecologists could help by studying diversity maintenance in larger fragments, but their research is likely to have more impact if it were focused on improving matrix management (Kupfer *et al.* 2006).

Similarly, despite substantial reservations about its relevance to conservation (Hanski 1997, Freckleton and Watkinson 2002), metapopulation theory has been invoked to support campaigns to connect forest fragments with habitat corridors (e.g., Laurance and Bierregaard 1997). Corridors make sense and sometimes work as predicted (Beier and Noss 1998, Haddad *et al.* 2003), but their likely biodiversity benefits should not be exaggerated (Harrison and Bruna 1999), especially when they are long (Simberloff *et al.* 1992) and the matrix surrounding fragments is already favorable for many species (e.g., Gascon *et al.* 1999). Despite these possible limitations, the theory has drawn attention to the value of small habitat patches for species survival, to the issue of species movement, and to the dynamic aspects of species maintenance. Research on increasing the retention of biodiversity in multifunctional landscapes consisting of mosaics of forest fragments (small to large, connected to isolated), production forests, plantations, and agricultural lands is urgently required (Zuidema and Sayer 2003).

A body of ecological theory that has not yet been fully utilized in tropical forest conservation concerns the capacity of disturbed ecosystems to return to pre-intervention states, the roles of diversity in this resilience, and the likelihood that further disturbance will precipitate dramatic changes in ecosystem structure and composition (e.g., Holling 2001; see www.resalliance.org). Resilience theory is broad and flexible enough to include human-induced as well as other sorts of perturbations, and accepts that responses to stresses and disturbances are often non-linear and sometimes completely unexpected. Such surprises are sometimes manifest in "phase shifts" (*sensu* Folke *et al.* 2004), such as when tropical forest is degraded to the point that it becomes savanna (e.g., Oyama and Nobre 2003). Unfortunately, models based on resilience theory are unavoidably only as good as the input data, which must include social as well as biophysical variables. For example, failure to recognize the importance of historical land-use practices and inappropriate extrapolations of ecological evidence resulted in mistaken impressions of the role of local people in forest destruction in both West and East Africa (Fairhead and Leach 1996, Bassett and Zuéli 2003). Given the complexity of tropical forests and the diversity of issues relevant to their management and conservation, the humbling occurrence of surprising responses to planned and unexpected interventions is and will remain common. In light of this uncertainty, a diversity of locally tailored approaches to conservation is warranted, with adjustments as justified from experience and frequent monitoring (Figure 28.1; Bormann and Kiester 2004).

CONSERVATION SOLUTIONS VARY IN SIZE AND LAND-USE INTENSITIES

While we might like to think of tropical forests as extensive and uninhabited wilderness areas, most do not fit this description (e.g., Denevan 1992, van Gemerden *et al.* 2003, Willis *et al.*

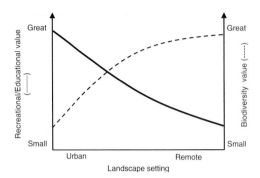

Figure 28.2 The direct value of forests to human societies is often highest for forests near urban centers, even if they are low in diversity and severely degraded. Forests with high biodiversity value are typically in remote areas, away from where many people live, and therefore of lower direct societal value.

2004, Baker *et al.* 2005, Mann 2005). Remaining tropical forests vary in size and, in some settings, even small and degraded forest patches can have large conservation values (Figure 28.2). Unfortunately, the following have typically failed to attract attention from ecologists concerned about tropical forest conservation:

• Small patches of forest in urban settings that are important for environmental education, recreation, and environmental services (e.g., noise abatement).

• Buffer zones and parks in the rapidly expanding suburbs around already huge and rapidly growing cities such as São Paulo, Accra, Jakarta, and others (e.g., Turner 1996, Turner and Corlett 1996).

• Larger forests in more rural settings that were mostly defaunated by hunters and otherwise degraded by repeated-entry logging, overgrazing, and wildfires.

• Extensive tracts of forests that are officially designated for production of timber and non-timber forest products (NTFPs; e.g., Brazil nuts, incense resins, and rattan palms).

• Equally extensive areas under the control of rural communities and intended for multiple uses including NTFP collection, hunting, subsistence farming, and timber stand management.

A major task in forest conservation will be to design ways to maintain ecosystem functions and maximize biodiversity conservation in landscapes that include this range of forest sizes, forest types, forest owners, use histories, neighborhoods, and benefits (Zuidema and Sayer 2003, Rudel 2005, Sayer and Maginnis 2005).

Although large protected areas are critical for maintaining the full complement of tropical species, forests under various sorts of commercial management are also of conservation value and are much more extensive in area (Figure 28.3). In fact, given the vast extents of populated, exploited, and managed areas (e.g., Asner *et al.* 2005), these "working forests" have a huge conservation potential, even if their biodiversity per unit area is lower than in protected areas (e.g., Zarin *et al.* 2004). It is also in these exploited or actively managed forests that ecologists could make their largest contributions to conservation by solving instead of just describing problems in ever-increasing detail (Putz 2004). In particular, while it is useful to know how different intensities of forest management influence the retention of biodiversity and the maintenance of ecosystem functions (e.g., hydrological processes and carbon sequestration), ecologists could help more in the development of management techniques that serve their purpose (e.g., promoting regeneration and tree growth to sustain yields; Peña-Claros *et al.* 2008a,b) while being financially viable, socially appropriate, and environmentally sound. Ecological insights would be particularly useful in the poorly stocked and weed-infested forests that suffered overexploitation due to short-term profit maximization. It is frustrating that uncontrolled natural resource exploitation and forest destruction continue, but this frustration should not be used as an excuse for not trying to solve management problems until available funds and political will are such that widespread protection is possible. Instead of waiting for this unlikely alignment, certifiers working with the Forest Stewardship Council (www.fsc.org) are addressing some of these problems, but without the assistance of the ecologists who prefer protected forests where they can ask purely ecological questions.

Given that substantial expansion of protected areas beyond the 18–23% of tropical forests

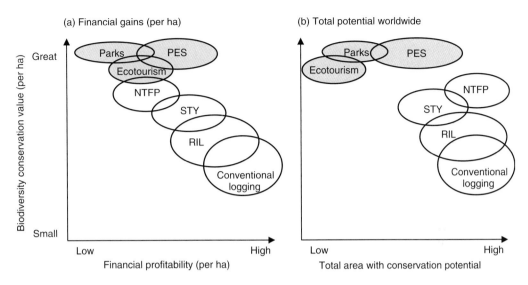

Figure 28.3 Biodiversity value of tropical forest conservation options versus (a) their financial gains on a per unit area basis, and (b) the total forest area with conservation potential (i.e., where conversion might be avoided) under these management regimes. Shaded options are low-intensity uses; white options are more intensive uses. Parks, protected areas; PES, payments for environmental services (e.g., carbon sequestration and hydrological function protection); NTFP, non-timber forest product harvesting; STY, sustained timber yield; RIL, reduced impact logging.

currently demarcated in preserves (Chape *et al.* 2005) is unlikely due to social, economic, and political conditions (Schwartzman *et al.* 2000, Balmford and Whitten 2003), ecologists need to conduct more research in degraded and managed landscapes to inform efforts at enhancing their value for conservation and development. NTFP harvesting, for example, can be carried out sustainably or not (e.g., Peres *et al.* 2003), but even where not sustainable, such harvesting generally has only minor direct impacts on forest structure and composition (Ticktin 2004). In contrast, if collectors of forest fruit, bark, and other products hunt for market purposes, widespread defaunation often results (Peres and Zimmerman 2001).

The selective logging that characterizes most tropical timber harvesting causes more forest damage than NTFP harvesting but is also often more lucrative (Putz *et al.* 2001, Chomitz 2007). Although uncontrolled logging by untrained and unsupervised crews paid solely on the basis of the volumes of timber they harvest can be

extremely damaging to soils and residual trees, substantial and often cost-saving improvements are possible through implementation of reduced impact logging (RIL) techniques (e.g., Dykstra and Heinrich 1996). For example, planning of skid trail locations, directional felling, and cutting of woody vines on trees to be harvested can reduce stand damage by 50% (Pinard *et al.* 2000, Putz *et al.* 2008). It should be noted, however, that even poorly logged forests support many species and supply many of the ecosystem services that society values (e.g., Chazdon 1998, Ter Steege 2003, Arets 2005, Azevedo-Ramos *et al.* 2005, Meijaard *et al.* 2005). In contrast, if logger-built roads open forest to hunters, render it fire-susceptible, and increase its accessibility to agricultural colonists, then the secondary impacts are substantial (e.g., Robinson and Bennett 2000, Fimbel *et al.* 2001, but see Blate 2005). Whether further forest degradation follows logging depends on the pressures on the area for conversion and on the effectiveness of governance (Chomitz 2007).

Where the objective of forest management is sustained timber yield (STY) of shade-tolerant tree species with abundant regeneration in the forest understory, RIL is tantamount to STY unless harvests exceed 60–80 m^3 ha^{-1} (e.g., Sist and Nyuyen-Thé 2002). In contrast, where the commercial tree species are light-demanding and regenerate only after substantial opening of the canopy, then silviculturalists intent on sustaining timber yields may call for increased harvest intensity, liberation of future crop trees from competition by poison-girdling their near neighbors, and even mechanical soil scarification to promote seed germination and seedling establishment (Fredericksen and Putz 2003, Peña-Claros et al. 2008b). Such treatments substantially and intentionally change pre-intervention forest structure and composition often beyond the assumed "historical" range (which is seldom based on more than 5–10 years of stand monitoring), but presumably the light-demanding trees that dominated the pre-logging canopy did not regenerate under the closed-canopy conditions to which they contributed later in their lives. Given that many light-demanding tropical trees of high commercial value can live 100–200 years (e.g., Brienen and Zuidema 2006), forests may reach equilibrium structure and composition only after more than 500 years following substantial natural or anthropogenic disturbance (e.g., van Gemerden et al. 2003, Worbes et al. 2003, Baker et al. 2005). It should therefore not be surprising that many of the tropical forests that are richest in tree species and wildlife benefited from centuries of husbandry by hunters, gatherers, and other traditional forest users (e.g., Gómez-Pompa et al. 1987, Balée 1994, 2000, Peters 2000, but see Parker 1992). The extent of historical humanization of tropical forests, particularly by pre-Columbian Amerindians in Amazonia, is hotly contested (e.g., Roosevelt 1991, Meggers 2001, Bush and Silman 2007), but widespread occurrence of anthropogenic soils (i.e., "terra preta do indio") suggests that many forests were indeed substantially modified before European diseases decimated human populations in the basin (e.g., Erickson 2003, Lehmann et al. 2004).

If we accept that the remaining tropical forests of the world cannot all be fenced off and otherwise protected in people-less preserves, then conservation values are maximized across the landscape where ecologically informed land-use plans are implemented responsibly. Although ease of access often overwhelms other considerations in determining how forest-lands end up being used (e.g., Kaimowitz and Angelsen 1998, Chomitz 2007), biodiversity value (e.g., species richness or the presence of rare or endemic taxa) as well as soil characteristics (e.g., fertility and erosion-proneness), costs of silvicultural management (versus timber mining), slope, and elevation should figure prominently in land-use planning. The importance of these ecological attributes notwithstanding, socioeconomic and political conditions such as contested land ownership, spontaneous and planned land colonization by people from other regions, and cultural traditions often coupled with the activities of smugglers, illegal loggers, wildlife poachers, and bands of guerrillas often determine how forests are treated. Even when conservation is approached in a quantitative manner by skilled ecologists, normative rather than technical issues typically prevail (McCool and Stankey 2004), which is to say that culture generally trumps ecology.

CONSERVATION SOLUTIONS REFLECT VARIOUS SOCIOECONOMIC CONTEXTS

In discussing the *tropical conservation problem*, many environmentalists disregard the fact that most tropical forests are in sovereign nations of which they are not citizens (Romero and Andrade 2004). Just as the government of the USA might object to international attempts to intervene in the cutting of the few remaining old-growth forests in the Pacific Northwest or the continued pollution of Everglades National Park with effluent from highly subsidized sugarcane plantations in Florida, tropical nations expect recognition of the legitimacy of their own political processes (Escobar 1998).

Deforestation and conversion of tropical forests to other land uses are often portrayed as simple consequences of population pressure and ignorance. As described by Corlett and Primack

(Chapter 26, this volume), the drivers of forest conversion vary from region to region, with poor people trying to survive and rich people trying to get richer being equally to blame. Unfortunately for forests and despite hopes to the contrary, logging is lucrative, especially when followed by forest conversion to cattle ranches or commodity crops such as soybeans and oil palm (e.g., Rice et al. 1997, Pearce et al. 2002, Niesten et al. 2004, Chomitz 2007). Although the profits from tropical forest-destroying activities are not shared equitably, entire nations can benefit from forest exploitation if natural capital is converted into social capital (Luckert and Williamson 2005). Malaysia, for example, is well known for having "cashed in" its forests. Although the government failed to capture much of the revenue it was due from timber companies (Repetto and Gillis 1988), logging profits fueled economic development but at the expense of the forest. In other words, the biological costs of this macroeconomic success were substantial. The rich forests that blanketed the country have mostly been replaced by oil palm plantations, some of which are now being cleared for sprawling suburbs and traffic-choked highways (F.E. Putz personal observation). At the other end of the development gradient, the mostly poor people in rural communities across the tropics now control at least 21% of the world's remaining tropical forests (White and Martin 2002). The combined effects of millions of small-scale farmers on tropical forests are substantial, but rural communities have also instituted forest protection programs that now cover an area equivalent to that which is included in nature preserves demarcated by central governments (Molnar et al. 2004). In Brazil alone, indigenous groups now have title to about 1×10^6 km^2 (Schwartzman and Zimmerman 2005). Whether these lands are protected and well managed depends in part on interactions between visiting conservation biologists and the property owners, which are sometimes good (e.g., Zimmerman et al. 2001) and sometimes not so good (e.g., Colchester 1996).

Many of the tropical forests where conservation needs to happen are in frontier areas characterized by poverty, insecure land tenure, and lawlessness (Rudel 2005). It is critical to keep in mind that armed insurrections and full-scale wars are currently underway in about two dozen places in the tropics (Álvarez 2003, McNeely 2003, Price 2003) and approximately 50% of tropical forest logging is illegal (e.g., Ravenel et al. 2004). Even where local people recognize the value of their forests as reliable long-term sources of food, medicines, and building materials, they are forced to address the challenge of short-term survival in what are sometimes destructive ways. In most places, local empowerment, improved governance, and poverty alleviation are prerequisites for conservation (Sanderson 2004) – necessary but not always sufficient. The message is that forests will be protected out of enlightened self-interest only where social, economic, and political conditions allow (Figure 28.1). Fortunately, while the beliefs upon which Western neoliberal economics are based are not universally held (e.g., the benefits of privatization and globalization; Wade 2004), there is some evidence that respect for nature is a characteristic shared by many cultures around the world (Selin 2003).

CONSERVATION OPPORTUNITIES

Before considering further how ecologists could contribute more substantially to the conservation of tropical forests, it might help to describe some of the approaches that are having real conservation impacts. The examples we discuss are mostly market-based and motivated by recognition that the costs of tropical forest conservation are often borne by local people whereas the benefits (e.g., protected biodiversity and sequestered carbon) are enjoyed globally or at least beyond the forest boundaries (e.g., maintained hydrological functions). Various ways of capturing these "externalities" (i.e., values that are not included in standard financial cost–benefit analyses) are being employed to make conservation a more economically attractive option to the people who determine forest fates (Wunder 2007).

Where forest protection entails relocation of forest-dwelling people or substantial restrictions on their forest-based activities, financial compensation for their lost livelihoods is ethically warranted and might promote conservation

(Wilshusen *et al.* 2002). The "conservation incentive agreement" (CIA) approach promoted by Conservation International (Gullison *et al.* 2001) is the best-known type of direct payment for tropical forest conservation. Advocates of this approach argue that the benefits of CIAs to rural communities generally exceed those that the communities would receive from loggers harvesting timber from the same lands. Opponents are concerned that negotiations between well-funded international conservation organizations and poor rural communities are unlikely to be fair, and that development of locally adapted conservation strategies is impeded by these outside interventions (Chapin 2004, Romero and Andrade 2004). Clearly there will only be long-term conservation benefits of agreements negotiated between wealthy and powerful outside groups and forest-rich but financially poor rural people if the latter remain satisfied with the negotiated agreements. In light of recent progress towards understanding community–company partnerships in the tropics (e.g., Mayers and Vermeulen 2002), there are good reasons to hope for fruitful marriages of conservation and development, at least when the relationship is approached honestly and in an informed way.

Another option to promote forest protection is to pay forest owners for the environmental services their forests provide (Landell-Mills and Porras 2002, Wunder 2007). For example, such payments for environmental services (PES) are being used to maintain the hydrological functions of the forested water catchments for the city of Quito, Ecuador, and PES from a national program in Costa Rica are being used to compensate farmers for profits lost as a result of protecting patches of forest important for wildlife populations and the ecotourists they attract (Pagiola *et al.* 2002, Scherr *et al.* 2004). Similarly, in the interest of reducing atmospheric concentrations of carbon dioxide, PES are being used to promote carbon sequestration through reforestation in Ecuador, Mexico, Uganda, Malaysia, and elsewhere in the tropics (e.g., de Jong 2004). Resource economists and sociologists, with the help of a few ecologists (e.g., Bass *et al.* 2000), are developing ways to use PES for biodiversity protection and to increase the effectiveness and equitability of PES projects.

More than any other approach to improving matrix management practices, voluntary third-party certification of forest products harvested in environmentally sound, socially appropriate, and economically viable manners has changed the ways tropical forests are treated (e.g., Dickinson *et al.* 2004, Nebel *et al.* 2005). By linking consumers concerned about the fates of the forests from which their flooring, furniture, Brazil nuts, or health-care products are derived with the harvesters of these products, the certification program of the FSC has stimulated substantial improvements in forest management practices (Nittler and Nash 1999), now covering millions of hectares of natural forest in the tropics (www.fsc.org). The costs of certification are both direct (i.e., paying for the forest audits) and indirect (i.e., modifying management practices so as to be eligible for certification), but for companies and communities interested in marketing their products to environmental and socially concerned consumers, the benefits are apparently sufficient to warrant the additional expenditures (Dickinson *et al.* 2004). Unfortunately, although the FSC is an international non-governmental and non-profit organization with many members from the USA, markets for FSC-certified products are much stronger in the UK and northern Europe.

The ecological foundations for these market-based mechanisms for promoting forest preservation and conservation through sustainable resource use require careful regulation and regular monitoring because markets have no conscience and least-cost options are typically preferred. For example, a PES for climate change mitigation might involve reforestation, a process about which most people have good feelings. Unfortunately, maximizing rates of carbon sequestration in plantations often entails narrow spacings of exotic trees, fertilizing heavily with nitrogen fixed by the fossil-fuel expensive Haber–Bosch process, and treating native species as weeds (e.g., Evans 1982). The undesirable ecological consequences of such an approach may seem obvious, but the alternatives also have disadvantages. It is often argued, for example, that by intensifying management in a small area of plantations, pressure is relieved on the more extensive areas of natural forest (reviewed by Cossalter and Pye-Smith 2003). Unfortunately, the high-grade cabinet woods and many other products harvested from natural forests are not likely to be grown in plantations and their harvesting will consequently continue. Perhaps if all the environmental costs are considered, intensively managed biomass plantations will be less attractive recipients of PES and funds will become available for improved natural forest management. Recent political discussions on including reducing emissions from deforestation and ecosystem degradation (REDD) as a mechanism for carbon sequestration in the follow-up of the Kyoto protocol represents a step in that direction. Inclusion of REDD in the Kyoto follow-up would imply that apart from the net sequestration of carbon in plantations and reforestation projects, carbon credits can also be obtained by securing the retention of carbon in existing forests.

CONCLUSIONS: ROLES FOR ECOLOGY AND ECOLOGISTS

Given the demonstrated and growing impacts of forest certification and the substantial but yet to be realized potential of environmental service payments to alter the fates of tropical forests, we suggest that ecologists should endeavour to reinforce the scientific basis for these approaches. Rather than just describing the problem of tropical forest loss in ever-increasing detail, bemoaning the fates of these forests, and condemning those most directly responsible for the destruction, ecologists should help to solve the complex problems that are the root cause of deforestation and forest degradation. Many of these problems are political and social in nature, but others are ecological and therefore appropriate for addressing with the theories, tools, and methods in the portfolios of ecologists (Sayer and Campbell 2004, Balmford and Bond 2005).

There is no "one-size-fits-all" method for tropical forest conservation (Figure 28.1; cf. Sayer and Maginnis 2005). Instead, the appropriate approaches are locally adapted, slow to develop, subject to change, and otherwise idiosyncratic if they are to resonate with local ecological, social, economic, and political conditions. There

are dangers in oversimplification and in advocating simple shortcuts, thereby disregarding the multidimensional space in which tropical conservation works. Conversely, there are also dangers in expecting deeply entrenched social problems to be solved before effective conservation can happen. In any event, although long-lasting solutions are not solely or even predominantly ecological, ecologists nevertheless have much to contribute.

Obviously, ecologists should avoid making recommendations that are socially inappropriate, economically unviable, politically impossible, and based on questionable science. For example, promoters of strict preservation as the only acceptable approach to conservation in all contexts fail to recognize the historical and continuing roles of humans in tropical forests and the fact that preservation is a luxury that typically cannot be afforded and may not be preferred by disenfranchised and desperately poor people (Neumann 1997, Wilshusen et al. 2002, Dove et al. 2003). Similarly, while explaining the coexistence of so many species of tropical trees is high on the list of research priorities of many ecologists from developed countries, this is not a central issue in the minds of most people, including those who determine the fates of tropical forests (Kaimowitz and Sheil 2007). Instead, research on issues such as how to secure the regeneration and promote the growth of merchantable species needs to figure prominently in the research agendas of those concerned with maintaining tropical forest diversity against the forces causing deforestation and widespread forest degradation. Carrying out real-world, problem-solving research need not imply any reduction in scientific rigor or less emphasis on ecological theory, but in addition to tallies of publications, citations, and research grants, the impacts of applied research are measured in hectares of forest well managed, tons of carbon sequestered, cubic meters of timber sustainably logged, profits accrued to appropriate stakeholders, management regulations adopted, and human capacity built. That said, given that most conservation problems are multidimensional in origin, effective solutions are likely to reflect sound multidisciplinary thinking. Applied ecologists intent on solving problems must therefore frame their research in more than an ecological context. Scientific rigor is also required in measuring the conservation impacts of the realized interventions (Ferraro and Pattanayak 2006).

Finally, if ecologists want to improve the fates of tropical forests they first need to specify the impact they want their research to have (e.g., behavioural, attitudinal, or political change; Spilsbury 2001). The next step is to select the appropriate target audience for causing the desired impact. Then a strategy is needed for reaching this audience, a goal that is more likely attained if audience members are involved in the project from its inception (Sayer and Campbell 2004). With a clear impact chain the likelihood of having the desired effect is increased, but the amount of required effort should not be underestimated. And while work in strictly protected nature preserves where traditional forest uses ceased centuries ago remains useful (e.g., see most of the other chapters in this volume), the research most likely to have conservation impacts will be conducted under the realistic and complex socioeconomic and political conditions that characterize most of the world's remaining tropical forests.

REFERENCES

Álvarez, M.D. (2003) Forests in the time of violence: conservation implications of the Colombian war. *Journal of Sustainable Forestry* 16, 49–70.

Andrade, G.I. (2005) Science and society at the World Parks Congress. *Conservation Biology* 19, 4–5.

Arets, E.J.M.M. (2005) *Responses of tree population dynamics and tree species composition to selective logging in a rain forest in Guyana.* PhD thesis, Utrecht University, The Netherlands. Tropenbos Guyana Series.

Asner, G.P., Knapp, D.E., Broadbent, E.N., Oliveira, P.J.C., Keller, M., and Silva, J.N. (2005) Selective logging in the Brazilian Amazon. *Science* 310, 480–482.

Azevedo-Ramos, C., de Carvalho, O. Jr., and Nasi, R. (2005) *Animal Indicators: A Tool to Assess Biotic Integrity after Logging Tropical Forests?* Instituto de Pesquisa Ambiental da Amazonia, Brasília. [available at: www.ipam.org.br]

Baker, P.J., Bunyavejchewin, S., Oliver, C.D., and Ashton, P.S. (2005) Disturbance history and historical stand dynamics of a seasonal tropical forest in western Thailand. *Ecological Monographs* 75, 317–343.

Balée, W. (1994) *Footprints in the Forest: Ka'apor Ethnobotany – The Historical Ecology of Plant Utilization by an Amazonian People.* Columbia University Press, New York.

Balée, W. (2000) Elevating the Amazonian landscape. *Forum for Applied Research and Public Policy* 15, 28–32.

Balmford, A. and Bond, W. (2005) Trends in the state of nature and their implications for human well-being. *Ecology Letters* 8, 1218–1234.

Balmford, A. and Whitten, T. (2003) Who should pay for tropical conservation, and how could the costs be met? *Oryx* 37, 238–250.

Barlow, J., Gardner, T.A., Araujo, I.S. *et al.* (2007) Quantifying the biodiversity value of tropical primary, secondary, and plantation forests. *Proceedings of the National Academy of Sciences* 104, 18555–18560.

Bass, S., Dubois, O., Moura Costa, P., Pinard, M., Tipper, R., and Wilson, C. (2000) *Rural Livelihoods and Carbon Management.* International Institute for Environment and Development, London.

Bassett, T.J. and Bi Zuéli, K. (2003) The Ivorian savanna: global narratives and local knowledge of environmental change. In K.S. Zimmerer and T.J. Bassett (eds), *Political Ecology: An Integrative Approach to Geography and Environment-Development Studies.* The Guilford Press, New York, pp. 115–136.

Beier, P. and Noss, R.F. (1998) Do habitat corridors provide connectivity? *Conservation Biology* 12, 1241–1252.

Blate, G.M. (2005) Modest trade-offs between timber management and fire susceptibility of a Bolivian semi-deciduous forest. *Ecological Applications* 15, 1649–1663.

Bormann, B.T. and Kiester, A.R. (2004) Options forestry: acting on uncertainty. *Journal of Forestry* 102, 22–27.

Bray, D.B. (2004) Community forestry as a strategy for sustainable management: perspectives from Quintana Roo, Mexico. In D. Zarin, F.E. Putz, J. Alavalapati, and M. Schmink (eds), *Working Forests in the Tropics.* Columbia University Press, New York, pp. 221–237.

Brienen, R.J.W. and Zuidema, P.A. (2006) The use of tree rings in tropical forest management: projecting timber yields of four Bolivian tree species. *Forest Ecology and Management* 226, 256–267.

Brosius, J.P. (1997) Endangered forest, endangered people: environmentalist representations of indigenous knowledge. *Human Ecology* 25, 47–70.

Bush, M.B. and Silman, M.R. (2007) Amazonian exploitation revisited: ecological asymmetry and the policy pendulum. *Frontiers in Ecology and the Environment* 5, 457–465.

Chape, S., Harrison, J., Spalding, M., and Lysenko, I. (2005) Measuring the extent and effectiveness of protected areas as an indicator for meeting global bio-diversity targets. *Philosophical Transactions of the Royal Society B-Biological Sciences* 360, 443–455.

Chapin, M. (2004) A challenge to conservationists. *World-Watch* November/December, 17–31.

Chazdon, R.L. (1998) Ecology – tropical forests – log 'em or leave 'em? *Science* 281, 1295–1296.

Chomitz, K. (2007) *At Loggerheads? Agricultural Expansion, Poverty Reduction, and Environment in the Tropical Forests.* The World Bank, Washington, DC.

Colchester, M. (1996) Beyond "participation": indigenous peoples, biological diversity conservation and protected area management. *Unasylva* 186, 33–39.

Colfer, C.J.P. (ed.) (2005) *The Equitable Forest: Diversity, Community, and Resource Management.* Resources for the Future, Washington, DC.

Cossalter, C. and Pye-Smith, C. (2003) *Fast-Wood Forestry: Myths and Realities.* Center for International Forestry Research, Bogor, Indonesia.

Croker, T.C. (1987) *Longleaf Pine: A History of Man and a Forest.* Forestry Report R8-FR7. USDA Forest Service, Atlanta, GA.

Denevan, W. (1992) The pristine myth: the landscape of the Americas in 1492. *Annals of the Association of American Geographers* 82, 369–385.

Dickinson, J.C., Forgach, J.M., and Wilson, T.E. (2004) The business of certification. In D. Zarin, F.E. Putz, J. Alavalapati, and M. Schmink (eds), *Working Forests in the Tropics.* Columbia University Press, New York, pp. 97–116.

Dove, M.R., Campos, M.T., Matthews, A.S. *et al.* (2003) The global mobilization of environmental concepts: re-thinking the Western–Non-Western divide. In H. Selin (ed.), *Nature Across Cultures: Views of Nature and the Environment in Non-Western Cultures.* Kluwer Academic Publishers, Dordrecht, The Netherlands, pp. 19–46.

Dykstra, D.P. and Heinrich, R. (1996) *FAO Model Code of Forest Harvesting Practice.* FAO, Rome.

Erickson, C. (2003) Historical ecology and future explorations. In J. Lehmann, D.C. Kern, B. Glaser, and W.I. Woods (eds), *Amazonian Dark Earths: Origin, Properties, Management.* Kluwer Academic Publishers, Dordrecht, The Netherlands, pp. 455–500.

Escobar, A. (1998) Whose knowledge, whose nature? Biodiversity, conservation, and the political ecology of social movements. *Journal of Political Ecology* 5, 53–82.

Evans, J. (1982) *Plantation Forestry in the Tropics.* Oxford University Press, Oxford.

Fairhead, J. and Leach, M. (1996) *Misreading the African Landscape: Society and Ecology in a Forest–Savanna Mosaic.* Cambridge University Press, Cambridge.

Ferraro, P.J. and Pattanayak, S.K. (2006) Money for nothing? A call for empirical evaluation of biodiversity conservation investments. *PLoS Biology* 4, 482–488 (e105).

Fimbel, R., Grajal, A., and Robinson, J. (eds) (2001) *The Cutting Edge: Conserving Wildlife in Logged Tropical Forests.* Columbia University Press, New York.

Folke, C., Carpenter, S., Walker, B. *et al.* (2004) Regime shifts, resilience, and biodiversity in ecosystem management. *Annual Review of Ecology and Systematics* 35, 557–581.

Freckleton, R.P. and Watkinson, A.R. (2002) Large-scale spatial dynamics of plants: metapopulations, regional ensembles and patchy populations. *Journal of Ecology* 90, 419–434.

Fredericksen, T.S. and Putz, F.E. (2003) Silvicultural intensification for tropical forest conservation. *Biodiversity and Conservation* 12, 1445–1453.

Gascon, C., Lovejoy, T.E., Bierregaard, R.O. *et al.* (1999) Matrix habitat and species richness in tropical forest remnants. *Biological Conservation* 91, 223–229.

Gómez-Pompa, A., Salvador Flores, J., and Sosa, V. (1987) The "Pet Kot": a man-made tropical forest of the Maya. *Interciencia* 12, 10–15.

Groom, M.J., Meffe, G.K., Carroll, C.R. *et al.* (2006) *Principles of Conservation Biology.* Sinauer, Sunderland, MA.

Gullison, T., Melnyk, M., and Wong, C. (2001) *Logging Off: Mechanisms to Stop or Prevent Industrial Logging in Forests of High Conservation Value.* Union of Concerned Scientists – Center for Tropical Forest Science, Smithsonian Institution, Washington, DC.

Haddad, N.M., Bowne, D.R., Cunningham, A. *et al.* (2003) Corridor use by diverse taxa. *Ecology* 84, 609–615.

Hanski, I. (1997) Habitat destruction and metapopulation dynamics. In S.T.A. Pickett, R.S. Ostfeld, M. Shachak, and G.E. Likens (eds), *The Ecological Basis of Conservation. Heterogeneity, Ecosystems, and Biodiversity.* Chapman and Hall, New York, pp. 217–227.

Harrison, S. and Bruna, E. (1999) Habitat fragmentation and large-scale conservation: what do we know for sure? *Ecography* 22, 225–232.

Holling, C.S. (2001) Understanding the complexity of economic, ecological, and social systems. *Ecosystems* 4, 390–405.

Hutton, J.M. and Leader-Williams, N. (2003) Sustainable use and incentive-driven conservation: realigning human and conservation interests. *Oryx* 37, 215–226.

Jong, B.H.J. de (2004) Carbon sequestration potential through forestry activities in tropical Mexico In

D. Zarin, F.E. Putz, J. Alavalapati, and M. Schmink (eds), *Working Forests in the Tropics*. Columbia University Press, New York, pp. 238–257.

Kaimowitz, D. and Angelsen, A. (1998) *Economic Models of Tropical Deforestation: A Review*. CIFOR, Bogor, Indonesia.

Kaimowitz, D. and Sheil, D. (2007) Conserving what and for whom? Why conservation should help meet basic human needs in the Tropics. *Biotropica* 39, 567–574.

Kupfer, J.A., Malanson, G.P. and Franklin, S.B. (2006) Not seeing the ocean for the islands: the mediating influence of matrix-based processes on forest fragmentation effects. *Global Ecology and Biogeography* 15, 8–20.

Landell-Mills, N. and Porras, I. (2002) *Silver Bullet or Fools' Gold?: A Global Review of Markets for Forest Environmental Services and Their Impact on the Poor*. International Institute for Environment and Development, London.

Laurance, W.F. (2000) Do edge effects occur over large spatial scales? *Trends in Ecology and Evolution* 15, 134–135.

Laurance, W.F. (2004) Forest–climate interactions in fragmented tropical landscapes. *Philosophical Transactions of the Royal Society, Series B* 359, 345–352.

Laurance, W.F. (2005) When bigger is better: the need for Amazonian mega-reserves. *Trends in Ecology and Evolution* 20, 645–648.

Laurance, W.F. and Bierregaard, R.O. (eds) (1997) *Tropical Forest Remnants: Ecology, Management, and Conservation of Fragmented Communities*. University of Chicago Press, Chicago.

Laurance, W.F., Lovejoy, T.E., Vasconcelos, H.L. *et al.* (2002) Ecosystem decay of Amazonian forest fragments: a 22-year investigation. *Conservation Biology* 16, 605–618.

Lehmann, J., Kern, D.C., Glaser, B., and Woods, W.I. (eds) (2004) *Amazonian Dark Earths: Origin, Properties, Management*. Kluwer Academic Publishers, Dordrecht, The Netherlands.

Leigh, E.G. Jr., Wright, S.J., Herre, E.A., and Putz, F.E. (1993) A null hypothesis concerning the diversity of tropical trees: a test and some implications. *Evolutionary Ecology* 7, 76–102.

Luckert, M.K. and Williamson, T. (2005) Should sustained yield be part of sustainable forest management? *Canadian Journal of Forest Research* 35, 356–364.

MacArthur, R.H. and Wilson, E.O. (1967) *The Theory of Island Biogeography*. Princeton University Press, Princeton, NJ.

Malcolm, J.R. (1994) Edge effects in central Amazonian forest fragments. *Ecology* 75, 2438–2445.

Mann, C. (2005) *1491: New Revelations of the Americas Before Columbus*. Alfred A. Knopf, New York.

Mayers, J. and Vermeulen, S. (2002) *Company–Community Partnerships: From Raw Deals to Mutual Gains*. International Institute for Environment and Development, London.

McCool, S.F. and Stankey, G.H. (2004) Indicators of sustainability: challenges and opportunities at the interface of science and policy. *Environmental Management* 33, 294–305.

McNeely, J.A. (2003) Conserving forest biodiversity in times of violent conflict. *Oryx* 37, 142–152.

Meggers, B.J. (2001) The continuing quest for El Dorado: round two. *Latin American Antiquity* 12, 304–325.

Meijaard, E., Sheil, D., Nasi, R. *et al.* (2005) *Life After Logging: Reconciling Wildlife Conservation and Production Forestry in Indonesian Borneo*. CIFOR and UNESCO, Bogor, Indonesia.

Millennium Ecosystem Assessment (2005) *Ecosystems and Human Well-being: Synthesis*. Island Press, Washington, DC.

Molnar, A., Scherr, S.J., and Khare, A. (2004) *Who Conserves the World's Forests? Community-Driven Strategies to Protect Forests and Respect Rights*. Forest Trends and Ecoagriculture Partners, Washington, DC.

Nebel, G., Quevedo, L., Jacobsen, J.B., and Helles, F. (2005) Development and economic significance of forest certification: the case of FSC in Bolivia. *Forest Policy and Economics* 7, 175–186.

Nepstad, D.C., Verissimo, A., Alencar, A. *et al.* (1999) Large-scale impoverishment of Amazonian forests by logging and fire. *Nature* 398, 505–508.

Neumann, R.P. (1997) Primitive ideas: Protected area buffer zones and the politics of land in Africa. *Development and Change* 28, 559–582.

Niesten, E.T., Rice, R.E., Ratay, S.M., and Paratore, K. (eds) (2004) *Commodities and Conservation: The Need for Greater Habitat Protection in the Tropics*. Center for Applied Biodiversity Science, Conservation International, Washington, DC.

Nittler, J.B. and Nash, D.W. (1999) The certification model for forestry in Bolivia. *Journal of Forestry* 97, 32–36.

Noss, R.F. and Beier, P. (2000) Arguing over little things: response to Haddad *et al. Conservation Biology* 14, 1546–1548.

Oyama, M.D. and Nobre, C.A. (2003) A new climate-vegetation equilibrium state for tropical South America. *Geophysical Research Letters* 30, 1–5.

Pagiola, S., Bishop, J., and Landell-Mills, N. (eds) (2002) *Selling Forest Environmental Services. Market-based Mechanisms for Conservation and Development.* Earthscan, London.

Parker, E. (1992) Forest islands and Kayapó resource management in Amazonia: a reappraisal of the apete. *American Anthropologist* 94, 406–428.

Pearce, D., Putz, F.E., and Vanclay, J. (2002) Sustainable forestry: panacea or pipedream? *Forest Ecology and Management* 172, 229–247.

Peña-Claros, M., Peters, E.M., Justiniano, M.J. *et al.* (2008a) Regeneration of commercial tree species following silvicultural treatments in a moist tropical forest. *Forest Ecology and Management*, in press. DOI: 10.1016/j.foreco.2007.10.033.

Peña-Claros, M., Fredericksen, T.S., Alarcón, A. *et al.* (2008b) Beyond reduced-impact logging: silvicultural treatments to increase growth rates of tropical trees. *Forest Ecology and Management*, in press. DOI: 10.1016/j.foreco.2007.11.013.

Peres, C.A. (2005) Why we need megareserves in Amazonia. *Conservation Biology* 19, 728–733.

Peres, C.A., Baider, C., Zuidema, P.A. *et al.* (2003) Demographic threats to the sustainability of Brazil nut exploitation. *Science* 302, 2112–2114.

Peres, C.A. and Lake, I.R. (2003) Extent of nontimber resource extraction in tropical forests: Accessibility to game vertebrates by hunters in the Amazon basin. *Conservation Biology* 17, 521–535.

Peres, C.A. and Zimmerman, B. (2001) Perils in parks or parks in peril? Reconciling conservation in Amazonian reserves with and without use. *Conservation Biology* 15, 793–797.

Peters, C.M. (2000) Pre-Columbian silviculture and indigenous management of tropical forests. In D.L. Lentz (ed.), *Imperfect Balance: Landscape Transformations in the Pre-Columbian Americas.* Columbia University Press, New York, pp. 203–223.

Phillips, O.L., Rose, S., Monteagudo Mendoza, A., and Núñez Vargas, P. (2006) Resilience of Amazon forest to anthropogenic edge effects. *Conservation Biology* 20, 1698–1710.

Pickett, S.T.A., Ostfeld, R.S., Shachak, M., and Likens, G.E. (eds) (1997) *The Ecological Basis of Conservation. Heterogeneity, Ecosystems, and Biodiversity.* Chapman and Hall, New York.

Pinard, M.A., Putz, F.E., and Tay, J. (2000) Lessons learned from the implementation of reduced-impact logging in hilly terrain in Sabah, Malaysia. *International Forestry Review* 2, 33–39.

Price, S.V. (ed.) (2003) *War and Tropical Forests: Conservation in Areas of Armed Conflict.* Haworth Press, Binghamton, NY.

Putz, F.E. (2000) Some roles for North American ecologists in land-use planning in the tropics. *Ecological Applications* 10, 676–679.

Putz, F.E. (2004) Are you a logging advocate or a conservationist? In D. Zarin, F.E. Putz, J. Alavalapati, and M. Schmink (eds), *Working Forests in the Tropics.* Columbia University Press, New York, pp. 15–30.

Putz, F.E., Blate, G.M., Redford, K.H., Fimbel, R. and Robinson, J.G. (2001) Biodiversity conservation in the context of tropical forest management. *Conservation Biology* 15, 7–20.

Putz, F.E., Sist, P., Fredericksen, T. and Dykstra, D. (2008) Reduced-impact logging: challenges and opportunities. *Forest Ecology and Management* (in press).

Ravenel, R.M., Granoff, I.M.E., and Magee, C.A. (eds) (2004) *Illegal Logging in the Tropics.* Haworth Press, New York.

Repetto, R. and Gillis, M. (eds) (1988) *Public Policies and the Misuse of Forest Resources.* Cambridge University Press, Cambridge.

Rice, R., Gullison, R., and Reed, J. (1997) Can sustainable management save tropical forests? *Scientific American* 276, 1246–1256.

Robbins, P. (2004) *Political Ecology.* Blackwell Publishing, Oxford.

Robinson, J.G. and Bennett, E.L. (eds) (2000) *Hunting for Sustainability in Tropical Forests.* Columbia University Press, New York.

Romero, C. and Andrade, G.I. (2004) International conservation organizations and the fate of local tropical forest conservation initiatives. *Conservation Biology* 18, 578–580.

Roosevelt, A.C. (1991) *Moundbuilders of the Amazon.* Academic Press, San Diego, CA.

Rudel, T.K. (2005) *Tropical Forests: Regional Paths of Destruction and Regeneration in the Late Twentieth Century.* Columbia University Press, New York.

Sanderson, S. (2004) Poverty and conservation: The new century's "peasant question?" *World Development* 33, 323–332.

Sayer, J. and Campbell, B. (2004) *The Science of Sustainable Development: Local Livelihoods and the Global Environment.* Cambridge University Press, Cambridge.

Sayer, J.A. and Maginnis, S. (eds) (2005) *Forests in Landscapes: Ecosystem Approaches to Sustainability.* Earthscan, London.

Scherr, S., White, A., and Khare, A. (2004) *For Services Rendered: The Current Status and Future Potential Markets for the Ecosystem Services Provided by Tropical Forests.* ITTO Technical Series 21, International Tropical Timber Organization, Yokohama, Japan.

Schwartzman, S., Moreira, A., and Nepstad, D. (2000) Rethinking tropical forest conservation: peril in parks. *Conservation Biology* 14, 1351–1358.

Schwartzman, S. and Zimmerman, B. (2005) Conservation alliances with indigenous peoples of the Amazon. *Conservation Biology* 19, 721–727.

Selin, H. (ed.) (2003) *Nature Across Cultures: Views of Nature and the Environment in Non-Western Cultures.* Kluwer Academic Publishers, Dordrecht, The Netherlands.

Sheil, D. and Lawrence, A. (2004) Tropical biologists, local people and conservation: new opportunities for collaboration. *Trends in Ecology and Evolution* 19, 634–638.

Silva Dias, M.A., Rutledge, S., Kabat, P. *et al.* (2002) Clouds and rain processes in a biosphere atmosphere interaction context in the Amazon Region. *Journal of Geophysical Research* 107, 8078.

Simberloff, D. (1997) Biogeographic approaches and the new conservation biology. In S.T.A. Pickett, R.S. Ostfeld, M. Shachak, and G.E. Likens (eds), *The Ecological Basis of Conservation. Heterogeneity, Ecosystems, and Biodiversity.* Chapman & Hall, New York, pp. 274–284.

Simberloff, D., Farr, J.A., Cox, J., and Mehlman, D.W. (1992) Movement corridors: conservation bargains or poor investments? *Conservation Biology* 6, 493–504.

Sist, P. and Nyuyen-Thé, N. (2002) Logging damage and the subsequent dynamics of a dipterocarp forest in East Kalimantan (1990–1996). *Forest Ecology and Management* 165, 85–103.

Slater, C. (ed.) (2003) *In Search of the Rain Forest.* Duke University Press, Durham, NC.

Spilsbury, M.J. (2001) From research to impact and the tricky part between. *FAO Regional Office for Asia and the Pacific, FORSPA Publication* 27, 15–22.

Tabarelli, M., Da Silva, M.J.C., and Gascon, C. (2004) Forest fragmentation, synergisms and the impoverishment of neotropical forests. *Biodiversity and Conservation* 13, 1419–1425.

Tabarelli, M. and Gascon, C. (2005) Lessons from fragmentation research: Improving management and policy guidelines for biodiversity conservation. *Conservation Biology* 19, 734–739.

Terborgh, J. (2000) The fate of tropical forests: a matter of stewardship. *Conservation Biology* 14, 1358–1361.

Terborgh, J. (2004) Reflections of a scientist on the World Parks Congress. *Conservation Biology* 18, 619–620.

Terborgh, J. (2005) Science and society at the World Parks Congress. *Conservation Biology* 19, 5–6.

Terborgh, J., Lopez, L., Tello, J., Yu, D. and Bruni, A.R. (1997) Transitory states in relaxing ecosystems of land bridge islands. In W.F. Laurance and R.O. Bierregaard, Jr. (eds), *Tropical Forest Remnants: Ecology, Management, and Conservation of Fragmented Communities.* University of Chicago Press, Chicago, pp. 256–274.

Ter Steege, H. (2003) *Long-Term Changes in Tropical Tree Diversity.* Studies from the Guiana Shield, Africa, Borneo and Melanesia. Tropenbos Series 22, Wageningen, The Netherlands.

Thomas, S.C. (2004) Ecological correlates of tree species persistence in tropical forest fragments. In E.C. Losos and E.G. Leigh (eds), *Forest Diversity and Dynamism: Findings from a Large-Scale Plot Network.* University of Chicago Press, Chicago, pp. 279–314.

Ticktin, T. (2004) The ecological implications of harvesting non-timber forest products. *Journal of Applied Ecology* 41, 11–21.

Turner, I.M. (1996) Species loss in fragments of tropical rain forests: a review of the evidence. *Journal of Applied Ecology* 33, 200–209.

Turner, I.M. and Corlett, R.T. (1996) The conservation value of small, isolated fragments of lowland tropical rain forest. *Trends in Ecology and Evolution* 11, 330–333.

van Gemerden, B.S., Olff, H., Parren, M.P.E., and Bongers, F. (2003) The pristine rain forest? Remnants of historical human impacts on current tree species composition and diversity. *Journal of Biogeography* 30, 1381–1390.

Wade, R.H. (2004) Is globalization reducing poverty and inequality? *World Development* 32, 567–589.

White, A. and Martin, A. (2002) *Who Owns the World's Forests?* Forest Trends, Washington, DC.

Willis, K.J., Gillson, L., and Brncic, T.M. (2004) How "virgin" is virgin rainforest. *Science* 304, 402–403.

Wilshusen, P.R., Brechin, S.R., Fortwangler, C.L., and West, P.C. (2002) Reinventing the square wheel: critique of a resurgent "protection paradigm" in international biodiversity conservation. *Society and Natural Resources* 15, 17–40.

Worbes, M., Staschel, R., Roloff, A., and Junk, W.J. (2003) Tree ring analysis reveals age structure, dynamics and wood production of a natural forest stand in Cameroon. *Forest Ecology and Management* 173, 105–123.

Wunder, S. (2007) The efficiency of payments for environmental services in tropical conservation. *Conservation Biology* 21, 48–58.

Zarin, D., Putz, F.E., Alavalapati, J., and Schmink, M. (eds) (2004) *Working Forests in the Tropics.* Columbia University Press, New York.

Zimmerman, B. Peres, C.A., Malcolm, J., and Turner, T. (2001) Conservation and development of alliances with the Kayapó of south-eastern Amazonia, a tropical forest indigenous peoples. *Environmental Conservation* 28, 10–22.

Zuidema, P.A. and Sayer, J.A. (2003) Tropical forests in multi-functional landscapes: the need for new approaches to conservation and research. In P.A. Zuidema (ed.), *Tropical Forests in Multi-functional Landscapes*. Prince Bernhard Centre for International Nature Conservation, Utrecht University, The Netherlands, pp. 9–19.

Zuidema, P.A., Sayer, J.A., and Dijkman, W. (1996) Forest fragmentation and biodiversity: the case for intermediate-sized conservation areas. *Environmental Conservation* 23, 290–297.

INDEX